Study Guide Volume I

to accompany

CALCULUS

Mustafa A. Munem Macomb County Community College

David J. Foulis University of Massachusetts

Prepared with the assistance of Hyla Gold Foulis

Worth Publishers, Inc.

STUDY GUIDE

Volume One

to accompany

CALCULUS

Printed in the United States of America
ISBN: 0-87901-091-6
First printing, February 1978

WORTH PUBLISHERS, INC.
444 Park Avenue South
New York, New York 10016

PREFACE

PURPOSE: This study guide accompanies *Calculus* by Mustafa Munem and David Foulis. It consists of three volumes—Volume I covers chapters 0 through 7, Volume II covers chapters 6 through 13, and Volume III covers chapters 14 through 17. Chapters 6 and 7 are included in both Volumes I and II to provide flexibility in the scheduling and the ordering of topics. The study guide is not an independent textbook but rather a workbook that helps students to master the material presented in Munem and Foulis, *Calculus*. The chapters conform section by section to those in the parent text. Volume I includes those topics usually covered in the first semester of the course.

ORGANIZATION: The study guide is written in a semiprogrammed format in that the correct answer to a question will guide the student to correctly answering the following question. Difficult concepts and problems are broken down into a sequence of steps. In supplying the correct answer at each step, the student can see the logic and reasoning that underlie the concepts and techniques of calculus. Questions that are shaded highlight definitions, properties, rules, and theorems. The tests at the end of each chapter have been selected from test questions used by the authors in class-testing the textbook and study guide. The test problems are similar to those in each chapter of the text and may be used by students to prepare for examinations.

HOW TO USE IT: We recommend that the workbook be used as follows:

1. Place a strip of paper over the left column to cover the answers.
2. Read the statement in the right column carefully and write the answers which complete the statements in the blanks provided.
3. After all of the answers for a problem are filled in, lower the strip of paper to compare your answers with the correct answers.
4. Review the material in the corresponding section of the textbook whenever you do not fully understand the answers given.
5. Continue on to the next problem and repeat the above procedure.

The use of this workbook obviously depends on the needs of the student; however, here are some suggestions:

1. An auxiliary learning resource to be used for each section covered in the textbook.
2. A supplementary learning resource to be used only when a student has difficulty mastering the material in a particular section of the textbook.
3. An auxiliary learning resource to be used in studying for examinations. Here the tests at the end of each chapter will be particularly useful.
4. A learning resource to be used to cover material missed because of absences from class.

From the enthusiastic response of our own students we feel confident that this workbook can provide a valuable resource in helping students to master the concepts and techniques presented in *Calculus*. We would like to thank our students for their many suggestions during the class-testing of the study guide. We also wish to thank Professor Karl Folley of Wayne State University for reading page proofs and the staff of Worth Publishers for their help and encouragement.

MUSTAFA A. MUNEM
DAVID J. FOULIS

CONTENTS

Chapter 0 PRECALCULUS REVIEW

The objective of the chapter is to present the mathematical background required for understanding calculus. In working the problems of this chapter, the student should be able to:

1 Understand real numbers and their properties.

2 Solve linear and absolute value inequalities.

3 Use the cartesian coordinate system.

4 Understand the geometry of lines and their slopes.

5 Understand functions and their graphs.

6 Recognize different types of functions.

7 Define and graph trigonometric functions.

8 Deal with the algebra of functions and the composition of functions.

9 Understand inverse functions.

1 Real Numbers

real numbers

$\mathbb{R}$

1 The set of all coordinates of points on a number scale or a number line or a coordinate axis is called the set of ______. This set is denoted by ______.

$x \div y$, 0

2 Two real numbers x and y in $\mathbb{R}$ can be combined by means of arithmetic operations to yield new real numbers $x + y$, xy, $x - y$, and ______ provided that $y \neq$ ______.

$y - x$ is a positive number

3 For any two real numbers x and y, x is less than y $(x < y)$, or, equivalently, y is greater than x $(y > x)$ if ______.

inequality

4 An assertion of the form $y > x$ or $x < y$ is called an ______.

$y < x$, $x = y$

5 *Trichotomy Principle:* If x and y are any real numbers, then one and only one of the following conditions will hold: $x < y$ or ______ or ______.

negative real number

positive real number

negative

6 If we put $y = 0$ in Problem 5, then one and only one of the following conditions is true:

(i) $x < 0$, in which case we call x a ______.

(ii) $x > 0$, in which case we call x a ______.

(iii) $x = 0$, in which case x is neither positive nor ______.

In Problems 7 through 22, list the rules for manipulating inequalities.

Answer		Problem
$b + c$	7	$a < b$ holds if and only if $a + c <$ ______.
$b - c$	8	$a < b$ holds if and only if $a - c <$ ______.
$b + d$	9	If $a < b$ and $c < d$, then $a + c <$ ______.
bc	10	If $a < b$ and $c > 0$, then $ac <$ ______.
$\frac{b}{c}$	11	If $a < b$ and $c > 0$, then $\frac{a}{c} <$ ______.
ac	12	If $a < b$ and $c < 0$, then $bc <$ ______.
$\frac{a}{c}$	13	If $a < b$ and $c < 0$, then $\frac{b}{c} <$ ______.
$-a$	14	$a < b$ holds if and only if $-b <$ ______.
bd	15	If $a > 0$, $b > 0$, $c > 0$, and $d > 0$, then $a < b$ and $c < d$ implies that $ac <$ ______.
$\frac{1}{b}$	16	If $ab > 0$, then $a < b$ if and only if $\frac{1}{a} >$ ______.
$b^2 > 0$	17	If $a > b > 0$, then $a^2 >$ ______.
$>$	18	If $a > b > 0$, then $\sqrt{a}$ ______ $\sqrt{b} > 0$.

Answer	Problem
> 0	**19** If $a \neq 0$, then a^2 ____.
positive, negative	**20** The condition $ab > 0$ is equivalent to the assertion that both a and b have the same algebraic sign; that is, either a and b are both ________ or they are both ________.
same algebraic sign	**21** If $b \neq 0$, then $\frac{a}{b} > 0$ if and only if a and b have the ________________.
$a < c$	**22** **If $a < b$ and $b < c$, then ________.**

In Problems 23 through 49, use the properties of real numbers, given above, to justify the following.

Answer	Problem
$b + 2$	**23** If $a < b$, then $a + 2 <$ ________.
$b - 11$	**24** If $a < b$, then $a - 11 <$ ________.
3	**25** If $a + 5 < 3 + 5$, then $a <$ ____.
6	**26** If $a - 7 < 6 - 7$, then $a <$ ____.
7	**27** If $x < 2$ and $y < 5$, then $x + y <$ ____.
11	**28** If $x > 3$ and $y > 8$, then $x + y >$ ____.
$7y$	**29** If $x < y$, then $7x <$ ____.
y	**30** If $9x < 9y$, then $x <$ ____.
$-5y$	**31** If $x < y$, then $-5x >$ ____.
y	**32** If $-3x < -3y$, then $x >$ ____.
$\frac{y}{3}$	**33** If $x < y$, then $\frac{x}{3} <$ ____.
$x < y$	**34** If $\frac{x}{7} < \frac{y}{7}$, then ________.
$\frac{y}{-2}$	**35** If $x < y$, then $\frac{x}{-2} >$ ____.
y	**36** If $\frac{x}{-7} < \frac{y}{-7}$, then $x >$ ____.
10	**37** If $x > 0$ and $x < 5$, then $2x <$ ____.
$\frac{1}{x}$	**38** If $x > 3$, then $\frac{1}{3} >$ ____.
$\frac{1}{x}$	**39** If $7x > 0$ and $x < 7$, then $\frac{1}{7} <$ ____.
4	**40** If $x > 2$, then $x^2 >$ ____.
3, -3	**41** If $x^2 > 9$, then $x >$ ____ or $x <$ ____.
$\frac{1}{y}$	**42** If $x > 0$ and $y > 0$ and $x < y$, then $\frac{1}{x} >$ ____.
y	**43** If $x > 0$ and $y > 0$ and $x^2 < y^2$, then $x <$ ____.
0	**44** If $x < 0$, then $x^2 >$ ____.
x	**45** If $0 < x < 1$, then $x^2 <$ ____.

Answer	Frame
z	**46** If $x < y < z$, then $\frac{x+y+z}{3} <$ ____.
$-\frac{1}{y}$	**47** If $x > 0$ and $y > 0$ and $x < y$, then $-\frac{1}{x} <$ ____.
y	**48** If $x < 7$ and $7 < y$, then $x <$ ____.
y	**49** If $x > 2$ and $2 > y$, then $x >$ ____.

2 Solution of Inequalities, Intervals, and Absolute Value

Answer	Frame
interval belongs	**50** An ________ is defined to be a set of real numbers such that, whenever a number x lies between two numbers that belong to the set, then x also ________ to the set.
degenerate	**51** The empty set is generally regarded as a ________ interval; so is a set that consists precisely of one real number.
$a < x < b$	**52** The open interval from a to b, denoted by (a, b), is defined to be the set of all real numbers x such that ________.
$a \leqslant x \leqslant b$	**53** The closed interval from a to b, denoted by $[a, b]$, is defined to be the set of all real numbers x such that ________.
$a \leqslant x < b$	**54** The half-open interval on the right from a to b, denoted by $[a, b)$, is defined to be the set of all real numbers x such that ________.
$a < x \leqslant b$	**55** The half-open interval on the left from a to b, denoted by $(a, b]$, is defined to be the set of all real numbers x such that ________.
$x > a$	**56** The open interval from a to $+\infty$, denoted by $(a, +\infty)$, is defined to be the set of all real numbers x such that ________.
$x < a$	**57** The open interval from $-\infty$ to a, denoted by $(-\infty, a)$, is defined to be the set of all real numbers x such that ________.
$x \leqslant a$	**58** The closed interval from $-\infty$ to a, denoted by $(-\infty, a]$, is defined to be the set of all real numbers x such that ________.
$x \geqslant a$	**59** The closed interval from a to $+\infty$, denoted by $[a, +\infty)$, is defined to be the set of all real numbers x such that ________.
real numbers	**60** The interval notation $(-\infty, +\infty)$ will be used to denote the set $\mathbb{R}$ of all ________.

In Problems 61 through 73, solve each inequality using interval notation to represent the solution. Also, show the solution on the number line.

Answer	Frame
	61 $3 < 7 - 4x \leqslant 19$.
12	$3 < 7 - 4x \leqslant 19$ is equivalent to $-4 < -4x \leqslant$ ____, so that
-12	$4 > 4x \geqslant$ ____. Dividing both sides by 4, we have
$1 > x \geqslant -3$, $-3 \leqslant x < 1$	________ or ________. Hence, the solution set is

$[-3, 1)$

__________. The graph is

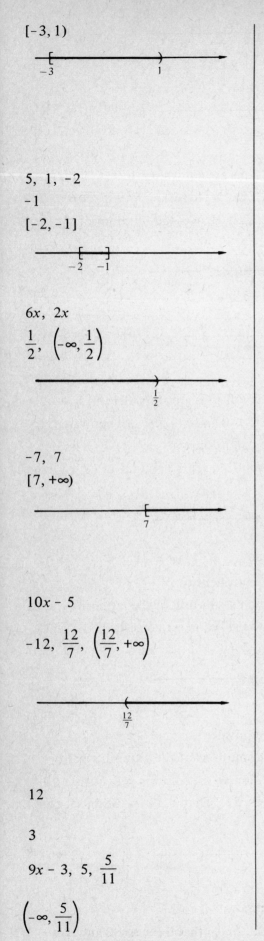

62 $-0.2 \leqslant \dfrac{2x+3}{5} \leqslant 0.2.$

5, 1, −2

−1

$[-2, -1]$

$-0.2 \leqslant \dfrac{2x+3}{5} \leqslant 0.2$ is equivalent to $5(-0.2) \leqslant 5\left(\dfrac{2x+3}{5}\right) \leqslant$ (____) (0.2), so that $-1 \leqslant 2x+3 \leqslant$ ____ or $-4 \leqslant 2x \leqslant$ ____. Dividing both sides by 2, we have $-2 \leqslant x \leqslant$ ____. Hence, the solution is __________. The graph is

63 $x + 2 > 7x - 1.$

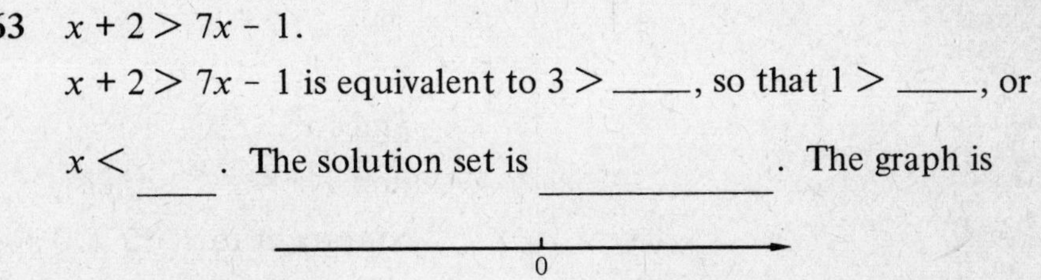

$6x$, $2x$

$\dfrac{1}{2}$, $\left(-\infty, \dfrac{1}{2}\right)$

$x + 2 > 7x - 1$ is equivalent to $3 >$ ____, so that $1 >$ ____, or $x <$ ____. The solution set is __________. The graph is

64 $3 - x \leqslant -4.$

−7, 7

$[7, +\infty)$

$3 - x \leqslant -4$ is equivalent to $-x \leqslant$ ____, so that $x \geqslant$ ____. The solution set is __________. The graph is

65 $\dfrac{3x+7}{5} < 2x - 1.$

$10x - 5$

-12, $\dfrac{12}{7}$, $\left(\dfrac{12}{7}, +\infty\right)$

$\dfrac{3x+7}{5} < 2x - 1$ is equivalent to $3x + 7 <$ ________, so that $-7x <$ ____ or $x >$ ____. The solution set is __________. The graph is

66 $\dfrac{5x-2}{3} < \dfrac{3x-1}{4}.$

12

3

$9x - 3$, 5, $\dfrac{5}{11}$

$\left(-\infty, \dfrac{5}{11}\right)$

Multiplying both sides by ____, we have $12\left(\dfrac{5x-2}{3}\right) < 12\left(\dfrac{3x-1}{4}\right)$, so that $4(5x-2) <$ (____) $(3x-1)$ or $20x - 8 <$ ________, so that $11x <$ ____ or $x <$ ____. Hence, the solution set is __________. The graph is

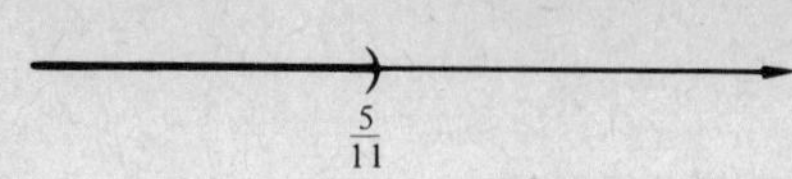

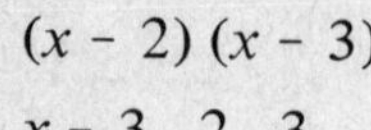

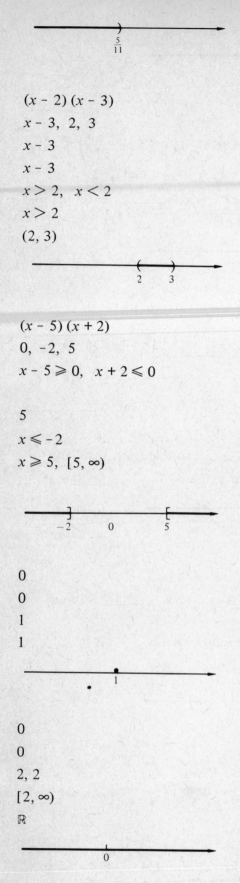

$(x-2)(x-3)$

$x-3,\ 2,\ 3$

$x-3$

$x-3$

$x > 2,\ \ x < 2$

$x > 2$

$(2, 3)$

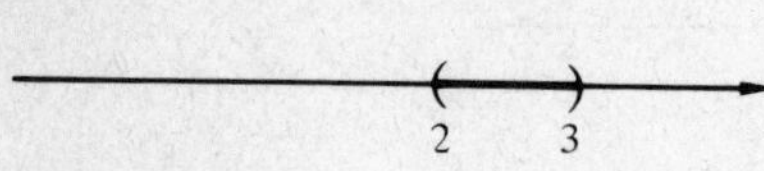

$(x-5)(x+2)$

$0,\ -2,\ 5$

$x-5 \geqslant 0,\ \ x+2 \leqslant 0$

5

$x \leqslant -2$

$x \geqslant 5,\ [5, \infty)$

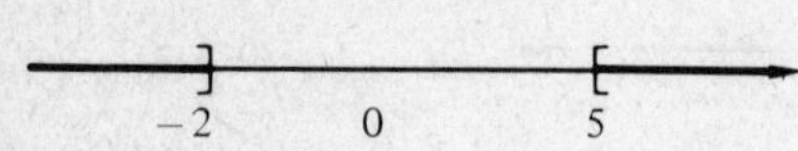

0

0

1

1

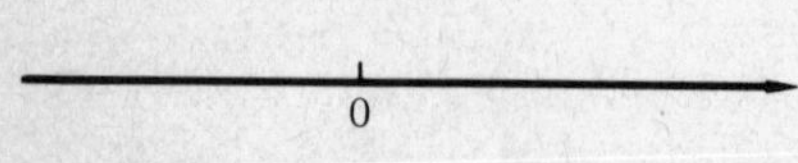

0

0

2, 2

$[2, \infty)$

$\mathbb{R}$

0

$x+1$

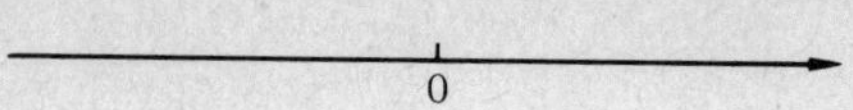

67 $x^2 - 5x + 6 < 0$.
Since $x^2 - 5x + 6 =$ ________________, so that $x - 2 = 0$ or ________ $= 0$, then $x =$ ____ or $x =$ ____. $x^2 - 5x + 6 < 0$ is equivalent to $x - 2 < 0$ and ________ > 0 or $x - 2 > 0$ and ________ < 0. These conditions are equivalent to $x < 3$ and ________ or $x > 3$ and ________. Therefore, $x^2 - 5x + 6 < 0$ holds when $x < 3$ and ________. Hence, the solution set is ________. The graph is

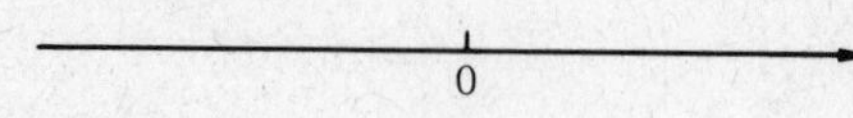

68 $x^2 - 2x - 10 \geqslant 0$.
Since $x^2 - 2x - 10 =$ ________________, so that $x + 2 = 0$ or $x - 5 =$ ____, then $x =$ ____ or $x =$ ____. $x^2 - 2x - 10 \geqslant 0$ is equivalent to $x + 2 \geqslant 0$ and ____________ or ____________ and $x - 5 \leqslant 0$. The first two conditions are equivalent to $x \geqslant -2$ and $x \geqslant$ ____. Similarly, $x + 2 \leqslant 0$ and $x - 5 \leqslant 0$ is equivalent to ______ and $x \leqslant 5$. Hence, the condition $x^2 - 2x - 10 \geqslant 0$ holds when $x \leqslant -2$ or ______. The solution set is $(-\infty, -2]$ or ______. The graph is

69 $x^2 + 1 \leqslant 2x$.
The condition $x^2 + 1 \leqslant 2x$ is equivalent to $x^2 - 2x + 1 \leqslant$ ____; that is, $(x-1)^2 \leqslant$ ____. The only solution of the latter inequality is $x =$ ____. Hence, the solution set consists of the real number ____. The graph is

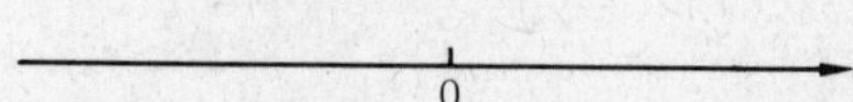

70 $x^2 + 4 \geqslant 4x$.
The condition $x^2 + 4 \geqslant 4x$ is equivalent to $x^2 - 4x + 4 \geqslant$ ____. That is, $(x-2)^2 \geqslant$ ____. The solutions of the latter inequality are $x \geqslant$ ____ and $x \leqslant$ ____. Hence, the solution set consists of the intervals ________ and $(-\infty, 2]$. That is, the solution set consists of ____. The graph is

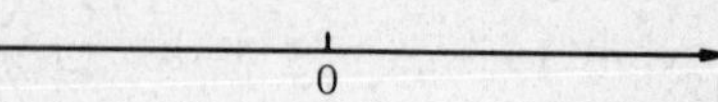

71 $2x^2 < x + 3$.
The condition $2x^2 < x + 3$ is equivalent to $2x^2 - x - 3 <$ ____. But $2x^2 - x - 3 = (2x - 3)($______$)$, so that $2x - 3 < 0$ and

$x+1>0,\ \ x+1<0$

$x>-1$

$x+1<0,\ \ x>\frac{3}{2}$

$x<\frac{3}{2},\ \left(-1,\frac{3}{2}\right)$

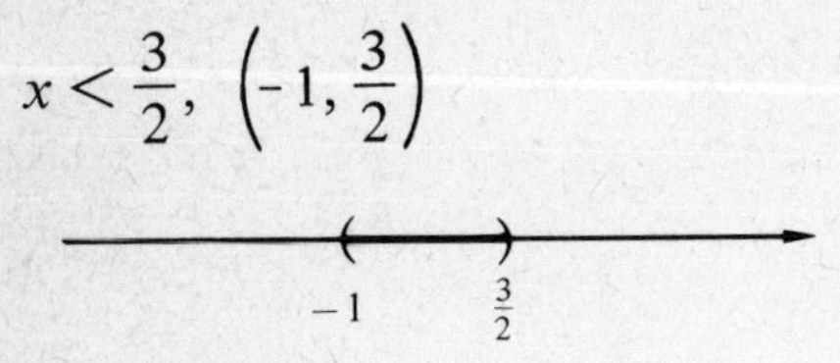

______ or $(2x-3)>0$ and ______. The latter two inequalities are equivalent to $x<\frac{3}{2}$ and ______. Similarly, $2x-3>0$ and ______ is equivalent to ______ and $x<-1$. Therefore, $2x^2-x-3<0$ holds when $x>-1$ and ______. The solution set is ______. The graph is

0

3, $-2x-5$

$2x+5$

$x+2>0,\ \ x+2<0$

$x>-2,\ (-2,\infty)$

$-\frac{5}{2}$ -2

72 $\dfrac{x+1}{x+2}\leqslant 3.$

$\dfrac{x+1}{x+2}\leqslant 3$ is equivalent to $\dfrac{x+1}{x+2}-$ ______ $\leqslant 0$, so that $\dfrac{\quad}{x+2}\leqslant 0$ or $\dfrac{\quad}{x+2}\geqslant 0$. But $0\leqslant\dfrac{2x+5}{x+2}$ holds if and only if $2x+5$ and $x+2$ have the same algebraic sign or $2x+5=0$; that is, $2x+5\geqslant 0$ and ______ or $2x+5<0$ and ______. Hence, $\dfrac{2x+5}{x+2}\geqslant 0$ holds if $x\leqslant-\frac{5}{2}$ and $x<-2$ or $x\geqslant-\frac{5}{2}$ and ______. The solution set is $\left(-\infty,-\frac{5}{2}\right]$ or ______. The graph is

0

$x-2<0$

$x>1$ and $x<2$

$1<x<2$

(1, 2)

1 2

73 $\dfrac{x-1}{x-2}<0.$

The inequality $\dfrac{x-1}{x-2}<0$ holds if and only if $x-1<0$ and and $x-2>0$ or $x-1>0$ and ______. These conditions are equivalent to $x<1$ and $x>2$ or ______. Therefore, $\dfrac{x-1}{x-2}<0$ holds only if ______. The solution set is ______. The graph is

0

2.1 Absolute Value

$x\geqslant 0$

$x<0$

74 If x is a real number, then the absolute value of x, denoted by $|x|$, is defined by

$$|x|=\begin{cases}x & \text{if } \_\_\_\_\_\_ \\ -x & \text{if } \_\_\_\_\_\_.\end{cases}$$

$x \geqslant 0$

$x < 0$

$|a - b|$

5, -3, 3

$\pm x$

$-\dfrac{x}{x}$

$x < 0$

$|b|$

6

6, 6

$|-b|$

75 Since $|x| = \sqrt{x^2}$, we have

$$\sqrt{x^2} = \begin{cases} x & \text{if } \_\_\_\_\_\_ \\ -x & \text{if } \_\_\_\_\_\_. \end{cases}$$

76 The distance between any two points a and b on the number line is expressed by ________ or $|b - a|$.

77 $|5| =$ ____ and $|-3| = -($____$) =$ ____.

78 If $x \neq 0$, determine $\dfrac{|x|}{x}$. By definition, $|x| =$ ____, so that

$$\frac{|x|}{x} = \begin{cases} \dfrac{x}{x} & \text{if } x \geqslant 0 \\ \_\_\_\_ & \text{if } x < 0. \end{cases}$$

Therefore,

$$\frac{|x|}{x} = \begin{cases} 1 & \text{if } x \geqslant 0 \\ -1 & \text{if } \_\_\_\_\_\_. \end{cases}$$

79 *Triangle Inequality:* If a and b are any real numbers, then $|a + b| \leqslant |a| +$ ____.

80 If $|x| < 2$ and $|y| < 4$, find $|x + y|$. Since $|x| < 2$ and $|y| < 4$, then $|x| + |y| <$ ____. Thus, by the triangle inequality, $|x + y| \leqslant |x| + |y| <$ ____, so that $|x + y| <$ ____.

81 Let a, b, and c be real numbers. Then $|a - b| \leqslant |a| +$ ____.

In Problems 82 through 99, list the basic properties of absolute value.

x

$-x$

$|x|$

$\pm y$

$x = \pm y$

$-y < x < y$

$-y \leqslant x \leqslant y$

$x > y, \; x < -y$

$x \leqslant -y$

$|x|$

x^2

$|x|$

$|y|$

$|y|$

$|y - z|$

$|y|$

1

$|x|$

82 $x \geqslant 0$ if and only if $|x| =$ ____.

83 $x < 0$ if and only if $|x| =$ ____.

84 $-|x| \leqslant x \leqslant$ ____.

85 $|x| = y$ if and only if $y \geqslant 0$ and $x =$ ____.

86 $|x| = |y|$ if and only if ________.

87 $|x| < y$ if and only if ________.

88 $|x| \leqslant y$ if and only if ________.

89 $|x| > y$ if and only if ________ or ________.

90 $|x| \geqslant y$ if and only if $x \geqslant y$ or ________.

91 $|-x| =$ ____.

92 $|x|^2 =$ ____.

93 $\sqrt{x^2} =$ ____.

94 $|x - y| \geqslant |x| -$ ____.

95 $\big||x| - |y|\big| \leqslant |x - y| \leqslant |x| +$ ____.

96 $|x - y| \leqslant |x - z| +$ ________.

97 $|xy| = |x|$ ____.

98 If $y \neq 0$, then $\left|\dfrac{1}{y}\right| = \dfrac{\_\_\_\_}{|y|}$.

99 If $y \neq 0$, then $\left|\dfrac{x}{y}\right| = \dfrac{\_\_\_\_}{|y|}$.

In Problems 100 through 105, solve each equation.

100 $|x| = 3$.
$|x| = 3$ is equivalent to $x =$ ____ or $x =$ ____.

$-3, 3$

101 $|5x - 1| = 9$.
$|5x - 1| = 9$ is equivalent to $5x - 1 = -9$ or $5x - 1 =$ ____, so that $5x =$ ____ or $5x =$ ____. Therefore, $x =$ ____ or $x =$ ____.

9

$-8, 10, -\frac{8}{5}, 2$

102 $|2x - 1| = 3$.
$|2x - 1| = 3$ is equivalent to $2x - 1 = -3$ or $2x - 1 =$ ____, so that $2x =$ ____ or $2x =$ ____. Therefore, $x =$ ____ or $x =$ ____.

3

$-2, 4, -1, 2$

103 $|2x - 3| = 7$.
$|2x - 3| = 7$ is equivalent to $2x - 3 = -7$ or $2x - 3 =$ ____, so that $2x =$ ____ or $2x =$ ____. Therefore, $x =$ ____ or $x =$ ____.

7

$-4, 10, -2, 5$

104 $|7x - 3| = |2x|$.
$|7x - 3| = |2x|$ is equivalent to $7x - 3 = -2x$ or $7x - 3 =$ ____, so that $9x =$ ____ or $5x =$ ____. Therefore, $x =$ ____ or $x =$ ____.

$2x$

$-3, 3, -\frac{1}{3}, \frac{3}{5}$

105 $|4x - 1| = |3x + 7|$.
$|4x - 1| = |3x + 7|$ is equivalent to $4x - 1 = -(3x + 7)$ or $4x - 1 =$ ________, so that $7x =$ ____ or $x =$ ____. Therefore, $x =$ ____ or $x =$ ____.

$3x + 7, -6, 8$

$-\frac{6}{7}, 8$

In Problems 106 through 111, solve each inequality.

106 $|x| < 2$.
$|x| < 2$ is equivalent to $-2 < x <$ ____. Hence, the solution set is the open interval ________. The graph is

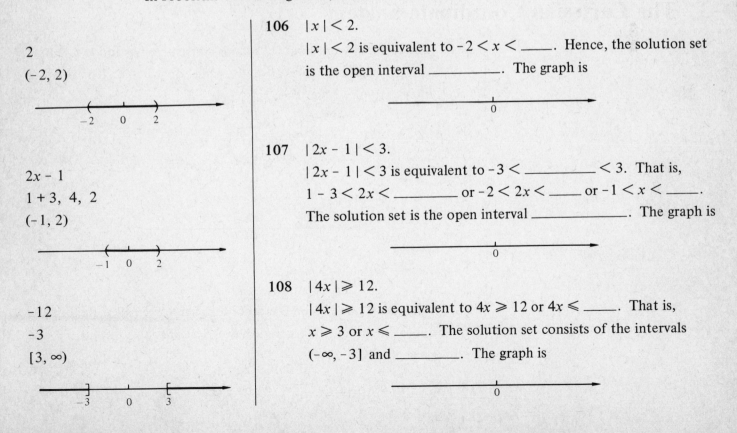

0

2

$(-2, 2)$

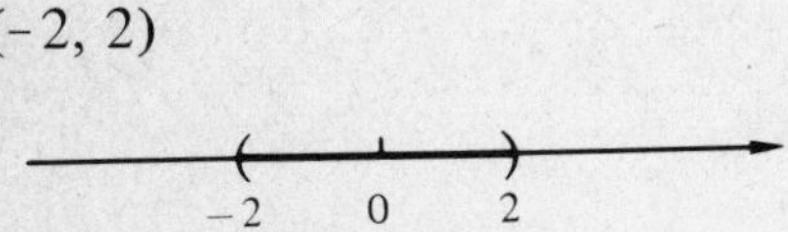

107 $|2x - 1| < 3$.
$|2x - 1| < 3$ is equivalent to $-3 <$ ________ < 3. That is, $1 - 3 < 2x <$ ________ or $-2 < 2x <$ ____ or $-1 < x <$ ____. The solution set is the open interval ____________. The graph is

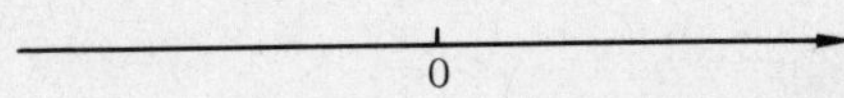

$2x - 1$

$1 + 3, 4, 2$

$(-1, 2)$

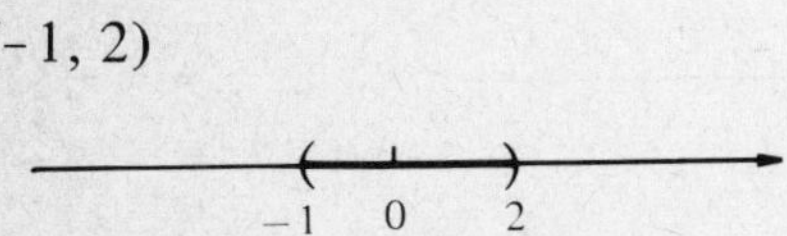

108 $|4x| \geqslant 12$.
$|4x| \geqslant 12$ is equivalent to $4x \geqslant 12$ or $4x \leqslant$ ____. That is, $x \geqslant 3$ or $x \leqslant$ ____. The solution set consists of the intervals $(-\infty, -3]$ and ________. The graph is

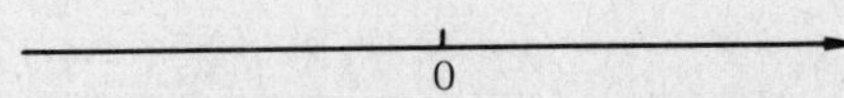

-12

-3

$[3, \infty)$

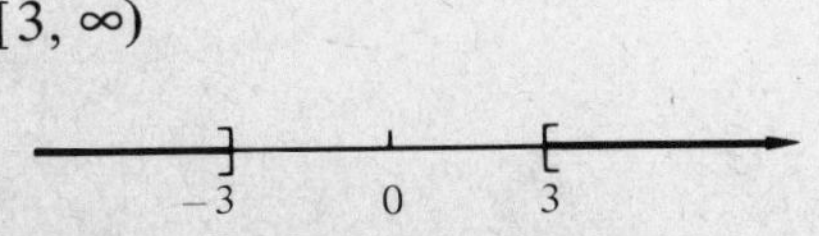

109 $|2x - 7| > 3$.
$|2x - 7| > 3$ is equivalent to $2x - 7 > 3$ or $2x - 7 <$ ____.
That is, $2x >$ ____ or $2x <$ ____, so that $x >$ ____ or $x <$ ____.
The solution set consists of the intervals $(5, \infty)$ and ________.
The graph is

0

-3
10, 4, 5, 2
$(-\infty, 2)$

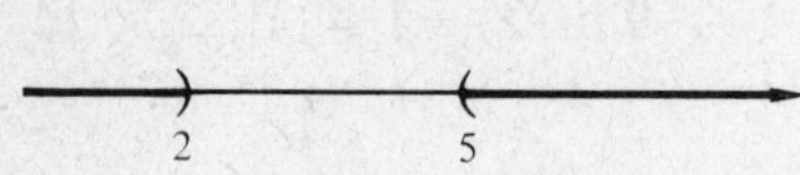

110 $|7x - 3| \geqslant 11$.
$|7x - 3| \geqslant 11$ is equivalent to $7x - 3 \geqslant +11$ or $7x - 3 \leqslant -11$.
That is, $7x \geqslant$ ____ or $7x \leqslant$ ____, so that $x \geqslant$ ____ or $x \leqslant$ ____.
The solution set consists of the intervals $\left(-\infty, -\frac{8}{7}\right]$ and ________. The graph is

0

14, -8, 2, $-\frac{8}{7}$
$[2, \infty)$

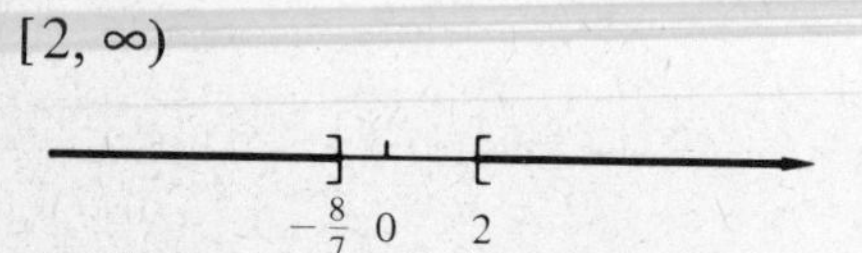

111 $|x^2 - 8| \leqslant 1$.
$|x^2 - 8| \leqslant 1$ is equivalent to $-1 \leqslant x^2 - 8 \leqslant$ ____ or $7 \leqslant x^2 \leqslant$ ____, so that $\sqrt{7} \leqslant x \leqslant$ ____. The solution set is the interval ________. The graph is

0

1
9, 3
$[\sqrt{7}, 3]$

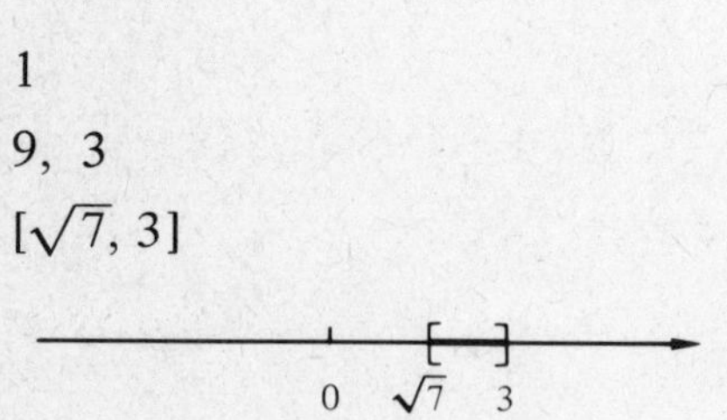

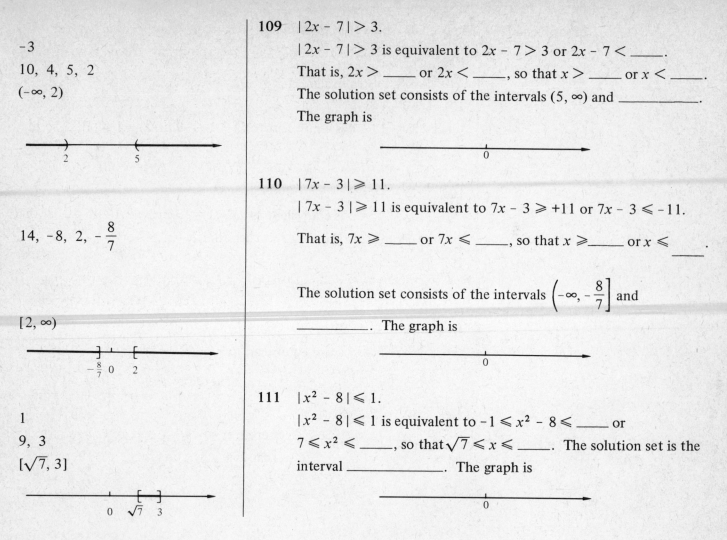

3 The Cartesian Coordinate System

112 The cartesian coordinate system is constructed as follows. First two perpendicular lines L_1 and L_2 are constructed. The point of intersection is called the ________.

origin

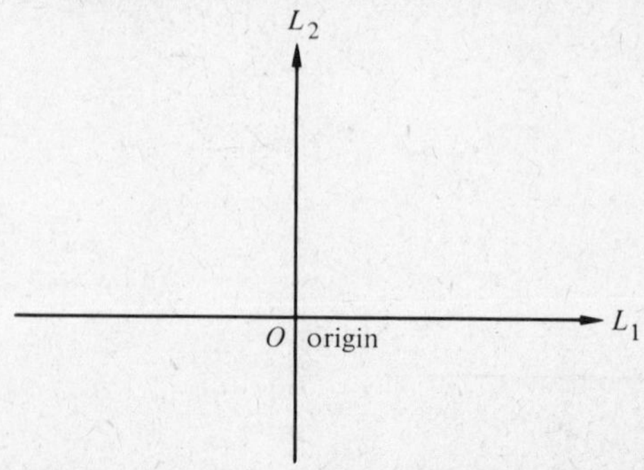

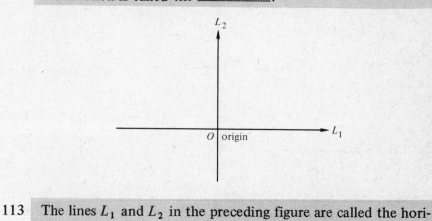

113 The lines L_1 and L_2 in the preceding figure are called the horizontal or the ____ axis, and the vertical or the ____ axis, respectively.

x, y

abscissa
ordinate
coordinates
(x, y)

114 Let P be any point in the plane and drop perpendiculars from P to the x and y axes. Let these perpendiculars intersect the x and the y axes at the points P_1 and P_2, respectively. The coordinate x of P_1 on the horizontal axis is called the ______________ of P, while the coordinate y of P_2 on the vertical axis is called the ______________ of P. The real numbers x and y are called the ______________ of P. Traditionally, these coordinates are written as an ordered pair ____________.

ordered pairs (x, y)

115 A cartesian coordinate system establishes a one-to-one correspondence between points P in the plane and the ______________ ______________ of real numbers.

quadrants
$y > 0$
$x < 0$
$y < 0$
$y < 0$

116 The coordinate axes divide the plane into four disjoint regions called ______________. The first quadrant includes all points (x, y) such that $x > 0$ and __________, the second quadrant includes all points (x, y) such that __________ and $y > 0$, the third includes all points (x, y) such that $x < 0$ and __________, and the fourth includes all points such that $x > 0$ and __________.

In Problems 117 and 118, plot the point $P = (3, -1)$ on the cartesian system and give the coordinates of the points.

(3, 1)

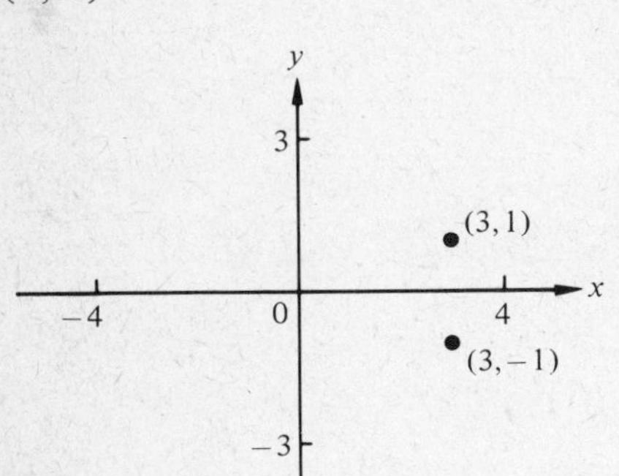

117 Q such that $\overline{PQ}$ is perpendicular to the x axis and is bisected by it.

$Q =$ ______________________

(−3, 1)

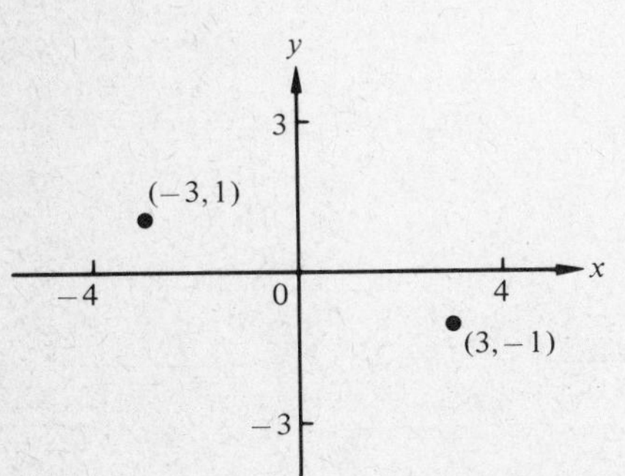

118 R such that the line segment $\overline{PR}$ is bisected by the origin.

$R =$ ______________________

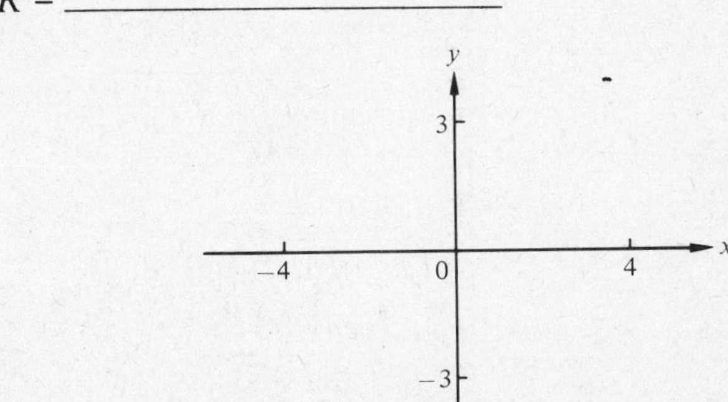

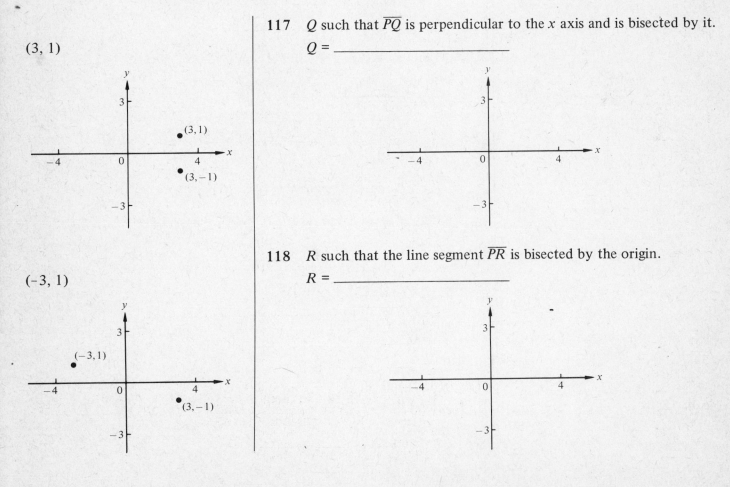

3.1 The Distance Formula

$\sqrt{(x_1 - x_2)^2 + (y_1 - y_2)^2}$

$\sqrt{(x_2 - x_1)^2 + (y_2 - y_1)^2}$

119 In the cartesian plane, the distance $d = |\overline{P_1P_2}|$ between the points $P_1 = (x_1, y_1)$ and $P_2 = (x_2, y_2)$ is given by

$d =$ __________ or

$d =$ __________.

In Problems 120 through 124, find the distance between each pair of points.

(3, 1), (7, −1)

1 + 1

$2\sqrt{5}$

120 (3, 1) and (7, −1).

Here, $P_1 = (x_1, y_1) =$ ________ and $P_2 = (x_2, y_2) =$ ________,

so that

$d = \sqrt{(x_1 - x_2)^2 + (y_1 - y_2)^2}$

$= \sqrt{(3 - 7)^2 + (\_\_\_\_)^2}$

$= \sqrt{20} =$ ____.

(−12, −5)

0 + 5, 169

13

121 (0, 0) and (−12, −5).

Here, $P_1 = (x_1, y_1) = (0, 0)$ and $P_2 = (x_2, y_2)$ ________,

so that

$d = \sqrt{(x_1 - x_2)^2 + (y_1 - y_2)^2}$

$= \sqrt{(0 + 12)^2 + (\_\_\_\_)^2} = \sqrt{\_\_\_\_}$

$=$ ____.

(−4, −1)

1, 37

$\sqrt{37}$

122 (2, −2) and (−4, −1).

Here, $P_1 = (x_1, y_1) = (2, -2)$ and $P_2 = (x_2, y_2) =$ ________,

so that

$d = \sqrt{(x_2 - x_1)^2 + (y_2 - y_1)^2}$

$= \sqrt{(-4 - 2)^2 + (-1 + 2)^2}$

$= \sqrt{(-6)^2 + (\_\_\_\_)^2} = \sqrt{\_\_\_\_}$

$=$ ____.

(−1, 2)

25

34

$\sqrt{34}$

123 (2, −3) and (−1, 2).

Here, $P_1 = (x_1, y_1) = (2, -3)$ and $P_2 = (x_2, y_2) =$ ________,

so that

$d = \sqrt{(x_2 - x_1)^2 + (y_2 - y_1)^2}$

$= \sqrt{(-1 - 2)^2 + (2 + 3)^2}$

$= \sqrt{9 + \_\_\_\_}$

$= \sqrt{\_\_\_\_}$

$=$ ____.

(4, 0), (8, 3)

5

124 (4, 0) and (8, 3).

Here, $P_1 = (x_1, y_1) =$ ________ and $P_2 = (x_2, y_2) =$ ________,

so that

$d = \sqrt{(x_2 - x_1)^2 + (y_2 - y_1)^2}$

$= \sqrt{(8 - 4)^2 + (3 - 0)^2} =$ ____.

4 Straight Lines and Their Slope

125 Consider the inclined line segment $\overline{AB}$.

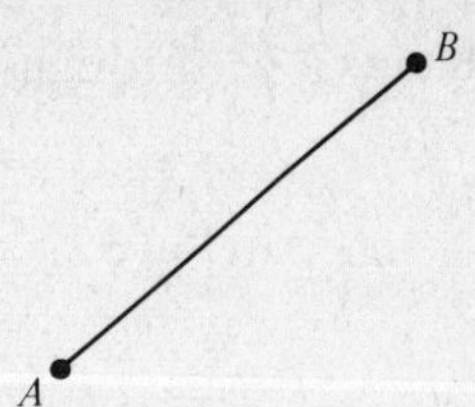

run
rise
slope

The horizontal distance between A and B is called the ____, while the vertical distance between A and B is called the ____. The ratio of rise to run is called the ______ of the line segment.

rise

Thus, slope of $\overline{AB} = \dfrac{\quad}{\text{run}}$.

$\dfrac{y_2 - y_1}{x_2 - x_1}, \dfrac{y_1 - y_2}{x_1 - x_2}$

126 Let $A = (x_1, y_1)$ and $B = (x_2, y_2)$ be any two points in the cartesian plane. Then the slope m of the line segment between A and B is given by $m = \dfrac{\quad}{\quad} = \dfrac{\quad}{\quad}$, provided that $x_1 \neq x_2$.

In Problems 127 through 130, find the slope of the line containing the given pair of points.

(1, 5), (-3, 2)

$2 - 5, \frac{3}{4}$

127 (1, 5) and (-3, 2).
Here, $P_1 = (x_1, y_1) =$ ____ and $P_2 = (x_2, y_2) =$ ____, so that
$m = \dfrac{y_2 - y_1}{x_2 - x_1} = \dfrac{\quad}{-3 - 1} = \underline{\quad}$.

(-1, 2), (2, -3)

$-3 - 2, -\frac{5}{3}$

128 (-1, 2) and (2, -3).
Here, $P_1 = (x_1, y_1) = ($____$)$ and $P_2 = (x_2, y_2) = ($____$)$, so that $m = \dfrac{y_2 - y_1}{x_2 - x_1} = \dfrac{\quad}{2 + 1} = \underline{\quad}$.

(-4, 3), (6, 3)

3 - 3, 0

129 (-4, 3) and (6, 3).
Here, $P_1 = (x_1, y_1) =$ ____ and $P_2 = (x_2, y_2) =$ ____, so that
$m = \dfrac{y_2 - y_1}{x_2 - x_1} = \dfrac{\quad}{6 + 4} =$ ____.

(-4, 2)

(-4, 6), 6 - 2, undefined

130 (-4, 2) and (-4, 6).
Here, $P_1 = (x_1, y_1) =$ ________ and $P_2 = (x_2, y_2) =$ ________, so that $m = \dfrac{y_2 - y_1}{x_2 - x_1} = \dfrac{\quad}{-4 + 4}$ is ________.

x intercept

131 Suppose that L is a nonvertical straight line whose equation is $Ax + By + C = 0$ with $A \neq 0$ and $B \neq 0$. The number $-\dfrac{C}{A}$ is called the ________ of L and the number $-\dfrac{C}{B}$ is called the

y intercept
general form

_____ of L. The equation of the line is called the _____ of the line.

In Problems 132 through 136, find the x intercept and the y intercept of the line.

132 $x - 2y + 3 = 0$.

$-3, \frac{3}{2}$

The x intercept is _____ and the y intercept is _____.

133 $3x + 7y - 11 = 0$.

$\frac{11}{3}, \frac{11}{7}$

The x intercept is _____ and the y intercept is _____.

134 $-3x + 8y - 31 = 0$.

$-\frac{31}{3}, \frac{31}{8}$

The x intercept is _____ and the y intercept is _____.

135 The point-slope form for the equation of the line containing the point $P_1 = (x_1, y_1)$ and having slope m is _____.

$y - y_1 = m(x - x_1)$

136 The slope-intercept form of the line L with slope m and intercept b is _____.

$y = mx + b$

In Problems 137 through 142, find the equation of the line with the given information.

137 $m = \frac{3}{2}$ and $(x_1, y_1) = (-1, 5)$.

Using the point-slope form, we have $y - y_1 = m(x - x_1)$, so that

$y - 5 = \frac{3}{2}(\_\_\_\_\_)$.

$x + 1$

138 $m = -6$ and $(x_1, y_1) = (4, -1)$.

Using the point-slope form, we have $y + 1 = -6(\_\_\_\_\_)$.

$x - 4$

139 $m = \frac{1}{6}$ and $b = 4$.

Using the slope-intercept form, $y = mx + b$, we have

$y = \frac{1}{6}x + \_\_\_\_\_ = \_\_\_\_\_$.

$4, \frac{1}{6}x + 4$

140 $m = -2$ and $b = -7$.

Using the slope-intercept form, $y = -2x + \_\_\_\_\_ = \_\_\_\_\_$.

$-7, -2x - 7$

141 $m = -\frac{3}{4}$ and $(x_1, y_1) = (3, 1)$.

The equation of the line is $y - 1 = -\frac{3}{4}(\_\_\_\_\_)$ or _____.

$x - 3$

$4y + 3x = 13$

$x - 5,\ y = 2$

142 $m = 0$ and $(x_1, y_1) = (5, 2)$.
The equation of the line is $y - 2 = 0($______$)$ or ______.

$m_1 = m_2$

143 **Two distinct straight lines L_1 and L_2, with slopes m_1 and m_2, respectively, are parallel if and only if ______.**

In Problems 144 and 145, indicate whether or not each pair of lines is parallel.

$\frac{2}{3}x - \frac{5}{3}$

$\frac{2}{3}$

$\frac{2}{3}x - \frac{8}{9},\ \frac{2}{3}$

parallel

144 $2x - 3y = 5$ and $6x - 9y = 8$.

$2x - 3y = 5$ is written as $y =$ ______, so that the slope of the line $y = \frac{2}{3}x - \frac{5}{3}$ is ____. Also, $6x - 9y = 8$ is written as $y =$ ______, so that the slope of the line $y = \frac{2}{3}x - \frac{8}{9}$ is ____.

Since $\frac{2}{3} = \frac{2}{3}$, the two lines are ______.

$\frac{5}{4}x + 7$

$\frac{10}{8}x + 1$

parallel

145 $5x - 4y + 28 = 0$ and $10x - 8y + 8 = 0$.
The equation of the line $5x - 4y + 28 = 0$ is written as $y =$ ______. Also, the equation of the line $10x - 8y + 8 = 0$ is written as $y =$ ______, so that $\frac{5}{4} = \frac{10}{8}$ and the two lines are ______.

In Problems 146 through 148, find the equation of the line with the given information.

$-3,\ -\frac{3}{4}x - \frac{1}{4}$

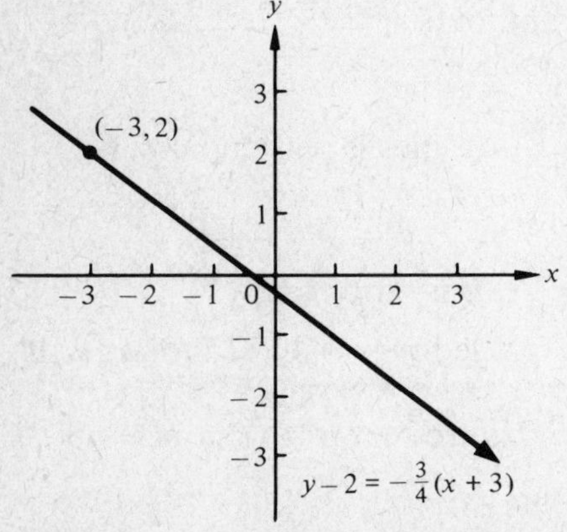

146 $m = -\frac{3}{4}$, $(x_1, y_1) = (-3, 2)$.

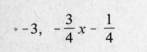

The equation is $y - 2 = -\frac{3}{4}[x - ($______$)]$ or $y =$ ______.

The graph is

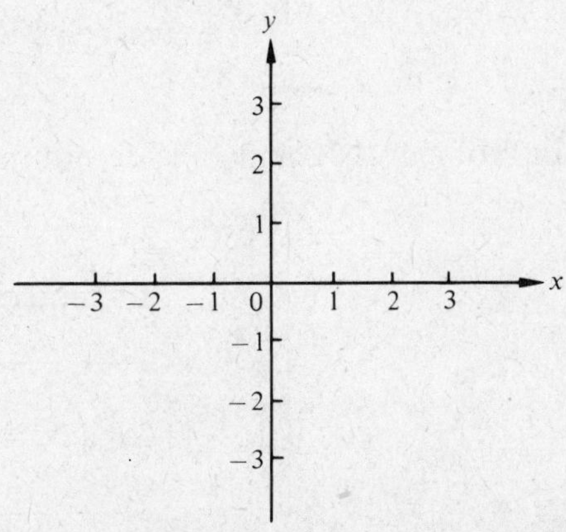

$-\frac{1}{2}$, $-\frac{1}{2}x + 4$

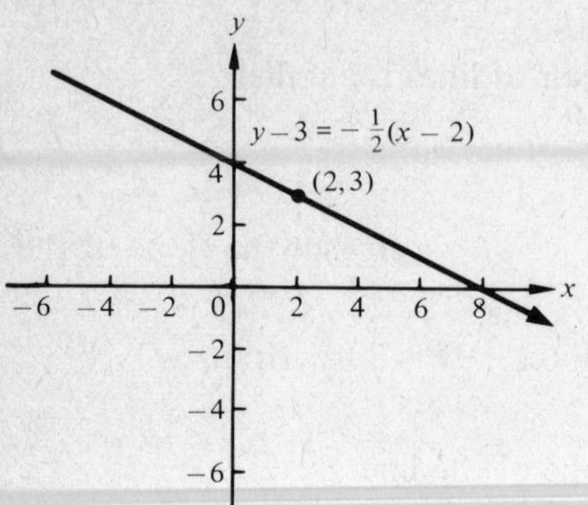

147 $m = -\frac{1}{2}$, $(x_1, y_1) = (2, 3)$.

The equation is $y - 3 =$ ____ $(x - 2)$ or $y =$ ______________ . The graph is

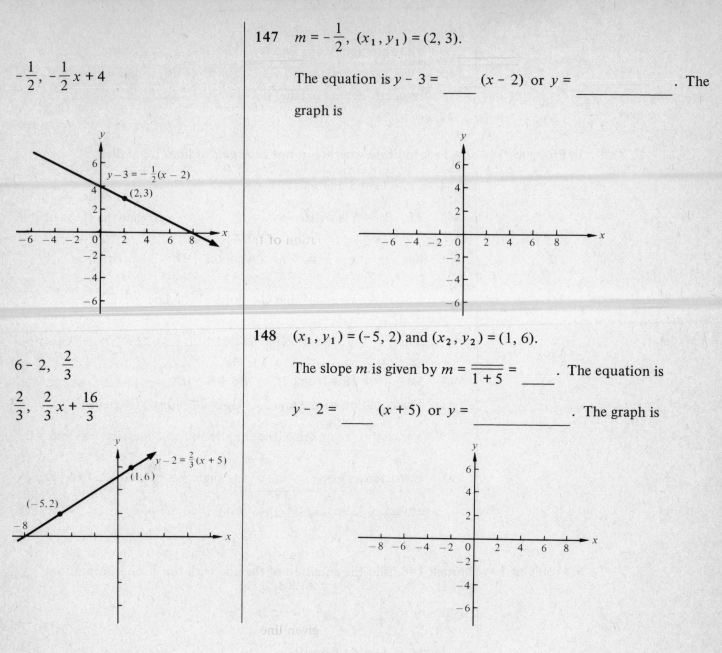

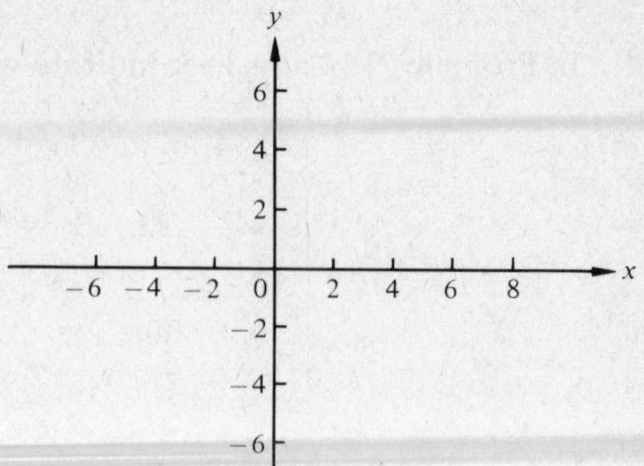

$6 - 2$, $\frac{2}{3}$

$\frac{2}{3}$, $\frac{2}{3}x + \frac{16}{3}$

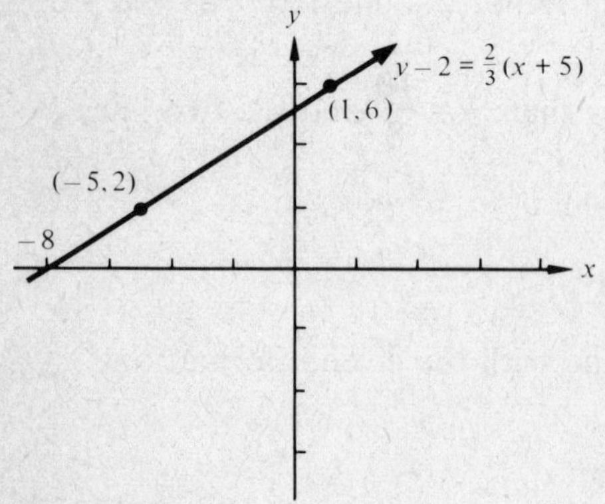

148 $(x_1, y_1) = (-5, 2)$ and $(x_2, y_2) = (1, 6)$.

The slope m is given by $m = \frac{\quad}{1 + 5} =$ ____ . The equation is

$y - 2 =$ ____ $(x + 5)$ or $y =$ ______________ . The graph is

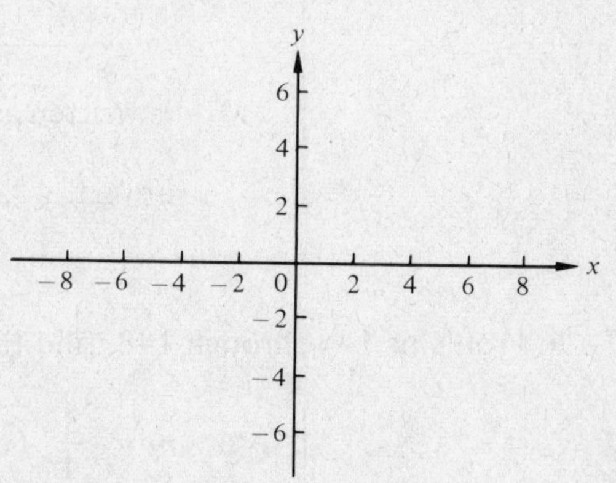

4.1 Perpendicular Lines

$m_1 m_2 = -1$

149 Two nonvertical straight lines L_1 and L_2, with slopes m_1 and m_2, respectively, are perpendicular if and only if ______________ .

In Problems 150 and 151, determine whether or not each pair of lines is perpendicular.

$3x + 5$

$-\frac{1}{3}x + \frac{5}{3}$

-1, perpendicular

150 $3x - y + 5 = 0$ and $x + 3y - 5 = 0$.

The equation of the line $3x - y + 5 = 0$ is written as $y =$ ________, whereas the equation of the line $x + 3y - 5 = 0$ is written as $y =$ ______________ . Since the product of the slopes $(3)\left(-\frac{1}{3}\right) =$ ____, the lines are ______________ .

151 $5x - 2y + 7 = 0$ and $2x + 5y - 3 = 0$.

The equation of the line $5x - 2y + 7 = 0$ is written as

$y =$ ________, whereas the equation of the line $2x + 5y - 3 = 0$

is written as $y =$ ________. Since the product of the slopes

$\left(\frac{5}{2}\right)\left(-\frac{2}{5}\right) =$ ____, the lines are __________.

$\frac{5}{2}x + \frac{7}{2}$

$-\frac{2}{5}x + \frac{3}{5}$

-1, perpendicular

In Problems 152 through 157, find the equation of the line using the given information.

152 Containing the point $(-1, 2)$ and is parallel to the line $3x - y - 1 = 0$.

The slope of the given line is ____. Since the line contains the point $(-1, 2)$ and the line is parallel to the given line, its equation in point-slope form is __________ or $y =$ ________.

3

$y - 2 = 3(x + 1)$

$3x + 5$

153 Containing the point $(-1, 2)$ and is perpendicular to the line $3x - y - 1 = 0$.

The slope of the given line is ____. Since the line contains the point $(-1, 2)$ and the line is perpendicular to the given line, its slope is ____ and its equation in point-slope form is __________ or $y =$ ________.

3

$-\frac{1}{3}$

$y - 2 = -\frac{1}{3}(x + 1)$, $-\frac{1}{3}x + \frac{5}{3}$

154 Containing the point $(3, 2)$ and is parallel to the line $2x - y + 7 = 0$.

The slope of the given line is ____. Since the line is parallel to the given line, its slope is ____ and its equation in point-slope form is __________ or $y =$ ________.

2

2

$y - 2 = 2(x - 3)$, $2x - 4$

155 Containing the point $(3, 2)$ and is perpendicular to the line $2x - y + 13 = 0$.

The slope of the given line is ____. Since the line is perpendicular to the given line, its slope is ____ and its equation in point-slope form is __________ or $y =$ ________.

2

$-\frac{1}{2}$

$y - 2 = -\frac{1}{2}(x - 3)$, $-\frac{1}{2}x + \frac{7}{2}$

156 Containing the point $(7, -1)$ and is parallel to the line $7x - y + 15 = 0$.

The slope of the given line is ____. Since the line is parallel to the given line, its slope is ____ and its equation in point-slope form is __________ or $y =$ ________.

7

7

$y + 1 = 7(x - 7)$, $7x - 50$

Answer	Problem
0 infinite $-1, \ x + 1 = 0$	**157** Containing the point $(-1, 4)$ and is perpendicular to the line $y = 4$. The slope of the given line is ____. Since the line is perpendicular to the given line, its slope is ____________ and its equation is $x =$ ____ or ____________.

5 Functions and Their Graphs

Answer	Problem
correspondence	**158** A function is a ________________ that assigns a unique numerical value to a variable y for each numerical value of a variable x.
dependent independent domain	**159** The variable y is called the ____________ variable, because its value depends on x. The variable x is called the ____________ variable, since it can take on any value in a certain set of numbers called the ________ of the function.
function	**160** To graph a function we simply display all points in the cartesian plane whose coordinates (x, y) belong to the ____________.

In Problems 161 through 172, find the domain and range of each function and sketch its graph.

Answer	Problem
$(-\infty, \infty)$ $\frac{y-2}{5}$ $(-\infty, \infty)$	**161** $y = 5x + 2$. The domain of the function is the interval __________. Solving this equation for x, we obtain $x =$ ________, so that the range is the interval __________. The graph is

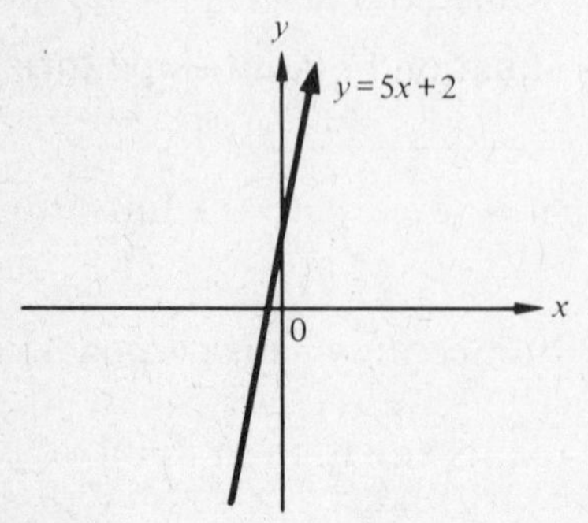

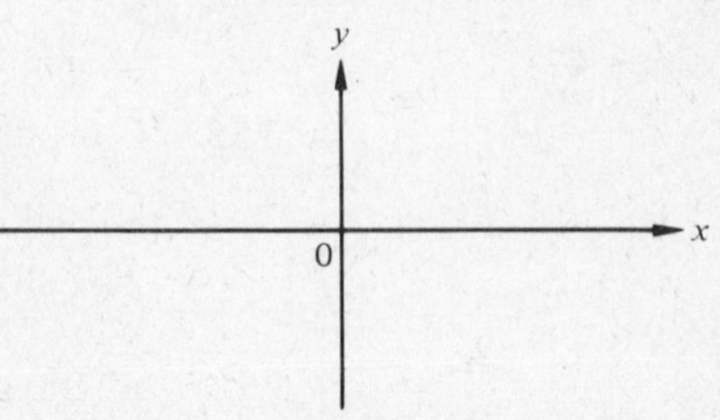

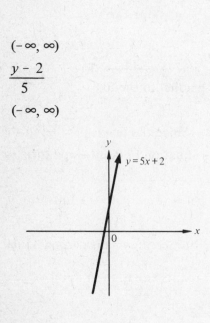

Answer	Problem
$(-\infty, 1]$ $1 - x \geqslant 0$ $1 - y^2$ $[0, \infty)$	**162** $y = \sqrt{1 - x}$. The domain of the function is the interval __________, since $\sqrt{1-x}$ is defined for ____________. Solving the equation for x, we obtain $x =$ ________, so that the range is the interval ________. The graph is

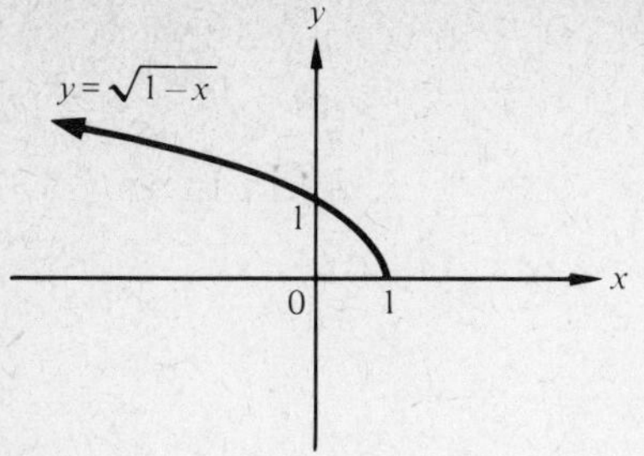

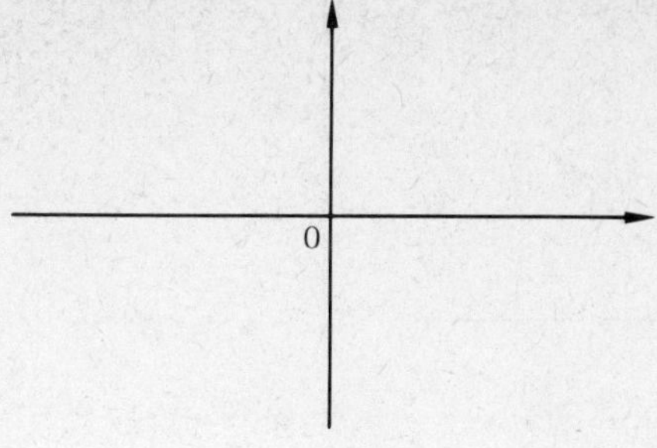

(-∞, 9]
$9 - y^2$
[0, ∞)

163 $y = \sqrt{9 - x}$.

The domain of the function is the interval ________. Solving the equation for x, we obtain $x =$ ________, so that the range of the function is the interval ________. The graph is

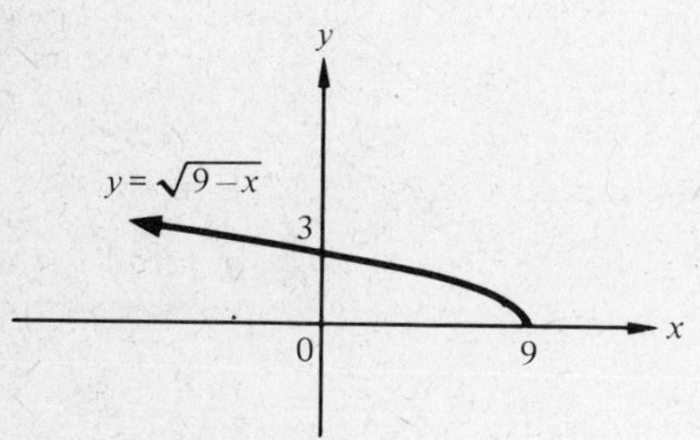

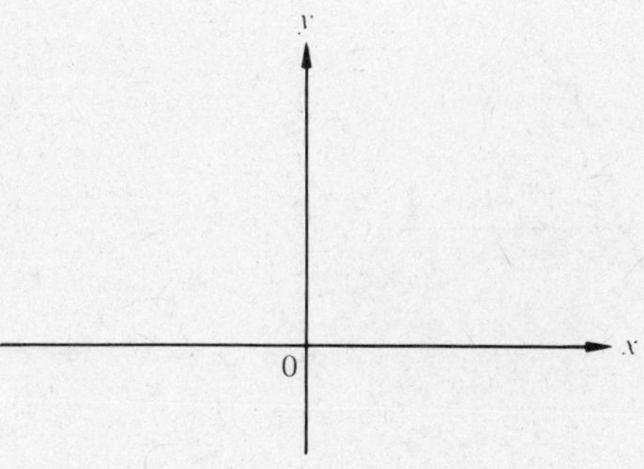

of real numbers ℝ
negative
[0, ∞]

164 $y = |2x|$.

The domain of the function is the set ____________________.
Since y cannot be ____________, the range of the function f is ________.

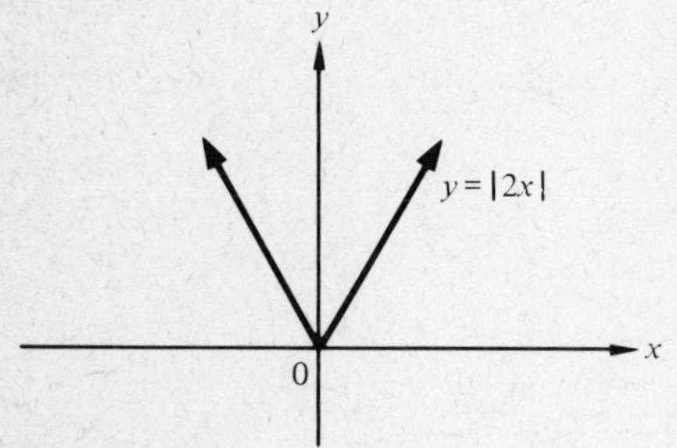

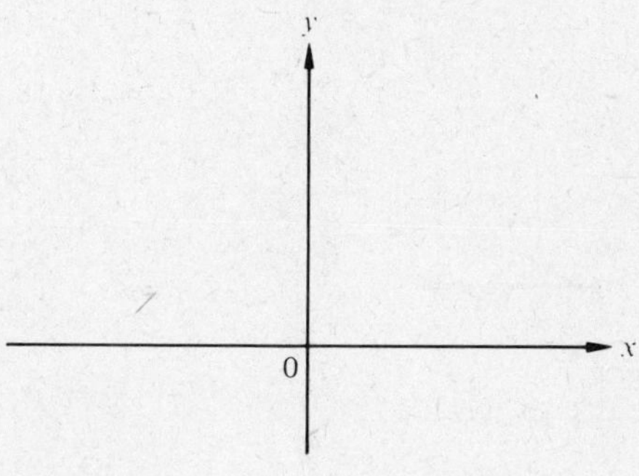

$|x|$
real numbers ℝ
nonnegative
[0, ∞)

165 $y = |-x|$.

$|-x| =$ ____. The domain of the function is the set of ________________. Since y cannot be negative but can take on any ________________ value, the range of the function is ________. The graph is

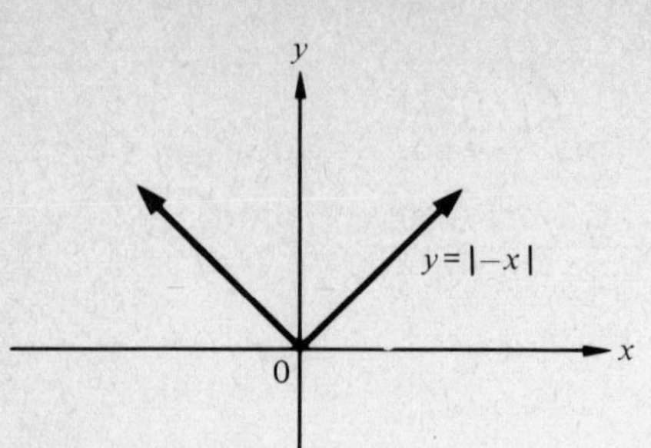

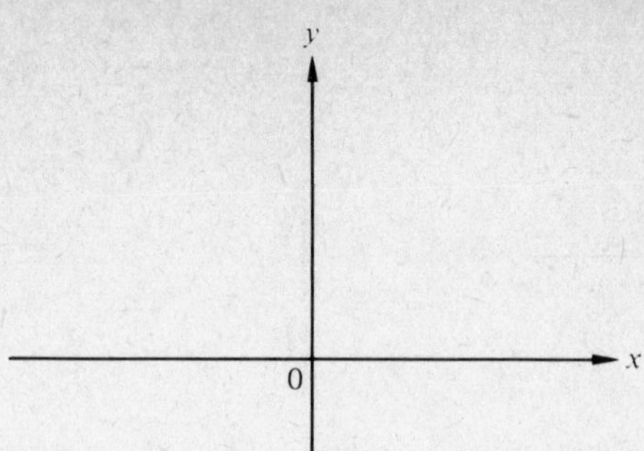

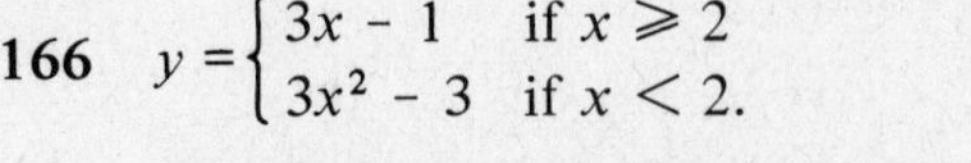

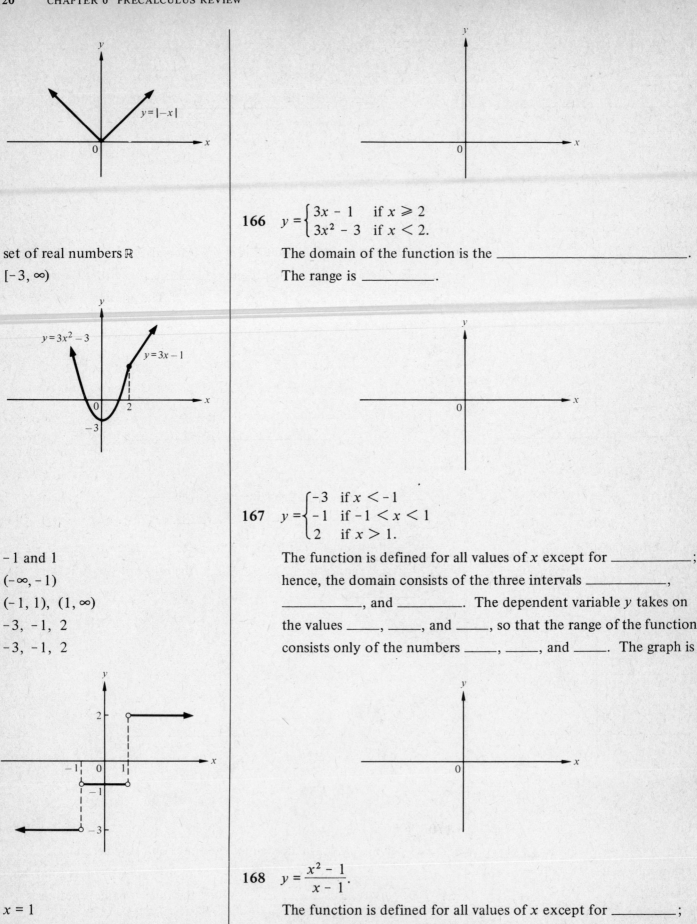

166 $y = \begin{cases} 3x - 1 & \text{if } x \geqslant 2 \\ 3x^2 - 3 & \text{if } x < 2. \end{cases}$

set of real numbers ℝ

$[-3, \infty)$

The domain of the function is the ______________________________.
The range is ____________.

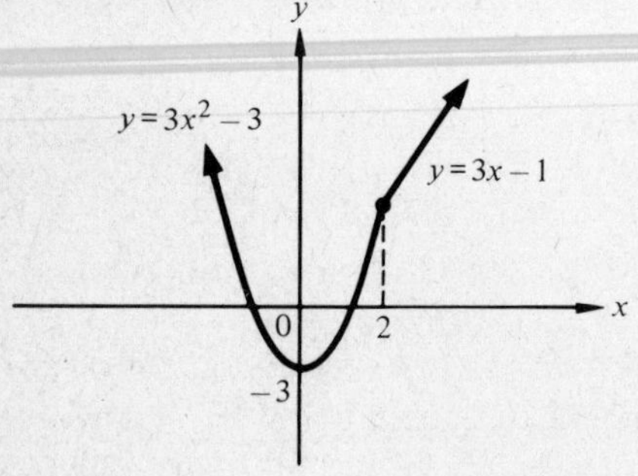

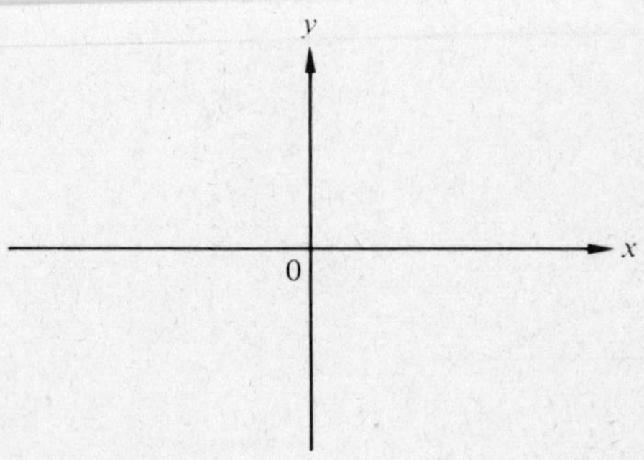

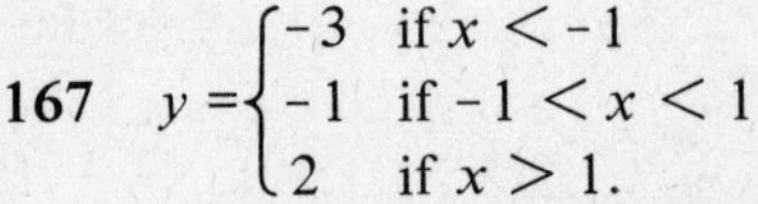

167 $y = \begin{cases} -3 & \text{if } x < -1 \\ -1 & \text{if } -1 < x < 1 \\ 2 & \text{if } x > 1. \end{cases}$

-1 and 1

$(-\infty, -1)$

$(-1, 1)$, $(1, \infty)$

-3, -1, 2

-3, -1, 2

The function is defined for all values of x except for ____________; hence, the domain consists of the three intervals ____________, ____________, and __________. The dependent variable y takes on the values _____, _____, and _____, so that the range of the function consists only of the numbers _____, _____, and _____. The graph is

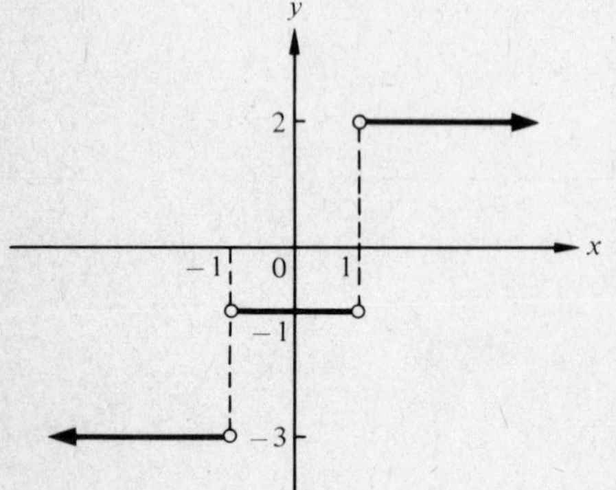

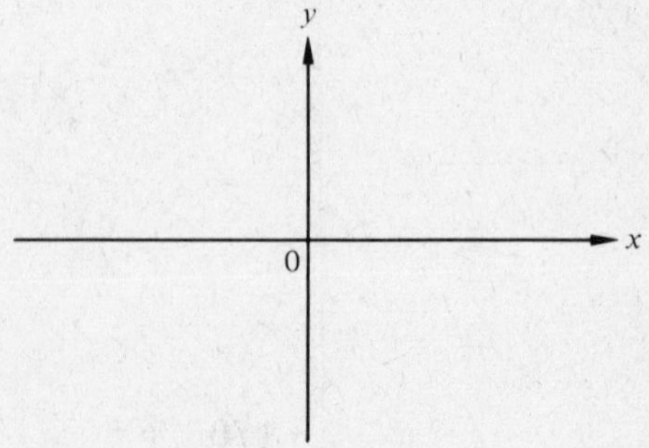

168 $y = \dfrac{x^2 - 1}{x - 1}.$

$x = 1$

$(-\infty, 1)$

The function is defined for all values of x except for __________; hence, the domain consists of the two intervals ______________

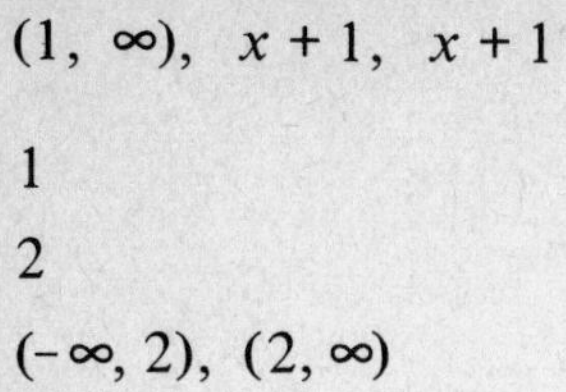

$(1, \infty)$, $x + 1$, $x + 1$

1

2

$(-\infty, 2)$, $(2, \infty)$

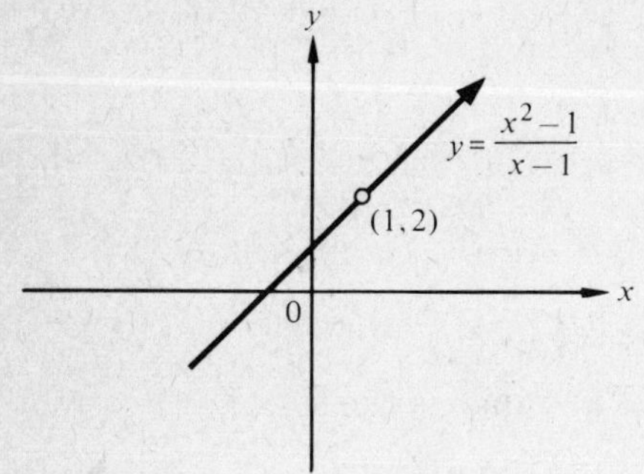

and ________. Since $\dfrac{x^2 - 1}{x - 1}$ = ________, y = ________ for $x \neq$ ____. The range of the function is all real numbers except for ____, so that the range consists of the two intervals ________ and ________. The graph is

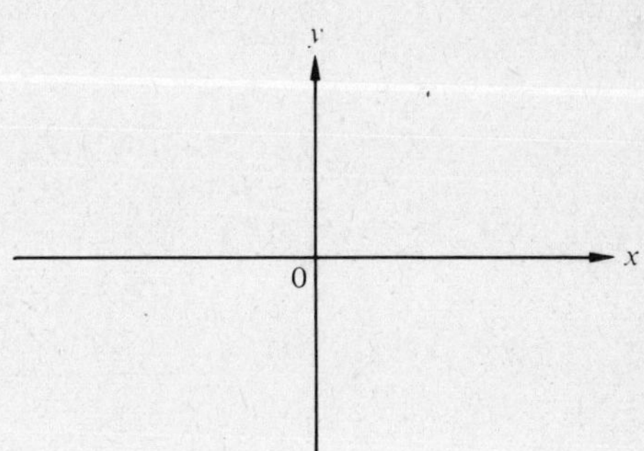

$x - 2$

$-1, -5$

$(-\infty, -5)$, $(-5, -1)$, $(-1, \infty)$

$x - 2$

$(-1, -3)$

$(-7, -3)$, $(-3, \infty)$

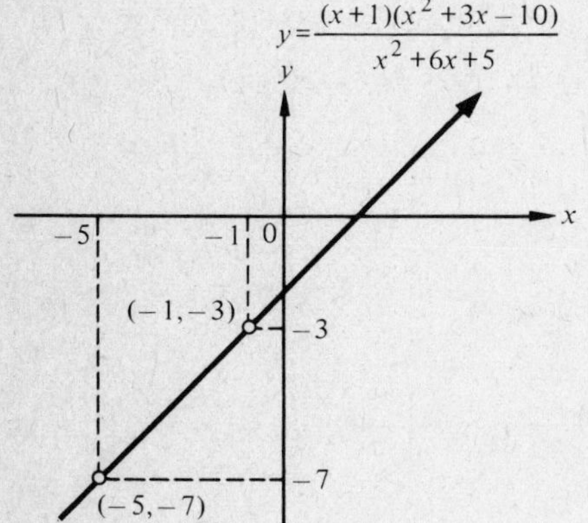

169 $y = \dfrac{(x + 1)(x^2 + 3x - 10)}{x^2 + 6x + 5}$.

Factoring the numerator and the denominator, we have $y = \dfrac{(x + 1)(x + 5)(\_\_\_\_\_\_)}{(x + 1)(x + 5)}$, so that the function is defined for all values of x except ____ and ____. The domain consists of the intervals ________, ________, and ________. Hence, the graph consists of all points on the line y = ________ except for the points $(-5, -7)$ and ________. The range consists of the intervals $(-\infty, -7)$, ________, and ________. The graph is

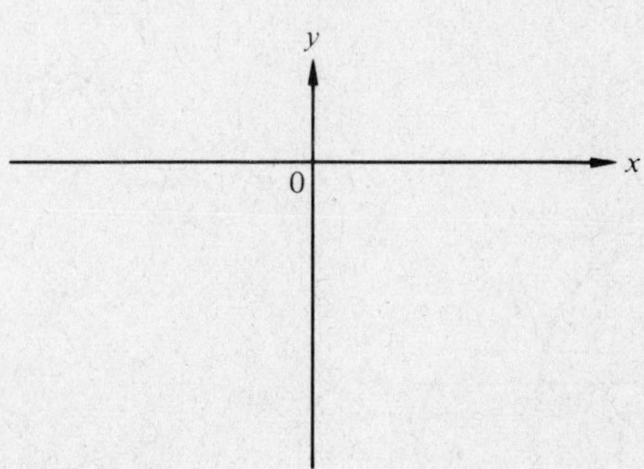

-1, 1

$(-1, 1)$, $(1, \infty)$

$(-\infty, 0]$, $(1, \infty)$

170 $y = \dfrac{x^2}{x^2 - 1}$.

The domain of the function is the set of all real numbers except ____ and ____. Therefore, it consists of the intervals $(-\infty, -1)$, ________, and ________. The range consists of the two intervals ________ and ________. The graph is

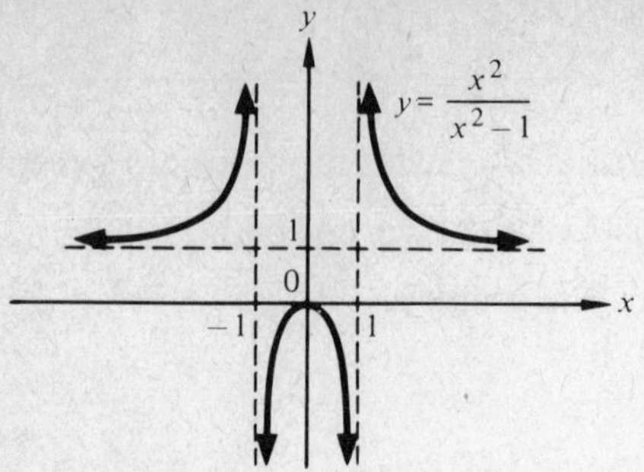

171 $y = |x - 2|$.

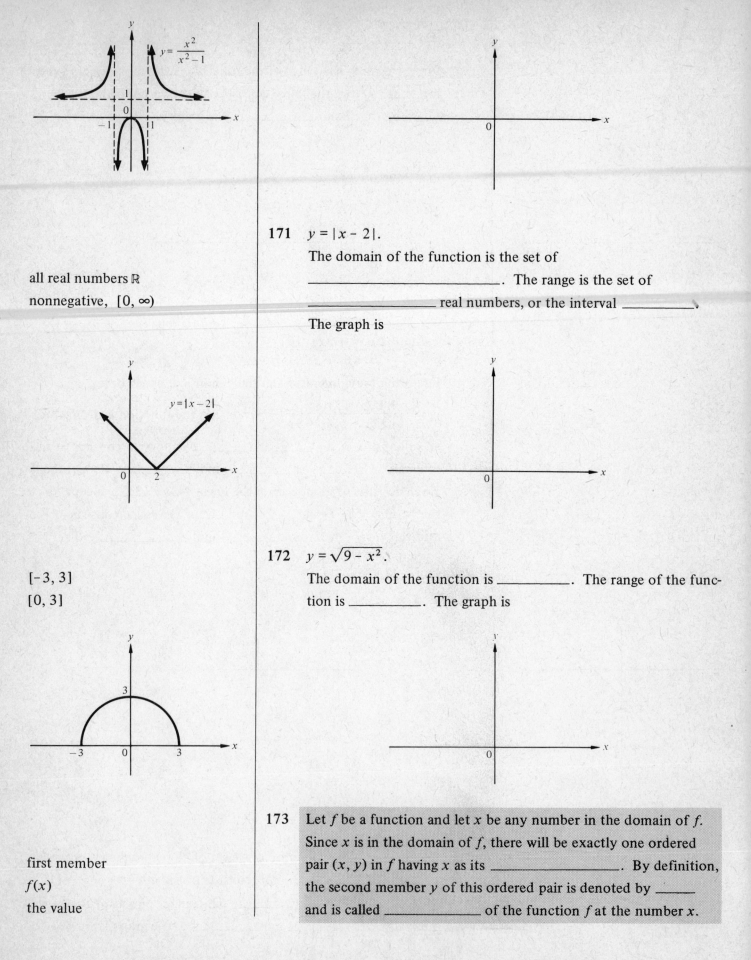

The domain of the function is the set of ______________________. The range is the set of ______________ real numbers, or the interval __________.

The graph is

all real numbers $\mathbb{R}$

nonnegative, $[0, \infty)$

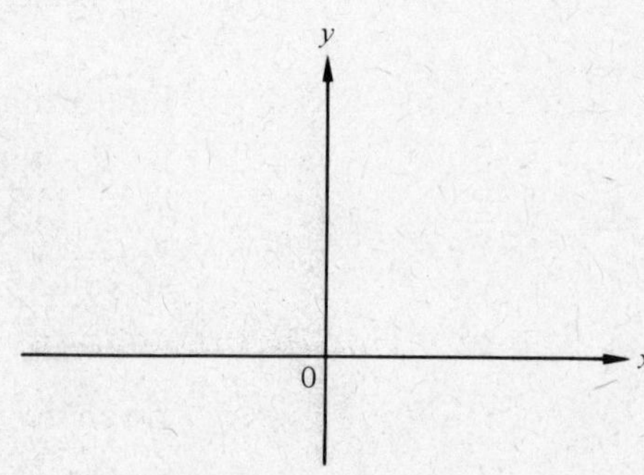

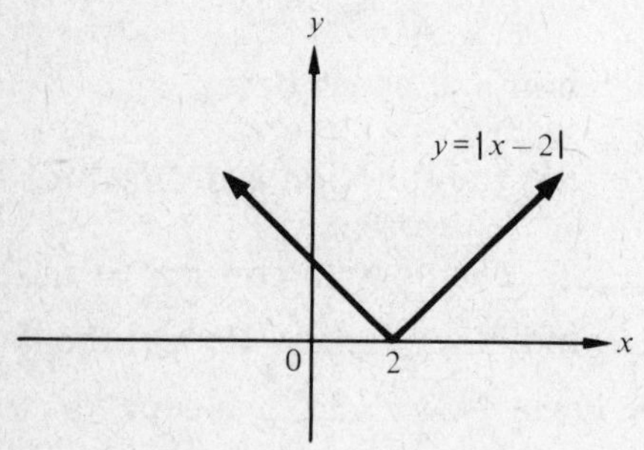

172 $y = \sqrt{9 - x^2}$.

The domain of the function is __________. The range of the function is __________. The graph is

$[-3, 3]$

$[0, 3]$

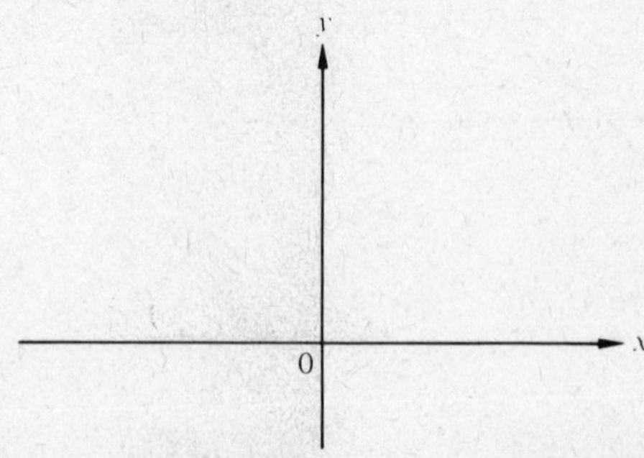

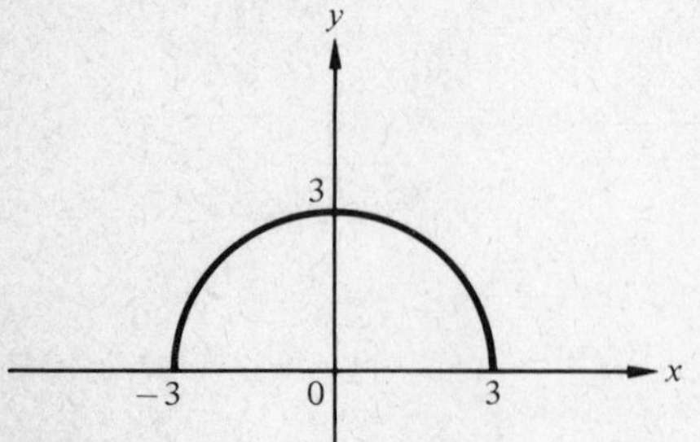

173 Let f be a function and let x be any number in the domain of f. Since x is in the domain of f, there will be exactly one ordered pair (x, y) in f having x as its ______________. By definition, the second member y of this ordered pair is denoted by _____ and is called ______________ of the function f at the number x.

first member

$f(x)$

the value

In Problems 174 through 183, find the values of each function and indicate the domain of the function.

174 $f(-2)$, $f(2)$, $f(0)$, $f(-1)$, and $f(1)$ if $f(x) = x^2$.

The domain of f is the set of ____________.

$f(-2) = (\_\_\_)^2 = \_\_\_$,

$f(2) = (\_\_\_)^2 = \_\_\_$,

$f(0) = (\_\_\_)^2 = \_\_\_$,

$f(-1) = (\_\_\_)^2 = \_\_\_$,

$f(1) = (\_\_\_)^2 = \_\_\_$.

real numbers ℝ

-2, 4

2, 4

0, 0

-1, 1

1, 1

175 $g(-4)$, $g(4)$, and $g(0)$ if $g(x) = \sqrt{16 - x^2}$.

The domain of g is ________.

$g(-4) = \sqrt{16 - (\_\_\_)^2} = \_\_\_$,

$g(4) = \sqrt{16 - (\_\_\_)^2} = \_\_\_$,

$g(0) = \sqrt{16 - (\_\_\_)^2} = \_\_\_$.

$[-4, 4]$

-4, 0

4. 0

0, 4

176 $f(-4)$, $f(2)$, and $f\left(\frac{1}{2}\right)$ if $f(x) = 3x^2 - 4$.

The domain of f is the set ____________.

$f(-4) = 3(\_\_\_)^2 - 4 = \_\_\_$,

$f(2) = 3(\_\_\_)^2 - 4 = \_\_\_$,

$f\left(\frac{1}{2}\right) = 3\left(\_\_\_\right)^2 - 4 = \_\_\_$.

of real numbers ℝ

-4, 44

2, 8

$\frac{1}{2}$, $\frac{-13}{4}$

177 $h(-2)$, $h(0)$, and $h(2)$ if $h(x) = \sqrt{2x^2 + 1}$.

The domain of h is the set of ____________.

$h(-2) = \sqrt{2(\_\_\_)^2 + 1} = \_\_\_$,

$h(0) = \sqrt{2(\_\_\_)^2 + 1} = \_\_\_$,

$h(2) = \sqrt{2(\_\_\_)^2 + 1} = \_\_\_$.

all real numbers ℝ

-2, 3

0, 1

2, 3

178 $g(10)$, $g(5)$, and $g(2)$ if $g(x) = \frac{1}{\sqrt{x - 1}}$.

The domain of g is ________.

$g(10) = \frac{1}{\sqrt{\_\_\_ - 1}} = \_\_\_$,

$g(5) = \frac{1}{\sqrt{\_\_\_ - 1}} = \_\_\_$,

$g(2) = \frac{1}{\sqrt{\_\_\_ - 1}} = \_\_\_$.

$(1, \infty)$

10, $\frac{1}{3}$

5, $\frac{1}{2}$

2, 1

179 $F(-3)$, $F(0)$, and $F(3)$ if $F(x) = |2x - 1|$.

The domain of F is the set of ____________.

$F(-3) = |2(\_\_\_) - 1| = \_\_\_$,

$F(0) = |2(\_\_\_) - 1| = \_\_\_$,

$F(3) = |2(\_\_\_) - 1| = \_\_\_$.

real numbers ℝ

-3, 7

0, 1

3, 5

180 $G(-1)$, $G\left(\frac{1}{2}\right)$, and $G(2a+3)$ if $G(x) = \sqrt{2x+3}$.

The domain of G is ______.

$G(-1) = \sqrt{2(\_\_\_) + 3} = \_\_\_$,

$G\left(\frac{1}{2}\right) = \sqrt{2(\_\_\_) + 3} = \_\_\_$,

$G(2a+3) = \sqrt{2(\_\_\_\_\_) + 3} = \_\_\_\_\_$.

$\left[-\frac{3}{2}, \infty\right)$

-1, 1

$\frac{1}{2}$, 2

$2a+3$, $\sqrt{4a+9}$

181 $f(1)$, $f(4)$, $f(a^2)$, and $f(a+b)$ if $f(x) = \sqrt{x}$.

The domain of f is ______,

$f(1) = \sqrt{\_\_\_} = \_\_\_$,

$f(4) = \sqrt{\_\_\_} = \_\_\_$,

$f(a^2) = \sqrt{\_\_\_} = \_\_\_$,

$f(a+b) = \sqrt{\_\_\_\_\_} = \_\_\_\_\_$.

$[0, \infty)$

1, 1

4, 2

a^2, $|a|$

$a+b$, $\sqrt{a+b}$

182 $f(2)$, $f(a^2)$, and $[f(a)]^2$ if $f(x) = \dfrac{3x-1}{1+2x}$.

The domain of f is the set of ______________ except $x =$ ____.

$f(2) = \dfrac{3(\_\_\_) - 1}{1 + 2(2)} = \_\_\_$,

$f(a^2) = \dfrac{3(\_\_\_) - 1}{1 + 2a^2} = \_\_\_\_\_$.

$[f(a)]^2 = \left(\_\_\_\_\_\right)^2 = \_\_\_\_\_$.

real numbers $\mathbb{R}$

$-\frac{1}{2}$

2, 1

a^2, $\dfrac{3a^2 - 1}{1 + 2a^2}$

$\dfrac{3a-1}{1+2a}$, $\dfrac{9a^2 - 6a + 1}{1 + 4a + 4a^2}$

183 **Let f be a function and suppose that x_0 is a fixed number in the domain of f. The difference quotient function for f at x_0 is the function q defined by the equation $q(h) = \dfrac{\_\_\_\_\_\_\_\_}{h}$.**

The formula $\dfrac{f(x_0 + h) - f(x_0)}{h}$ is called the ______________ of f at x_0.

$f(x_0 + h) - f(x_0)$

difference quotient

In Problems 184 through 187, write the difference quotient of f at x.

184 $f(x) = 7x + 11$.

$$\frac{f(x+h) - f(x)}{h} = \frac{[7(x+h) + 11] - (\_\_\_\_\_\_)}{h}$$

$$= \frac{\_\_\_}{h} = \_\_\_.$$

$7x + 11$

$7h$, 7

185 $f(x) = -x^2$.

$$\frac{f(x+h) - f(x)}{h} = \frac{-(x+h)^2 - \_\_\_}{h}$$

$-x^2$

x^2

$-2xh - h^2,\ -2x - h$

$x + 2$

$\sqrt{x+h+2} + \sqrt{x+2}$

h

$\sqrt{x+h+2} + \sqrt{x+2}$

$\dfrac{3}{x}$

$3(x+h)$

$-3h,\ \dfrac{-3}{x(x+h)}$

$\dfrac{\sqrt{3}}{2}x$

$\dfrac{\sqrt{3}}{2}x,\ \dfrac{\sqrt{3}}{4}x^2$

$36,\ \sqrt{36 - x^2}$

$2y,\ \sqrt{36 - x^2}$

$r,\ \dfrac{1}{3}\pi r^3$

$-x^2,\ \dfrac{-1 + \sqrt{1 + 4x^2}}{2}$

$$= \frac{-x^2 - 2xh - h^2 + \underline{\qquad}}{h}$$

$$= \frac{\underline{\qquad\qquad}}{h} = \underline{\qquad\qquad}.$$

186 $f(x) = \sqrt{x+2}$.

$$\frac{f(x+h) - f(x)}{h} = \frac{\sqrt{x+h+2} - \sqrt{\underline{\qquad}}}{h}$$

$$= \frac{(\sqrt{x+h+2} - \sqrt{x+2})(\underline{\qquad\qquad\qquad})}{h(\sqrt{x+h+2} + \sqrt{x+2})}$$

$$= \frac{\underline{\qquad}}{h(\sqrt{x+h+2} + \sqrt{x+2})}$$

$$= \frac{1}{\underline{\qquad\qquad\qquad}}.$$

187 $f(x) = \dfrac{3}{x}$.

$$\frac{f(x+h) - f(x)}{h} = \frac{\frac{3}{x+h} - \left(\underline{\quad}\right)}{h}$$

$$= \frac{3x - (\underline{\qquad\qquad})}{hx(x+h)}$$

$$= \frac{\underline{\qquad}}{hx(x+h)} = \underline{\qquad\qquad}.$$

188 Express the area A of an equilateral triangle as a function of its side of length x.

Let x be the base of the triangle. Then its height is $\underline{\qquad}$, so that $A = \dfrac{1}{2}(x)\left(\underline{\qquad}\right) = \underline{\qquad}$.

189 A rectangle of dimensions $2x$ by $2y$ is inscribed in a circle of radius 6 inches. Express the area A of the rectangle as a function of x.

$x^2 + y^2 = \underline{\quad}$, so that $y = \underline{\qquad\qquad}$. Thus,

$A = (2x)(\underline{\quad}) = 4x(\underline{\qquad\qquad})$.

190 A lateral circular cone has altitude equal to r, the radius of the base. Express the volume V of the cone as a function of r.

$V = \dfrac{1}{3}\pi r^2(\underline{\quad}) = \underline{\qquad}$.

191 For each x, let y be the nonnegative root of the quadratic equation $y^2 + y - x^2 = 0$. Express y as a function of x.

$$y = \frac{-1 + \sqrt{1 - 4(\underline{\quad})}}{2} = \underline{\qquad\qquad}.$$

5.1 Functions as Sets of Ordered Pairs

ordered
first
domain
range

192 A function is a set of ________ pairs in which no two different ordered pairs have the same ____ member. The set of the first members of the ordered pairs in the function is the ________ of the function, while the set of all second members of these ordered pairs is the ________ of the function.

In Problems 193 through 201, indicate which of the following is a function.

193 The set of all ordered pairs with $y = 3x + 1$.

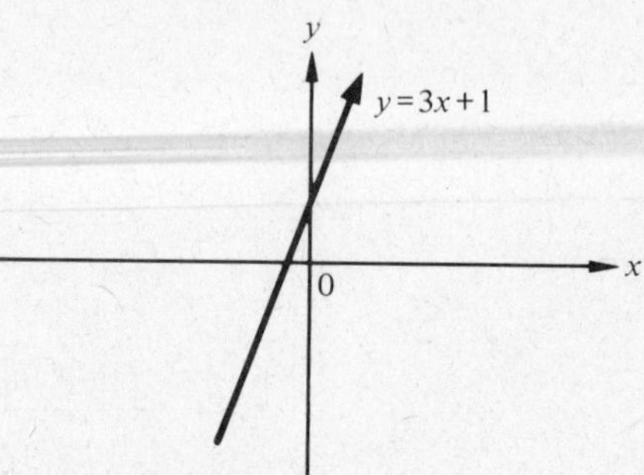

same abscissa
function

We see from the graph of $y = 3x + 1$ that no two different points on the graph can have the ________________. Therefore, the set of all ordered pairs with $y = 3x + 1$ is a ____________.

194 The set of all ordered pairs with $y = -x^2$.

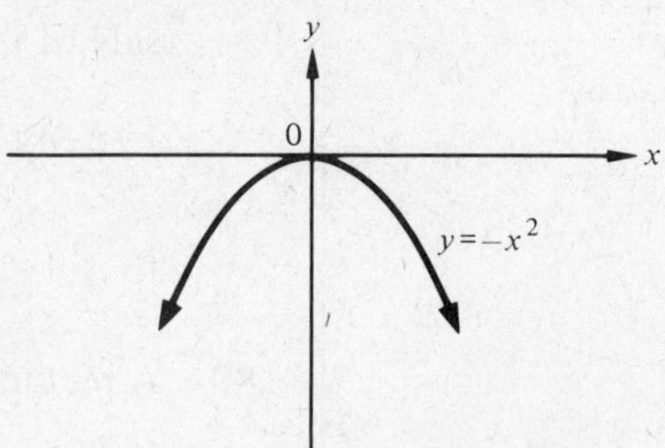

two different points
abscissa
function

We see from the graph of $y = -x^2$ that no ________________ on the graph can have the same ____________.
Therefore, the set of all ordered pairs with $y = -x^2$ is a ________.

195 The set of ordered pairs with $x = 3y^2$.

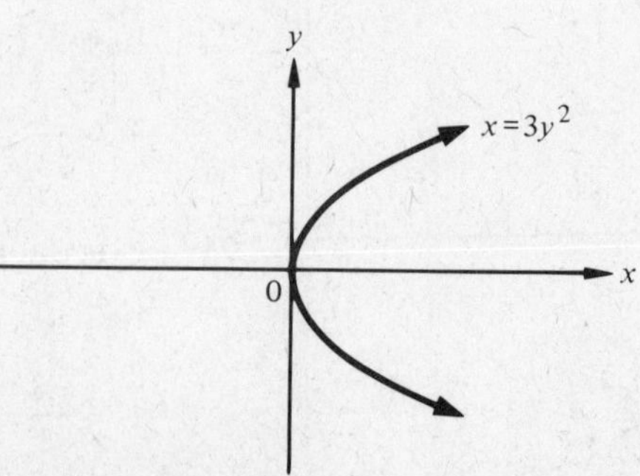

From the graph of $x = 3y^2$, we notice that there are two different points on the curve that have ________________, so that the curve cannot be the graph of a ________.

the same abscissa
function

196 The set of ordered pairs with $x = -2y^2$.

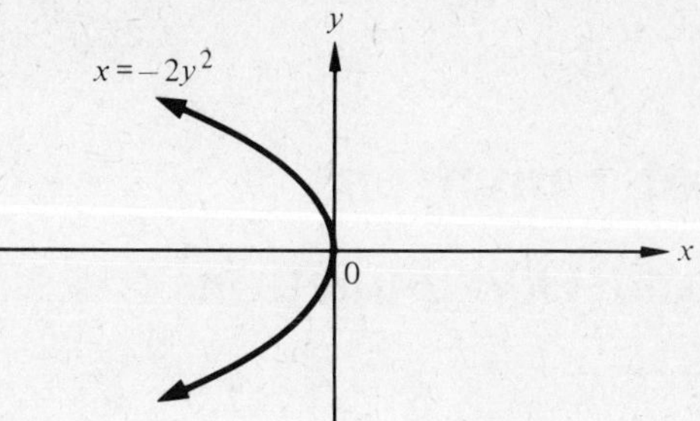

From the graph of $x = -2y^2$, we notice that there are two ________ on the curve having ________________, so that the curve cannot be the graph of a ________.

different points, the same abscissa
function

197 $5x + 13$.
Since this expression is not a set of ________, then it cannot be a ________.

ordered pairs
function

198 The set consisting of $(0, 3)$ and $(0, -3)$.
Since the two ordered pairs have the ________ first member, this set is not ________.

same
a function

199 The set consisting of $(-1, 5)$ and $(1, 5)$.
Since the two ordered pairs do not have the ________ first member, this set is ________.

same
a function

200 The set consisting of $(3, 1)$ and $(5, 2)$.
Since the two ordered pairs do not have the ________ first member, this set is ________.

same
a function

201 The set of all ordered pairs with $y \leqslant 3x$.
This set is ________________, since we can find two points with the ________________.

not a function
same abscissa

In Problems 202 through 206, state which equations define a function.

202 $x^2 + y^2 = 9$.
This equation ________ define a function, since we can find two points on the circle with the same ________.

does not
abscissa

203 $3x^2 + 4y^2 = 12$.
This equation ________ define a function, since we can find two points on its graph with the same ________.

does not
abscissa

204 $y = x^2 - 2$.
The equation defines ________, since we cannot find two points on its graph with the same ________.

a function
abscissa

over, under

205 The graph of a function cannot pass ______ or ______ itself.

graph
abscissas
ordinates

206 The domain and range of a function can be found from its ______. Thus, the domain of a function f is the set of all ______ of points on its graph, while the range of a function f is the set of all ______ of points on its graph.

6 Types of Functions

6.1 Even and Odd Functions

$f(-x) = f(x)$

207 A function f is said to be an even function if, for every number x in the domain of f, $-x$ is also in the domain of f and ______.

y

208 If f is an even function, then the graph of f is symmetric about the ______ axis.

$-f(x)$

209 A function f is said to be an odd function if, for every number x in the domain of f, $-x$ is also in the domain of f and $f(-x) =$ ______.

origin

210 If f is an odd function, then the graph of f is symmetric about the ______.

In Problems 211 through 221, determine whether each function is even or odd; also discuss the symmetry in each case.

$-x$, $5x^2$, even
y

211 $f(x) = 5x^2$.
$f(-x) = 5$ (______$)^2 =$ ______ $= f(x)$, so that f is an ______ function. The graph of f is symmetric about the ______ axis.

$-x$, $x^2 + 8$, even
y

212 $g(x) = x^2 + 8$.
$g(-x) = ($______$)^2 + 8 =$ ______ $= g(x)$, so that g is an ______ function. The graph of g is symmetric about the ______ axis.

$-x$, $-3x^4$, even
y

213 $h(x) = -3x^4$.
$h(-x) = -3($______$)^4 =$ ______ $= h(x)$, so that h is an ______ function. The graph of h is symmetric about the ______ axis.

$-x$, $|5x|$, $F(x)$
even, y

214 $F(x) = |-5x|$.
$F(-x) = |-5($______$)| =$ ______ $=$ ______, so that F is an ______ function. The graph of F is symmetric about the ______ axis.

$-x$, $-8x$, $f(x)$, odd
origin

215 $f(x) = 8x$.
$f(-x) = 8($______$) =$ ______ $= -($______$)$, so that f is an ______ function. The graph of f is symmetric about the ______.

$-x$, $-x^3|x|$, odd
origin

216 $g(x) = x^3|x|$.
$g(-x) = (-x)^3|$______$| =$ ______ $= -g(x)$, so that g is an ______ function. The graph of g is symmetric about the ______.

$-x,\ -x,\ x(|2x| - x^2)$
$h(x)$, odd
origin

217 $h(x) = x(|2x| - x^2)$.
$h(-x) = (\_\_\_\_)(|-2x| - (\_\_\_\_)^2 = - \_\_\_\_\_\_\_\_\_\_ =$
$-(\_\_\_\_)$, so that h is an ________ function. The graph of h is symmetric about the ________.

$-x,\ x + \frac{1}{x},\ f(x)$
odd, origin

218 $f(x) = x + \frac{1}{x}$.
$f(-x) = \_\_\_\_ - \frac{1}{x} = -\left(\_\_\_\_\_\_\right) = - \_\_\_\_\_$, so that f is an ____ function. The graph of f is symmetric about the ________.

$-x,\ -x + 5,\ f(x)$
even, $-f(x)$
odd

219 $f(x) = x + 5$.
$f(-x) = \_\_\_\_ + 5 = \_\_\_\_\_\_\_ \neq \_\_\_\_\_\_$, so that f is not an ________ function. Also, $f(-x) \neq$ ________. Therefore, f is not an ____ function.

$-x,\ g(x)$, even
$-g(x)$, odd

220 $g(x) = \sqrt{4x}$.
$g(-x) = \sqrt{4(\_\_\_\_)} \neq$ ______, so that g is not an ________ function. Also, $g(-x) \neq$ ________. Therefore, g is not an ______ function.

$-x,\ h(x)$, even
$h(x)$
function

221 $h(x) = x^2 - 5x$.
$h(-x) = (\_\_\_\_)^2 + 5x \neq$ ________, so that h is not an ________ function. Also, $h(-x) \neq -$ ________. Therefore, h is not an odd ________.

6.2 Polynomial Functions

polynomial
degree
coefficients

222 A function f which is expressed by an equation of the form
$f(x) = a_n x^n + a_{n-1}x^{n-1} + a_{n-2}x^{n-2} + \cdots + a_1 x + a_0$,
where n is a nonnegative integer, $a_n, a_{n-1}, a_{n-2}, \ldots, a_1, a_0$ are real numbers, and $a_n \neq 0$ is called a ____________ function of ________ n in the variable x. The numbers $a_n, a_{n-1}, a_{n-2}, \ldots, a_1, a_0$ are called the ____________ of the polynomial function.

undefined

223 A function f defined by the equation $f(x) = 0$ is a polynomial function whose degree is ____________.

constant

224 A function f defined by an equation of the form $f(x) = a$, where a is a fixed real number, is called a ____________ function.

$\mathbb{R}$, 2
2, 2, 2, 2

225 Let $f(x) = 2$. Find $f(-2)$, $f(-1)$, $f(0)$, $f(1)$, and $f(2)$; also find the domain and range of f.
The domain of f is ____; the range of f is the set $\{\_\_\_\_\}$.
$f(-2) = \_\_\_\_$, $f(-1) = \_\_\_\_$, $f(0) = \_\_\_\_$, $f(1) = \_\_\_\_$, and

2

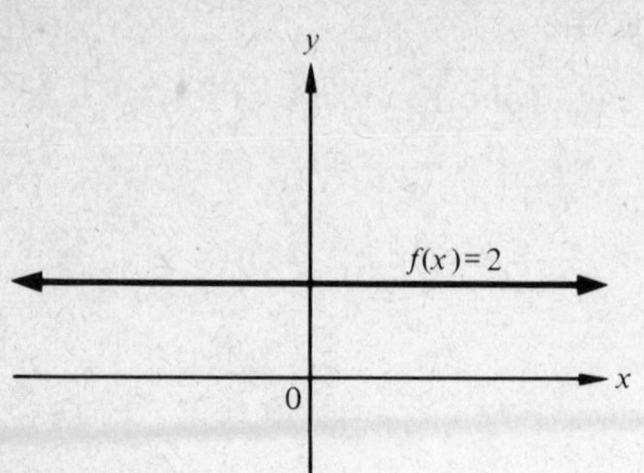

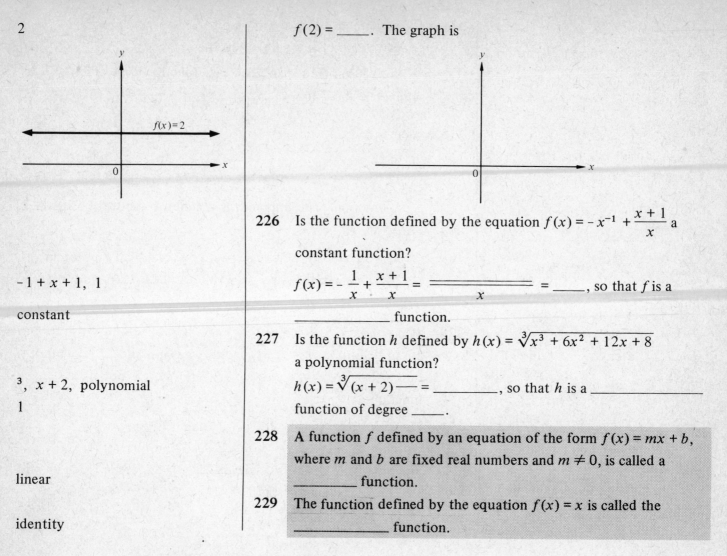

$f(2) =$ _____. The graph is

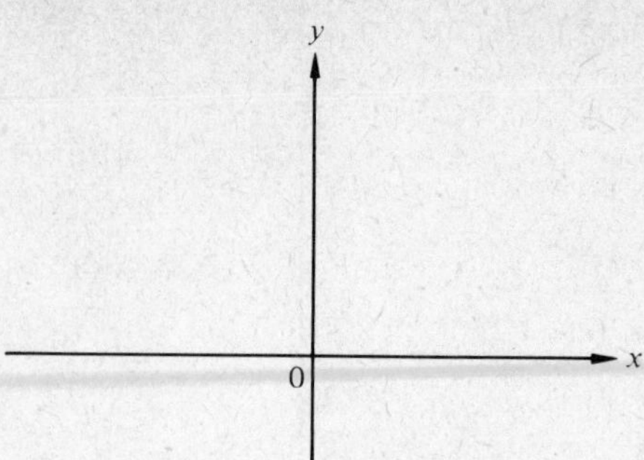

226 Is the function defined by the equation $f(x) = -x^{-1} + \dfrac{x+1}{x}$ a constant function?

$f(x) = -\dfrac{1}{x} + \dfrac{x+1}{x} = \dfrac{}{x} =$ _____, so that f is a __________ function.

$-1 + x + 1,\ 1$

constant

227 Is the function h defined by $h(x) = \sqrt[3]{x^3 + 6x^2 + 12x + 8}$ a polynomial function?

$h(x) = \sqrt[3]{(x+2)^{\_\_}} =$ _______, so that h is a __________ function of degree _____.

^{3}, $x + 2$, polynomial

1

228 **A function f defined by an equation of the form $f(x) = mx + b$, where m and b are fixed real numbers and $m \neq 0$, is called a ________ function.**

linear

229 **The function defined by the equation $f(x) = x$ is called the __________ function.**

identity

In Problems 230 through 233, find the domain and range of each function; also, find the values indicated for each function.

230 $f(-2)$, $f(-1)$, $f(0)$, $f(1)$, and $f(2)$ if $f(x) = 5x - 1$.

The domain of f is _____ and the range of f is _____.

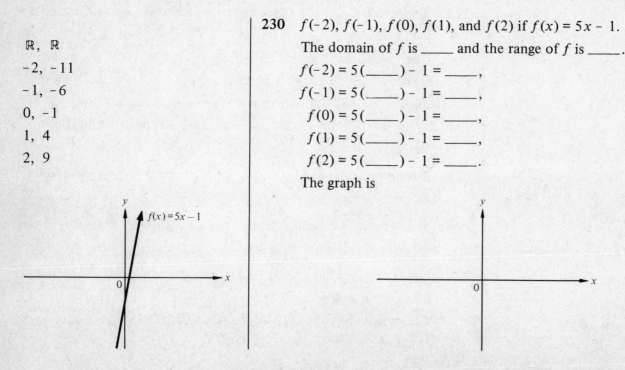

$f(-2) = 5($_____$) - 1 =$ _____,

$f(-1) = 5($_____$) - 1 =$ _____,

$f(0) = 5($_____$) - 1 =$ _____,

$f(1) = 5($_____$) - 1 =$ _____,

$f(2) = 5($_____$) - 1 =$ _____.

The graph is

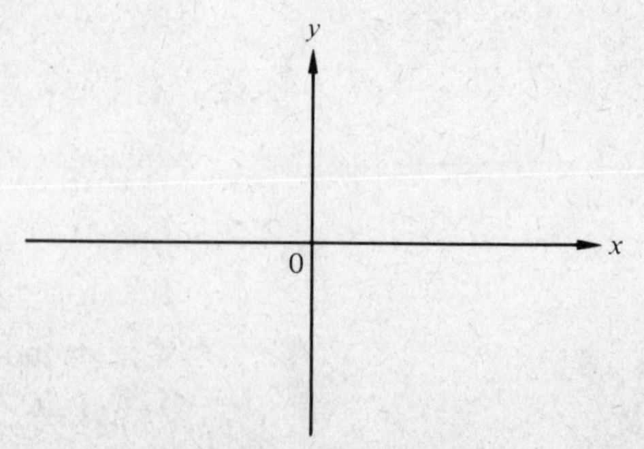

$\mathbb{R}$, $\mathbb{R}$

$-2, -11$

$-1, -6$

$0, -1$

$1, 4$

$2, 9$

y
$f(x) = 5x - 1$
0
x

231 $f(-1)$, $f(0)$, and $f(1)$ if $f(x) = -2x + 3$.

The domain of f is ____ and the range is ____.

$f(-1) = -2(\_\_\_\_) + 3 = \_\_\_\_,$

$f(0) = -2(\_\_\_\_) + 3 = \_\_\_\_,$

$f(1) = -2(\_\_\_\_) + 3 = \_\_\_\_.$

The graph is

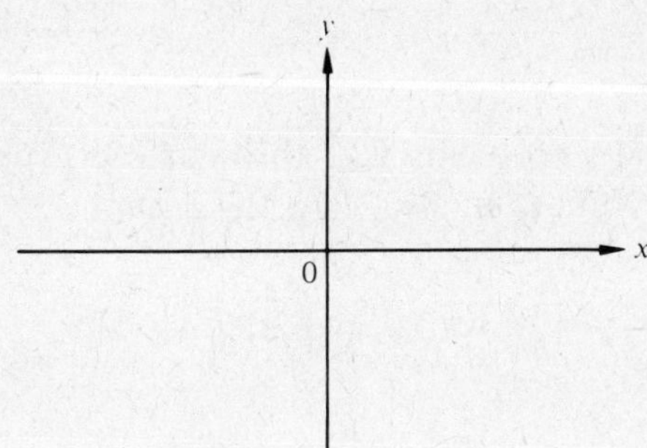

Answers:

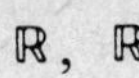
ℝ, ℝ

$-1, 5$

$0, 3$

$1, 1$

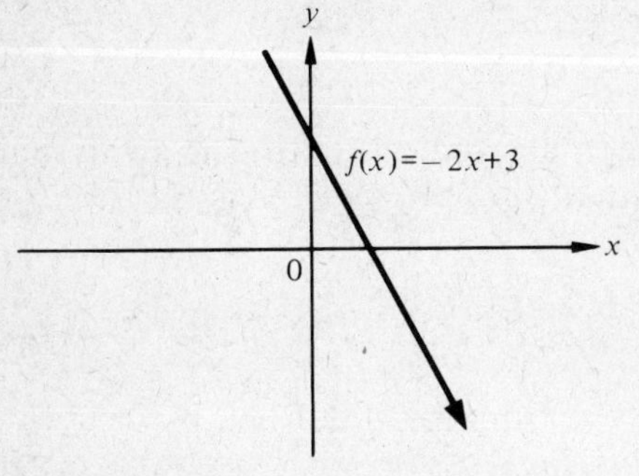

232 $f(-1)$, $f(0)$, and $f(1)$ if $f(x) = -x$.

The domain of f is ____ and the range of f is ____.

$f(-1) = -(\_\_\_\_) = \_\_\_\_,$

$f(0) = -(\_\_\_\_) = \_\_\_\_,$

$f(1) = -(\_\_\_\_) = \_\_\_\_.$

The graph is

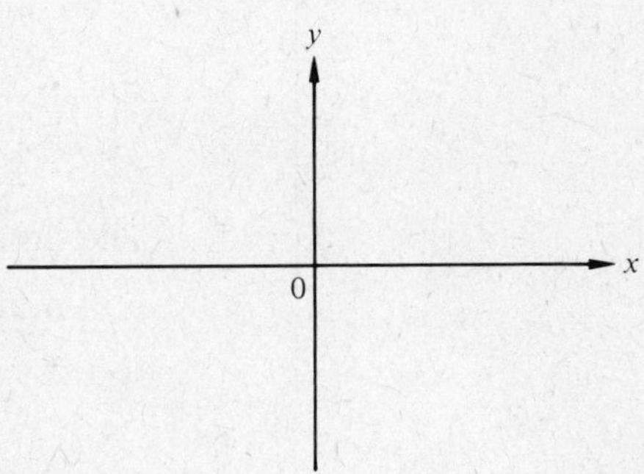

Answers:

ℝ, ℝ

$-1, 1$

$0, 0$

$1, -1$

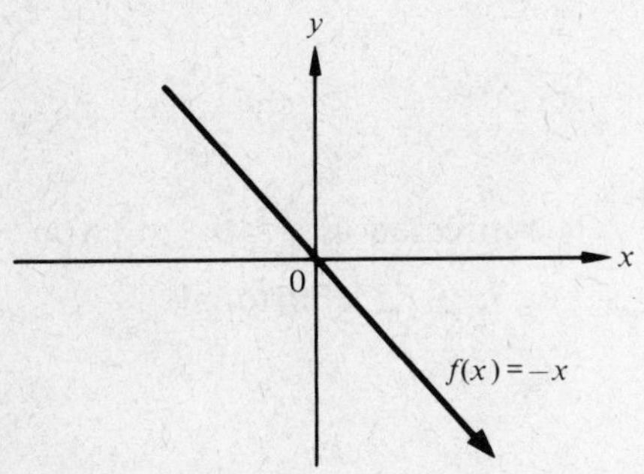

233 Given a linear function f so that $f(1) = 4$, and $f(2) = 5$, write the equation that determines f.

Since f is linear, $f(x) = \_\_\_\_\_\_$. $f(1) = m(\_\_\_\_) + b = \_\_\_\_$, while $f(2) = m(\_\_\_\_) + b = \_\_\_\_$. Solving the equations simultaneously for m and b, we obtain $m = \_\_\_\_$ and $b = \_\_\_\_$, so that $f(x) = \_\_\_\_\_\_$.

Answers:

$mx + b$, 1, 4

2, 5

1, 3

$x + 3$

6.3 Rational Functions and Algebraic Functions

234 A function f which is defined by an equation of the form $f(x) = \frac{\_\_\_\_}{q(x)}$, where $p(x)$ and $q(x)$ are polynomial functions and $q(x)$ is not the constant zero function, is called a ______ function.

Answers:

$p(x)$

rational

235 The function f defined by the equation $f(x) = |x|$ is called the ____________ function.

absolute value

236 If a function f can be formed by a finite number of algebraic operations (that is, addition, subtraction, multiplication, division, and extraction of positive integral roots), starting with the identity function and constant functions, it is called an elementary ____________.

algebraic function

In Problems 237 through 239, specify whether or not each algebraic function is a rational function; also, find the domain.

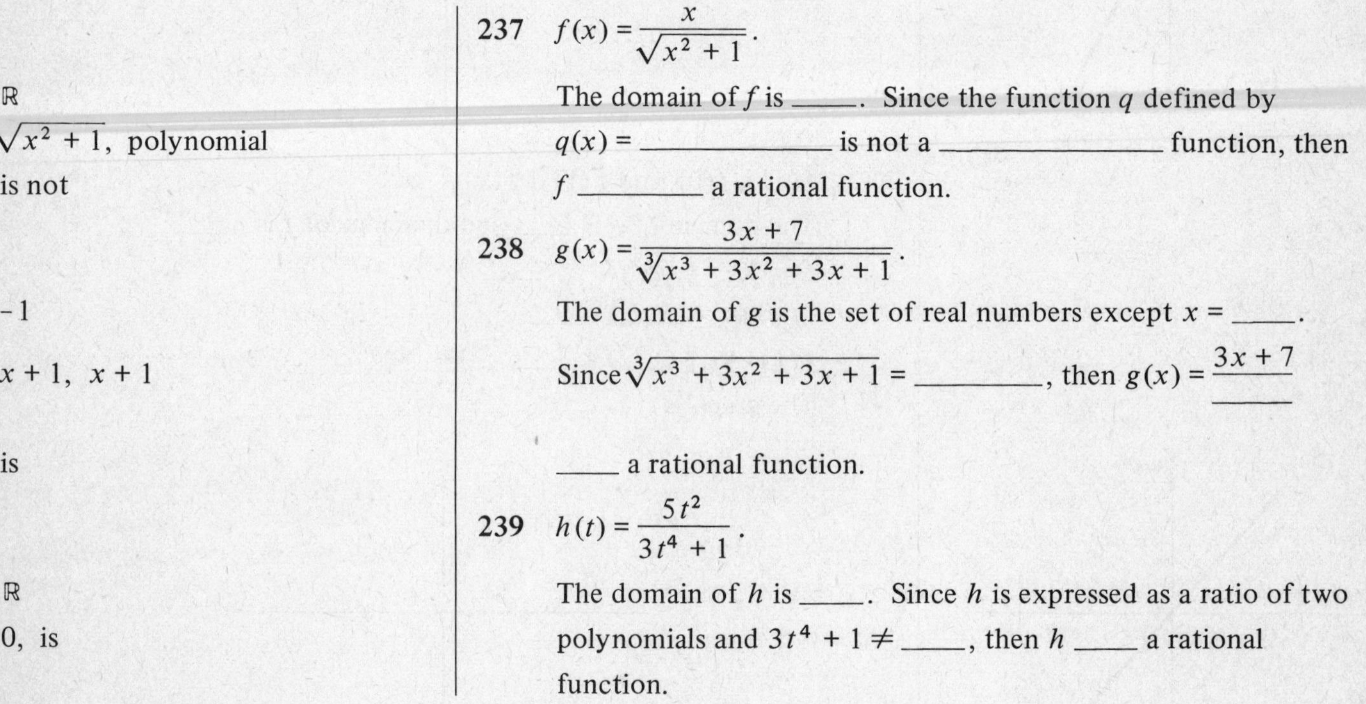

237 $f(x) = \dfrac{x}{\sqrt{x^2 + 1}}$.

The domain of f is ____. Since the function q defined by $q(x) =$ ________ is not a ________ function, then f ______ a rational function.

$\mathbb{R}$

$\sqrt{x^2 + 1}$, polynomial

is not

238 $g(x) = \dfrac{3x + 7}{\sqrt[3]{x^3 + 3x^2 + 3x + 1}}$.

The domain of g is the set of real numbers except $x =$ ____.

Since $\sqrt[3]{x^3 + 3x^2 + 3x + 1} =$ ______, then $g(x) = \dfrac{3x + 7}{\_\_\_\_}$

____ a rational function.

-1

$x + 1$, $x + 1$

is

239 $h(t) = \dfrac{5t^2}{3t^4 + 1}$.

The domain of h is ____. Since h is expressed as a ratio of two polynomials and $3t^4 + 1 \neq$ ____, then h ____ a rational function.

$\mathbb{R}$

0, is

6.4 Discontinuous Functions

240 If x is a real number, then the symbol $[\![x]\!]$ denotes the greatest integer not exceeding x. Thus, $[\![x]\!]$ is an integer and $[\![x]\!] \leqslant$ ____ $\leqslant [\![x]\!] + 1$. The function defined by the equation $f(x) = [\![x]\!]$ is called the ____________ function. Its domain is the set ____ and the range is the set of ________.

x

greatest integer

$\mathbb{R}$, integers

In Problems 241 through 243, indicate the domain and range of each function.

241 $f(x) = [\![3x^2]\!]$.

The domain of f is ____ and the range of f is the set of ____________.

$\mathbb{R}$

nonnegative integers

242 $f(x) = [\![x + 3]\!]$.

The domain of f is ____ and the range is the set of ________.

$\mathbb{R}$, integers

243 $f(x) = \dfrac{[\![x]\!]}{x}$.

0 | The domain of f is the set of real numbers except $x =$ ____, and

$\left(\frac{1}{2}, 1\right]$, $[1, \infty)$ | the range consists of 0, ________, and ________.

7 Trigonometric Functions

244 If P is a point on a unit circle and $P(t) = (x, y)$, then the values at t of the six trigonometric functions are defined as follows:

x, $\frac{1}{x}$ | $\cos t =$ ____, $\sec t =$ ____,

y, $\frac{1}{y}$ | $\sin t =$ ____, $\csc t =$ ____,

$\frac{y}{x}$, $\frac{x}{y}$ | $\tan t =$ ____, $\cot t =$ ____.

In Problems 245 through 268, evaluate the six trigonometric functions at the given value of t.

245 $t = \dfrac{\pi}{6}$.

30° | $\frac{\pi}{6} =$ ____; the terminal side of a 30° angle contains the point

$\left(\frac{\sqrt{3}}{2}, \frac{1}{2}\right)$ | ________, so that

$\frac{\sqrt{3}}{2}$, $\frac{2}{\sqrt{3}}$ | $\cos \frac{\pi}{6} =$ ____, $\sec \frac{\pi}{6} =$ ____,

$\frac{1}{2}$, 2 | $\sin \frac{\pi}{6} =$ ____, $\csc \frac{\pi}{6} =$ ____,

$\frac{1}{\sqrt{3}}$, $\sqrt{3}$ | $\tan \frac{\pi}{6} =$ ____, $\cot \frac{\pi}{6} =$ ____.

246 $t = \dfrac{\pi}{3}$.

60° | $\frac{\pi}{3} =$ ____; the terminal side of a 60° angle contains the point

$\left(\frac{1}{2}, \frac{\sqrt{3}}{2}\right)$ | ________, so that

$\frac{1}{2}$, 2 | $\cos \frac{\pi}{3} =$ ____, $\sec \frac{\pi}{3} =$ ____,

$\frac{\sqrt{3}}{2}, \frac{2}{\sqrt{3}}$

$\sqrt{3}, \frac{1}{\sqrt{3}}$

$\frac{\pi}{4}$

$\left(\frac{\sqrt{2}}{2}, \frac{\sqrt{2}}{2}\right)$

$\frac{\sqrt{2}}{2}, \sqrt{2}$

$\frac{\sqrt{2}}{2}, \sqrt{2}$

1, 1

$\frac{4\pi}{3}$

$\left(-\frac{1}{2}, -\frac{\sqrt{3}}{2}\right)$

$-\frac{1}{2}, -2$

$-\frac{\sqrt{3}}{2}, -\frac{2}{\sqrt{3}}$

$\sqrt{3}, \frac{1}{\sqrt{3}}$

$\left(-\frac{\sqrt{2}}{2}, -\frac{\sqrt{2}}{2}\right)$

$-\frac{\sqrt{2}}{2}, -\sqrt{2}$

$-\frac{\sqrt{2}}{2}, -\sqrt{2}$

1, 1

$\sin\frac{\pi}{3} =$ ____, $\csc\frac{\pi}{3} =$ ____,

$\tan\frac{\pi}{3} =$ ____, $\cot\frac{\pi}{3} =$ ____.

247 $t = \frac{17\pi}{4}$.

$\frac{17\pi}{4} = 4\pi +$ ____. The terminal side of the angle $\frac{17\pi}{4}$ contains the point ________, so that

$\cos\frac{17\pi}{4} =$ ____, $\sec\frac{17\pi}{4} =$ ____,

$\sin\frac{17\pi}{4} =$ ____, $\csc\frac{17\pi}{4} =$ ____,

$\tan\frac{17\pi}{4} =$ ____, $\cot\frac{17\pi}{4} =$ ____.

248 $t = \frac{82\pi}{3}$.

$\frac{82\pi}{3} = 26\pi +$ ____. The terminal side of the angle $\frac{82\pi}{3}$ contains the point ________, so that

$\cos\frac{82\pi}{3} =$ ____, $\sec\frac{82\pi}{3} =$ ____,

$\sin\frac{82\pi}{3} =$ ____, $\csc\frac{82\pi}{3} =$ ____,

$\tan\frac{82\pi}{3} =$ ____, $\cot\frac{82\pi}{3} =$ ____.

249 $t = -\frac{91\pi}{4}$.

$-\frac{91\pi}{4} = -22\pi - \frac{3\pi}{4}$. The terminal side of the angle $-\frac{91\pi}{4}$ contains the point ________, so that

$\cos\left(-\frac{91\pi}{4}\right) =$ ____, $\sec\left(-\frac{91\pi}{4}\right) =$ ____,

$\sin\left(-\frac{91\pi}{4}\right) =$ ____, $\csc\left(-\frac{91\pi}{4}\right) =$ ____,

$\tan\left(-\frac{91\pi}{4}\right) =$ ____, $\cot\left(-\frac{91\pi}{4}\right) =$ ____.

Answer	Problem
$\sin t$	**250** $\tan t = \dfrac{\_\_\_\_\_\_}{\cos t}$.
$\cos t$	**251** $\cot t = \dfrac{\_\_\_\_\_\_}{\sin t}$.
$\cos t$	**252** $\sec t = \dfrac{1}{\_\_\_\_\_\_}$.
$\sin t$	**253** $\csc t = \dfrac{1}{\_\_\_\_\_\_}$.
1	**254** $\cos^2 t + \sin^2 t = \_\_\_\_$.
$\sec^2 t$	**255** $1 + \tan^2 t = \_\_\_\_\_\_$.
$\csc^2 t$	**256** $1 + \cot^2 t = \_\_\_\_\_\_$.
$\sin b$	**257** $\sin(a + b) = \sin a \cos b + \_\_\_\_\_\_ \cos a$.
$\sin a \sin b$	**258** $\cos(a + b) = \cos a \cos b - \_\_\_\_\_\_$.
$\sin b \cos a$	**259** $\sin(a - b) = \sin a \cos b - \_\_\_\_\_\_$.
$\sin a \sin b$	**260** $\cos(a - b) = \cos a \cos b + \_\_\_\_\_\_$.
$\dfrac{\pi}{2} - t$	**261** $\sin t = \cos\left(\_\_\_\_\_\_\right)$.
$\dfrac{\pi}{2} - t$	**262** $\cos t = \sin\left(\_\_\_\_\_\_\right)$.
$\tan b$	**263** $\tan(a + b) = \dfrac{\tan a + \_\_\_\_\_\_}{1 - \tan a \tan b}$.
$\tan b$	**264** $\tan(a - b) = \dfrac{\tan a - \_\_\_\_\_\_}{1 + \tan a \tan b}$.
$2 \sin t \cos t$	**265** $\sin 2t = \_\_\_\_\_\_$.
$\cos^2 t - \sin^2 t$	**266** $\cos 2t = \_\_\_\_\_\_$.
$1 - \cos t$	**267** $\sin^2 \dfrac{t}{2} = \dfrac{1}{2}(\_\_\_\_\_\_)$.
$1 + \cos t$	**268** $\cos^2 \dfrac{t}{2} = \dfrac{1}{2}(\_\_\_\_\_\_)$.

In Problems 269 through 279, use the preceding identities to simplify each expression.

Answer	Problem
	269 $7 \sin^2 \dfrac{\pi}{6} + 7 \cos^2 \dfrac{\pi}{6}$.
$\sin^2 \dfrac{\pi}{6} + \cos^2 \dfrac{\pi}{6}$	$7 \sin^2 \dfrac{\pi}{6} + 7 \cos^2 \dfrac{\pi}{6} = 7\left(\_\_\_\_\_\_\right)$
1	$= 7(\_\_\_\_)$
7	$= \_\_\_\_$.
	270 $\csc^2 \dfrac{81\pi}{5} - \cot^2 \dfrac{81\pi}{5}$.
1	Using Problem 256, we have $\csc^2 t - \cot^2 t = \_\_\_\_$, so that
1	$\csc^2 \dfrac{81\pi}{5} - \cot^2 \dfrac{81\pi}{5} = \_\_\_\_$.

271 $\tan^2 t \csc^2 t - \tan^2 t$.

$\tan^2 t \csc^2 t - \tan^2 t = \tan^2 t$ (________)

$= \tan^2 t$ (________) = ______.

$\csc^2 t - 1$

$\cot^2 t,\ 1$

272 $\cos^4 3t - \sin^4 3t$.

$\cos^4 3t - \sin^4 3t = (\cos^2 3t - \sin^2 3t)$ (______)

= ________.

1

$\cos 6t$

273 $\cos(\pi - t)$.

$\cos(\pi - t) = \cos\pi \cos t + \sin\pi$ (______)

= (______) $\cos t$ + (______) $\sin t$

= ________.

$\sin t$

$-1,\ 0$

$-\cos t$

274 $\sin\left(\frac{3\pi}{2} - t\right)$.

$\sin\left(\frac{3\pi}{2} - t\right) = \sin\frac{3\pi}{2}\cos t - \sin t$ ________

= (______) $\cos t - \sin t$ (______)

= ________.

$\cos\frac{3\pi}{2}$

$-1,\ 0$

$-\cos t$

275 $\cos\left(\frac{3\pi}{2} + t\right)$.

$\cos\left(\frac{3\pi}{2} + t\right) = \cos\frac{3\pi}{2}\cos t - \sin\frac{3\pi}{2}$ (______)

= (______) $\cos t$ − (______) $\sin t$

= ________.

$\sin t$

$0,\ -1$

$\sin t$

276 $\sin\left(t + \frac{\pi}{6}\right) + \sin\left(t - \frac{\pi}{6}\right)$.

$\sin\left(t + \frac{\pi}{6}\right) + \sin\left(t - \frac{\pi}{6}\right)$

$= \sin t \cos\frac{\pi}{6} + \sin\frac{\pi}{6}\cos t + \sin t \cos\frac{\pi}{6} -$ (________)

= ________ $= 2\sin t$(______)

= ________.

$\sin\frac{\pi}{6}\cos t$

$2\sin t\cos\frac{\pi}{6},\ \frac{\sqrt{3}}{2}$

$\sqrt{3}\sin t$

277 $\cot\left(t + \frac{\pi}{4}\right) + \tan\left(t + \frac{\pi}{4}\right)$.

$\cot\left(t + \frac{\pi}{4}\right) + \tan\left(t + \frac{\pi}{4}\right)$

$= \dfrac{1}{\_\_\_\_\_\_} + \tan\left(t + \frac{\pi}{4}\right)$

$= \dfrac{1 + (\_\_\_\_\_\_)}{\tan\left(t + \frac{\pi}{4}\right)} = \dfrac{\_\_\_\_\_\_}{\tan\left(t + \frac{\pi}{4}\right)}$

$= \dfrac{1}{\_\_\_\_\_\_}$.

$\tan\left(t + \frac{\pi}{4}\right)$

$\tan^2\left(t + \frac{\pi}{4}\right),\ \sec^2\left(t + \frac{\pi}{4}\right)$

$\sin\left(t + \frac{\pi}{4}\right)\cos\left(t + \frac{\pi}{4}\right)$

278 $\cot t \sin 2t - \cos 2t$.

$\cot t \sin 2t - \cos 2t$

$= \cot t$ (__________) - (__________) — $2 \sin t \cos t,\ 2\cos^2 t - 1$

$= 2\cos^2 t - 2\cos^2 t +$ ____ — 1

$=$ ____. — 1

279 $(\sin t - \cos t)^2 + \sin 2t$.

$(\sin t - \cos t)^2 + \sin 2t$

$= \sin^2 t + \cos^2 t -$ __________ $+ \sin 2t$ — $2 \sin t \cos t$

$=$ ____ - ________ $+ \sin 2t$ — $1,\ \sin 2t$

$=$ ____. — 1

In Problems 280 through 283, find the exact value of each expression.

280 $\cos \frac{7\pi}{12}$.

$\cos \frac{7\pi}{12} = \cos$(________) — $\frac{\pi}{4} + \frac{\pi}{3}$

$= \cos\frac{\pi}{4}\cos\frac{\pi}{3} -$ ________ — $\sin\frac{\pi}{4}\sin\frac{\pi}{3}$

$= \left(\frac{\sqrt{2}}{2}\right)\left(\frac{1}{2}\right) - \left(\frac{\sqrt{2}}{2}\right)$(____) — $\frac{\sqrt{3}}{2}$

$= \frac{\sqrt{2}}{4} -$ ____. — $\frac{\sqrt{6}}{4}$

281 $\sin \frac{5\pi}{12}$.

$\sin \frac{5\pi}{12} = \sin$(______) — $\frac{\pi}{4} + \frac{\pi}{6}$

$= \sin\frac{\pi}{4}\cos\frac{\pi}{6} + \sin\frac{\pi}{6}$(______) — $\cos\frac{\pi}{4}$

$= \left(\frac{\sqrt{2}}{2}\right)\left(\frac{\sqrt{3}}{2}\right) +$ (____)(____) — $\frac{1}{2},\ \frac{\sqrt{2}}{2}$

$= \frac{\sqrt{6}}{4} +$ ____. — $\frac{\sqrt{2}}{4}$

282 $\cos \frac{11\pi}{12}$.

$\cos \frac{11\pi}{12} = \cos$(________) — $\frac{2\pi}{3} + \frac{\pi}{4}$

$= \cos\frac{2\pi}{3}\cos\frac{\pi}{4} - \sin\frac{2\pi}{3}$(______) — $\sin\frac{\pi}{4}$

$\frac{\sqrt{3}}{2}, \frac{\sqrt{2}}{2}$

$\frac{\sqrt{6}}{4}$

$= \left(-\frac{1}{2}\right)\left(\frac{\sqrt{2}}{2}\right) - (\_\_\_)(\_\_\_)$

$= -\frac{\sqrt{2}}{4} - \_\_\_\_.$

30°, $1 - \cos t$

$\cos 30°, \frac{\sqrt{3}}{2}$

$\frac{2-\sqrt{3}}{4}, \frac{\sqrt{2-\sqrt{3}}}{2}$

283 sin 15°.

$15° = \frac{1}{2}(\_\_\_\_)$ and $\sin^2 \frac{t}{2} = \frac{1}{2}(\_\_\_\_\_\_\_\_)$, so that

$\sin^2 15° = \frac{1}{2}(1 - \_\_\_\_\_\_) = \frac{1}{2}\left(1 - \_\_\_\right)$. Therefore,

$\sin 15° = \sqrt{\_\_\_\_\_} = \_\_\_\_\_\_.$

8 Algebra of Functions and Composition of Functions

8.1 Algebra of Functions

284 Let f and g be two functions with domains that intersect. We define functions $f + g$, $f - g$, $f \cdot g$, and f/g, called respectively the sum, the difference, the product, and the quotient of f and g, as follows:

$(f + g)(x) = \_\_\_\_\_\_\_\_$,

$(f - g)(x) = \_\_\_\_\_\_\_\_$,

$(f \cdot g)(x) = \_\_\_\_\_\_\_\_$,

$\left(\frac{f}{g}\right)(x) = \frac{\_\_\_\_}{g(x)}$ if $g(x) \neq 0$.

$f(x) + g(x)$

$f(x) - g(x)$

$f(x) \cdot g(x)$

$f(x)$

In Problems 285 through 288, find $f + g$, $f - g$, $f \cdot g$, and f/g.

285 $f(x) = x^2 - 4$ and $g(x) = 2x + 7$.

$(f + g)(x) = f(x) + g(x) = \_\_\_\_\_\_\_\_$,

$(f - g)(x) = f(x) - g(x) = \_\_\_\_\_\_\_\_$,

$(f \cdot g)(x) = f(x) \cdot g(x) = \_\_\_\_\_\_\_\_\_\_\_\_$,

$\left(\frac{f}{g}\right)(x) = \frac{f(x)}{g(x)} = \_\_\_\_\_\_.$

$x^2 + 2x + 3$

$x^2 - 2x - 11$

$2x^3 + 7x^2 - 8x - 28$

$\frac{x^2 - 4}{2x + 7}$

286 $f(x) = x^3$ and $g(x) = -3x^2 + 1$.

$(f + g)(x) = f(x) + g(x) = \_\_\_\_\_\_\_\_$,

$(f - g)(x) = f(x) - g(x) = \_\_\_\_\_\_\_\_$,

$(f \cdot g)(x) = f(x) \cdot g(x) = \_\_\_\_\_\_\_\_$,

$\left(\frac{f}{g}\right)(x) = \frac{f(x)}{g(x)} = \_\_\_\_\_\_.$

$x^3 - 3x^2 + 1$

$x^3 + 3x^2 - 1$

$-3x^5 + x^3$

$\frac{x^3}{-3x^2 + 1}$

287 f consists of the ordered pairs (2, 2), (3, 3), (4, 2), and (5, 1). g consists of the ordered pairs (2, 4), (3, 9), (4, 6), and (5, 7). The domain of f is the set consisting of the numbers ________. The domain of g is the set consisting of the numbers ________. Thus, $f + g$ consists of the ordered pairs (2, 6), ________; $f - g$ consists of the ordered pairs (2, −2), (3, −6), ________; $f \cdot g$ consists of the ordered pairs ________, (4, 12), (5, 7); and $\frac{f}{g}$ consists of the ordered pairs $\left(2, \frac{1}{2}\right)$, $\left(3, \frac{1}{3}\right)$, ________, ________.

Answers: 2, 3, 4, 5
(3, 12), (4, 8), (5, 8)
(4, −4), (5, −6)
(2, 8), (3, 27)
$\left(4, \frac{1}{3}\right)$, $\left(5, \frac{1}{7}\right)$

288 $f(x) = -3x + 7$ and $g(x) = 3x - 7$.
$(f + g)(x) = f(x) + g(x) =$ ____,
$(f - g)(x) = f(x) - g(x) =$ ________,
$(f \cdot g)(x) = f(x) \cdot g(x) =$ ________,
$\left(\frac{f}{g}\right)(x) = \frac{f(x)}{g(x)} = \frac{\_\_\_\_}{3x - 7} =$ ____.

Answers: 0
$-6x + 14$
$-9x^2 + 42x - 49$
$-3x + 7$, -1

8.2 Composition of Functions

289 Let f and g be two functions satisfying the condition that at least one number in the range of g belongs to the domain of f. Then the composition of f and g (or the composition of g by f), in symbols $f \circ g$, is the function defined by the equation $(f \circ g)(x) =$ ________.

Answer: $f[g(x)]$

In Problems 290 through 297, let $f(x) = 5x + 2$, $g(x) = x^2$, $p(x) = \sin x$, and $k(x) = \frac{1}{5}(x - 2)$. Evaluate each composition.

290 $(f \circ g)(3)$.
$(f \circ g)(3) = f[g(3)] = f(\_\_\_\_)$
$= 5(\_\_\_\_) + 2 =$ ____.

Answers: 9
9, 47

291 $(g \circ f)(3)$.
$(g \circ f)(3) = g[f(3)] = g(\_\_\_\_)$
$= (\_\_\_\_)^2 =$ ____.

Answers: 17
17, 289

292 $(f \circ g)(x)$.
$(f \circ g)(x) = f[g(x)] = f(\_\_\_\_)$
$= 5(\_\_\_\_) + 2 =$ ________.

Answers: x^2
x^2, $5x^2 + 2$

293 $(f \circ p)\left(\frac{\pi}{2}\right)$.

$(f \circ p)\left(\frac{\pi}{2}\right) = f\left[p\left(\frac{\pi}{2}\right)\right] = f(\_\_\_\_)$

$= 5(\_\_\_\_) + 2 = \_\_\_\_.$

1

1, 7

294 $(p \circ g)(x)$.

$(p \circ g)(x) = p[g(x)] = p(\_\_\_\_) = \sin(\_\_\_\_)$

$= \_\_\_\_\_\_.$

x^2, x^2

$\sin x^2$

295 $(f \circ k)(x)$ and $(k \circ f)(x)$.

$(f \circ k)(x) = f[k(x)] = f(\_\_\_\_\_\_\_\_)$

$= 5(\_\_\_\_\_\_\_\_) + 2 = \_\_\_\_,$

$\frac{1}{5}(x - 2)$

$\frac{1}{5}(x - 2),\ x$

$(k \circ f)(x) = k[f(x)] = k(\_\_\_\_\_\_)$

$= \frac{1}{5}[(\_\_\_\_\_\_) - 2] = \_\_\_\_.$

$5x + 2$

$5x + 2,\ x$

296 $(f \circ f)(x)$ and $(p \circ p)(x)$.

$(f \circ f)(x) = f[\_\_\_\_\_\_] = f(\_\_\_\_\_\_)$

$= 5(\_\_\_\_\_\_) + 2 = \_\_\_\_\_\_\_\_,$

$(p \circ p)(x) = p[p(x)] = p(\_\_\_\_\_\_)$

$= \sin(\_\_\_\_\_\_) = \_\_\_\_\_\_\_\_\_\_.$

$f(x),\ 5x + 2$

$5x + 2,\ 25x + 12$

$\sin x$

$\sin x,\ \sin(\sin x)$

297 $[f \circ (g + k)](x)$.

$[f \circ (g + k)](x) = f[(g + k)(x)]$

$= f(\_\_\_\_\_\_\_\_\_\_)$

$= 5\left(x^2 + \frac{1}{5}x - \frac{2}{5}\right) + \_\_\_\_$

$= \_\_\_\_\_\_\_\_.$

$x^2 + \frac{1}{5}x - \frac{2}{5}$

2

$5x^2 + x$

298 Suppose that f is an even function and g is an odd function. Is $f \circ f$ even or odd? Is $f \circ g$ even or odd?

f is even, so that $f(-x) = \_\_\_\_$. g is odd, so that $g(-x) = -\ \_\_\_\_$.

$(f \circ f)(-x) = f[f(-x)] = f(\_\_\_\_)$

$= \_\_\_\_\_\_\_\_\_\_.$

Therefore, $f \circ f$ is an ____ function.

$(f \circ g)(-x) = f[g(-x)] = f(\_\_\_\_\_\_)$

$= f(\_\_\_\_\_\_) = \_\_\_\_\_\_\_\_\_\_.$

Therefore, $f \circ g$ is an ____ function.

$f(x),\ g(x)$

$f(x)$

$(f \circ f)(x)$

even

$-g(x)$

$g(x),\ (f \circ g)(x)$

even

299 Let $f(x) = 3x - 7$ and $g(x) = 2x + k$. Find k so that $(f \circ g)(x) = (g \circ f)(x)$.

$(f \circ g)(x) = f[g(x)] = f(\_\_\_\_\_\_)$

$= 3(\_\_\_\_\_\_) - 7 = \_\_\_\_\_\_\_\_\_\_,$

$(g \circ f)(x) = g[\_\_\_\_] = g(\_\_\_\_\_\_)$

$2x + k$

$2x + k,\ 6x + 3k - 7$

$f(x),\ 3x - 7$

$3x - 7$, $6x - 14 + k$

-7, $-\frac{7}{2}$

$= 2(\_\_\_\_\_\_) + k = \_\_\_\_\_\_\_\_\_\_$,

so that $6x + 3k - 7 = 6x - 14 + k$ or $2k = \_\_\_\_$ or $k = \_\_\_\_$.

9 Inverse Functions

domain; x; domain; x; $y = x$

300 Two functions f and g are said to be inverses of each other if the following four conditions hold:

(i) The range of g is contained in the ________ of f.

(ii) For every number x in the domain of g, $(f \circ g)(x) = \_\_\_\_$.

(iii) The range of f is contained in the ________ of g.

(iv) For every number x in the domain of f, $(g \circ f)(x) = \_\_\_\_$.

301 Suppose that f is an invertible function. Then we define the inverse of the function f, in symbols f^{-1}, to be the function whose graph is the mirror image of the graph of f across the straight line ________.

In Problems 302 through 305, show that f is the inverse of g by showing that $(f \circ g)(x) = (g \circ f)(x) = x$. Observe that $\mathbb{R}$ is the domain and range of each function.

$\frac{x}{9}$, x; $9x$, x; x, inverses

302 $f(x) = 9x$ and $g(x) = \frac{x}{9}$.

$(f \circ g)(x) = f[g(x)] = f\left(\frac{x}{9}\right) = 9\left(\_\_\_\_\right) = \_\_\_\_$,

$(g \circ f)(x) = g[f(x)] = g(9x) = \frac{1}{9}(\_\_\_\_) = \_\_\_\_$.

Since $(f \circ g)(x) = (g \circ f)(x) = \_\_\_\_$, f and g are __________ of each other.

$\frac{x-2}{7}$, x; $7x + 2$, x; inverses

303 $f(x) = 7x + 2$ and $g(x) = \frac{x - 2}{7}$.

$(f \circ g)(x) = f[g(x)] = f\left(\_\_\_\_\_\_\right) = \_\_\_\_$,

$(g \circ f)(x) = g[f(x)] = g(\_\_\_\_\_\_) = \_\_\_\_$.

Therefore, f and g are __________ of each other.

$\frac{x}{2} - 1$, x; $2x + 2$, x; inverses

304 $f(x) = 2x + 2$ and $g(x) = \frac{x}{2} - 1$.

$(f \circ g)(x) = f[g(x)] = f\left(\_\_\_\_\_\_\right) = \_\_\_\_$,

$(g \circ f)(x) = g[f(x)] = g(\_\_\_\_\_\_) = \_\_\_\_$.

Therefore, f and g are __________ of each other.

$\frac{x+4}{3}$, x

305 $f(x) = 3x - 4$ and $g(x) = \frac{x + 4}{3}$.

$(f \circ g)(x) = f[g(x)] = f\left(\_\_\_\_\_\_\right) = \_\_\_\_$,

$3x - 4$, x

inverses

$(g \circ f)(x) = g[f(x)] = g(\_\_\_\_\_\_) = \_\_\_\_.$
Therefore, f and g are ________ of each other.

In Problems 306 through 310, find f^{-1} and verify that $(f \circ f^{-1})(x) = x$ and $(f^{-1} \circ f)(x) = x$.

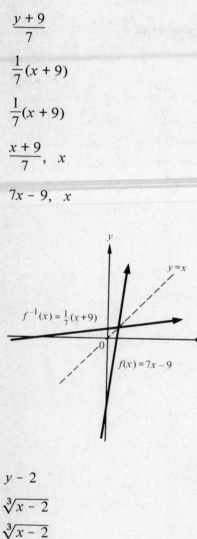

$\frac{y+9}{7}$

$\frac{1}{7}(x+9)$

$\frac{1}{7}(x+9)$

$\frac{x+9}{7}$, x

$7x - 9$, x

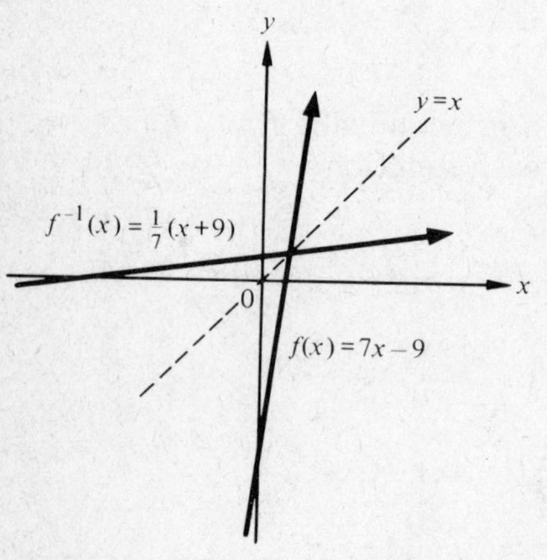

306 $f(x) = 7x - 9$.

Let $y = 7x - 9$. Then $x =$ ________. Now interchange x and y in the latter equation to get $y =$ ________. Hence, f^{-1} is defined by the equation $f^{-1}(x) =$ ________.

$(f \circ f^{-1})(x) = f(f^{-1}(x)) = f(\_\_\_\_\_\_) = \_\_\_\_.$

$(f^{-1} \circ f)(x) = f^{-1}(f(x)) = f^{-1}(\_\_\_\_\_\_) = \_\_\_\_.$

The graph is

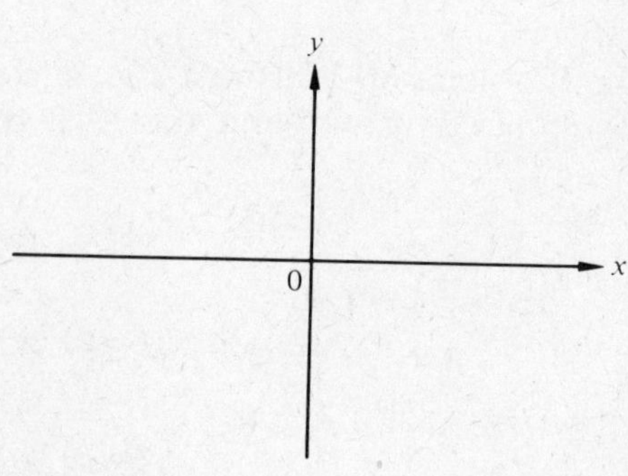

$y - 2$

$\sqrt[3]{x-2}$

$\sqrt[3]{x-2}$

$\sqrt[3]{x-2}$, x

$x^3 + 2$, x

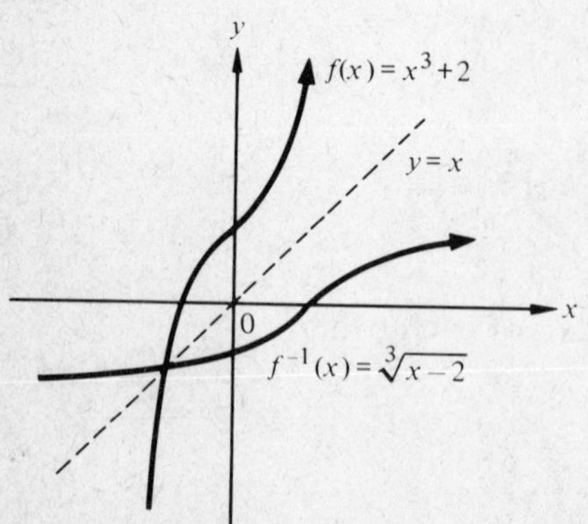

307 $f(x) = x^3 + 2$.

Let $y = x^3 + 2$. Then $x = \sqrt[3]{\_\_\_\_\_\_}$. Now interchange x and y in the latter equation to get $y =$ ________. Hence, $f^{-1}(x) =$ ________.

$(f \circ f^{-1})(x) = f(f^{-1}(x)) = f(\_\_\_\_\_\_) = \_\_\_\_,$

$(f^{-1} \circ f)(x) = f^{-1}(f(x)) = f^{-1}(\_\_\_\_\_\_) = \_\_\_\_.$

The graph is

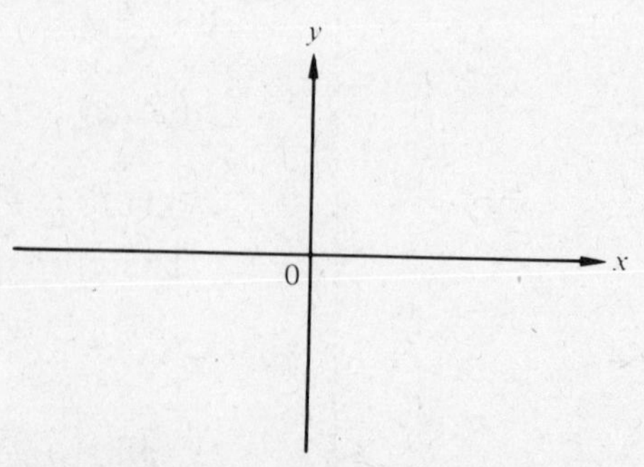

$y + 1$

$x + 1$

$\sqrt{x+1}$

$\sqrt{x+1},\ x$

$x^2 - 1,\ x$

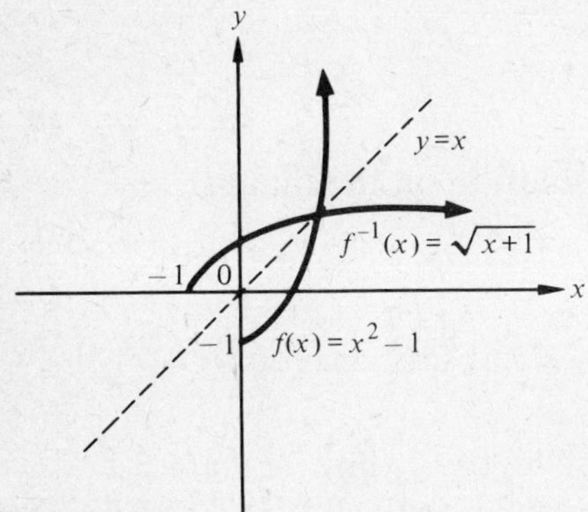

308 $f(x) = x^2 - 1$ for $x \geqslant 0$.

Let $y = x^2 - 1$, so that $x = \sqrt{\_\_\_\_\_\_}$. Interchange x and y in the latter equation to get $y = \sqrt{\_\_\_\_\_\_}$, so that $f^{-1}(x) =$ ______.

$(f \circ f^{-1})(x) = f(f^{-1}(x)) = f(\_\_\_\_\_\_) = \_\_\_\_$,
$(f^{-1} \circ f)(x) = f^{-1}(f(x)) = f^{-1}(\_\_\_\_\_\_) = \_\_\_\_$.

The graph is

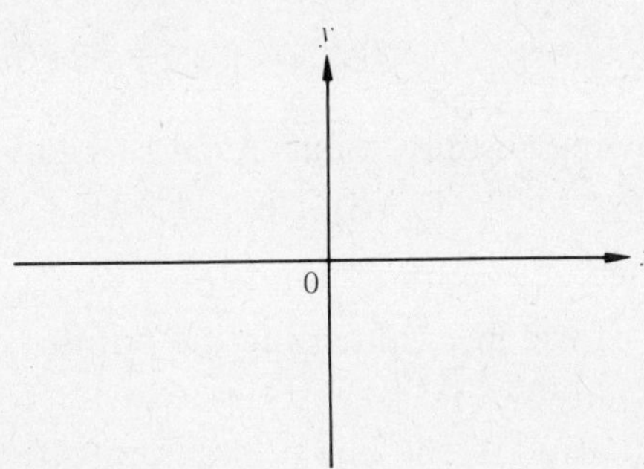

$y + 4$

$x + 4$

$-\sqrt{x+4}$

$-\sqrt{x+4},\ x$

$x^2 - 4,\ x$

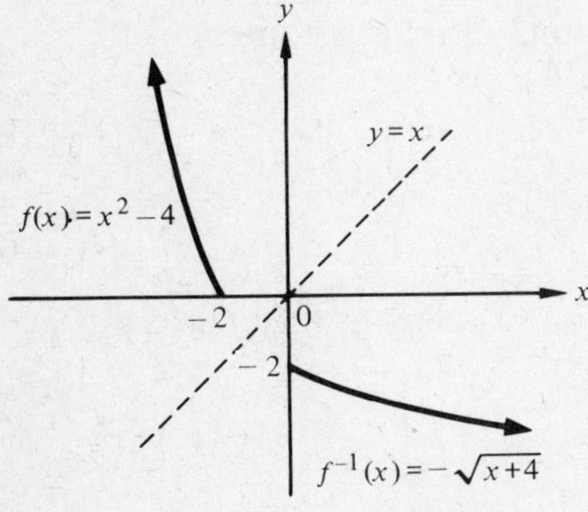

309 $f(x) = x^2 - 4$ for $x \leqslant -2$.

Let $y = x^2 - 4$, so that $x = -\sqrt{\_\_\_\_\_\_}$. Interchange x and y in the latter equation to get $y = -\sqrt{\_\_\_\_\_\_}$, so that $f^{-1}(x) =$ ______.

$(f \circ f^{-1})(x) = f(f^{-1}(x)) = f(\_\_\_\_\_\_) = \_\_\_\_$,
$(f^{-1} \circ f)(x) = f^{-1}(f(x)) = f^{-1}(\_\_\_\_\_\_) = \_\_\_\_$.

The graph is

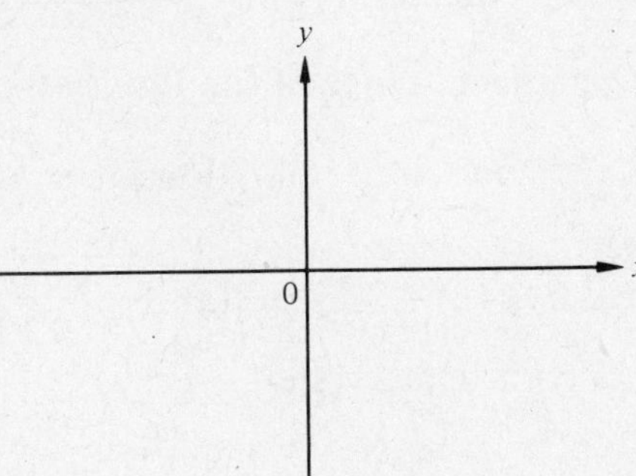

$\dfrac{y-2}{2y-1}$

$\dfrac{x-2}{2x-1}$

$\dfrac{x-2}{2x-1}$

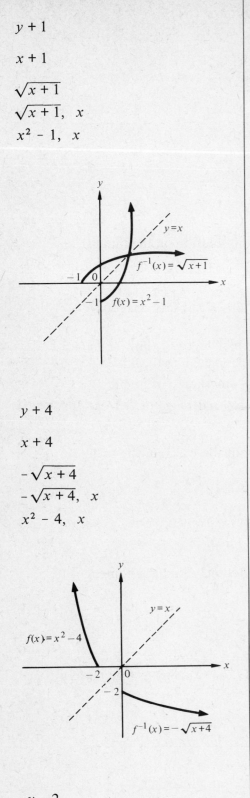

310 $f(x) = \dfrac{x-2}{2x-1}$.

Let $y = \dfrac{x-2}{2x-1}$, so that $x =$ ______.

Interchange x and y in the latter equation to get $y =$ ______, so that $f^{-1}(x) =$ ______.

$\frac{x-2}{2x-1}$, x | $(f \circ f^{-1})(x) = f(f^{-1}(x)) = f(\_\_\_\_\_\_) = \_\_\_\_.$

$\frac{x-2}{2x-1}$, x | $(f^{-1} \circ f)(x) = f^{-1}(f(x)) = f^{-1}(\_\_\_\_\_\_) = \_\_\_\_.$

Chapter Test

1 Solve the following inequalities and show the solution on the number line.

(a) $x + 3 < 5x - 1$ (b) $x^2 + 3x + 2 \geqslant 0$ (c) $\frac{5+x}{5-x} < 2$

2 Solve the following absolute value equations and inequalities. Show the solution on the number line.

(a) $|3x - 2| = 5$ (b) $|3x - 2| \leqslant 4$ (c) $|3x + 2| > 5$

3 Find the distance between the points $P_1 = (3, 2)$ and $P_2 = (-1, 2)$.

4 Find the slope of the line containing the points $P_1 = (8, -2)$ and $P_2 = (3, 7)$, and find the equation of the line containing P_1 and P_2.

5 Find (a) the equation of the line that contains the point $(-3, 7)$ and is parallel to the line $7x - y + 3 = 0$, and (b) the equation of the line that contains the point $(-3, 7)$ and is perpendicular to the line $7x - y + 3 = 0$.

6 Find the domain and range of the function f defined by $f(x) = \sqrt{1 - x}$, and sketch its graph.

7 Let f be defined by the equation $f(x) = x^2 + 3x - 4$. Find the domain and range of f. Also find $f(-2)$, $f(0)$, $f(2)$, $f(2a)$, and $f(a + b)$ and sketch its graph.

8 Is f, defined by $f(x) = \frac{x^2 - 4}{x^2 + 4}$, even or odd? Why? Is it symmetric about the y axis or the origin?

9 Let f and g be defined by $f(x) = \sqrt{x}$ and $g(x) = x^2 - 1$. Find:

(a) $(f + g)(x)$ (b) $(f \cdot g)(x)$ (c) $\left(\frac{f}{g}\right)(x)$

(d) $(f \circ f)(16)$ (e) $(f \circ g)(x)$ (f) $(g \circ f)(4)$.

10 Is f, defined by the equation $f(x) = \frac{2x + 1}{\sqrt[3]{x^3 - 3x^2 + 3x - 1}}$, a rational function? Find the domain of f.

11 Indicate the domain and range of the function defined by $f(x) = [\![7x^2]\!]$.

12 Evaluate the six trigonometric functions at $\frac{5\pi}{6}$.

13 Let f be defined by $f(x) = 7x - 3$. (a) Find $\frac{f(a + h) - f(a)}{h}$. (b) Find f^{-1}. (c) Sketch the graphs of f and f^{-1} on the same coordinate system.

Answers

1 (a) $(1, \infty)$
(b) $(-\infty, -2]$ or $[-1, \infty)$
(c) $\left(-\infty, \frac{5}{3}\right)$ or $(5, \infty)$

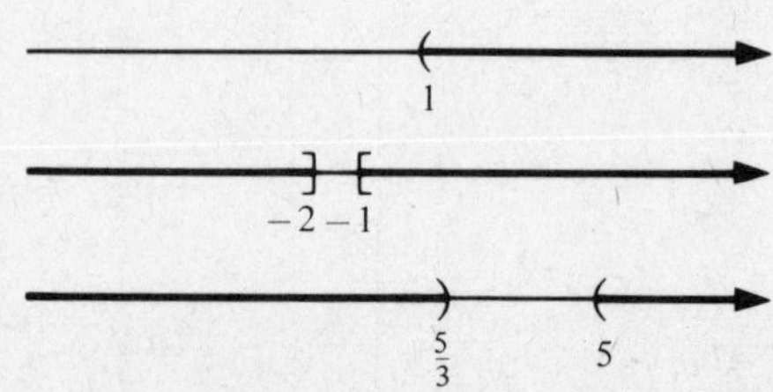

2 (a) $x = -1$ or $x = \frac{7}{3}$

(b) 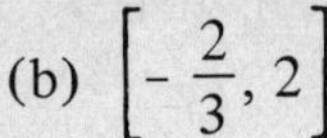 $\left[-\frac{2}{3}, 2\right]$

(c) $(1, \infty)$ or $\left(-\infty, -\frac{7}{3}\right)$

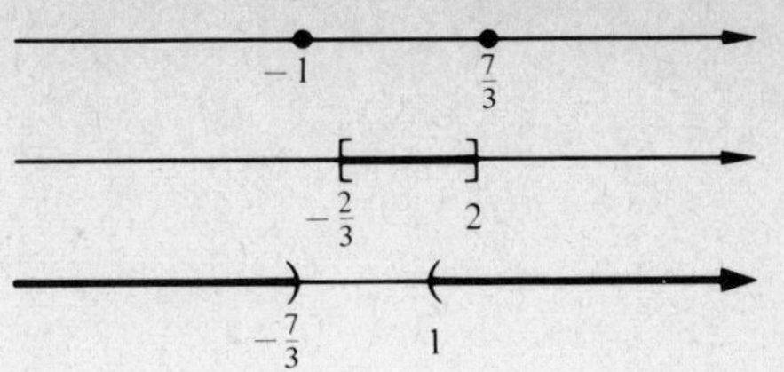

3 4

4 slope $= -\frac{9}{5}$, $y = -\frac{9}{5}x + \frac{62}{5}$

5 (a) $y = 7x + 28$ (b) $y = -\frac{1}{7}x + \frac{46}{7}$

6 The domain of f is the interval $(-\infty, 1]$. The range of f is the interval $[0, \infty)$.

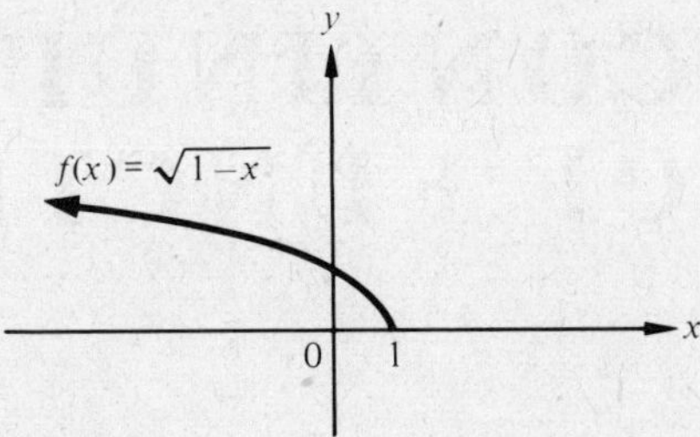

7 The domain of f is $\mathbb{R}$. The range of f is $\left[-\frac{25}{4}, \infty\right)$. $f(-2) = -6$, $f(0) = -4$, $f(2) = 6$, $f(2a) = 4a^2 + 6a - 4$, and $f(a+b) = a^2 + (2b+3)a + (b^2 + 3b - 4)$.

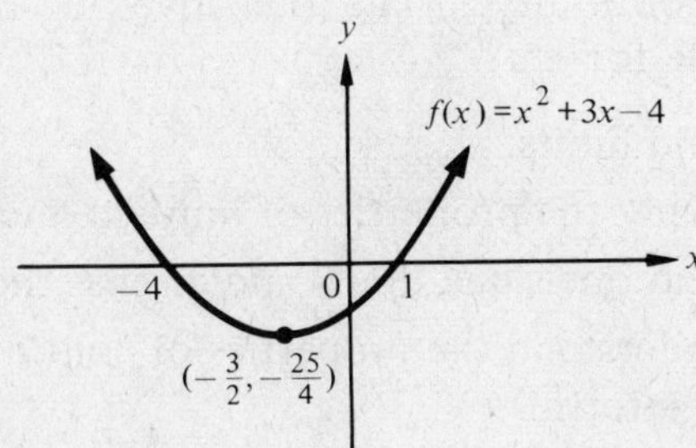

8 Even, since $f(-x) = f(x)$; the y axis.

9 (a) $\sqrt{x} + x^2 - 1$ (b) $x^{5/2} - x^{1/2}$ (c) $\frac{\sqrt{x}}{x^2 - 1}$

(d) 2 (e) $\sqrt{x^2 - 1}$ (f) 3

10 Yes; the domain of f is the set of real numbers except $x = 1$.

11 The domain of f is $\mathbb{R}$; the range of f is the set of nonnegative integers.

12 $\sin\frac{5\pi}{6} = \frac{1}{2}$, $\csc\frac{5\pi}{6} = 2$, $\cos\frac{5\pi}{6} = -\frac{\sqrt{3}}{2}$, $\sec\frac{5\pi}{6} = -\frac{2}{\sqrt{3}}$, $\tan\frac{5\pi}{6} = -\frac{1}{\sqrt{3}}$, $\cot\frac{5\pi}{6} = -\sqrt{3}$.

13 (a) 7 (b) $\frac{x+3}{7}$ (c)

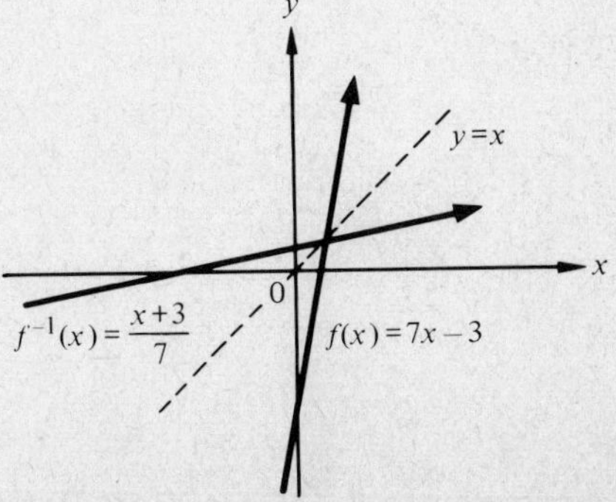

Chapter 1 LIMITS AND CONTINUITY OF FUNCTIONS

Some of the most useful concepts in mathematics are function, limit, and continuity. The objectives of the chapter are that the student be able to:

1 Find limits.

2 Apply the properties of limits to evaluate limits of functions.

3 Find one-sided limits and discuss the continuity of functions.

4 Understand the properties of continuous functions and continuity on an interval.

5 Evaluate limits involving infinity.

6 Graph functions and identify their horizontal and vertical asymptotes.

7 Understand some of the proofs of the basic properties of limits and continuity of functions.

1 Limits and Continuity

4

4

1 Consider the graph of $f(x) = x^2$. The function values, $f(x)$, approach ____ as a limit as x approaches 2; in symbols, we have $\lim_{x \to 2} f(x) = \lim_{x \to 2} x^2 =$ ____. The graph is

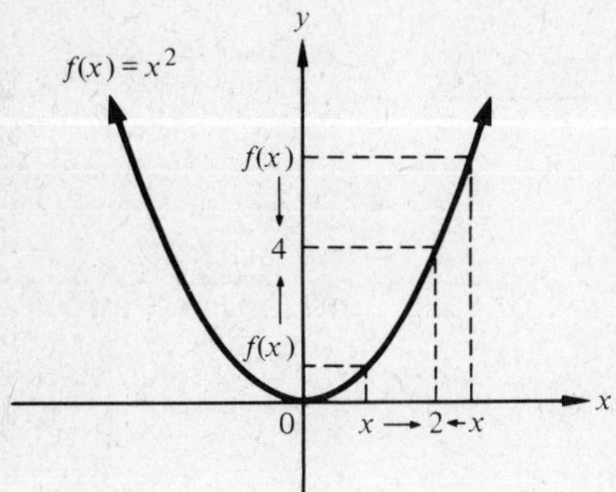

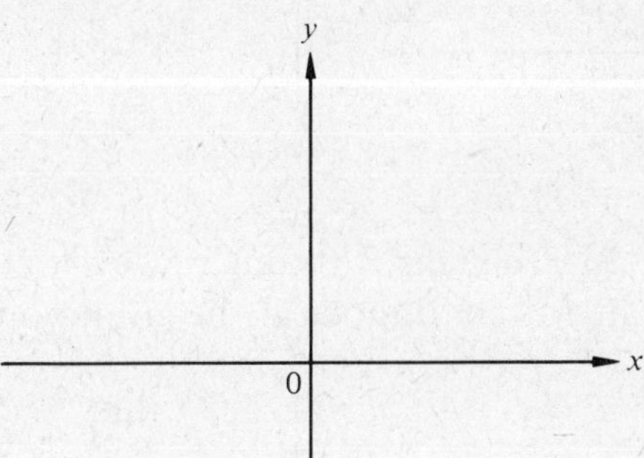

near a

2 In finding $\lim_{x \to a} f(x)$, it does not matter how f is defined at a. The only thing that does matter is how f is defined for values ________. In fact, if $\lim_{x \to a} f(x) = L$, we can distinguish three possible cases:

continuous

Case 1: f is defined at a and $f(a) = L$. In this case, we say f is ________ at a. The graph is

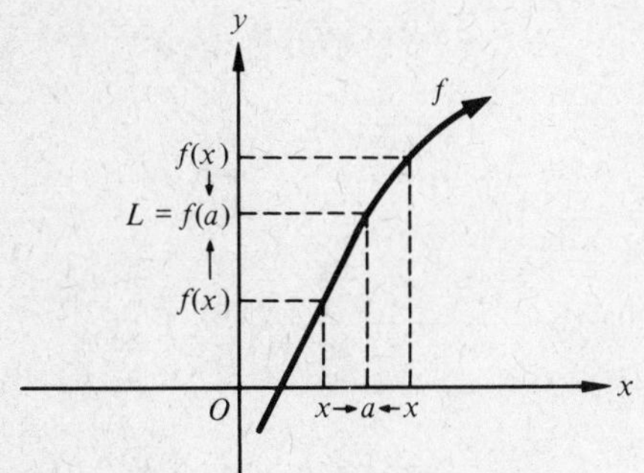

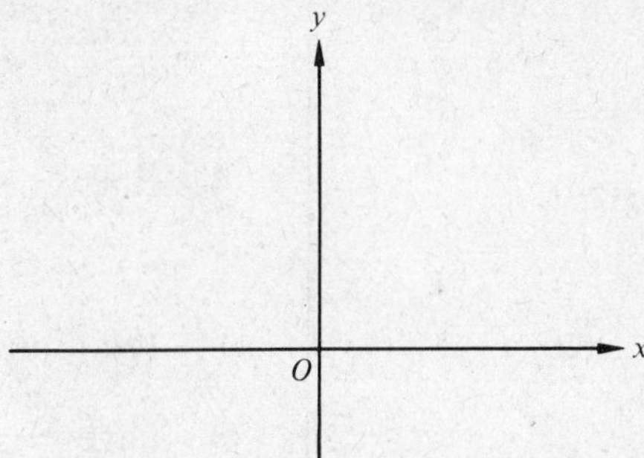

Case 2: f is not defined at a. The graph is

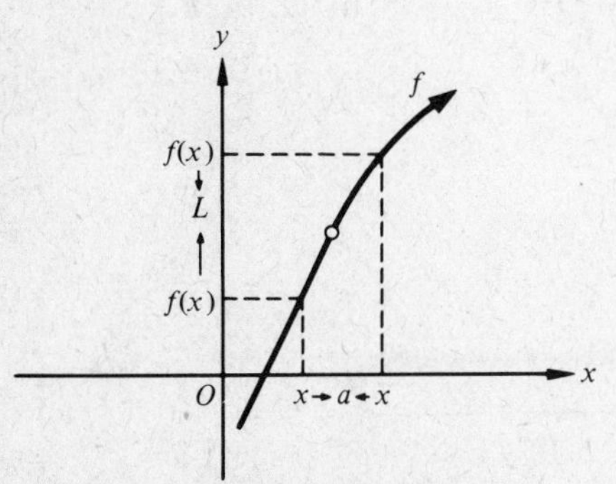

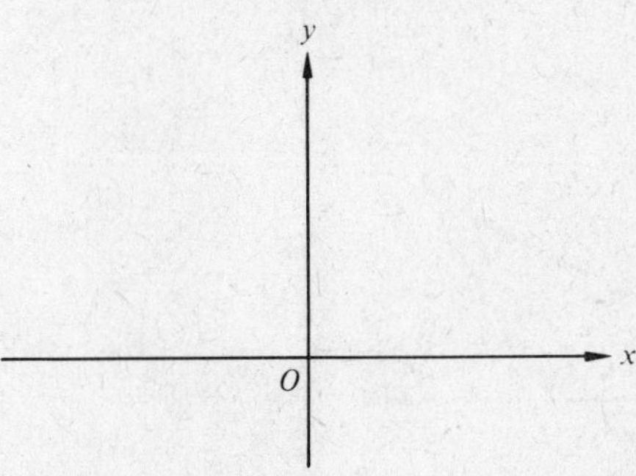

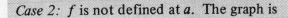

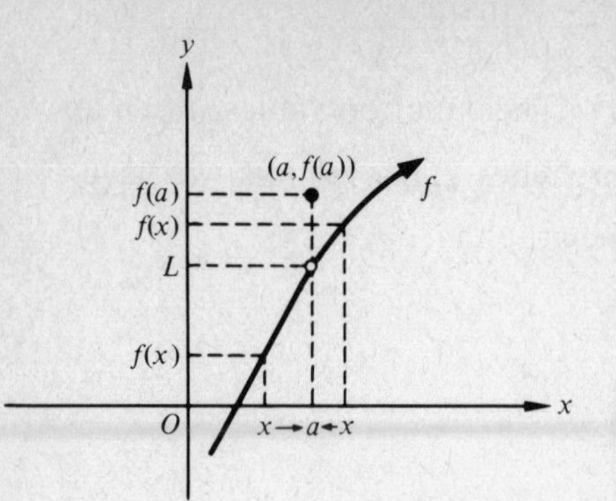

Case 3: f is defined at a, but $f(a) \neq L$. The graph is

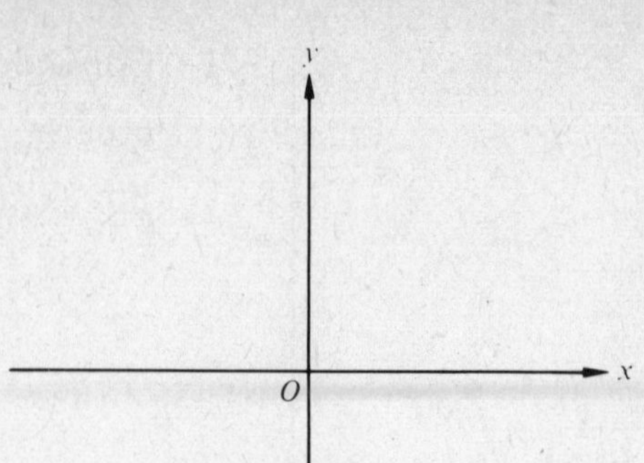

In Problems 3 through 5, evaluate the limit, sketch the graph of each function, and indicate if f is continuous at the given point.

3 $\lim_{x \to 1} 5x$.

1, 5 | As x approaches 1, $5x$ approaches (5) (____) = ____. Thus,

5 | $\lim_{x \to 1} 5x =$ ____. The graph is

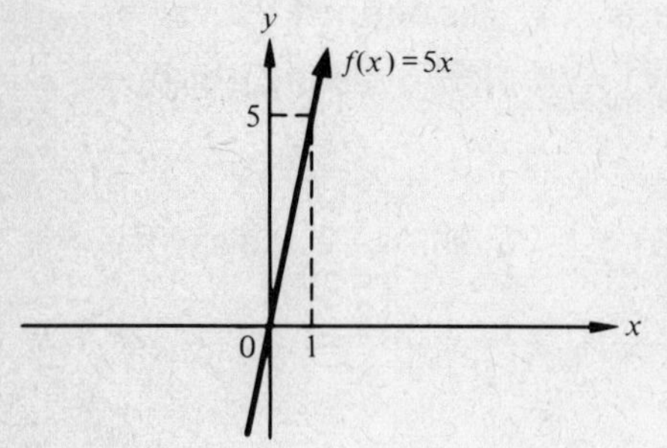

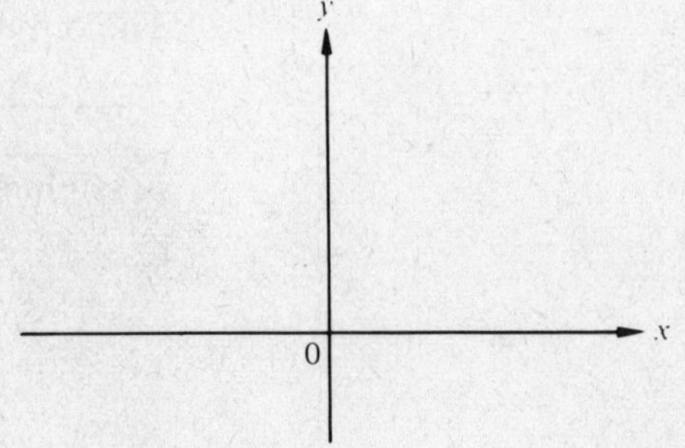

continuous | f is __________ at $x = 1$.

4 $\lim_{x \to 3} \dfrac{x^2 - 9}{x - 3}$.

undefined | At $x = 3$, the function in question is ________, since both

zero | the numerator and the denominator are _____. But for $x \neq 3$, we

$x + 3$ | have $\dfrac{x^2 - 9}{x - 3} = \dfrac{(x - 3)(x + 3)}{x - 3} =$ ______; hence, when x ap-

$x + 3$ | proaches 3, but is different from 3, $\dfrac{x^2 - 9}{x - 3} =$ ______ will

6, 6 | approach ____. Thus, $\lim_{x \to 3} \dfrac{x^2 - 9}{x - 3} =$ ____. The graph is

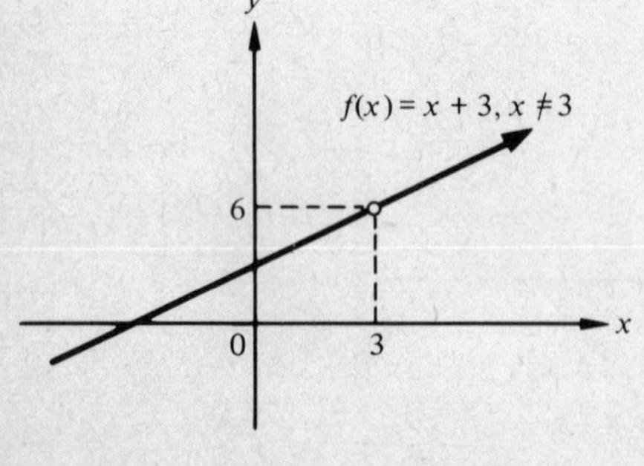

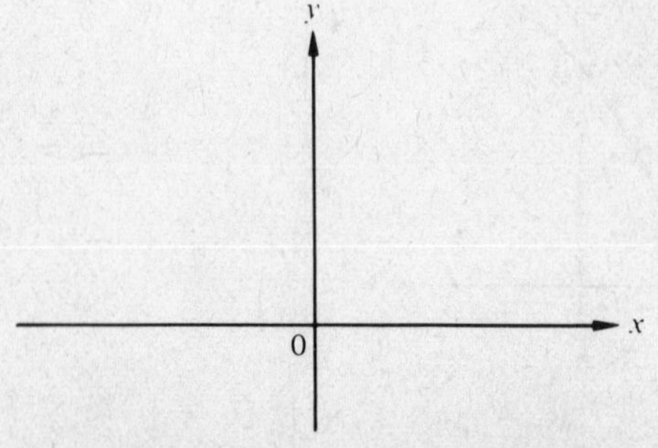

discontinuous

f is ________________ at $x = 3$.

5 $\lim_{x \to 1} f(x)$, where $f(x) = \begin{cases} x + 3 & \text{if } x \neq 1 \\ 6 & \text{if } x = 1. \end{cases}$

When x approaches 1 but is different from 1, $f(x) = x + 3$ approaches ____. Thus, $\lim_{x \to 1} f(x) =$ ____. The graph is

4, 4

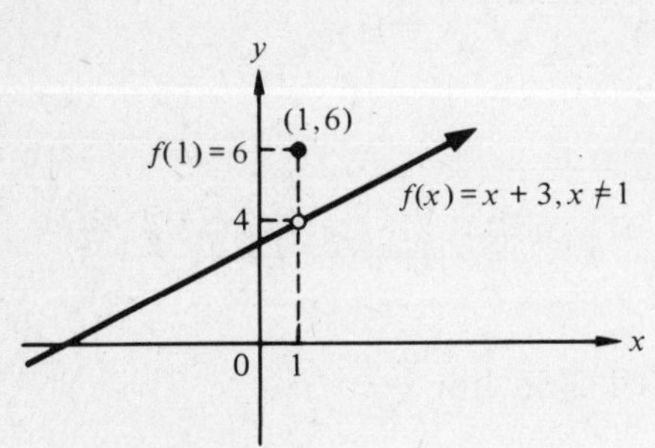

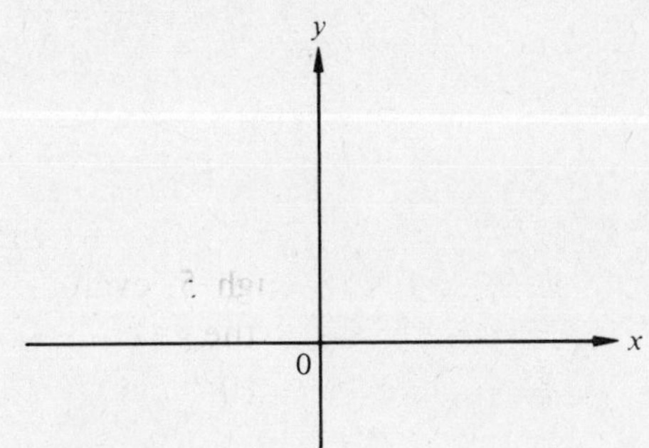

discontinuous

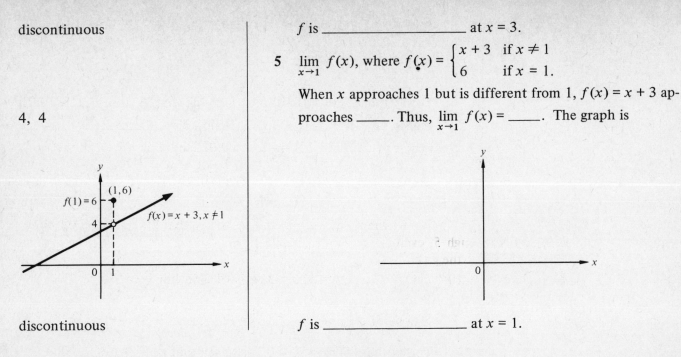

f is ________________ at $x = 1$.

1.1 The Formal Definition of Limit

6 **Let f be a function defined on some open interval, containing a, except possibly at the number a itself. We say that the limit as x approaches a of $f(x)$ is L and write ____________ if, for each positive number ϵ, no matter how small, there exists a positive number δ such that ________________ holds whenever $0 < |x - a| < \delta$. The graph is**

$\lim_{x \to a} f(x) = L$

$|f(x) - L| < \epsilon$

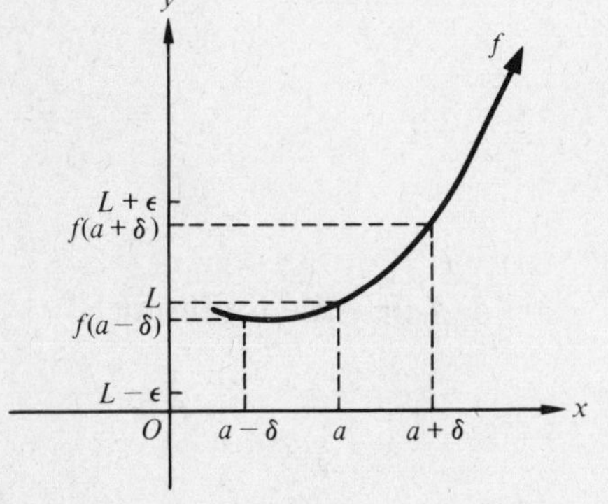

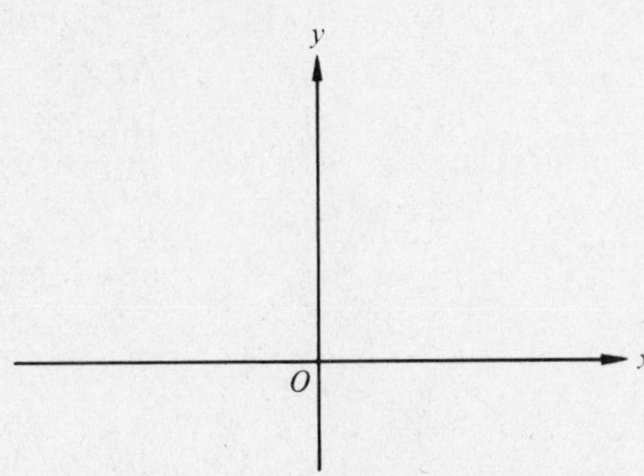

In Problems 7 through 10, for a and the given ϵ, determine a positive δ so that $|f(x) - L| < \epsilon$ will hold whenever $0 < |x - a| < \delta$.

7 $f(x) = 2x - 1$, $L = 3$, $a = 2$, $\epsilon = 0.01$, and $\lim_{x \to 2} (2x - 1) = 3$.

$|f(x) - 3| = |2x - 1 -$ ____$| = |2x - 4|$

$= 2|$________$|$.

To say that $|f(x) - 2| < 0.01$ is to say that $2|$________$| < 0.01$, that is, $|x - 2| <$ ________. Hence, $\delta =$ ________.

3

$x - 2$

$x - 2$

0.005, 0.005

8 $f(x) = 7x + 3$, $L = 10$, $a = 1$, $\epsilon = 0.01$, and $\lim_{x\to 1} (7x + 3) = 10$.

10

$|f(x) - 10| = |7x + 3 - \_\_\_\_| = |7x - 7|$

$x - 1$

$= 7|\_\_\_\_\_\_|$.

$x - 1$

To say that $|f(x) - 10| < 0.01$ is to say that $7|\_\_\_\_\_\_| < 0.01$,

0.0014, 0.0014

that is, $|x - 1| < \_\_\_\_\_\_$. Hence, $\delta = \_\_\_\_\_\_$.

9 $f(x) = 5 - 2x$, $L = 7$, $a = -1$, $\epsilon = 0.02$, and $\lim_{x\to -1} (5 - 2x) = 7$.

7

$|f(x) - 7| = |5 - 2x - \_\_\_\_| = |-2x - 2|$

$x + 1$, $x + 1$

$= |-2(\_\_\_\_\_\_)| = 2|\_\_\_\_\_\_|$.

To say that $|f(x) - 7| < 0.02$ is to say that $2|x + 1| < 0.02$, that

0.01, 0.01

is, $|x + 1| < \_\_\_\_$. Hence, $\delta = \_\_\_\_$.

10 $f(x) = \dfrac{x^2 - 4}{x - 2}$, $L = 5$, $a = 3$, $\epsilon = 0.1$, and $\lim_{x\to 3} \dfrac{x^2 - 4}{x - 2} = 5$.

5, $x - 3$

$|f(x) - 5| = \left|\dfrac{x^2 - 4}{x - 2} - \_\_\_\_\right| = |\_\_\_\_\_\_|$.

0.1

To say that $|f(x) - 5| < 0.1$ is to say that $|x - 3| < \_\_\_\_$ and

0.1

$\delta = \_\_\_\_$.

2 Properties of Limits of Functions

$\lim_{x\to a} g(x)$

11 $\lim_{x\to a} [f(x) + g(x)] = \lim_{x\to a} f(x) + \_\_\_\_\_\_$.

$\lim_{x\to a} g(x)$

12 $\lim_{x\to a} [f(x) - g(x)] = \lim_{x\to a} f(x) - \_\_\_\_\_\_$.

$\lim_{x\to a} f(x)$

13 $\lim_{x\to a} [cf(x)] = c\left(\_\_\_\_\_\_\right)$.

$\lim_{x\to a} g(x)$

14 $\lim_{x\to a} [f(x) \cdot g(x)] = \lim_{x\to a} f(x) \cdot \_\_\_\_\_\_$.

$\lim_{x\to a} f(x)$

15 If $\lim_{x\to a} g(x) \neq 0$, then $\lim_{x\to a} \dfrac{f(x)}{g(x)} = \dfrac{\_\_\_\_\_\_}{\lim_{x\to a} g(x)}$.

$\lim_{x\to a} f(x)$

16 $\lim_{x\to a} [f(x)]^n = \left(\_\_\_\_\_\_\right)^n$.

$\lim_{x\to a} f(x)$

17 $\lim_{x\to a} \sqrt[n]{f(x)} = \sqrt[n]{\_\_\_\_\_\_}$ provided that the limit exists.

$\lim_{x\to a} f(x)$

18 $\lim_{x\to a} |f(x)| = |\_\_\_\_\_\_|$.

c

19 $\lim_{x\to a} c = \_\_\_\_$.

a

20 $\lim_{x\to a} x = \_\_\_\_$.

21 If h is a function such that $f(x) = h(x)$ holds for all values of x in some open interval containing a except possibly $x = a$, then

$\lim_{x\to a} f(x)$

$\lim_{x\to a} h(x) = \_\_\_\_\_\_$.

22 $\lim_{x\to a} [f_1(x) \pm f_2(x) \pm \cdots \pm f_n(x)] =$

$\lim_{x\to a} f_n(x)$

$\lim_{x\to a} f_1(x) \pm \lim_{x\to a} f_2(x) \pm \cdots \pm \_\_\_\_\_\_$.

In Problems 23 through 30, let $\lim_{x\to 2} f(x) = 4$ and $\lim_{x\to 2} g(x) = 25$. Use properties of limits to find each expression.

Answers	Problems
	23 $\lim_{x\to 2} [f(x) + g(x)]$.
$\lim_{x\to 2} g(x)$	$\lim_{x\to 2} [f(x)+g(x)] = \lim_{x\to 2} f(x) +$ ________
25, 29	$= 4 +$ ____ $=$ ____.
	24 $\lim_{x\to 2} [3f(x) + 5g(x)]$.
$\lim_{x\to 2} g(x)$	$\lim_{x\to 2} [3f(x) + 5g(x)] = 3 \lim_{x\to 2} f(x) + 5($________$)$
4, 25	$= 3($____$) + 5($____$)$
125, 137	$= 12 +$ ____ $=$ ____.
	25 $\lim_{x\to 2} [f(x) - g(x)]$.
$\lim_{x\to 2} g(x)$	$\lim_{x\to 2} [f(x) - g(x)] = \lim_{x\to 2} f(x) -$ ________
25, -21	$= 4 -$ ____ $=$ ____.
	26 $\lim_{x\to 2} [5f(x) - 2g(x)]$.
$\lim_{x\to 2} g(x)$	$\lim_{x\to 2} [5f(x) - 2g(x)] = 5 \lim_{x\to 2} f(x) - 2($________$)$
4, 25	$= 5($____$) - 2($____$)$
50	$= 20 -$ ____
-30	$=$ ____.
	27 $\lim_{x\to 2} [f(x) \cdot g(x)]$.
$\lim_{x\to 2} g(x)$	$\lim_{x\to 2} [f(x) \cdot g(x)] = \lim_{x\to 2} f(x) \cdot$ ________
25, 100	$= 4 \cdot$ ____ $=$ ____.
	28 $\lim_{x\to 2} [f(x)]^3$.
$\lim_{x\to 2} f(x)$, 4, 64	$\lim_{x\to 2} [f(x)]^3 = ($________$)^3 = ($____$)^3 =$ ____.
	29 $\lim_{x\to 2} \sqrt{f(x) \cdot g(x)}$.
$\lim_{x\to 2} [f(x) \cdot g(x)]$	$\lim_{x\to 2} \sqrt{f(x) \cdot g(x)} = \sqrt{\_\_\_\_\_\_\_\_}$
25, 10	$= \sqrt{4(\_\_\_\_)} =$ ____.
	30 $\lim_{x\to 2} \left\lvert\frac{f(x)}{g(x)}\right\rvert$.
	$\lim_{x\to 2} \left\lvert\frac{f(x)}{g(x)}\right\rvert = \left\lvert\lim_{x\to 2} \frac{f(x)}{g(x)}\right\rvert$
$\lim_{x\to 2} f(x)$	$= \frac{\lvert\_\_\_\_\_\_\_\_\rvert}{\lvert\lim_{x\to 2} g(x)\rvert}$
$\frac{4}{25}$	$= \frac{\_\_\_\_}{\_\_\_\_}$.

In Problems 31 through 43, find each limit.

31 $\lim_{x \to 2} (2x^2 - 5x + 3)$.

$\lim_{x \to 2} (2x^2 - 5x + 3) = \lim_{x \to 2} 2x^2 - \lim_{x \to 2} 5x +$ ______

$= 8 -$ ____ $+$ ____

$=$ ____.

Answer: $\lim_{x \to 2} 3$; 10, 3; 1

32 $\lim_{x \to 3} \dfrac{x^2 + 2x}{2x + 3}$.

$\lim_{x \to 3} \dfrac{x^2 + 2x}{2x + 3} = \dfrac{\text{\_\_\_\_\_\_}}{\lim_{x \to 3} (2x + 3)} = \dfrac{15}{\text{\_\_\_\_}}$.

$=$ ____.

Answer: $\lim_{x \to 3} (x^2 + 2x)$, 9; $\frac{5}{3}$

33 $\lim_{x \to 1} \dfrac{3 - x}{x + 2}$.

$\lim_{x \to 1} \dfrac{3 - x}{x + 2} = \dfrac{\text{\_\_\_\_\_\_}}{\lim_{x \to 1} (x + 2)} = \dfrac{2}{\text{\_\_\_\_}} =$ ____.

Answer: $\lim_{x \to 1} (3 - x)$, 3, $\frac{2}{3}$

34 $\lim_{x \to 1} \sqrt[3]{\dfrac{2x^2 + x + 5}{3x^2 + 19x + 5}}$.

$\lim_{x \to 1} \sqrt[3]{\dfrac{2x^2 + x + 5}{3x^2 + 19x + 5}} = \sqrt[3]{\lim_{x \to 1} \dfrac{\text{\_\_\_\_\_\_}}{3x^2 + 19x + 5}}$

$= \sqrt[3]{\dfrac{\lim_{x \to 1} (2x^2 + x + 5)}{\lim_{x \to 1} (3x^2 + 19x + 5)}}$

$= \sqrt[3]{\dfrac{\text{\_\_\_\_}}{27}} =$ ____.

Answer: $2x^2 + x + 5$; 8, $\frac{2}{3}$

35 $\lim_{x \to 2} \sqrt{\dfrac{x^3 + 2x + 4}{x^2 + 5}}$.

$\lim_{x \to 2} \sqrt{\dfrac{x^3 + 2x + 4}{x^2 + 5}} = \sqrt{\lim_{x \to 2} \dfrac{\text{\_\_\_\_\_\_}}{x^2 + 5}}$

$= \sqrt{\dfrac{\lim_{x \to 2} (x^3 + 2x + 4)}{\lim_{x \to 2} (x^2 + 5)}}$

$= \sqrt{\dfrac{\text{\_\_\_\_}}{9}} =$ ____.

Answer: $x^3 + 2x + 4$; 16, $\frac{4}{3}$

36 $\lim_{x \to 2} \dfrac{x^3 - 8}{x - 2}$.

$\dfrac{x^3 - 8}{x - 2} = \dfrac{(x - 2)(\text{\_\_\_\_\_\_})}{x - 2} =$ ______,

so that

$\lim_{x \to 2} \dfrac{x^3 - 8}{x - 2} = \lim_{x \to 2}$ ______ $=$ ____.

Answer: $x^2 + 2x + 4$, $x^2 + 2x + 4$; $x^2 + 2x + 4$, 12

Answers	Problems
	37 $\lim\limits_{x\to 3} \dfrac{2x^2 - 5x - 3}{3 - x}$.
$(x - 3),\ 2x + 1$	$\dfrac{2x^2 - 5x - 3}{3 - x} = \dfrac{(\underline{\hspace{2cm}})(2x + 1)}{3 - x} = -\underline{\hspace{2cm}}$, so that
$2x + 1,\ -7$	$\lim\limits_{x\to 3} \dfrac{2x^2 - 5x + 3}{3 - x} = -\lim\limits_{x\to 3} \underline{\hspace{2cm}} = \underline{\hspace{1cm}}$.
	38 $\lim\limits_{h\to 0} \dfrac{\sqrt{9 + h} - 3}{h}$.
$\sqrt{9 + h} + 3$	$\dfrac{\sqrt{9 + h} - 3}{h} = \dfrac{(\sqrt{9 + h} - 3)(\underline{\hspace{3cm}})}{h(\sqrt{9 + h} + 3)}$
	$= \dfrac{9 + h - 9}{h(\sqrt{9 + h} + 3)}$
h	$= \dfrac{\underline{\hspace{1cm}}}{h(\sqrt{9 + h} + 3)}$
$\dfrac{1}{\sqrt{9 + h} + 3}$	$= \underline{\hspace{2cm}}$,
	so that
$\sqrt{9 + h} + 3,\ \dfrac{1}{6}$	$\lim\limits_{h\to 0} \dfrac{\sqrt{9 + h} - 3}{h} = \lim\limits_{h\to 0} \dfrac{1}{\underline{\hspace{2cm}}} = \underline{\hspace{1cm}}$.
	39 $\lim\limits_{x\to 0} \dfrac{1}{x}\left(\dfrac{6 + x}{3 + 2x} - 2\right)$.
$6 + x - 6 - 4x$	$\lim\limits_{x\to 0} \dfrac{1}{x}\left(\dfrac{6 + x}{3 + 2x} - 2\right) = \lim\limits_{x\to 0} \dfrac{\underline{\hspace{3cm}}}{x(3 + 2x)}$
-3	$= \lim\limits_{x\to 0} \dfrac{\underline{\hspace{1cm}}}{3 + 2x}$
-1	$= \underline{\hspace{1cm}}$.
	40 $\lim\limits_{h\to 0} \dfrac{(1 + h)^3 - 1}{h}$.
$1 + 3h + 3h^2 + h^3$	$\lim\limits_{h\to 0} \dfrac{(1 + h)^3 - 1}{h} = \lim\limits_{h\to 0} \dfrac{(\underline{\hspace{3cm}}) - 1}{h}$
$3h + 3h^2 + h^3$	$= \lim\limits_{h\to 0} \dfrac{\underline{\hspace{3cm}}}{h}$
$3 + 3h + h^2$	$= \lim\limits_{h\to 0} \underline{\hspace{3cm}}$
3	$= \underline{\hspace{1cm}}$.
	41 $\lim\limits_{x\to 0} \dfrac{1}{x}\left[1 - \dfrac{1}{(1 + x)^2}\right]$.
$(1 + x)^2$	$\lim\limits_{x\to 0} \dfrac{1}{x}\left[1 - \dfrac{1}{(1 + x)^2}\right] = \lim\limits_{x\to 0} \dfrac{\underline{\hspace{2cm}} - 1}{x(1 + x)^2}$
$2x + x^2$	$= \lim\limits_{x\to 0} \dfrac{\underline{\hspace{2cm}}}{x(1 + x)^2}$
$2 + x,\ 2$	$= \lim\limits_{x\to 0} \dfrac{\underline{\hspace{2cm}}}{(1 + x)^2} = \underline{\hspace{1cm}}$.

42 $\lim_{x \to 0} \dfrac{\sqrt{1+x+x^2}-1}{x}$.

$\dfrac{\sqrt{1+x+x^2}-1}{x} = \dfrac{(\sqrt{1+x+x^2}-1)(\_\_\_\_\_\_)}{x(\sqrt{1+x+x^2}+1)}$ — $\sqrt{1+x+x^2}+1$

$= \dfrac{\_\_\_\_}{x(\sqrt{1+x+x^2}+1)}$ — $x + x^2$

$= \dfrac{\_\_\_\_}{\sqrt{1+x+x^2}+1}$, — $1 + x$

so that

$\lim_{x \to 0} \dfrac{\sqrt{1+x+x^2}-1}{x} = \lim_{x \to 0} \dfrac{\_\_\_\_}{\sqrt{1+x+x^2}+1}$ — $1 + x$

$= \_\_\_\_$. — $\frac{1}{2}$

43 $\lim_{x \to 0} \dfrac{\sqrt[3]{x+1}-1}{x}$.

$\dfrac{\sqrt[3]{x+1}-1}{x} = \dfrac{(\sqrt[3]{x+1}-1)(\_\_\_\_\_\_)}{x[(x+1)^{2/3}+(x+1)^{1/3}+1]}$ — $(x+1)^{2/3} + (x+1)^{1/3} + 1$

$= \dfrac{\_\_\_\_}{x[(x+1)^{2/3}+(x+1)^{1/3}+1]}$ — x

$= \dfrac{\_\_\_\_}{(x+1)^{2/3}+(x+1)^{1/3}+1}$, — 1

so that

$\lim_{x \to 0} \dfrac{\sqrt[3]{x+1}-1}{x} = \lim_{x \to 0} \dfrac{\_\_\_\_}{(x+1)^{2/3}+(x+1)^{1/3}+1}$ — 1

$= \_\_\_\_$. — $\frac{1}{3}$

3 Continuity—One-Sided Limits

44 A function f is said to be continuous at the number a if

(i) $f(a)$ is defined,

(ii) $\lim_{x \to a} f(x)$ ______, and — exists

(iii) ______ — $\lim_{x \to a} f(x) = f(a)$

In Problems 45 through 47, determine if the function is continuous at the given point.

45 $f(x) = x^2 - 2$ at $a = 0$.

$f(0) = 0^2 - 2 =$ ____, so that $f(0)$ is defined. Also, $\lim_{x \to 0} f(x) =$ — -2

$\lim_{x \to 0} (x^2 - 2) =$ ____ $= f(0)$. Therefore, f is ______ at — -2, continuous

____. — 0

46 $g(x) = |x + 1|$ at $a = -1$.

$g(-1) = |-1 + 1| =$ ____, so that $g(-1)$ is defined. Also, $\lim_{x \to -1} g(x) = \lim_{x \to -1} |x + 1| =$ ____ $= g(-1)$. Therefore, g is __________ at -1.

0
0
continuous

47 $f(x) = \sqrt{x^2 + 16}$ at $a = -3$.

$f(-3) = \sqrt{9 + 16} =$ ____, so that $f(-3)$ is defined. Also, $\lim_{x \to -3} f(x) = \lim_{x \to -3} \sqrt{x^2 + 16} =$ ____ $= f(-3)$. Therefore, f is __________ at -3.

5
5
continuous

48 If the function f is not continuous at the number a, then we say that f is __________ at a.

discontinuous

In Problems 49 through 57, find $\lim_{x \to a^+} f(x)$, $\lim_{x \to a^-} f(x)$, and determine if $\lim_{x \to a} f(x)$ exists; also, determine if f is continuous at a.

49 $f(x) = \begin{cases} 3x + 2 & \text{if } x < 1 \\ 7 - 2x & \text{if } x \geqslant 1 \end{cases}; a = 1.$

$\lim_{x \to 1^+} f(x) = \lim_{x \to 1^+} (7 - 2x) =$ ____,

$\lim_{x \to 1^-} f(x) = \lim_{x \to 1^-}$ (______) = ____.

Since $\lim_{x \to 1^+} f(x) = \lim_{x \to 1^-} f(x)$, then $\lim_{x \to 1} f(x)$ ______ and is equal to ____. $f(1) = 7 - 2($____$) = 5$, so that $\lim_{x \to 1} f(x) =$ ____ = ______ and f is __________ at 1. The graph is

5
$3x + 2$, 5
exists
5, 1
5, $f(1)$, continuous

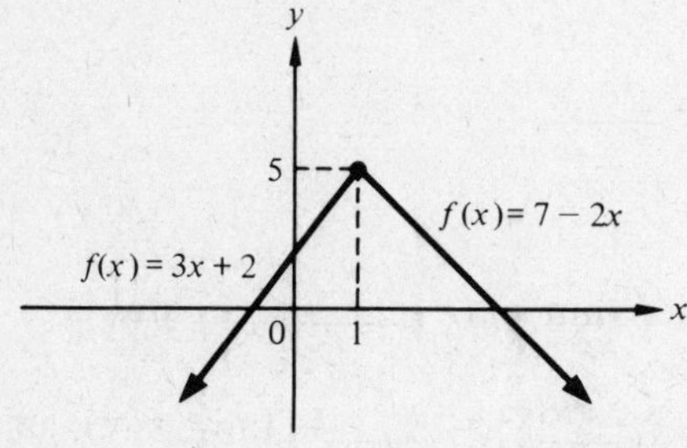

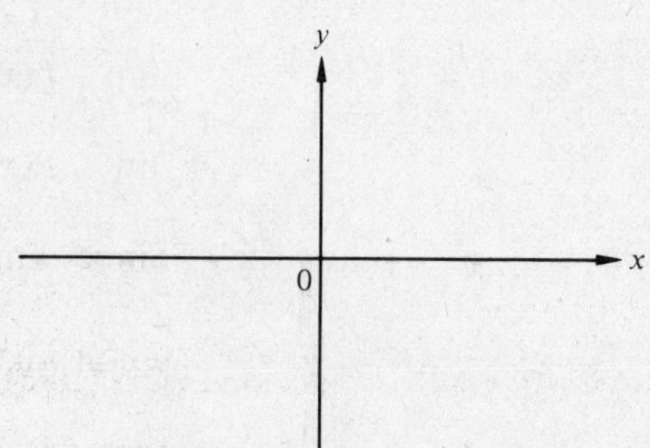

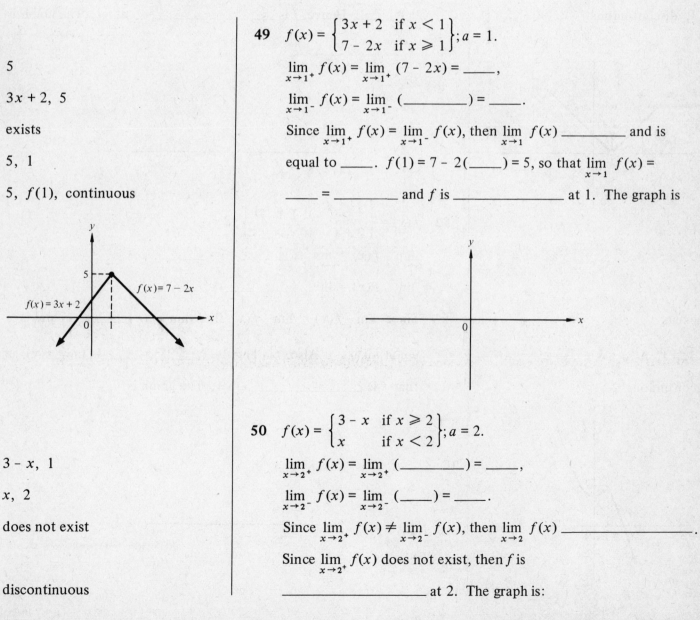

50 $f(x) = \begin{cases} 3 - x & \text{if } x \geqslant 2 \\ x & \text{if } x < 2 \end{cases}; a = 2.$

$\lim_{x \to 2^+} f(x) = \lim_{x \to 2^+}$ (______) = ____,

$\lim_{x \to 2^-} f(x) = \lim_{x \to 2^-}$ (____) = ____.

Since $\lim_{x \to 2^+} f(x) \neq \lim_{x \to 2^-} f(x)$, then $\lim_{x \to 2} f(x)$ __________.

Since $\lim_{x \to 2^+} f(x)$ does not exist, then f is __________ at 2. The graph is:

$3 - x$, 1
x, 2
does not exist
discontinuous

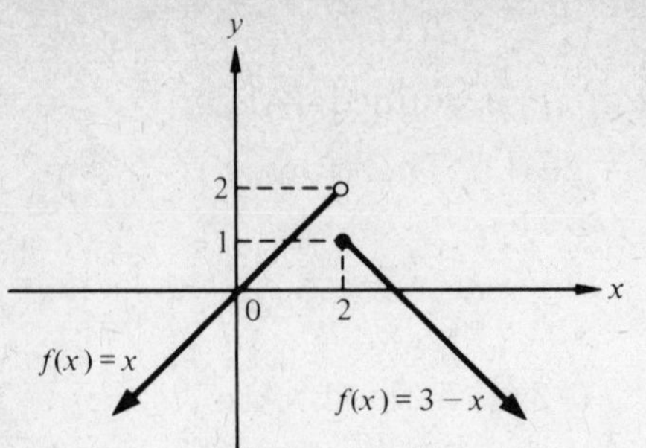

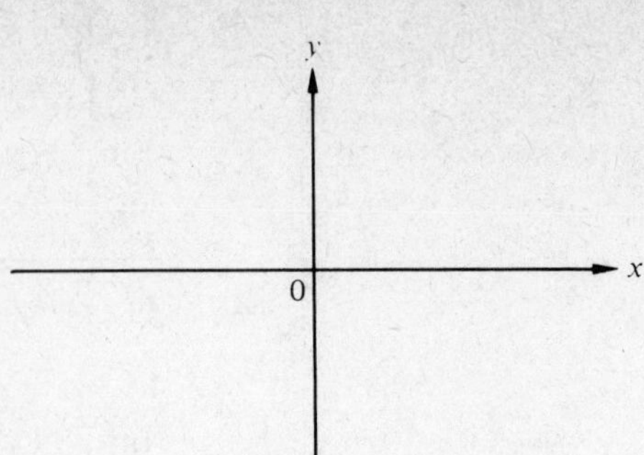

51 $f(x) = \begin{cases} |x+1| & \text{if } x \neq -1 \\ 1 & \text{if } x = -1 \end{cases}$; $a = -1$.

$\lim_{x\to -1} f(x) = \lim_{x\to -1} |x+1| = |\_\_\_\_\_\_\_\_\_\_| = \_\_\_\_$.

Therefore, $\lim_{x\to -1^+} f(x) = \_\_\_\_\_\_\_\_\_\_ = \lim_{x\to -1} f(x) = \_\_\_\_$

and $\lim_{x\to -1} f(x)$ ________. $f(-1) = \_\_\_\_$, so that $\lim_{x\to -1} f(x) =$

$0 \neq \_\_\_\_$. Hence, f is ____________ at −1. The graph is

$\lim_{x\to -1}(x+1)$, 0

$\lim_{x\to -1^-} f(x)$, 0

exists, 1

1, discontinuous

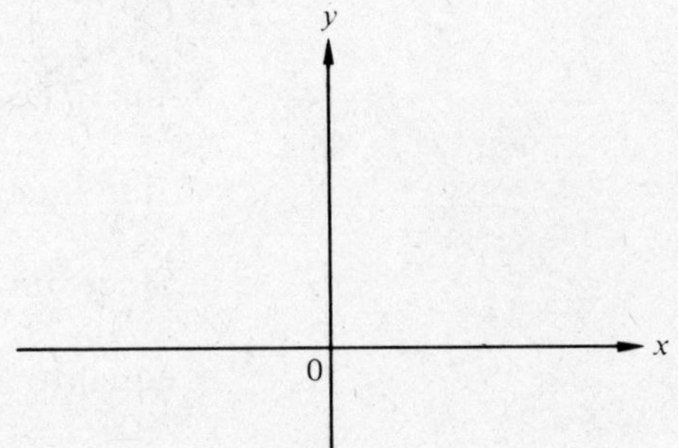

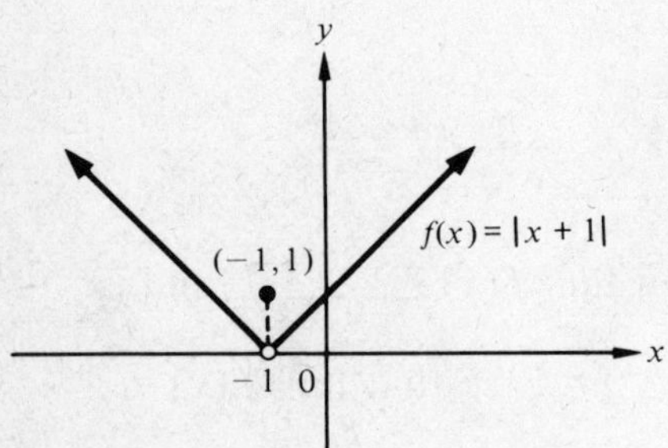

52 $f(x) = \begin{cases} 4 - x^2 & \text{if } x \leqslant -1 \\ 2 + x^2 & \text{if } x > -1 \end{cases}$; $a = -1$.

$\lim_{x\to -1^+} f(x) = \lim_{x\to -1^+} (\_\_\_\_\_\_\_\_) = \_\_\_\_$,

$\lim_{x\to -1^-} f(x) = \lim_{x\to -1^-} (\_\_\_\_\_\_\_\_) = \_\_\_\_$.

Since $\lim_{x\to -1^+} f(x) = \lim_{x\to -1^-} f(x)$, then $\lim_{x\to -1} f(x)$ ________ and is

equal to ____. Also, $f(-1) = 4 - (\_\_\_\_)^2 = \_\_\_\_ = \lim_{x\to -1} f(x)$, so

that f is ____________ at −1. The graph is

$2 + x^2$, 3

$4 - x^2$, 3

exists

3, −1, 3

continuous

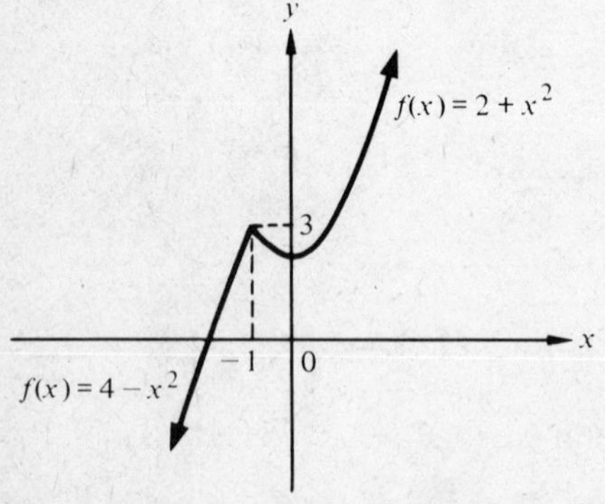

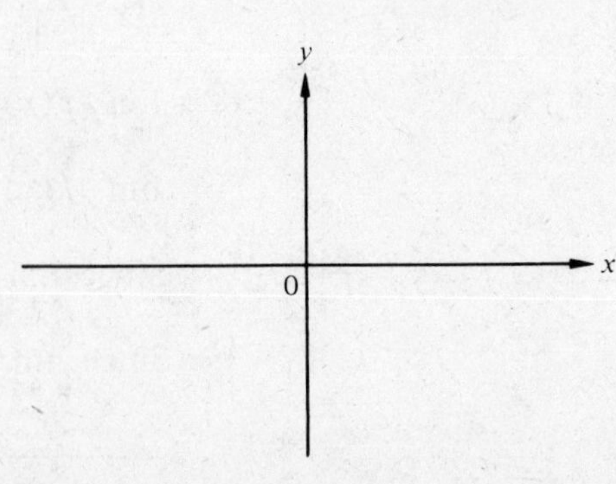

$x - 3$

$x + 3$

6

6, 6

6, discontinuous

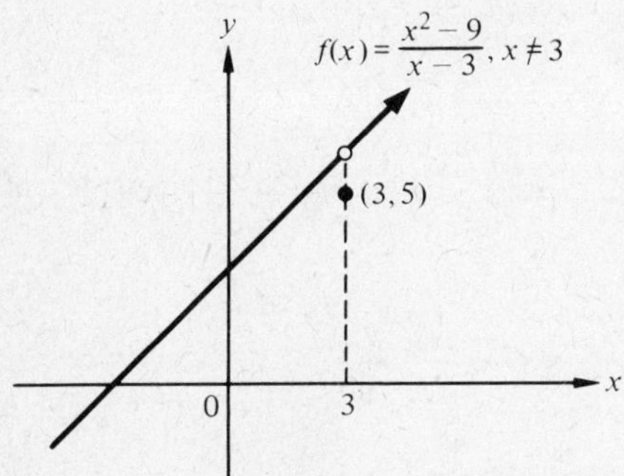

53 $f(x) = \begin{cases} \dfrac{x^2 - 9}{x - 3} & \text{if } x \neq 3 \\ 5 & \text{if } x = 3 \end{cases}$; $a = 3$.

For $x \neq 3$, $\dfrac{x^2 - 9}{x - 3} = \dfrac{(\_\_\_\_\_\_\_\_)(x + 3)}{x - 3}$

$= \_\_\_\_\_\_\_\_$;

hence, $\lim_{x \to 3^-} f(x) = \lim_{x \to 3^-} (x + 3) =$ _____ and $\lim_{x \to 3^+} f(x) =$ $\lim_{x \to 3^+} (x + 3) =$ _____. Therefore, $\lim_{x \to 3} f(x) =$ _____. Since $\lim_{x \to 3} f(x) =$ _____ $\neq 5 = f(3)$, then f is __________________ at 3.

The graph is

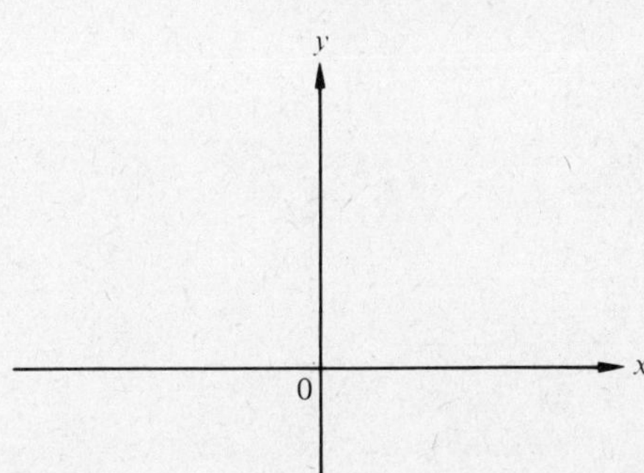

$\dfrac{x + 1}{|x + 1|}$

1, 1

$\dfrac{x + 1}{|x + 1|}$

−1, −1

exist

discontinuous

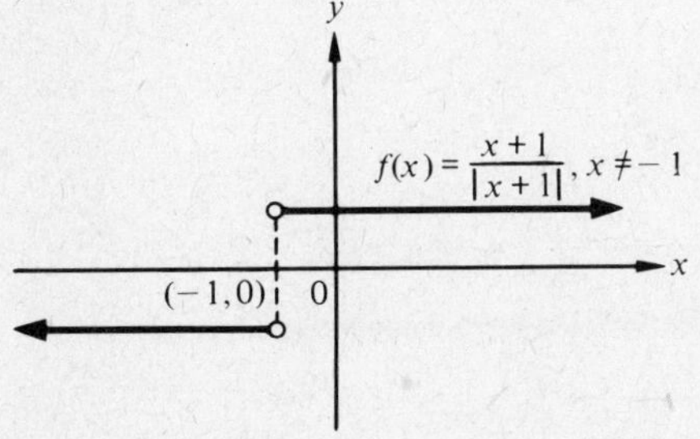

54 $f(x) = \begin{cases} \dfrac{x + 1}{|x + 1|} & \text{if } x \neq -1 \\ 0 & \text{if } x = -1 \end{cases}$; $a = -1$.

$\lim_{x \to -1^+} f(x) = \lim_{x \to -1^+} \left(\_\_\_\_\_\_\_\_\right) = \lim_{x \to -1^+} \dfrac{x + 1}{x + 1}$

$= \lim_{x \to -1^+} (\_\_\_\_) = \_\_\_\_$.

$\lim_{x \to -1^-} f(x) = \lim_{x \to -1^-} \left(\_\_\_\_\_\_\_\_\right) = \lim_{x \to -1^-} \dfrac{x + 1}{-(x + 1)}$

$= \lim_{x \to -1^-} (\_\_\_\_) = \_\_\_\_$.

$\lim_{x \to -1} f(x)$ does not ________, since $\lim_{x \to -1^+} f(x) \neq \lim_{x \to -1^-} f(x)$.

Therefore, f is __________________ at −1. The graph is

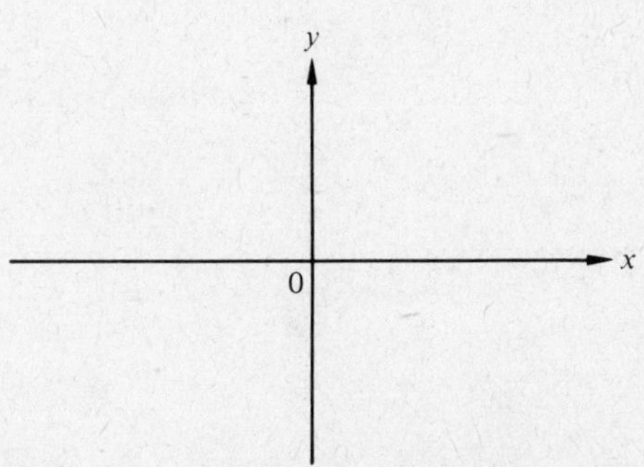

55 $f(x) = [\![x]\!] + [\![3 - x]\!]$; $a = 2$.

Suppose that $2 < x < 3$; then $[\![x]\!] =$ ____. Also, $0 < 3 - x < 1$, so that $[\![3 - x]\!] =$ ____. Then $f(x) = 2 +$ ____ $= 2$ for $2 < x < 3$; hence, $\lim_{x \to 2^+} f(x) = \lim_{x \to 2^+} 2 =$ ____. Suppose that $1 < x < 2$; then $[\![x]\!] =$ ____. Also, $1 < 3 - x < 2$ implies that $[\![3 - x]\!] =$ ____, so that $f(x) = 1 +$ ____ $=$ ____ for $1 < x < 2$. Then $\lim_{x \to 2^-} f(x) = \lim_{x \to 2^-} 2 =$ ____. $\lim_{x \to 2^+} f(x) = \lim_{x \to 2^-} f(x) =$ ____, so that $\lim_{x \to 2} f(x)$ exists and is equal to ____. $f(2) = [\![2]\!] + [\![3 - ($____$)]\!] = 2 + 1 =$ ____. Thus, $\lim_{x \to 2} f(x) = 2 \neq f(2)$, and f is ____________ at 2. The graph is

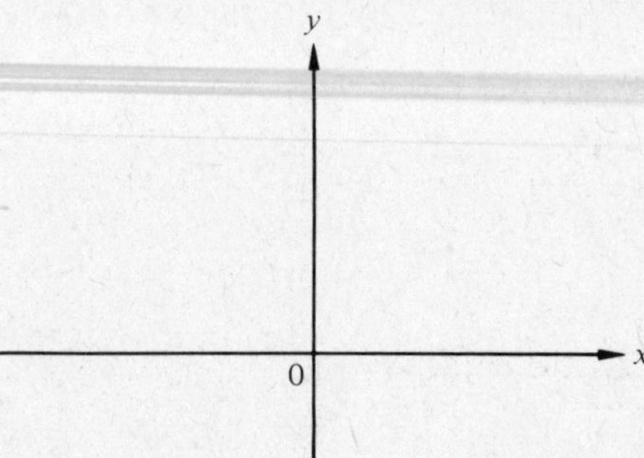

Answers: 2; 0, 0; 2; 1, 1; 1, 2; 2, 2; 2, 2; 3, discontinuous

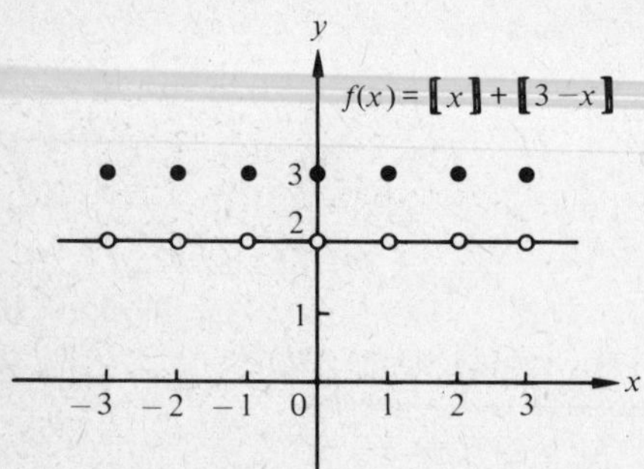

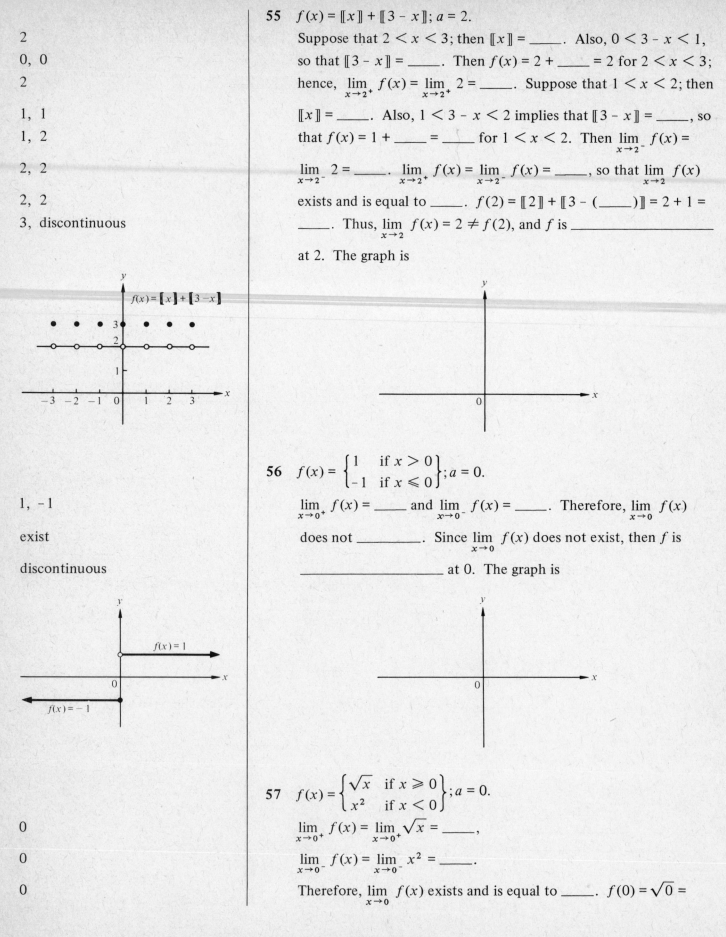

56 $f(x) = \begin{cases} 1 & \text{if } x > 0 \\ -1 & \text{if } x \leqslant 0 \end{cases}$; $a = 0$.

$\lim_{x \to 0^+} f(x) =$ ____ and $\lim_{x \to 0^-} f(x) =$ ____. Therefore, $\lim_{x \to 0} f(x)$ does not ______. Since $\lim_{x \to 0} f(x)$ does not exist, then f is ____________ at 0. The graph is

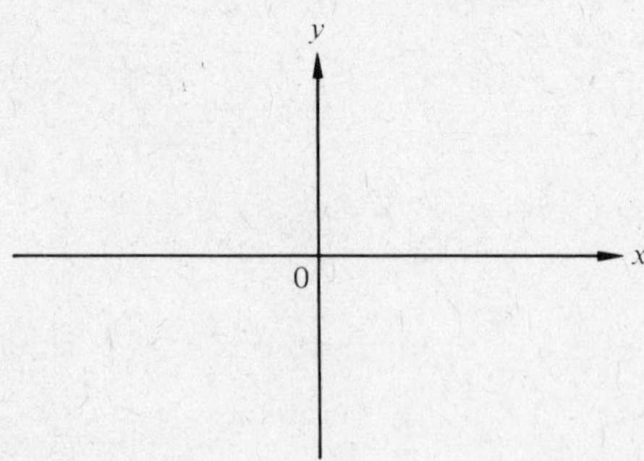

Answers: 1, −1; exist; discontinuous

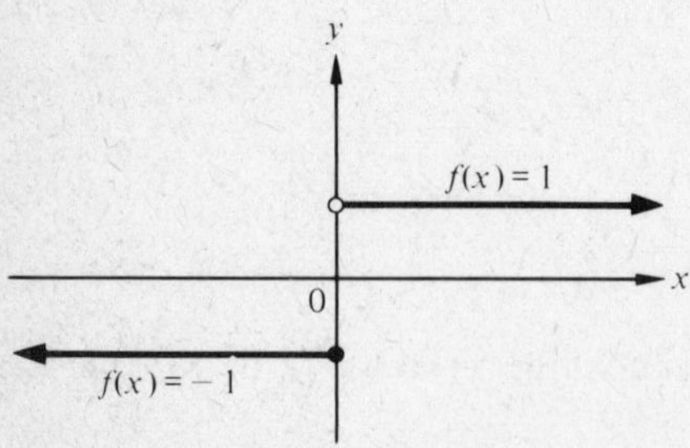

57 $f(x) = \begin{cases} \sqrt{x} & \text{if } x \geqslant 0 \\ x^2 & \text{if } x < 0 \end{cases}$; $a = 0$.

$\lim_{x \to 0^+} f(x) = \lim_{x \to 0^+} \sqrt{x} =$ ____,

$\lim_{x \to 0^-} f(x) = \lim_{x \to 0^-} x^2 =$ ____.

Therefore, $\lim_{x \to 0} f(x)$ exists and is equal to ____. $f(0) = \sqrt{0} =$

Answers: 0; 0; 0

0, continuous

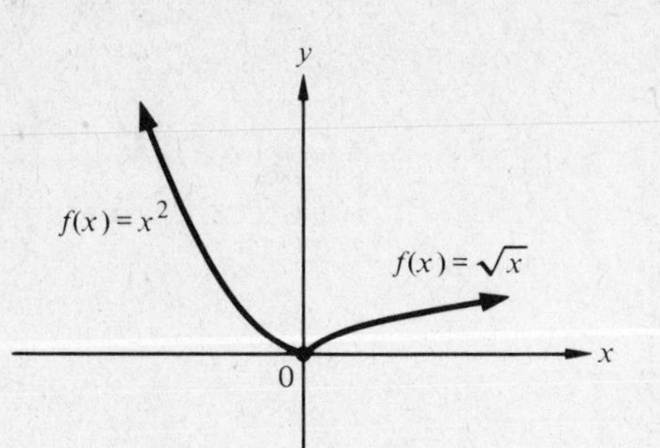

_____ $= \lim_{x \to 0} f(x)$. Hence, f is _____________ at 0. The graph is

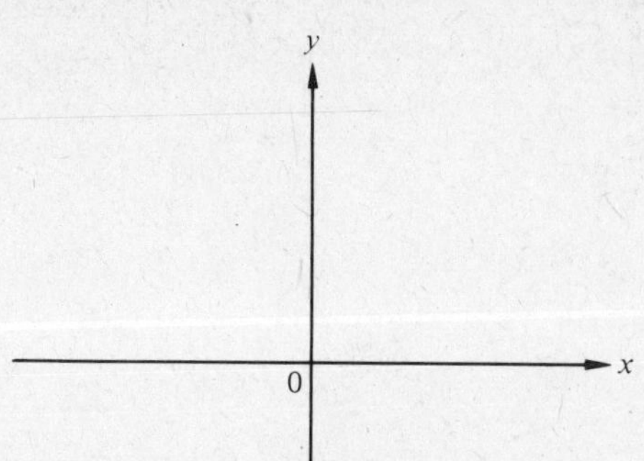

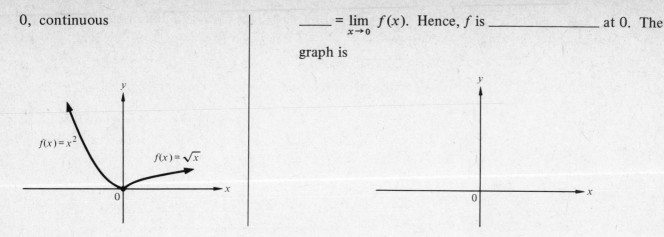

3.1 Formal Definitions of One-Sided Limits

$\lim_{x \to a^+} f(x) = L$

$0 < x - a < \delta$

58 Let f be a function defined at least in some open interval (a, b). We say that L is the limit of $f(x)$ as x approaches a from the right and we write _______________ if, for each positive number ϵ, there exists a positive number δ such that $|f(x) - L| < \epsilon$ will hold whenever _______________.

$\lim_{x \to a^-} f(x) = L$

$0 < a - x < \delta$

59 Let f be a function defined at least in some open interval (c, a). We say that L is the limit of $f(x)$ as x approaches a from the left and we write _______________ if, for each positive number ϵ, there exists a positive number δ such that $|f(x) - L| < \epsilon$ will hold whenever _______________.

exists

L

60 $\lim_{x \to a} f(x)$ exists and is equal to L if and only if both of the one-sided limits $\lim_{x \to a^+} f(x)$ and $\lim_{x \to a^-} f(x)$ _________ and their common value is _____.

4 Properties of Continuous Functions

continuous

61 If f and g are continuous at a, then $f + g$, $f - g$, and $f \cdot g$ are _____________ at a.

continuous

62 If f and g are continuous at a, and $g(a) \neq 0$, then $\frac{f}{g}$ is _____________ at a.

continuous

63 If g is continuous at a and f is continuous at $g(a)$, then $f \circ g$ is _____________ at a.

every number

64 A polynomial function is continuous at _____________.

continuous

65 A rational function is _____________ at every number for which it is defined.

In Problems 66 through 71, sketch the graph of each function and determine the numbers a at which the function is continuous.

66 $f(x) = |3x|$.

The graph is

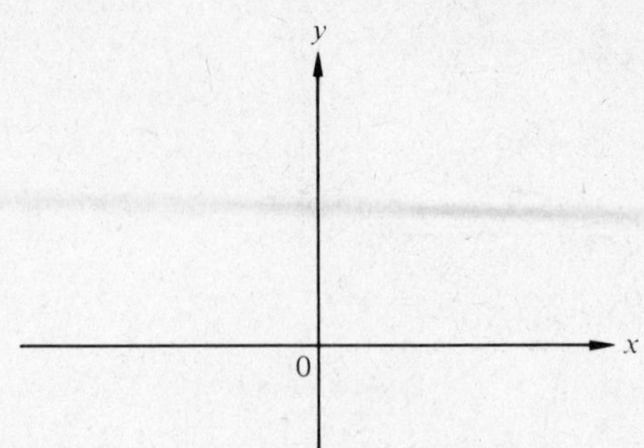

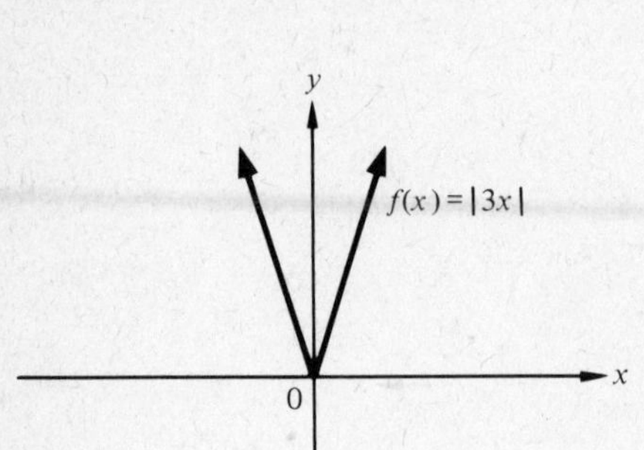

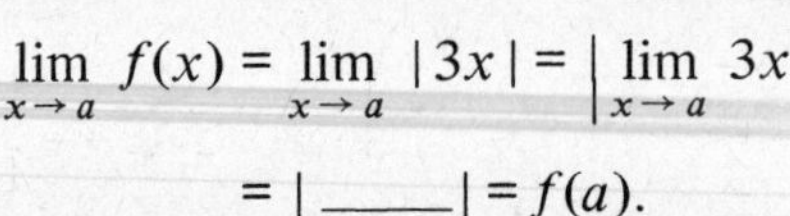

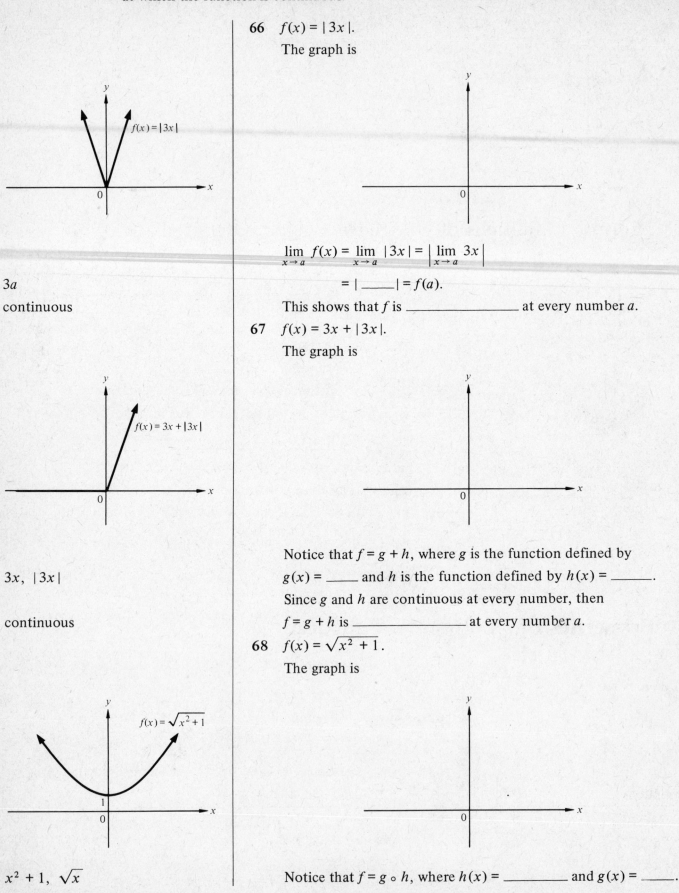

$$\lim_{x \to a} f(x) = \lim_{x \to a} |3x| = \left| \lim_{x \to a} 3x \right|$$

$= |\_\_\_\_| = f(a)$.

This shows that f is ________________ at every number a.

$3a$

continuous

67 $f(x) = 3x + |3x|$.

The graph is

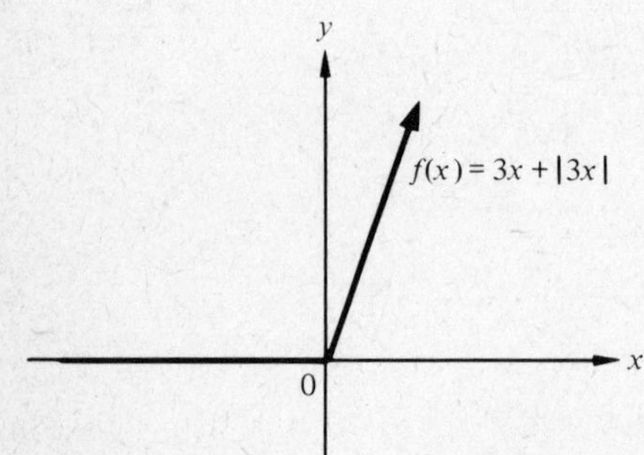

Notice that $f = g + h$, where g is the function defined by $g(x) =$ _____ and h is the function defined by $h(x) =$ ______. Since g and h are continuous at every number, then $f = g + h$ is ________________ at every number a.

$3x$, $|3x|$

continuous

68 $f(x) = \sqrt{x^2 + 1}$.

The graph is

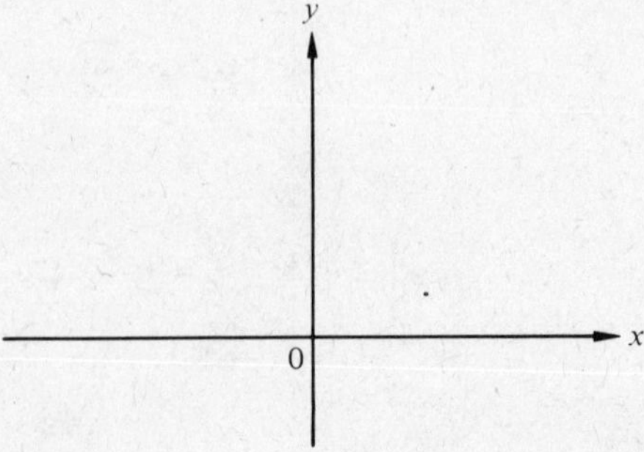

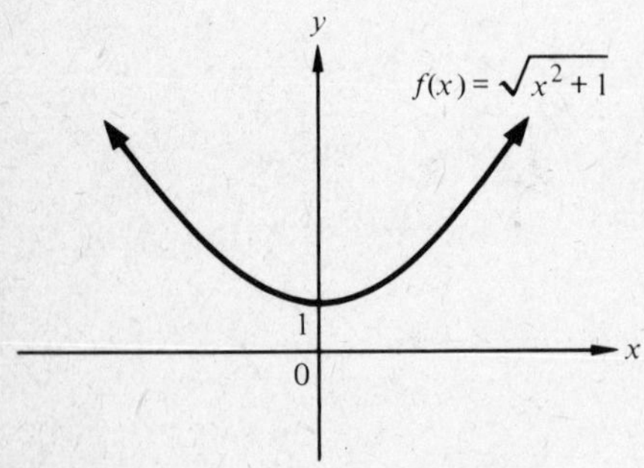

Notice that $f = g \circ h$, where $h(x) =$ _________ and $g(x) =$ _____.

$x^2 + 1$, $\sqrt{x}$

continuous

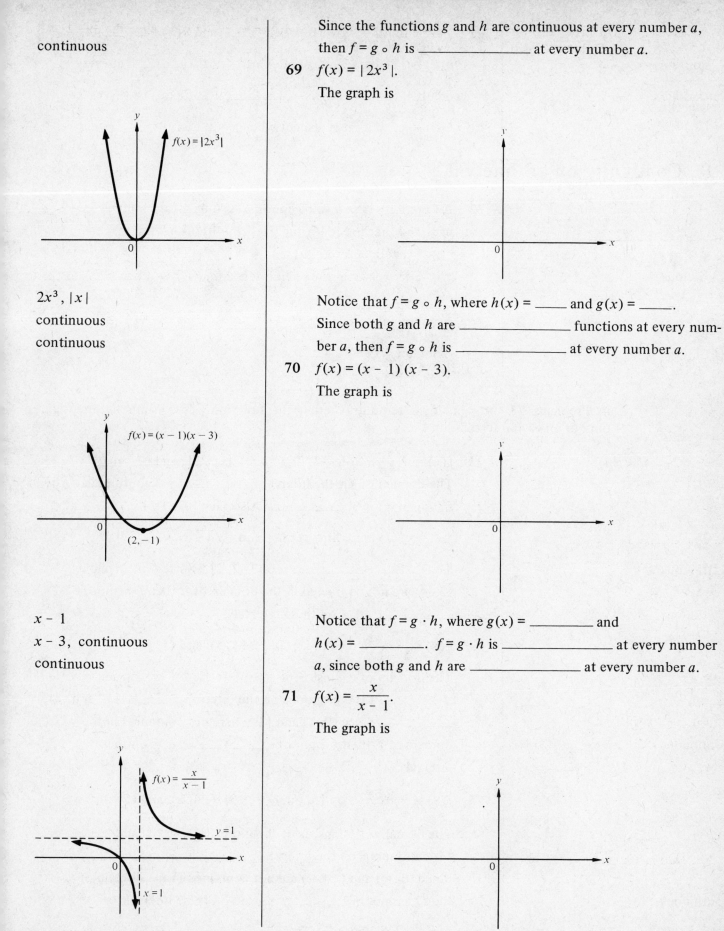

Since the functions g and h are continuous at every number a, then $f = g \circ h$ is ______________ at every number a.

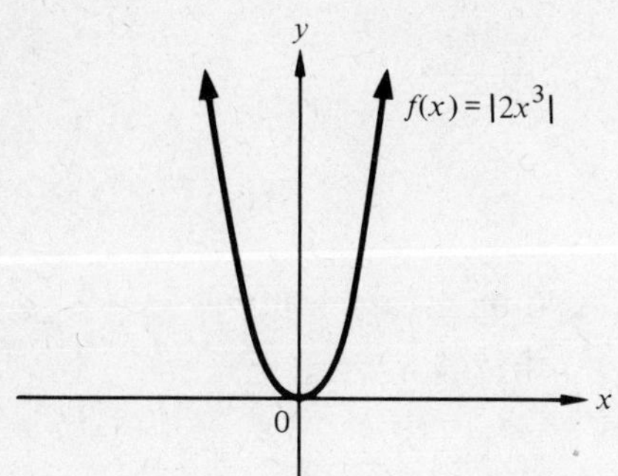

69 $f(x) = |2x^3|$.

The graph is

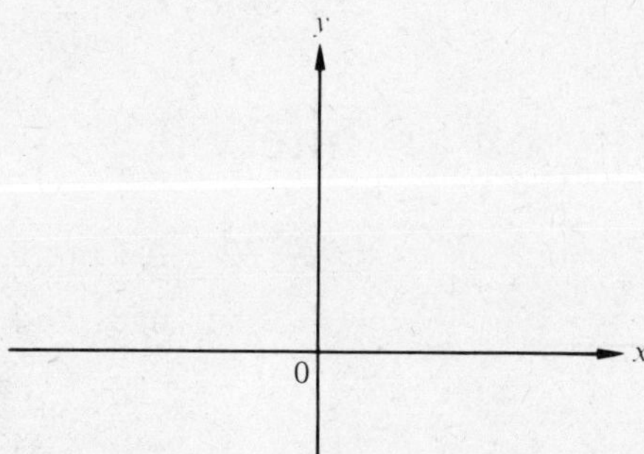

$2x^3$, $|x|$
continuous
continuous

Notice that $f = g \circ h$, where $h(x) =$ _____ and $g(x) =$ _____. Since both g and h are ______________ functions at every number a, then $f = g \circ h$ is ______________ at every number a.

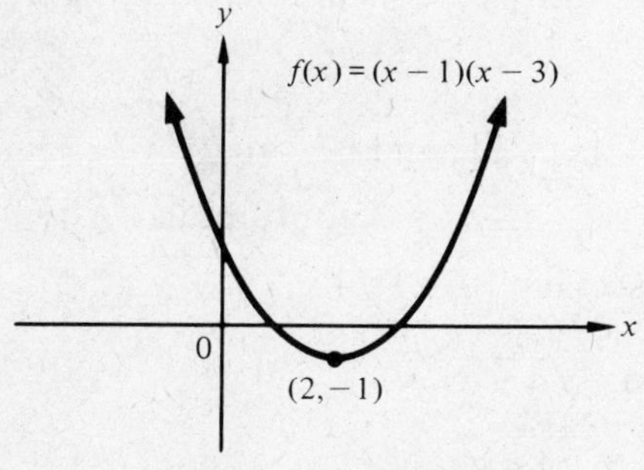

70 $f(x) = (x - 1)(x - 3)$.

The graph is

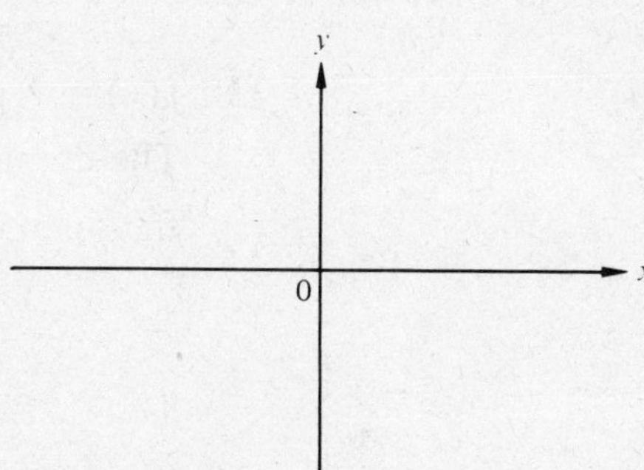

$x - 1$
$x - 3$, continuous
continuous

Notice that $f = g \cdot h$, where $g(x) =$ _________ and $h(x) =$ _________. $f = g \cdot h$ is ______________ at every number a, since both g and h are ______________ at every number a.

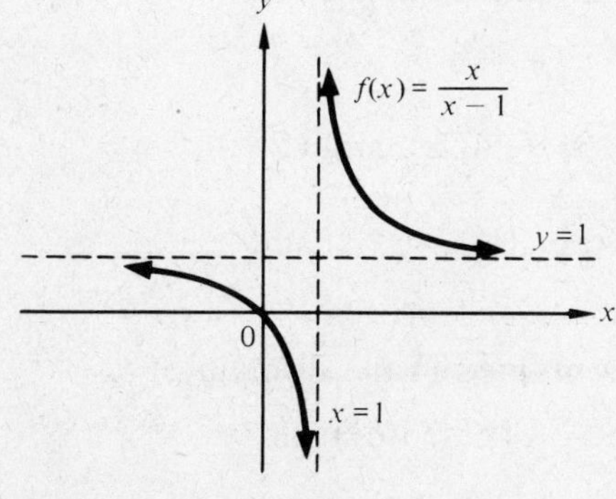

71 $f(x) = \dfrac{x}{x - 1}$.

The graph is

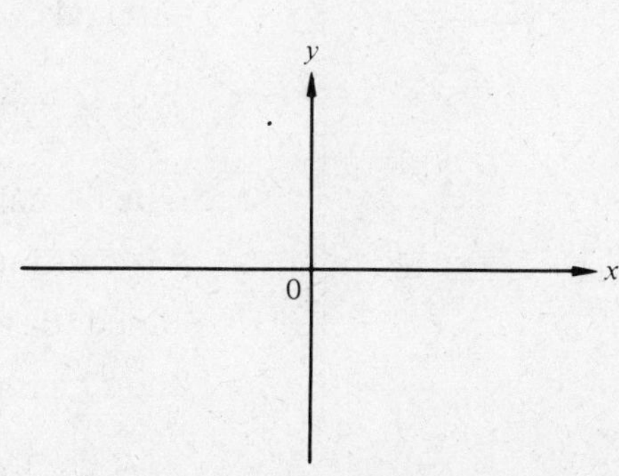

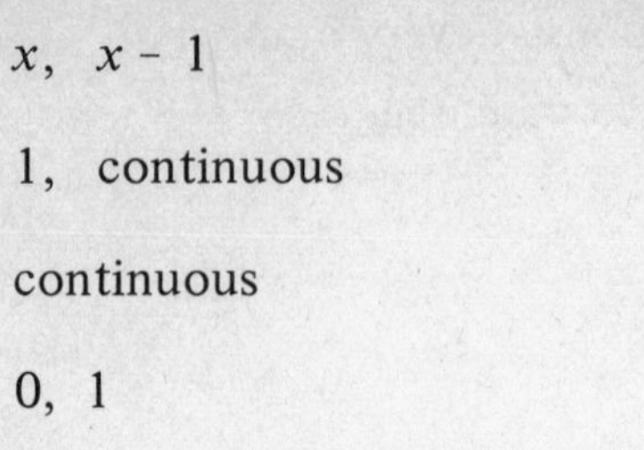
x, $x - 1$
1, continuous
continuous
0, 1

Notice that $f = \frac{g}{h}$, where $g(x) =$ ____ and $h(x) =$ ________. $h(x) = 0$ at $x =$ ____. g and h are ____________ at every number, so that $f = \frac{g}{h}$ is ____________ except when $h(x) =$ ____, that is, except when $x =$ ____.

4.1 Continuity on an Interval

continuous
$f(a)$
$f(b)$

72 A function f is said to be continuous on a closed interval $[a, b]$ provided that the following four conditions hold:

(i) f is defined on $[a, b]$.

(ii) For every c in the open interval (a, b), f is ____________ at c.

(iii) $\lim_{x \to a^+} f(x) =$ ______.

(iv) $\lim_{x \to b^-} f(x) =$ ______.

In Problems 73 through 76, determine whether the function f is continuous or discontinuous on each interval.

$[-2, 2]$
$(-2, 2)$
0, 0
continuous
continuous

73 $f(x) = \sqrt{4 - x^2}$ on $(-3, 2)$, $(-2, 2)$, $[-2, 2]$, and $(2, \infty)$.
The domain of f is the interval ________. f is continuous on the open interval ________. Also, $\lim_{x \to 2^-} f(x) = \lim_{x \to 2^-} \sqrt{4 - x^2} =$ ____ $= f(2)$ and $\lim_{x \to -2^+} f(x) = \lim_{x \to -2^+} \sqrt{4 - x^2} =$ ____ $= f(2)$.
Thus, f is ____________ on $[-2, 2]$. Since $(-3, 2)$ and $(2, \infty)$ are not contained in the domain of f, then f cannot be ____________ on these intervals.

$(-\infty, 1)$
$(1, \infty)$, $(-\infty, -2)$
$(1, \infty)$
continuous

74 $g(x) = \dfrac{2}{x - 1}$ on $(-\infty, -2)$, $(0, 2)$, $[1, 3)$, and $(1, \infty)$.
The domain of g consists of the intervals ________ and ________. f is continuous on the intervals ________ and ________. Since $(0, 2)$ and $[1, 3)$ are not contained in the domain of f, then f cannot be ____________ on these intervals.

$(3, \infty)$
$(-\infty, -3)$, $(5, \infty)$
continuous

75 $f(x) = \dfrac{x + 3}{\sqrt{x^2 - 9}}$ on $[3, \infty)$, $(-\infty, -3)$, $(-4, 2)$, and $(5, \infty)$.
The domain of f consists of the intervals $(-\infty, -3)$ and ________. f is continuous on the intervals ________ and ________. Since $[3, \infty)$ and $(-4, 2)$ are not contained in the domain of f, then f cannot be ____________ on these intervals.

$(-\infty, 3)$
$(3, \infty)$
continuous
$[-1, 1]$

76 $g(x) = \dfrac{|x - 3|}{x - 3}$ on $[-1, 1]$, $(-3, \infty)$, $(-\infty, 3]$, and $[-4, 5]$.

The domain of g consists of the intervals ________ and ________. Since $(-3, \infty)$, $(-\infty, 3]$, and $[-4, 5]$ are not contained in the domain of g, then g cannot be ______________ on these intervals. g is continuous on ________.

5 Limits Involving Infinity

77 **In finding one-sided infinite limits, there are four possibilities:**

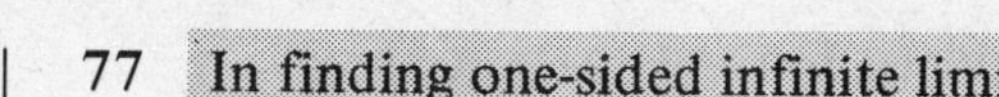

(i) $\lim_{x \to a^+} f(x) = +\infty$. The graph is:

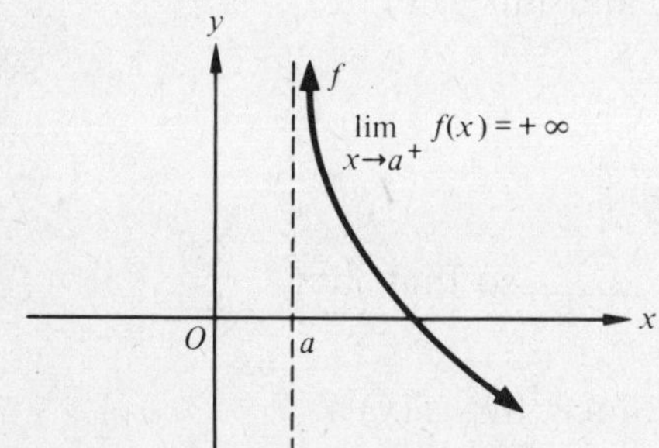

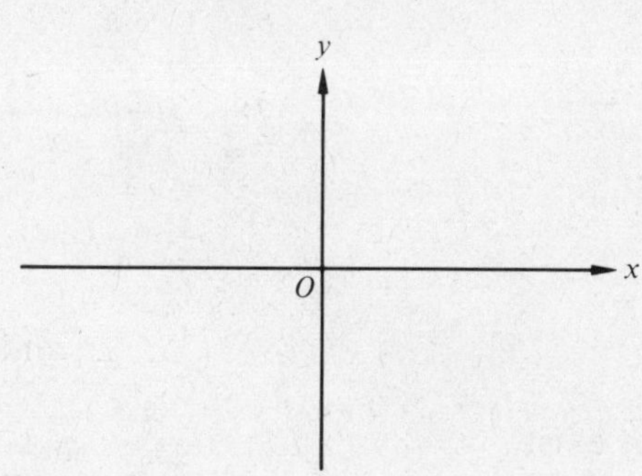

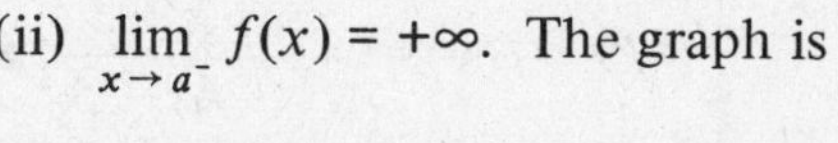

(ii) $\lim_{x \to a^-} f(x) = +\infty$. The graph is

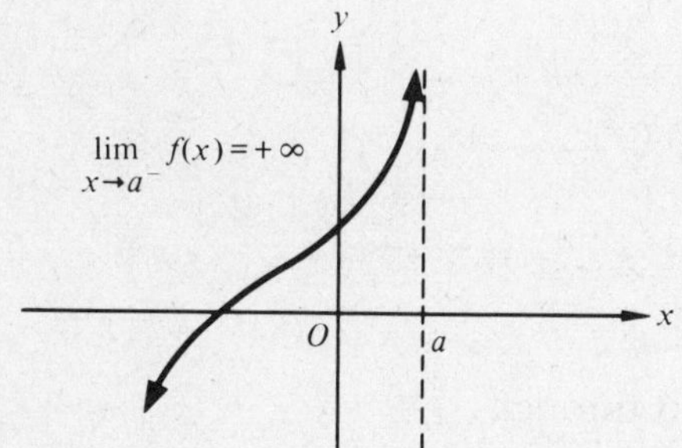

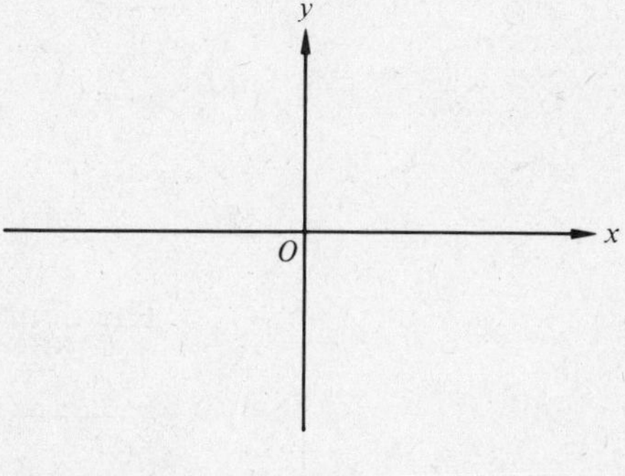

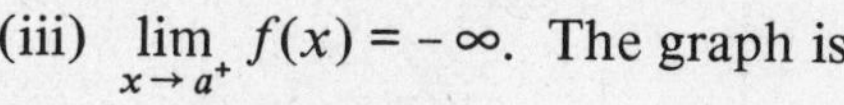

(iii) $\lim_{x \to a^+} f(x) = -\infty$. The graph is

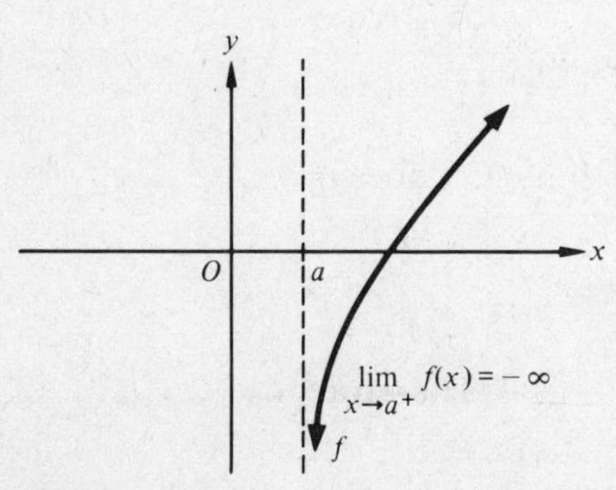

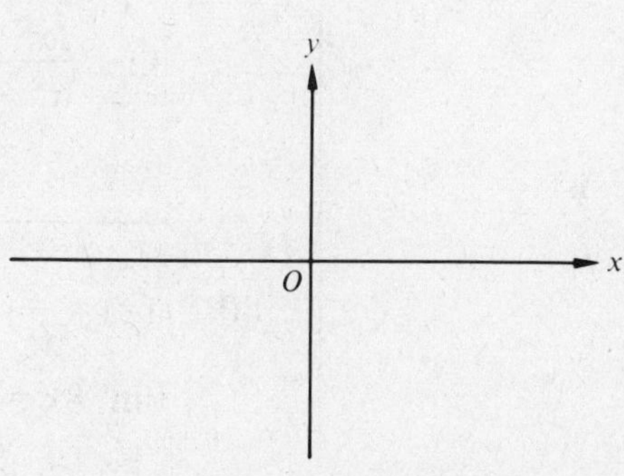

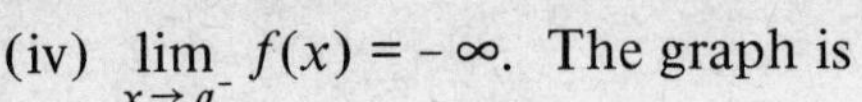

(iv) $\lim_{x \to a^-} f(x) = -\infty$. The graph is

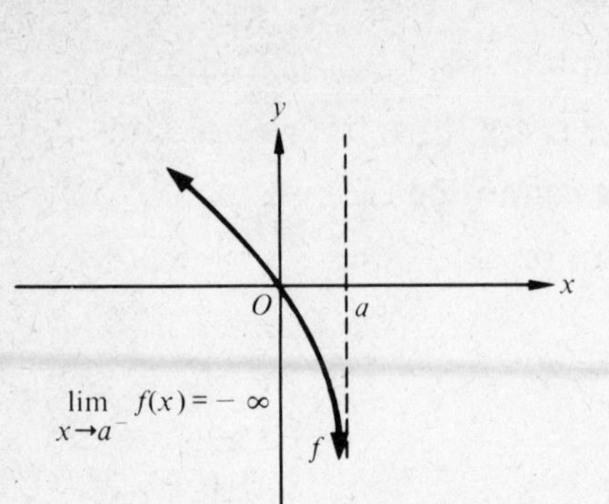

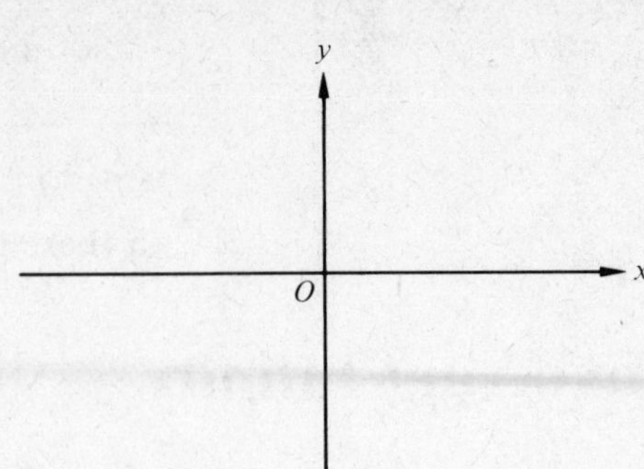

In Problems 78 through 81, determine $\lim_{x \to a^+} f(x)$, $\lim_{x \to a^-} f(x)$, and $\lim_{x \to a} f(x)$.

78 $f(x) = \dfrac{3x}{x-2}$, $a = 2$.

$\lim_{x \to 2} (3x) =$ ____. $\lim_{x \to 2^+} (x - 2) =$ ____, so that $\lim_{x \to 2^+} \dfrac{3x}{x-2} =$ ____, and $\lim_{x \to 2^-} \dfrac{3x}{x-2} =$ ____. Since $\lim_{x \to 2^+} f(x) \neq$ ________, then $\lim_{x \to 2} f(x)$ ____________.

6, 0

$+\infty$, $-\infty$

$\lim_{x \to 2^-} f(x)$, does not exist

79 $f(x) = \dfrac{x^2 + x + 3}{x^2 - 4x - 5}$, $a = 5$.

$\lim_{x \to 5} (x^2 + x + 3) =$ ____.

$\lim_{x \to 5^+} (x^2 - 4x - 5) = \lim_{x \to 5^+} (x - 5)($____$)$

$= \lim_{x \to 5^+} (x - 5) \cdot \lim_{x \to 5^+}$ ____

$= 0 \cdot$ ____ $=$ ____.

The denominator is approaching 0 through ____________. Therefore, $\lim_{x \to 5^+} \dfrac{x^2 + x + 3}{x^2 - 4x - 5} =$ ____. $\lim_{x \to 5^-} (x^2 - 4x - 5) =$ ____. The denominator is approaching 0 through ____________, so that $\lim_{x \to 5^-} \dfrac{x^2 + x + 3}{x^2 - 4x - 5} =$ ____. Since $\lim_{x \to 5^+} \dfrac{x^2 + x + 3}{x^2 - 4x - 5} \neq$ ____________, then $\lim_{x \to 5} f(x)$ does not ____.

33

$x + 1$

$x + 1$

6, 0

positive values

$+\infty$, 0

negative values

$-\infty$

$\lim_{x \to 5^-} \dfrac{x^2 + x + 3}{x^2 - 4x - 5}$, exist

80 $f(x) = \dfrac{8x}{(x-3)^2}$, $a = 3$.

$\lim_{x \to 3} 8x =$ ____. $\lim_{x \to 3} (x - 3)^2 =$ ____. Therefore,

$\lim_{x \to 3^+} \dfrac{8x}{(x-3)^2} =$ ____,

24, 0

$+\infty$

$+\infty$

$$\lim_{x\to 3^-} \frac{8x}{(x-3)^2} = \_\_\_\_,$$

$+\infty$

$$\lim_{x\to 3} \frac{8x}{(x-3)^2} = \_\_\_\_.$$

81 $f(x) = \dfrac{-3}{(x+2)^2}$, $a = -2$.

0

$\lim_{x\to -2} (x+2)^2 = \_\_\_\_$, so that

$-\infty$

$$\lim_{x\to -2^+} \frac{-3}{(x+2)^2} = \_\_\_\_,$$

$-\infty$

$$\lim_{x\to -2^-} \frac{-3}{(x+2)^2} = \_\_\_\_,$$

$-\infty$

$$\lim_{x\to -2} \frac{-3}{(x+2)^2} = \_\_\_\_.$$

5.1 Limits at Infinity

82 **In calculating limits at infinity, it is often useful to notice that, for any positive integer p,**

x^p, 0

(i) $\lim_{x\to +\infty} \left(\frac{1}{x}\right)^p = \lim_{x\to +\infty} \frac{1}{\_\_\_\_} = \_\_\_\_.$

x^p, 0

(ii) $\lim_{x\to -\infty} \left(\frac{1}{x}\right)^p = \lim_{x\to -\infty} \frac{1}{\_\_\_\_} = \_\_\_\_.$

In Problems 83 through 88, evaluate each limit.

83 $\lim_{x\to +\infty} \dfrac{x^2}{7x^2+3}$.

We divide the numerator and denominator by

the highest power, x^2

____________ of x, namely ____, so that

$\frac{3}{x^2}$

$$\lim_{x\to +\infty} \frac{x^2}{7x^2+3} = \lim_{x\to +\infty} \frac{1}{7 + \_\_\_\_}$$

0, $\frac{1}{7}$

$$= \frac{1}{7 + \_\_\_\_} = \_\_\_\_$$

84 $\lim_{x\to +\infty} \dfrac{5x^3}{1+2x-4x^3}$.

x^3

Divide the numerator and the denominator by ____, so that

$\frac{2}{x^2}$

$$\lim_{x\to +\infty} \frac{5x^3}{1+2x-4x^3} = \lim_{x\to +\infty} \frac{5}{\frac{1}{x^3} + \_\_\_\_ - 4}$$

$-\frac{5}{4}$

$$= \frac{5}{0+0-4} = \_\_\_\_.$$

85 $\lim_{x\to-\infty} \frac{3x^4 - 12x^2 + 7}{11x^4 + 13}$.

Divide the numerator and the denominator by ____, so that

$$\lim_{x\to-\infty} \frac{3x^4 - 12x^2 + 7}{11x^4 + 13} = \lim_{x\to-\infty} \frac{3 - \frac{12}{x^2} + \frac{7}{x^4}}{11 + \underline{\qquad}}$$

$$= \frac{3 - 0 + 0}{11 + \underline{\qquad}} = \underline{\qquad}.$$

86 $\lim_{y\to+\infty} \frac{5y^2 + 4}{3y + 7}$.

$$\lim_{y\to+\infty} \frac{5y^2 + 4}{3y + 7} = \lim_{y\to+\infty} \frac{5 + \frac{4}{y^2}}{\frac{3}{y} + \underline{\qquad}}$$

$$= \frac{5 + \underline{\qquad}}{0 + 0} = \underline{\qquad}.$$

87 $\lim_{x\to+\infty} \frac{7x}{\sqrt[3]{8x^3 + 1}}$.

$$\lim_{x\to+\infty} \frac{7x}{\sqrt[3]{8x^3 + 1}} = \lim_{x\to+\infty} \frac{7}{\frac{1}{x}\sqrt[3]{\underline{\qquad}}}$$

$$= \lim_{x\to+\infty} \frac{7}{\sqrt[3]{\underline{\qquad}}}$$

$$= \lim_{x\to+\infty} \frac{7}{\sqrt[3]{8 + \frac{1}{x^3}}} = \underline{\qquad}.$$

88 $\lim_{x\to+\infty} \frac{\sqrt{x^2 + 13}}{6x + 5}$.

$$\lim_{x\to+\infty} \frac{\sqrt{x^2 + 13}}{6x + 5} = \lim_{x\to+\infty} \frac{\frac{1}{x}\sqrt{x^2 + 13}}{6 + \underline{\qquad}}$$

$$= \lim_{x\to+\infty} \frac{\sqrt{1 + \frac{13}{x^2}}}{6 + \underline{\qquad}} = \frac{\sqrt{1 + 0}}{6 + 0} = \underline{\qquad}.$$

Answers (margin): 85: x^4; $\frac{13}{x^4}$; $0, \frac{3}{11}$. 86: $\frac{7}{y^2}$; $0, +\infty$. 87: $8x^3 + 1$; $\frac{1}{x^3}(8x^3 + 1)$; $\frac{7}{2}$. 88: $\frac{5}{x}$; $\frac{5}{x}, \frac{1}{6}$.

In Problems 89 through 97, let c denote any number such that $-\infty < c < \infty$. Then we "calculate" with the symbols $+\infty$ and $-\infty$ as follows:

89 $(+\infty) \pm c = (\pm c) + (+\infty) = (+\infty) + (+\infty) = \underline{\qquad}$.

90 $(-\infty) + c = c + (-\infty) = \underline{\qquad} + \underline{\qquad} = -\infty$.

Answers (margin): 89: $+\infty$. 90: $-\infty, -\infty$.

Answer	Problem
$+\infty$	**91** $c(+\infty) = (+\infty)\,c =$ ____ if $c > 0$.
$-\infty$	**92** $c(+\infty) = (+\infty)\,c =$ ____ if $c < 0$.
$-\infty$	**93** $c(-\infty) = (-\infty)\,c =$ ____ if $c > 0$.
$+\infty$	**94** $c(-\infty) = (-\infty)\,c =$ ____ if $c < 0$.
$+\infty$	**95** $(+\infty)(+\infty) = (-\infty)(-\infty) =$ ____.
$-\infty$	**96** $(+\infty)(-\infty) = (-\infty)(+\infty) =$ ____.
0	**97** $\dfrac{c}{+\infty} = \dfrac{c}{-\infty} =$ ____.

In Problems 98 through 100, evaluate each limit.

98 $\lim\limits_{x\to 2}\left(\dfrac{1}{x-2} - \dfrac{4}{x^2-4}\right)$.

$x + 2 - 4,\ x - 2$

$$\frac{1}{x-2} - \frac{4}{x^2-4} = \frac{\underline{\qquad}}{x^2-4} = \frac{\underline{\qquad}}{x^2-4}$$

$$= \frac{x-2}{(x-2)(x+2)}$$

$x + 2$

$$= \frac{1}{\underline{\qquad}},$$

so that

$x + 2$

$$\lim_{x\to 2}\left(\frac{1}{x-2} - \frac{4}{x^2-4}\right) = \lim_{x\to 2}\frac{1}{\underline{\qquad}}$$

$\dfrac{1}{4}$

$$= \underline{\qquad}.$$

99 $\lim\limits_{t\to+\infty}(t^2 - 7t)$.

$t - 7$

$$\lim_{t\to+\infty}(t^2 - 7t) = \lim_{t\to+\infty} t(\underline{\qquad})$$

$\lim\limits_{t\to+\infty}(t - 7)$

$$= \lim_{t\to+\infty} t\left(\underline{\qquad\qquad}\right)$$

$+\infty$

$$= (+\infty)\cdot\underline{\qquad}.$$

$+\infty$

$$= \underline{\qquad}.$$

100 $\lim\limits_{x\to+\infty}(\sqrt{x^2+x} - x)$.

$\sqrt{x^2+x} + x$

$$\sqrt{x^2+x} - x = \frac{(\sqrt{x^2+x} - x)(\underline{\qquad\qquad})}{\sqrt{x^2+x} + x}$$

x

$$= \frac{\underline{\qquad}}{\sqrt{x^2+x} + x}.$$

$$\lim_{x\to+\infty}(\sqrt{x^2+x} - x) = \lim_{x\to+\infty}\frac{x}{\sqrt{x^2+x} + x}$$

1

$$= \lim_{x\to+\infty}\frac{1}{\sqrt{1 + \frac{1}{x}} + \underline{\qquad}}$$

$\dfrac{1}{2}$

$$= \frac{1}{1+1} = \underline{\qquad}.$$

5.2 Formal Definitions of Limits Involving Infinity

101 $\lim_{x \to a^+} f(x) = +\infty$ means that f is defined at least in some open interval (a, b), and that, given any positive number M, no matter how large, there exists a positive number δ such that $f(x) > M$ holds whenever $0 < x - a <$ ____.

δ

102 $\lim_{x \to a^-} f(x) = +\infty$ means that f is defined at least in some open interval (c, a), and that, given any positive number M, no matter how large, there exists a positive number δ such that ________ holds whenever $0 < a - x <$ ____.

$f(x) > M$

δ

103 $\lim_{x \to a^+} f(x) = -\infty$ means that f is defined at least in some open interval (a, b), and that, given any positive number M, no matter how large, there exists a positive number δ such that $f(x) < -M$ holds whenever ________.

$0 < x - a < \delta$

104 $\lim_{x \to a^-} f(x) = -\infty$ means that f is defined at least in some open interval (c, a), and that, given any positive number M, no matter how large, there exists a positive number δ such that ________ holds whenever ________.

$f(x) < -M$, $0 < a - x < \delta$

105 We take $\lim_{x \to a} f(x) = +\infty$ to mean that $\lim_{x \to a^+} f(x) =$ ________ $= +\infty$. Similarly, we take $\lim_{x \to a} f(x) = -\infty$ to mean that $\lim_{x \to a^+} f(x) = \lim_{x \to a^-} f(x) =$ ____.

$\lim_{x \to a^-} f(x)$

$-\infty$

106 $\lim_{x \to +\infty} f(x) = L$ means that f is defined at least on an unbounded open interval (a, ∞), and that, for each positive number ϵ, no matter how small, there exists a positive number N such that $|f(x) - L| <$ ____ holds whenever $x >$ ____. Likewise, $\lim_{x \to -\infty} f(x) = L$ means that f is defined at least on an unbounded open interval $(-\infty, a)$, and that, for each positive number ϵ, no matter how small, there exists a positive number N such that $|f(x) - L| <$ ____ holds whenever $x <$ ____.

ϵ, N

ϵ, $-N$

6 Horizontal and Vertical Asymptotes

107 The vertical straight line $x = a$ is called a ________ of the graph of the function f if at least one of the following conditions holds:

(i) $\lim_{x \to a^+} f(x) = +\infty$.

(ii) $\lim_{x \to a^+} f(x) =$ ____.

vertical asymptote

$-\infty$

$+\infty$

$-\infty$

horizontal asymptote

b

b

(iii) $\lim_{x \to a^-} f(x) =$ ____.

(iv) $\lim_{x \to a^-} f(x) =$ ____.

108 The horizontal straight line $y = b$ is called a ____________________ of the graph of the function f if at least one of the following conditions holds:

(i) $\lim_{x \to +\infty} f(x) =$ ____.

(ii) $\lim_{x \to -\infty} f(x) =$ ____.

In Problems 109 through 115, find the horizontal and vertical asymptotes of the graph of the function f and sketch the graph.

3

3

$+\infty$, vertical

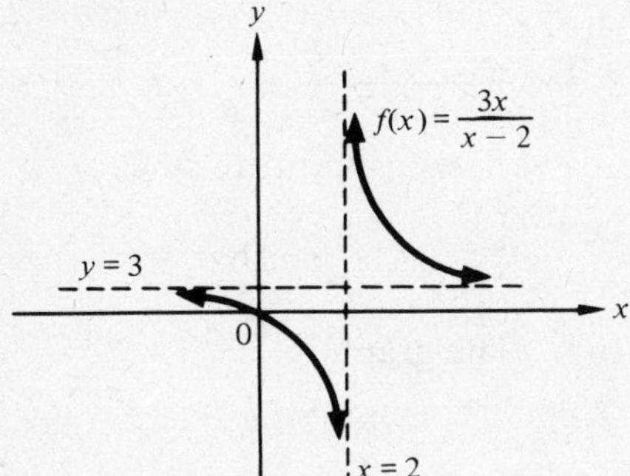

109 $f(x) = \dfrac{3x}{x - 2}$.

To find the horizontal asymptotes, consider

$$\lim_{x \to +\infty} f(x) = \lim_{x \to +\infty} \frac{3x}{x - 2} = \_\_\_\_.$$

Thus, $y =$ ____ is a horizontal asymptote. Since $\lim_{x \to 2^+} \dfrac{3x}{x - 2} =$ ____, then $x = 2$ is a __________ asymptote. The graph is

y

x

0

2, 2

asymptote

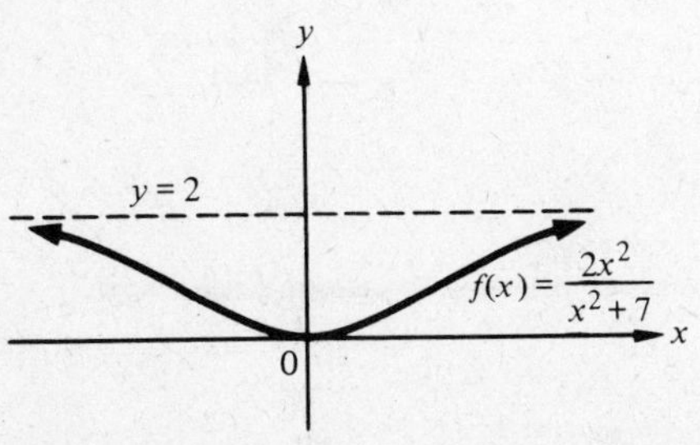

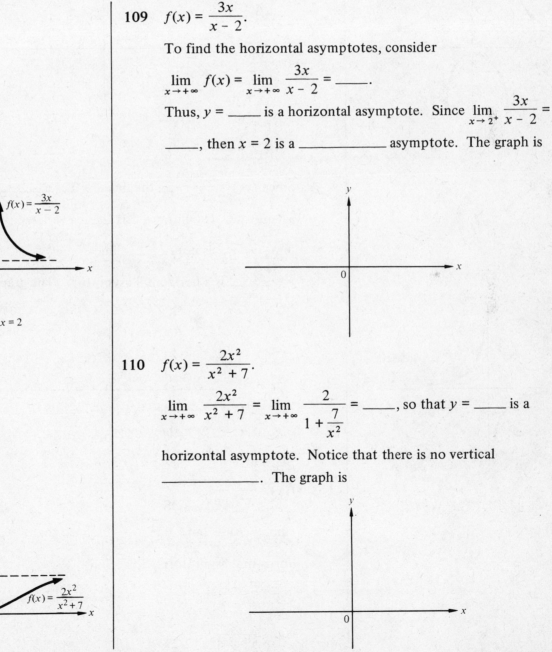

110 $f(x) = \dfrac{2x^2}{x^2 + 7}$.

$$\lim_{x \to +\infty} \frac{2x^2}{x^2 + 7} = \lim_{x \to +\infty} \frac{2}{1 + \frac{7}{x^2}} = \_\_\_\_,$$ so that $y =$ ____ is a horizontal asymptote. Notice that there is no vertical __________. The graph is

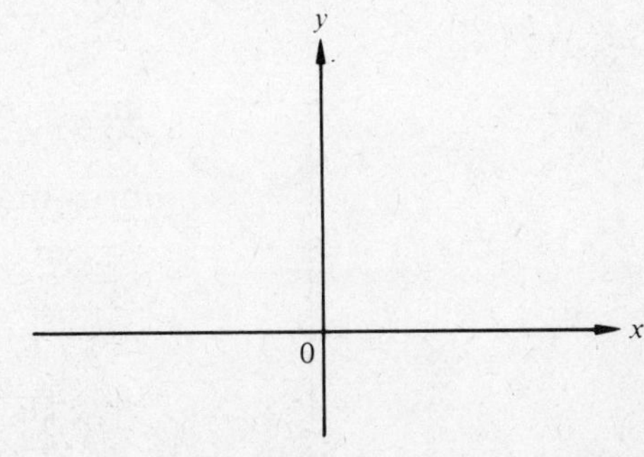

111 $f(x) = \dfrac{x^2 + 2}{x^2 + x - 6}$.

1, 1

$\lim\limits_{x \to +\infty} f(x) = \lim\limits_{x \to +\infty} \dfrac{x^2 + 2}{x^2 + x - 6} =$ ____, so that $y =$ ____ is a horizontal asymptote. Since

$-\infty$

$\lim\limits_{x \to -3^+} \dfrac{x^2 + 2}{(x + 3)(x - 2)} =$ ____,

$+\infty$

$\lim\limits_{x \to 2^+} \dfrac{x^2 + 2}{(x + 3)(x - 2)} =$ ____,

$-3, 2$

then $x =$ ____ and $x =$ ____ are vertical asymptotes. The graph is

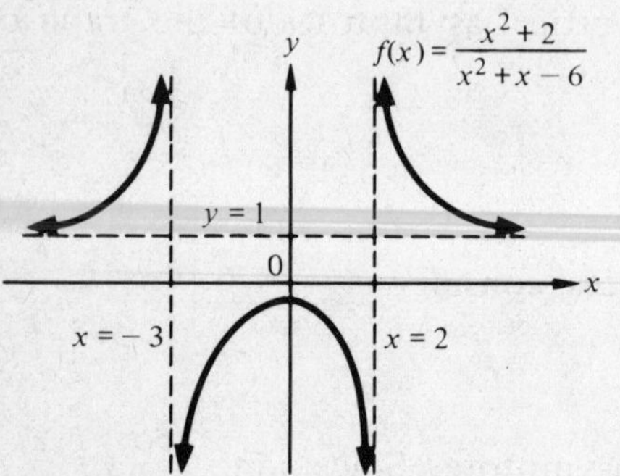

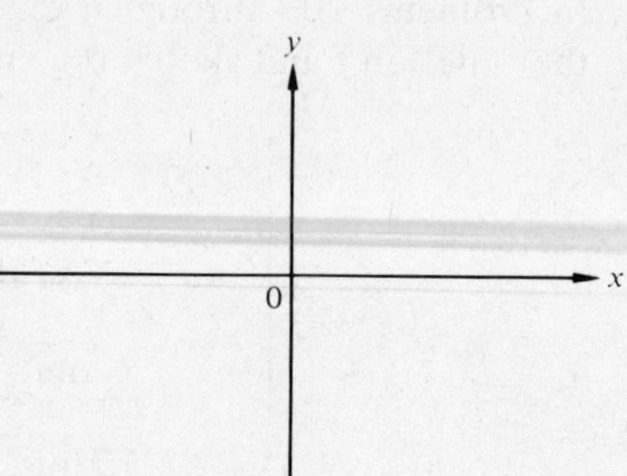

112 $f(x) = \dfrac{x - 1}{x^2 - x}$.

vertical

Since $f(x) = \dfrac{x - 1}{x(x - 1)}$, the line $x = 0$ is a ______

not

asymptote. The line $x = 1$ is ____ a vertical asymptote since

$\pm\infty$, 0

$\lim\limits_{x \to 1^+} \dfrac{x - 1}{x^2 - x}$ is not ____. $\lim\limits_{x \to +\infty} \dfrac{x - 1}{x^2 - x} =$ ____, so that

0

$y =$ ____ is a horizontal asymptote. The graph is

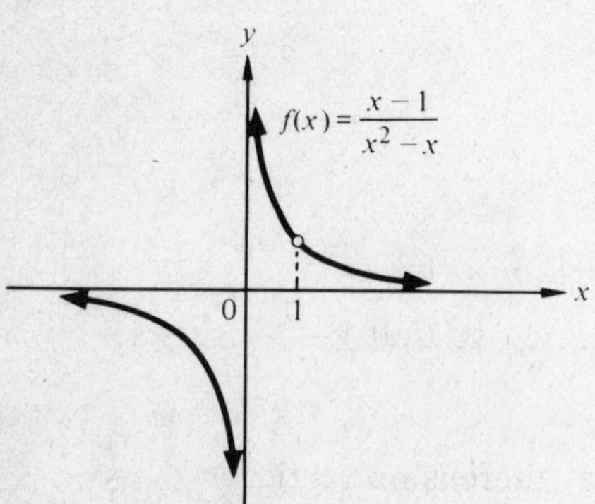

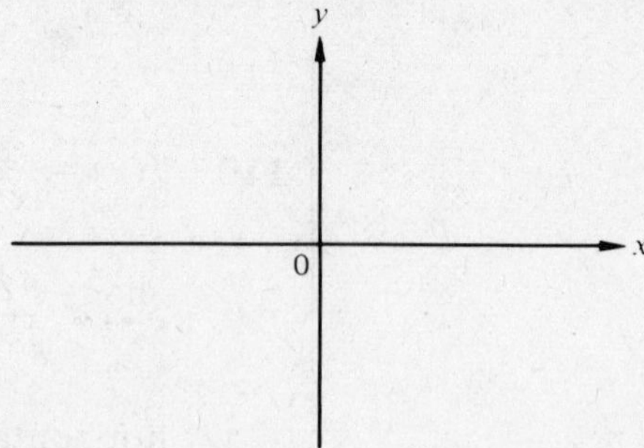

113 $f(x) = \dfrac{2}{x^2 - 4}$.

0, 0

$\lim\limits_{x \to +\infty} f(x) = \lim\limits_{x \to +\infty} \dfrac{2}{x^2 - 4} =$ ____, so that $y =$ ____ is a horizontal asymptote. Since

$+\infty$

$\lim\limits_{x \to 2^+} \dfrac{2}{x^2 - 4} =$ ____,

$-\infty$

$\lim\limits_{x \to -2^+} \dfrac{2}{x^2 - 4} =$ ____,

vertical

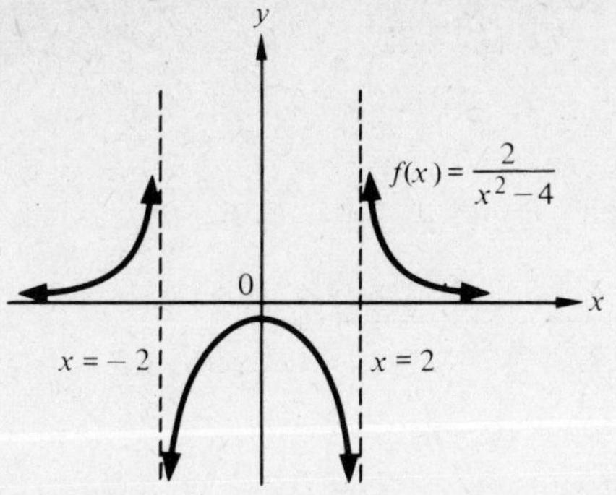

then $x = 2$ and $x = -2$ are __________ asymptotes. The graph is

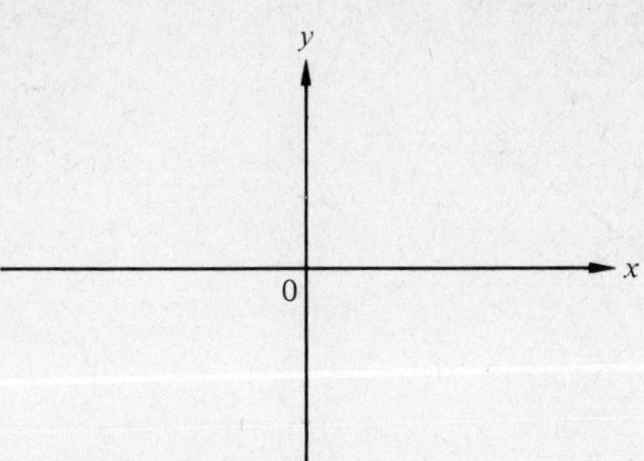

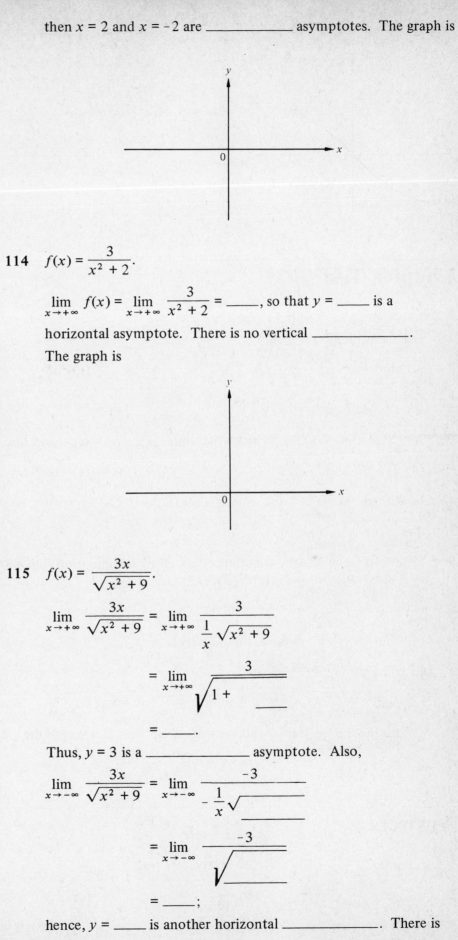

114 $f(x) = \dfrac{3}{x^2 + 2}$.

0, 0

$\lim\limits_{x \to +\infty} f(x) = \lim\limits_{x \to +\infty} \dfrac{3}{x^2 + 2} =$ ____, so that $y =$ ____ is a horizontal asymptote. There is no vertical __________.

asymptote

The graph is

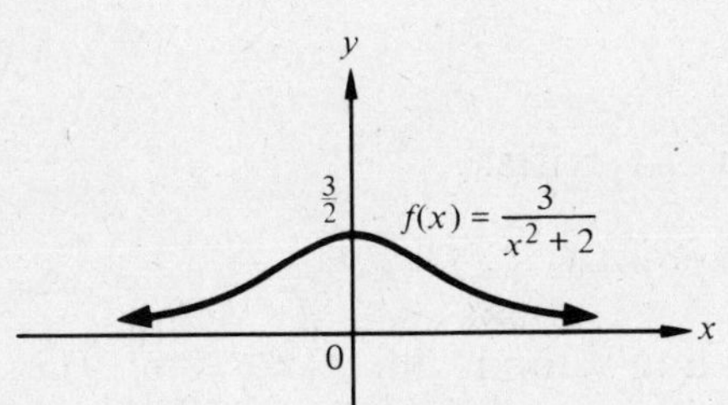

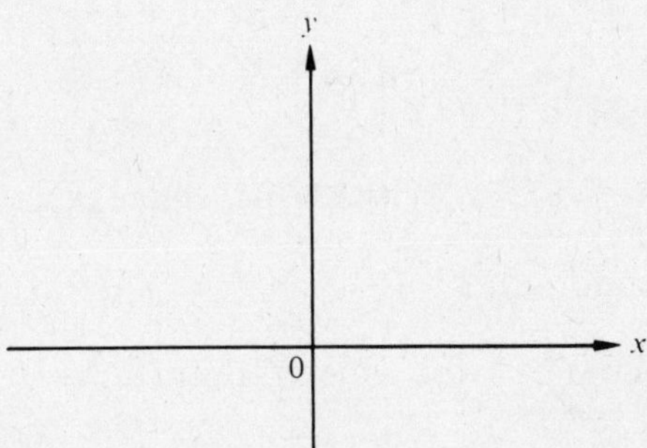

115 $f(x) = \dfrac{3x}{\sqrt{x^2 + 9}}$.

$$\lim_{x \to +\infty} \frac{3x}{\sqrt{x^2 + 9}} = \lim_{x \to +\infty} \frac{3}{\frac{1}{x}\sqrt{x^2 + 9}}$$

$\dfrac{9}{x^2}$

$$= \lim_{x \to +\infty} \frac{3}{\sqrt{1 + \_\_\_\_}}$$

3

$$= \_\_\_\_.$$

horizontal

Thus, $y = 3$ is a __________ asymptote. Also,

$x^2 + 9$

$$\lim_{x \to -\infty} \frac{3x}{\sqrt{x^2 + 9}} = \lim_{x \to -\infty} \frac{-3}{-\frac{1}{x}\sqrt{\_\_\_\_}}$$

$1 + \dfrac{9}{x^2}$

$$= \lim_{x \to -\infty} \frac{-3}{\sqrt{\_\_\_\_}}$$

-3

$$= \_\_\_\_;$$

-3, asymptote

hence, $y =$ ____ is another horizontal __________. There is vertical asymptote | no ______________. The graph is

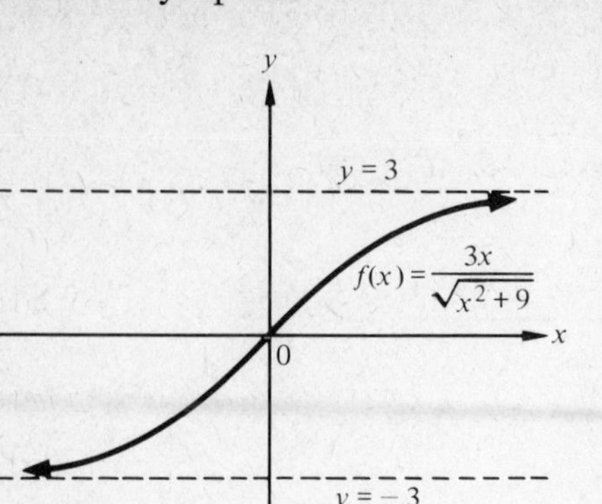

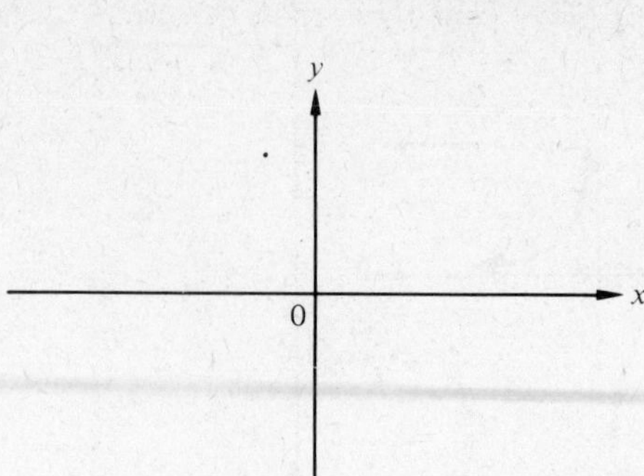

Chapter Test

1 Let $f(x) = 5x - 3$. Find a positive δ such that $|f(x) - 2| < \epsilon$ whenever $0 < |x - 1| < \delta$ if $\epsilon = 0.01$.

2 Evaluate the following limits.

(a) $\lim\limits_{x \to 1} \dfrac{x^2 - 5x + 6}{x - 2}$ (b) $\lim\limits_{t \to 0} \dfrac{t^2 + 7t + 3}{t + 13}$ (c) $\lim\limits_{x \to 2} \dfrac{x^4 - 16}{x - 2}$ (d) $\lim\limits_{h \to 0} \dfrac{\sqrt{1 + h} - 1}{h}$

3 Let f be defined by $f(x) = \begin{Bmatrix} 3x + 2 & \text{if } x \geqslant 1 \\ 6 - x & \text{if } x < 1 \end{Bmatrix}$. Let $a = 1$.

(a) Find $\lim\limits_{x \to a^+} f(x)$. (b) Find $\lim\limits_{x \to a^-} f(x)$. (c) Find $\lim\limits_{x \to a} f(x)$ if it exists.

(d) Is f continuous at a? (e) Sketch the graph of f.

4 Find those values of x for which f, defined by $f(x) = \dfrac{x - 1}{\sqrt{x^2 - 1}}$, is continuous. Find $\lim\limits_{x \to +\infty} f(x)$, $\lim\limits_{x \to -\infty} f(x)$, $\lim\limits_{x \to 1^+} f(x)$, and $\lim\limits_{x \to -1^-} f(x)$.

5 Determine whether the function f defined by the equation $f(x) = \sqrt{1 - x^2}$ is continuous or discontinuous on each of the following intervals: $(-2, 1)$, $[-1, 1]$, $(1, 2)$, $(-1, 1)$, and $(1, \infty)$.

6 Given $f(x) = \dfrac{2x^2 + 5x + 1}{x^2 - x - 6}$, evaluate:

(a) $\lim\limits_{x \to 3^+} f(x)$ (b) $\lim\limits_{x \to 3^-} f(x)$.

7 Evaluate the following limits.

(a) $\lim\limits_{x \to +\infty} \dfrac{3 + x - 7x^2}{4 - x + 5x^2}$ (b) $\lim\limits_{x \to +\infty} (x - \sqrt{x^2 + 1})$

8 Determine the horizontal and vertical asymptotes and sketch the graph of the functions.

(a) $f(x) = \dfrac{x^2}{4 - x^2}$ (b) $f(x) = \dfrac{x^2 - 3x + 2}{x^2 + 2x - 3}$

Answers

1 $\delta = 0.002$

2 (a) -2 (b) $\dfrac{3}{13}$ (c) 32 (d) $\dfrac{1}{2}$

3 (a) 5 (b) 5 (c) It exists and is 5. (d) Yes

(e)

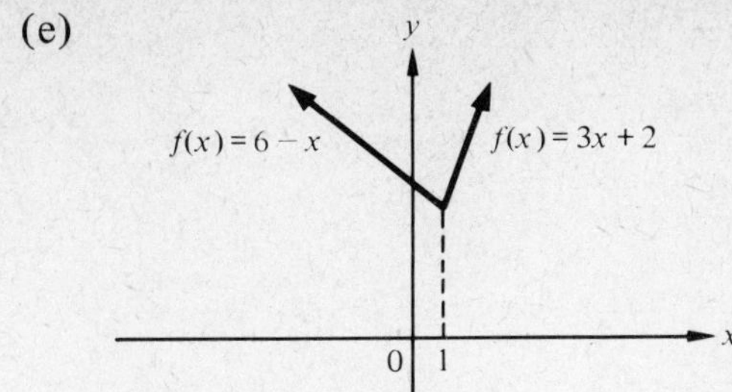

4 f is continuous on the intervals $(-\infty, -1)$ and $(1, \infty)$. $\lim_{x \to +\infty} f(x) = 1$, $\lim_{x \to -\infty} f(x) = -1$, $\lim_{x \to 1^+} f(x) = 0$, $\lim_{x \to -1^-} f(x) = -\infty$.

5 f is continuous on $[-1, 1]$ and discontinuous on $(-2, 1)$, $(1, 2)$, and $(1, \infty)$.

6 (a) $+\infty$ (b) $-\infty$

7 (a) $-\frac{7}{5}$ (b) 0

8 (a) Vertical asymptotes $x = -2$ and $x = 2$; horizontal asymptote $y = -1$.

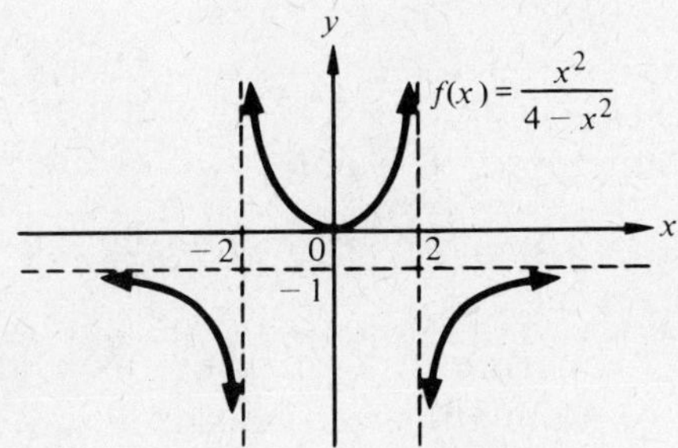

(b) Vertical asymptote $x = -3$; horizontal asymptote $y = 1$.

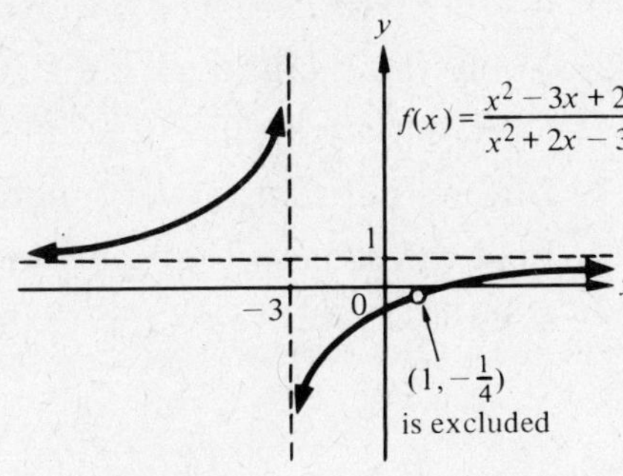

Chapter 2 THE DERIVATIVE

The chapter will be devoted primarily to differentiation of algebraic functions. The objectives of the chapter are that the student be able to:

1 Distinguish between instantaneous rate of change and average rate of change.

2 Interpret the geometric meaning of the slope of the tangent line to a graph at a given point.

3 Define the derivative, and discuss continuity and differentiability of a function.

4 Use the basic rules for differentiation.

5 Apply the chain rule for differentiation of composite functions.

6 Differentiate inverse functions.

7 Differentiate the power functions with rational exponents.

8 Find the equations of tangent and normal lines.

9 Use the derivative to approximate function values.

1 Rates of Change and Slopes of Tangent Lines

1.1 Speed of an Automobile

1 Suppose that an automobile travels over a straight road from city A at the rate of $g(t)$ miles per hour. At time t, it is $f(t)$ miles from city A, and after an interval of time h, it is at a distance ______ miles from A and its speed is ______ miles per hour. The average speed of the automobile during the interval of time h is $\dfrac{\quad\quad}{h}$ miles per hour.

$f(t+h)$, $g(t+h)$

$f(t+h) - f(t)$

2 The instantaneous speed $g(t)$ of the automobile is given by

$$g(t) = \lim_{h\to 0} \frac{\_\_\_\_\_\_ - f(t)}{h}.$$

$f(t+h)$

1.2 Instantaneous Rate of Change in General

3 Let x_1 and x_2 be two different particular values of x and suppose that $y_1 = f(x_1)$ and $y_2 = f(x_2)$ are the corresponding values of y. Thus, as x changes from x_1 to x_2, y will change from y_1 to y_2. Denote the change in x by $\Delta x =$ ______ and the corresponding change in y by $\Delta y =$ ______ $=$ ______ $= f(x_1 + \Delta x) - f(x_1)$. $\dfrac{\Delta y}{\Delta x} = \dfrac{\quad\quad}{\Delta x}$ is called the average rate of change of y per unit change in x over the interval from x_1 to $x_2 =$ ______.

$x_2 - x_1$

$y_2 - y_1$, $f(x_2) - f(x_1)$

$f(x_1 + \Delta x) - f(x_1)$

$x_1 + \Delta x$

4 Find the average rate of change of the surface area S of a cube with respect to the length x of a side as x increases from 2 inches to 2.25 inches. The surface area of the cube is given by $S =$ ____, so that the average rate of change of the surface area S with respect to a length x of a side as x increases from 2 inches to 2.25 inches is

$6x^2$

$$\frac{\Delta S}{\Delta x} = \frac{6(x_1+\Delta x)^2 - \_\_\_\_}{\Delta x}$$

$6x_1^2$

$$= \frac{6(2.25)^2 - \_\_\_\_}{0.25}$$

24

$$= \frac{\quad\quad}{0.25}$$

6.375

$=$ _____ square inches per inch of edge length.

25.5

5 Find the average rate of change of the volume V of a cylinder with respect to its height h if the radius r is constant and equal to 0.5 centimeter, and the height h increases from 1 centimeter to 1.25 centimeters.

$\frac{\Delta V}{\Delta h}$

The average rate of change of the volume V with respect to the height h is ____, so that

$\frac{\pi}{4}(1+\Delta h)^2$

$$\frac{\Delta V}{\Delta h} = \frac{(\_\_\_\_\_\_) - \frac{\pi}{4}}{\Delta h}$$

1.5625

$$= \frac{\frac{\pi}{4}(\_\_\_\_\_\_) - \frac{\pi}{4}}{0.25}$$

1.7671

$\approx$ ________ cubic centimeters per centimeter of height.

6 The pressure P of a gas depends upon its volume V according to Boyle's law $P = \frac{C}{V}$, where C is a constant. Suppose that $C = 1000$, P is in pounds per square inch, and V is in cubic inches. Find the average rate of change of P with respect to V as V increases from 100 cubic inches to 110 cubic inches.

$\frac{\Delta P}{\Delta V}$

The average rate of change of P with respect to V is ____, so that

10

$$\frac{\Delta P}{\Delta V} = \frac{\frac{1000}{110} - \frac{\_\_\_\_}{}}{10}$$

$-\frac{1}{11}$

$=$ ____ pound per square inch per cubic inch.

$f(x_1 + x) - f(x_1)$

7 **The instantaneous rate of change of y with respect to x at the instant when $x = x_1$ is $\lim_{\Delta x \to 0} \frac{\Delta y}{\Delta x} = \lim_{\Delta x \to 0} \frac{\_\_\_\_\_\_}{\Delta x}$.**

8 Find the instantaneous rate of change of the area A of a square with respect to a side x when the length of the side is 10 inches.

x^2

The area A of a square of side x is given by $A =$ ____, so that the instantaneous rate of change is given by

$(10 + \Delta x)^2$

$$\lim_{\Delta x \to 0} \frac{\Delta A}{\Delta x} = \lim_{\Delta x \to 0} \frac{\_\_\_\_\_\_ - 100}{\Delta x}$$

$20(\Delta x) + (\Delta x)^2$

$$= \lim_{\Delta x \to 0} \frac{\_\_\_\_\_\_}{\Delta x}$$

20

$=$ ____ square inches per inch of side length.

9 Find the instantaneous rate of change of the volume V of a sphere with respect to its radius r when the radius is 3 centimeters.

$\frac{4}{3}\pi r^3$

The volume V of a sphere is given by $V =$ ________, so that the instantaneous rate of change is given by

$\frac{4}{3}\pi(3 + \Delta r)^3$

$$\lim_{\Delta r \to 0} \frac{\Delta V}{\Delta r} = \lim_{\Delta r \to 0} \frac{\_\_\_\_\_\_ - 36\pi}{\Delta r}$$

$36\pi\Delta r + 36\pi(\Delta r)^2 + \frac{4}{3}\pi(\Delta r)^3$

$$= \lim_{\Delta r \to 0} \frac{\rule{6cm}{0.4pt}}{\Delta r}$$

$36\pi + 36\pi(\Delta r) + \frac{4}{3}\pi(\Delta r)^2$

$$= \lim_{\Delta r \to 0} \rule{6cm}{0.4pt}$$

36π

$=$ ____ cubic centimeters per centimeter of radius length.

10 Find the instantaneous rate of change of the volume V of a right circular cylinder of constant altitude 10 feet with respect to the radius r of a right section when the radius of the right section is 2 feet.

$\pi r^2(10)$

$10\pi r^2$

The volume V of the right circular cylinder is $V =$ __________ $=$ ______, so that the instantaneous rate of change is given by

$10\pi(2 + \Delta r)^2$

$$\lim_{\Delta r \to 0} \frac{\Delta V}{\Delta r} = \lim_{\Delta r \to 0} \frac{\rule{3cm}{0.4pt} - 40\pi}{\Delta r}$$

$40\pi(\Delta r) + 10\pi(\Delta r)^2$

$$= \lim_{\Delta r \to 0} \frac{\rule{4cm}{0.4pt}}{\Delta r}$$

40π

$=$ ____ cubic feet per foot of radius length.

11 Suppose that a particle P is moving along a straight line and is a distance s from its starting point A at time t. If $s = f(t)$, then the speed of the particle at the instant when $t = t_1$ is given by

$f(t_1 + \Delta t) - f(t_1)$

$$\lim_{\Delta t \to 0} \frac{\Delta s}{\Delta t} = \lim_{\Delta t \to 0} \frac{\rule{4cm}{0.4pt}}{\Delta t}.$$

In Problems 12 through 15, suppose that a particle is moving along a straight line according to the given equation of motion, where s is the distance in feet from the starting point at the end of t seconds. Find the instantaneous speed when t has the value t_1.

12 $s = 3t^2$; $t_1 = 2$.

$f(t_1)$

$$\lim_{\Delta t \to 0} \frac{\Delta s}{\Delta t} = \lim_{\Delta t \to 0} \frac{f(t_1 + \Delta t) - \rule{2cm}{0.4pt}}{\Delta t}$$

$3t_1^2$

$$= \lim_{\Delta t \to 0} \frac{3(t_1 + \Delta t)^2 - \rule{1.5cm}{0.4pt}}{\Delta t}$$

$3t_1^2$

$$= \lim_{\Delta t \to 0} \frac{3t_1^2 + 6t_1\Delta t + 3(\Delta t)^2 - \rule{1.5cm}{0.4pt}}{\Delta t}$$

$6t_1\Delta t + 3(\Delta t)^2$, $6t_1 + 3\Delta t$

$$= \lim_{\Delta t \to 0} \frac{\rule{4cm}{0.4pt}}{\Delta t} = \lim_{\Delta t \to 0} \rule{3cm}{0.4pt}$$

$6t_1$

$=$ ____.

2, 12

In particular, for $t_1 = 2$ seconds, we get $6($____$) =$ ____ feet per second.

13 $s = t^3 - t$; $t_1 = 2$.

$f(t_1 + \Delta t) - f(t_1)$

$$\lim_{\Delta t \to 0} \frac{\Delta s}{\Delta t} = \lim_{\Delta t \to 0} \frac{\rule{4cm}{0.4pt}}{\Delta t}$$

6

$$= \lim_{\Delta t \to 0} \frac{[(2+\Delta t)^3 - (2+\Delta t)] - \_\_\_\_}{\Delta t}$$

6

$$= \lim_{\Delta t \to 0} \frac{8 + 12\Delta t + 6(\Delta t)^2 + (\Delta t)^3 - 2 - \Delta t - \_\_\_\_}{\Delta t}$$

$11\Delta t + 6(\Delta t)^2 + (\Delta t)^3$

$$= \lim_{\Delta t \to 0} \frac{\_\_\_\_\_\_\_\_\_\_\_\_\_\_\_\_}{\Delta t}$$

11

$= \_\_\_\_$ feet per second.

14 $s = \dfrac{1}{t}$; $t_1 = 2$.

$\dfrac{f(t_1 + \Delta t) - f(t_1)}{\Delta t}$

$$\lim_{\Delta t \to 0} \frac{\Delta s}{\Delta t} = \lim_{\Delta t \to 0} \_\_\_\_\_\_\_\_\_\_\_\_\_\_\_\_$$

$\dfrac{1}{2}, -\Delta t$

$$= \lim_{\Delta t \to 0} \frac{\dfrac{1}{2+\Delta t} - \_\_\_\_}{\Delta t} = \lim_{\Delta t \to 0} \frac{\_\_\_\_\_\_\_\_}{2\Delta t(2+t)}$$

$\dfrac{-1}{2(2+\Delta t)}, -\dfrac{1}{4}$

$= \lim\limits_{\Delta t \to 0} \_\_\_\_\_\_\_\_\_\_\_\_ = \_\_\_\_$ foot per second.

15 $s = \sqrt{t}$; $t_1 = 4$.

$\dfrac{f(t_1 + \Delta t) - f(t_1)}{\Delta t}$

$$\lim_{\Delta t \to 0} \frac{\Delta s}{\Delta t} = \lim_{\Delta t \to 0} \_\_\_\_\_\_\_\_\_\_\_\_\_\_\_\_$$

2

$$= \lim_{\Delta t \to 0} \frac{\sqrt{4+\Delta t} - \_\_\_\_}{\Delta t}$$

$(\sqrt{4+\Delta t} + 2)$

$$= \lim_{\Delta t \to 0} \frac{(\sqrt{4+\Delta t} - 2)(\sqrt{4+\Delta t} + 2)}{\Delta t(\_\_\_\_\_\_\_\_\_\_\_\_)}$$

$\dfrac{1}{4}$

$= \lim\limits_{\Delta t \to 0} \dfrac{\Delta t}{\Delta t(\sqrt{4+\Delta t} + 2)} = \_\_\_\_$ foot per second.

16 A projectile is fired upward (vertically), and its position above the ground t seconds after being fired is given by $s = 50t - 16t^2$ feet.

(a) Find the instantaneous speed of the projectile t_1 seconds after being fired.

(b) How long does it take for the projectile to reach its maximum height?

(c) What is the maximum height achieved by the projectile?

(a) $\lim\limits_{\Delta t \to 0} \dfrac{\Delta s}{\Delta t}$

$50(t_1 + \Delta t) - 16(t_1 + \Delta t)^2$

$$= \lim_{\Delta t \to 0} \frac{(\_\_\_\_\_\_\_\_\_\_\_\_\_\_\_\_) - (50t_1 - 16t_1^2)}{\Delta t}$$

$50t_1 + 50(\Delta t) - 16t_1^2 - 32t_1\Delta t - 16(\Delta t)^2$

$$= \lim_{\Delta t \to 0} \frac{(\_\_\_\_\_\_\_\_\_\_\_\_\_\_\_\_) - (50t_1 - 16t_1^2)}{\Delta t}$$

$50(\Delta t) - 32t_1\Delta t - 16(\Delta t)^2$

$$= \lim_{\Delta t \to 0} \frac{\_\_\_\_\_\_\_\_\_\_\_\_\_\_\_\_}{\Delta t}$$

$50 - 32t_1 - 16(\Delta t)$

$$= \lim_{\Delta t \to 0} \_\_\_\_\_\_\_\_\_\_\_\_\_\_\_\_$$

$50 - 32t_1$

$= \_\_\_\_\_\_\_\_$.

0, 0

$\frac{25}{16}$, $\frac{25}{16}$

(b) At the instant when the projectile reaches maximum height, its speed will be ____, so that $50 - 32t_1 =$ ____ or $t_1 =$ ____. Therefore ____ seconds after being fired, the projectile will reach its maximum height.

$16t^2$

(c) The projectile will be s feet above the ground t seconds after being fired, where $s = 50t -$ ____; hence, $\frac{25}{16}$ seconds after being fired, the projectile will be at its maximum height of

$\frac{25}{16}$, $\frac{(25)^2}{(16)^2}$, $\frac{625}{16}$, 39.06

50(____) - 16(____) = ____ ≈ _______ feet.

1.3 Slope of the Tangent Line to a Graph at a Point

tangent line

17 Let f be a function defined at least in some open interval containing the number x_1 and let $y_1 = f(x_1)$. If the limit

$$m = \lim_{\Delta x \to 0} \frac{f(x_1 + \Delta x) - f(x_1)}{\Delta x}$$

exists and is finite, we say that the straight line in the xy plane containing the point (x_1, y_1) and having slope m is the ____________ to the graph of f at (x_1, y_1).

graph of f

(x_1, y_1)

18 If $\lim_{\Delta x \to 0} \left| \frac{f(x_1 + \Delta x) - f(x_1)}{\Delta x} \right| = +\infty$, we say that the vertical line $x = x_1$ in the xy plane is the *tangent line* to the ____________ at the point __________.

no tangent

19 If neither of the limits in Problems 17 and 18 exists, we say that the graph of f has __________ line at the point (x_1, y_1).

In Problems 20 through 25, find the slope of the tangent line to the graph of f at the given point.

$2(1 + \Delta x)^2 + 1$

$4(\Delta x) + 2(\Delta x)^2$, $4 + 2\Delta x$

4

20 $f(x) = 2x^2 + 1$ at $(1, 3)$.

The slope of the tangent line is given by

$$m = \lim_{\Delta x \to 0} \frac{f(1 + \Delta x) - f(1)}{\Delta x} = \lim_{\Delta x \to 0} \frac{\_\_\_\_\_\_\_\_ - 3}{\Delta x}$$

$$= \lim_{\Delta x \to 0} \frac{\_\_\_\_\_\_\_\_}{\Delta x} = \lim_{\Delta x \to 0} \_\_\_\_\_\_$$

$= \_\_\_\_.$

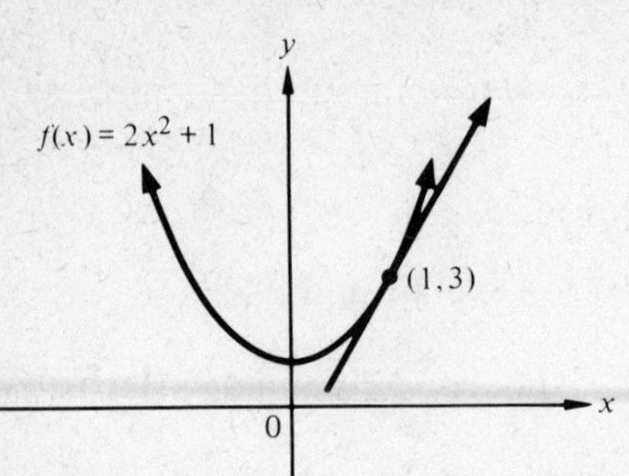

The graph is

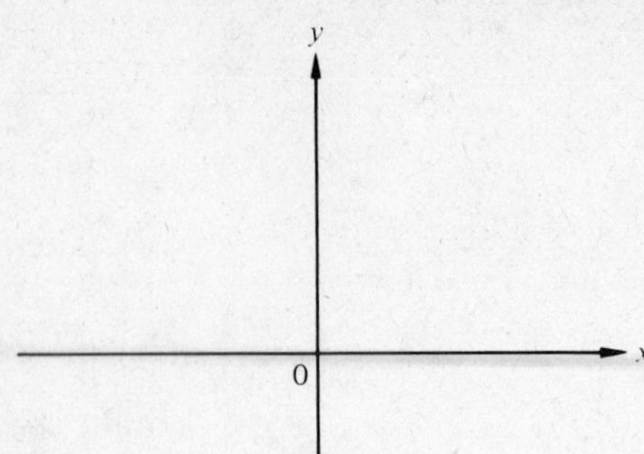

21 $f(x) = -3x^2$ at $(-2, -12)$.

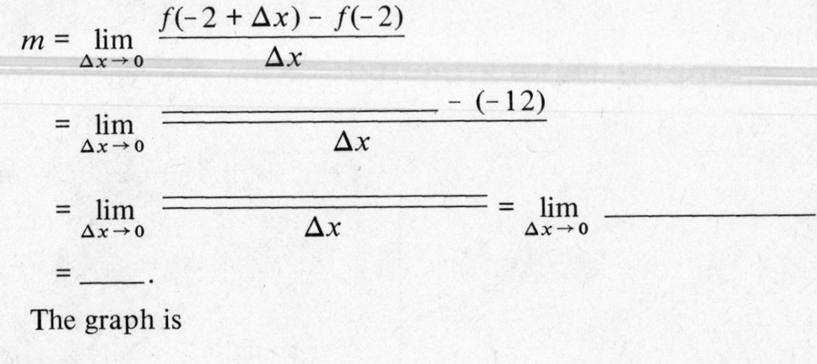

$$m = \lim_{\Delta x \to 0} \frac{f(-2 + \Delta x) - f(-2)}{\Delta x}$$

$$= \lim_{\Delta x \to 0} \frac{\rule{3cm}{0.4pt} - (-12)}{\Delta x}$$

$$= \lim_{\Delta x \to 0} \frac{\rule{3cm}{0.4pt}}{\Delta x} = \lim_{\Delta x \to 0} \rule{3cm}{0.4pt}$$

$= \rule{1cm}{0.4pt}$.

The graph is

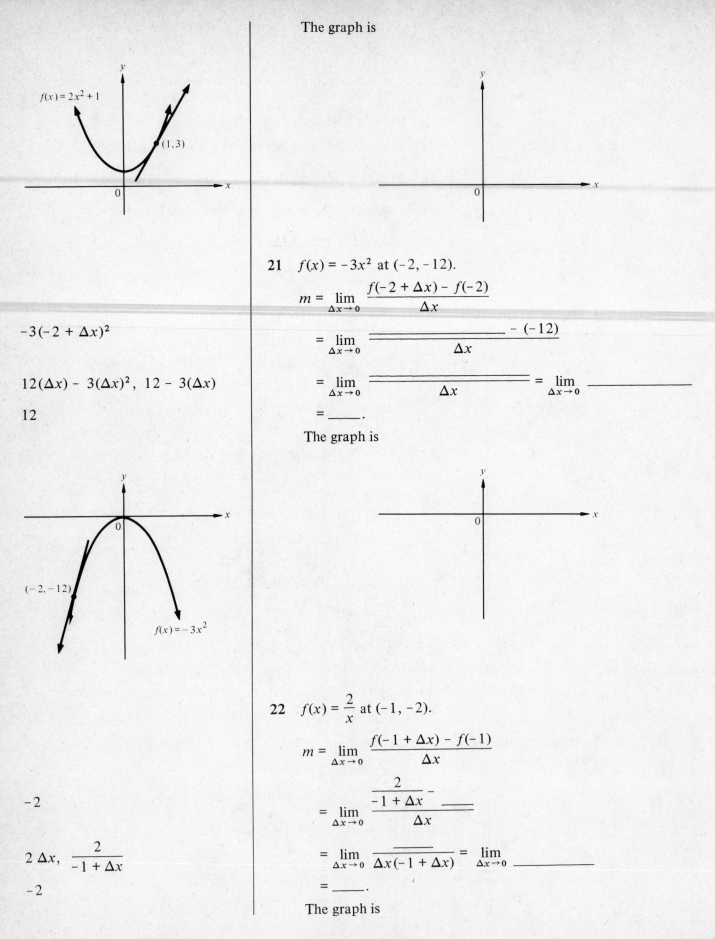

$-3(-2 + \Delta x)^2$

$12(\Delta x) - 3(\Delta x)^2$, $12 - 3(\Delta x)$

12

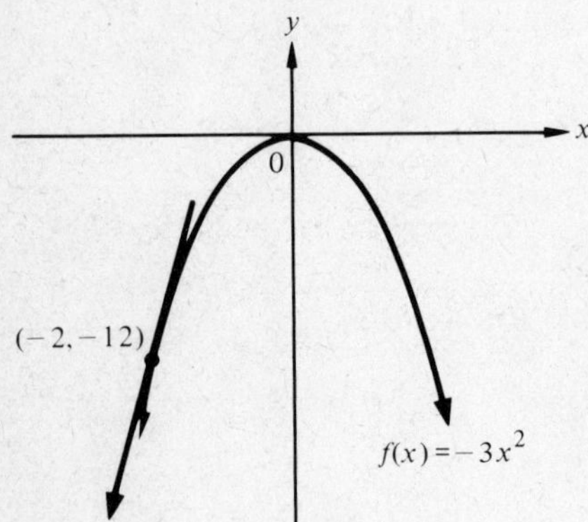

22 $f(x) = \dfrac{2}{x}$ at $(-1, -2)$.

$$m = \lim_{\Delta x \to 0} \frac{f(-1 + \Delta x) - f(-1)}{\Delta x}$$

$$= \lim_{\Delta x \to 0} \frac{\dfrac{2}{-1 + \Delta x} - \rule{1cm}{0.4pt}}{\Delta x}$$

$$= \lim_{\Delta x \to 0} \frac{\rule{1.5cm}{0.4pt}}{\Delta x(-1 + \Delta x)} = \lim_{\Delta x \to 0} \rule{3cm}{0.4pt}$$

$= \rule{1cm}{0.4pt}$.

The graph is

-2

$2\,\Delta x$, $\dfrac{2}{-1 + \Delta x}$

-2

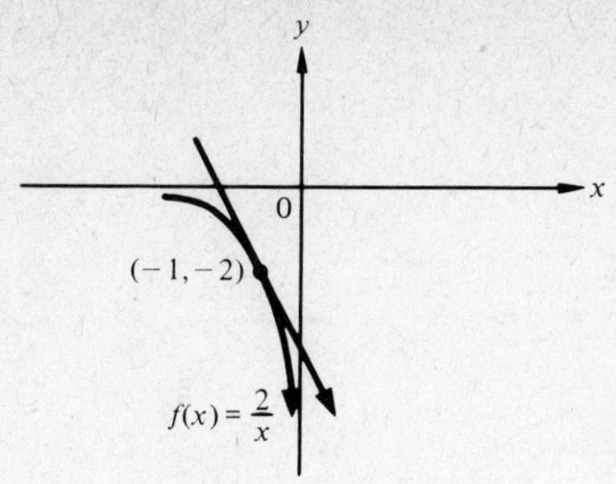

4

$\sqrt{3(5+\Delta x)+1}+4$

16

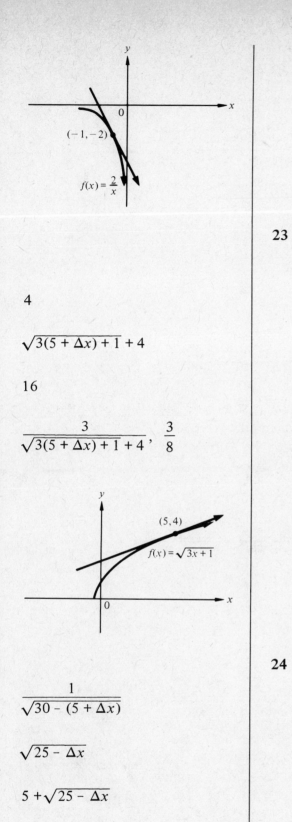

$\dfrac{3}{\sqrt{3(5+\Delta x)+1}+4}$, $\dfrac{3}{8}$

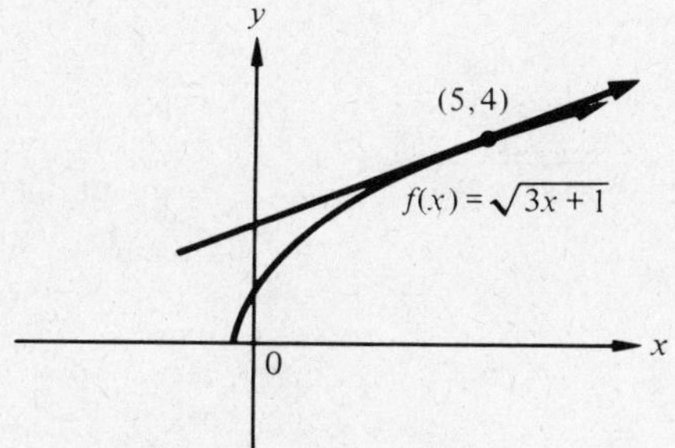

$\dfrac{1}{\sqrt{30-(5+\Delta x)}}$

$\sqrt{25-\Delta x}$

$5+\sqrt{25-\Delta x}$

Δx

1

250, $\dfrac{1}{250}$

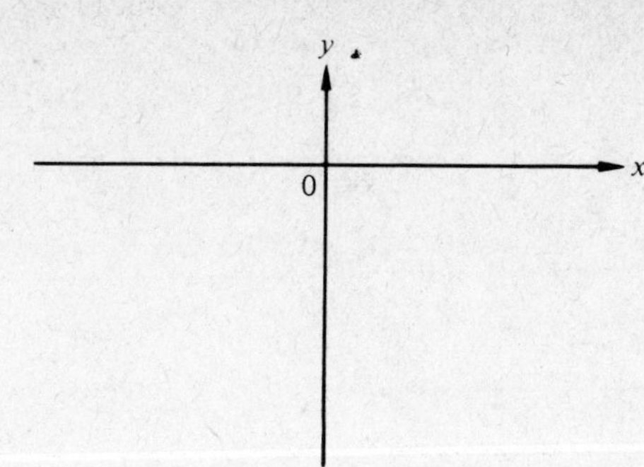

23 $f(x)=\sqrt{3x+1}$ at (5, 4).

$$m = \lim_{\Delta x \to 0} \frac{f(5+\Delta x)-f(5)}{\Delta x}$$

$$= \lim_{\Delta x \to 0} \frac{\sqrt{3(5+\Delta x)+1} - \underline{\qquad}}{\Delta x}$$

$$= \lim_{\Delta x \to 0} \frac{[\sqrt{3(5+\Delta x)+1}-4]\,\underline{\qquad\qquad\qquad}}{\Delta x[\sqrt{3(5+\Delta x)+1}+4]}$$

$$= \lim_{\Delta x \to 0} \frac{3(5+\Delta x)+1-\underline{\qquad}}{\Delta x[\sqrt{3(5+\Delta x)+1}+4]}$$

$$= \lim_{\Delta x \to 0} \underline{\qquad\qquad\qquad} = \underline{\qquad}$$

The graph is

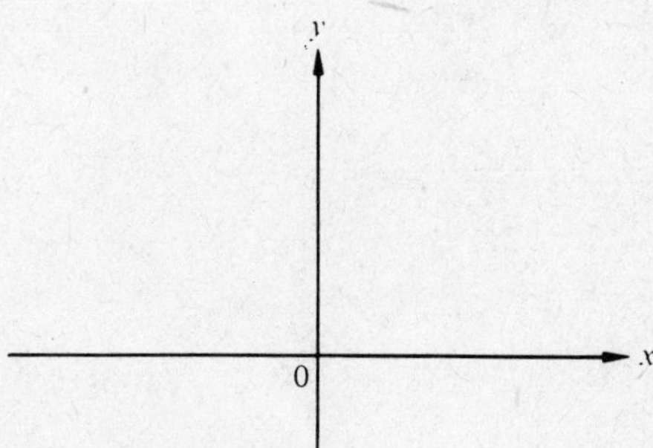

24 $f(x)=\dfrac{1}{\sqrt{30-x}}$ at $\left(5, \dfrac{1}{5}\right)$.

$$m = \lim_{\Delta x \to 0} \frac{f(5+\Delta x)-f(5)}{\Delta x} = \lim_{\Delta x \to 0} \frac{\underline{\qquad\qquad} - \frac{1}{5}}{\Delta x}$$

$$= \lim_{\Delta x \to 0} \frac{5-\underline{\qquad\qquad}}{5\,\Delta x\sqrt{25-\Delta x}}$$

$$= \lim_{\Delta x \to 0} \frac{[5-\sqrt{25-\Delta x}]\,\underline{\qquad\qquad}}{(5\,\Delta x\sqrt{25-\Delta x})(5+\sqrt{25-\Delta x})}$$

$$= \lim_{\Delta x \to 0} \frac{\underline{\qquad}}{(5\,\Delta x\sqrt{25-\Delta x})(5+\sqrt{25-\Delta x})}$$

$$= \lim_{\Delta x \to 0} \frac{\underline{\qquad}}{(5\sqrt{25-\Delta x})(5+\sqrt{25-\Delta x})}$$

$$= \frac{1}{\underline{\qquad}} = \underline{\qquad}.$$

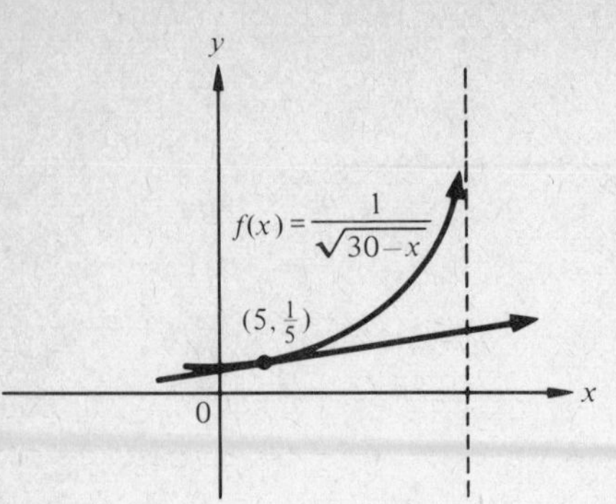

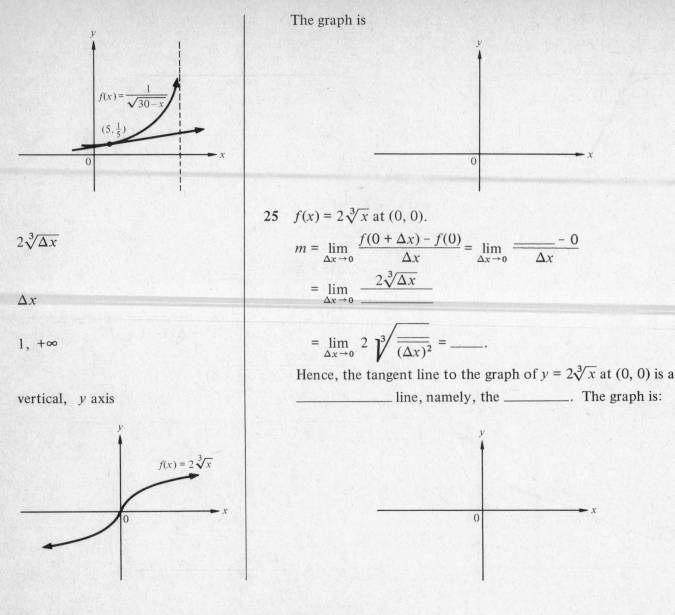

The graph is

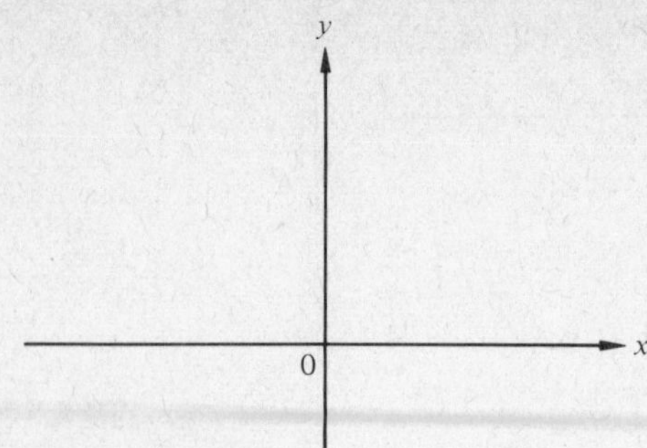

$2\sqrt[3]{\Delta x}$

Δx

1, $+\infty$

vertical, y axis

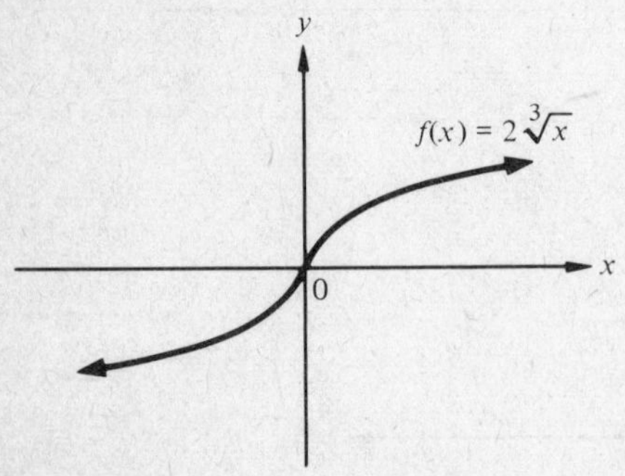

25 $f(x) = 2\sqrt[3]{x}$ at (0, 0).

$$m = \lim_{\Delta x \to 0} \frac{f(0 + \Delta x) - f(0)}{\Delta x} = \lim_{\Delta x \to 0} \frac{\_\_\_\_ - 0}{\Delta x}$$

$$= \lim_{\Delta x \to 0} \frac{2\sqrt[3]{\Delta x}}{\_\_\_\_}$$

$$= \lim_{\Delta x \to 0} 2\sqrt[3]{\frac{\_\_\_\_}{(\Delta x)^2}} = \_\_\_\_.$$

Hence, the tangent line to the graph of $y = 2\sqrt[3]{x}$ at (0, 0) is a ________ line, namely, the ________. The graph is:

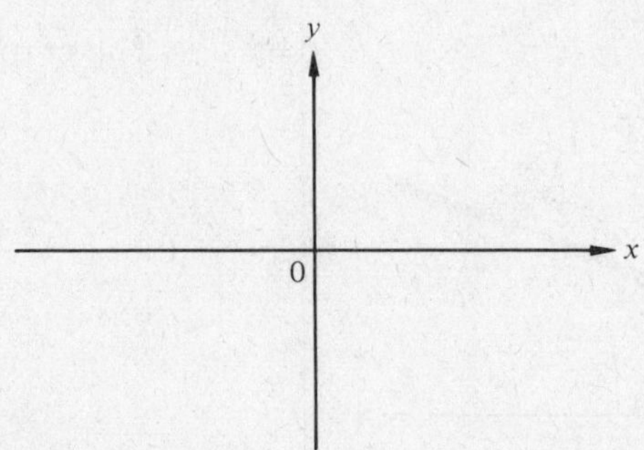

2 The Derivative of a Function

difference quotient

derivative

limit

constant

26 A quotient of the form $\frac{\Delta y}{\Delta x} = \frac{f(x + \Delta x) - f(x)}{\Delta x}$ is called a ________________.

27 Define $f'(x) = \lim_{\Delta x \to 0} \frac{f(x + \Delta x) - f(x)}{\Delta x} = \lim_{\Delta x \to 0} \frac{\Delta y}{\Delta x}$. The function f' is called the __________ of f, since it is derived from the function of f. The domain of f' is the set of all numbers x in the domain of f for which the ______ of the difference quotient exists.

28 In calculating the value of a derivative $f'(x)$ directly from the definition, one must be careful to treat x as a __________ while computing the limit $\lim_{\Delta x \to 0} \frac{f(x + \Delta x) - f(x)}{\Delta x}$.

In Problems 29 through 33, find $f'(x)$ for each function f directly from the definition.

29 $f(x) = (x - 1)^2$.

$$f'(x) = \lim_{\Delta x \to 0} \frac{\underline{\qquad\qquad\qquad}}{\Delta x}$$

$$= \lim_{\Delta x \to 0} \frac{(\underline{\qquad\qquad\qquad}) - (x-1)^2}{\Delta x}$$

$$= \lim_{\Delta x \to 0} \frac{(\underline{\qquad\qquad\qquad\qquad\qquad}) - (x^2 - 2x + 1)}{\Delta x}$$

$$= \lim_{\Delta x \to 0} \frac{\underline{\qquad\qquad\qquad\qquad}}{\Delta x}$$

$$= \lim_{\Delta x \to 0} \underline{\qquad\qquad\qquad}$$

$$= \underline{\qquad\qquad}.$$

$f(x + \Delta x) - f(x)$

$(x + \Delta x - 1)^2$

$x^2 + (\Delta x)^2 + 1 + 2x(\Delta x) - 2x - 2(\Delta x)$

$(\Delta x)^2 + 2x(\Delta x) - 2(\Delta x)$

$\Delta x + 2x - 2$

$2x - 2$

30 $f(x) = x^3$.

$$f'(x) = \lim_{\Delta x \to 0} \frac{\underline{\qquad\qquad\qquad}}{\Delta x}$$

$$= \lim_{\Delta x \to 0} \frac{(x + \Delta x)^3 - \underline{\qquad}}{\Delta x}$$

$$= \lim_{\Delta x \to 0} \frac{\underline{\qquad\qquad\qquad\qquad\qquad}}{\Delta x}$$

$$= \lim_{\Delta x \to 0} \underline{\qquad\qquad\qquad\qquad\qquad}$$

$$= \underline{\qquad}.$$

$f(x + \Delta x) - f(x)$

x^3

$3x^2(\Delta x) + 3x(\Delta x)^2 + (\Delta x)^3$

$3x^2 + 3x(\Delta x) + (\Delta x)^2$

$3x^2$

31 $f(x) = \frac{1}{x + 2}$.

$$f'(x) = \lim_{\Delta x \to 0} \frac{\frac{1}{x + \Delta x + 2} - \underline{\qquad\qquad}}{\Delta x}$$

$$= \lim_{\Delta x \to 0} \frac{\underline{\qquad}}{\Delta x(x + 2)(x + \Delta x + 2)}$$

$$= \lim_{\Delta x \to 0} \frac{\underline{\qquad}}{(x + 2)(x + \Delta x + 2)}$$

$$= \underline{\qquad\qquad\qquad}.$$

$\frac{1}{x + 2}$

$-\Delta x$

-1

$\frac{-1}{(x + 2)^2}$

32 $f(x) = \frac{1}{\sqrt{x}}$.

$$f'(x) = \lim_{\Delta x \to 0} \frac{\left(\underline{\qquad\qquad}\right) - \frac{1}{\sqrt{x}}}{\Delta x}$$

$$= \lim_{\Delta x \to 0} \frac{\underline{\qquad\qquad\qquad\qquad}}{\Delta x \sqrt{x}\sqrt{x + \Delta x}}$$

$$= \lim_{\Delta x \to 0} \frac{(\sqrt{x} - \sqrt{x + \Delta x})(\underline{\qquad\qquad\qquad})}{\Delta x \sqrt{x}\sqrt{x + \Delta x}(\sqrt{x} + \sqrt{x + \Delta x})}$$

$\frac{1}{\sqrt{x + \Delta x}}$

$\sqrt{x} - \sqrt{x + \Delta x}$

$\sqrt{x} + \sqrt{x + \Delta x}$

-1	$= \lim_{\Delta x \to 0} \frac{\_\_\_\_}{\sqrt{x}\sqrt{x+\Delta x}(\sqrt{x}+\sqrt{x+\Delta x})}$
$\frac{-1}{2x\sqrt{x}}$	$= \_\_\_\_\_\_\_\_.$
	33 $f(x) = \sqrt{1-x^2}$.
$\sqrt{1-x^2}$	$f'(x) = \lim_{\Delta x \to 0} \frac{\sqrt{1-(x+\Delta x)^2} - (\_\_\_\_\_\_\_\_)}{\Delta x}$
$\sqrt{1-(x+\Delta x)^2} + \sqrt{1-x^2}$	$= \lim_{\Delta x \to 0} \frac{[\sqrt{1-(x+\Delta x)^2} - \sqrt{1-x^2}\,][\_\_\_\_\_\_\_\_\_\_\_\_]}{\Delta x[\sqrt{1-(x+\Delta x)^2} + \sqrt{1-x^2}\,]}$
$1 - x^2 - 2x(\Delta x) - (\Delta x)^2 - 1 + x^2$	$= \lim_{\Delta x \to 0} \frac{\_\_\_\_\_\_\_\_\_\_\_\_}{\Delta x[\sqrt{1-(x+\Delta x)^2} + \sqrt{1-x^2}\,]}$
$-2x - \Delta x$	$= \lim_{\Delta x \to 0} \frac{\_\_\_\_\_\_\_\_\_\_\_\_}{\sqrt{1-(x+\Delta x)^2} + \sqrt{1-x^2}}$
$\frac{-x}{\sqrt{1-x^2}}$	$= \_\_\_\_\_\_\_\_.$

2.1 The Derivative Notations

$\frac{dy}{dx}$ $D_x y$ $\dot{y}$	34 Let $y = f(x)$. Then the value of the derivative at a given number is written in Leibniz notation as $\frac{\_\_\_}{\_\_\_}$, and in differentiation operator notation as ____. Also, in Newton's notation it is written as ____.
differentiation	35 The operation of finding the derivative f' of a given function f is called ____________.

In Problems 36 through 40, express the derivative of the functions in Problems 29 through 33 in Leibniz notation and in operator notation.

$2x - 2$	36 From Problem 29, we have $f(x) = (x-1)^2$, so that $f'(x) = \frac{d}{dx}(x-1)^2 = D_x(x-1)^2 = \_\_\_\_\_\_\_\_.$
$3x^2$	37 From Problem 30, we have $f(x) = x^3$, so that $f'(x) = \frac{d}{dx}x^3 = D_x x^3 = \_\_\_\_.$
$D_x\left(\frac{1}{x+2}\right)$, $\frac{-1}{(x+2)^2}$	38 From Problem 31, we have $f(x) = \frac{1}{x+2}$, so that $f'(x) = \frac{d}{dx}\left(\frac{1}{x+2}\right) = \_\_\_\_\_\_\_\_ = \_\_\_\_\_\_\_\_.$
$\frac{d}{dx}\left(\frac{1}{\sqrt{x}}\right)$, $\frac{-1}{2x\sqrt{x}}$	39 From Problem 32, we have $f(x) = \frac{1}{\sqrt{x}}$, so that $f'(x) = \_\_\_\_\_\_\_\_ = D_x\left(\frac{1}{\sqrt{x}}\right) = \_\_\_\_\_\_\_\_.$

$\frac{d}{dx}(\sqrt{1-x^2})$, $D_x(\sqrt{1-x^2})$

40 From Problem 33, we have $f(x) = \sqrt{1-x^2}$, so that $f'(x) =$ ________ = ________ $= \frac{-x}{\sqrt{1-x^2}}$.

2.2 Differentiability and Continuity

$f(x_1 + \Delta x) - f(x_1)$

$\lim_{\Delta x \to 0^-} \frac{f(x_1 + \Delta x) - f(x_1)}{\Delta x}$

$f'_-(x_1)$

41 Let f be a function defined at the number x_1, and assume that f is defined in some open interval (x_1, b). Then we define the derivative from the right of f at x_1 by $f'_+(x_1) = \lim_{\Delta x \to 0^+}$ ________, provided that the limit exists.

42 Let f be a function defined at the number x_1, and assume that f is defined in some open interval (a, x_1). Then we define the derivative from the left of f at x_1 by $f'_-(x_1) =$ ________, provided that the limit exists.

43 The derivative $f'(x_1)$ exists and has the value A if and only if both $f'_+(x_1)$ and $f'_-(x_1)$ exist and have the common value A, that is, if and only if $f'_+(x_1) =$ ________ $= A$.

In Problems 44 through 47, find $f'_+(x_1)$ and $f'_-(x_1)$; then indicate if $f'(x_1)$ exists.

3

$6(\Delta x)$, 6

3

$6(\Delta x) + 3(\Delta x)^2$, 6

exists

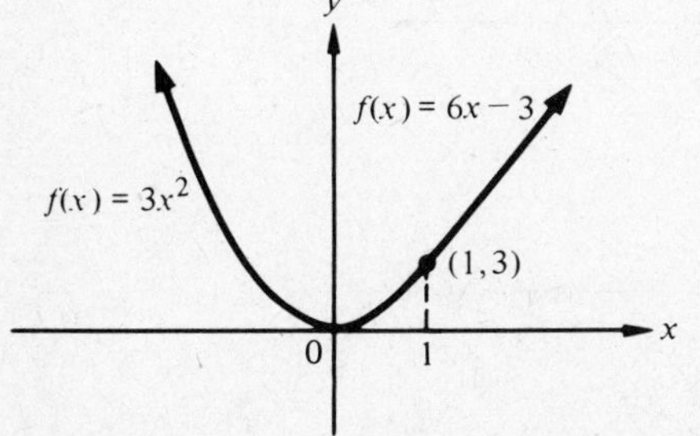

44 $f(x) = \begin{cases} 3x^2 & \text{if } x < 1 \\ 6x - 3 & \text{if } x \geq 1 \end{cases}$; $x_1 = 1$.

$f'_+(1) = \lim_{\Delta x \to 0^+} \frac{[6(1 + \Delta x) - 3] - \_\_\_\_}{\Delta x}$

$= \lim_{\Delta x \to 0^+} \frac{\_\_\_\_\_\_}{\Delta x} = \_\_\_\_,$

whereas

$f'_-(1) = \lim_{\Delta x \to 0^-} \frac{3(1 + \Delta x)^2 - \_\_\_\_}{\Delta x}$

$= \lim_{\Delta x \to 0^-} \frac{\_\_\_\_\_\_}{\Delta x} = \_\_\_\_.$

Therefore, $f'_+(1) = f'_-(1)$ and we conclude that $f'(1)$ ________.

The graph is

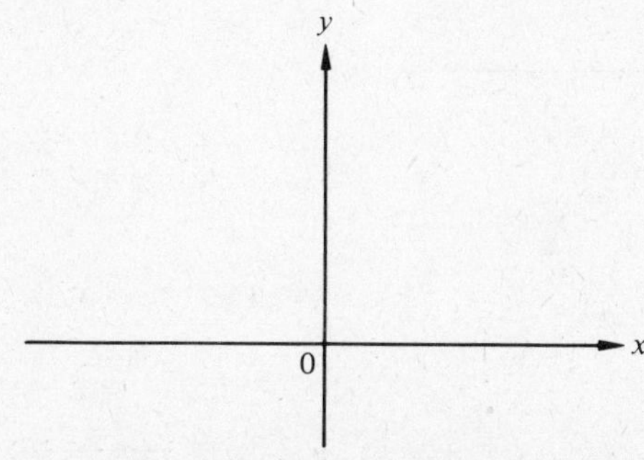

5

$-(\Delta x)$, -1

5

$2(\Delta x)$, 2

not exist

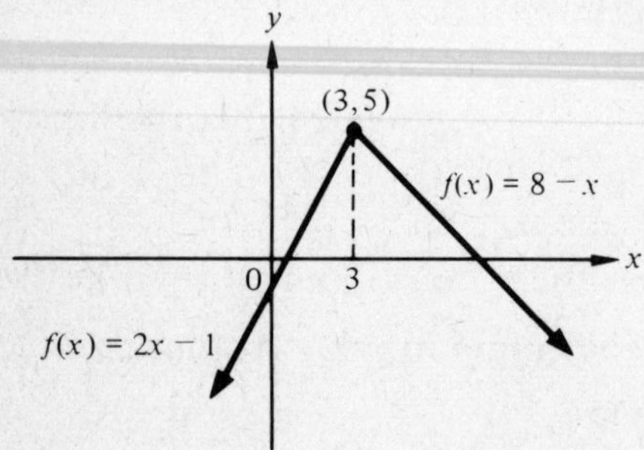

45 $f(x) = \begin{cases} 8 - x & \text{if } x \geq 3 \\ 2x - 1 & \text{if } x < 3 \end{cases}$; $x_1 = 3$.

$$f'_+(3) = \lim_{\Delta x \to 0^+} \frac{[8 - (3 + \Delta x)] - \_\_\_\_}{\Delta x}$$

$$= \lim_{\Delta x \to 0^+} \frac{\_\_\_\_}{\Delta x} = \_\_\_\_.$$

$$f'_-(3) = \lim_{\Delta x \to 0^-} \frac{[2(3 + \Delta x) - 1] - \_\_\_\_}{\Delta x}$$

$$= \lim_{\Delta x \to 0^-} \frac{\_\_\_\_}{\Delta x} = \_\_\_\_.$$

Since $f'_+(3) \neq f'_-(3)$, we conclude that $f'(3)$ does ________.
The graph is

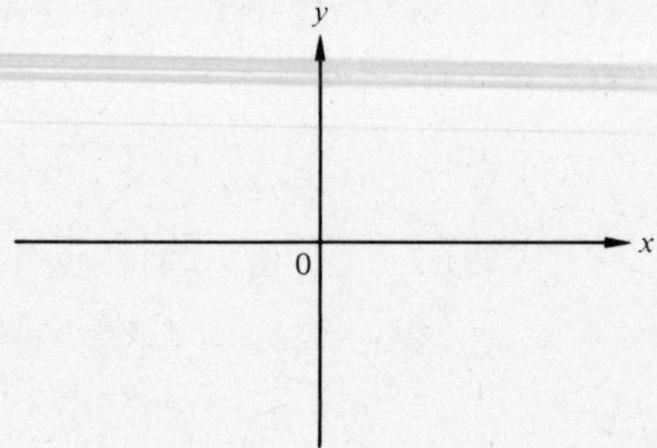

0

$2(\Delta x)$, 2

0

$-(\Delta x)^2$, 0

not exist

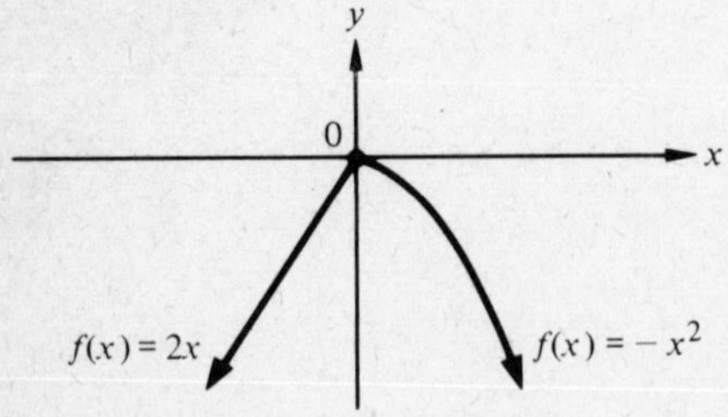

46 $f(x) = \begin{cases} 2x & \text{if } x \leq 0 \\ -x^2 & \text{if } x > 0 \end{cases}$; $x_1 = 0$.

$$f'_-(0) = \lim_{\Delta x \to 0^-} \frac{2(0 + \Delta x) - \_\_\_\_}{\Delta x}$$

$$= \lim_{\Delta x \to 0^-} \frac{\_\_\_\_}{\Delta x} = \_\_\_\_,$$

$$f'_+(0) = \lim_{\Delta x \to 0^+} \frac{-(0 + \Delta x)^2 - \_\_\_\_}{\Delta x}$$

$$= \lim_{\Delta x \to 0^+} \frac{\_\_\_\_}{\Delta x} = \_\_\_\_,$$

so that $f'_+(0) \neq f'_-(0)$ and we conclude that $f'(0)$ does ________.
The graph is

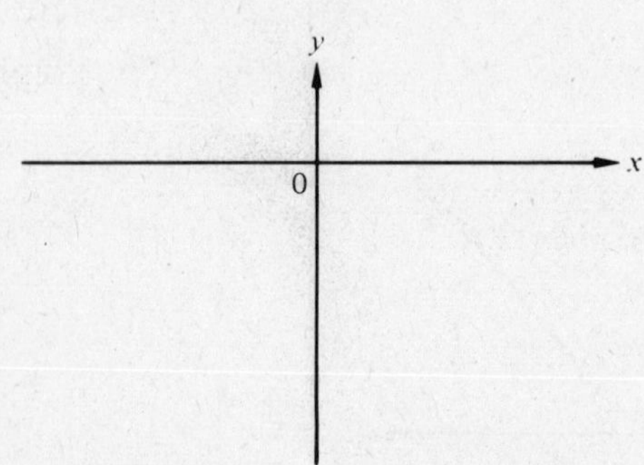

47 $f(x) = \begin{Bmatrix} x^2 - 9 & \text{if } x < 3 \\ \sqrt{x - 3} & \text{if } x \geqslant 3 \end{Bmatrix}$; $x_1 = 3$.

0

$f'_-(3) = \lim_{\Delta x \to 0^-} \dfrac{[(3 + \Delta x)^2 - 9] - \_\_\_\_}{\Delta x}$

6(Δx) + (Δx)², 6

$= \lim_{\Delta x \to 0^-} \dfrac{\_\_\_\_\_\_\_\_}{\Delta x} = \_\_\_\_.$

0

$f'_+(3) = \lim_{\Delta x \to 0^+} \dfrac{\sqrt{3 + \Delta x - 3} - \_\_\_\_}{\Delta x}$

$\sqrt{\Delta x}$, $+\infty$

$= \lim_{\Delta x \to 0^+} \dfrac{\_\_\_\_}{\Delta x} = \_\_\_\_.$

not exist

so that $f'_+(3) \neq f'_-(3)$ and we conclude that $f'(3)$ does ________.
The graph is

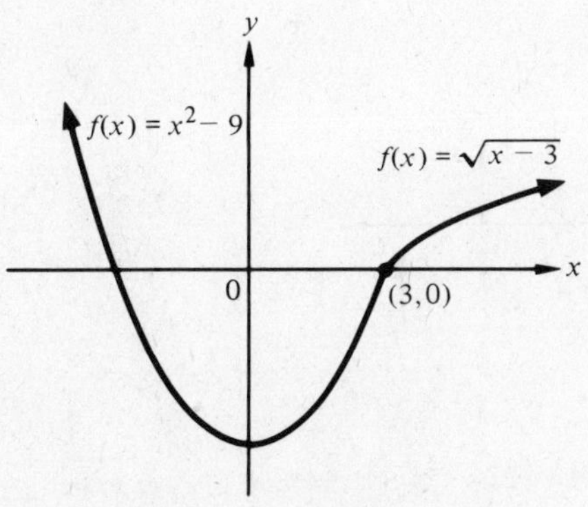

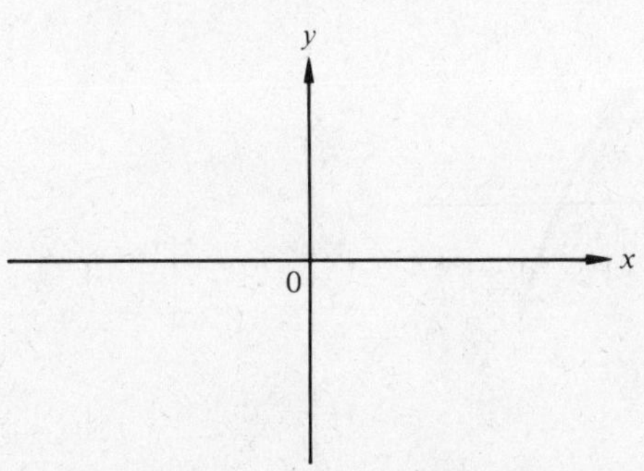

$f'(x)$

$f'_-(x)$

48 A function f is said to be differentiable at the number x if f is defined on some open interval containing x and ________ exists and is finite. This is equivalent to the statement that $f'_+(x)$ and $f'_-(x)$ both exist, are finite, and $f'_+(x) =$ ________.

differentiable

differentiable

49 A function f which is defined on an open interval (a, b) is said to be ____________ on the open interval (a, b) if it is differentiable at each number in this interval. If a function f is differentiable at each number belonging to its domain, it is called a ____________ function.

continuous

50 If a function f is differentiable at the number x, it is ____________ at the number x.

differentiable

51 Problem 50 provides a new way for checking the continuity of functions. That is, in order to show that the function f is continuous at the number x, it is enough to show that f is ____________ at x.

differentiable

continuous

52 If f is a continuous function at the number x, it is not necessarily ____________ at the number x; and if f is not differentiable at the number x, then f is not ____________ at the number x.

In Problems 53 through 58, indicate whether f is continuous at x_1 and whether it is differentiable at x_1.

53 $f(x) = 4 - x^2$, $x_1 = 1$.

Since f is a polynomial function, it is ______ at every number; hence, it is continuous at ____.

$$f'(1) = \lim_{\Delta x \to 0} \frac{[4 - (1 + \Delta x)^2] - \_\_\_\_}{\Delta x}$$

$$= \lim_{\Delta x \to 0} \frac{\_\_\_\_\_\_}{\Delta x} = \_\_\_\_,$$

so that f is ______ at 1. The graph is

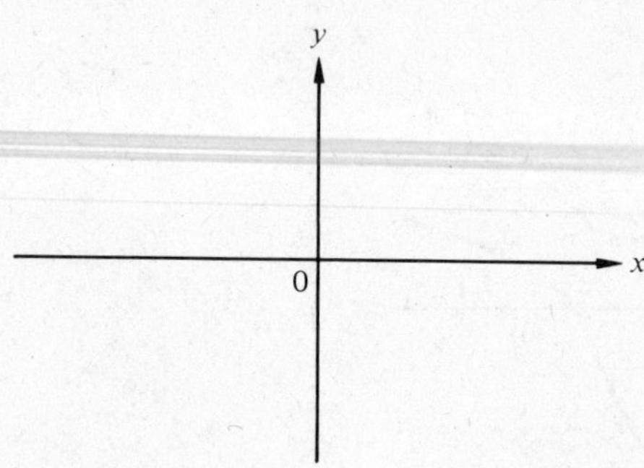

continuous

1

3

$-2(\Delta x) - (\Delta x)^2, \ -2$

differentiable

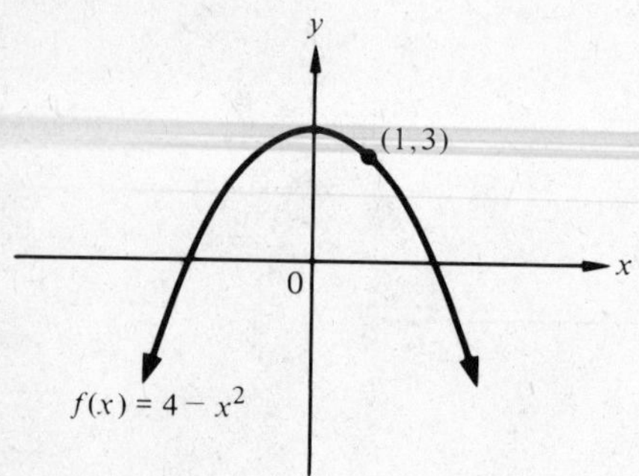

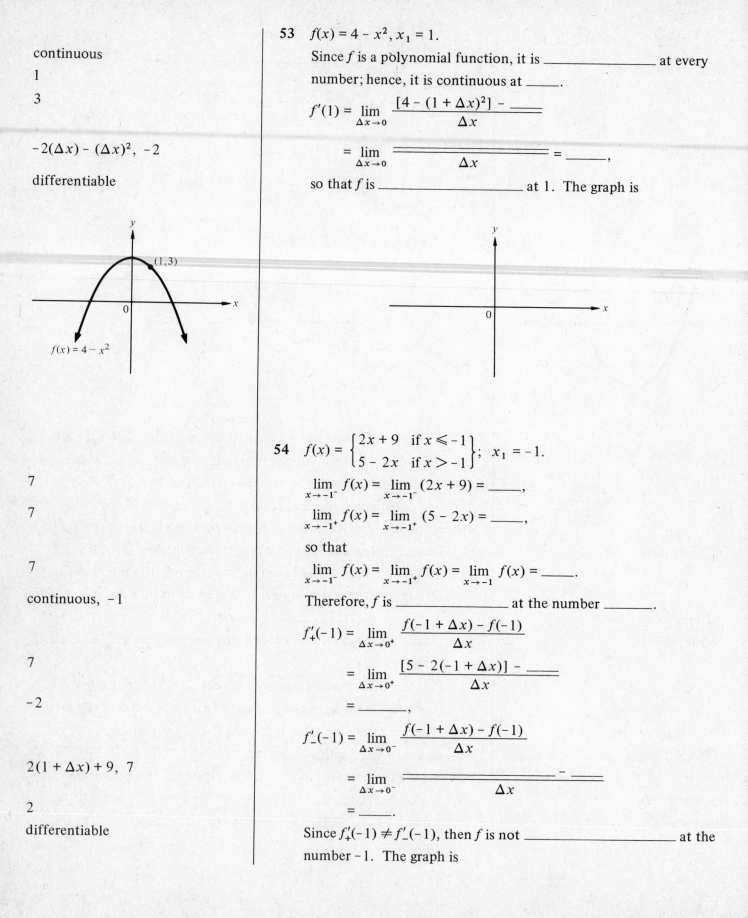

54 $f(x) = \begin{cases} 2x + 9 & \text{if } x \leqslant -1 \\ 5 - 2x & \text{if } x > -1 \end{cases}; \quad x_1 = -1.$

$$\lim_{x \to -1^-} f(x) = \lim_{x \to -1^-} (2x + 9) = \_\_\_\_,$$

$$\lim_{x \to -1^+} f(x) = \lim_{x \to -1^+} (5 - 2x) = \_\_\_\_,$$

so that

$$\lim_{x \to -1^-} f(x) = \lim_{x \to -1^+} f(x) = \lim_{x \to -1} f(x) = \_\_\_\_.$$

Therefore, f is ______ at the number ____.

$$f'_+(-1) = \lim_{\Delta x \to 0^+} \frac{f(-1 + \Delta x) - f(-1)}{\Delta x}$$

$$= \lim_{\Delta x \to 0^+} \frac{[5 - 2(-1 + \Delta x)] - \_\_\_\_}{\Delta x}$$

$$= \_\_\_\_,$$

$$f'_-(-1) = \lim_{\Delta x \to 0^-} \frac{f(-1 + \Delta x) - f(-1)}{\Delta x}$$

$$= \lim_{\Delta x \to 0^-} \frac{\_\_\_\_\_\_ - \_\_\_\_}{\Delta x}$$

$$= \_\_\_\_.$$

Since $f'_+(-1) \neq f'_-(-1)$, then f is not ______ at the number -1. The graph is

7

7

7

continuous, -1

7

-2

$2(1 + \Delta x) + 9, \ 7$

2

differentiable

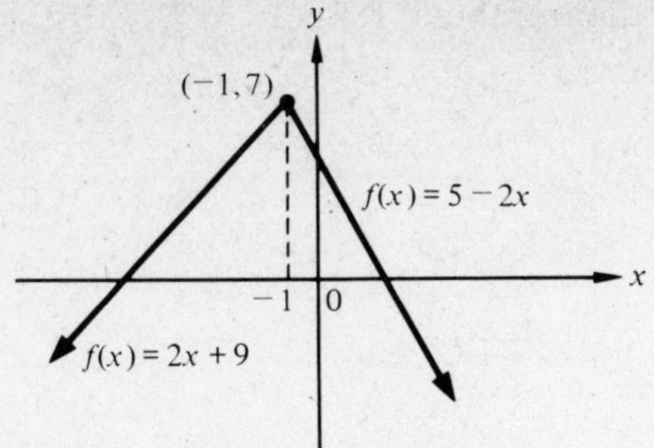

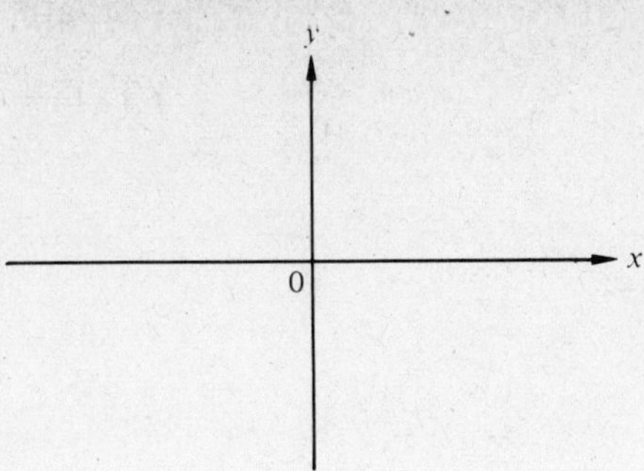

55 $f(x) = \begin{cases} 9 - x^2 & \text{if } x > 2 \\ 2x + 1 & \text{if } x \leq 2 \end{cases}$; $x_1 = 2$.

$\lim_{x \to 2^+} f(x) = \lim_{x \to 2^+} (9 - x^2) = 5$,

$\lim_{x \to 2^-} f(x) = \lim_{x \to 2^-} (2x + 1) = 5$,

continuous

so that f is ______________ at the number 2.

$$f'_+(2) = \lim_{\Delta x \to 0^+} \frac{f(2 + \Delta x) - f(2)}{\Delta x}$$

5

$$= \lim_{\Delta x \to 0^+} \frac{[9 - (2 + \Delta x)^2] - \_\_\_\_}{\Delta x}$$

$-4(\Delta x) - (\Delta x)^2$, -4

$$= \lim_{\Delta x \to 0^+} \frac{\_\_\_\_\_\_\_\_\_\_\_\_}{\Delta x} = \_\_\_\_,$$

$$f'_-(2) = \lim_{\Delta x \to 0^-} \frac{f(2 + \Delta x) - f(2)}{\Delta x}$$

5

$$= \lim_{\Delta x \to 0^-} \frac{[2(2 + \Delta x) + 1] - \_\_\_\_}{\Delta x}$$

$2(\Delta x)$, 2

$$= \lim_{\Delta x \to 0^-} \frac{\_\_\_\_\_\_\_\_}{\Delta x} = \_\_\_\_.$$

differentiable

Since $f'_+(2) \neq f'_-(2)$, then f is not ____________________ at the number 2. The graph is

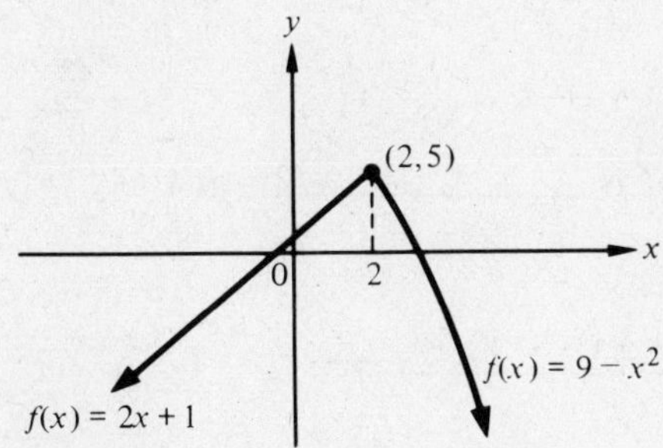

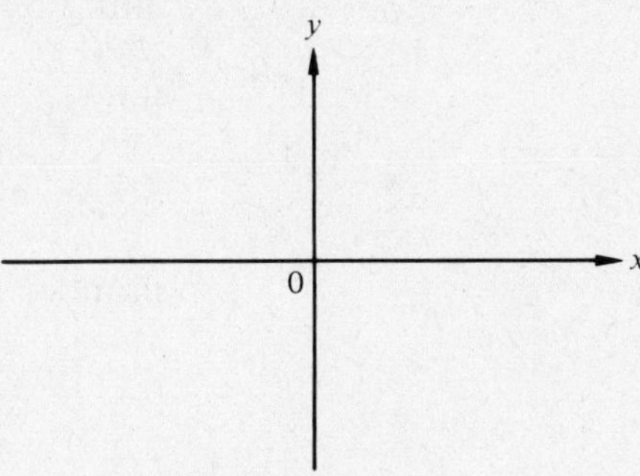

56 $f(x) = |x + 1|$, $x_1 = -1$.

0

$\lim_{x \to -1^-} f(x) = \lim_{x \to -1^-} |x + 1| = (\_\_\_\_)$,

0

$\lim_{x \to -1^+} f(x) = \lim_{x \to -1^+} |x + 1| = (\_\_\_\_)$, continuous | so that f is ________ at the number -1.

$$f'_+(-1) = \lim_{\Delta x \to 0^+} \frac{|-1 + \Delta x + 1|}{\Delta x}$$

1 | $$= \lim_{\Delta x \to 0^+} \frac{|\Delta x|}{\Delta x} = \_\_\_\_,$$

$$f'_-(-1) = \lim_{\Delta x \to 0^-} \frac{|-1 + \Delta x + 1|}{\Delta x} = \lim_{\Delta x \to 0^-} \frac{|\Delta x|}{\Delta x}$$

-1 | $$= \_\_\_\_.$$

differentiable | Since $f'_+(-1) \neq f'_-(-1)$, then f is not ________ at the number -1. The graph is

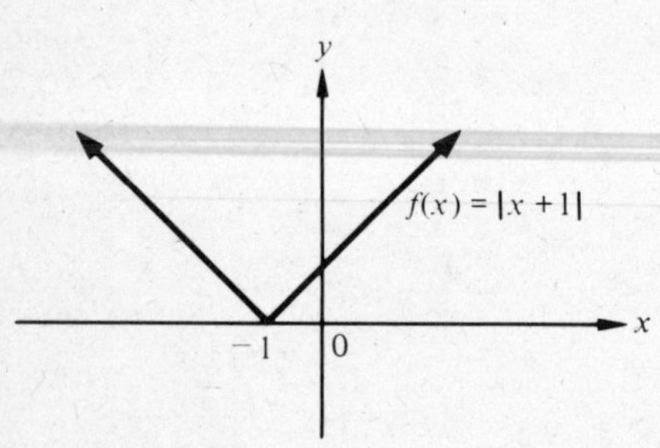

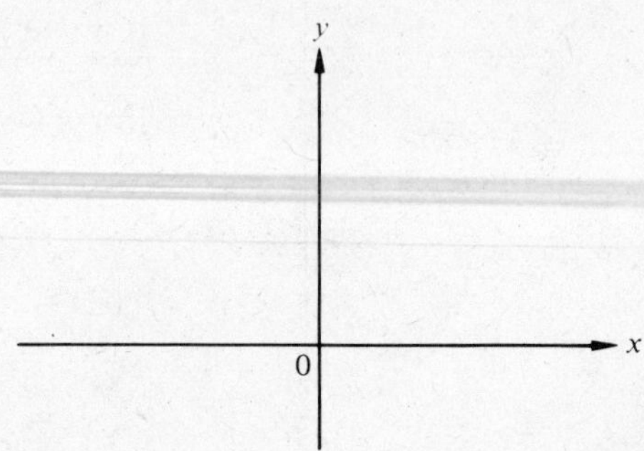

57 $f(x) = x - [\![x]\!]$, $x_1 = 0$.

$$\lim_{x \to 0^+} f(x) = \lim_{x \to 0^+} \{x - [\![x]\!]\} = 0,$$

1 | $$\lim_{x \to 0^-} f(x) = \lim_{x \to 0^-} \{x - [\![x]\!]\} = \_\_\_\_,$$

continuous | so that f is not ________ at the number 0.

differentiable | By Problem 52, f is not ________ at the number 0.

58 $f(x) = (x - 1)[\![x - 1]\!]$, $x_1 = 1$.

0 | $$\lim_{x \to 1^+} f(x) = \lim_{x \to 1^+} (x - 1)[\![x - 1]\!] = \_\_\_\_,$$

0 | $$\lim_{x \to 1^-} f(x) = \lim_{x \to 1^-} (x - 1)[\![x - 1]\!] = \_\_\_\_.$$

continuous | Since $\lim_{x \to 1^+} f(x) = \lim_{x \to 1^-} f(x)$, then f is ________ at the number 1.

0 | $$f'_+(1) = \lim_{\Delta x \to 0^+} \frac{\{(1 + \Delta x - 1)[\![1 + \Delta x - 1]\!]\} - \_\_\_\_}{\Delta x}$$

0 | $$= \_\_\_\_,$$

0 | $$f'_-(1) = \lim_{\Delta x \to 0^-} \frac{\{(1 + \Delta x - 1)[\![1 + \Delta x - 1]\!]\} - \_\_\_\_}{\Delta x}$$

-1 | $$= \_\_\_\_,$$

1 | so that $f'_+(1) \neq f'_-(1)$ and f is not differentiable at the number ____.

3 Basic Rules for Differentiation

59 **If c is a constant number and if f is the function defined by the equation $f(x) = c$ for all values of x, then f is differentiable at every x and f' is defined by the equation $f'(x) =$ ____.**

0

60 If $f(x) = 7$, then $f'(x) =$ ____.

0

61 $\frac{d}{dx}(3) =$ ____.

0

62 $D_x(7 + \sqrt{2}) =$ ____.

0

63 $\frac{d}{dx}(\pi^2) =$ ____.

0

64 $\frac{d}{dx}(1 + \sqrt{17}) =$ ____.

0

65 $D_x(3 - \sqrt{2}) =$ ____.

0

66 Let n be a fixed positive integer and let f be the function defined by $f(x) = x^n$ for all values of x; then $f'(x) =$ ____.

nx^{n-1}

67 If $f(x) = x^9$, then $f'(x) = 9x^{(\_\_\_)} =$ ____.

^{8}, $9x^8$

68 If $f(x) = x^{17}$, then $f'(x) = (\_\_\_)\, x^{16} =$ ____.

17, $17x^{16}$

69 $\frac{d}{dx}(x^{71}) = 71x^{(\_\_\_)} =$ ____.

^{70}, $71x^{70}$

70 $\frac{d}{dx}(x^{37}) = (\_\_\_)\, x^{36} =$ ____.

37, $37x^{36}$

71 $D_x(x^{15}) = (\_\_\_)\, x^{14} =$ ____.

15, $15x^{14}$

72 $D_x(x^{89}) = 89x^{(\_\_\_)} =$ ____.

^{88}, $89x^{88}$

73 $D_x(x^{113}) = 113x^{(\_\_\_)} =$ ____.

^{112}, $113x^{112}$

74 **Let f be a function which is differentiable at the number x, and let c be a constant number. Let the function g be defined by $g(x) = cf(x)$. Then g is ____ and $g'(x) =$ ____. In operator notation $D_x(cf(x)) = c(\_\_\_\_)$. In Leibniz notation $\frac{d}{dx}(cf(x)) = c\left(\_\_\_\_\right)$.**

differentiable, $cf'(x)$

$D_x f(x)$

$\frac{df(x)}{dx}$

75 If $f(x) = \frac{5}{6}x^6$, then $f'(x) = (\_\_\_)\, x^5 =$ ____.

5, $5x^5$

76 If $y = \frac{2}{3}x^3$, then $y' = (\_\_\_)\, x^2 =$ ____.

2, $2x^2$

77 If $y = \frac{3}{7}x^{21}$, then $y' = (\_\_\_)\, x^{20} =$ ____.

9, $9x^{20}$

78 If $y = \frac{2}{9}x^{36}$, then $y' = (\_\_\_)\, x^{35} =$ ____.

8, $8x^{35}$

79 $\frac{d}{dx}(13x^4) = 52x^{(\_\_\_)} =$ ____.

^{3}, $52x^3$

80 $\frac{d}{dx}\left(\frac{3}{17}x^{34}\right) = 6x^{(\_\_\_)} =$ ____.

^{33}, $6x^{33}$

81 $\frac{d}{dx}\left(\frac{1}{12}x^{72}\right) = (\_\_\_\_)\,x^{71} = \_\_\_\_$.

$6,\ 6x^{71}$

82 $D_x\left(\frac{5}{13}x^{39}\right) = \_\_\_\_$.

$15x^{38}$

83 $D_x\left(-\frac{\pi}{3}x^3\right) = \_\_\_\_$.

$-\pi x^2$

84 $D_x\left(-\frac{3}{5}x^{15}\right) = \_\_\_\_$.

$-9x^{14}$

85 **Let f and g be functions both of which are differentiable at the number x_1, and let the function h be defined by $h(x) = f(x) + g(x)$. Then h is also ____ at the number x_1 and $h'(x_1) =$ ____. In operator notation $D_x[f(x) + g(x)] =$ ____ $+ D_x g(x)$. In Leibniz notation $\frac{d}{dx}[f(x) + g(x)] =$ ____.**

differentiable

$f'(x_1) + g'(x_1)$

$D_x f(x)$

$\frac{df(x)}{dx} + \frac{dg(x)}{dx}$

86 If $f(x) = x^5 + x^3$, then $f'(x) = (\_\_\_\_)\,x^4 + (\_\_\_\_)\,x^2 =$ ____.

5, 3

$5x^4 + 3x^2$

87 If $f(x) = \frac{5}{6}x^6 + 9x^4$, then $y' = (\_\_\_\_)\,x^5 + (\_\_\_\_)\,x^3 =$ ____.

5, 36

$5x^5 + 36x^3$

88 If $y = \frac{1}{2}x^{10} + \frac{2}{5}x^5$, then $y' = (\_\_\_\_)\,x^9 + 2x^{(\_\_\_\_)} =$ ____.

5, 4

$5x^9 + 2x^4$

89 If $f(x) = \frac{2}{3}x^3$ and $g(x) = \frac{1}{2}x^2$, then $(f+g)'(x) =$ ____ $=$ ____.

$f'(x) + g'(x),\ 2x^2 + x$

90 $\frac{d}{dx}(3x^2 + 17x + 2) = \_\_\_\_$.

$6x + 17$

91 $\frac{d}{dx}\left(8x^4 - \frac{1}{3}x^3 + 7\right) = \_\_\_\_$.

$32x^3 - x^2$

92 $\frac{d}{dx}(3x^4 + 2x^3 + 111) = \_\_\_\_$.

$12x^3 + 6x^2$

93 $D_x(2x^{100} - 21x^3 - 3x^2 + 9) = \_\_\_\_$.

$200x^{99} - 63x^2 - 6x$

94 $D_x\left(\sqrt{2}x^3 - \frac{1}{3}x^3 + \frac{2}{5}x + 21\right) = \_\_\_\_$.

$3\sqrt{2}x^2 - x^2 + \frac{2}{5}$

In Problems 95 through 97, let f and g be functions for which $f'(2) = -8$ and $g'(2) = 7$. Find each expression.

95 $(f+g)'(2)$.

$(f+g)'(2) =$ ____ $+ g'(2)$

$=$ ____ $+ 7 =$ ____.

$f'(2)$

$-8,\ -1$

96 $(2f+g)'(2)$.

$2f'(2),\ g'(2)$
7
-9

$(2f+g)'(2) = \_\_\_\_\_ + \_\_\_\_\_$
$= 2(-8) + \_\_\_\_$
$= \_\_\_\_.$

$3f'(2)$
$-8,\ 7$
35
-59

97 $(3f-5g)'(2)$.
$(3f-5g)'(2) = \_\_\_\_\_ - 5g'(2)$
$= 3(\_\_\_\_) - 5(\_\_\_\_)$
$= -24 - \_\_\_\_$
$= \_\_\_\_.$

$f_1'(x)+f_2'(x)+f_3'(x)$

98 Let f_1, f_2, and f_3 be differentiable functions. Then by the sum rule, $(f_1+f_2+f_3)'(x) = \_\_\_\_\_\_\_\_\_\_$.

differentiable

99 Using all the rules studied up to this point, we can see that all polynomial functions are __________, and we can differentiate such functions by these rules.

differentiable
$f(x_1)\cdot g'(x_1)+f'(x_1)\cdot g(x_1)$

$D_x[f(x)]\ g(x)$

100 Let f and g be functions both of which are differentiable at the number x_1 and let the function h be defined by $h(x)=f(x)\cdot g(x)$. Then h is also __________ at x_1 and $h'(x_1) = \_\_\_\_\_\_\_\_\_\_$. In operator notation, we have $D_x[f(x)\cdot g(x)] = f(x)\cdot D_x g(x) + \_\_\_\_\_\_\_\_\_\_$.

$D_x(x^2+2)$
$2x$
$2x^3+4x$
$4x^3+6x$

101 If $f(x) = (x^2+2)(x^2+1)$, then
$f'(x) = (x^2+2)\,D_x(x^2+1) + (\_\_\_\_\_\_\_\_)(x^2+1)$
$= (x^2+2)(\_\_\_\_) + 2x(x^2+1)$
$= \_\_\_\_\_\_\_\_ + 2x^3 + 2x$
$= \_\_\_\_\_\_\_\_.$

$D_x(x^2-2)$
$6x^2,\ 2x$
$10x^4 - 12x^2 + 14x$

102 If $f(x) = (x^2-2)(2x^3+7)$, then
$f'(x) = (x^2-2)\,D_x(2x^3+7) + (\_\_\_\_\_\_\_\_)(2x^3+7)$
$= (x^2-2)(\_\_\_\_) + (\_\_\_\_)(2x^3+7)$
$= \_\_\_\_\_\_\_\_\_\_.$

$\frac{d}{dx}(2x-1)$
$2x$
$6x^2-2x+6$

103 $\frac{d}{dx}[(2x-1)(x^2+3)]$
$= (2x-1)\frac{d}{dx}(x^2+3) + \left(\_\_\_\_\_\_\_\_\right)(x^2+3)$
$= (2x-1)(\_\_\_\_) + 2(x^2+3)$
$= \_\_\_\_\_\_\_\_.$

$6y,\ 4y^3+2$
$6y^5+12y^2,\ 12y^5+20y^3+6y^2+10$
$18y^5+20y^3+18y^2+10$

104 $\frac{d}{dy}[(y^4+2y)(3y^2+5)]$
$= (y^4+2y)(\_\_\_\_) + (\_\_\_\_\_\_\_\_)(3y^2+5)$
$= (\_\_\_\_\_\_\_\_) + (\_\_\_\_\_\_\_\_\_\_)$
$= \_\_\_\_\_\_\_\_\_\_.$

105 $D_t[(3t+7)(t^2-5t+4)]$

$2t - 5,\ 3$

$6t^2 - t - 35,\ 3t^2 - 15t + 12$

$9t^2 - 16t - 23$

$= (3t + 7)(\_\_\_\_\_\_) + (\_\_\_\_)(t^2 - 5t + 4)$

$= (\_\_\_\_\_\_\_\_\_\_) + (\_\_\_\_\_\_\_\_\_\_\_\_)$

$= \_\_\_\_\_\_\_\_\_\_\_\_.$

$g'(3),\ g(3)$

$-\frac{1}{5},\ 2$

$-1,\ -8$

-9

106 If f and g are differentiable functions such that $f(3) = 5, f'(3) = 2$, $g(3) = -4$, and $g'(3) = -\frac{1}{5}$, then

$(fg)'(3) = f(3)(\_\_\_\_\_\_) + f'(3)(\_\_\_\_\_)$

$= (5)\left(\_\_\_\right) + (\_\_\_\_)(-4)$

$= (\_\_\_\_) + (\_\_\_\_)$

$= \_\_\_\_.$

differentiable

$\frac{-g'(x_1)}{[g(x_1)]^2}$

107 Let g be a function which is differentiable at x_1. Suppose that $g(x_1) \neq 0$. Let h be the function defined by the equation $h(x) = \frac{1}{g(x)}$ for all numbers x in the domain of g for which $g(x) \neq 0$. Then h is ____________ at x_1 and $h'(x_1) =$ ________.

x^2

108 If $f(x) = \frac{3}{x}$, then $f'(x) = -\frac{3}{\_\_\_\_}$.

$2x$

109 If $y = \frac{1}{x^2 + 7}$, then $\frac{dy}{dx} = -\frac{\_\_\_\_\_\_}{(x^2 + 7)^2}$.

$12x^2$

110 Let $y = \frac{4}{x^3 + 5}$; then $D_x y = -\frac{\_\_\_\_\_\_}{(x^3 + 5)^2}$.

$30x^5,\ -\frac{30}{x^7}$

111 $\frac{d}{dx}\left(\frac{5}{x^6}\right) = \frac{-(\_\_\_\_\_)}{x^{12}} = \_\_\_\_\_.$

$4x^3,\ -\frac{4}{x^5}$

112 $\frac{d}{dx}(x^{-4}) = \frac{-(\_\_\_\_)}{x^8} = \_\_\_\_\_.$

$36x,\ -\frac{36x}{(3x^2 - 11)^2}$

113 $D_x\left(\frac{6}{3x^2 - 11}\right) = \frac{-(\_\_\_\_)}{(3x^2 - 11)^2} = \_\_\_\_\_\_\_\_\_\_.$

differentiable

$\frac{g(x_1) f'(x_1) - f(x_1) g'(x_1)}{[g(x_1)]^2}$

114 Let f and g be functions which are differentiable at the number x_1 and suppose that $g(x_1) \neq 0$. Let $h = \frac{f(x)}{g(x)}$ for all x in the common portion of the domains of f and g such that $g(x) \neq 0$. Then h is ____________ at x_1 and $h'(x_1) =$ ______________.

$D_x(3x),\ D_x(x^2 + 8)$

$3x^2 + 24 - 6x^2,\ \frac{-3x^2 + 24}{(x^2 + 8)^2}$

115 Let $f(x) = \frac{3x}{x^2 + 8}$; then

$f'(x) = \frac{(x^2 + 8)(\_\_\_\_\_\_) - 3x(\_\_\_\_\_\_\_\_)}{(x^2 + 8)^2}$

$= \frac{\_\_\_\_\_\_\_\_\_\_\_\_}{(x^2 + 8)^2} = \_\_\_\_\_\_\_\_.$

116 If $f(x) = \dfrac{x^2 - 1}{x^2 + 1}$, then

$D_x(x^2 - 1),\ D_x(x^2 + 1)$

$$f'(x) = \frac{(x^2 + 1)(\_\_\_\_\_\_) - (x^2 - 1)(\_\_\_\_\_\_)}{(x^2 + 1)^2}$$

$2x^3 - 2x,\ 4x$

$$= \frac{2x^3 + 2x - (\_\_\_\_\_\_)}{(x^2 + 1)^2} = \frac{\_\_\_\_\_\_}{(x^2 + 1)^2}.$$

117 Let $y = \dfrac{x^5 - 6x^2}{x^7 + 3}$; then

$\dfrac{d}{dx}(x^5 - 6x^2)$

$$\frac{dy}{dx} = \frac{(x^7 + 3)\left(\_\_\_\_\_\_\right) - (x^5 - 6x^2)\dfrac{d}{dx}(x^7 + 3)}{(x^7 + 3)^2}$$

$5x^4 - 12x,\ 7x^6$

$$= \frac{(x^7 + 3)(\_\_\_\_\_\_) - (x^5 - 6x^2)(\_\_\_)}{(x^7 + 3)^2}$$

$5x^{11} - 12x^8 + 15x^4 - 36x$

$$= \frac{(\_\_\_\_\_\_) - (7x^{11} - 42x^8)}{(x^7 + 3)^2}$$

$-2x^{11} + 30x^8 + 15x^4 - 36x$

$$= \frac{\_\_\_\_\_\_}{(x^7 + 3)^2}.$$

118 Let $y = \dfrac{t^3 - 2}{t^2 + 9}$; then

$D_t(t^2 + 9)$

$$D_t y = \frac{(t^2 + 9)\, D_t(t^3 - 2) - (t^3 - 2)(\_\_\_\_\_\_)}{(t^2 + 9)^2}$$

$3t^2$

$$= \frac{(t^2 + 9)(\_\_\_) - (t^3 - 2)(2t)}{(t^2 + 9)^2}$$

$2t^4 - 4t$

$$= \frac{3t^4 + 27t^2 - (\_\_\_\_\_\_)}{(t^2 + 9)^2}$$

$t^4 + 27t^2 + 4t$

$$= \frac{\_\_\_\_\_\_}{(t^2 + 9)^2}.$$

119 $\dfrac{d}{ds}\left(\dfrac{s^2}{1 - s}\right)$

$\dfrac{d}{ds}(s^2),\ 1 - s$

$$= \frac{(1 - s)\left(\_\_\_\_\_\_\right) - s^2 \dfrac{d}{ds}(\_\_\_\_\_\_)}{(1 - s)^2}$$

-1

$$= \frac{(1 - s)(2s) - s^2(\_\_\_)}{(1 - s)^2}$$

$2s - s^2$

$$= \frac{\_\_\_\_\_\_}{(1 - s)^2}.$$

120 $\dfrac{d}{dt}\left(\dfrac{t^2 + 2t + 7}{2t^2 - 3t - 5}\right)$

$t^2 + 2t + 7,\ 2t^2 - 3t - 5$

$$= \frac{(2t^2 - 3t - 5)\dfrac{d}{dt}(\_\_\_\_\_\_) - (t^2 + 2t + 7)\dfrac{d}{dt}(\_\_\_\_\_\_)}{(2t^2 - 3t - 5)^2}$$

$2t + 2,\ 4t - 3$

$$= \frac{(2t^2 - 3t - 5)(\_\_\_\_\_\_) - (t^2 + 2t + 7)(\_\_\_\_\_\_)}{(2t^2 - 3t - 5)^2}$$

$4t^3 - 2t^2 - 16t - 10$

$$= \frac{(\underline{\qquad\qquad\qquad\qquad}) - (4t^3 + 5t^2 + 22t - 21)}{(2t^2 - 3t - 5)^2}$$

$-7t^2 - 38t + 11$

$$= \frac{\underline{\qquad\qquad\qquad}}{(2t^2 - 3t - 5)^2}.$$

121 Let $y = \dfrac{2x+1}{x+1}(x+3)$; then

$\dfrac{2x+1}{x+1}$

$$\frac{dy}{dx} = \frac{2x+1}{x+1}\frac{d}{dx}(x+3) + (x+3)\frac{d}{dx}\left(\underline{\qquad\qquad}\right)$$

1, 1

$$= \frac{2x+1}{x+1}(\underline{\qquad}) + (x+3)\frac{(x+1)(2) - (2x+1)(\underline{\qquad})}{(x+1)^2}$$

1

$$= \frac{2x+1}{x+1} + (x+3)\frac{\underline{\qquad\qquad}}{(x+1)^2}$$

$2x^2 + 4x + 4$

$$= \frac{\underline{\qquad\qquad\qquad}}{(x+1)^2}.$$

122 Given two differentiable functions f and g so that $f(4) = 2$, $f'(4) = 5$, $g(4) = 6$, and $g'(4) = 3$, we have

$g'(4)$

$$\left(\frac{f}{g}\right)'(4) = \frac{g(4)f'(4) - f(4)(\underline{\qquad\qquad})}{g(4)^2}$$

2

$$= \frac{(6)(5) - (\underline{\qquad})(3)}{(6)^2}$$

24, $\dfrac{2}{3}$

$$= \frac{\underline{\qquad}}{36} = \underline{\qquad}.$$

123 Given that $\dfrac{d}{dx}(\sin x) = \cos x$ and $\dfrac{d}{dx}(\cos x) = -\sin x$, we have

$$\frac{d}{dx}(\cot x) = \frac{d}{dx}\left(\frac{\cos x}{\sin x}\right)$$

$\cos x$, $\sin x$

$$= \frac{(\sin x)\dfrac{d}{dx}(\underline{\qquad\qquad}) - (\cos x)\dfrac{d}{dx}(\underline{\qquad\qquad})}{\sin^2 x}$$

$-\sin x$, $\cos x$

$$= \frac{\sin x(\underline{\qquad\qquad}) - \cos x(\underline{\qquad\qquad})}{\sin^2 x}$$

$\sin^2 x + \cos^2 x$, 1

$$= \frac{-(\underline{\qquad\qquad\qquad\qquad})}{\sin^2 x} = \frac{-(\underline{\qquad})}{\sin^2 x}$$

$-\csc^2 x$

$$= \underline{\qquad\qquad\qquad}.$$

$\cos x$

124 $D_x(\sec x) = D_x\left(\dfrac{1}{\cos x}\right) = -\dfrac{D_x(\underline{\qquad\qquad})}{\cos^2 x}$

$\sin x$, $\cos x$

$$= \frac{\underline{\qquad\qquad}}{\cos^2 x} = \frac{\sin x}{\cos x}\cdot\frac{1}{\underline{\qquad\qquad}}$$

$\tan x \sec x$

$$= \underline{\qquad\qquad\qquad\qquad}.$$

125 **If the function f is defined by $f(x) = x^n$, where n is an integer, then $f'(x) = \underline{\qquad\qquad}$.**

nx^{n-1}

126 If $f(x) = x^{-4}$, then $f'(x) = -4x^{(\underline{\quad})} = \underline{\qquad\qquad}$.

$^{-5}$, $-4x^{-5}$

127 Let $y = \frac{2}{x^3} + \frac{5}{x^2}$; then $y = 2x^{-3} + 5x^{-2}$, so that $\frac{dy}{dx} =$ $-6x^{(\_\_\_)} - 10x^{(\_\_\_)} =$ ________.

$^{-4}$, $^{-3}$, $-6x^{-4} - 10x^{-3}$

128 $\frac{d}{dx}\left(\frac{\sqrt{7}}{x^3} - \frac{5}{x^7}\right) = \frac{d}{dx}(\sqrt{7}x^{-3} - 5x^{-7})$ $= (\_\_\_)\,x^{-4} + (\_\_\_)\,x^{-8} =$ ________.

$-3\sqrt{7}$, 35, $-3\sqrt{7}x^{-4} + 35x^{-8}$

129 $D_x\left(\frac{2}{x^{100}} - 3x^4\right) = D_x(2x^{-100} - 3x^4)$ $= (\_\_\_)\,x^{-101} - (\_\_\_)\,x^3 =$ ________.

-200, 12, $-200x^{-101} - 12x^3$

130 $D_x(x^{-21} - 3x^{-19} + 5x^2 + 8) = -21x^{(\_\_\_)} + 57x^{(\_\_\_)} + 10x$.

$^{-22}$, $^{-20}$

4 The Chain Rule

131 **If y is a differentiable function of u and if u is a differentiable function of x, then y is a differentiable function of ___ and $\frac{dy}{dx} =$ ________. In operator notation, we have $D_x y =$ ________.**

x

$\frac{dy}{du} \cdot \frac{du}{dx}$

$D_u y \cdot D_x u$

132 By direct calculation from the definition of the derivative, if $y = \sqrt{u}$, then $\frac{dy}{du} =$ ______. Therefore, if $y = \sqrt{x^4 + 3x^2 + 1}$, then $\frac{dy}{dx} = \frac{dy}{du} \cdot$ ______ gives

$$\frac{dy}{dx} = \frac{1}{2\sqrt{x^4 + 3x^2 + 1}} \frac{d}{dx}(\_\_\_\_\_\_\_\_)$$

$$= \frac{1}{2\sqrt{x^4 + 3x^2 + 1}}(\_\_\_\_\_\_\_\_).$$

$\frac{1}{2\sqrt{u}}$

$\frac{du}{dx}$

$x^4 + 3x^2 + 1$

$4x^3 + 6x$

133 If $y = \sqrt{x^2 + (2x - 1)^4}$, then $D_x y = D_u y \cdot$ ______ gives

$$D_x y = \frac{1}{2\sqrt{x^2 + (2x - 1)^4}} D_x(\_\_\_\_\_\_\_\_)$$

$$= \frac{1}{2\sqrt{x^2 + (2x - 1)^4}}(\_\_\_\_\_\_\_\_).$$

$D_x u$

$x^2 + (2x - 1)^4$

$2x + 8(2x - 1)^3$

134 If $y = \sqrt{\frac{x + 5}{x - 5}}$, then

$$\frac{dy}{dx} = \frac{1}{2\sqrt{\frac{x+5}{x-5}}} \frac{d}{dx}\left[\frac{(x - 5)(1) - (\_\_\_\_\_\_\_\_)}{(x - 5)^2}\right]$$

$$= \frac{1}{2\sqrt{\frac{x+5}{x-5}}} \frac{\_\_\_\_}{(x - 5)^2}.$$

$(x + 5)(1)$

-10

135 **The chain rule is often used to calculate derivatives of the form $D_x u^n$, where u is a differentiable function of x and n is an integer. The formula is $D_x u^n = nu^{n-1}(\_\_\_)$.**

$D_x u$

136 If $y = (6x^2 - 3x + 2)^{20}$, then

$$\frac{dy}{dx} = 20(\underline{\qquad\qquad\qquad})^{19}(\underline{\qquad\qquad})$$
$$= 60(6x^2 - 3x + 2)^{19}(\underline{\qquad\qquad})$$
$$= \underline{\qquad\qquad\qquad\qquad\qquad\qquad}.$$

$6x^2 - 3x + 2,\ 12x - 3$

$4x - 1$

$60(4x - 1)(6x^2 - 3x + 2)^{19}$

137 If $y = \left(\frac{3x^2 - 4}{x^2}\right)^5$, then

$$D_x y = 5\left(\frac{3x^2 - 4}{x^2}\right)^{(\underline{\quad})} D_x\left(\underline{\qquad\qquad}\right)$$
$$= 5\left(\frac{3x^2 - 4}{x^2}\right)^4\left(\underline{\qquad}\right).$$

$4,\ \frac{3x^2 - 4}{x^2}$

$\frac{8}{x^3}$

138 If $y = (4x^{-2} + 3x^{-1} + 7)^{14}$, then

$$D_x y = 14(4x^{-2} + 3x^{-1} + 7)^{(\underline{\quad})} D_x(\underline{\qquad\qquad\qquad\qquad})$$
$$= 14(4x^{-2} + 3x^{-1} + 7)^{13}(\underline{\qquad\qquad\qquad}).$$

$13,\ 4x^{-2} + 3x^{-1} + 7$

$-8x^{-3} - 3x^{-2}$

139 If $y = \left(\frac{2x + 1}{2x - 1}\right)^{10}$, then

$$\frac{dy}{dx} = 10\left(\underline{\qquad\qquad}\right)^9 \frac{d}{dx}\left(\underline{\qquad\qquad}\right)$$
$$= 10\left(\frac{2x + 1}{2x - 1}\right)^9 \frac{(2x - 1)(2) - (\underline{\qquad\qquad})(2)}{(2x - 1)^2}$$
$$= 10\left(\frac{2x + 1}{2x - 1}\right)^9 \frac{\underline{\qquad\qquad}}{(2x - 1)^2}.$$

$\frac{2x + 1}{2x - 1},\ \frac{2x + 1}{2x - 1}$

$2x + 1$

-4

140 If $f(x) = [(x^4 - 1)(2x^4 + 1)]^3$, then

$$f'(x) = 3[(x^4 - 1)(2x^4 + 1)]^{(\underline{\quad})} D_x(\underline{\qquad\qquad\qquad\qquad})$$
$$= 3[(x^4 - 1)(2x^4 + 1)]^2[(x^4 - 1)(8x^3) + \underline{\qquad\qquad\qquad}]$$
$$= 3[(x^4 - 1)(2x^4 + 1)]^2(\underline{\qquad\qquad\qquad}).$$

$2,\ (x^4 - 1)(2x^4 + 1)$

$4(2x^4 + 1)x^3$

$16x^7 - 4x^3$

141 **If y is a function of v, v is a function of u, and u is a function of x, then $\frac{dy}{dx} = \underline{\qquad\qquad\qquad}$.**

$\frac{dy}{dv} \cdot \frac{dv}{du} \cdot \frac{du}{dx}$

142 **Let f and g be differentiable functions and let $h = f \circ g$. Then h is $\underline{\qquad\qquad\qquad}$ and we write the chain rule as $h'(x) = \underline{\qquad\qquad\qquad} = f'[g(x)](\underline{\qquad\qquad})$.**

differentiable

$(f \circ g)'(x),\ g'(x)$

143 Let $f(u) = u^{10}$ and $g(x) = x^3 - 3x + 1$. To find $(f \circ g)'(x)$, we have $f'(u) = \underline{\qquad}$ and $g'(x) = \underline{\qquad\qquad}$, so that

$$(f \circ g)'(x) = f'[g(x)](\underline{\qquad\qquad})$$
$$= 10[g(x)]^{(\underline{\quad})} g'(x)$$
$$= 10(\underline{\qquad\qquad\qquad})^9(3x^2 - 3)$$
$$= \underline{\qquad\qquad\qquad\qquad\qquad\qquad}.$$

$10u^9,\ 3x^2 - 3$

$g'(x)$

9

$x^3 - 3x + 1$

$30(x^2 - 1)(x^3 - 3x + 1)^9$

$8u^7$, $6x$ $g(x)$ $6x$ $48x(3x^2 - 7)^7$	**144** Let $f(u) = u^8$ and $u = g(x) = 3x^2 - 7$. To find $(f \circ g)'(x)$, we have $f'(u) =$ ____ and $g'(x) =$ ____, so that $(f \circ g)'(x) = f'(\_\_\_\_)\, g'(x)$ $= 8u^7(\_\_\_\_)$ $=$ __________.
  $g'(-1)$ 3, 2	**145** Suppose that f and g are differentiable functions so that $g(-1) = 5$, $g'(-1) = 3$, $f(5) = 7$ and $f'(5) = \frac{2}{3}$. Find $(f \circ g)'(-1)$. $(f \circ g)'(-1) = f'[g(-1)]\,[\_\_\_\_\_\_]$ $= f'(5)(\_\_\_\_) = \frac{2}{3}(3) =$ ____.
 $(5x + 2)^2$ $(5x + 2)^2$ $2(5x + 2)\,5$ $5(5x + 2)$	**146** Use the fact that $\lvert u\rvert = \sqrt{u^2}$ to find $\frac{d}{dx}\lvert 5x + 2\rvert$. $\frac{d}{dx}\lvert 5x + 2\rvert = \frac{d}{dx}(\sqrt{\_\_\_\_\_\_})$ $= \frac{1}{2\sqrt{(5x + 2)^2}}\frac{d}{dx}(\_\_\_\_\_\_)$ $= \frac{\_\_\_\_\_\_}{2\sqrt{(5x + 2)^2}}$ $= \frac{\_\_\_\_\_\_}{\lvert 5x + 2\rvert}$.

In Problems 147 through 159, assume that $D_x(\sin x) = \cos x$, $D_x(\cos x) = -\sin x$, $D_x(\tan x) = \sec^2 x$, $D_x(\cot x) = -\csc^2 x$, $D_x(\sec x) = \sec x \tan x$, and $D_x(\csc x) = -\csc x \cot x$.

$7x + 9$ $7x + 9$ $\cos(7x + 9)$, $7\cos(7x + 9)$	**147** If $f(x) = \sin(7x + 9)$, then using the chain rule with $u =$ ______, we have $f'(x) = (\cos u)\frac{d}{dx}(\_\_\_\_\_\_)$, so that $f'(x) = (\_\_\_\_\_\_)(7) =$ ________.
$1 - 13x$ $-\sin u$ -13 $13\sin(1 - 13x)$	**148** If $g(x) = \cos(1 - 13x)$, then if we let $u =$ ______, we have $g'(x) = (\_\_\_\_\_\_)\, D_x(1 - 13x)$ $= -\sin(1 - 13x)(\_\_\_\_)$ $=$ ________.
 $\sec^2 u$ $2x$, $2x\sec^2 x^2$	**149** If $h(x) = \tan x^2$, then if we let $u = x^2$, we have $h'(x) = (\_\_\_\_\_\_)\frac{d}{dx}(x^2)$ $= (\sec^2 u)(\_\_\_\_) =$ ________.
 $-\csc^2(3t + 11)$, $(3t + 11)$ 3 $-3\csc^2(3t + 11)$	**150** If $y = \cot(3t + 11)$, then $\frac{dy}{dt} = (\_\_\_\_\_\_)\frac{d}{dt}(\_\_\_\_\_\_)$ $= -\csc^2(3t + 11)(\_\_\_\_)$ $=$ ________.

151 If $y = \sec(8x)$, then

$D_x y = (\_\_\_\_\_\_\_\_\_\_\_\_)\, D_x(8x)$

$= (\sec 8x \tan 8x)(\_\_\_\_)$

$= \_\_\_\_\_\_\_\_\_\_\_\_.$

$\sec 8x \tan 8x$

8

$8 \sec 8x \tan 8x$

152 If $y = \csc(25x + 3)$, then

$D_x y = -\csc(25x + 3)\cot(25x + 3)\, D_x(\_\_\_\_\_\_\_\_\_\_\_\_)$

$= \_\_\_\_\_\_\_\_\_\_\_\_.$

$25x + 3$

$-25 \csc(25x + 3) \cot(25x + 3)$

153 If $y = \sin^2 x$, then if we let $u = \sin x$, we have $y = \_\_\_\_$ and

$\frac{dy}{dx} = (2u)\left(\_\_\_\_\right)$, so that

$\frac{dy}{dx} = 2 \sin x \frac{d}{dx}(\_\_\_\_\_\_)$

$= \_\_\_\_\_\_\_\_\_\_\_\_.$

u^2

$\frac{du}{dx}$

$\sin x$

$2 \sin x \cos x$

154 If $y = \cos^6 t$, then if we let $u = \_\_\_\_\_\_$, we have $y = \_\_\_\_$ and

$\frac{dy}{dt} = (\_\_\_\_)\frac{du}{dt}$, so that

$\frac{dy}{dt} = 6(\_\_\_\_\_\_)\frac{d}{dt}(\_\_\_\_\_\_)$

$= 6 \cos^5 t(\_\_\_\_\_\_)$

$= \_\_\_\_\_\_\_\_\_\_\_\_.$

$\cos t,\ u^6$

$6u^5$

$\cos^5 t,\ \cos t$

$-\sin t$

$-6 \cos^5 t \sin t$

155 If $f(\theta) = \sqrt{\sin \theta}$, then if we let $u = \_\_\_\_\_\_$, we have

$f'(\theta) = \frac{1}{2\sqrt{\sin \theta}} D_\theta(\_\_\_\_\_\_)$

$= \frac{1}{2\sqrt{\sin \theta}}(\_\_\_\_\_\_)$

$= \frac{\qquad}{\_\_\_\_\_\_\_\_}.$

$\sin \theta$

$\sin \theta$

$\cos \theta$

$\frac{\cos \theta}{2\sqrt{\sin \theta}}$

156 If $g(y) = \csc^4 7y$, then

$g'(y) = 4 \csc^3 7y\, D_y(\_\_\_\_\_\_)$

$= 4 \csc^3 7y(-\csc 7y \cot 7y)\, D_y(\_\_\_\_)$

$= -28(\_\_\_\_\_\_\_\_\_\_\_\_).$

$\csc 7y$

$7y$

$\csc^4 7y \cot 7y$

157 If $h(t) = t^2 \cot 2t$, then

$h'(t) = t^2 D_t(\_\_\_\_\_\_) + 2t \cot 2t$

$= t^2(\_\_\_\_\_\_) + 2t \cot 2t$

$= \_\_\_\_\_\_\_\_\_\_\_\_.$

$\cot 2t$

$-2 \csc^2 2t$

$-2t^2 \csc^2 2t + 2t \cot 2t$

158 If $f(x) = \frac{\cos x}{x^2 + 3}$, then

$f'(x) = \frac{(x^2 + 3)D_x(\_\_\_\_\_\_) - \cos x\, D_x(\_\_\_\_\_\_)}{(x^2 + 3)^2}$

$= \frac{(x^2 + 3)(\_\_\_\_\_\_) - \cos x(\_\_\_\_)}{(x^2 + 3)^2}$

$= \frac{\_\_\_\_\_\_\_\_\_\_\_\_}{(x^2 + 3)^2}.$

$\cos x,\ x^2 + 3$

$-\sin x,\ 2x$

$-x^2 \sin x - 3 \sin x - 2x \cos x$

159 If $g(\theta) = \sec\sqrt{\theta}$, then

$g'(\theta) = \sec\sqrt{\theta}\,\tan\sqrt{\theta}\,D_\theta(\_\_\_\_)$

$= \sec\sqrt{\theta}\,\tan\sqrt{\theta}\,\dfrac{1}{\_\_\_\_}$

$= \dfrac{\_\_\_\_\_\_\_\_}{2\sqrt{\theta}}.$

$\sqrt{\theta}$

$2\sqrt{\theta}$

$\sec\sqrt{\theta}\,\tan\sqrt{\theta}$

160 If $h(t) = (3t^2+1)^2(2t^2+5)^3$, then

$h'(t) = (3t^2+1)^2 D_t(2t^2+5)^3 + D_t(\_\_\_\_\_\_\_\_)(2t^2+5)^3$

$= (3t^2+1)^2(3)(2t^2+5)^2(\_\_\_\_) + 2(3t^2+1)(\_\_\_\_)(2t^2+5)^3$

$= 12t(3t^2+1)^2(2t^2+5)^2 + (\_\_\_\_)(3t^2+1)(2t^2+5)^3$

$= (12t)(3t^2+1)(2t^2+5)^2(\_\_\_\_\_\_\_\_).$

$(3t^2+1)^2$

$4t,\ 6t$

$12t$

$5t^2+6$

161 If $y = (3x+1)^{-3}(5x-3)^4$, then

$y' = (3x+1)^{-3}(4)(5x-3)^3(5) + (\_\_\_\_)(3x+1)^{-4}(\_\_\_\_)(5x-3)^4$

$= \dfrac{20(\_\_\_\_\_\_\_\_)}{(3x+1)^3} - \dfrac{9(\_\_\_\_\_\_\_\_)}{(3x+1)^4}$

$= \dfrac{20(\_\_\_\_\_\_\_\_)(5x-3)^3 - (\_\_\_\_)(5x-3)^4}{(3x+1)^4}$

$= \dfrac{(5x-3)^3(\_\_\_\_\_\_\_\_\_\_\_\_)}{(3x+1)^4}$

$= \dfrac{(5x-3)^3(\_\_\_\_\_\_\_\_)}{(3x+1)^4}.$

$-3,\ 3$

$(5x-3)^3,\ (5x-3)^4$

$3x+1,\ 9$

$60x+20-45x+27$

$15x+47$

162 Let $w = xy$, $x = t^2+5$, $y = \dfrac{3}{u}$, and $u = 5t+7$. Find $\dfrac{dw}{dt}$.

$\dfrac{dw}{dt} = x\dfrac{dy}{dt} + y\left(\_\_\_\_\right)$

$= x\left(\_\_\_\_\_\_\_\_\right) + y(\_\_\_\_)$

$= x\left(-\dfrac{3}{u^2}\right)(\_\_\_\_) + y(2t)$

$= \dfrac{-15(\_\_\_\_\_\_)}{(5t+7)^2} + \dfrac{3(2t)}{5t+7}$

$= \dfrac{-15t^2 - 75 + (\_\_\_\_\_\_\_\_)}{(5t+7)^2}$

$= \dfrac{\_\_\_\_\_\_\_\_}{(5t+7)^2}.$

$\dfrac{dx}{dt}$

$\dfrac{dy}{du}\cdot\dfrac{du}{dt},\ 2t$

5

t^2+5

$30t^2+42t$

$15t^2+42t-75$

5 The Inverse Function Rule and the Rational Power Rule

163 If $\dfrac{dy}{dx} \neq 0$, then the inverse-function rule using Leibniz notation is given by $\dfrac{dx}{dy} = \_\_\_\_\_\_\_\_.$

$\dfrac{1}{\dfrac{dy}{dx}}$

In Problems 164 and 165, use the Leibniz notation and the inverse function rule to find the value of $\frac{dx}{dy}$ when y has the given value.

164 $y = x^3$ when $y = 8$.

When $y = 8$, then $x =$ ____. Since $\frac{dy}{dx} =$ _____, then $\frac{dx}{dy} = \frac{1}{\frac{dy}{dx}} =$ _____. Hence, when $y = 8$, then $x =$ ____ and $\frac{dx}{dy} =$ _____.

2, $3x^2$

$\frac{1}{3x^2}$, 2, $\frac{1}{12}$

165 $y = \frac{x+1}{x-1}$ when $y = 4$.

When $y = 4$, then $4 = \frac{x+1}{x-1}$, $4(x - 1) = x + 1$, $3x =$ ____, so $x =$ ____. Since

$$\frac{dy}{dx} = \frac{(x-1)(\_\_\_\_) - (x+1)(\_\_\_\_)}{(x-1)^2}$$

$$= -\frac{(\_\_\_\_)}{(x-1)^2},$$

then $\frac{dx}{dy} = -\frac{(x-1)^2}{\_\_\_\_}$. When $y = 4$, then $x =$ ____ and $\frac{dx}{dy} =$ _____.

5

$\frac{5}{3}$

1, 1

2

2, $\frac{5}{3}$, $-\frac{2}{9}$

5.1 The Inverse Function Theorem

166 Let f be a function whose domain is an open interval I, assume that f is differentiable on I, and suppose that $f'(c) \neq 0$ for every number c in I. Then f has an inverse g, g is __________, and $g'(x) = \frac{1}{\_\_\_\_\_\_\_\_}$ holds for every number x in the domain of g.

differentiable

$f'[g(x)]$

167 If a function f has an inverse function g, then we write $g =$ _____, and we can rewrite the inverse function theorem as

$$(f^{-1})'(x) = \frac{1}{\_\_\_\_\_\_\_\_}.$$

f^{-1}

$f'[f^{-1}(x)]$

168 Let $f(x) = 2x + 1$ and $g(x) = \frac{x-1}{2}$.

Here $(f \circ g)(x) = f[g(x)] = f\left(\frac{x-1}{2}\right) =$ ____ $(g \circ f)(x)$, so f and g are inverses of each other. Thus, $g'(x) = \frac{1}{\_\_\_\_\_\_\_\_} = \frac{1}{\_\_\_\_}$.

x

$f'[g(x)]$, $\frac{1}{2}$

5.2 The Power Rule for Rational Exponents

$\frac{1}{n}x^{(1/n)-1}$

169 If n is a positive integer, then $D_x x^{1/n} = D_x \sqrt[n]{x} =$ ____________

0

for all values of x for which $\sqrt[n]{x}$ is defined, except $x =$ ____.

170 Let $f(x) = \sqrt[4]{x}$; then

$-3/4$, x^3, $\sqrt[4]{x}$

$$f'(x) = \frac{1}{4}x^{(\_\_\_)} = \frac{\_\_\_}{4\sqrt[4]{\_\_\_}} = \frac{\_\_\_}{4x}.$$

$1/7$, $-6/7$, $\frac{1}{7}x^{-6/7}$

171 If $y = \sqrt[7]{x}$, then $y = x^{(\_\_\_)}$, so that $\frac{dy}{dx} = \frac{1}{7}x^{(\_\_\_)} =$ ________.

$1/9$, $-8/9$, $x^{-8/9}$

172 If $y = \sqrt[9]{x}$, then $y = x^{(\_\_\_)}$, so that $\frac{dy}{dx} = \frac{1}{9}x^{(\_\_\_)} = \frac{1}{9}(\_\_\_\_)$.

173 Let $r = \frac{m}{n}$ be a rational number, reduced to lowest terms, n a positive integer. Then $D_x x^r =$ ________ holds for all values of x for which $x^r = (x^{1/n})^m$ is defined, except possibly for $x = 0$. It also holds for $x = 0$, provided that n is ______ and $m > n$.

rx^{r-1}

odd

$5/2$, $x^{5/2}$

174 If $y = x^{7/2}$, then $D_x y = \frac{7}{2}x^{(\_\_\_)} = \frac{7}{2}(\_\_\_\_)$.

175 If $y = \frac{2}{3}x^{3/2} - \frac{3}{4}x^{4/3}$, then

1, 1

$$\frac{dy}{dx} = (\_\_\_)x^{1/2} - (\_\_\_)x^{1/3}$$

$x^{1/2} - x^{1/3}$

$$= \_\_\_\_\_\_\_\_\_\_\_\_.$$

$1/3$

176 If $y = \sqrt[3]{1 + x^2}$, then $y = (1 + x^2)^{(\_\_\_)}$, so that

$-2/3$, $1 + x^2$

$$\frac{dy}{dx} = \frac{1}{3}(1 + x^2)^{(\_\_\_)} \cdot \frac{d}{dx}(\_\_\_\_\_\_)$$

$2x$

$$= \frac{1}{3}(1 + x^2)^{-2/3}(\_\_\_)$$

$2x$

$$= \frac{\_\_\_}{3\sqrt[3]{(1 + x^2)^2}}.$$

177 If $y = (2 + x^3)^{4/5}$, then

$-1/5$, $2 + x^3$

$$D_x y = \frac{4}{5}(2 + x^3)^{(\_\_\_)} D_x(\_\_\_\_\_)$$

$3x^2$

$$= \frac{4}{5}(2 + x^3)^{-1/5}(\_\_\_)$$

$3x^2$

$$= \frac{4(\_\_\_)}{5\sqrt[5]{2 + x^3}}$$

$12x^2$

$$= \frac{\_\_\_}{5\sqrt[5]{2 + x^3}}.$$

$1/2$

178 If $y = \sqrt{\sqrt[3]{x} + \sqrt[4]{x}}$, then $y = (x^{1/3} + x^{1/4})^{(\_\_\_)}$, so that

$-1/2,\ x^{1/3} + x^{1/4}$

$$\frac{dy}{dx} = \frac{1}{2}(x^{1/3} + x^{1/4})^{(\_\_\_\_)}\frac{d}{dx}(\_\_\_\_\_\_\_\_\_\_)$$

$\frac{1}{3}x^{-2/3} + \frac{1}{4}x^{-3/4}$

$$= \frac{1}{2}(x^{1/3} + x^{1/4})^{-1/2}\left(\_\_\_\_\_\_\_\_\_\_\_\_\_\_\right).$$

179 Assume that $\frac{d}{dx}\sin x = \cos x$. Then, if $y = \sin^{5/3} x$, we have

$2/3$

$$\frac{dy}{dx} = \frac{5}{3}\sin^{(\_\_\_\_)} x \frac{d}{dx}\sin x$$

$\cos x$

$$= \frac{5}{3}\sin^{2/3} x(\_\_\_\_\_\_\_\_).$$

180 If $f(x) = \dfrac{\sqrt{x}}{(x^3 - 5)^{1/3}}$, then

$x^{1/2},\ (x^3 - 5)^{1/3}$

$$f'(x) = \frac{(x^3 - 5)^{1/3} D_x(\_\_\_\_\_\_) - x^{1/2} D_x(\_\_\_\_\_\_\_\_\_\_)}{[(x^3 - 5)^{1/3}]^2}$$

$\frac{1}{2}x^{-1/2},\ x^2(x^3 - 5)^{-2/3}$

$$= \frac{(x^3 - 5)^{1/3}\left(\_\_\_\_\_\_\_\_\right) - x^{1/2}(\_\_\_\_\_\_\_\_\_\_)}{(x^3 - 5)^{2/3}}$$

$-x^3 - 5$

$$= \frac{\_\_\_\_\_\_\_\_\_\_}{2\sqrt{x}\sqrt[3]{(x^3 - 5)^4}}.$$

181 If $y = \left(\dfrac{x^2}{2x + 5}\right)^{11/7}$, then

$4/7,\ \dfrac{x^2}{2x + 5}$

$$\frac{dy}{dx} = \frac{11}{7}\left(\frac{x^2}{2x + 5}\right)^{(\_\_\_\_)}\frac{d}{dx}\left(\_\_\_\_\_\_\_\_\right)$$

$2x,\ 2$

$$= \frac{11}{7}\left(\frac{x^2}{2x + 5}\right)^{4/7}\frac{(2x + 5)(\_\_\_\_) - x^2(\_\_\_\_)}{(2x + 5)^2}$$

$2x^2 + 10x$

$$= \frac{11}{7}\left(\frac{x^2}{2x + 5}\right)^{4/7}\frac{\_\_\_\_\_\_\_\_\_\_}{(2x + 5)^2}.$$

182 If $y = x^2\sqrt[4]{3x^5 + 7}$, then $y = x^2(3x^5 + 7)^{1/4}$, so that

$(3x^5 + 7)^{1/4},\ x^2$

$$\frac{dy}{dx} = x^2\frac{d}{dx}(\_\_\_\_\_\_\_\_\_\_\_\_) + (3x^5 + 7)^{1/4}\frac{d}{dx}(\_\_\_\_)$$

$2x$

$$= x^2 \cdot \frac{1}{4}(3x^5 + 7)^{-3/4}\frac{d}{dx}(3x^5 + 7) + (3x^5 + 7)^{1/4}(\_\_\_\_)$$

$15x^4,\ 2x$

$$= \frac{x^2(\_\_\_\_\_\_)}{4\sqrt[4]{(3x^5 + 7)^3}} + \sqrt[4]{3x^5 + 7}\cdot(\_\_\_\_)$$

$3x^5 + 7$

$$= \frac{15x^6 + 8x(\_\_\_\_\_\_\_\_)}{4\sqrt[4]{(3x^5 + 7)^3}}$$

$39x^6 + 56x$

$$= \frac{\_\_\_\_\_\_\_\_\_\_}{4\sqrt[4]{(3x^5 + 7)^3}}.$$

183 If $g(t) = \sqrt[3]{t + 1}\sqrt[7]{t + 3}$, then

$D_t(t + 1)^{1/3}$

$$g'(t) = (t + 1)^{1/3} D_t(t + 3)^{1/7} + (\_\_\_\_\_\_\_\_\_\_\_\_)(t + 3)^{1/7}$$

$(t + 3)^{-6/7}$

$$= (t + 1)^{1/3}\left(\frac{1}{7}\right)(\_\_\_\_\_\_\_\_)$$

$(t+1)^{-2/3}$

$(t+3)^{1/7}$

$+\frac{1}{3}(\_\_\_\_\_\_)(t+3)^{1/7}$

$=\frac{(t+1)^{1/3}}{7\sqrt[7]{(t+3)^6}}+\frac{\_\_\_\_\_\_}{3\sqrt[3]{(t+1)^2}}.$

$^{1/2}$, $(3-6t^{1/3})^{1/4}+t$

$-2t^{-2/3}$

184 If $f(t)=[(3-6t^{1/3})^{1/4}+t]^{3/2}$, then

$f'(t)=\frac{3}{2}[(3-6t^{1/3})^{1/4}+t]^{(\_\_\_)}D_t(\_\_\_\_\_\_)$

$=\frac{3}{2}[(3-6t^{1/3})^{1/4}+t]^{1/2}\left[\frac{1}{4}(3-6t^{1/3})^{-3/4}(\_\_\_\_\_\_)+1\right].$

5.3 Further Examples of the Inverse Function Theorem

$f^{-1}(5)$, 2

$\frac{1}{7}$

185 Suppose that $f(2)=5$ and $f'(2)=7$. Then

$(f^{-1})'(5)=\frac{1}{f'(\_\_\_\_\_\_)}=\frac{1}{f'(\_\_\_)}$

$=\_\_\_.$

$f^{-1}(8)$, 5

$\frac{1}{3}$

186 Suppose that $f(5)=8$ and $f'(5)=3$. Then

$(f^{-1})'(8)=\frac{1}{f'(\_\_\_\_\_\_)}=\frac{1}{f'(\_\_\_)}$

$=\_\_\_.$

$f^{-1}(9)$

$\frac{1}{6}$

187 Suppose that $f(-2)=9$ and $f'(-2)=6$. Then

$(f^{-1})'(9)=\frac{1}{f'(\_\_\_\_\_\_)}=\frac{1}{f'(-2)}$

$=\_\_\_.$

$f^{-1}\left(\frac{5}{3}\right)$

$\frac{7}{2}$

188 Suppose that $f\left(\frac{1}{4}\right)=\frac{5}{3}$ and $f'\left(\frac{1}{4}\right)=\frac{2}{7}$. Then

$(f^{-1})'\left(\frac{5}{3}\right)=\frac{1}{f'\left(\_\_\_\_\_\_\right)}=\frac{1}{f'\left(\frac{1}{4}\right)}$

$=\_\_\_.$

$5x^4$

f^{-1}

189 Let $f(x)=x^5+2$ for $x>0$. Since $f'(x)=\_\_\_\_\_\_\neq 0$ for $x>0$, it follows that ______ exists.

3

$f(1)$

1

190 Let $f(x)=x^5+2$. We wish to find $(f^{-1})'(3)$. Since $f(1)=\_\_\_\_$, then

$f^{-1}(3)=f^{-1}(\_\_\_\_\_\_)=(f^{-1}\circ f)(1)$

$=\_\_\_\_.$

By the inverse function theorem,

$f^{-1}(3)$, $f'(1)$

$$(f^{-1})'(3) = \frac{1}{f'(\_\_\_\_\_\_)} = \frac{1}{\_\_\_\_\_\_}$$

5

$$= \frac{1}{\_\_\_\_}.$$

6 The Equations of Tangent and Normal Lines

$f'(x_1)(x - x_1)$

191 The equation of the tangent line to the graph of a function f at the point (x_1, y_1), where $y_1 = f(x_1)$ is given by $y - y_1 =$ ________________.

perpendicular

192 The normal line to the graph of f at (x_1, y_1) is a line ________________ to the tangent line at (x_1, y_1). This is illustrated by

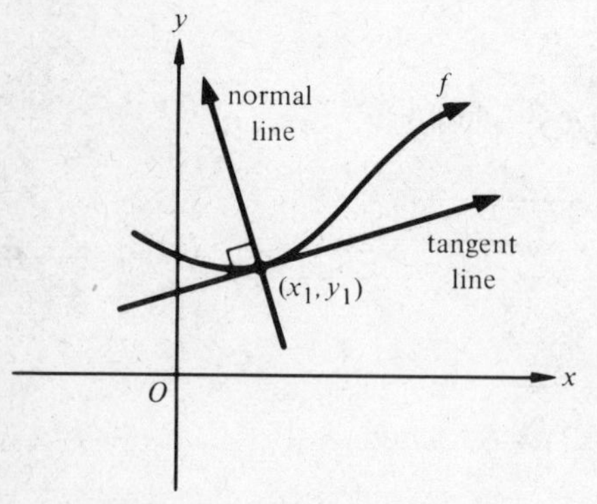

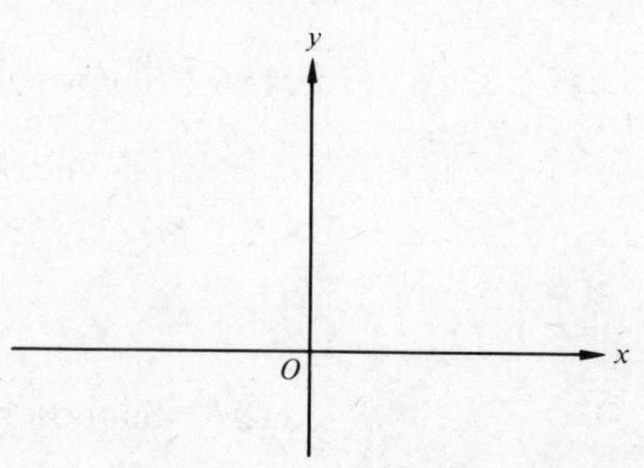

$-\frac{1}{f'(x_1)}$

$y - y_1 = -\frac{1}{f'(x_1)}(x - x_1)$

193 The slope of the normal line to the graph of f at (x_1, y_1) is __________ and the equation of the normal line to the graph of f at (x_1, y_1) is ________________________.

In Problems 194 through 198, find the equations of the tangent and the normal lines to the graph of the function f at the given point.

$6x$, 6

$6(x - 1)$

$6x - 3$

$-\frac{1}{6}(x - 1)$, $-\frac{1}{6}x + \frac{19}{6}$

194 $f(x) = 3x^2$ at $(1, 3)$.
Here, $f'(x) =$ ____, so that $f'(1) =$ ____. Hence, the equation of the tangent line to the graph of f at $(1, 3)$ is $y - 3 =$ __________ or $y =$ __________, while the equation of the normal line is $y - 3 =$ __________ or $y =$ __________. The graph is

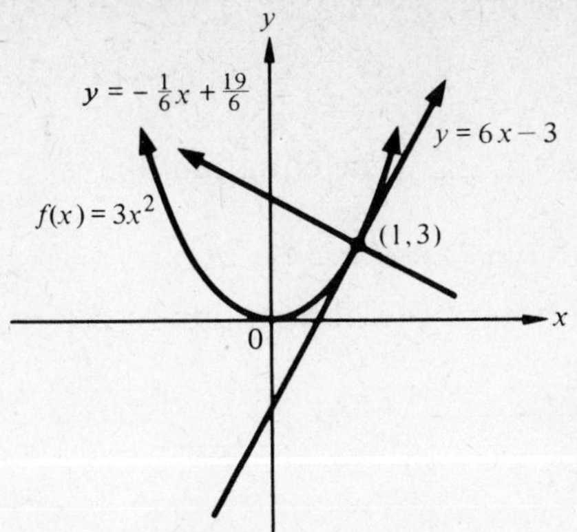

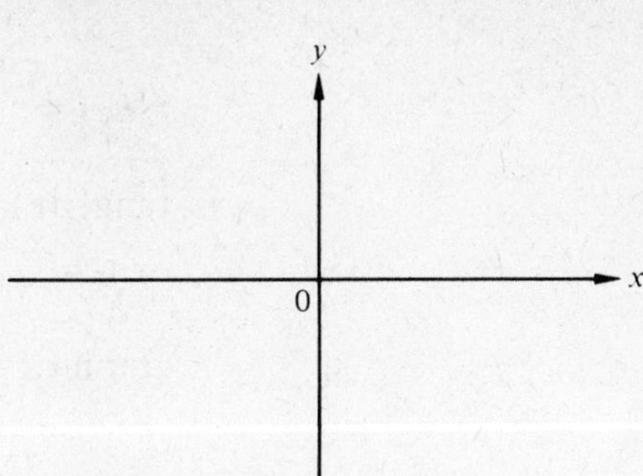

195 $f(x) = x^2 - 2$ at $(-1, -1)$.

Here, $f'(x) =$ ____, so $f'(-1) =$ ____. Hence, the equation of the tangent line to the graph of f at $(-1, -1)$ is $y + 1 =$ ________ or $y = -2x +$ ____. The equation of the normal line to the graph of f at $(-1, -1)$ is $y + 1 =$ ________ or $y = \frac{1}{2}x - (\_\_\_\_)$.

The graph is

$2x$, -2

$-2(x + 1)$

-3

$\frac{1}{2}(x + 1)$, $\frac{1}{2}$

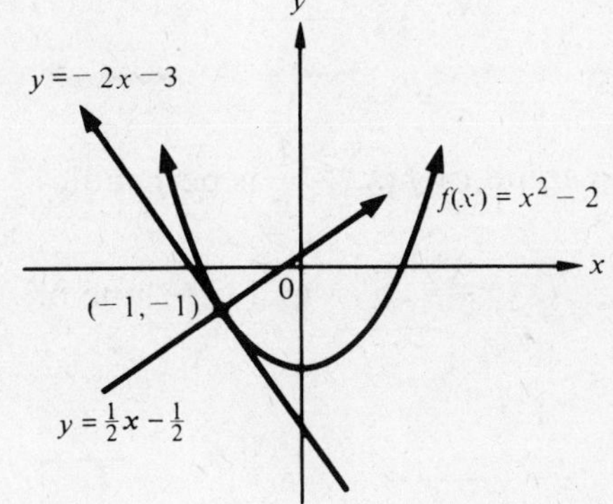

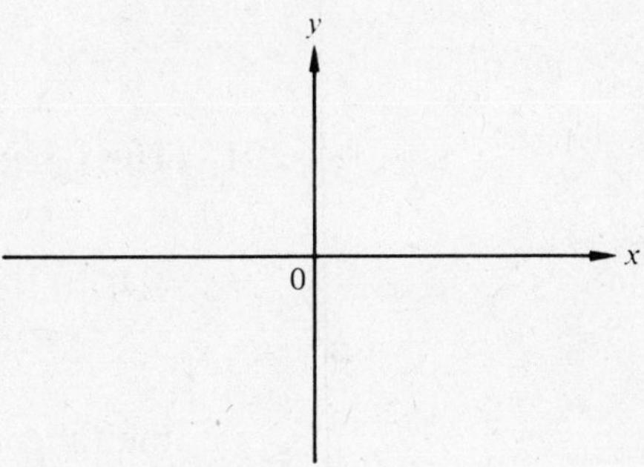

196 $f(x) = 2x^2 + 1$ at $(1, 3)$.

$f'(x) =$ ____, so $f'(1) =$ ____ and the equation of the tangent line to the graph of f at $(1, 3)$ is $y - 3 = 4($________$)$ or $y =$ ________.

The equation of the normal line to the graph of f at $(1, 3)$ is

$y - 3 = (\_\_\_\_)(x - 1)$ or $y =$ ____________.

$4x$, 4

$x - 1$, $4x - 1$

$-\frac{1}{4}$, $-\frac{1}{4}x + \frac{13}{4}$

197 $f(x) = -3x^2$ at $(-2, -12)$.

$f'(x) =$ _____, so $f'(-2) =$ ____. The equation of the tangent line to the graph of f at $(-2, -12)$ is $y + 12 = 12($________$)$ or $y =$ ____________.

The equation of the normal line to the graph of f at $(-2, -12)$ is

$y + 12 = -\frac{1}{12}($________$)$ or $y =$ ____________.

$-6x$, 12

$x + 2$

$12x + 12$

$x + 2$, $-\frac{1}{12}x - \frac{73}{6}$

198 $f(x) = \frac{2}{x}$ at $(-1, -2)$.

$f'(x) =$ ________, so that $f'(-1) =$ ____. The equation of the tangent line to the graph of f at $(-1, -2)$ is $y + 2 = -2($________$)$ or $y =$ ________. The equation of the normal line to the graph of f at $(-1, -2)$ is $y + 2 = \frac{1}{2}($______$)$ or $y =$ ______.

Answers: $-\frac{2}{x^2}$, -2; $x + 1$; $-2x - 4$; $x + 1$, $\frac{x}{2} - \frac{3}{2}$

In Problems 199 through 202, find a point on the curve that satisfies the given conditions.

199 The slope of the tangent line at (x_1, y_1) to the graph of $f(x) = x^2 + 8$ is 16. $f'(x_1) =$ ____ $= 16$, so that $x_1 =$ ____. Therefore, the point (x_1, y_1) is ________.

Answers: $2x_1$, 8; $(8, 72)$

200 The tangent line at (x_1, y_1) to the graph of $f(x) = \sqrt{x}$ is parallel to the line $y = \frac{1}{2}x + 3$. Here, $f'(x_1) =$ ______ and the slope of the line $y = \frac{1}{2}x + 3$ is ____, so that $\frac{1}{2\sqrt{x_1}} =$ ____ and so $x_1 =$ ____. The point (x_1, y_1) is ________.

Answers: $\frac{1}{2\sqrt{x_1}}$; $\frac{1}{2}$, $\frac{1}{2}$; 1, $(1, 1)$

201 The tangent line at (x_1, y_1) to the graph of $f(x) = \frac{1}{x}$ is perpendicular to the line $y = 4x + 1$. Here, $f'(x_1) =$ ______ and the slope of the tangent line is ______, so that $-\frac{1}{x_1^2} =$ ____, so that $x_1 =$ ____ or $x_1 =$ ____. Therefore, the point (x_1, y_1) is ________ or ________.

Answers: $-\frac{1}{x_1^2}$; $-\frac{1}{4}$, $-\frac{1}{4}$, -2; 2, $\left(-2, -\frac{1}{2}\right)$; $\left(2, \frac{1}{2}\right)$

202 The normal line at (x_1, y_1) to the graph of $f(x) = 2x^2 - 3x$ is parallel to the line $x + 9y - 5 = 0$. Here, $f'(x_1) =$ ________. The slope of the normal line is ____, so the slope of the tangent line is ____. Thus, $4x_1 - 3 =$ ____ or $4x_1 =$ ____. Therefore, $x_1 =$ ____. The point (x_1, y_1) is ________.

Answers: $4x_1 - 3$; $-\frac{1}{9}$; 9, 9, 12; 3, $(3, 9)$

7 The Use of Derivatives to Approximate Function Values

203 Suppose that the function f is differentiable at the number x_1 and that $y_1 = f(x_1)$ is known. Then, for values of x near x_1, an approximation to the value of $f(x)$ is given by

$f(x) \approx$ ______________________; that is,

$f(x) \approx$ ______________________.

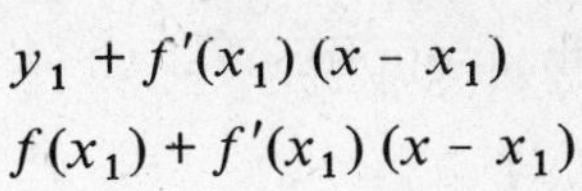

$y_1 + f'(x_1)(x - x_1)$

$f(x_1) + f'(x_1)(x - x_1)$

204 The approximation procedure given in Problem 203 is called a __________ approximation procedure because it is based on the use of ______________________, which is a straight line.

linear

the tangent line

In Problems 205 through 210, use the linear approximation procedure to find an approximate value of each expression.

205 $(1.04)^2$.

Let f be defined by $f(x) =$ ____. In the linear approximation procedure of Problem 203, put $x_1 =$ ____ and $x =$ _____, so that $f'(x_1) =$ _____ and $f'(1) =$ ____. Using

$f(x) \approx$ ______________________, we have

$f(1.04) = (1.04)^2 \approx 1 + 2($_____ $- 1) =$ _____.

x^2

1, 1.04

$2x_1$, 2

$f(x_1) + f'(x_1)(x - x_1)$

1.04, 1.08

206 $(1.02)^3$.

Let f be defined by $f(x) =$ ____. In the linear approximation procedure, put $x_1 =$ ____ and $x =$ _____. Here, $f'(x_1) =$ _____, so that $f'(1) =$ ____. Therefore,

$f(x) \approx$ ______________________, so that

$(1.02)^3 = f(1.02) \approx 1 + 3($_____ $- 1) =$ _____.

x^3

1, 1.02

$3x_1^2$, 3

$f(x_1) + f'(x_1)(x - x_1)$

1.02, 1.06

207 $\dfrac{1}{1.02}$.

Let f be defined by $f(x) = \overline{\quad}$. In the linear approximation procedure, put $x_1 = 1$ and $x =$ _____, so $f'(x_1) =$ _____ and $f'(1) =$ ____. Hence, $f(x) \approx ($______________________$)$,

so that $\dfrac{1}{1.02} = f(1.02) \approx 1 - 1($_____ $- 1) =$ _____.

$\dfrac{1}{x}$

1.02, $-\dfrac{1}{x_1^2}$

-1, $f(x_1) + f'(x_1)(x - x_1)$

1.02, 0.98

208 $(0.98)^2$.

Let f be defined by $f(x) =$ ____. In the linear approximation procedure, put $x_1 = 1$ and $x = 0.98$. Here, $f'(1) =$ ____. Therefore, $f(x) \approx$ ______________________, and so

$(0.98)^2 = f(0.98) \approx 1 + 2(0.98 -$ ____$) = 1 -$ _____ $=$ _____.

x^2

2

$f(x_1) + f'(x_1)(x - x_1)$

1, 0.04, 0.96

209 $\sqrt{1.03}$.

Let f be defined by $f(x) =$ ____. In the linear approximation procedure, put $x_1 = 1$ and $x =$ _____. Here, $f'(1) =$ ____.

Therefore, $f(x) \approx$ ______________________, so that $\sqrt{1.03} = f(1.03) = 1 + \frac{1}{2}($_____$) = 1 +$ _____ $=$ _____.

$\sqrt{x}$

1.03, $\dfrac{1}{2}$

$f(x_1) + f'(x_1)(x - x_1)$

0.03, 0.015, 1.015

210 $\sqrt{0.96}$.

Let f be defined by $f(x) =$ _____. In the linear approximation procedure, put $x_1 = 1$ and $x =$ _____. Here, $f'(1) =$ ____.

Therefore, $f(x) \approx$ ____________________, so that

$$\sqrt{0.96} = f(0.96) \approx 1 + \frac{1}{2}(\_\_\_\_ - 1)$$

$$= 1 + \frac{1}{2}(\_\_\_\_) = 1 - \_\_\_\_$$

$$= \_\_\_\_.$$

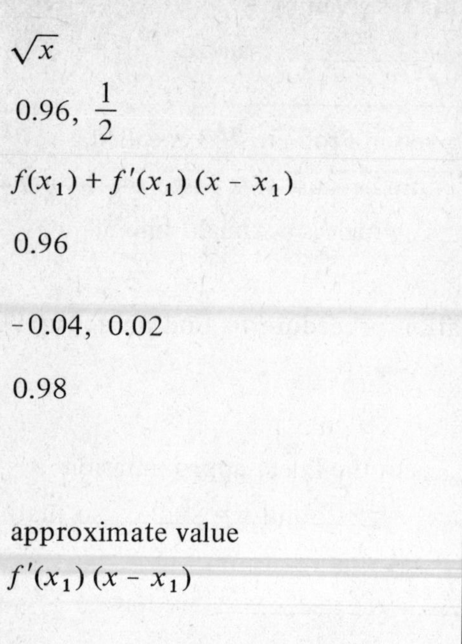

$\sqrt{x}$

$0.96, \frac{1}{2}$

$f(x_1) + f'(x_1)(x - x_1)$

0.96

$-0.04,\ 0.02$

0.98

211 The error $E(x)$ involved in the linear approximation procedure is given by

$E(x) =$ true value − ____________________

$= f(x) - f(x_1) -$ ____________________.

approximate value

$f'(x_1)(x - x_1)$

212 Let f be differentiable at the number x_1. Then, there is a function $\mathcal{E}$ with the same domain as f such that

(i) $f(x) = f(x_1) + f'(x_1)(x - x_1) +$ ________________.

(ii) $\mathcal{E}(x_1) =$ ____.

(iii) $\lim_{x \to x_1} \mathcal{E}(x) =$ ____.

$\mathcal{E}(x)(x - x_1)$

0

0

In Problems 213 through 216, use the linear approximation theorem in Problem 212 to write each function value as the sum of a linear term and an error term near the given value of x_1.

213 $f(x) = x^3 - 2x - 1$ near $x_1 = 2$. Calculating as in Section 3, we have $f'(2) =$ ____. Also, $f(2) =$ ____. Using Problem 212, we have $f(x) = f(2) + f'(2)(x - 2) +$ ________________, that is, $x^3 - 2x - 1 = 3 + 10($________$) + \mathcal{E}(x)(x - 2)$, where

$$\mathcal{E}(x) = \frac{\_\_\_\_\_\_\_\_\_\_\_\_}{x - 2} = \_\_\_\_\_\_\_\_\_\_\_\_.$$

Thus, $x^3 - 2x - 1 = 3 + 10(x - 2) +$ ________________, where $\lim_{x \to 2} \mathcal{E}(x) =$ ____.

$10,\ 3$

$\mathcal{E}(x)(x - 2)$

$x - 2$

$x^3 - 12x + 16,\ x^2 + 2x - 8$

$(x - 2)(x^2 + 2x - 8)$

0

214 $f(x) = \frac{1}{x}$ near $x_1 = 1$.

Since $f(1) =$ ____ and $f'(1) =$ ____, then

$$\frac{1}{x} = f(1) + f'(1)(x - 1) + \_\_\_\_\_\_\_\_\_\_\_\_ \text{ or}$$

$$\frac{1}{x} = 1 - 1(x - 1) + \mathcal{E}(x)(\_\_\_\_), \text{ where } \mathcal{E}(x) = \frac{\frac{1}{x} + x - 2}{x - 1} =$$

______ and $\lim_{x \to 1} \mathcal{E}(x) =$ ____.

$1,\ -1$

$\mathcal{E}(x)(x - 1)$

$x - 1$

$x - \frac{1}{x},\ 0$

215 $f(x) = x^3$ near $x_1 = 1$.

Since $f(1) =$ ____ and $f'(1) =$ ____, then

$x^3 = f(1) + f'(1)(x - 1) +$ ________________, so that

$x^3 = 1 + 3(x - 1) +$ ________________, where

$\&(x) =$ ________________ and $\lim_{x \to 1} \&(x) =$ ____.

Answers: 1, 3; $\&(x)(x - 1)$; $\&(x)(x - 1)$; $x^2 + x - 2$, 0

216 $f(x) = \sqrt{x}$ near $x_1 = 1$.

Since $f(1) =$ ____ and $f'(1) =$ ____, then $\sqrt{x} =$

$1 + \frac{1}{2}(x - 1) + \&(x)$(________), so that

$\sqrt{x} = \frac{1}{2} + \frac{1}{2}x +$ ________________, where

$$\&(x) = \begin{cases} \dfrac{\rule{3cm}{0.4pt}}{x - 1} & \text{if } x \neq 1, x > 0 \\ 0 & \text{if } x = 1 \end{cases}$$

and $\lim_{x \to 1} \&(x) =$ ____.

Answers: $1, \frac{1}{2}$; $x - 1$; $\&(x)(x - 1)$; $\sqrt{x} - \frac{1}{2} - \frac{1}{2}x$; 0

217 Let f be a function defined at least on an open interval (a, b) containing the number x_1. If there is a function $\&$ defined on the open interval (a, b) and if there are constants m and c such that $f(x) = mx + c + \&(x)(x - x_1)$ for $a < x < b$ and $\lim_{x \to x_1} \&(x) = 0$, then f is ________________ at x_1, $m =$ ________, and $c =$ ________________.

Answers: differentiable, $f'(x_1)$; $f(x_1) - f'(x_1)x_1$

In Problems 218 through 220, use the given identity and the converse of the linear approximation theorem (Problem 217) to find the indicated derivative.

218 $\frac{d}{dx}(x^3)$ at $x = 1$.

Let $f(x) =$ ____ and $\&(x) = x^2 + x - 2$. Then $\&(x)(x - 1) =$ ________________, so that $f(x) = mx + c + \&(x)(x - 1)$, where $m =$ ____ and $c =$ ____. Since $\lim_{x \to 1} \&(x) =$ ____, then $f'(1) = m =$ ____; that is, the value of $\frac{d}{dx}(x^3)$ at $x = 1$ is ____.

Answers: x^3; $x^3 - 3x + 2$; 3, -2, 0; 3, 3

219 $\frac{d}{dx}(x^2)$ at $x = -1$.

Since $(x + 1)^2 =$ ________________, then $x^2 =$ ________________. Define f and $\&$ by $f(x) =$ ____ and $\&$ by $\&(x) = x + 1$, so that $f(x) = -2x - 1 + \&(x)$(________). Since $\lim_{x \to -1} \&(x) =$ ____, then $f'(-1) =$ ____; that is, the value of $\frac{d}{dx}(x^2)$ at $x = -1$ is ____.

Answers: $x^2 + 2x + 1$; $-2x - 1 + (x + 1)^2$; x^2; $x + 1$, 0; -2, -2

	220 $\frac{d}{dx}(x^3)$ at $x = -1$.
x^3	Define f by $f(x) =$ ____ and $\mathcal{E}$ by $\mathcal{E}(x) = x^2 - x - 2$. Then
$x^3 - 3x - 2$	$\mathcal{E}(x)(x+1) =$ ________________, so that
3, 2	$f(x) = mx + c + \mathcal{E}(x)(x-1)$, where $m =$ ____ and $c =$ ____. Since
0, 3	$\lim_{x \to -1} \mathcal{E}(x) =$ ____, then $f'(-1) = m =$ ____; that is, the value
3	of $\frac{d}{dx}(x^3)$ at $x = -1$ is ____.

Chapter Test

1 Suppose that a particle moves along a straight line according to the equation $s = 8t^2 - 3t$, where s is the distance measured in feet of the particle from its starting point at the end of t seconds. Find the instantaneous speed of the particle when $t = 2$.

2 Find the instantaneous rate of change of the volume of a sphere with respect to its radius when the radius is 6 inches.

3 Let f be defined by $f(x) = \begin{cases} 3x - 1 & \text{if } x < 2 \\ 7 - x & \text{if } x \geqslant 2. \end{cases}$

(a) Is f continuous at 2?
(b) Find $f'_-(2)$ and $f'_+(2)$.
(c) Is f differentiable at 2?
(d) Sketch the graph of f.

4 (a) Find the slope of the tangent line to the graph of f at the point $(1, f(1))$.
(i) $f(x) = x^2 + 5$ (ii) $f(x) = \sqrt{16x}$
(b) Find the equations of the tangent and normal lines to the graph of f in part (a) at the point $(1, f(1))$.

5 Find $\frac{dy}{dx}$.

(a) $y = 3x^6 - x^3 + 5$
(b) $y = (x^2 + 9)(2x^3 + 11)$
(c) $y = \frac{5}{x^2 + 1}$
(d) $y = \frac{2x + 9}{2x - 9}$

6 Let f and g be differentiable functions so that $f(7) = 1$, $f'(7) = 5$, $g(7) = 3$, and $g'(7) = -2$. Find:

(a) $(f + g)'(7)$ (b) $(2f - g)'(7)$ (c) $(fg)'(7)$ (d) $\left(\frac{f}{g}\right)'(7)$ (e) $\left(\frac{f + g}{2f - g}\right)'(7)$

7 Use the chain rule to find $D_x y$.

(a) $y = (3x^2 + x + 11)^{25}$
(b) $y = \sqrt{3x^4 + x^2 + 5}$
(c) $y = \sin(7x + 2)$
(d) $y = x \cos 5x$

8 Let f and g be two functions so that $f(5) = 7$, $f'(5) = 3$, $g(2) = 5$, and $g'(2) = -1$. Find $(f \circ g)'(2)$.

9 Indicate which of the following are true statements of the chain rule equation.

(a) $D_x y = D_u y D_x u$
(b) $\frac{dy}{ds} = \frac{ds}{dt} \cdot \frac{dt}{dy}$
(c) $(f \circ g)'(x) = f'[g(x)]g'(x)$
(d) $\frac{dv}{dx} = \frac{du}{dv} \cdot \frac{dv}{dx}$
(e) $\frac{dy}{dt} = \frac{dy}{dx} \cdot \frac{dt}{dx}$

10 Find $\frac{dy}{dx}$.

(a) $y = x|x|$ (b) $y = \sqrt[3]{3x + 7}$ (c) $y = (x - 1)^4(2x + 1)^3$
(d) $y = x^{3/2} - 5x^{3/5}$ (e) $y = \frac{x}{\sqrt{x + 1}}$

11 Let $f(x) = x^3 + 7$ for $x > 0$.

(a) Show that f^{-1} exists. (b) Find $(f^{-1})'(8)$.

12 Use the linear approximation procedure to find an approximate value of $(0.99)^2$.

Answers

1 29 ft/sec

2 144π

3 (a) Yes (b) 3, −1 (c) No (d) The graph is

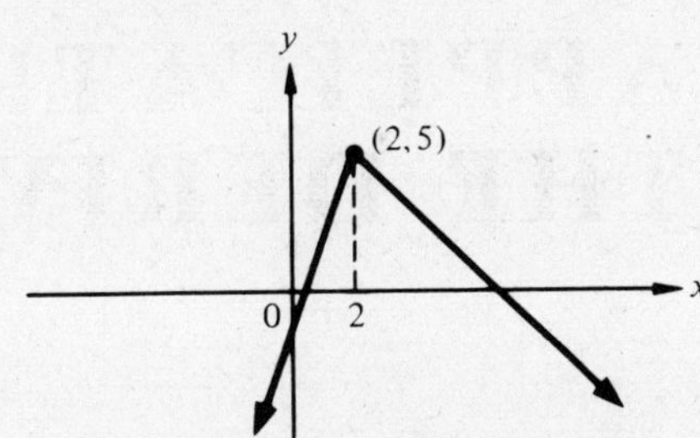

4 (a) (i) 2 (ii) 2

(b) (i) Tangent line is $y - 6 = 2(x - 1)$; normal line is $y - 6 = -\frac{1}{2}(x - 1)$.

(ii) Tangent line is $y - 4 = 2(x - 1)$; normal line is $y - 4 = -\frac{1}{2}(x - 1)$.

5 (a) $\frac{dy}{dx} = 18x^5 - 3x^2$ (b) $\frac{dy}{dx} = 10x^4 + 54x^2 + 22x$

(c) $\frac{dy}{dx} = \frac{-10x}{(x^2 + 1)^2}$ (d) $\frac{dy}{dx} = \frac{-36}{(2x - 9)^2}$

6 (a) 3 (b) 12 (c) 13 (d) $\frac{17}{9}$ (e) −51

7 (a) $D_x y = 25(3x^2 + x + 11)^{24}(6x + 1)$ (b) $D_x y = \frac{6x^3 + x}{\sqrt{3x^4 + x^2 + 5}}$

(c) $D_x y = 7 \cos (7x + 2)$ (d) $D_x y = -5x \sin 5x + \cos 5x$

8 −3

9 (a) T (b) F (c) T (d) F (e) F

10 (a) $\frac{dy}{dx} = 2|x|$ (b) $\frac{dy}{dx} = \frac{1}{\sqrt[3]{(3x + 7)^2}}$

(c) $\frac{dy}{dx} = (x - 1)^3(2x + 1)^2(14x - 2)$ (d) $\frac{dy}{dx} = \frac{3}{2}x^{1/2} - 3x^{-2/5}$

(e) $\frac{dy}{dx} = \frac{x + 2}{2(x + 1)^{3/2}}$

11 (a) $f'(x) > 0$ and f' is continuous for $x > 0$. (b) $\frac{1}{3}$

12 0.98

Chapter 3 APPLICATIONS OF THE DERIVATIVE

In Chapter 2 laws were developed which allow straightforward calculation of the derivatives of algebraic functions. The objectives of this chapter are that the student be able to:

1 Apply the intermediate value theorem and the mean value theorem to functions.

2 Compute higher-order derivatives of functions.

3 Find intervals where the graph of f is increasing, decreasing, concave upward, and concave downward.

4 Find the relative extrema of f and sketch the graph of f.

5 Find absolute extrema and solve maximum–minimum problems.

6 Differentiate implicitly.

7 Deal with related rates.

1 The Intermediate Value Theorem and the Mean Value Theorem

1.1 The Intermediate Value Theorem

k

1 **Let the function f be continuous on the closed interval $[a, b]$ and suppose that $f(a) \neq f(b)$. Then, if k is any number strictly between $f(a)$ and $f(b)$, there is at least one number c strictly between a and b such that $f(c) =$ _____. Graphically,**

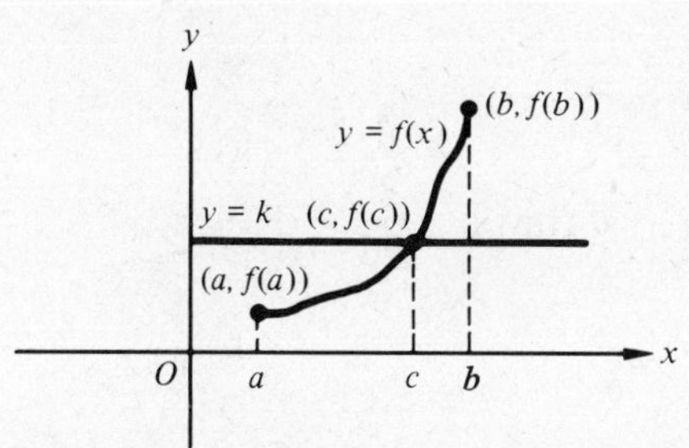

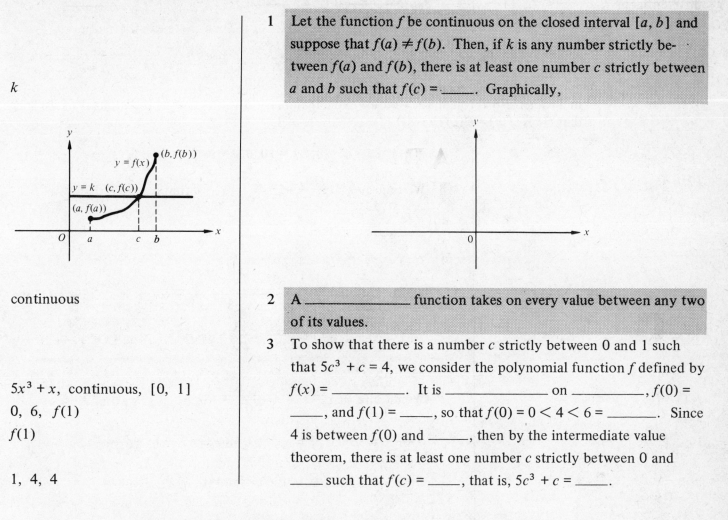

continuous

2 **A _____________ function takes on every value between any two of its values.**

$5x^3 + x$, continuous, $[0, 1]$
0, 6, $f(1)$
$f(1)$

1, 4, 4

3 To show that there is a number c strictly between 0 and 1 such that $5c^3 + c = 4$, we consider the polynomial function f defined by $f(x) =$ __________. It is ____________ on __________, $f(0) =$ _____, and $f(1) =$ _____, so that $f(0) = 0 < 4 < 6 =$ _______. Since 4 is between $f(0)$ and _____, then by the intermediate value theorem, there is at least one number c strictly between 0 and _____ such that $f(c) =$ _____, that is, $5c^3 + c =$ _____.

In Problems 4 through 6, use the intermediate value theorem to verify that each function f has a zero (root) on the interval (a, b).

-1.79, 0.61, opposite signs
continuous
root, $(1.1, 1.9)$

4 $f(x) = x^2 - 3$; $(a, b) = (1.1, 1.9)$.
Since $f(1.1) =$ _______ and $f(1.9) =$ _______ have _____________ _______, and the polynomial function f is ____________, it has a _______ in the open interval ____________.

-1, 8
continuous, root
$(1, 2)$

5 $f(x) = x^3 + 2x - 4$; $(a, b) = (1, 2)$.
Since $f(1) =$ _____, and $f(2) =$ _____ have opposite signs and the polynomial function f is ____________, it has a _______ in the open interval ________.

-3, 3
continuous, root
$(0, 1)$

6 $f(x) = x^5 + 5x^4 - 3$; $(a, b) = (0, 1)$.
Since $f(0) =$ _____ and $f(1) =$ _____ have opposite signs and the polynomial function f is ____________, it has a _______ in the open interval ________.

1.2 The Mean Value Theorem

continuous
differentiable
(a, b)
$a < c < b,\ f(b) - f(a)$

7 **Let the function f be defined and ______ on the closed interval $[a, b]$ and let f be ______ on the open interval ______. Then there exists at least one number c with ______ and $f'(c) = \dfrac{\_\_\_\_\_\_}{b - a}$.**

In Problems 8 through 13, find an explicit numerical value of c in the interval (a, b) such that $f(b) - f(a) = f'(c)(b - a)$.

$f(0),\ 0$

$2 - 2c,\ 3 - 2c,\ \frac{3}{2}$

8 $f(x) = 2x - x^2$; $[a, b] = [0, 3]$.

We require that $\dfrac{f(3) - \_\_\_\_}{3 - 0} = f'(c)$, that is, $\dfrac{-3 - \_\_\_\_}{3 - 0} =$ ______, so that ______ $= 0$ or $c =$ ____.

$f(9) - f(1),\ f'(c)$

$3,\ 2,\ 2,\ 4$

9 $f(x) = \sqrt{x}$; $[a, b] = [1, 9]$.

We require that $\dfrac{\_\_\_\_\_\_}{9 - 1} =$ ______, so that $\dfrac{\_\_\_\_ - 1}{9 - 1} = \dfrac{1}{2\sqrt{c}}$ or $\dfrac{\_\_\_\_}{8} = \dfrac{1}{2\sqrt{c}}$ or $\sqrt{c} =$ ____ or $c =$ ____.

$f(1) - f(0),\ f'(c)$

$0,\ \frac{1}{3},\ \frac{1}{\sqrt{3}}$

not

10 $f(x) = x^3$; $[a, b] = [0, 1]$.

We require that $\dfrac{\_\_\_\_\_\_}{1 - 0} =$ ____, so that $\dfrac{1 - \_\_\_\_}{1 - 0} = 3c^2$ or $c^2 =$ ____ and $c =$ ____. We reject $c = -\dfrac{1}{\sqrt{3}}$, since $-\dfrac{1}{\sqrt{3}}$ is ____ in the interval $(0, 1)$.

$f(3),\ f'(c),\ -2$

$\frac{2}{c^2},\ 3,\ \sqrt{3}$

does not belong to $(1, 3)$

11 $f(x) = -\dfrac{2}{x}$; $[a, b] = [1, 3]$.

We require that $\dfrac{\_\_\_\_ - f(1)}{3 - 1} =$ ______, so that $\dfrac{-\frac{2}{3} - \_\_\_\_}{2} =$ ____, so that $c^2 =$ ____ and $c =$ ____. We reject $c = -\sqrt{3}$, since $-\sqrt{3}$ ____________.

$f(10),\ f'(c),\ 3$

$\sqrt{c - 1},\ \sqrt{c - 1},\ 2$

5

12 $f(x) = \sqrt{x - 1}$; $[a, b] = [2, 10]$.

We require that $\dfrac{\_\_\_\_ - f(2)}{10 - 2} =$ ______, so that $\dfrac{\_\_\_\_ - 1}{8} = \dfrac{1}{2(\_\_\_\_\_\_)}$, that is, $\dfrac{2}{8} = \dfrac{1}{2(\_\_\_\_\_\_)}$ or $\sqrt{c - 1} =$ ____. Then $c =$ ____.

$f(2) - f(1),\ f'(c)$

13 $f(x) = \dfrac{1}{x^2}$; $[a, b] = [1, 2]$.

We require that $\dfrac{\_\_\_\_\_\_}{2 - 1} =$ ______, so that

$1, \frac{-2}{c^3}, 8, \sqrt[3]{\frac{8}{3}}$

$\frac{\frac{1}{4} - \_\_\_\_}{2-1} = \_\_\_\_$ or $3c^3 = \_\_\_\_$. Then $c = \_\_\_\_$.

In Problems 14 through 16, find an explicit numerical value of c such that $a < c < b$ and the tangent line to the graph of f at $(c, f(c))$ is parallel to the chord containing the points $(a, f(a))$ and $(b, f(b))$.

14 $f(x) = x^2 - 6x + 5; a = 1, b = 4$.

The slope of the tangent line at $(c, f(c))$ is $f'(c) = \_\_\_\_$. The slope of the chord containing $(1, f(1))$ and $(4, f(4))$ is $\frac{\_\_\_\_}{4-1}$. Since the tangent line is parallel to the chord we have $\frac{\_\_\_\_}{4-1} = \_\_\_\_$, so that $\_\_\_\_ = 2c - 6$ or $2c = \_\_\_\_$. Therefore, $c = \_\_\_\_$. The graph is

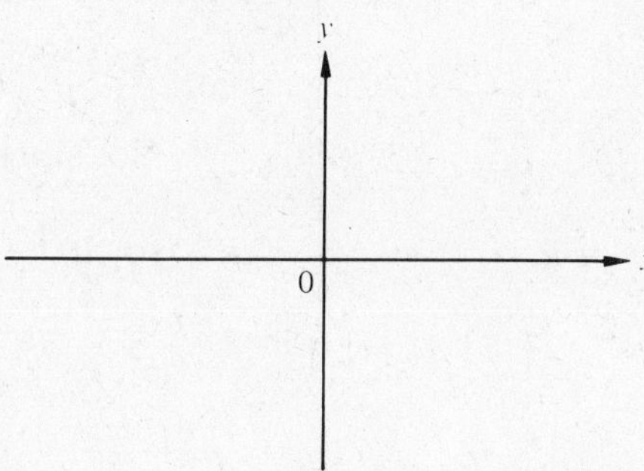

$2c - 6$

$f(4) - f(1)$

$f(4) - f(1),\ 2c - 6,\ -1$

$5, \frac{5}{2}$

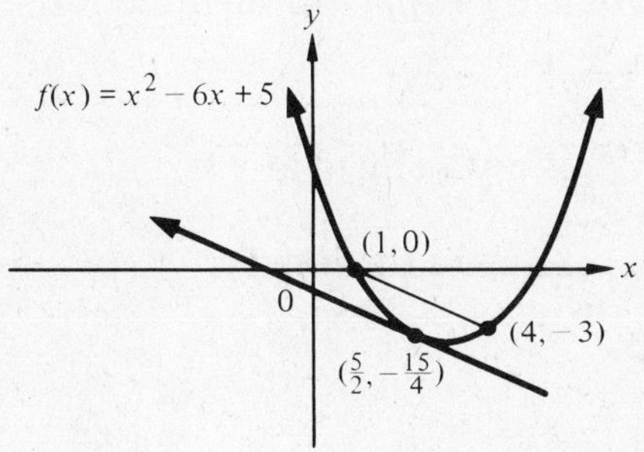

15 $f(x) = (2x + 1)^3; a = 1, b = 2$.

We require that $\frac{\_\_\_\_}{2-1} = \_\_\_\_$, so that $\frac{125 - \_\_\_\_}{2-1} = \_\_\_\_$; hence, $\frac{98}{1} = \_\_\_\_$ or $\frac{98}{6} = (2c+1)^2$. Then $2c + 1 = \_\_\_\_ = \frac{\pm 7}{\sqrt{3}}$, so that $2c = \_\_\_\_$. In the given interval $c = \_\_\_\_$. The graph is

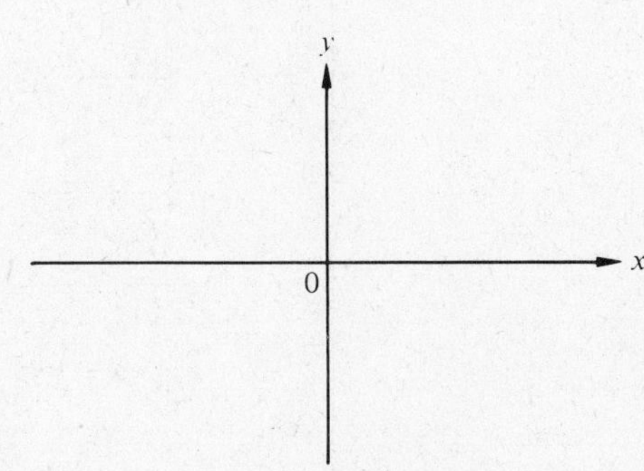

$f(2) - f(1),\ f'(c)$

$27,\ 6(2c+1)^2,\ 6(2c+1)^2$

$\sqrt{\frac{98}{6}}$

$\frac{7}{\sqrt{3}} - 1,\ \frac{7}{2\sqrt{3}} - \frac{1}{2}$

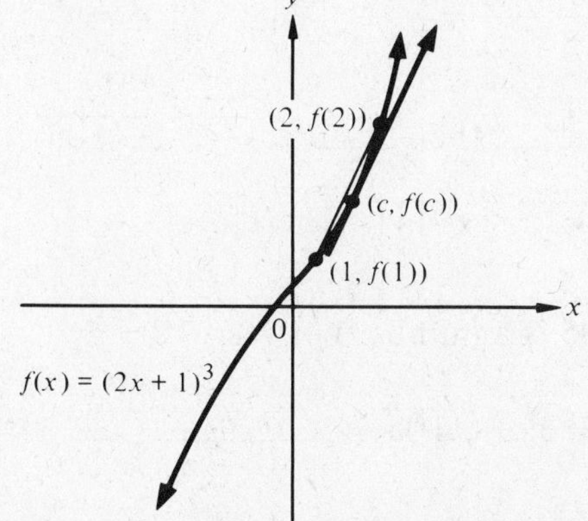

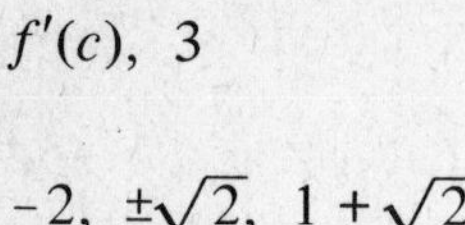

$f'(c)$, 3

-2, $\pm\sqrt{2}$, $1+\sqrt{2}$

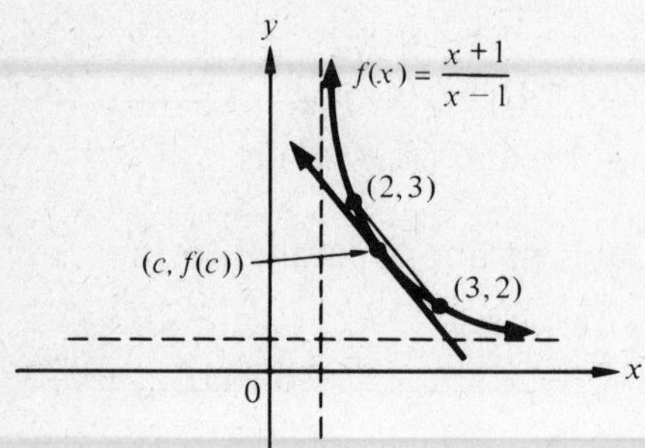

16 $f(x) = \dfrac{x+1}{x-1}$; $a = 2$, $b = 3$.

We require that $\dfrac{f(3) - f(2)}{3 - 2} =$ ________, so that $\dfrac{2 - \text{\_\_\_\_}}{3 - 2} =$ $\dfrac{\text{\_\_\_\_}}{(c-1)^2}$ or $c - 1 =$ ______. In the given interval $c =$ ________.

The graph is

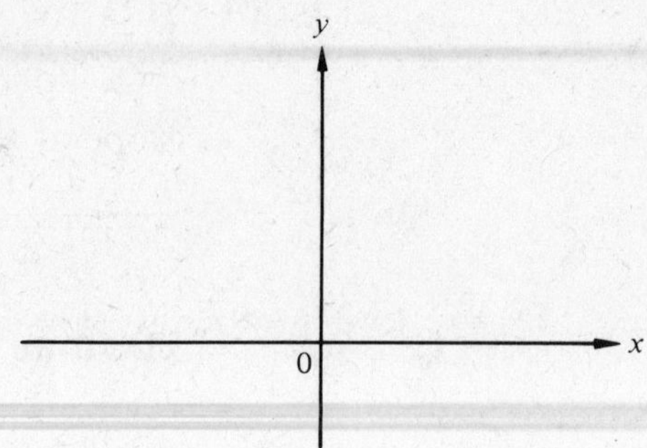

$x + 1$, -1

differentiable

$(-2, 1)$

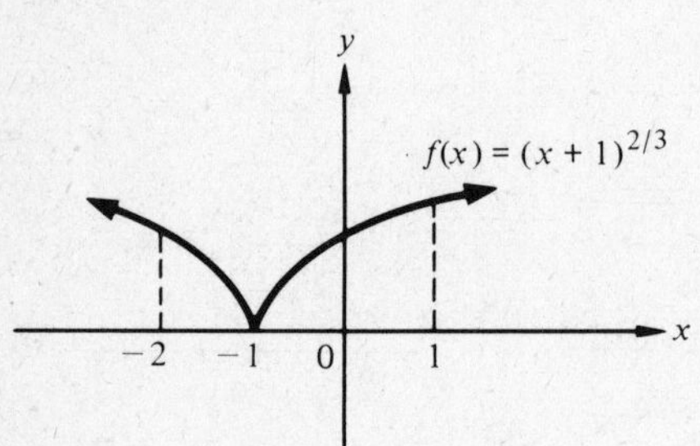

17 Does the mean value theorem (Problem 7) apply to the function $f(x) = (x + 1)^{2/3}$ in the interval $[-2, 1]$?

No, since $f'(x) = \dfrac{2}{3}(\text{\_\_\_\_\_\_\_\_})^{-1/3}$ is not defined at $x =$ ____, so that f is not ________________ in the interval ____________. The graph is

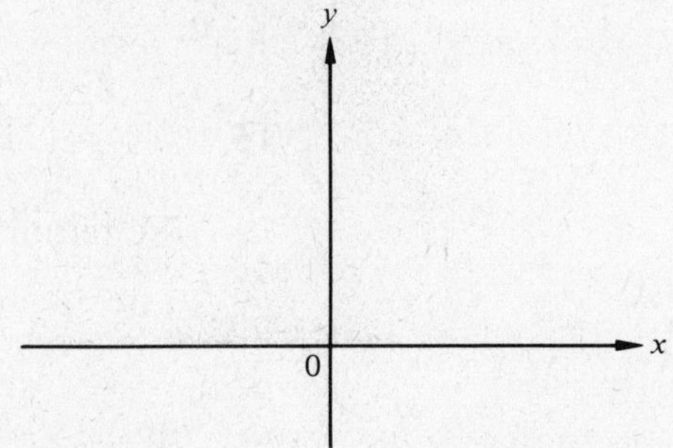

$\sqrt{x}$, 100, 105, 1

$\dfrac{1}{2\sqrt{c}}$

10, 105, 10

20

20

$10 + \dfrac{5}{20}$, 10.24695

18 Approximate $\sqrt{105}$.

Let $f(x) =$ ______, $a =$ _____ and $b =$ _____, $f'(c) = \dfrac{\text{\_\_\_\_}}{2\sqrt{c}}$, so that $\dfrac{f(105) - f(100)}{105 - 100} =$ _____ or $\dfrac{\sqrt{105} - 10}{5} = \dfrac{1}{2\sqrt{c}}$. Then $\sqrt{105} =$ $\dfrac{5}{2\sqrt{c}} +$ _____. Since $100 < c <$ _____, then _____ $< \sqrt{c} < \sqrt{105}$ or _____ $< 2\sqrt{c} < 2\sqrt{105} < \sqrt{420} < \sqrt{441}$. Therefore, $\dfrac{1}{21} < \dfrac{1}{2\sqrt{c}} < \dfrac{1}{\text{\_\_\_\_}}$, so that $\sqrt{105}$ is a number between $10 + \dfrac{5}{21}$ and ____________. The actual value of $\sqrt{105}$ is ____________, to five decimal places.

In Problems 19 through 22, indicate which hypotheses of the mean value theorem fail to hold on each interval.

19 $f(x) = 3x^{2/3}$; $[-2, 2]$.

f is continuous on ________ and $f'(x) =$ ________; therefore, f is ________________ on $(-2, 2)$, since $f'(x)$ is not defined at ____. The graph is

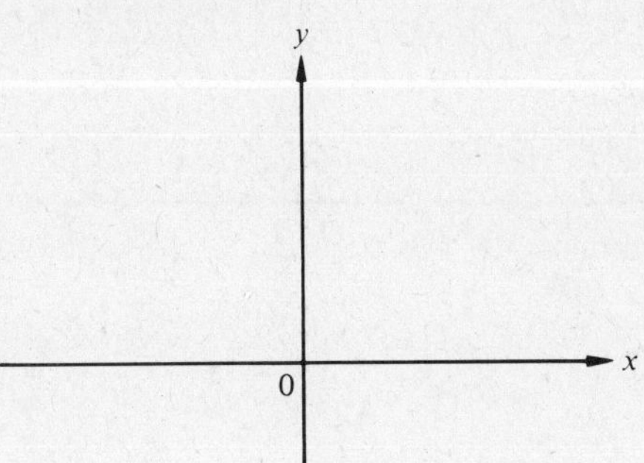

$[-2, 2]$, $2x^{-1/3}$

not differentiable

0

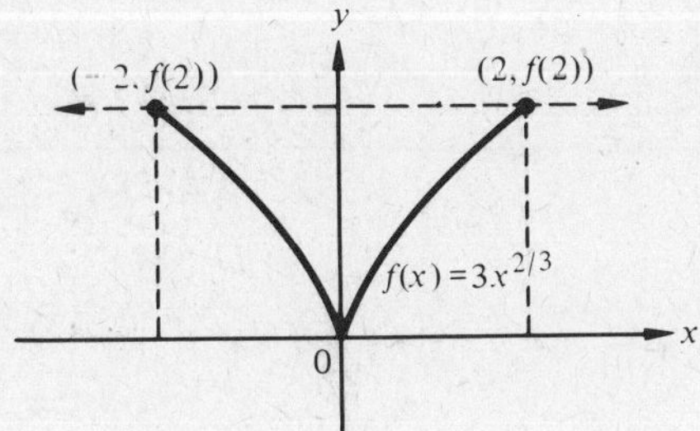

20 $f(x) = 3|x|$; $[-4, 4]$.

f is continuous on ________, but f fails to be ________ at the number ____ in the open interval ________. Hence, no tangent line to the graph of f is parallel to the line containing $(-4, f(-4))$ and ________. The graph is

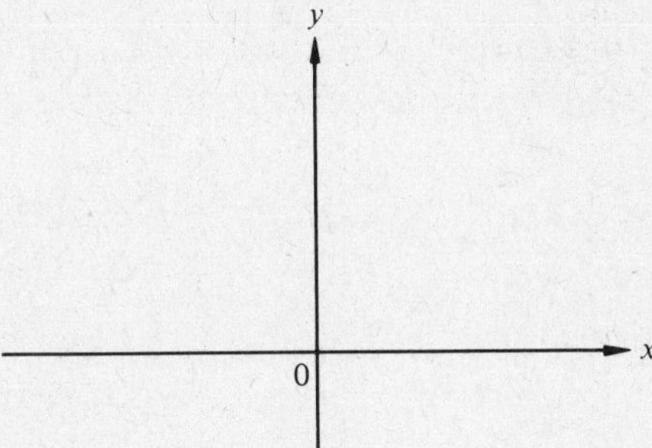

$[-4, 4]$, differentiable

0, $(-4, 4)$

$(4, f(4))$

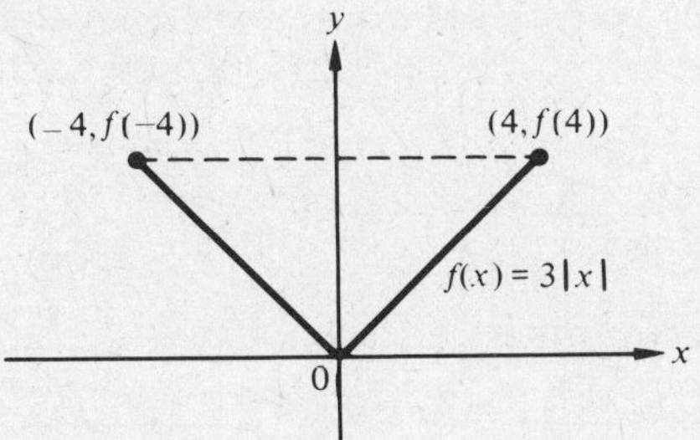

21

$$f(x) = \begin{cases} \frac{1}{x} & \text{if } x > 0 \\ 0 & \text{if } x = 0 \end{cases}; \ [0, 4].$$

Here f is ________________ on the open interval $(0, 4)$; however, it fails to be ________ on the closed interval $[0, 4]$, since $\lim_{x \to 0^+} f(x) =$ ____ $\neq f(0) =$ ____. The graph is

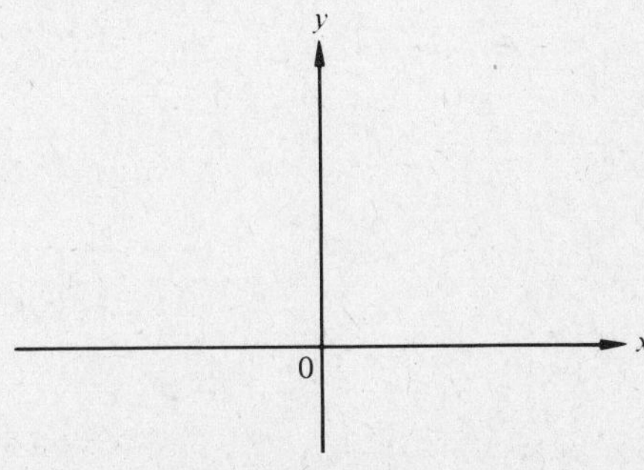

differentiable

continuous

$+\infty$, 0

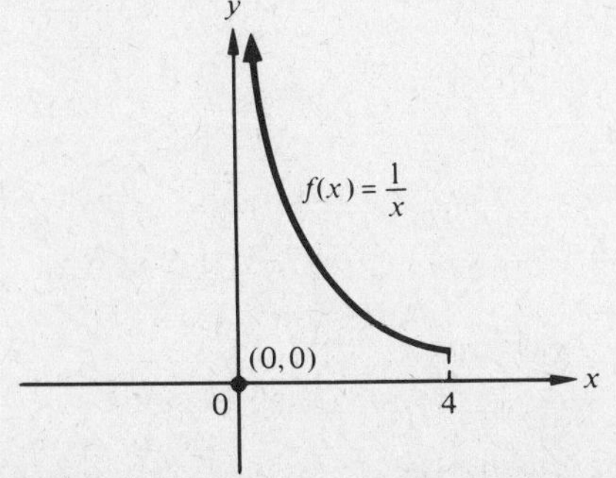

22 Let f be the function whose graph is

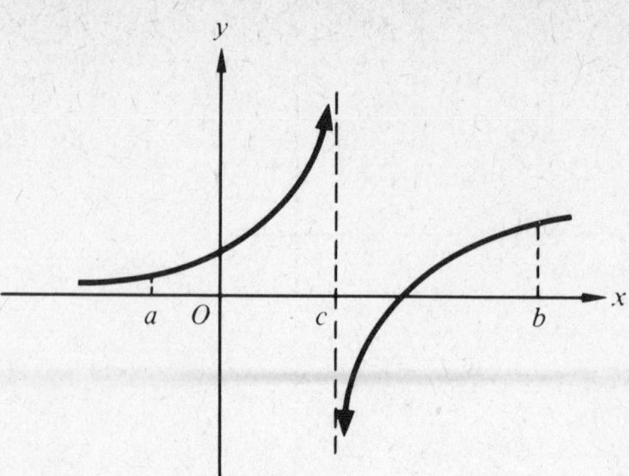

Then f is not ________ on the closed interval $[a, b]$ since it is not defined at ____.

defined

c

1.3 Rolle's Theorem

23 **Suppose that the function f is ________ on the closed interval $[a, b]$, that f is differentiable on the open interval (a, b), and that $f(a) =$ ______. Then there is at least one number c in the open interval (a, b) such that $f'(c) =$ ____.**

continuous

$f(b)$

0

In Problems 24 through 27, find a number c (if possible) on the open interval (a, b) that satisfies Rolle's theorem (Problem 23).

24 $f(x) = x^2 - 2x$; $[a, b] = [0, 2]$.
Since f is a polynomial function, then f is ________ and ________ at every number. Clearly, $f(0) =$ ____ and $f(2) =$ ____. Also, $f'(x) =$ ______ and $f'(c) =$ ______. Thus, $f'(c) = 2c - 2 =$ ____ when $c =$ ____. The graph is

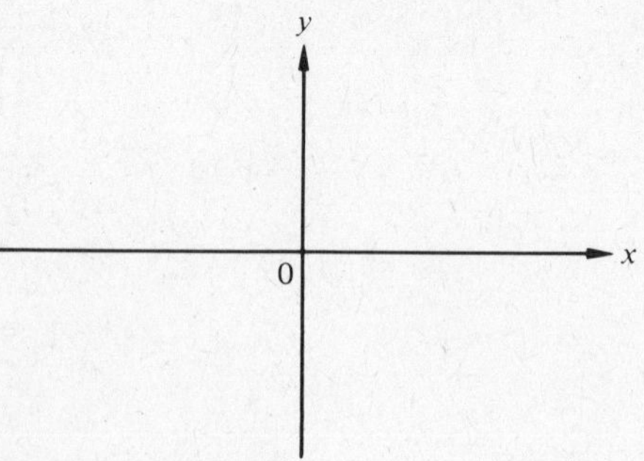

continuous

differentiable, 0

0, $2x - 2$, $2c - 2$

0, 1

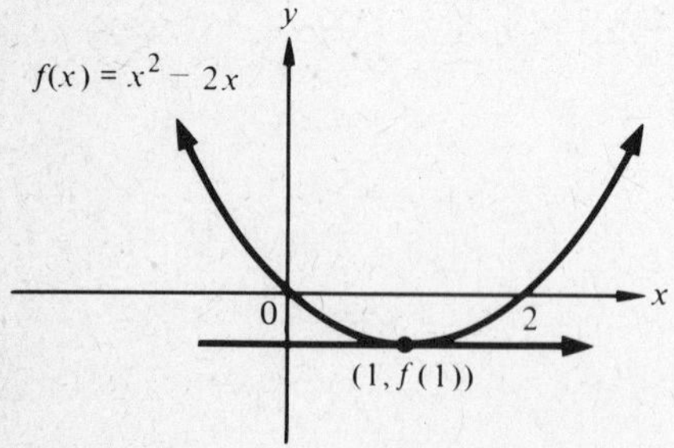

25 $f(x) = x^3 - 3x$, $[a, b] = [0, \sqrt{3}]$.
Since f is a polynomial function, then f is ________ and ________ at every number. Clearly, $f(0) =$ ____ and $f(\sqrt{3}) =$ ____. Also, $f'(x) =$ ______ and $f'(c) =$ ______. Then, $f'(c) = 3c^2 - 3 =$ ____; $c =$ ____ because we reject $c = -1$ since -1 does not belong to ______.

continuous

differentiable, 0

0, $3x^2 - 3$

$3c^2 - 3$, 0, 1

$(0, \sqrt{3})$

continuous
differentiable, 0
0, $12c - 3c^2$, 0
0, 4

26 $f(x) = 6x^2 - x^3$, $[a, b] = [0, 6]$.
Since f is a polynomial function, then f is ________ and ________ at every number. Clearly, $f(0) =$ ____ and $f(6) =$ ____. Also, $f'(c) =$ ________ $=$ ____, when $c =$ ____ or $c =$ ____.

continuous
differentiable, 0
$(-1, 1)$

27. $f(x) = 1 - |x|$; $[a, b] = [-1, 1]$.
f is ________ in the interval $[-1, 1]$; however, f is not ________ at ____. Hence, f is not differentiable on the open interval ________ and the hypotheses of Rolle's theorem fail to hold. The graph is

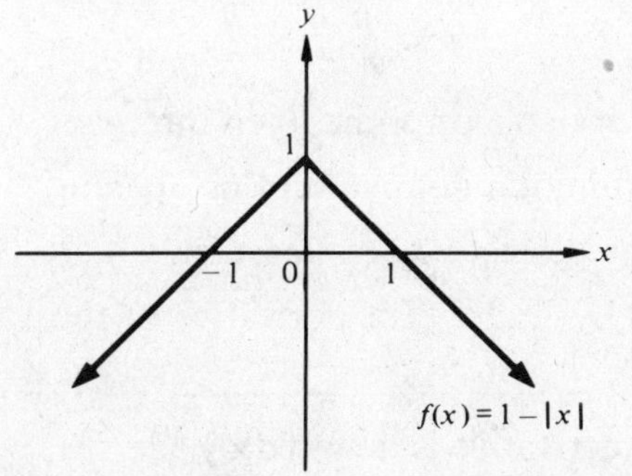

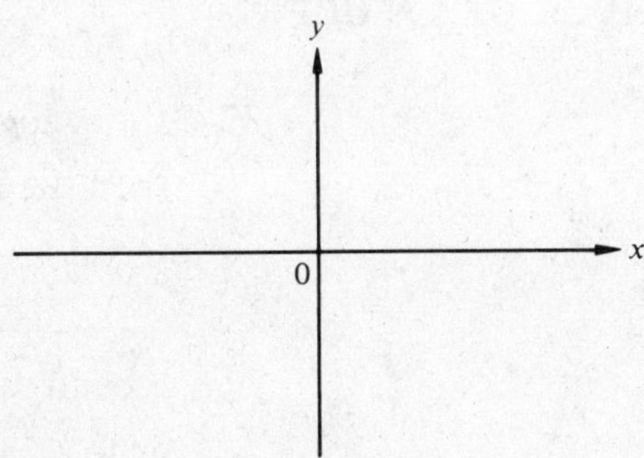

2 Higher-Order Derivatives

$\frac{ds}{dt}$
acceleration
$\frac{dv}{dt}$, $\frac{d^2s}{dt^2}$, $D_t^2 s$

28 Let $s = f(t)$, where s is the directed distance from the origin to a particle after t seconds have elapsed. The instantaneous velocity is given by $v =$ ____ and the instantaneous rate of change of velocity with respect to time is called ________ a; hence, $a =$ ____ $=$ ____. In operator notation, $a = D_t v =$ ____.

In Problems 29 through 32, a particle moves along a linear scale according to the given law of motion $s = f(t)$, where s is in feet and t is in seconds. Find the velocity v and the acceleration a at the given time t.

$6t + 5$, 17 ft/sec
$\frac{d^2s}{dt^2}$, 6 ft/sec^2

29 $s = 3t^2 + 5t$; $t = 2$.
$v = \frac{ds}{dt} =$ ________, so that at $t = 2$, $v =$ ________.
$a = \frac{dv}{dt} =$ ____, so that at $t = 2$, $a =$ ________.

$100 - 32t$, 68 ft/sec
$D_t^2 s$, -32 ft/sec^2

30 $s = 100t - 16t^2$; $t = 1$.
$v = D_t s =$ ________, so that v at $t = 1$ is ________.
$a = D_t v =$ ____, so that a at $t = 1$ is ________.

31 $s = t^3 - 3t^2 + 3t + 4$; $t = 1$.

$3t^2 - 6t + 3$, 0 ft/sec

0 ft/sec^2

$14t + \frac{1}{2}(5t - 4)^{-1/2}(5)$

$56\frac{5}{8}$ ft/sec, $\frac{d^2s}{dt^2}$, $14 - \frac{25}{4}(5t - 4)^{-3/2}$

$13\frac{231}{256}$ ft/sec^2

$v = D_t s =$ ________, so that v at $t = 1$ is ________.

$a = 6t - 6$, so that a at $t = 1$ is ________.

32 $s = 7t^2 + \sqrt{5t - 4}$; $t = 4$.

$v = \frac{ds}{dt} =$ ________, so that v at $t = 4$ is ________. $a =$ ____ $=$ ________, so that a at $t = 4$ is ________.

2.1 Derivatives of Order n

second derivative, f''

third derivative

f'''

f'''' or $f^{(iv)}$

$f^{(n)}$, y'

y'', $y^{(n)}$

$\frac{d^2 f}{dx^2}$, $\frac{d^3 f}{dx^3}$

$\frac{d^n f}{dx^n}$

$D_x^2 f$, $D_x^3 f$

$D_x^n f$

33 If f is any function that is differentiable on some open interval, then the derivative f' is again a function defined on this open interval. Also, the derivative of the derivative is called the ________ and is denoted by ____. The derivative of the second derivative is called the ________ and it is denoted by ____. The fourth derivative is denoted by ________. In general, the nth-order derivative, or nth derivative is denoted by ____. If $y = f(x)$, then $f'(x) =$ ____, $f''(x) =$ ____, and in general, $f^{(n)}(x) =$ ____.

34 The higher-order derivatives are denoted by Liebniz notation as follows: $f''(x) =$ ____, $f'''(x) =$ ____, and in general, $f^{(n)}(x) =$ ____.

35 The higher-order derivatives are denoted in operator notation as follows: $f''(x) =$ ____, $f'''(x) =$ ____, and in general $f^{(n)}(x) =$ ____.

In Problems 36 through 43, find the first, second, and the third derivatives.

$15x^2 - 6$

$30x$, 30

$-15x^2 - \frac{1}{x^2}$, $-30x + \frac{2}{x^3}$

$-30 - \frac{6}{x^4}$

$-\frac{3}{(x+1)^2}$, $\frac{6}{(x+1)^3}$

$-\frac{18}{(x+1)^4}$

36 If $f(x) = 5x^3 - 6x + 15$, then $f'(x) =$ ________, $f''(x) =$ ____, and $f'''(x) =$ ____.

37 If $y = -5x^3 + \frac{1}{x}$, then $\frac{dy}{dx} =$ ________, $\frac{d^2y}{dx^2} =$ ________, and $\frac{d^3y}{dx^3} =$ ________.

38 If $y = \frac{3}{x+1}$, then $D_x y =$ ________, $D_x^2 y =$ ________, and $D_x^3 y =$ ________.

39 If $s = \sqrt{4 - t^2}$, then $\dfrac{ds}{dt} = \dfrac{1}{2}(4 - t^2)^{-1/2}$ (_____) = __________.

$$\frac{d^2s}{dt^2} = -\frac{(\sqrt{4 - t^2})(\_\_\_\_) - \frac{t}{2}(\_\_\_\_\_\_\_\_\_\_)(-2t)}{4 - t^2}$$

$$= -\frac{4 - t^2 + \_\_\_\_}{(4 - t^2)^{3/2}}$$

$$= \frac{\_\_\_\_}{(4 - t^2)^{3/2}}.$$

$$\frac{d^3s}{dt^3} = \frac{d}{dt}[-4(4 - t^2)^{-3/2}] = 6(\_\_\_\_\_\_)^{-5/2}(-2t)$$

$$= \frac{\_\_\_\_}{(4 - t^2)^{5/2}}.$$

$-2t$, $\dfrac{-t}{\sqrt{4 - t^2}}$

1, $(4 - t^2)^{-1/2}$

t^2

-4

$4 - t^2$

$-12t$

40 If $f(x) = 9x^{7/3}$, then

$$f'(x) = 9\left(\frac{7}{3}\right)x^{(\_\_\_\_)} = \_\_\_\_\_\_,$$

$$f''(x) = 21\left(\frac{4}{3}\right)x^{(\_\_\_\_)} = \_\_\_\_\_\_,$$

$$f'''(x) = \frac{28}{3}(\_\_\_\_\_\_).$$

$4/3$, $21x^{4/3}$

$1/3$, $28x^{1/3}$

$x^{-2/3}$

41 If $f(x) = \dfrac{x + 1}{x - 1}$, then

$$f'(x) = \frac{(x - 1)(\_\_\_\_) - (x + 1)(\_\_\_\_)}{(x - 1)^2} = \frac{\_\_\_\_}{(x - 1)^2},$$

$$f''(x) = 4(x - 1)^{(\_\_\_\_)} = \frac{4}{(x - 1)^3},$$

$$f'''(x) = -12(\_\_\_\_\_\_)^{-4} = \frac{\_\_\_\_}{(x - 1)^4}.$$

1, 1, -2

-3

$x - 1$, -12

42 If $y = \sin(4x + 7)$, then $y' = 4$(__________), $y'' =$ ____________, and $y''' =$ ____________.

$\cos(4x + 7)$

$-16\sin(4x + 7)$, $-64\cos(4x + 7)$

43 If $f(x) = x - \dfrac{1}{x^3}$, then $f'(x) = 1 +$ ____, $f''(x) =$ ____, and $f'''(x) =$ ____.

$\dfrac{3}{x^4}$, $-\dfrac{12}{x^5}$

$\dfrac{60}{x^6}$

44 If $y = (x + 2)^{-1/2}$, then

$$D_x y = -\frac{1}{2}(x + 2)^{(\_\_\_\_)},$$

$$D_x^2 y = \left(\_\_\_\_\right)(x + 2)^{-5/2},$$

$$D_x^3 y = \left(\_\_\_\_\right)(\_\_\_\_\_\_)^{-7/2},$$

$$D_x^4 y = \frac{105}{16}(\_\_\_\_\_\_)^{-9/2}.$$

$-3/2$

$\dfrac{3}{4}$

$-\dfrac{15}{8}$, $x + 2$

$x + 2$

45 Let f and g be differentiable functions such that $f(1) = 2, f'(1) = 3, g(1) = 5, g'(1) = 1, f''(1) = -1$, and $g''(1) = 4$. To find $(f \cdot g)''(1)$, we have $(f \cdot g)' = f' \cdot g + (\_\_\_\_)$, so that

$$(f \cdot g)'' = (\_\_\_\_)' + (f \cdot g')'$$
$$= f'' \cdot g + (\_\_\_\_) + (\_\_\_\_) + f \cdot g''$$
$$= f'' \cdot g + 2(\_\_\_\_) + f \cdot g''.$$

Therefore,

$$(f \cdot g)''(1) = f''(1)\,(\_\_\_\_) + 2(\_\_\_\_) + \_\_\_\_$$
$$= (-1)\,(\_\_\_\_) + 2(\_\_\_\_) + \_\_\_\_$$
$$= \_\_\_\_.$$

$f \cdot g'$

$f' \cdot g$

$f' \cdot g', \; f' \cdot g'$

$f' \cdot g'$

$g(1), \; f'(1)\,g'(1)$

$f(1)\,g''(1)$

5, 3, 8

9

46 If $y = (2x + 1)^2 (x + 1)^3$, then

$$\frac{dy}{dx} = (2x + 1)^2 (3)\,(\_\_\_\_) + (\_\_\_\_)\,(x + 1)^3$$
$$= (2x + 1)\,(x + 1)^2\,(\_\_\_\_).$$

Therefore,

$$\frac{d^2y}{dx^2} = (\_\_\_\_)\,(x + 1)^2 (10x + 7)$$
$$+ (2x + 1)\,(\_\_\_\_)\,(10x + 7)$$
$$+ (2x + 1)\,(x + 1)^2 (\_\_\_\_).$$

$(x + 1)^2, \; 4(2x + 1)$

$10x + 7$

2

$2(x + 1)$

10

47 Suppose that g is a twice-differentiable function such that $g(3) = -2, g'(3) = 3, g''(3) = 5$. If f is defined by $f(x) = [1 + g(x)]^3$, find $f''(3)$. $f'(3) = 3[1 + g(3)]^2\,(\_\_\_\_)$, so that

$$f''(3) = 3\,[1 + g(3)]^2\,(\_\_\_\_)$$
$$+ 3[2[1 + g(3)]]\,(\_\_\_\_).$$

Then $f''(3) = 3[1 + (-2)]^2 (5) + 6[1 + (-2)]\,(\_\_\_\_)$, so that $f''(3) = \_\_\_\_$.

$g'(3)$

$g''(3)$

$2[g'(3)]^2$

9

-39

48 Let $f(x) = \begin{cases} \dfrac{x^2}{3} & \text{if } x \geqslant 0 \\ -\dfrac{x^2}{3} & \text{if } x < 0. \end{cases}$ Then

$$f'(x) = \begin{cases} \dfrac{2x}{3} & \text{if } x \geqslant 0 \\ \_\_\_\_ & \text{if } x < 0, \end{cases}$$

$$f''(x) = \begin{cases} \_\_\_\_ & \text{if } x > 0 \\ -\dfrac{2}{3} & \text{if } x < 0. \end{cases}$$

Notice that $f''(0)$ ____.

$-\dfrac{2x}{3}$

$\dfrac{2}{3}$

does not exist

3 Geometric Properties of Graphs of Functions—Increasing and Decreasing Functions and Concavity of Graphs

3.1 Increasing and Decreasing Functions

$f(b)$

$a < b$

49 A function f is said to be increasing on the interval I if f is defined on I and $f(a) <$ ______ holds whenever a and b are two numbers in I with ______.

$f(a) > f(b)$

$a < b$

50 A function f is said to be decreasing on the interval I if f is defined on I and __________ holds whenever a and b are two numbers in I with ______.

In Problems 51 through 53, determine whether each function f is increasing or decreasing in the indicated interval.

increasing

$\sqrt{b}$, $f(b)$

51 $f(x) = \sqrt{x}$; $[0, 4]$.

f is an __________ function on the interval $[0, 4]$, since, if $0 \leqslant a < b \leqslant 4$, then $f(a) = \sqrt{a} <$ ____ = _____. The graph is

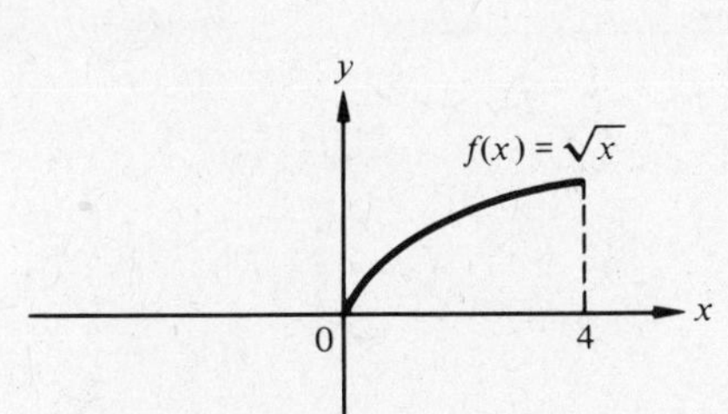

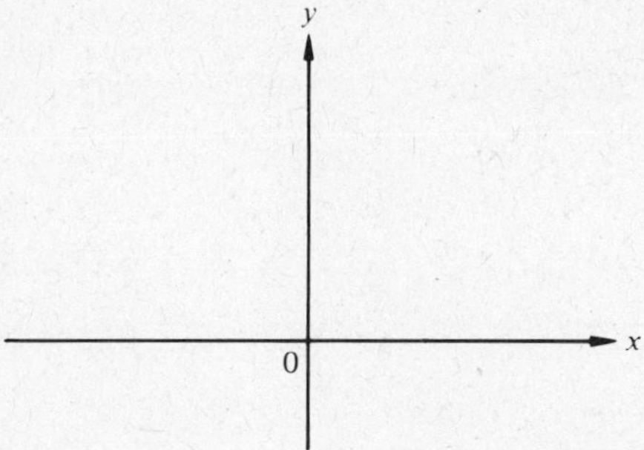

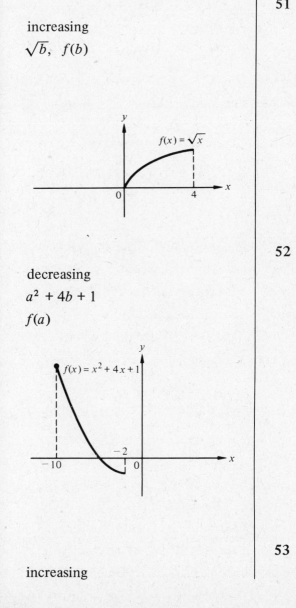

decreasing

$a^2 + 4b + 1$

$f(a)$

52 $f(x) = x^2 + 4x + 1$; $[-10, -2]$.

f is a __________ function on the interval $(-\infty, 2]$ since if $-10 \leqslant a < b \leqslant -2$, then $b^2 + 4b + 1 <$ __________, so that $f(b) = b^2 + 4b + 1 < a^2 + 4a + 1 =$ _____. The graph is

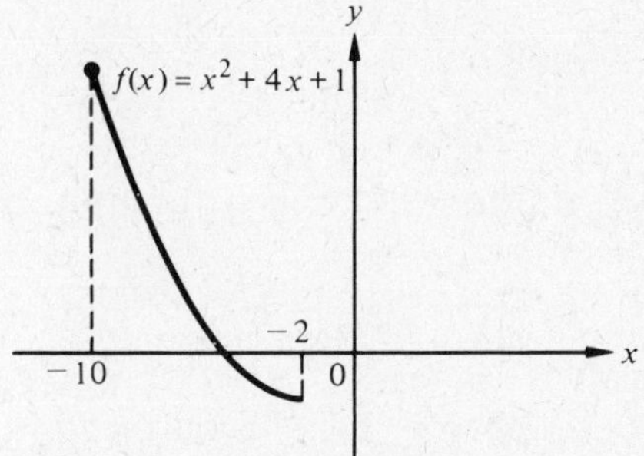

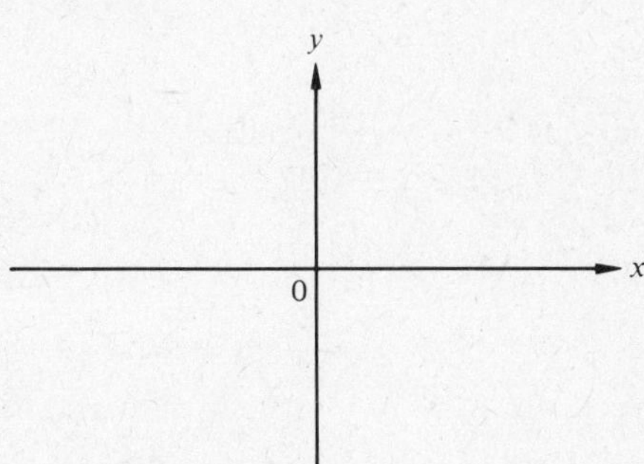

increasing

53 $f(x) = |x + 1|$; $[-1, 5]$.

f is an __________ function on the interval $[-1, 5]$; since, if

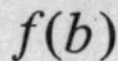

$-1 \leqslant a < b \leqslant 5$, then $f(a) = |a+1| < |b+1| =$ ______. The graph is

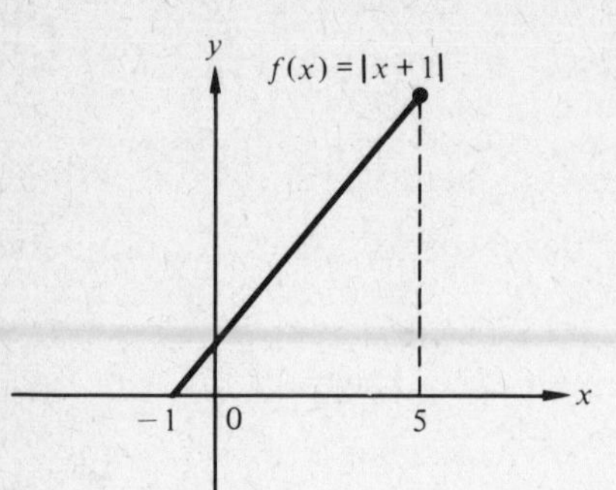

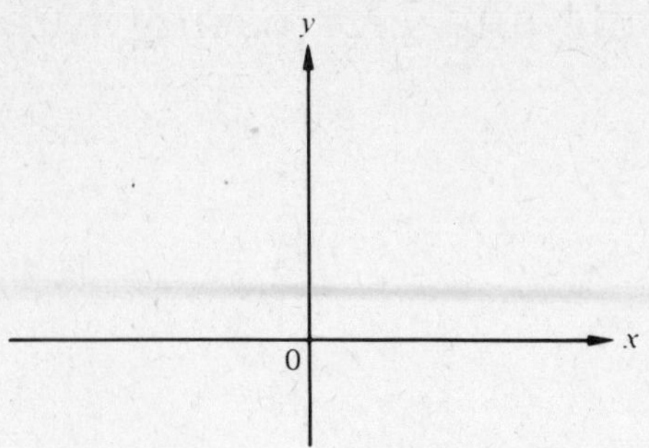

54 $f(x) = -x^2$; $[-4, 4]$.

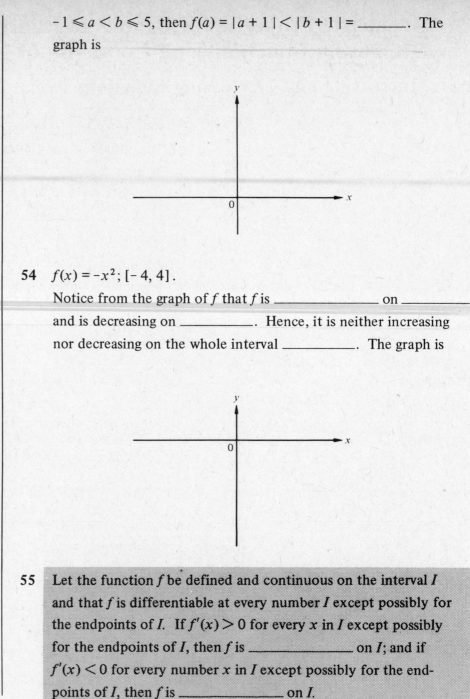

Notice from the graph of f that f is ______ on ______ and is decreasing on ______. Hence, it is neither increasing nor decreasing on the whole interval ______. The graph is

increasing, $[-4, 0]$
$[0, 4]$
$[-4, 4]$

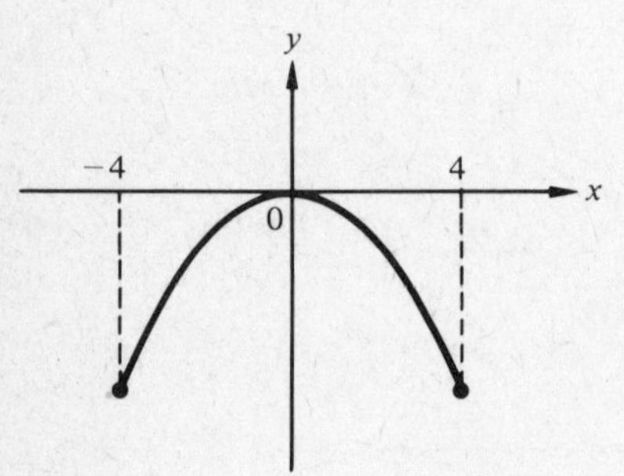

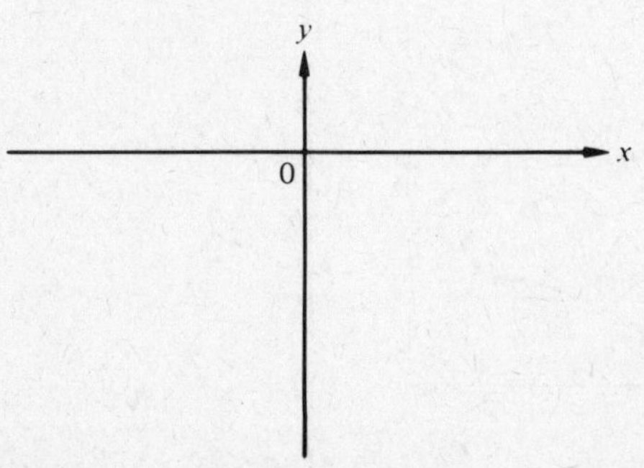

55 **Let the function f be defined and continuous on the interval I and that f is differentiable at every number I except possibly for the endpoints of I. If $f'(x) > 0$ for every x in I except possibly for the endpoints of I, then f is ______ on I; and if $f'(x) < 0$ for every number x in I except possibly for the endpoints of I, then f is ______ on I.**

increasing

decreasing

In Problems 56 through 61, use Problem 55 to find intervals on which each function is increasing or decreasing.

56 $f(x) = 2x^3 - 9x^2 + 12x - 3$.

$f'(x) =$ ______ $= 6(x-1)(x-2)$. f' is defined and continuous at every number in the interval ______. Here, $f'(x)$ is ______ when $x < 1$ or $x >$ ____ and $f'(x)$ is negative when ____ $< x <$ ____. Hence, f is increasing on ______ and ______, and f is decreasing on ______. The graph is

$6x^2 - 18x + 12$
$(-\infty, \infty)$
positive, 2
1, 2, $(-\infty, 1]$
$[2, \infty)$, $[1, 2]$

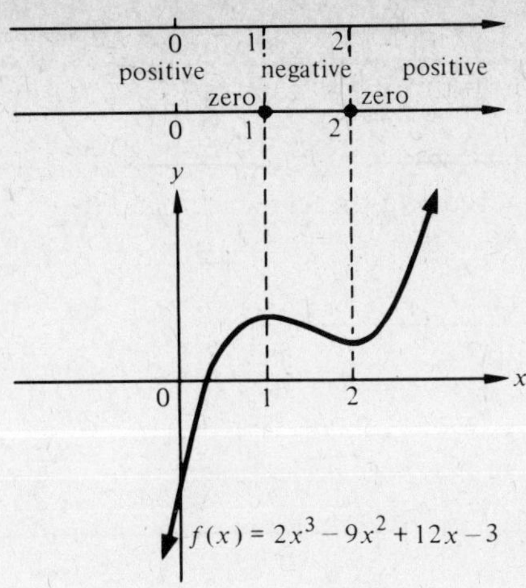

$x^2 - 4x + 3$, $(x - 1)(x - 3)$

$x < 1$, $x > 3$

3

$[3, \infty)$, $[1, 3]$

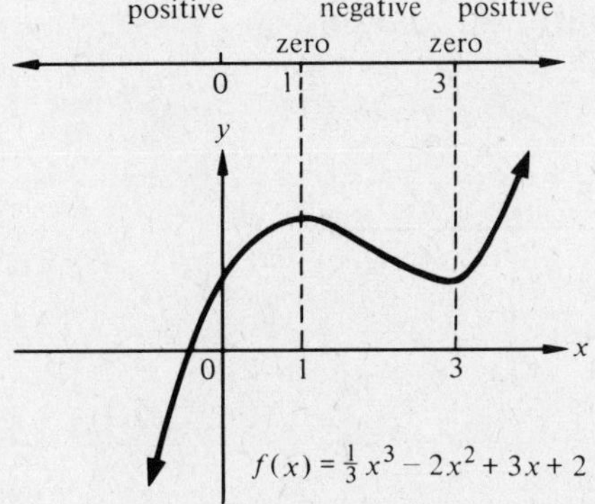

$2x - \frac{2}{x^2}$, $x^3 - 1$

$x > 1$, $x < 1$

$[1, \infty)$, $(0, 1]$

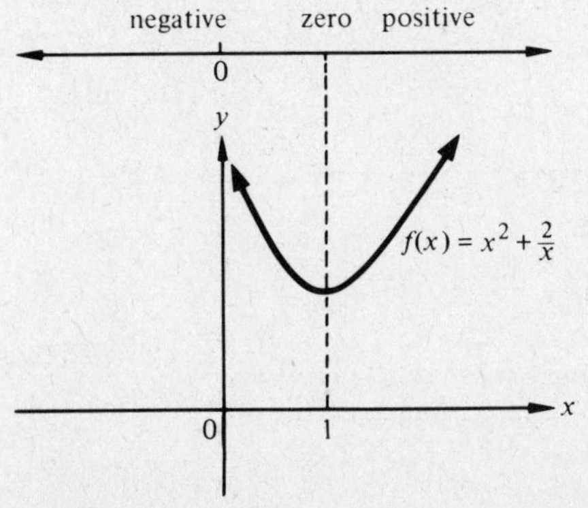

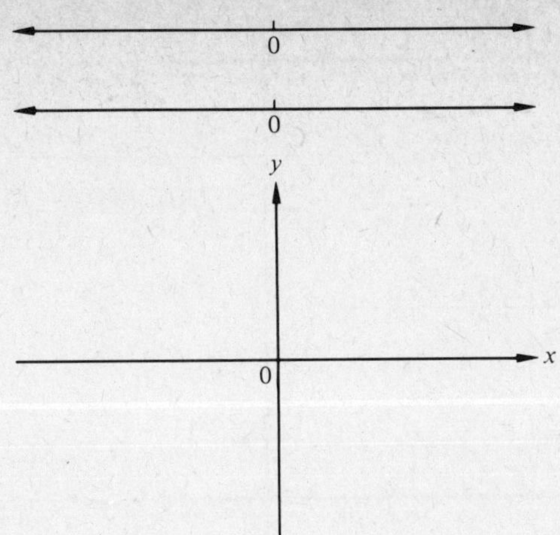

57 $f(x) = \frac{1}{3}x^3 - 2x^2 + 3x + 2$.

$f'(x) =$ ____________ $=$ ______________. Therefore, $f'(x)$ is positive when ______ and ______, and $f'(x)$ is negative when x is between 1 and ____. Hence, f is increasing on $(-\infty, 1)$ and ________, and is decreasing on ________. The graph is

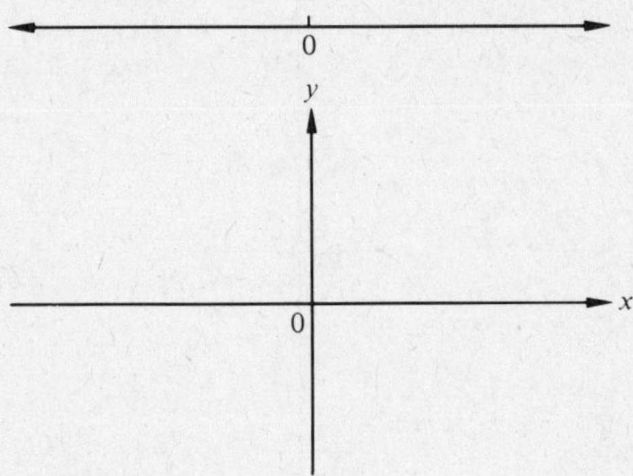

58 $f(x) = x^2 + \frac{2}{x}$ for $x > 0$.

$f'(x) = \frac{}{} = \frac{2}{x^2}$ (________). Therefore, $f'(x)$ is positive when ______ and $f'(x)$ is negative when ______. Hence, f is increasing on ________ and f is decreasing on ______. The graph is

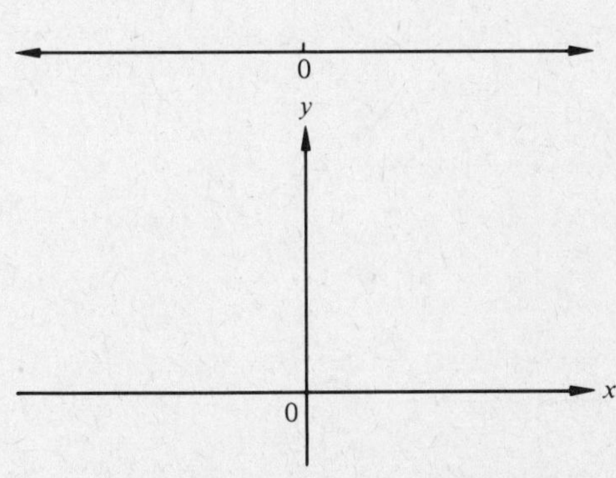

$6x + 12x^{1/2}$

nonnegative real numbers

$x > 0$, $[0, \infty)$

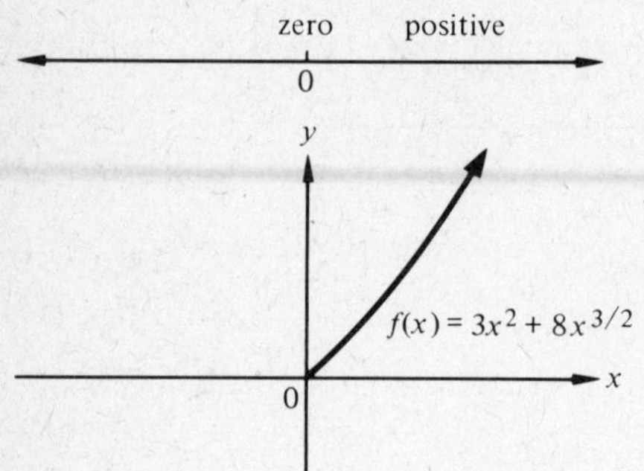

59 $f(x) = 3x^2 + 8x^{3/2}$.

$f'(x) =$ ____________. The domain of f is the set of ________________________. Therefore, $f'(x)$ is positive for _____; hence, f is increasing on ______. The graph is

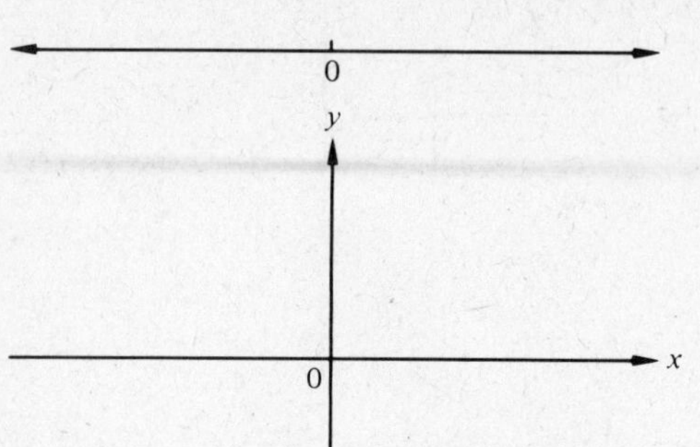

$1 - \frac{1}{2}x^{-1/2}$, $2\sqrt{x} - 1$

$x > \frac{1}{4}$, $0 < x < \frac{1}{4}$

$\left[\left(\frac{1}{4}, \infty\right), \left[0, \frac{1}{4}\right]\right]$

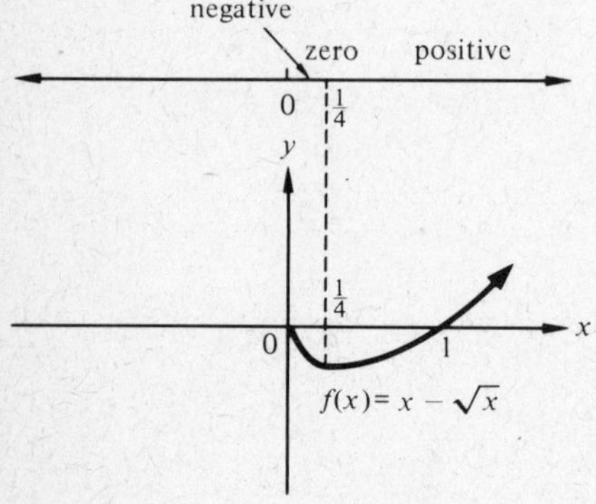

60 $f(x) = x - \sqrt{x}$.

$f'(x) =$ ____________ $= \dfrac{\_\_\_\_\_\_}{2\sqrt{x}}$. Therefore, $f'(x)$ is positive when ______ and $f'(x)$ is negative when ________.

Hence, f is increasing on ____ and f is decreasing on ____.

The graph is

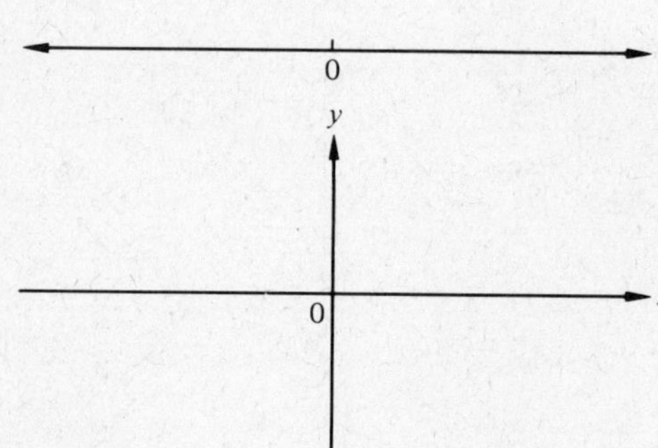

$x^{-2/3} - x^{-4/5}$, $x^{2/15} - 1$

$x < -1$, $x > 1$

$-1 < x < 1$

$[-1, 1]$

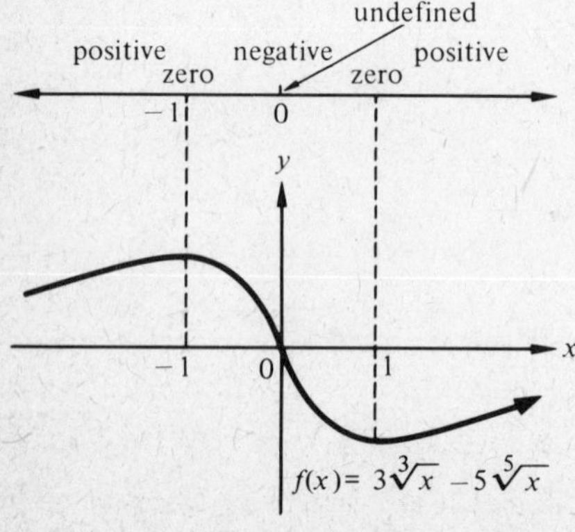

61 $f(x) = 3\sqrt[3]{x} - 5\sqrt[5]{x}$.

$f'(x) =$ ____________ $= x^{-4/5}($__________$)$, $x \neq 0$.

Therefore, $f'(x)$ is positive when ______ and ______, and $f'(x)$ is negative when __________, so that f is increasing on $(-\infty, -1]$ and $[1, \infty)$ and f is decreasing on _______. The graph is

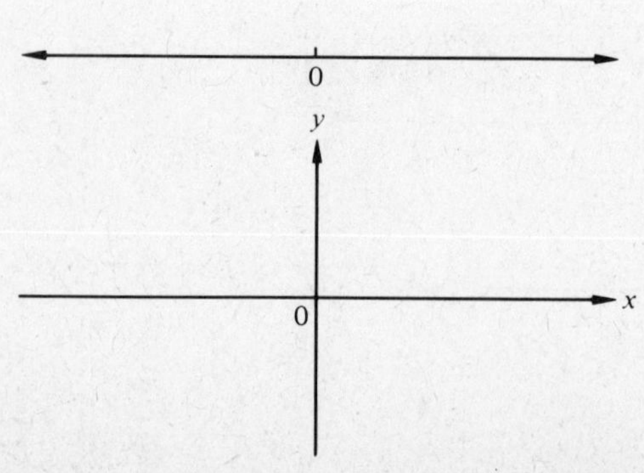

62 A function that is either increasing or else decreasing on an interval is said to be ________ on this interval.

monotone

63 If f is defined and continuous on the interval I, then f is invertible if and only if f is ________.

monotone

64 To determine if f^{-1} exists when $f(x) = 5x^3 + 3x$, we see that $f'(x) =$ ________ > 0 for all real numbers x. Thus, f is an ________ function on its entire domain. By Problem 63, it follows that f^{-1} ________.

$15x^2 + 3$
increasing
exists

65 To determine if f^{-1} exists when $f(x) = -x^3(x^2 + 2)$, we first notice that $f'(x) =$ ________ $<$ ____ for all real numbers $x \neq 0$. Then f is a ________ function on its entire domain. Using Problem 63, it follows that f^{-1} ________.

$-5x^4 - 6x^2$, 0
decreasing
exists

66 To determine if f^{-1} exists when $f(x) = \dfrac{x+2}{x-2}$ for $x > 2$, we first notice that $f'(x) =$ ________ $<$ ____, so that f is a ________ function on the interval ________. By Problem 63, it follows that f^{-1} ________.

$-\dfrac{4}{(x-2)^2}$, 0
decreasing, $(2, \infty)$
exists

67 To determine if f^{-1} exists when $f(x) = x^4 + 3x^2 + 7$ and $f'(x) =$ ________, notice that $f'(x) > 0$ if $x >$ ____ and $f'(x) < 0$ if $x <$ ____. Therefore, f is neither increasing nor decreasing and f^{-1} ________.

$4x^3 + 6x$, 0
0
does not exist

3.2 Concavity of the Graph of a Function

68 Let f be a differentiable function on the open interval I. The graph of f is said to be ________ on I if f' is an increasing function on I. Geometrically, this means that as a point P on this graph moves to the right, the tangent line at P turns ________ and its slope ________. Graphically,

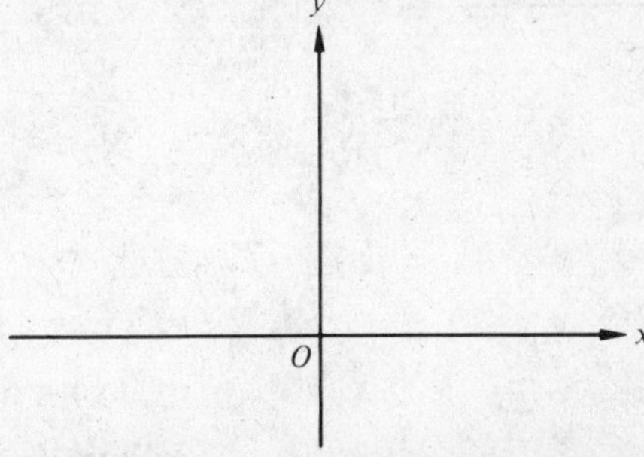

concave upward

counterclockwise, increases

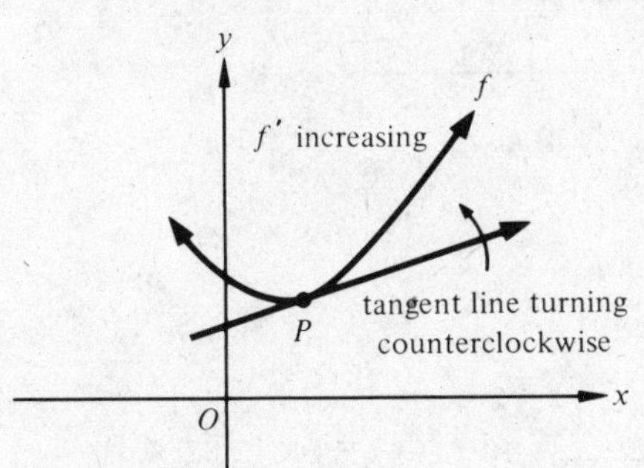

69 Let f be a differentiable function on the open interval I. The graph of f is said to be ________ on I if f' is a decreasing function on I. Geometrically, this means that as a point P on this graph moves to the right, the tangent line at P

concave downward

clockwise, decreases

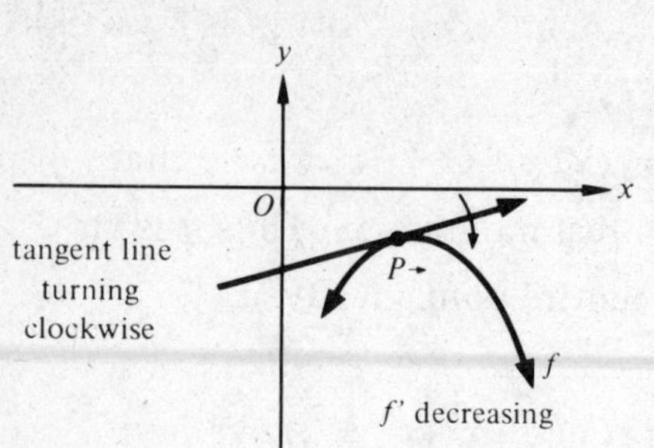

turns __________ and its slope __________.
Graphically,

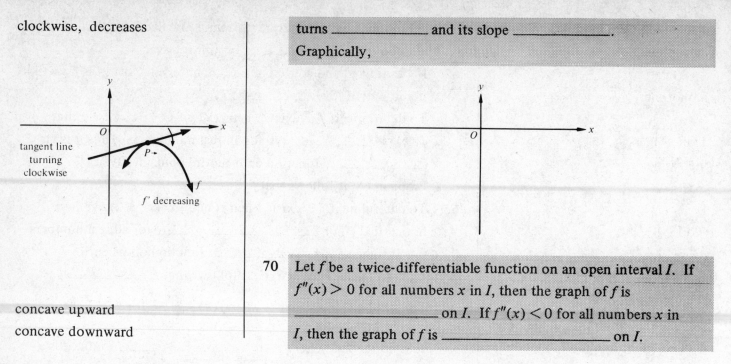

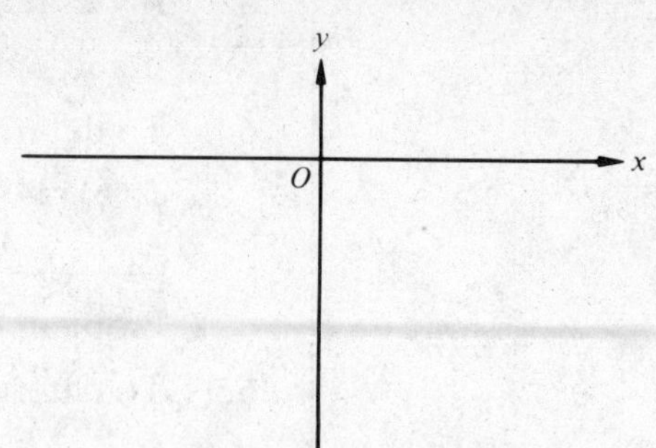

concave upward
concave downward

70 Let f be a twice-differentiable function on an open interval I. If $f''(x) > 0$ for all numbers x in I, then the graph of f is __________ on I. If $f''(x) < 0$ for all numbers x in I, then the graph of f is __________ on I.

In Problems 71 through 79, find intervals where the graph of f is concave upward or downward.

$3x^2 - 12x + 9$, $6x - 12$
$x > 2$
$x < 2$
$(2, \infty)$, $(-\infty, 2)$

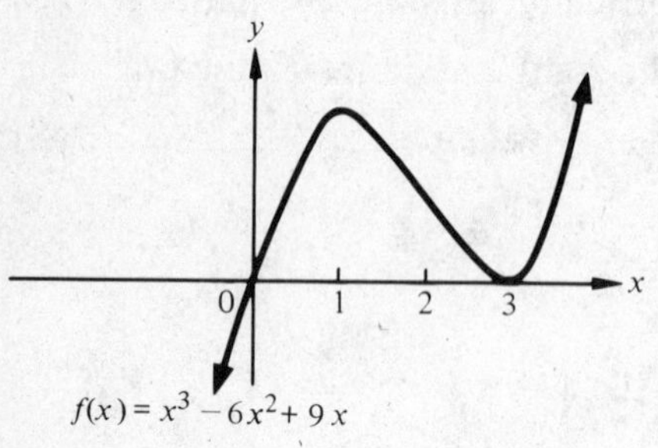

71 $f(x) = x^3 - 6x^2 + 9x$.
$f'(x) =$ __________, so that $f''(x) =$ __________. Since $f''(x) = 6x - 12 > 0$ when ______ and $f''(x) = 6x - 12 < 0$ when ______, then the graph of f is concave upward on the interval ______ and is concave downward on the interval ______.
The graph is

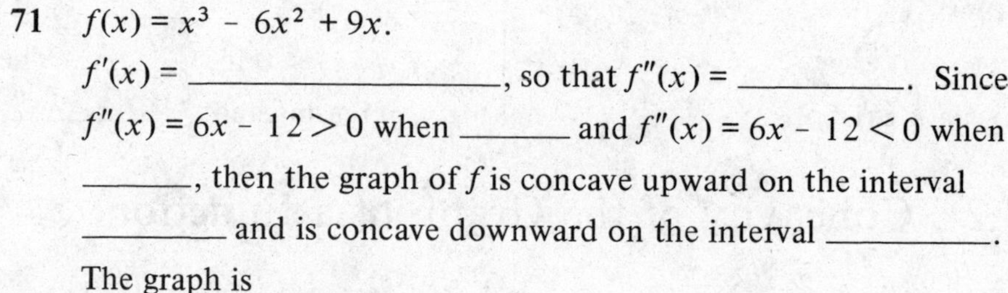

$3x^2 - 12x$, $6x - 12$
$x > 2$
$x < 2$
$(2, \infty)$, $(-\infty, 2)$

72 $f(x) = x^3 - 6x^2 + 12$.
$f'(x) =$ __________ and $f''(x) =$ __________.
$f''(x) = 6x - 12 > 0$ when ______ and $f''(x) = 6x - 12 < 0$ when ______. The graph of f is concave upward on the interval ______ and is concave downward on the interval ______.
The graph is

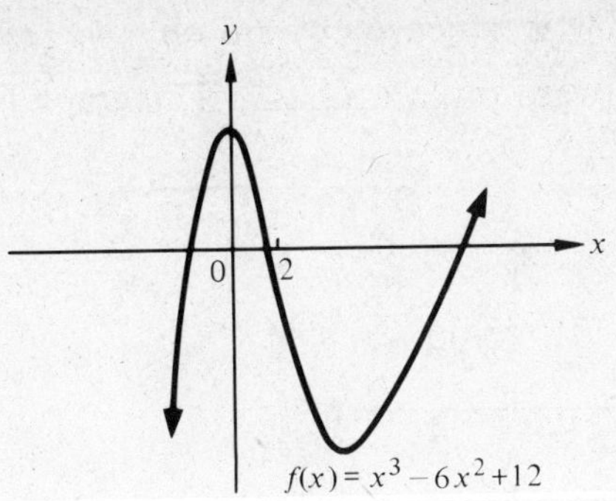

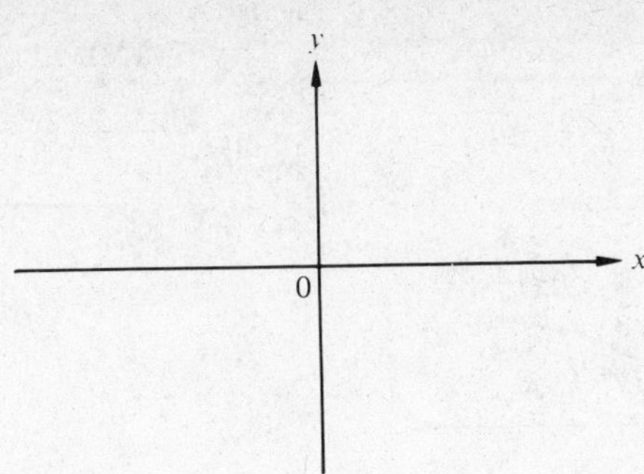

$15x^4 + 20x^3$, $60x^3 + 60x^2$

$x > -1$

$x < -1$

$(-1, \infty)$

$(-\infty, -1)$

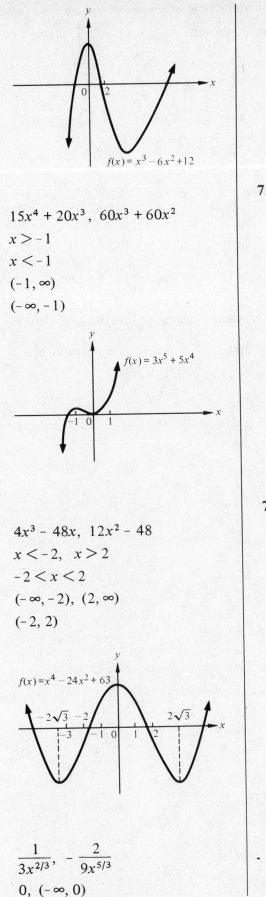

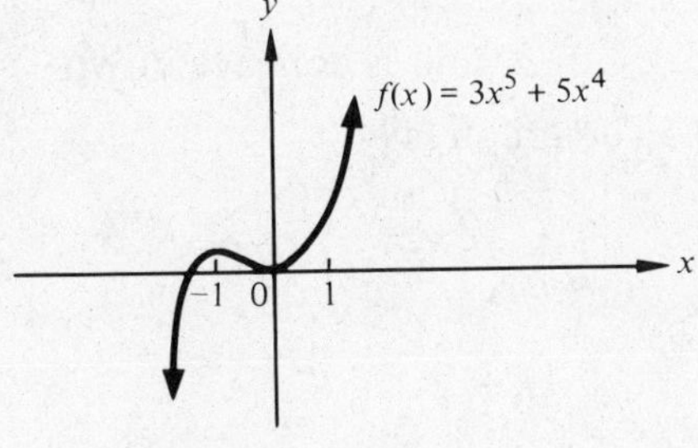

73 $f(x) = 3x^5 + 5x^4$.
$f'(x) =$ ________ and $f''(x) =$ ________.
$f''(x) = 60x^3 + 60x^2 > 0$ when ______ and
$f''(x) = 60x^3 + 60x^2 < 0$ when ______. The graph of f is
concave upward on the interval ______ and is concave downward on the interval ______. The graph is

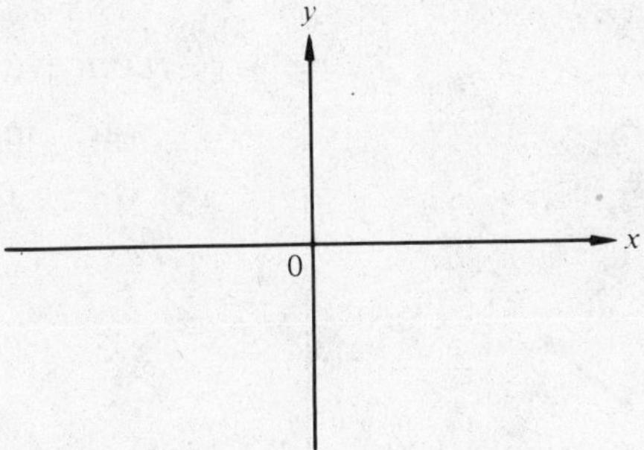

$4x^3 - 48x$, $12x^2 - 48$

$x < -2$, $x > 2$

$-2 < x < 2$

$(-\infty, -2)$, $(2, \infty)$

$(-2, 2)$

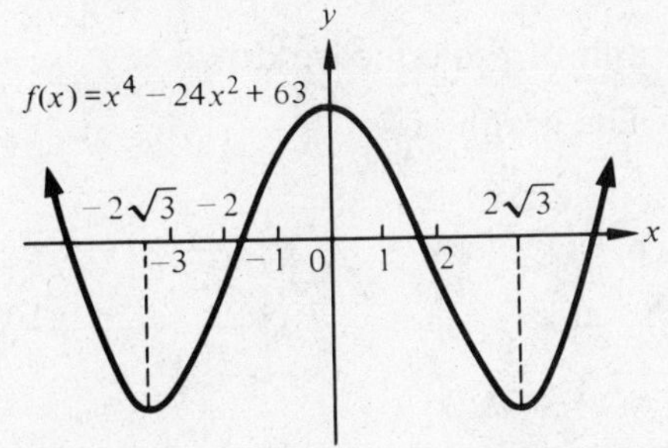

74 $f(x) = x^4 - 24x^2 + 63$.
$f'(x) =$ ________ and $f''(x) =$ ________.
$f''(x) = 12x^2 - 48 > 0$ when ______ or ______ and
$f''(x) = 12x^2 - 48 < 0$ when ______. The graph of f
is concave upward on the intervals ______ and ______,
and is concave downward on ______. The graph is

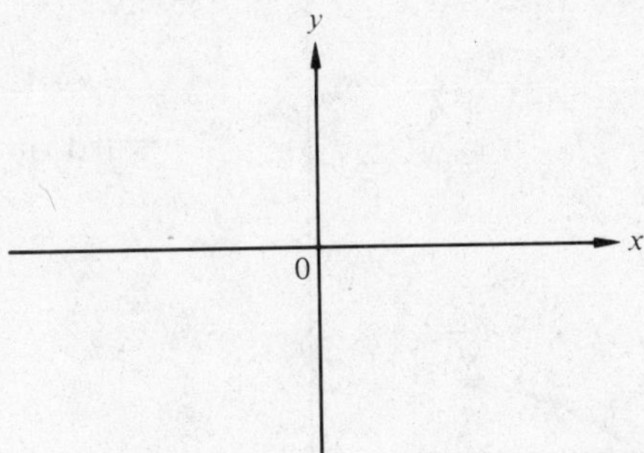

$\dfrac{1}{3x^{2/3}}$, $-\dfrac{2}{9x^{5/3}}$

0, $(-\infty, 0)$

75 $f(x) = \sqrt[3]{x}$.
$f'(x) =$ ______ and $f''(x) =$ ______. f is not differentiable at
____. $f''(x) > 0$ on the interval ______ and $f''(x) < 0$ on the

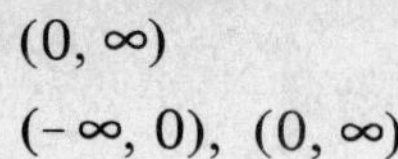

$(0, \infty)$

$(-\infty, 0)$, $(0, \infty)$

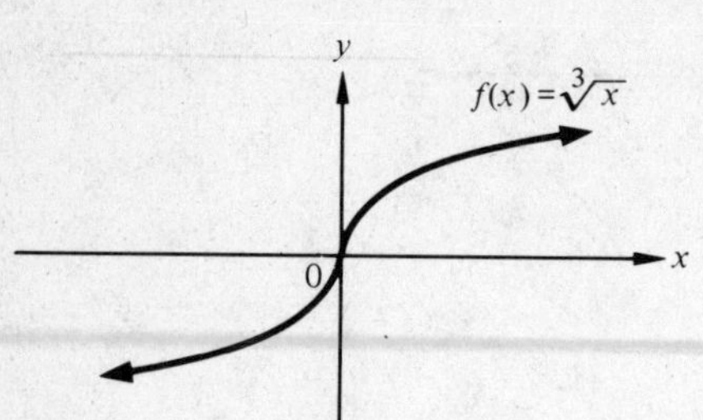

$1 - \frac{1}{x^2}$, $\frac{2}{x^3}$

$x > 0$, $x < 0$

$(0, \infty)$

$(-\infty, 0)$

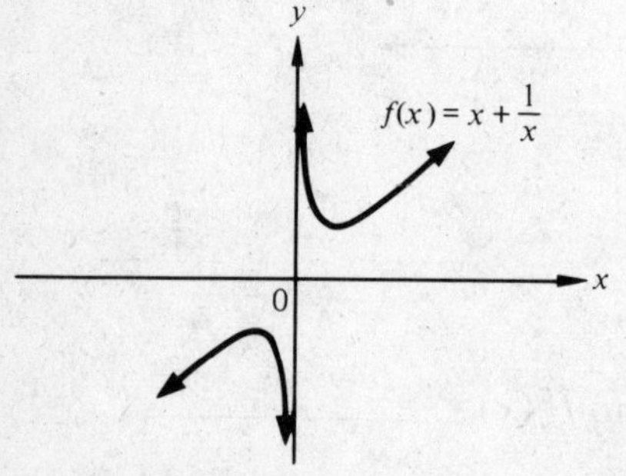

$\frac{1}{2\sqrt{x}} + \frac{1}{2\sqrt{x^3}}$, $-\frac{1}{4\sqrt{x^3}} - \frac{3}{4\sqrt{x^5}}$

$x > 0$

$(0, \infty)$

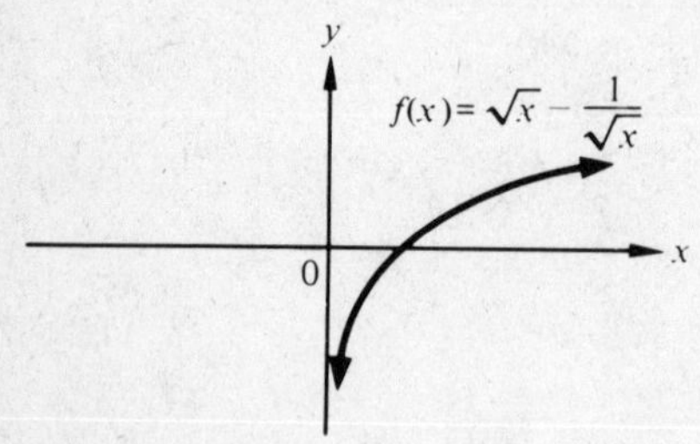

interval ________. The graph of f is concave upward on ________ and is concave downward on ________. The graph is

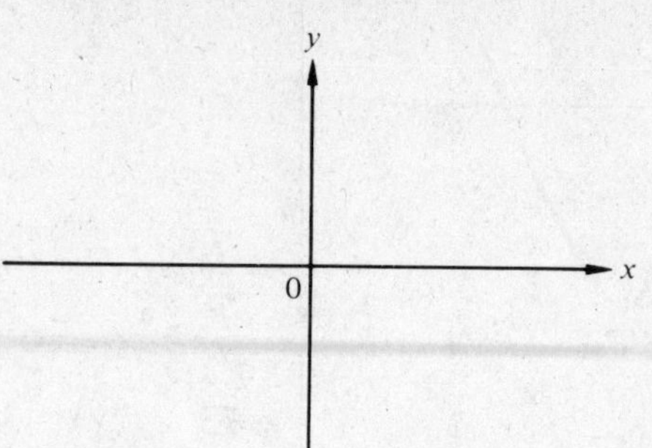

76 $f(x) = x + \frac{1}{x}$, $x \neq 0$.

$f'(x) =$ ________ and $f''(x) =$ ________. $f''(x) = \frac{2}{x^3} > 0$ when ________ and $f''(x) = \frac{2}{x^3} < 0$ when ________, so that the graph of f is concave upward on the interval ________ and is concave downward on the interval ________. The graph is

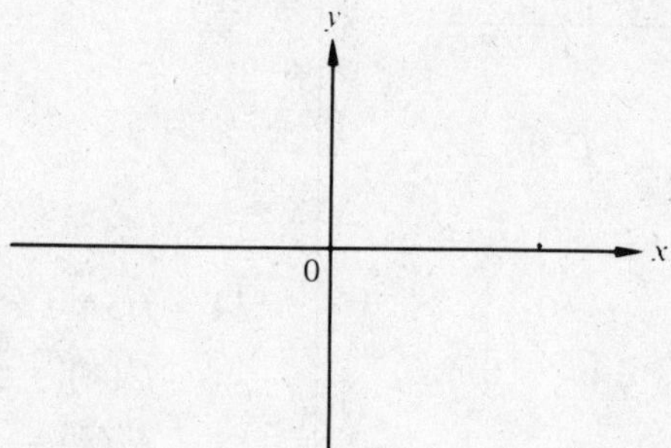

77 $f(x) = \sqrt{x} - \frac{1}{\sqrt{x}}$, $x > 0$.

$f'(x) =$ ________ and $f''(x) =$ ________.

$f''(x) < 0$ for ________, and the graph of f is concave downward on the interval ________. The graph is

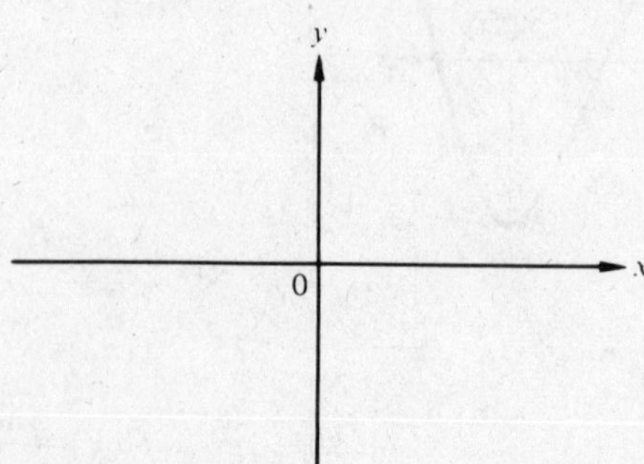

78 $f(x) = \frac{1}{x^2 + 1}$.

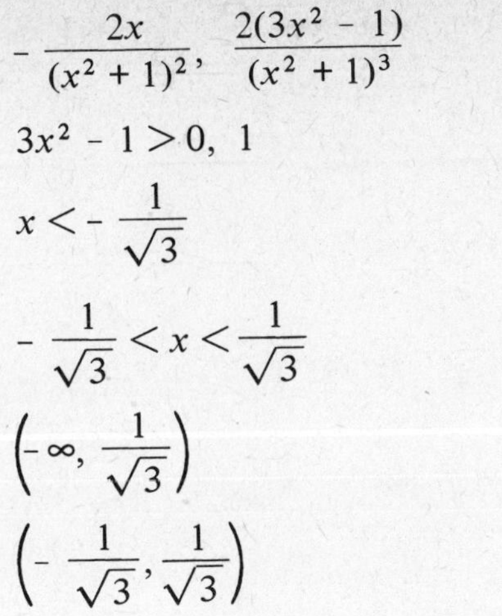

$-\frac{2x}{(x^2+1)^2}, \frac{2(3x^2-1)}{(x^2+1)^3}$

$3x^2 - 1 > 0,\ 1$

$x < -\frac{1}{\sqrt{3}}$

$-\frac{1}{\sqrt{3}} < x < \frac{1}{\sqrt{3}}$

$\left(-\infty, -\frac{1}{\sqrt{3}}\right)$

$\left(-\frac{1}{\sqrt{3}}, \frac{1}{\sqrt{3}}\right)$

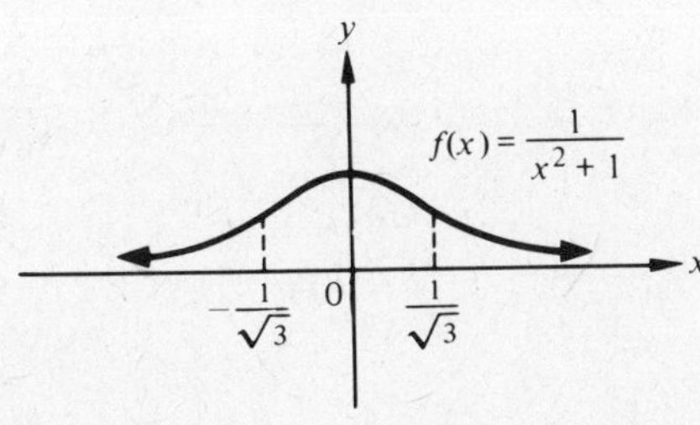

$\frac{2x}{(2x^2+1)^2}, \frac{2-12x^2}{(2x^2+1)^3}$

$2 - 12x^2 > 0,\ -\frac{1}{\sqrt{6}} < x < \frac{1}{\sqrt{6}}$

$2 - 12x^2 < 0$

$x < -\frac{1}{\sqrt{6}}$

$\left(-\frac{1}{\sqrt{6}}, \frac{1}{\sqrt{6}}\right)$

$\left(\frac{1}{\sqrt{6}}, \infty\right)$

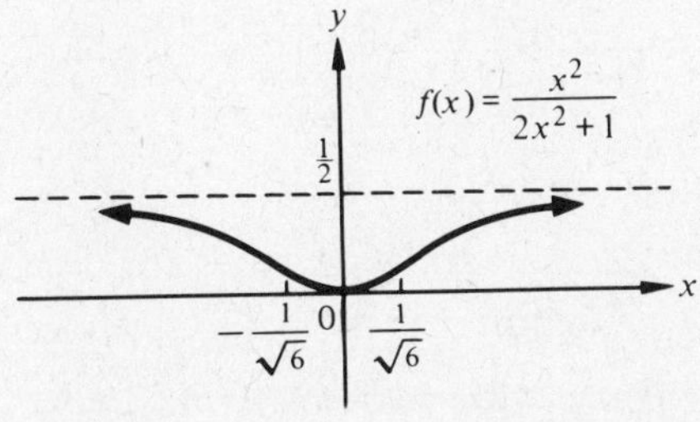

inflection

$f'(x) =$ ________ and $f''(x) =$ ________, so that $f''(x) > 0$ if ________ or $3x^2 >$ ____; that is, $x > \frac{1}{\sqrt{3}}$ or ________; and $f''(x) < 0$ if $3x^2 - 1 < 0$ or ________. Therefore, the graph of f is concave upward on the intervals $\left(\frac{1}{\sqrt{3}}, \infty\right)$ and ________ and is concave downward on ________. The graph is

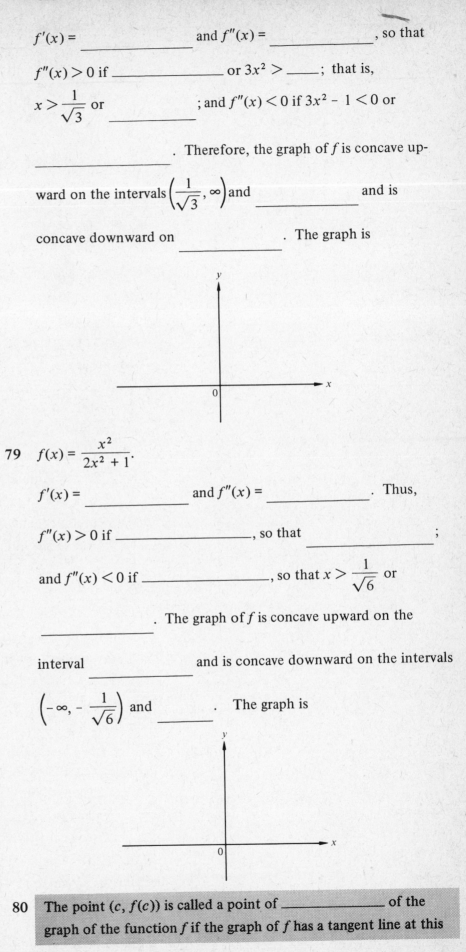

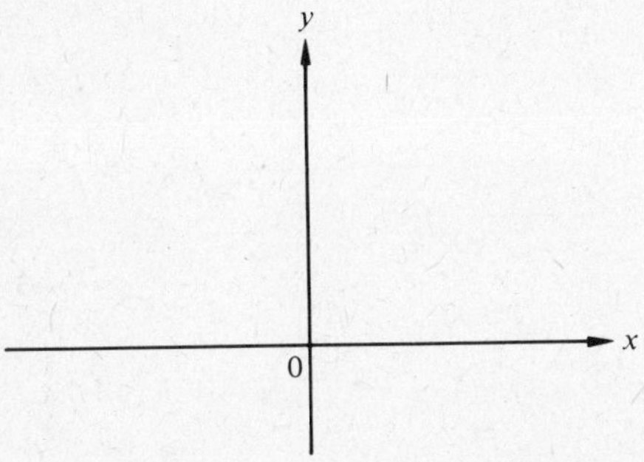

79 $f(x) = \frac{x^2}{2x^2+1}$.

$f'(x) =$ ________ and $f''(x) =$ ________. Thus, $f''(x) > 0$ if ________, so that ________; and $f''(x) < 0$ if ________, so that $x > \frac{1}{\sqrt{6}}$ or ________. The graph of f is concave upward on the interval ________ and is concave downward on the intervals $\left(-\infty, -\frac{1}{\sqrt{6}}\right)$ and ________. The graph is

80 The point $(c, f(c))$ is called a point of ________ of the graph of the function f if the graph of f has a tangent line at this

point and if there is an interval I containing the number c such that for every pair of numbers a and b in I with $a < c < b, f''(a)$ and $f''(b)$ ______ and have ______ algebraic signs.

exist, opposite

3.3 Graph Sketching

81 The procedure for determining the intervals on which f' is positive or negative consists of the following steps:

(1) Determine the numbers at which f' is undefined, or ______, or takes the value ____, and arrange in increasing order: $x_1, x_2, x_3, \ldots, x_n$.

(2) On each open interval $(x_1, x_2), (x_2, x_3), \ldots, (x_{n-1}, x_n)$ the values $f'(x)$ will have ______ algebriac signs. The same is true for an open interval contained in the domain of f' to the left of ____ or to the right of ____. To determine the sign for any one such interval, say (x_2, x_3), just select any number a in (x_2, x_3) and evaluate $f'(a)$. If ______, then $f'(x) > 0$ for all x in (x_2, x_3); and if ______, then $f'(x) < 0$ for all x in ______.

discontinuous, 0

constant

x_1, x_n

$f'(a) > 0$

$f'(a) < 0$

(x_2, x_3)

82 After carrying out the procedure of Problem 81, we can use Problem 55 to determine the intervals where f is ______ or ______.

increasing, decreasing

83 The procedure of Problem 81, with f' replaced by f'', can be used to determine the intervals on which the graph of f is ______ or ______ downward.

concave upward, concave

84 Let $f(x) = x^3 - 9x^2 + 15x - 5$.
Here $f'(x) =$ ______ $= 3($____$)(x - 1)$. f' is defined and ______ at every number x. Arranging the numbers at which $f'(x) = 0$ in increasing order, we have: 1, ____. Hence, f' has constant algebraic signs on each of the intervals $(-\infty, 1)$, ______, and ______. We select a number in each interval, say, 0, 2, and 6, respectively. Here, $f'(0) > 0$, $f'(2)$ ____, and $f'(6)$ ____; it follows that f' is ______ on $(-\infty, 1)$, f' is ______ on $(1, 5)$, and f' is ______ on $(5, \infty)$. Hence, f is ______ on $(-\infty, 1]$ and $[5, \infty)$ and f is ______ on $[1, 5]$. Now, $f''(x) =$ ______, so that $f''(x) < 0$ for ____ and $f''(x) > 0$ for ____. Hence, the graph of f is concave ______ on the interval $(-\infty, 3)$ and concave upward on the interval ______. Using Problem 80, we conclude that $(3, f(3)) = (3, -14)$ is a ______ of the graph of f. Plotting the

$3x^2 - 18x + 15$, $x - 5$

continuous

5

$(1, 5)$, $(5, \infty)$

< 0, > 0

positive, negative

positive, increasing

decreasing

$6x - 18$, $x < 3$

$x > 3$, downward

$(3, \infty)$

point of inflection

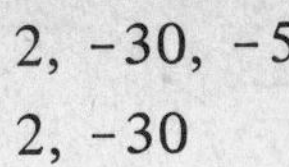
2, −30, −5
2, −30

points (1, ____), (3, −14), (5, ____), and, say, (0, ____), (7, ____), (−1, ____), and using the information obtained, we have the graph

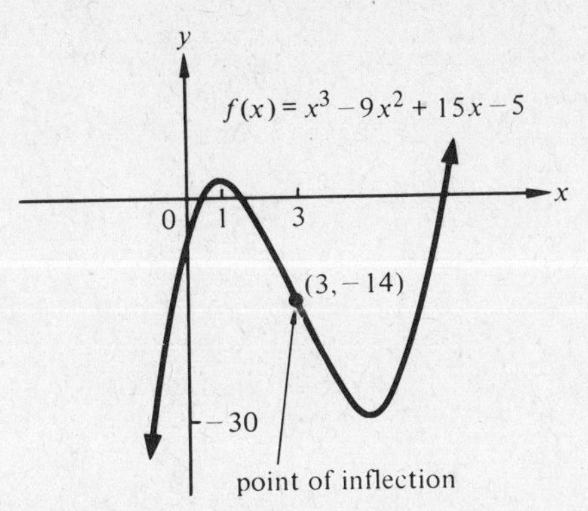

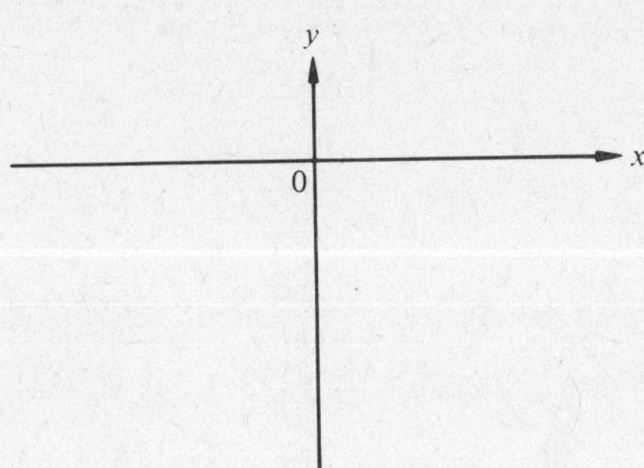

4 Relative Maximum and Minimum Values of Functions

$f(c) \geqslant f(x)$

85 A function f is said to have a relative (or a local) maximum at a number c if there is an open interval I containing c such that f is defined on I and ____________ holds for every x in I.

$f(c) \leqslant f(x)$

86 A function f is said to have a relative (or local) minimum at a number c if there is an open interval I containing c such that ____________ holds for every x in I.

extremum

87 If a function f has either a relative maximum or a relative minimum at a number c, we shall say that f has a relative ____________ at c.

$f'(c) = 0$

88 Suppose that the function f has a relative extremum at the number c. Then, if f is differentiable at the number c, it is necessary that ____________.

critical number

differentiable, $f'(c) = 0$

89 A number c is called a ____________ for the function f provided that f is defined at c, but either f is not ____________ at c or else ____________.

critical number

90 If a function f has a relative extremum at a number c, then c must be a ____________ for the function f.

In Problems 91 through 94, find all critical numbers at which f has a relative extremum.

$(-\infty, \infty)$, $-2x$
$(-\infty, \infty)$
0, critical number
maximum, 0, $> f(x)$

91 $f(x) = 9 - x^2$.
f is defined on ________ and $f'(x) =$ ____ is also defined on ________. There is only one solution to $f'(x) = 0$, namely $x =$ ____, and $x = 0$ is a ____________. f has a relative ________ at $x =$ ____, since $f(0)$ ________ for $x \neq 0$. The graph is

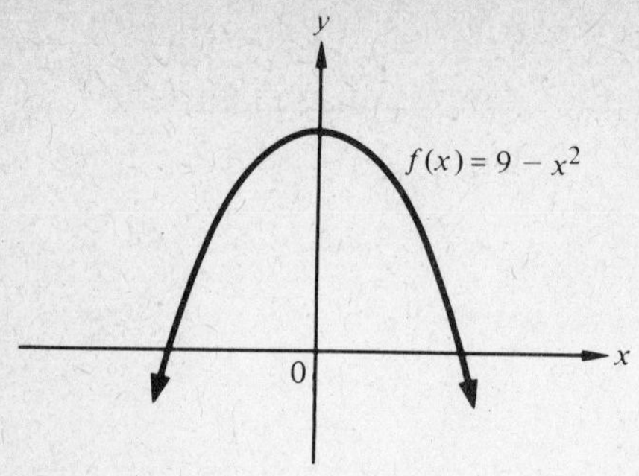

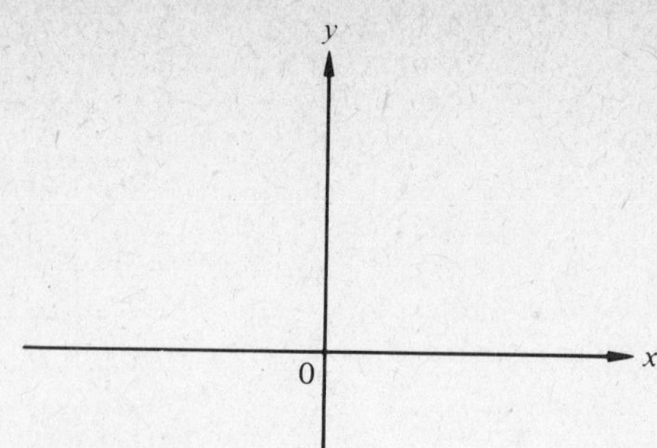

$(-\infty, \infty)$, $2x$
$(-\infty, \infty)$
0, critical
minimum, $f(x) > f(0)$

92 $f(x) = x^2 - 4$.
f is defined on ________ and $f'(x) =$ _____ is also defined on ________. There is only one solution to $f'(x) = 0$, namely $x =$ _____ and $x = 0$ is a __________ number. f has a relative __________, since ______________ for $x \neq 0$. The graph is

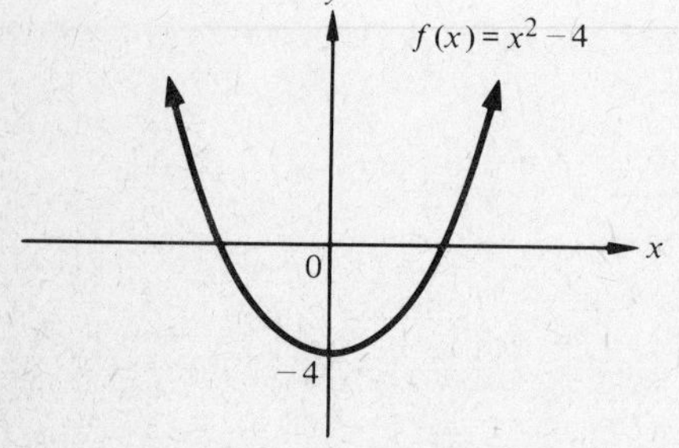

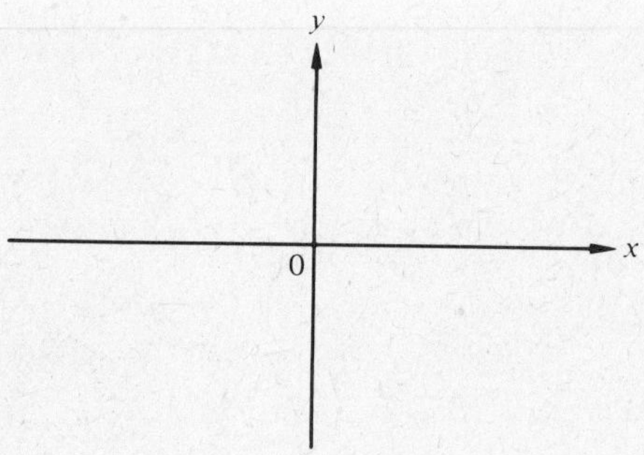

$-2x^2$, 0
$< f(0)$

$f(0)$, > 0

critical number
extrema

93 $f(x) = -\dfrac{2}{3}x^3$.

$f'(x) =$ _____, so that $x =$ _____ is the only critical number. For values of x such that $x > 0$, $f(x)$ ________, whereas if $x < 0$, we have $f(x) >$ _____, since $x < 0$ implies that $-\dfrac{2}{3}x^3$ _____. Since f does not have other critical numbers and since a relative extremum must occur at a ______________, there are no relative ________. The graph is

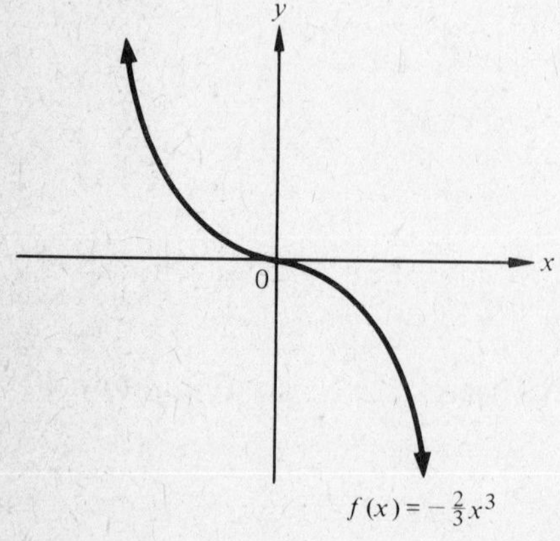

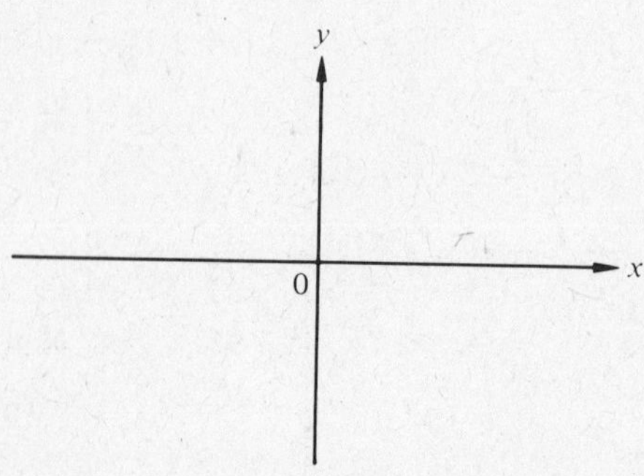

94 $f(x) = \begin{cases} 7x - 9 & \text{if } x \leqslant 2 \\ 7 - x & \text{if } x > 2. \end{cases}$

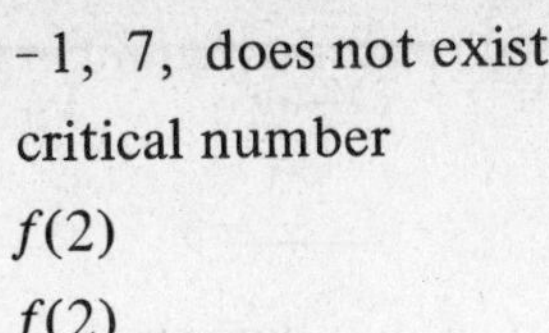

-1, 7, does not exist
critical number
$f(2)$

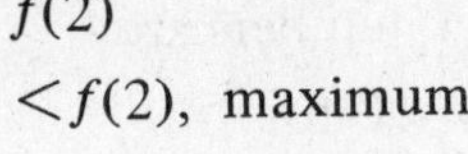

$f(2)$
$< f(2)$, maximum

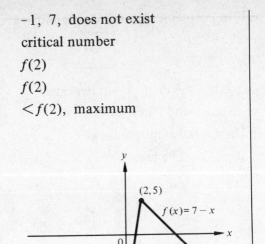

$f'_+(2) =$ ____, while $f'_-(2) =$ ____. Hence, $f'(2)$ ________. Thus, 2 is the only ________. If $x < 2$, then $f(x) = 7x - 9 < 7(2) - 9 =$ ____, whereas if $x > 2$, then $f(x) = 7 - x < 7 - 2 = 5 =$ ________. In any case, $x \neq 2$ implies that $f(x)$ ________. Therefore, f has a relative ________ at 2. The graph is

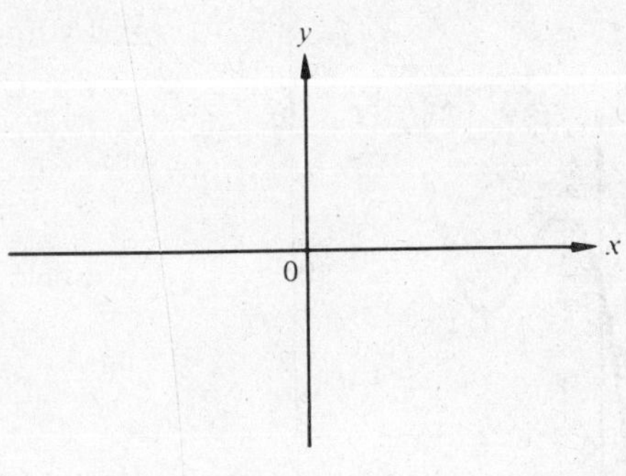

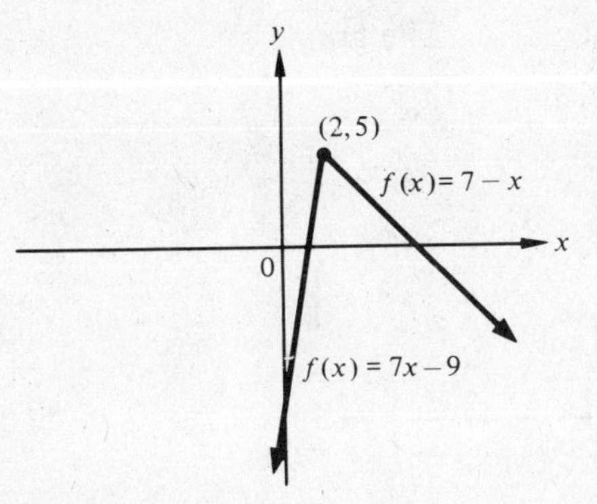

4.1 First and Second Derivative Tests for Relative Extrema

95 Let f be defined and continuous on the open interval (a, b) and suppose that the number c belongs to (a, b) and that f is differentiable at every number in (a, b) except possibly at c. Then if $f'(x) > 0$ for every number x in (a, c) and $f'(x) < 0$ for every number x in (c, b), then f has a ________ at c. If $f'(x) < 0$ for every x in (a, c) and $f'(x) > 0$ for every x in (c, b), then f has a ________ at c.

relative maximum

relative minimum

96 Let the function f be differentiable on the open interval I, and suppose that c is a number in I such that $f'(c) = 0$ and $f''(c)$ exists. If $f''(c) > 0$, then f has a ________ at c; if $f''(c) < 0$, then f has a ________ at c; if $f''(c) = 0$, then the test is ________.

relative minimum
relative maximum
inconclusive

4.2 Procedure for Finding Relative Extrema

97 To find all relative extrema of f, carry out the following steps.

(1) Find f'.

(2) Find the critical numbers for f, that is, all c such that $f'(c)$ does not exist or ________.

(3) Test each of the critical numbers to see whether it corresponds to a relative maximum, a relative minimum, or ________. The first or second derivative tests can be used.

$f'(c) = 0$

neither

In Problems 98 through 106, use the first derivative test (in Problem 95) to find all numbers at which each function f has a relative extremum.

98 $f(x) = x^3 - 6x^2 + 9x + 5$.

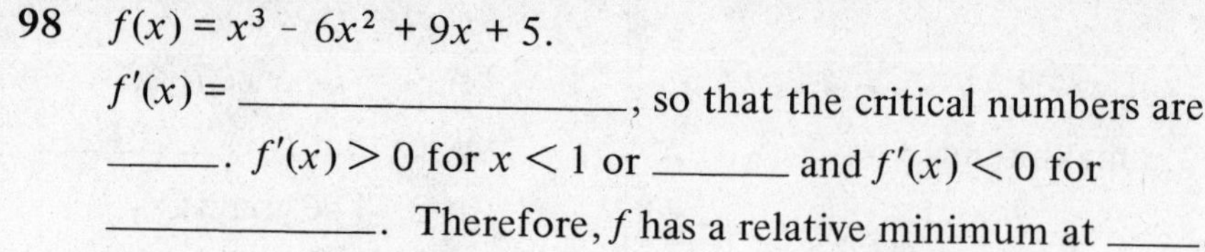

$f'(x) =$ ________________, so that the critical numbers are ______. $f'(x) > 0$ for $x < 1$ or ______ and $f'(x) < 0$ for ________. Therefore, f has a relative minimum at ____, and f has a relative maximum at ____. The graph is

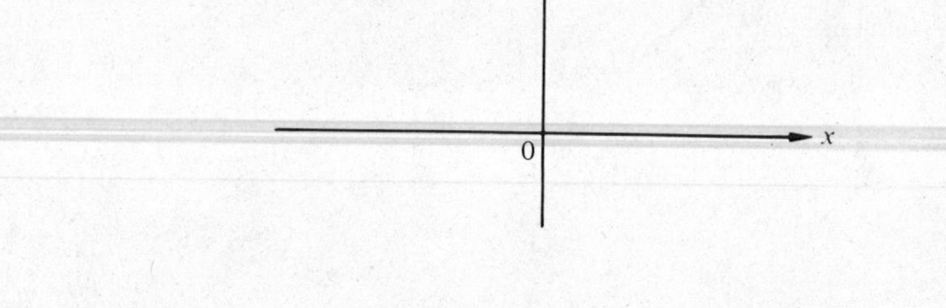

$3x^2 - 12x + 9$

1, 3, $x > 3$

$1 < x < 3$, 3

1

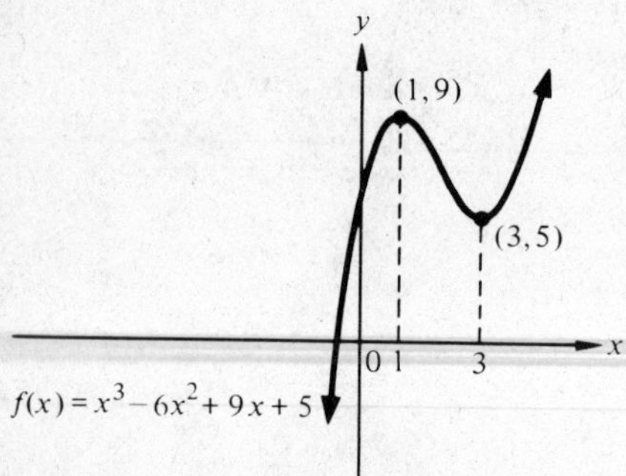

99 $f(x) = x^3 + 3x - 2$.

$f'(x) =$ ________, so that $f'(x) > 0$ for ________________. Thus, there is no critical number for f and therefore f has no ________________. The graph is

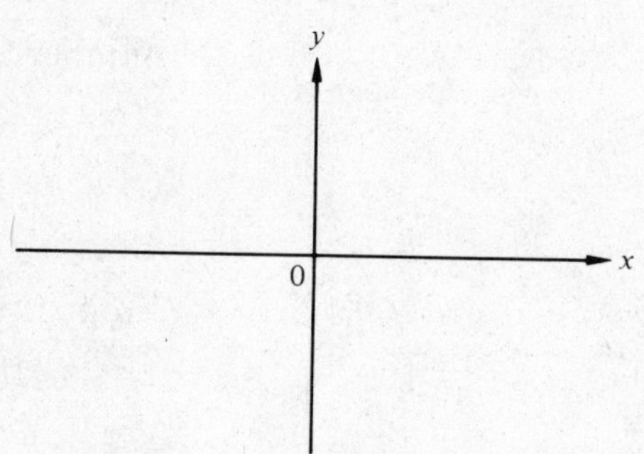

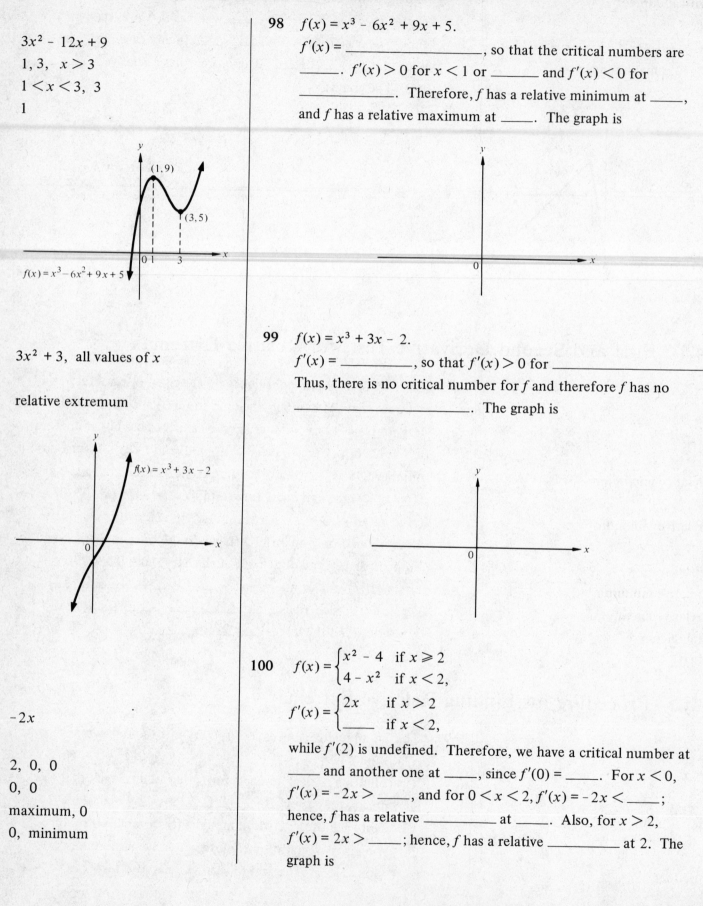

$3x^2 + 3$, all values of x

relative extremum

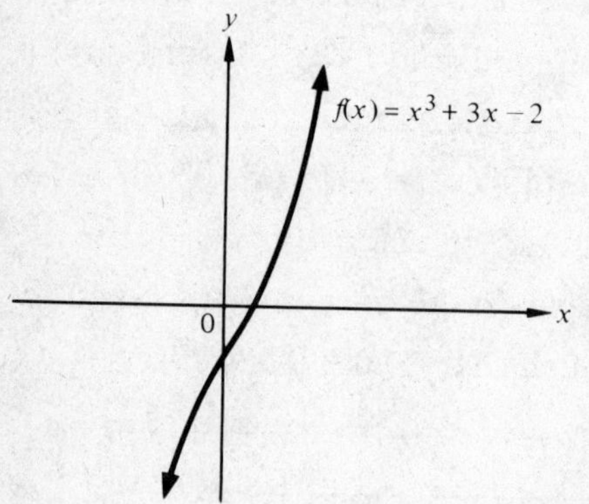

100 $f(x) = \begin{cases} x^2 - 4 & \text{if } x \geqslant 2 \\ 4 - x^2 & \text{if } x < 2, \end{cases}$

$f'(x) = \begin{cases} 2x & \text{if } x > 2 \\ \_\_\_\_ & \text{if } x < 2, \end{cases}$

while $f'(2)$ is undefined. Therefore, we have a critical number at ____ and another one at ____, since $f'(0) =$ ____. For $x < 0$, $f'(x) = -2x >$ ____, and for $0 < x < 2$, $f'(x) = -2x <$ ____; hence, f has a relative ________ at ____. Also, for $x > 2$, $f'(x) = 2x >$ ____; hence, f has a relative ________ at 2. The graph is

$-2x$

2, 0, 0

0, 0

maximum, 0

0, minimum

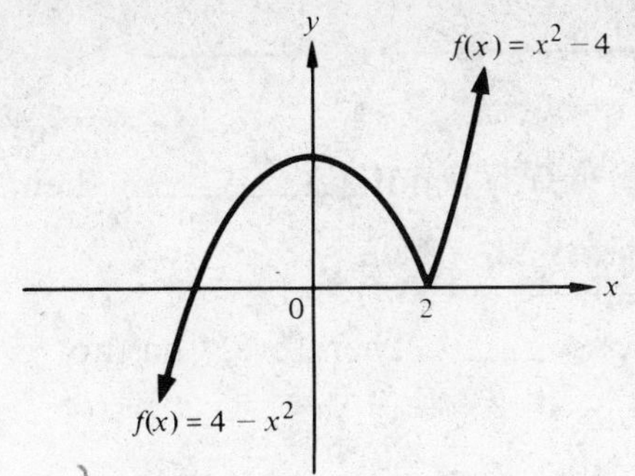

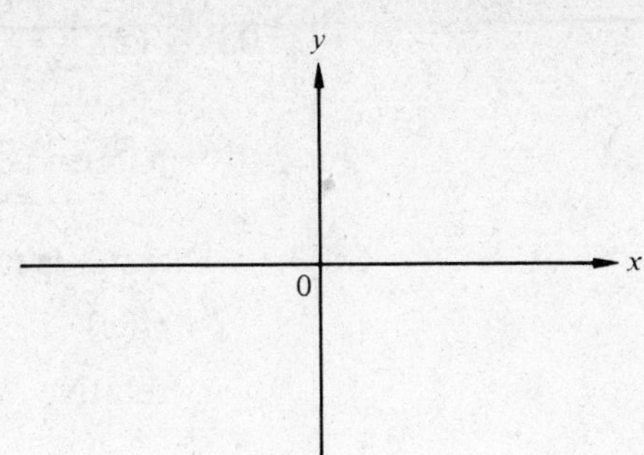

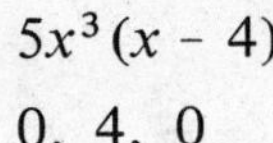
$5x^3(x - 4)$

0, 4, 0

0, >

4

0

101 $f(x) = x^4(x - 5)$.

$f'(x) =$ ____________, so that the critical numbers of f are ____ and ____. For $x < 0$, $f'(x) >$ ____ and for $0 < x < 4$, $f'(x) <$ ____. Also, for $x > 4$, $f'(x)$ ____ 0. Hence, f has a relative minimum at ____ and f has a relative maximum at ____. The graph is

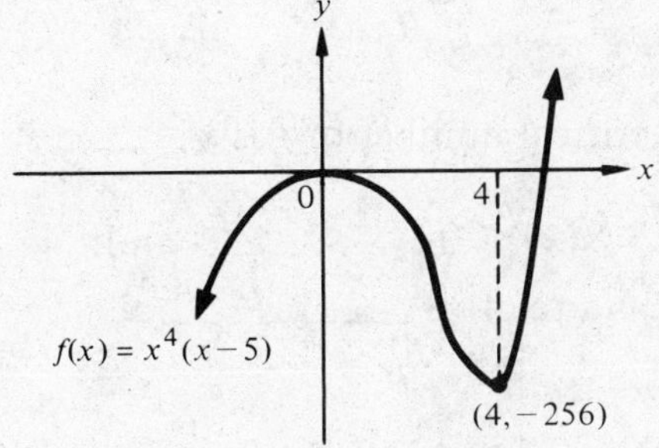

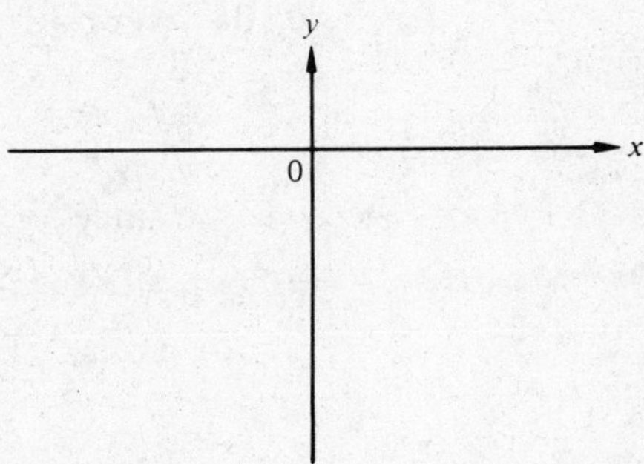

0

1

1

does not exist

1

relative minimum

$x < 1$, relative minimum

102 $f(x) = \begin{cases} 6 & \text{if } x \leqslant 1 \\ 5 + x & \text{if } x > 1, \end{cases}$

$f'(x) = \begin{cases} \text{____} & \text{if } x < 1 \\ \text{____} & \text{if } x > 1. \end{cases}$

Therefore, every number $x <$ ____ is a critical number; also, 1 is a critical number since $f'(1)$ ____________. Since the function values $f(x)$ are constant for $x <$ ____, then f has both a relative maximum and a ____________ at each number ________. At 1, f has only a ____________.

The graph is

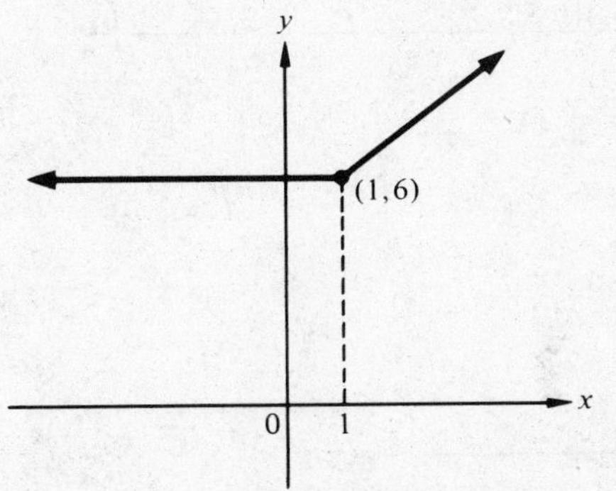

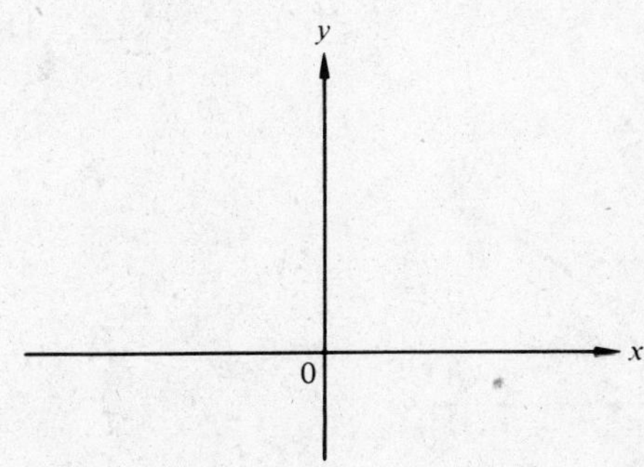

$-\dfrac{1}{(x-3)^2}$, defined

critical number

0, 0

extrema

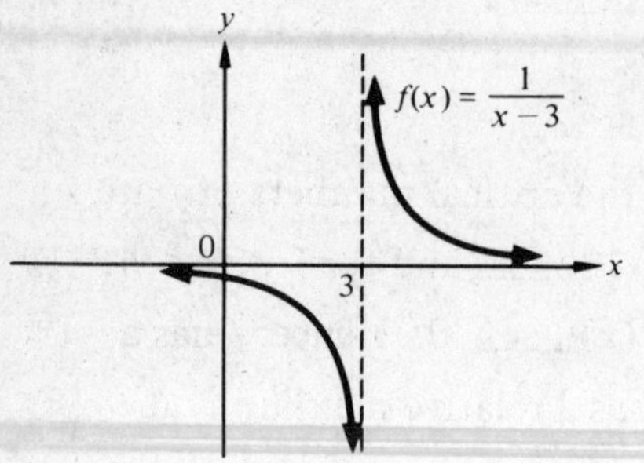

103 $f(x) = \dfrac{1}{x-3}$.

$f'(x) =$ ________. Since $f(3)$ is not ________, then $x = 3$ is not a ________. If $x < 3$, then $f'(x) <$ ____, and for $x > 3$, $f'(x) <$ ____. Therefore, f has no relative ________. The graph is

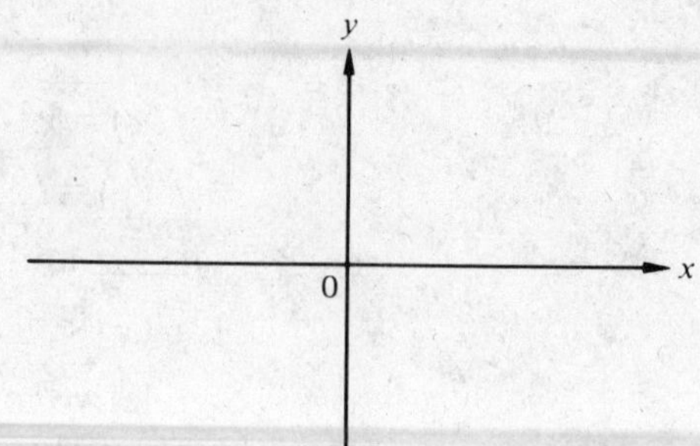

$\dfrac{2}{3x^{1/3}}$, 0

0, 0

$x > 0$, minimum

0

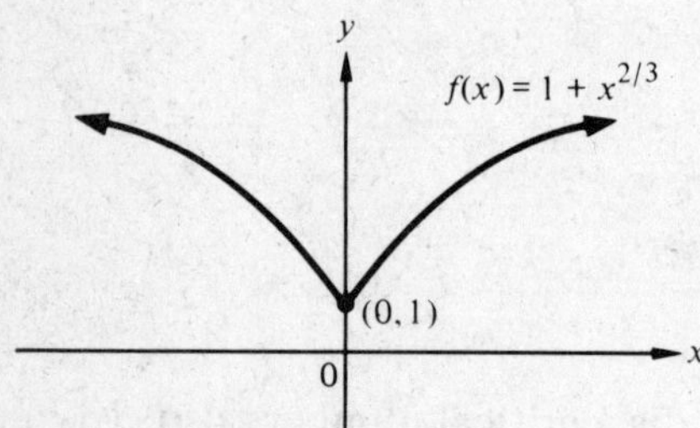

104 $f(x) = 1 + x^{2/3}$.

$f'(x) =$ ____, so that the only critical number of f is $x =$ ____, since $f'(x)$ is not defined at ____. $f'(x) < 0$ if $x <$ ____, and $f'(x) > 0$ if ________. Hence, f has a relative ________ at ____. The graph is

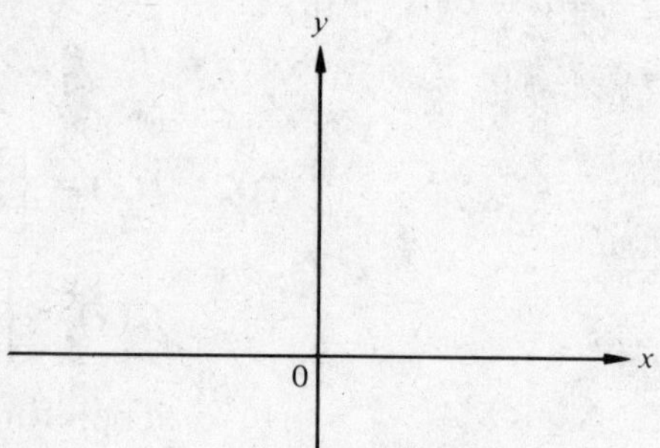

$\dfrac{2}{3}(x-1)^{-1/3}$

1, $x < 1$

$x > 1$, relative minimum

1

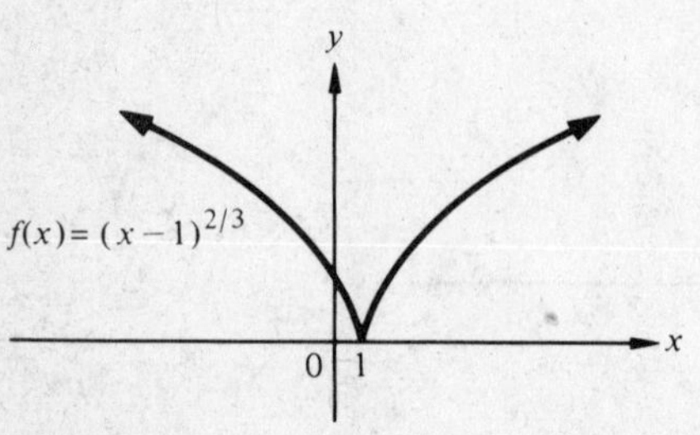

105 $f(x) = (x-1)^{2/3}$.

$f'(x) =$ ________, so that the critical number of f is $x =$ ____. $f'(x) < 0$ when ________ and $f'(x) > 0$ when ________. Therefore, f has a ________ at ____. The graph is

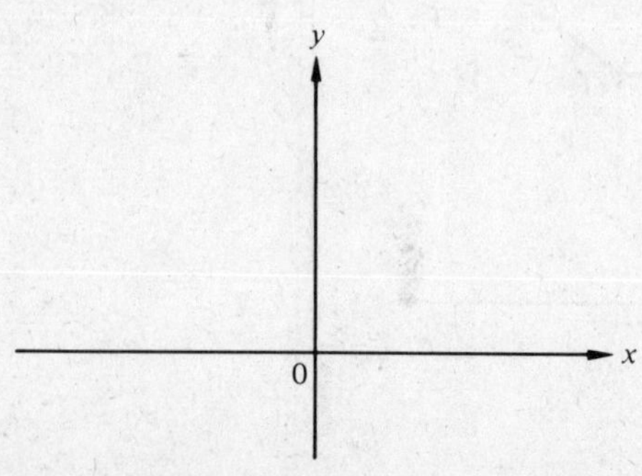

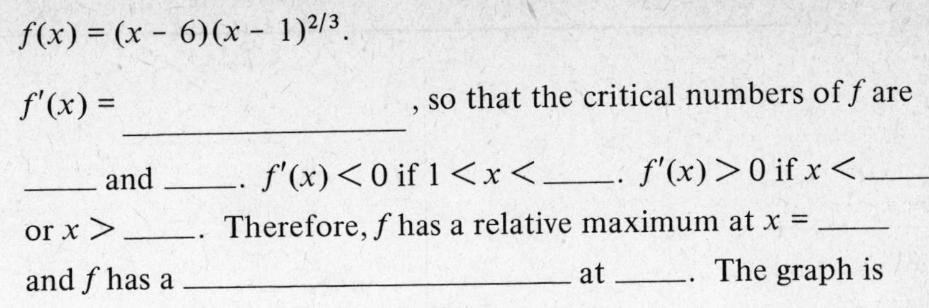

106 $f(x) = (x - 6)(x - 1)^{2/3}$.

$f'(x) =$ ________________, so that the critical numbers of f are _____ and _____. $f'(x) < 0$ if $1 < x <$ _____. $f'(x) > 0$ if $x <$ _____ or $x >$ _____. Therefore, f has a relative maximum at $x =$ _____ and f has a ____________________ at _____. The graph is

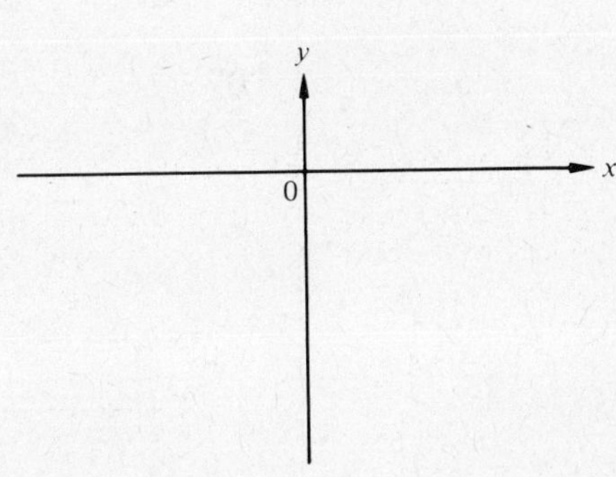

$\dfrac{5(x - 3)}{3(x - 1)^{1/3}}$

1, 3, 3, 1

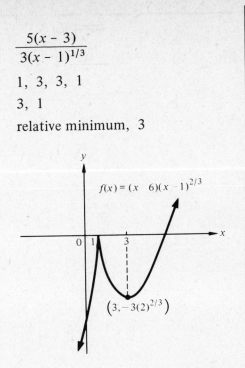

3, 1

relative minimum, 3

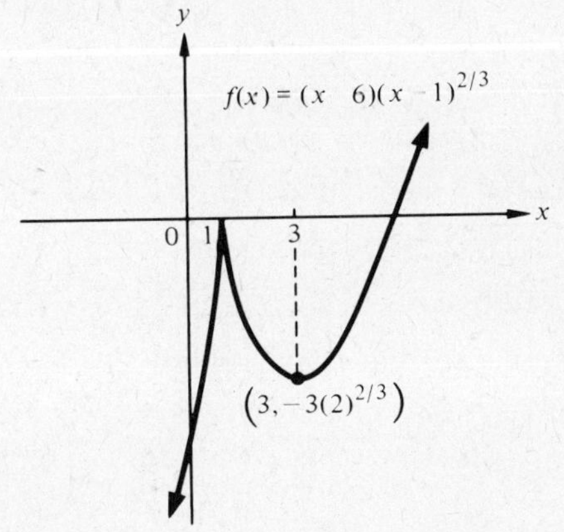

In Problems 107 through 117, apply the second derivative test (in Problem 96) to find the relative extrema of each function.

107 $f(x) = -2x^3 - 3x^2 + 12x + 10$.

$f'(x) =$ ____________________ $= -6(x + 2)($________$)$ and $f''(x) =$ ____________. The roots of $f'(x) = 0$ are $x =$ _____ and $x =$ _____. Since $f'(-2) = 0$ and $f''(-2) >$ _____, then f has a ____________________ at _____. Since $f'(1) = 0$ and $f''(1) <$ _____, then f has a ____________________ at _____. The graph is

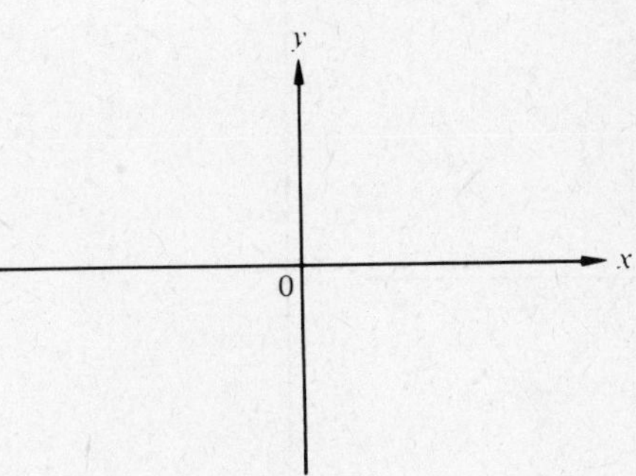

$-6x^2 - 6x + 12, \quad x - 1$

$-12x - 6, \quad -2$

1, 0

relative minimum, -2

0, relative maximum, 1

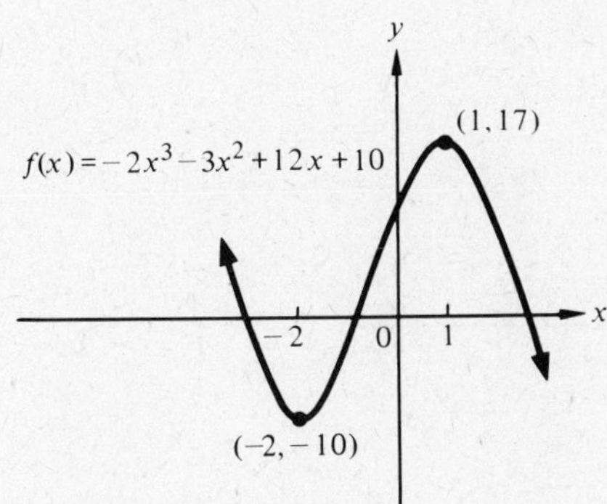

108 $f(x) = x^3 - 3x^2 + 3x + 1$.

$f'(x) =$ ________________ $= 3($________$)^2$ and $f''(x) =$ ________. The only root of $f'(x) = 0$ is ________. Since $f'(1) = 0$ and $f''(1) =$ _____, then the second derivative test is inconclusive. The graph is

$3x^2 - 6x + 3, \quad x - 1$

$6x - 6, \quad 1$

0

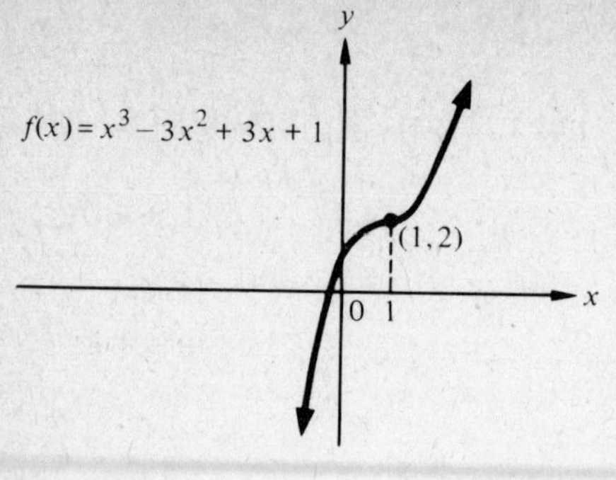

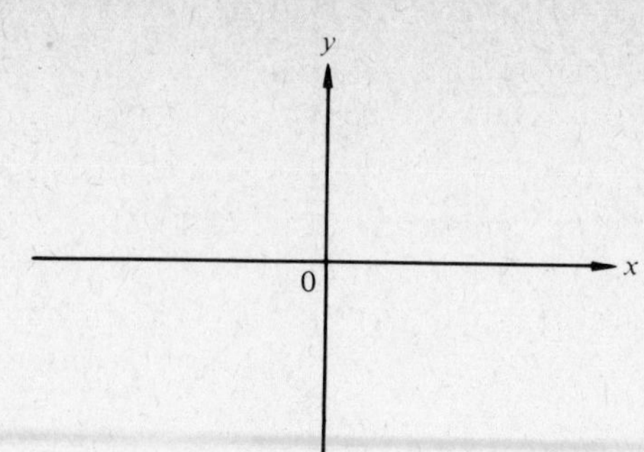

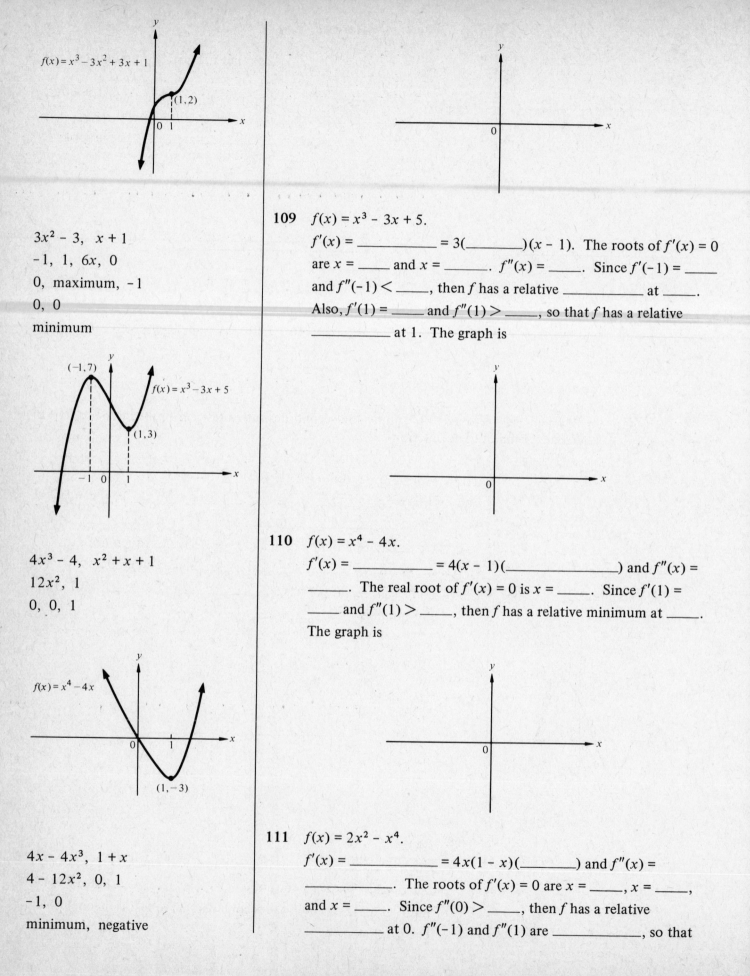

$3x^2 - 3$, $x + 1$
-1, 1, $6x$, 0
0, maximum, -1
0, 0
minimum

109 $f(x) = x^3 - 3x + 5$.
$f'(x) =$ ____________ $= 3($________$)(x - 1)$. The roots of $f'(x) = 0$ are $x =$ _____ and $x =$ ______. $f''(x) =$ _____. Since $f'(-1) =$ _____ and $f''(-1) <$ _____, then f has a relative ____________ at _____. Also, $f'(1) =$ _____ and $f''(1) >$ _____, so that f has a relative ____________ at 1. The graph is

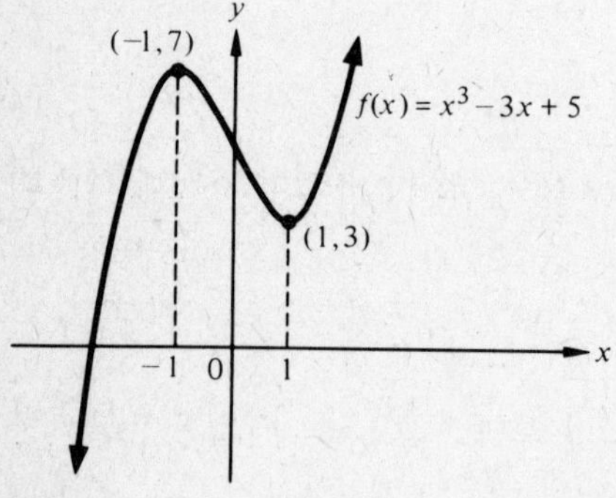

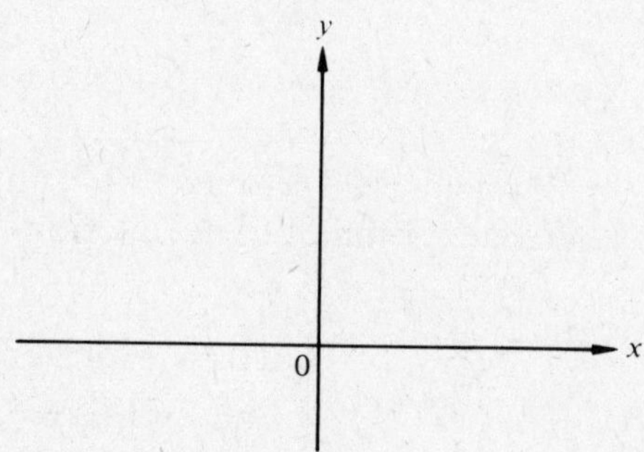

$4x^3 - 4$, $x^2 + x + 1$
$12x^2$, 1
0, 0, 1

110 $f(x) = x^4 - 4x$.
$f'(x) =$ ____________ $= 4(x - 1)($________________$)$ and $f''(x) =$ ______. The real root of $f'(x) = 0$ is $x =$ _____. Since $f'(1) =$ _____ and $f''(1) >$ _____, then f has a relative minimum at _____. The graph is

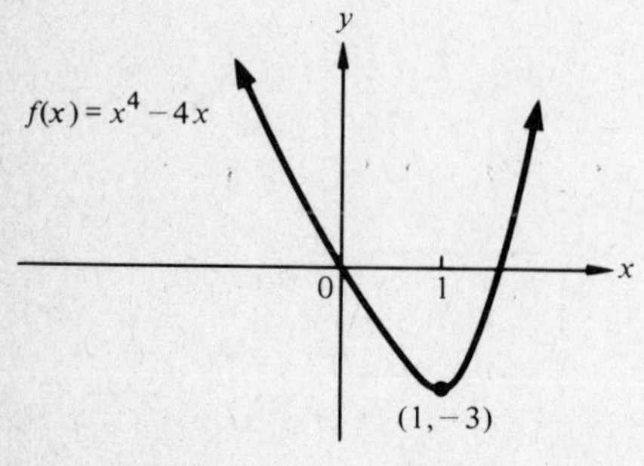

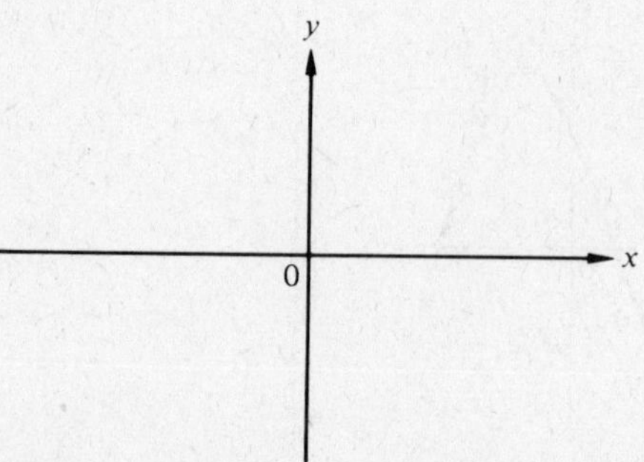

$4x - 4x^3$, $1 + x$
$4 - 12x^2$, 0, 1
-1, 0
minimum, negative

111 $f(x) = 2x^2 - x^4$.
$f'(x) =$ ______________ $= 4x(1 - x)($________$)$ and $f''(x) =$ ______________. The roots of $f'(x) = 0$ are $x =$ _____, $x =$ _____, and $x =$ _____. Since $f''(0) >$ _____, then f has a relative ____________ at 0. $f''(-1)$ and $f''(1)$ are ______________, so that

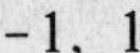

$-1,\ 1$

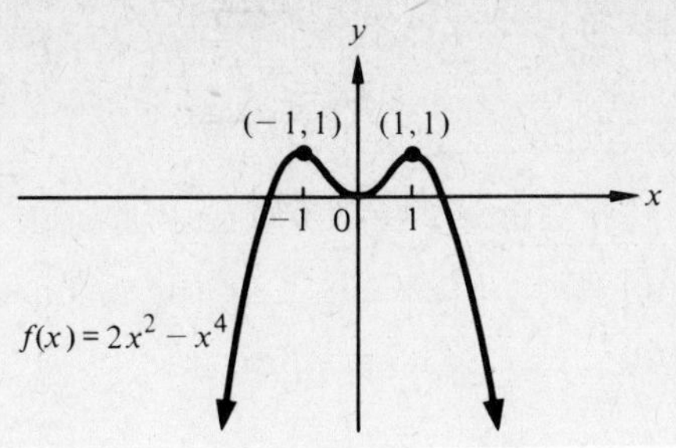

f has a relative maximum at ____ and ____. The graph is

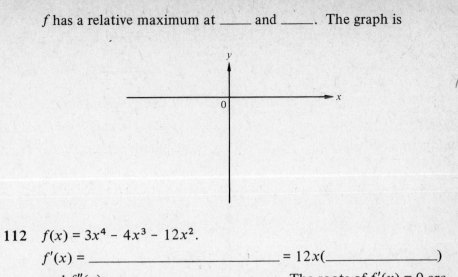

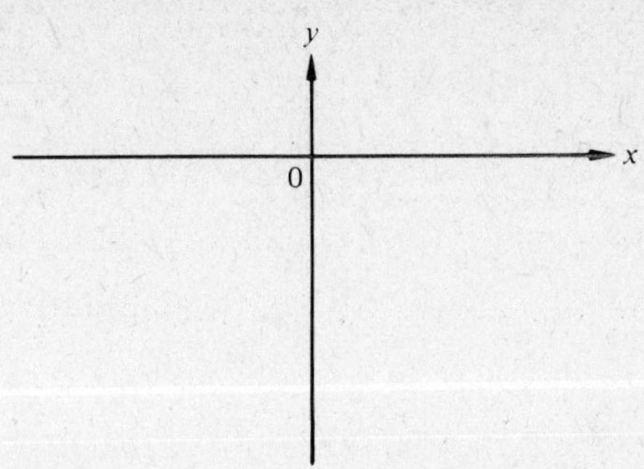

$12x^3 - 12x^2 - 24x,\ \ x^2 - x - 2$
$36x^2 - 24x - 24$
$-1,\ 0,\ 2,\ 0$
relative maximum, 0
positive, minimum, -1
2

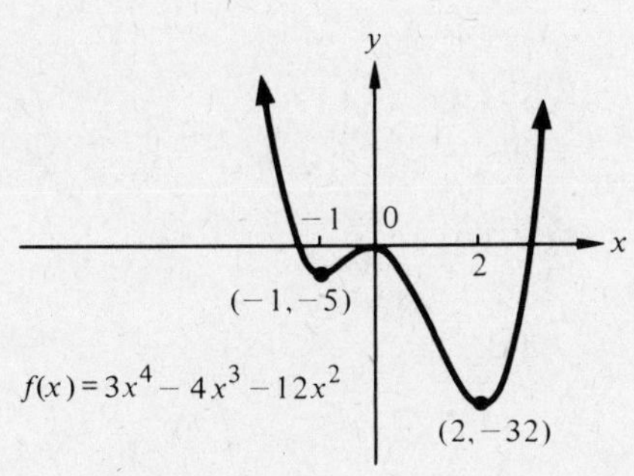

112 $f(x) = 3x^4 - 4x^3 - 12x^2$.
$f'(x) =$ ________________ $= 12x($____________$)$
and $f''(x) =$ ________________. The roots of $f'(x) = 0$ are $x =$ ____, $x =$ ____, and $x =$ ____. Since $f''(0) <$ ____, then f has a ________________ at ____. Also, $f''(-1)$ and $f''(2)$ are __________, so that f has a relative __________ at ____ and ____. The graph is

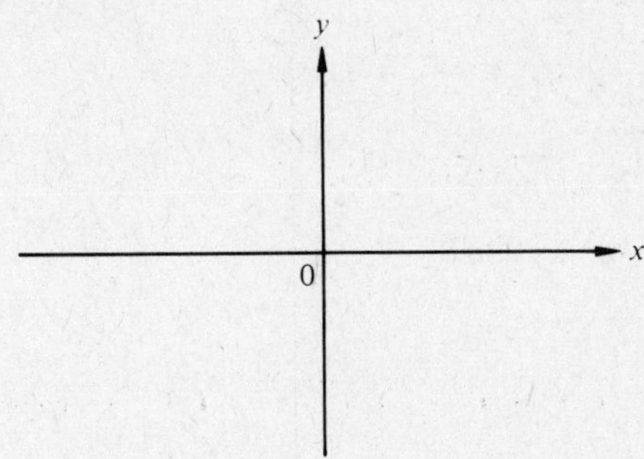

$2x - \dfrac{2}{x^2},\ \ 2 + \dfrac{4}{x^3}$

$1,\ 0$
minimum, 1

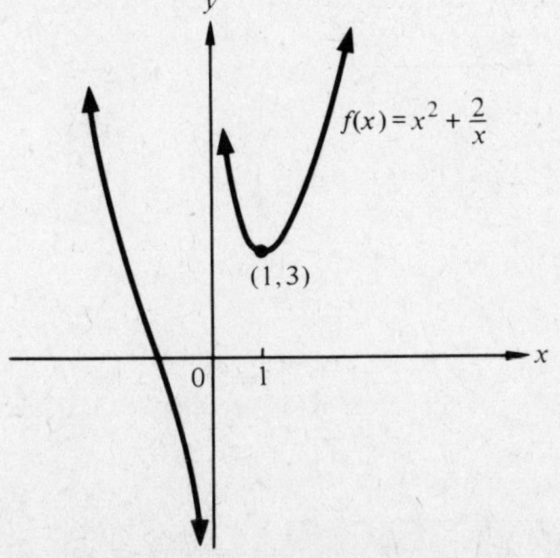

113 $f(x) = x^2 + \dfrac{2}{x}$.

$f'(x) =$ __________ and $f''(x) =$ __________. The only root of $f'(x) = 0$ is $x =$ ____. Since $f'(1) = 0$ and $f''(1) >$ ____, f has a relative __________ at ____. The graph is

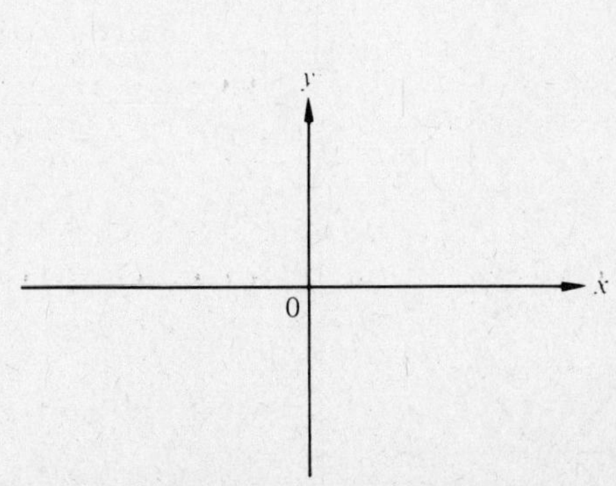

$2x - \frac{2}{x^3}$, $2 + \frac{6}{x^4}$

-1, 1

positive

minimum, -1, 1

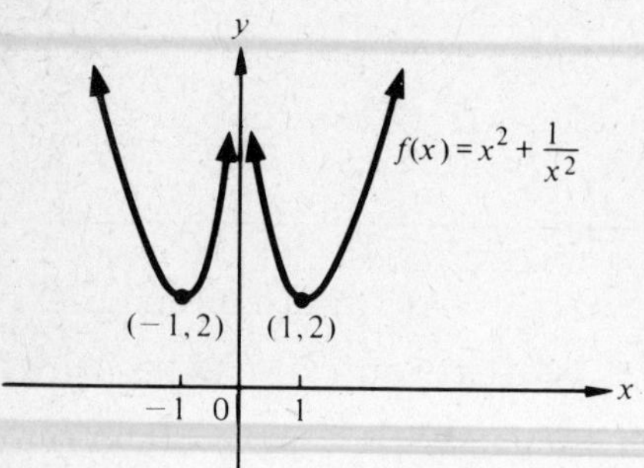

114 $f(x) = x^2 + \dfrac{1}{x^2}$.

$f'(x) =$ ________ and $f''(x) =$ ________ . The roots of $f'(x) = 0$ are $x =$ ____ and $x =$ ____. Since $f'(-1) = 0$ and $f'(1) = 0$, and since $f''(-1)$ and $f''(1)$ are ________, then f has a relative ________ at ____ and ____. The graph is

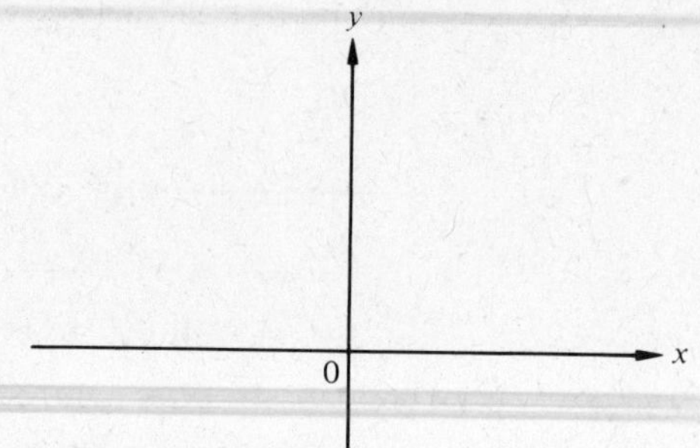

$\frac{1 - x^2}{(x^2 + 1)^2}$, $\frac{2x^3 - 6x}{(x^2 + 1)^3}$

-1, 1

0, minimum

0, maximum, 1

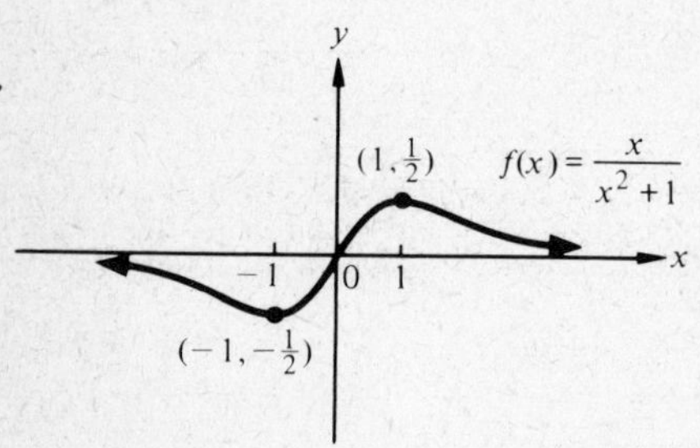

115 $f(x) = \dfrac{x}{x^2 + 1}$.

$f'(x) =$ ________ and $f''(x) =$ ________ . The roots of $f'(x) = 0$ are $x =$ ____ and $x =$ ____. Since $f'(-1) = 0$ and $f''(-1) >$ ____, f has a relative ________ at -1. Also, since $f'(1) = 0$ and $f''(1) <$ ____, f has a relative ________ at ____. The graph is

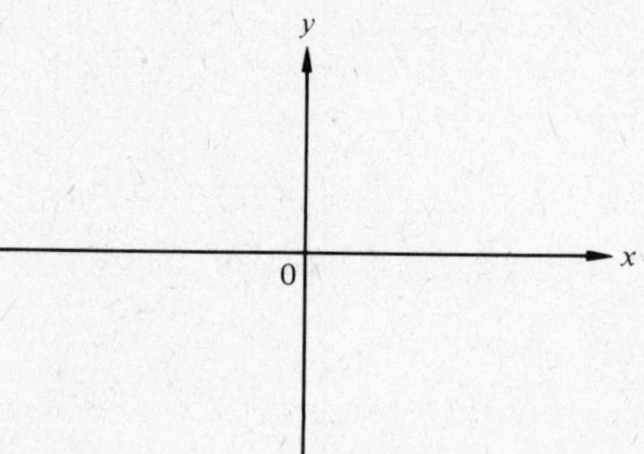

$36 - 24x + 3x^2$, $6x - 24$

2, 6

$<$, relative maximum, 2

$>$

relative minimum, 6

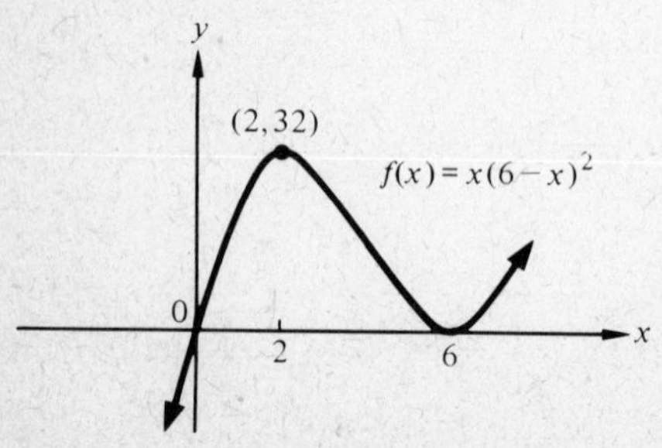

116 $f(x) = x(6 - x)^2$.

$f'(x) =$ ____________ and $f''(x) =$ ________. The roots of $f'(x) = 0$ are $x =$ ____ and $x =$ ____. Since $f'(2) = 0$ and $f''(2)$ ____ 0, then f has a ____________ at ____. Since $f'(6) = 0$ and $f''(6)$ ____ 0, then f has a ____________ at ____. The graph is

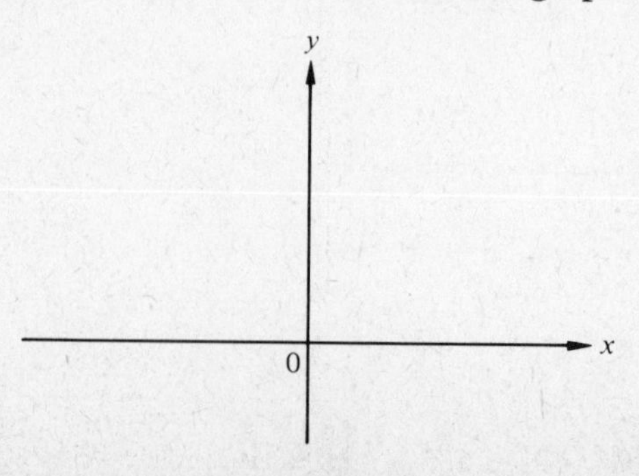

$3x^2 + 2ax + b,\ 6x + 2a$

$\frac{1}{3}(-a + \sqrt{a^2 - 3b})$

$\frac{1}{3}(-a - \sqrt{a^2 - 3b})$

$>$

minimum

$<$, maximum

117 $f(x) = x^3 + ax^2 + bx + c$, where $a^2 > 3b$. $f'(x) =$ ________ and $f''(x) =$ ________. The roots of $f'(x) = 0$ are given by the quadratic formula as $x =$ ________ and $x =$ ________. When $x = \frac{1}{3}(-a + \sqrt{a^2 - 3b})$, then $f''(x) = 2\sqrt{a^2 - 3b}$ ____ 0, and we have a relative ________. When $x = \frac{1}{3}(-a - \sqrt{a^2 - 3b})$, then $f''(x) = -2\sqrt{a^2 - 3b}$ ____ 0, and we have a relative ________.

5 Absolute Extrema

absolute maximum

absolute minimum

absolute extremum

$[a, b]$, absolute minimum

endpoint, extremum

endpoint

critical

c

$f(b)$

absolute maximum

118 Let f be a function that is defined (at least) on the interval I and let c be a number in I. If $f(c) \geqslant f(x)$ holds for all numbers x in I, then we say that on the interval I, the function f takes on its ________ value $f(c)$ at the number c. If $f(c) \leqslant f(x)$ holds for all numbers x in I, then we say that, on the interval I, the function f takes on its ________ value $f(c)$ at the number c.

119 If f takes on either an absolute maximum or an absolute minimum value at c, then we say that f takes on an ________ at c.

120 If a function f is defined and continuous on a closed interval $[a, b]$, then f takes on an absolute maximum value at some number on ________ and f takes on an ________ at some number in $[a, b]$.

121 If a function f takes an absolute extremum at a number c on an interval I, one of the following two conditions must hold:

(i) c is not an ________ of I and f has a relative ________ at c; or else

(ii) c is an ________ of I.

122 To find the absolute extrema for a continuous function f on a closed interval $[a, b]$, we carry out the following steps:

(1) Find all ________ numbers c for the function f on the open interval (a, b).

(2) Calculate the function values $f(c)$ for each number ____ obtained in Step 1.

(3) Calculate $f(a)$ and ____.

(4) The largest of all numbers calculated in Steps 2 and 3 is the ________ of f on $[a, b]$, and the smallest

absolute minimum

of these numbers is the ______________ of f on $[a, b]$.

In Problems 123 through 134, find absolute extrema of each function f on the given interval I.

123 $f(x) = 2x$ on $I = (0, 1]$.

Notice the graph of f.

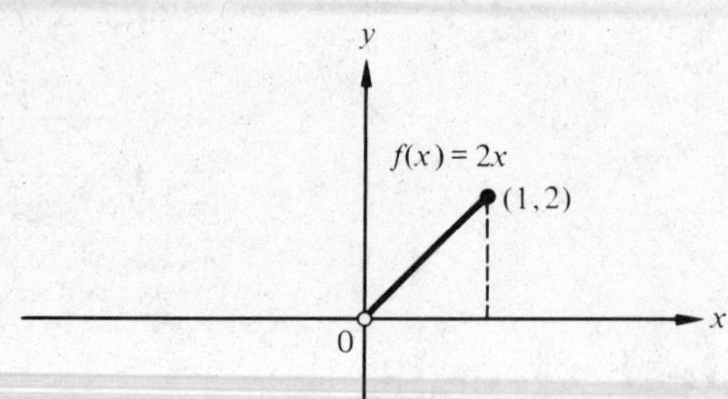

closed

maximum, 2, 1

minimum

Here I is not a ________ interval, but we notice that f takes on an absolute __________, $f(1)$ = _____, at the number _____ on $(0, 1]$. f does not take on any absolute __________ on I.

3x^2, 0

1

1, −1

−1

124 $f(x) = x^3$ on $I = [-1, 1]$.

$f'(x)$ = _____, so that x = _____ is a critical number on $[-1, 1]$. Notice that f takes an absolute maximum $f(1)$ = _____ at the number _____ and f takes an absolute minimum $f(-1)$ = _____ at the number _____. The graph is

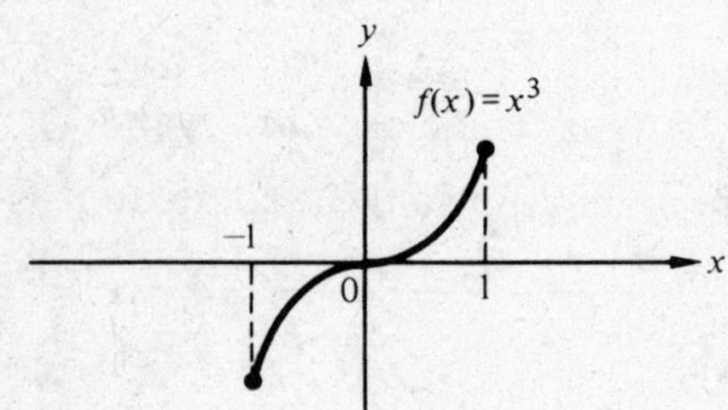

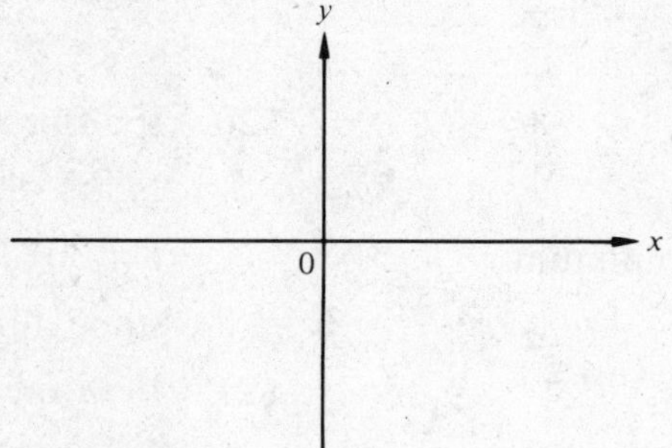

125 $f(x) = \sqrt{4 - x^2}$ on $I = [-2, 2)$.

Notice from the graph that

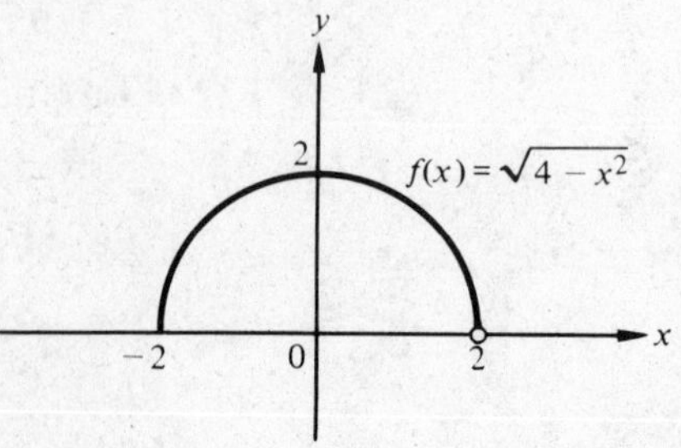

2, 0

absolute maximum, 2

$f(0)$ = _____, $f(-2)$ = _____, so that f takes on an ______________ value $f(0)$ = _____ at the number

0, absolute minimum
0, -2

_____ and f takes on an _____________. $f(-2)$ = _____ at the number _____.

$x^2 - 4x + 3$, 1
3
$\frac{26}{3}$
$-\frac{1}{24}$

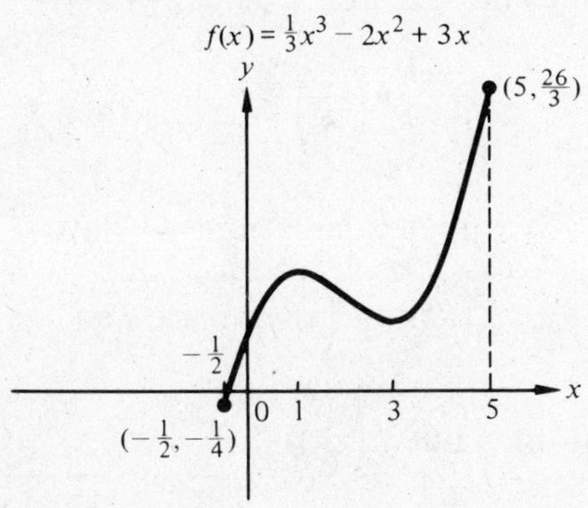

126 $f(x) = \frac{1}{3}x^3 - 2x^2 + 3x + 2$ on $I = \left[-\frac{1}{2}, 5\right]$.

$f'(x)$ = _____________, so that the critical numbers are _____ and _____. Notice that f takes on an absolute maximum value $f(5)$ = _____ at 5 on I and f takes on an absolute minimum value $f\left(-\frac{1}{2}\right)$ = _______ at $-\frac{1}{2}$ on I. The graph is

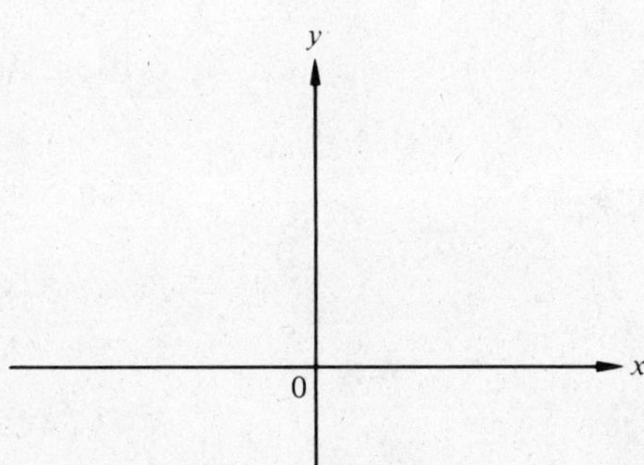

$3x^2 - 27$, continuous
$3x^2 - 27$
-3, 3
49
-59

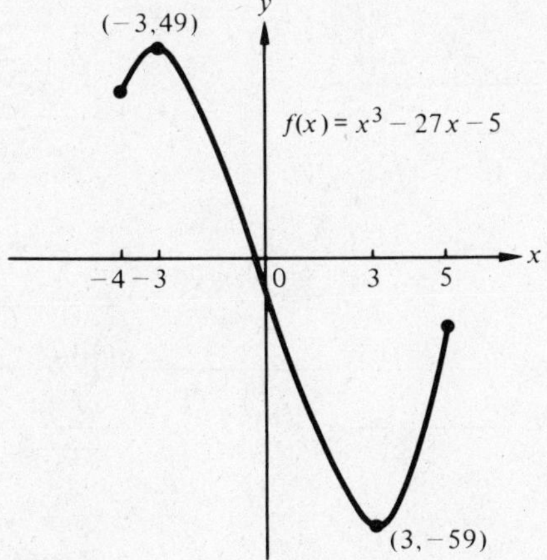

127 $f(x) = x^3 - 27x - 5$ on $I = [-4, 5]$.

$f'(x)$ = ___________ and f is ___________ on I. The critical numbers of f are found by setting ___________ = 0, so that x = _____ and x = _____. f takes on an absolute maximum value $f(-3)$ = _____ at -3 on I and f takes on an absolute minimum value $f(3)$ = _____ on I. The graph is

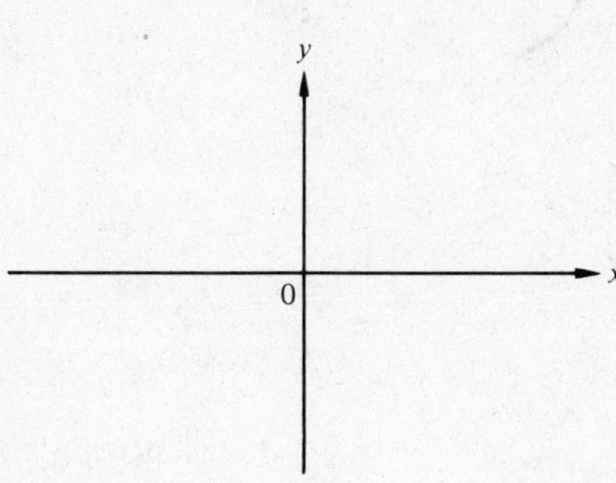

[−3, 3]

128 $f(x) = \begin{cases} x + 2 & \text{if } x < 0 \\ -x^2 + x + 2 & \text{if } x \geq 0 \end{cases}$ on $I = [-3, 3]$.

Here f is continuous on the closed interval _________.

$f'(x) = \begin{cases} 1 & \text{if } -3 < x < 0 \\ -2x + 1 & \text{if } 0 < x < 3, \end{cases}$

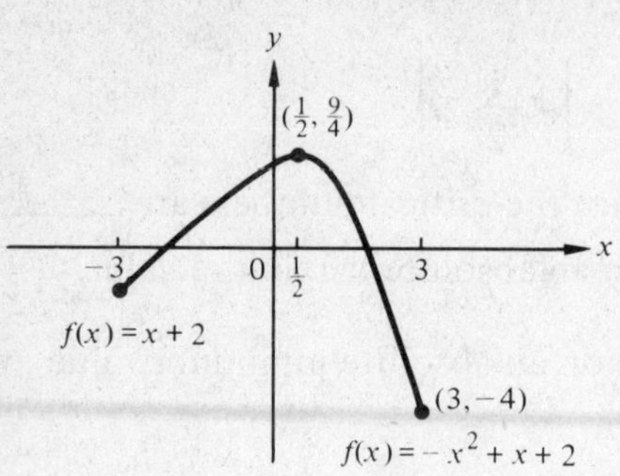

The graph is

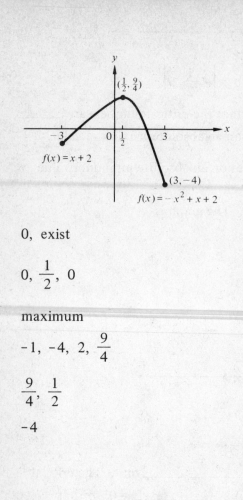

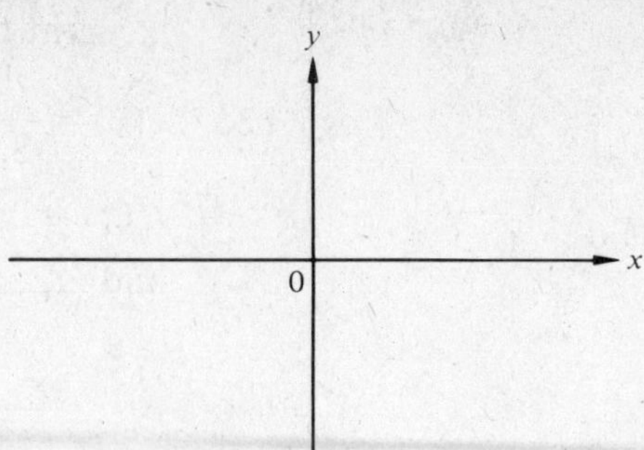

0, exist

0, $\frac{1}{2}$, 0

maximum

-1, -4, 2, $\frac{9}{4}$

$\frac{9}{4}$, $\frac{1}{2}$

-4

so that $f'\left(\frac{1}{2}\right) =$ ____ and $f'(0)$ does not ______. Thus, we have critical numbers at ____ and ____. Since $f''\left(\frac{1}{2}\right) <$ ____, then f has a relative __________ at $\frac{1}{2}$. By direct calculation, we have $f(-3) =$ ____, $f(3) =$ ____, $f(0) =$ ____, and $f\left(\frac{1}{2}\right) =$ ____, so that f takes on an absolute maximum value $f\left(\frac{1}{2}\right) =$ ____ at ____ on I and f takes on an absolute minimum value $f(3) =$ ____ at 3 on I.

129 $f(x) = \dfrac{1}{x - 2}$ on $I = [1, 4]$.

The graph is

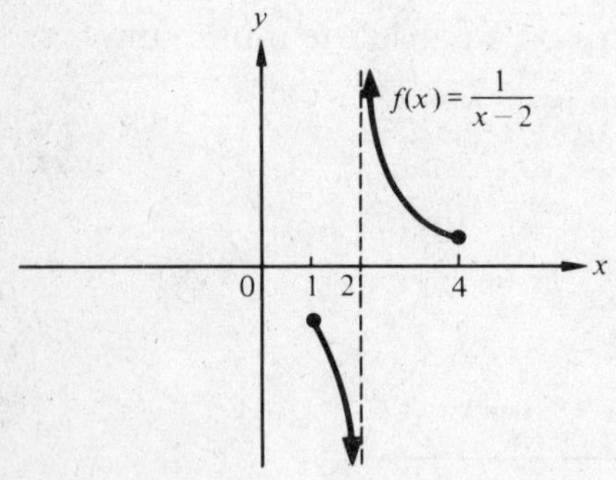

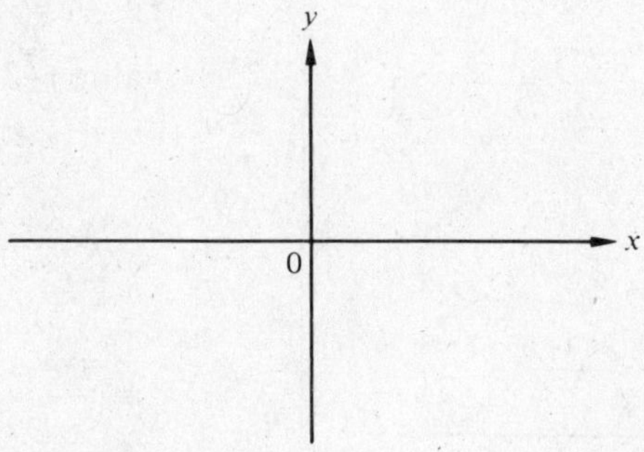

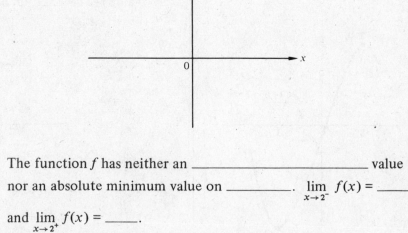

absolute maximum

[1, 4], $-\infty$

$+\infty$

The function f has neither an ______________ value nor an absolute minimum value on ________. $\lim_{x \to 2^-} f(x) =$ ____ and $\lim_{x \to 2^+} f(x) =$ ____.

130 $f(x) = x^2 + 4$ on $I = (-\infty, +\infty)$.

$2x$, 2, increasing

$[0, \infty)$, $(-\infty, 0]$

upward, minimum

0, maximum

$f'(x) =$ ____ and $f''(x) =$ ____, so that f is __________ on ________ and f is decreasing on ________, while the graph of f is concave ________. Thus, f takes on an absolute ________ value at ____ on I, but f has no absolute __________. The

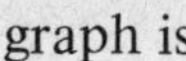

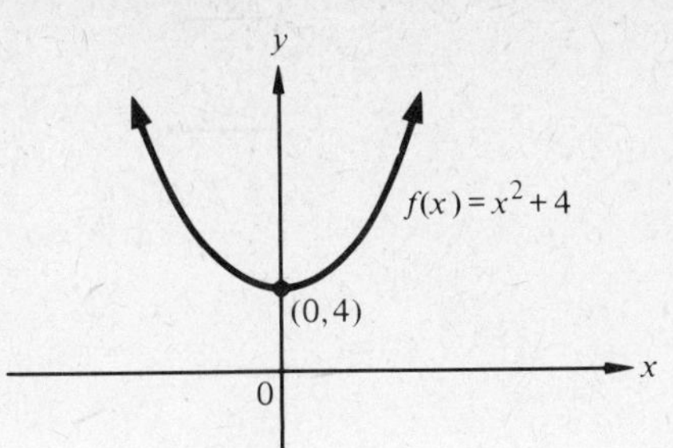

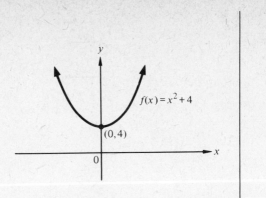

$\frac{1 - x^2}{(x^2+1)^2}$, $\frac{2x^3 - 6x}{(x^2+1)^3}$

0, $\left[-\frac{1}{2}, \frac{3}{2}\right]$

1, 0

$-\frac{1}{2}$, maximum, 1

$-\frac{2}{5}$, $\frac{6}{13}$, $\frac{1}{2}$

maximum, 1

$-\frac{2}{5}$

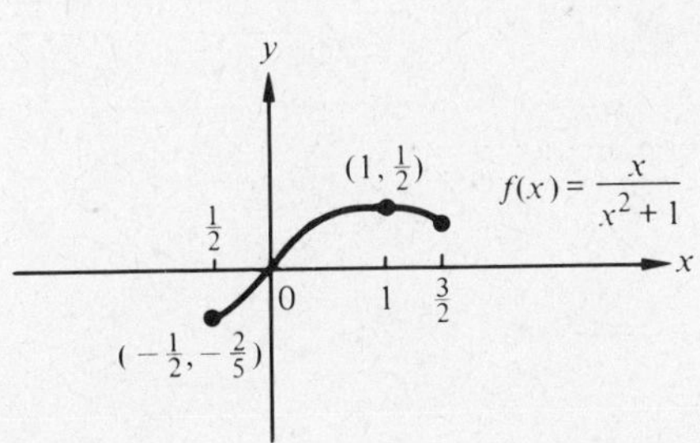

$\frac{2}{3}(x-1)^{-1/3}$, $-\frac{2}{9}(x-1)^{-4/3}$

0

1, 1, 0, $\sqrt[3]{9}$

1

4

graph is

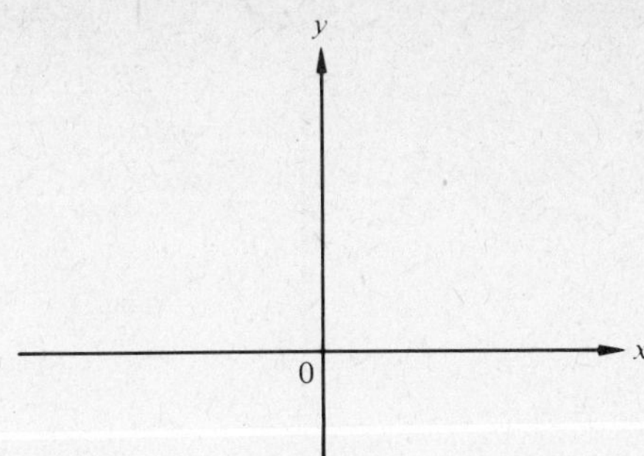

131 $f(x) = \frac{x}{x^2+1}$ on $I = \left[-\frac{1}{2}, \frac{3}{2}\right]$.

$f'(x) =$ ________ and $f''(x) =$ ________. The denominator is never ____ for x in ________. Hence, the only critical number for f in $\left(-\frac{1}{2}, \frac{3}{2}\right)$ is ____. Since $f'(1) =$ ____ and $f''(1) =$ ____, then f has a relative ________ at ____. Since $f\left(-\frac{1}{2}\right) =$ ____, $f\left(\frac{3}{2}\right) =$ ____, and $f(1) =$ ____, then f takes on an absolute ________ $\frac{1}{2}$ at ____ on I, and f takes on an absolute minimum ____ at $-\frac{1}{2}$ on I. The graph is

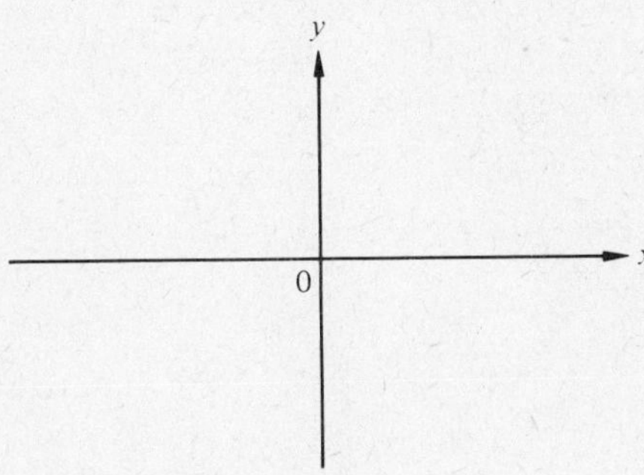

132 $f(x) = (x-1)^{2/3}$ on $I = [0, 4]$.

$f'(x) =$ ________ and $f''(x) =$ ________.

There is no value x for which $f'(x) =$ ____, but f has a critical number at ____. $f(0) =$ ____, $f(1) =$ ____, and $f(4) =$ ____. f takes on an absolute minimum at ____ on I and f takes on an absolute maximum at ____ on I. The graph is

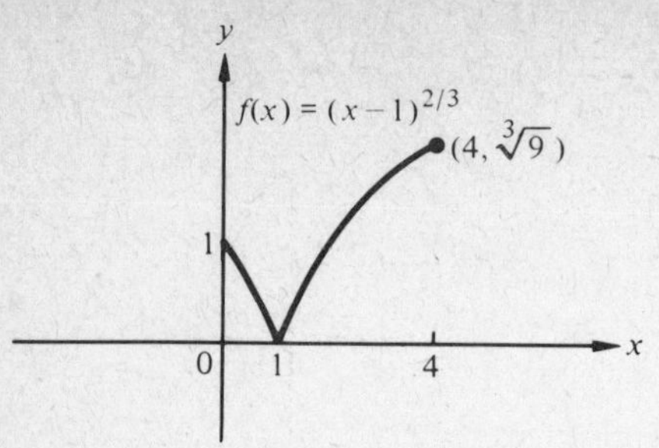

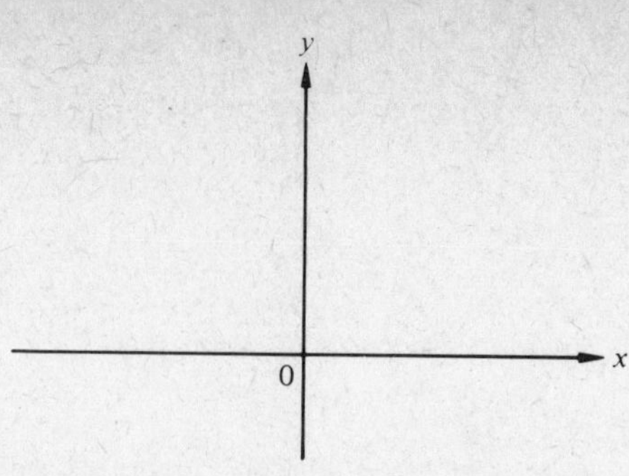

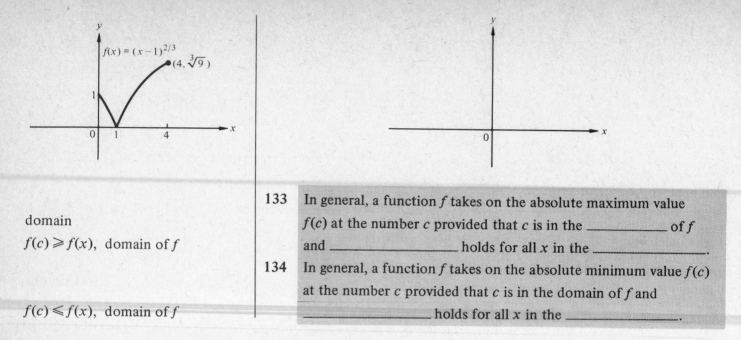

133 In general, a function f takes on the absolute maximum value $f(c)$ at the number c provided that c is in the ______ of f and ______ holds for all x in the ______.

domain

$f(c) \geqslant f(x)$, domain of f

134 In general, a function f takes on the absolute minimum value $f(c)$ at the number c provided that c is in the domain of f and ______ holds for all x in the ______.

$f(c) \leqslant f(x)$, domain of f

In Problems 135 through 140, sketch the graph of f.

135 $f(x) = \dfrac{x+1}{(x-2)^2}$.

$f'(x) =$ ______ and $f''(x) =$ ______. The only critical number of f is ____. $f'(-4) =$ ____ and $f''(-4) =$ ____.

$\dfrac{-x-4}{(x-2)^3}$, $\dfrac{2x+14}{(x-2)^4}$

-4, 0, $\dfrac{3}{648}$

By the second derivative test, f has a relative ______ at ____. The vertical asymptote of f is $x =$ ____ and the horizontal asymptote is $y =$ ____. Notice that $f'(x) > 0$ for ______ and $f'(x) < 0$ for ______ or ______, so that f is ______ on ______ and f is ______ on ______ and on ______. The graph is

minimum

-4, 2

0, $-4 < x < 2$

$x < -4$, $x > 2$, increasing

$[-4, 2)$, decreasing, $(-\infty, -4]$

$(2, \infty)$

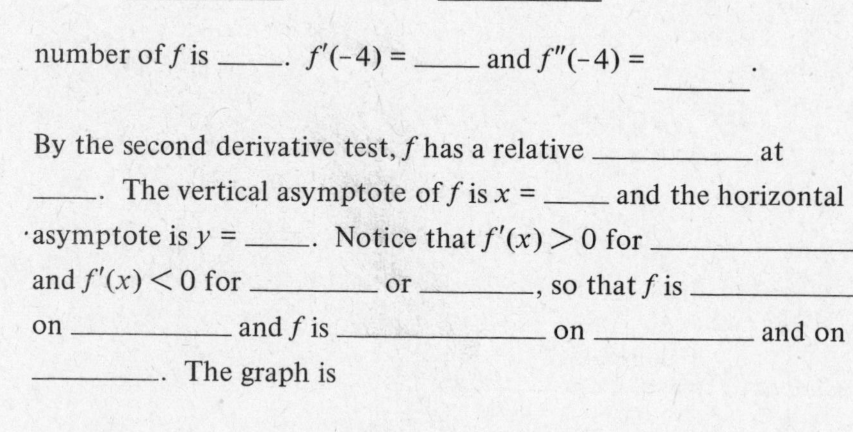

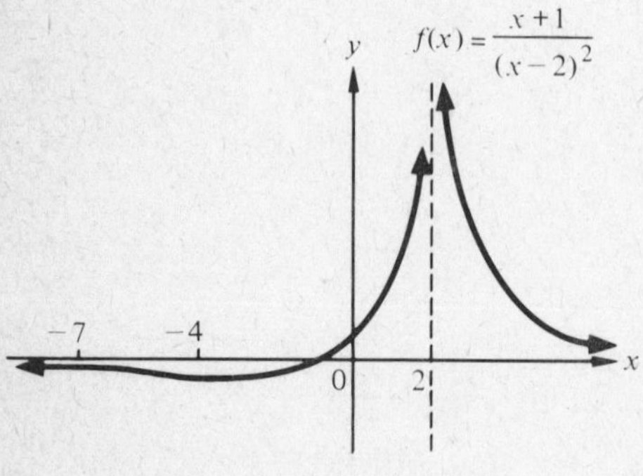

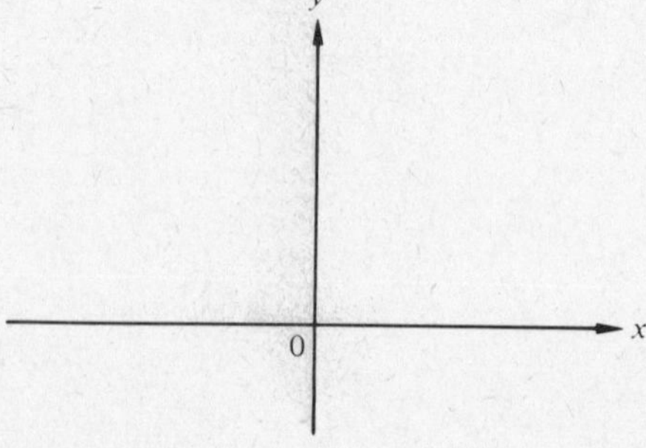

136 $f(x) = \dfrac{1}{x^2+1}$.

$f'(x) =$ ______ and $f''(x) =$ ______, so that

$\dfrac{-2x}{(x^2+1)^2}$, $\dfrac{2(3x^2-1)}{(x^2+1)^3}$

0, −2

maximum, 0

0, horizontal

0

0, decreasing

$[0, \infty)$, increasing, $(-\infty, 0]$

1

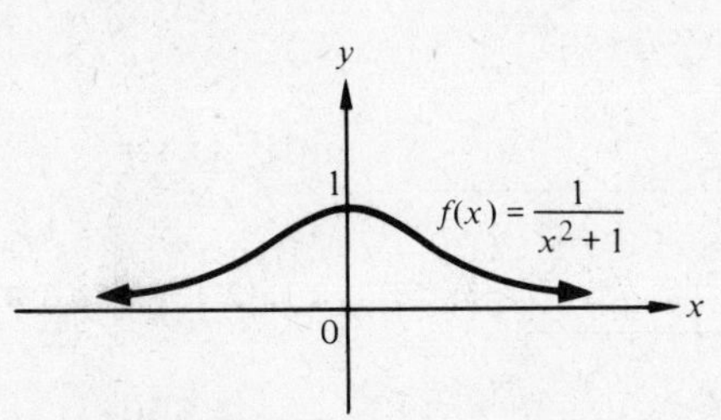

$f'(0) =$ ____ and $f''(0) =$ ____. By the second derivative test, f has a relative ________ at ____. Also, f is an even function and $\lim_{x \to +\infty} \frac{1}{x^2+1} =$ ____, so that $y = 0$ is a __________ asymptote of the graph of f. Notice that $f'(x) < 0$ if $x >$ ____ and $f'(x) > 0$ for $x <$ ____, so that f is __________ on ________ and f is __________ on ________. The graph has an absolute maximum of ____. The graph is

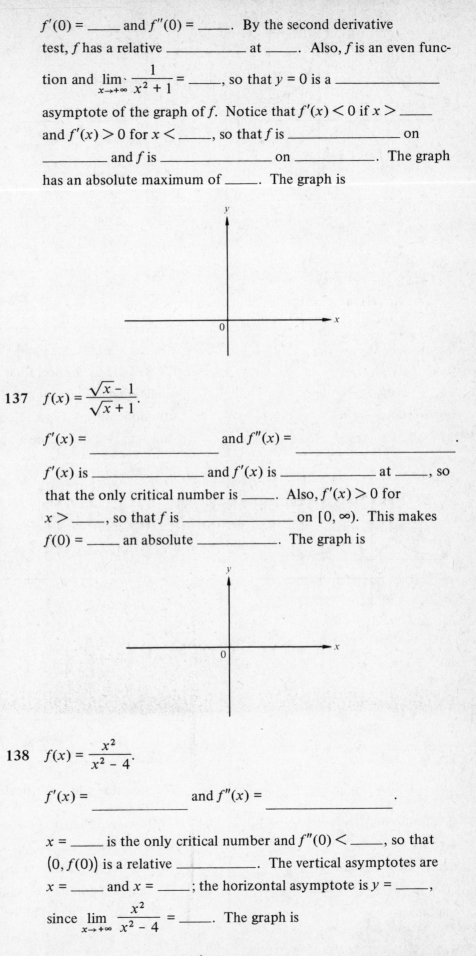

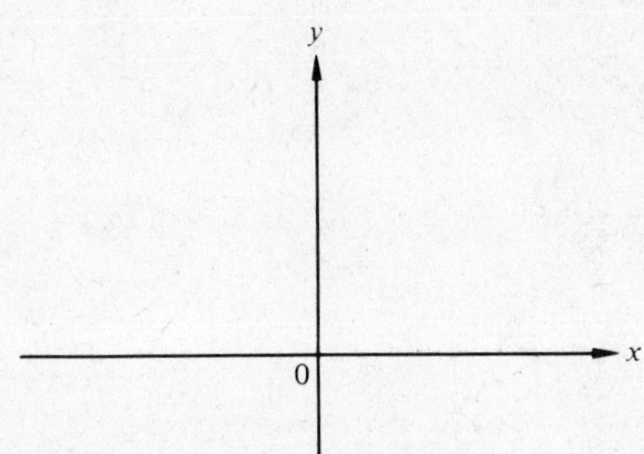

$\frac{1}{\sqrt{x}(\sqrt{x}+1)^2}$, $-\frac{3\sqrt{x}+1}{2x^{3/2}(\sqrt{x}+1)^3}$

never zero, undefined, 0

0

0, increasing

−1, minimum

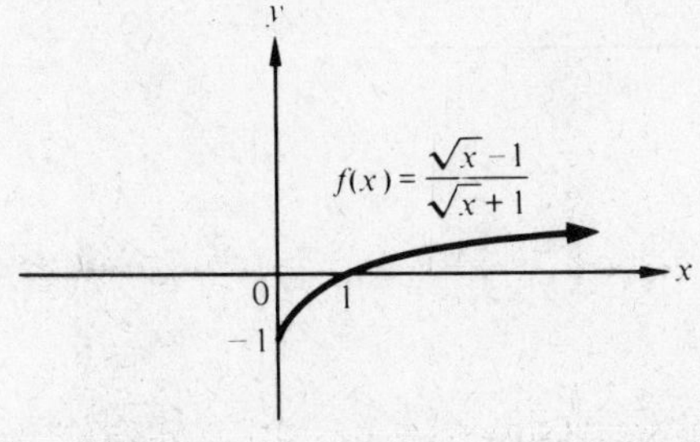

137 $f(x) = \frac{\sqrt{x}-1}{\sqrt{x}+1}$.

$f'(x) =$ ____________ and $f''(x) =$ ____________.

$f'(x)$ is __________ and $f'(x)$ is __________ at ____, so that the only critical number is ____. Also, $f'(x) > 0$ for $x >$ ____, so that f is __________ on $[0, \infty)$. This makes $f(0) =$ ____ an absolute ________. The graph is

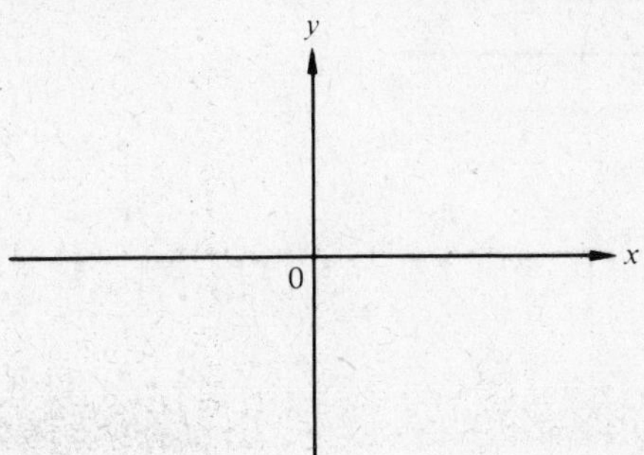

$\frac{-8x}{(x^2-4)^2}$, $\frac{8(3x^2+4)}{(x^2-4)^3}$

0, 0

maximum

−2, 2, 1

1

138 $f(x) = \frac{x^2}{x^2-4}$.

$f'(x) =$ ____________ and $f''(x) =$ ____________.

$x =$ ____ is the only critical number and $f''(0) <$ ____, so that $(0, f(0))$ is a relative ________. The vertical asymptotes are $x =$ ____ and $x =$ ____; the horizontal asymptote is $y =$ ____, since $\lim_{x \to +\infty} \frac{x^2}{x^2-4} =$ ____. The graph is

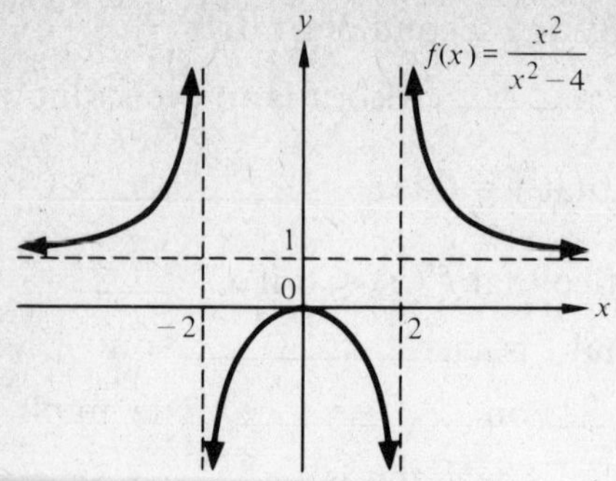

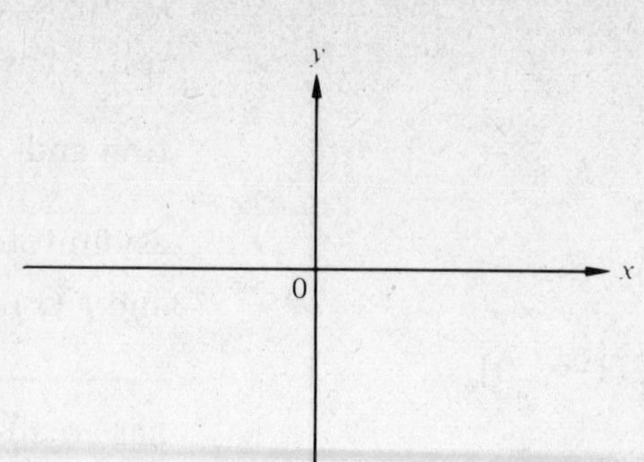

139 $f(x) = \dfrac{x^2 + 4}{\sqrt{x^4 + 4}}$.

$f'(x) =$ __________ $=$ __________ and $f''(x) =$ __________. We have $f'(x) = 0$ when $x =$ __________, and $f''(0) >$ ____, $f''(-1) <$ ____, and $f''(1) <$ ____. Hence, f has a relative minimum at ____ and f has a relative maximum at $x =$ ____ and at $x =$ ____. f is an ______ function and $y = 1$ is a __________ asymptote, since $\lim_{x \to +\infty} f(x) =$ ____. The graph is

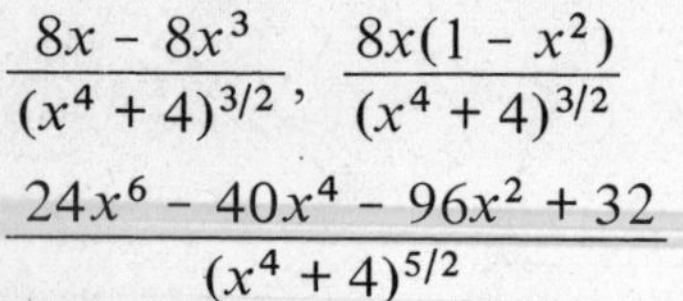

$\dfrac{8x - 8x^3}{(x^4 + 4)^{3/2}}$, $\dfrac{8x(1 - x^2)}{(x^4 + 4)^{3/2}}$

$\dfrac{24x^6 - 40x^4 - 96x^2 + 32}{(x^4 + 4)^{5/2}}$

0, −1 or 1, 0

0, 0

−1, 1, even

horizontal

1

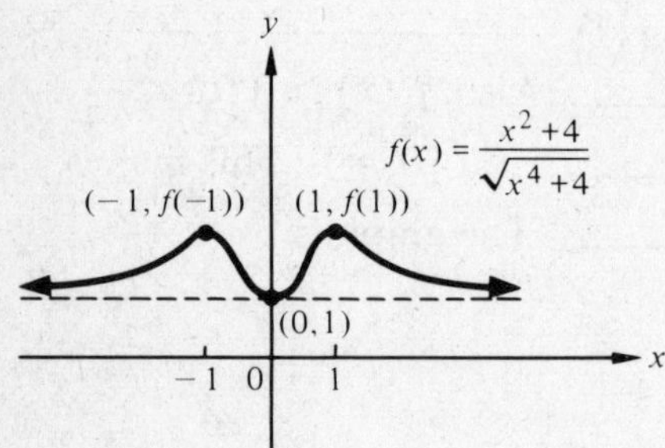

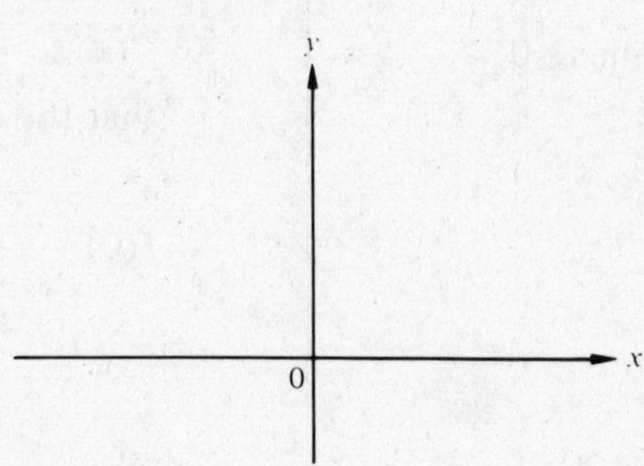

140 $f(x) = x + \dfrac{1}{\sqrt{x - 4}}$.

$f'(x) =$ __________ and $f''(x) =$ __________. The domain of f is the interval ______.

The function f has a critical number $c =$ ______ and $f''(c) >$ ____; hence, f has a relative ______ at c. Since $f'(x) < 0$ for ________ and $f'(x) >$ ____ for $x >$ ____, then f is ________ on $[4, c]$ and ________ on $[c, \infty)$. The graph of f has an absolute minimum at ____, but no absolute ______. The graph is

$1 - \dfrac{1}{2\sqrt{(x - 4)^3}}$

$\dfrac{3}{4\sqrt{(x - 4)^5}}$, $(4, \infty)$

$4 + \dfrac{1}{\sqrt[3]{4}}$

0, minimum

$4 < x < c$, 0, c

decreasing, increasing

c

maximum

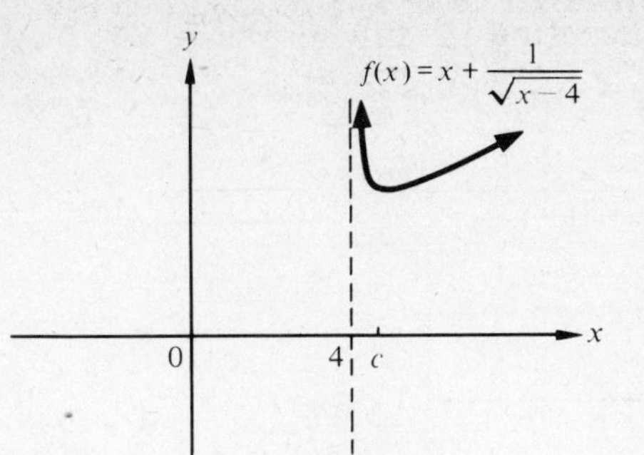

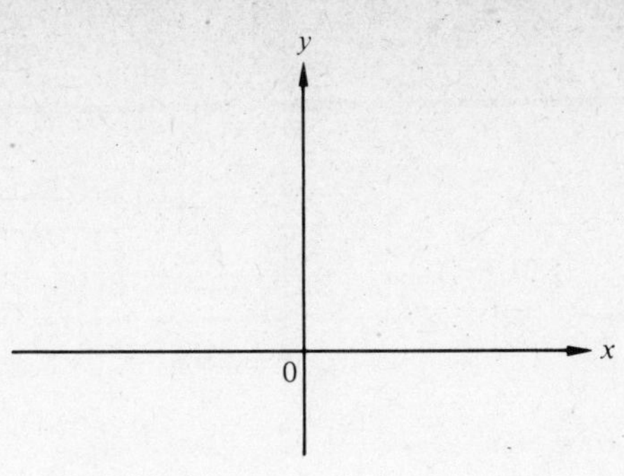

6 Maximum and Minimum—Applications to Geometry

6.1 Geometric Applications

141 Find the dimensions of the rectangle whose perimeter is 40 inches and whose area is a maximum.

Let l be the length of the rectangle and w be its width. Then $2l + 2w =$ ____. If A represents the area, then $A =$ ____, so that

40, lw

$A = (20 - w)w =$ ________. $\frac{dA}{dw} =$ ________ $=$ ____, so

$20w - w^2$, $20 - 2w$, 0

that $w =$ ____ and $l =$ ____. Therefore, the dimensions of the rectangle of maximum area are ____ inches by ____ inches.

10, 10

10, 10

142 A person wishes to fence off a rectangular swimming pool adjacent to one side of a house. The area enclosed is to be 288 square feet and no fence is required on the side next to the house. What should be the dimensions of the fence in order that the amount of fencing required is a minimum?

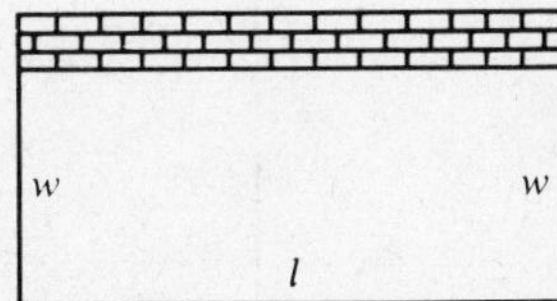

Let l ft and w ft represent the length and the width of the fence, respectively. Then $lw =$ ____. Let P ft represent the amount of

288

fencing so that $P =$ ________. But $l = \frac{}{\text{____}}$, so $P =$

$l + 2w$, $\frac{288}{w}$

$\frac{288}{w} +$ ____. $D_w P = \frac{}{\text{________}} =$ ____, so that $w^2 =$ ____

$2w$, $-\frac{288}{w^2} + 2$, 0, 144

or $w =$ ____. Then $l = \frac{288}{12} =$ ____. Therefore, the length and the

12, 24

width of the fence are ____ ft and ____ ft, respectively.

24, 12

143 A rectangular lot adjacent to a highway is to be enclosed by a fence. If the fencing costs \$2.50 per foot along the highway and

\$1.50 per foot on the other sides, find the dimensions of the largest lot that can be fenced off for \$720.

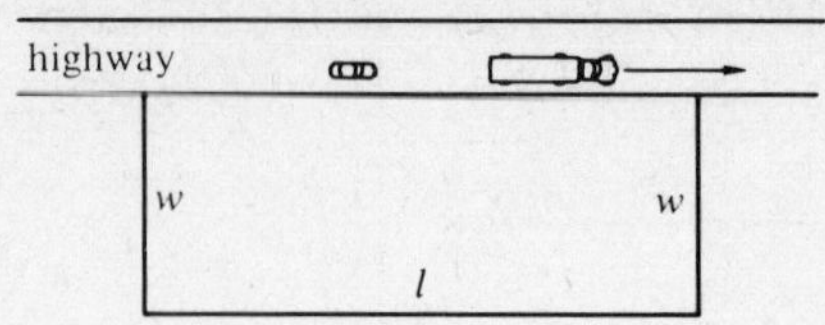

Let l ft and w ft represent the length and the width of the lot, respectively. Let \$$c$ represent the cost of the fence.

$4l + 3w$, $\dfrac{720 - 3w}{4}$

$720 = (2.50)\,l + (1.50)\,(l + 2w) =$ ________, $l =$ ________.

lw, $\dfrac{720w - 3w^2}{4}$

The area $A =$ ____ $=$ ________, so that $D_w A =$

$\dfrac{720 - 6w}{4}$, 0, 120, 90

________ $=$ ____. Then $w =$ ____ and $l =$ ____. The length

90, 120

and width of the lot are ____ ft and ____ ft.

144 A triangular lot is bounded by two streets intersecting at right angles. The lengths of the street frontage for the lot are 80 and 120 feet. Find the dimensions of the largest rectangular building that can be placed on the lot facing one of the streets.

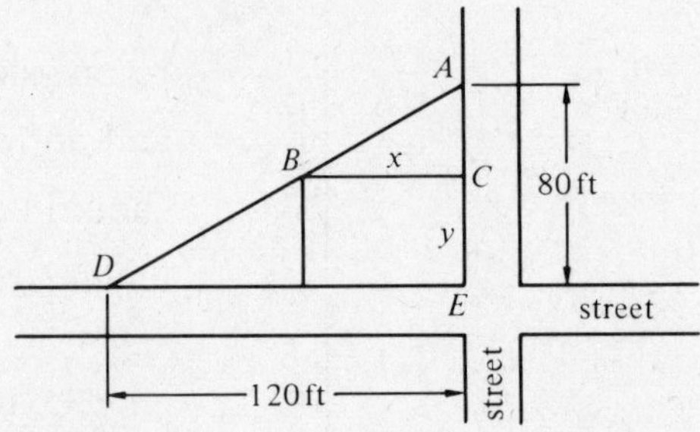

Let x ft and y ft represent the dimensions of the building. The

$80 - y$

triangles ABC and ADE are similar so that $\dfrac{x}{120} = \dfrac{}{80}$, and

$80 - y$, $9600 - 120y$

so $80x = 120($________$) =$ ________. Therefore,

9600, 240

$80x + 120y =$ ____ or $2x + 3y =$ ____. The area A of the build-

xy, $80 - \dfrac{2}{3}x$, $80x - \dfrac{2}{3}x^2$

ing is $A =$ ____, so that $A = x\Big($________$\Big) =$ ________.

$80 - \dfrac{4x}{3}$, 60, 40

Then $D_x A =$ ________ $= 0$, so that $x =$ ____ and $y =$ ____.

The dimensions of the building are 60 ft and 40 ft.

145 An open box is formed from a piece of cardboard 12 inches square by cutting equal squares out of the corners and turning up the sides. Find the volume of the largest box that can be made in this way.

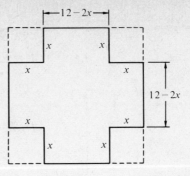

The height of the box will be ____ inches, the width will be ________ inches, and the length will also be ________ inches. Thus, $V =$ ____________ $=$ ____________, so that $D_x V =$ ____________ $=$ ____ and $x =$ ____ or $x =$ ____; we discard $x =$ ____. The maximum volume $V = 2(8)^2 =$ ____________.

x

$12 - 2x,\ 12 - 2x$

$x(12 - 2x)^2,\ 144x - 48x^2 + 4x^3$

$144 - 96x + 12x^2,\ 0,\ 2$

6, 6

128 cubic inches

146 A closed box, whose length is twice its width, is to have a surface of 192 square centimeters. Find the dimensions of the box when the volume is maximum.

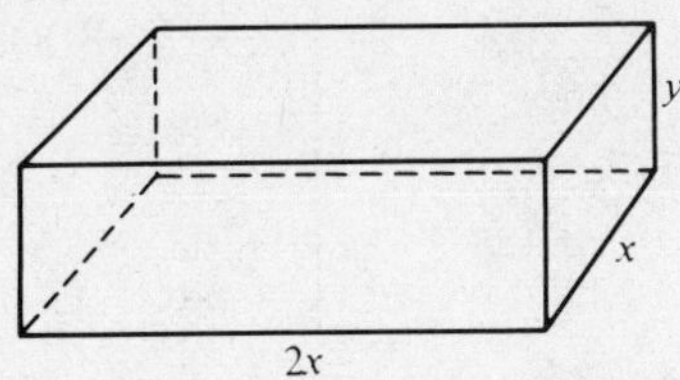

Let x and $2x$ centimeters represent the width and the length of the base and let y centimeters represent the height of the box. $2(x)(2x) + 2($____$) + 2(2x)y =$ ____, so that ____________ $=$ 192. $V = x(2x)(y) =$ ____, so that $V = 2x^2($____________$) =$ ____________. Then $D_x V =$ ____________ $=$ ____, so that ____________

$x =$ ____ or $x =$ ____. We take $x =$ ____, $y = \frac{32}{4} - \frac{2}{3}(4) =$ ____.

Therefore, the dimensions of the box are 4 by 8 by ____ centimeters.

$xy,\ 192,\ 4x^2 + 6xy$

$2x^2y,\ \frac{32}{x} - \frac{2}{3}x$

$64x - \frac{4}{3}x^3,\ 64 - 4x^2,\ 0$

$-4,\ 4,\ 4,\ \frac{16}{3}$

$\frac{16}{3}$

147 What are the dimensions of a closed cylindrical tin can which is to hold 16π cubic centimeters if the amount of tin used is to be a minimum?

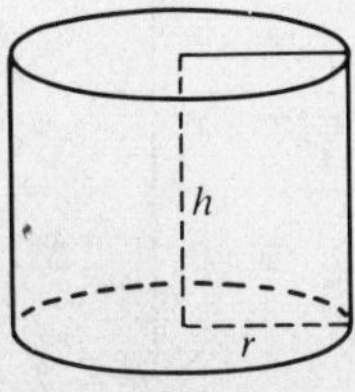

Let r and h centimeters represent the radius and the height of the cylinder respectively, and let V represent the volume. Then

$\pi r^2 h,\ 16\pi$

$V =$ ____ $=$ ____. If S represents the surface area of the

$2\pi r^2 + 2\pi rh,\ \dfrac{32\pi}{r}$

cylinder, then $S =$ ________ $= 2\pi r^2 +$ ____.

$4\pi r - \dfrac{32\pi}{r^2},\ 0,\ 2$

$D_rS =$ ________ $=$ ____, so that $r =$ ____ cm and

4

$h =$ ____ cm.

148 What is the least amount of material needed to make an open right circular cylindrical can of volume 8π cubic inches?

Let r, h, V, and S represent the radius, the height, the volume and

$\pi r^2 h,\ 8\pi,\ \dfrac{8}{r^2}$

the surface area, respectively. $V =$ ____ $=$ ____, so that $h =$ ____

$\pi r^2 + 2\pi rh,\ \pi r^2 + \dfrac{16\pi}{r},\ \dfrac{16\pi}{r^2}$

$S =$ ________ $=$ ________. Then $D_rS = 2\pi r -$ ____ $=$

$0,\ 2,\ 2$

____, so that $r =$ ____ and $h =$ ____. Therefore, the radius and

$2,\ 2$

and the height of the cylinder are ____ inches and ____ inches.

149 Find the dimensions of the right circular cone whose volume is 3000 cubic centimeters if its lateral surface area is a minimum.

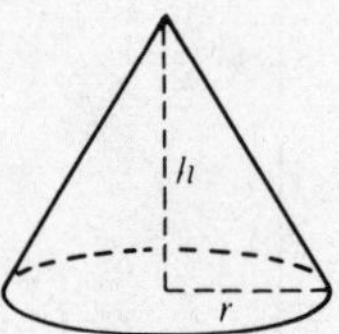

Let r, h, V, and S represent the radius, the height, the volume,

$\dfrac{1}{3}\pi r^2 h,\ 3000$

and the surface area of the cone. $V =$ ____ $=$ ____, so that

$\dfrac{9000}{\pi r^2},\ \pi r\sqrt{r^2 + \left(\dfrac{9000}{\pi r^2}\right)^2}$

$h =$ ____. $S =$ ________ $=$

$4\pi^2 r^3 - \dfrac{162{,}000{,}000}{r^3}$

$\sqrt{\pi^2 r^4 + \dfrac{81{,}000{,}000}{r^2}}$, so that $D_rS = \dfrac{\rule{3em}{0.4pt}}{2\sqrt{\pi^2 r^4 + \dfrac{81{,}000{,}000}{r^2}}}$

$0,\ 0,\ 81{,}000{,}000$

$=$ ____, and so $4\pi^2 r^6 - 162{,}000{,}000 =$ ____ or $r^6 = \dfrac{\rule{3em}{0.4pt}}{2\pi^2}$.

$81{,}000{,}000,\ 10\sqrt[6]{\dfrac{81}{2\pi^2}}$

Then $r = \sqrt[6]{\dfrac{\rule{3em}{0.4pt}}{2\pi^2}} =$ ________ and

$\dfrac{90}{\pi}\sqrt[3]{\dfrac{2\pi^2}{81}}$

$h =$ ________. Therefore, the radius and the height of the cone are $10\sqrt[6]{\dfrac{81}{2\pi^2}}$ cm and $\dfrac{90}{\pi}\sqrt[3]{\dfrac{2\pi^2}{81}}$ cm.

150 A sphere of radius a is inscribed in a right circular cone. Find the dimensions of the cone if its volume is minimum.

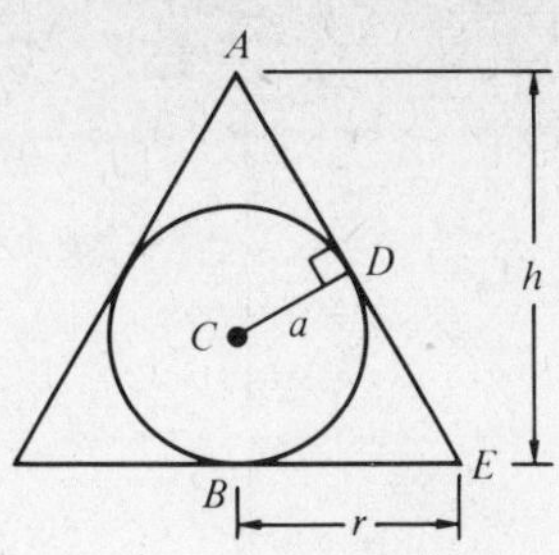

Let r, h, and V represent the radius, the height, and the volume of the cone, so that $V =$ ________ . ΔADC is similar to ΔABE, so

$$\frac{|\overline{AE}|}{|\overline{BE}|} = \frac{|\overline{AC}|}{|\overline{DC}|} \text{ or } \frac{\sqrt{h^2 + r^2}}{r} = \frac{\qquad}{a} \text{ and } r^2 = \underline{\qquad}.$$

Then $V =$ ________ . $D_h V =$ ____________ = ____, so that $h =$ ____ and $r =$ ____.

$\frac{1}{3}\pi r^2 h$

$h - a, \quad \frac{a^2 h}{h - 2a}$

$\frac{\pi a^2 h^2}{3(h - 2a)}, \quad \frac{a^2 \pi h}{3} \cdot \frac{h - 4a}{(h - 2a)^2}$

$0, \ 4a, \ a\sqrt{2}$

6.2 General Procedure for Solving Maximum-Minimum Problems

151 In order to work applied maximum–minimum problems, carry out the following procedure:

(1) Select a suitable symbol, say Q, for the quantity to be ________ or ________. Determine the remaining quantities upon which Q depends and select suitable symbols for these variables. Draw a ________ (if possible) and label its parts.

(2) Express Q in terms of a ________ involving the ________ upon which it depends. Find relationships among these variables and eliminate all but one of them from the formula.

(3) We now have $Q = f(x)$, where x denotes the single variable on which Q was found to depend in Step 2 and f is the ________ determined by this dependence. If there are constraints on the quantity x imposed by the physical nature of the problem, then specify these constraints explicitly.

(4) Find the ____________ of $f(x)$ subject to the imposed constraints on x.

maximized, minimized

diagram

formula, variables

function

absolute extrema

152 Find two numbers whose sum is 10 and the sum of the squares is minimum. Let x and y represent the ________. Then $x + y =$ ____ and $s = x^2 + y^2$, so that $s = x^2 +$ ________ = ____________. Then $D_x s =$ ________ = ____, so that $x =$ ____ gives a ________. Therefore, the numbers are ____ and ____.

numbers

$10, \ (10 - x)^2$

$2x^2 - 20x + 100, \ 4x - 20, \ 0$

5, minimum

5, 5

153 Find two positive numbers whose product is 64 and whose sum is a minimum.

Let x and y represent the numbers. Let s represent the sum. Then

$s =$ ______. Also, $xy =$ ____, so that $y =$ ____. Therefore,

$x + y,\ 64,\ \frac{64}{x}$

$s =$ ______. $D_x s =$ ______ $=$ ____, so that $x =$ ____; then

$x + \frac{64}{x},\ 1 - \frac{64}{x^2},\ 0,\ 8$

$y =$ ____. The two numbers are ____ and ____.

8, 8, 8

154 Divide the number 60 into two parts such that the product p of one part and the square of the other is a maximum.

Let x be the first part, and let y be the second part, so that

$x + y =$ ____ and $p = x^2$ (____). $y =$ ______, so that

60, y, $60 - x$

$p = x^2$ (______) $=$ ______. $\frac{dp}{dx} =$ ______ $= 0$,

$60 - x,\ 60x^2 - x^3,\ 120x - 3x^2$

so that $x =$ ____ or $x =$ ____. $\frac{d^2p}{dx^2} = 120 -$ ____, so that the

0, 40, $6x$

maximum product results from $x =$ ____ and $y =$ ____.

40, 20

7 Maximum and Minimum—Applications to Physics, Engineering, Business, and Economics

7.1 Applications to Physics and Engineering

155 Find the altitude of the cylinder of maximum volume that can be inscribed in a right circular cone of radius r and height h.

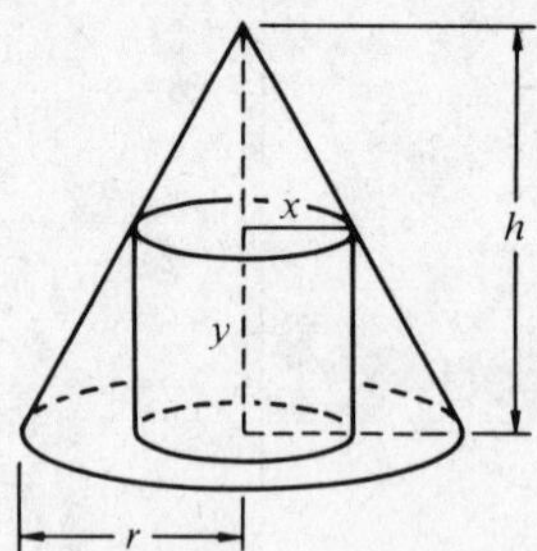

Let x and y represent the radius and the height of the cylinder, respectively. Then its volume $V =$ ______. To express the volume in terms of a single variable, we observe from the figure that

$\pi x^2 y$

$\frac{x}{r} = \frac{\text{______}}{h}$, so that $y =$ ______. Thus,

$h - y,\ \frac{h}{r}(r - x)$

$V =$ ______ and $\frac{dV}{dx} =$ ______ $=$ ____.

$\frac{\pi h}{r}(rx^2 - x^3),\ \frac{\pi h}{r}x(2r - 3x),\ 0$

Therefore, $x =$ ____ and $y =$ ____.

$\frac{2}{3}r,\ \frac{1}{3}h$

156 The height s of a ball t seconds after it is thrown straight up is expressed by the equation $s = -16t^2 + 40t$. When does it reach its maximum height?

$-32t + 40,\ 0,\ \frac{5}{4}$ $D_t s =$ ________ $=$ ____ and $t = \overline{}$. The ball reaches a

1.25 maximum after ____ seconds.

157 A man in a rowboat 6 miles from shore desires to reach a point on the shore at a distance of 10 miles from his present position. If he can walk 4 miles per hour and row 2 miles per hour, where should he land in order to reach his destination in the shortest possible time?

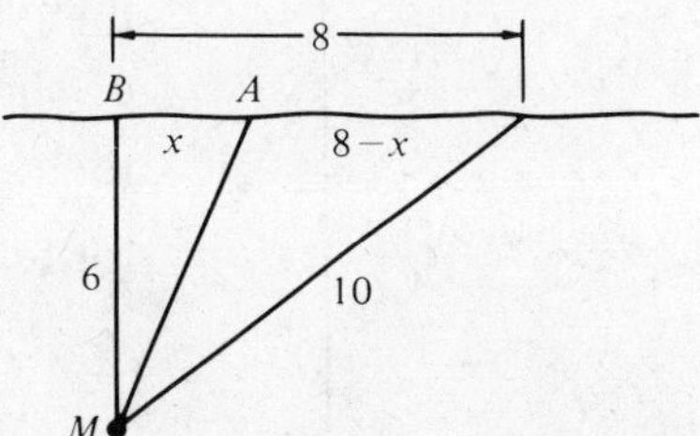

Suppose that the man lands at point A. Let T be the time in hours

$8 - x$ required for his journey. Then $T = \frac{\sqrt{x^2 + 36}}{2} + \frac{\rule{2em}{0.4pt}}{4}$, so that

$-\frac{1}{4},\ 0$ $\frac{dT}{dx} = \frac{\frac{1}{2}(x^2 + 36)^{-1/2}}{2}(2x) + \left(\rule{2em}{0.4pt}\right) =$ ____ and

$\frac{1}{4},\ \frac{1}{2}$ $\frac{x}{2\sqrt{x^2 + 36}} - \rule{2em}{0.4pt} = 0$ or $\frac{x}{\sqrt{x^2 + 36}} - \rule{2em}{0.4pt} = 0$. Then $2x =$

$\sqrt{x^2 + 36},\ x^2 + 36,\ 36$ ________ or $4x^2 =$ ________. Thus, $3x^2 =$ ____ or

$12,\ 2\sqrt{3}$ $x^2 =$ ____, so that $x =$ ____. He should land approximately 3.46 miles from point B down the shore.

158 A military courier is located on a desert 6 miles from a point P, which is the point on a long, straight road nearest to him. He is ordered to get to a point Q on the road. Assuming that he travels 14 miles per hour on the desert and 50 miles per hour on the road, find the point R where he should reach the road in order to get to Q in the least possible time when Q is 3 miles from P.

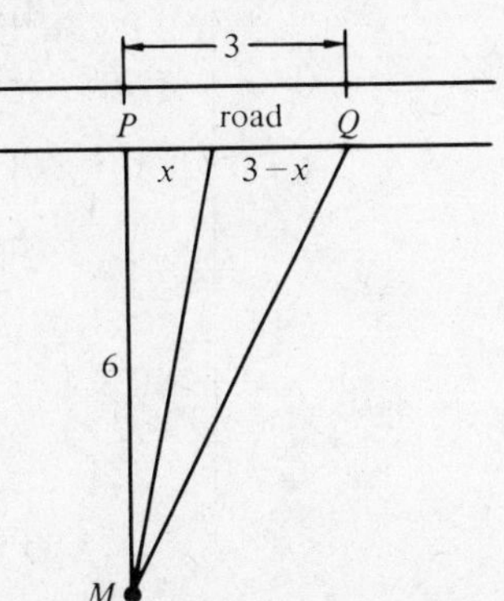

Let R be the point on the road and put $|\overline{PR}| = x$, so that $|\overline{RQ}| =$

$3 - x,\ x^2 + 36$

________ and $|\overline{MR}| = \sqrt{\_\_\_\_\_\_\_\_}$. Let T hours be the time

$3 - x$

it takes to get to Q, so that $T = \dfrac{\sqrt{x^2+36}}{14} + \dfrac{\_\_\_\_}{50}$.

$\dfrac{x}{14\sqrt{x^2+36}},\ 0$

$\dfrac{dT}{dx} =$ ________ $- \dfrac{1}{50} =$ ____, so that

$\dfrac{x^2}{14^2(x^2+36)},\ \dfrac{49 \times 36}{576},\ 1.75$

________ $= \dfrac{1}{50^2}$, $x^2 =$ ________, or $x =$ ______. He

1.75

should reach the point R that is ______ miles from P.

159 A ship A is docked 6 miles from shore, and opposite a point 12 miles farther along the shore another ship B is docked 18 miles offshore. A boat from the first ship is to land a passenger and then proceed to the other ship. What is the length of the shortest route that the boat can take?

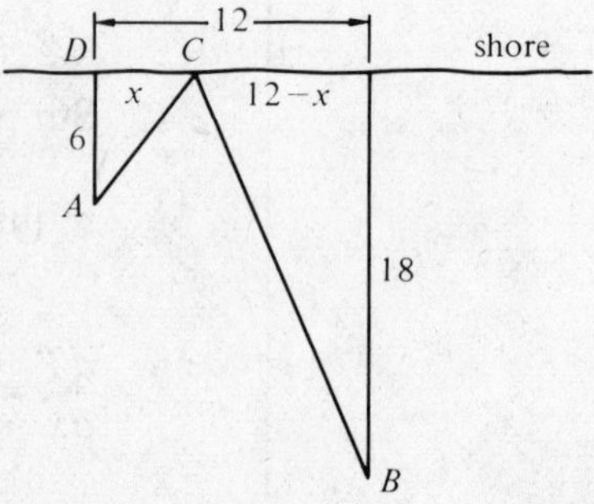

Let the passenger land at a point C which is x miles from D. Then the total distance from A to B is given by $y = |\overline{AC}| + |\overline{CB}| =$

324

$\sqrt{x^2+36} + \sqrt{(12-x)^2 + \_\_\_\_\_\_}$, so that

$2x$

$\dfrac{dy}{dx} = \dfrac{1}{2}(x^2+36)^{-1/2}\ (\_\_\_\_\_\_) + \dfrac{1}{2}[(12-x)^2 + 324]^{-1/2}(-2)(12-x)$

$12 - x$

$= \dfrac{x}{\sqrt{x^2+36}} - \dfrac{\_\_\_\_\_\_}{\sqrt{(12-x)^2+324}} = 0,$

$x^2 + 3x - 18 = 0$

and so $288x^2 + 864x - 5184 = 0$ or ________________;

$3,\ 12\sqrt{5},\ 26.8$

hence, $x =$ _____. Then $y =$ ______ $\approx$ ______, so that the shortest

26.8

distance the boat can take will be approximately ______ miles.

160 A truck is to be driven 300 miles on a freeway at a constant speed of x miles per hour. Speed laws require $30 \leqslant x \leqslant 60$. Assume that the fuel costs 60 cents per gallon and is consumed at the rate of $2 + \dfrac{x^2}{600}$ gallons per hour. If the driver's wages are 8 dollars per hour and if he obeys all speed laws, find the most economical speed.

$\frac{300}{x}$

$\frac{360}{x} + (0.3)x$

$\frac{2400}{x}$

$\frac{2760}{x} + (0.3)x$

$-\frac{2760}{x^2} + 0.3$

$\sqrt{9200}$, 96

[30, 60]

101, 64

60

$64t^2 + (100 - 5t)^2$

$128t - 10(100 - 5t)$, 0, $\frac{500}{89}$, 5.6

5171.06

7191.01

84.8

84.8

The time required for the journey is $T =$ ____ hours, the fuel cost is $(0.60)\left(2 + \frac{x^2}{600}\right)\left(\frac{300}{x}\right) =$ ____________ dollars, the driver's total wages are $\$8T = \$$ ____, and the total cost for the journey is $C = \frac{360}{x} + (0.3)x + \frac{2400}{x} =$ ____________ dollars. Setting $\frac{dC}{dx} =$ ____________ $= 0$, we obtain the critical number $x =$ ________ $\approx$ ____ miles per hour. Since this critical number is outside the interval ____________, we reject it. When $x = 30$, then $C = \$$____, and when $x = 60$, then $C = \$$____; hence, for minimum cost, $x =$ ____ mph.

161 As a man starts across a 200-foot bridge, a ship passes directly beneath the center of the bridge. If the ship is moving at the rate of 8 ft/sec and the man at the rate of 5 ft/sec, what is the shortest horizontal distance between them?

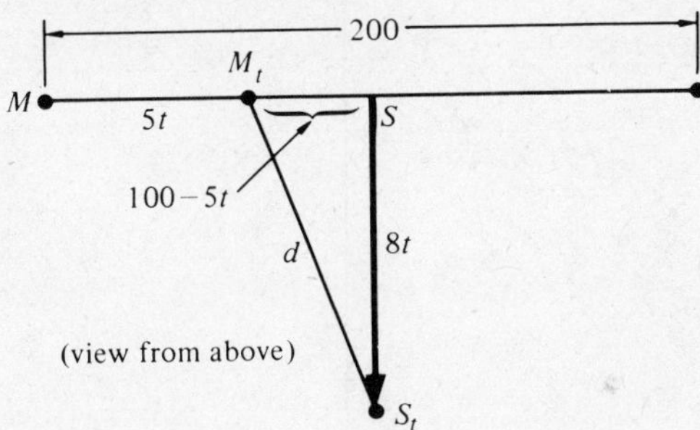

Let M and S be the initial locations of the man and ship, respectively, and let M_t and S_t be their positions at t seconds. The distance d is given by $d = \sqrt{(8t)^2 + (100 - 5t)^2}$, so that $y = d^2 =$ ____________. Then $\frac{dy}{dt} =$ ____________ $=$ ____ so that $t =$ ____ $\approx$ ____ sec. Then

$$d = \sqrt{\left[8\left(\frac{500}{89}\right)\right]^2 + \left[100 - 5\left(\frac{500}{89}\right)\right]^2}$$

$$\approx \sqrt{2019.95 + \_\_\_\_\_\_}$$

$$= \sqrt{\_\_\_\_\_\_}$$

$$\approx \_\_\_\_\_\_.$$

The minimum horizontal distance between them is approximately ____ feet.

162 The electric potential at a point (x, y) on the line extending from $(0, 3)$ to $(2, 0)$ is given by $P = 3x^2 + 2y^2$. At what point on this segment is the potential a maximum?

The equation of the line segment from $(0, 3)$ to $(2, 0)$ is ____________, so that $y =$ ________. Then

$P =$ ____________ $=$ ____________

for $0 \leqslant x \leqslant 2$. $\dfrac{dP}{dx} =$ ____________ $=$ ______, so that $x =$ ______

and $y =$ ______. Note that $\dfrac{d^2P}{dx^2} =$ ______ $>$ ______. Hence, the point ________ gives a minimum potential.

$3x + 2y = 6,\ \dfrac{6 - 3x}{2}$

$3x^2 + 2\left(\dfrac{6 - 3x}{2}\right)^2,\ \dfrac{3}{2}(5x^2 - 12x + 12)$

$3(5x - 6),\ 0,\ \dfrac{6}{5}$

$\dfrac{6}{5},\ 15,\ 0$

$\left(\dfrac{6}{5}, \dfrac{6}{5}\right)$

163 The strength s of a rectangular beam is given by the equation $s = xy^2$, where x and y in centimeters are the width and the depth, respectively. Find the dimensions of the strongest beam that can be cut from a circular log of diameter 15 centimeters.

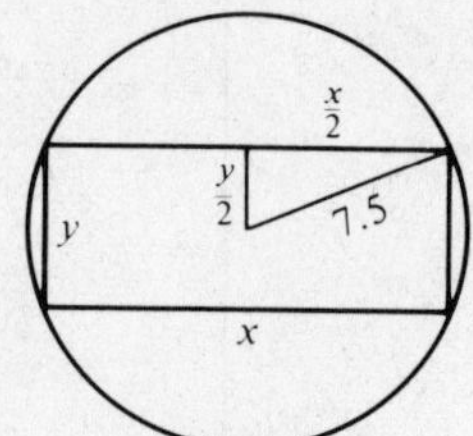

$s = xy^2$, but $\left(\dfrac{x}{2}\right)^2 + \left(\dfrac{y}{2}\right)^2 =$ ________, so that $x^2 + y^2 =$ ______.

Then $s = x($____________$) =$ ____________, so that $\dfrac{ds}{dx} =$

____________ $=$ ______. $x^2 =$ ______ or $x \approx$ ______, and

$y = \sqrt{225 - \text{\_\_\_\_}} = \sqrt{\text{\_\_\_\_}} \approx$ ________, so that the dimensions of the beam are approximately 8.66 cm and 12.25 cm.

$\left(\dfrac{15}{2}\right)^2,\ 225$

$225 - x^2,\ 225x - x^3$

$225 - 3x^2,\ 0,\ 75,\ 8.66$

$75,\ 150,\ 12.25$

164 A man wishes to heat his basement by plugging a long extension chord connected to a number of heaters into a 120-volt socket in the house. The power P in watts radiated by the 30-ohm heaters is given by $P = \dfrac{432{,}000n}{(30 + 9n)^2}$, where n is the number of heaters plugged in. What would be the maximum amount of heat?

$$\frac{dP}{dn} = \frac{(30 + 9n)^2(\text{\_\_\_\_\_\_\_\_}) - (432{,}000n)(2)(\text{\_\_\_\_\_\_\_\_})(9)}{(30 + 9n)^4}.$$

Setting $\dfrac{dP}{dn} =$ ______, we get ____________________ $= 0$.

$432{,}000,\ 30 + 9n$

$0,\ 432{,}000\,(30 - 9n)$

3

3, 400 watts

Then $n = \frac{10}{3}$ or $n =$ ____ (to the nearest whole number). Therefore, $P = \frac{(432{,}000)(\_\_\_\_)}{(30 + 27)^2} \approx$ ________.

120 - 10*I*,

0, 12

165 The power P generated by a 120-volt generator of internal resistance 5 ohms is given by $P = 120I - 5I^2$ watts, where I is the current in amperes. For what current does the generator produce maximum power? $P = 120I - 5I^2$, so that $\frac{dP}{dI} =$ ________ = ____. Then $I =$ ____. The generator produces maximum power when $I = 12$ amperes.

7.2 Applications to Business and Economics

$C'(x)$

the marginal cost function

166 If C is a cost function, so that $C(x)$ is the cost of producing x units of a certain commodity, then ____ is called the marginal cost of producing x units and C' is called ________.

$R'(x)$

167 If $R(x)$ denotes the total revenue obtained when x units of a commodity are demanded, then ____ is called the marginal revenue.

$R(x + 1) - R(x)$

168 If x is large, then the marginal revenue $R'(x)$ would be a good approximation to the additional revenue ________ generated by one additional unit of demand.

$6x^2 - 18x + 12$

12

169 A manufacturer sells a certain commodity. The total cost of producing a quantity x of the commodity is given by the equation $C(x) = 2x^3 - 9x^2 + 12x$. Find the marginal cost when $x = 3$. $C(x) = 2x^3 - 9x^2 + 12x$, so that $C'(x) =$ ________. Then $C'(3) =$ ____.

$x - 4$

$-80x + 640$

0, 8

\$8

170 A retailer of electric alarm clocks pays \$4 for each clock. If he charges x dollars per clock, he will be able to sell $480 - 40x$ clocks. How much should he charge per clock so as to maximize his total profit, P? $P(x) = (480 - 40x)(\_\_\_\_) = -40x^2 + 640x - 1920$, so that $P'(x) =$ ________. To maximize the total profit, we set $P'(x) =$ ____, so that $x =$ ____. Therefore, his profit will be maximum if he charges ____ per clock.

171 A manufacturer of a certain product knows that at x dollars per ton he can sell $(240 - x)$ tons and his production cost will be $\left(\frac{12{,}000}{240 - x} + 30\right)$ dollars per ton. How much should he charge per ton so as to maximize his total profit P? At \$$x$ per ton, the manufacturer's cost of production will be given by $\left(\frac{12{,}000}{240 - x} + 30\right)$ (________) = ________. Also, his total revenue will be ________; so $P(x)$ = revenue − cost = (________) − (________) = ________. Therefore, $P'(x)$ = ________ = 0, so that x = ____ dollars per ton.

$240 - x$, $8400 - 30x$

$(240 - x)x$

$240x - x^2$, $8400 - 30x$

$-x^2 + 270x - 8400$

$-2x + 270$, 135

172 A firm that manufactures a certain commodity estimates that the total cost in dollars for making x items is given by the equation $C(x) = \frac{x^3}{3} - 5x^2 + 16x + 150$. Determine the minimum cost.

$C(x) = \frac{x^3}{3} - 5x^2 + 16x + 150$, so that $C'(x)$ = ________.

To minimize the cost, we set $C'(x)$ = ____, so that $x^2 - 10x + 16$ = ____. Then x = ____ or x = ____. $C''(x)$ = ________ and $C''(8) >$ ____. The minimum cost will be $\frac{\quad}{\quad} \approx$ ____ dollars.

$x^2 - 10x + 16$

0

0, 2, 8, $2x - 10$

0, $\frac{386}{3}$, 128.67

173 A company estimates that if x is the number of units of some commodity produced, then the total cost $C(x)$ and the total revenue $R(x)$ will be given by the equations $C(x) = \frac{x^3}{3} - 4x^2 + 10x$ and $R(x) = 10x - 2x^2$. If $P(x)$ is the profit, find the number of units produced that yields the maximum profit.

$P(x)$ = ________, so that $P(x)$ = $(10x - 2x^2) - \left(\_\_\_\_\_\_\_\_\right) = 2x^2 - \frac{x^3}{3}$. Then $P'(x)$ = ________. To maximize the profit, set $P'(x)$ = ____, so that $4x - x^2$ = ____; then x = ____ or x = ____. $P''(x) = 4 - 2x$, so that $P''(0) >$ ____ and $P''(4) <$ ____. Then x = ____ yields the maximum profit.

$R(x) - C(x)$

$\frac{x^3}{3} - 4x^2 + 10x$

$4x - x^2$, 0

0, 0, 4

0, 0, 4

174 A toy manufacturer knows that the total cost C in dollars of making x thousands of toys is given by the equation $C(x) = 600 + 3x$ and that the corresponding sales revenue R in dollars is given by $R(x) = 4x - 0.0002x^2$. Find the number of toys (in thousands) that will maximize the manufacturer's profit P.

$P(x)$ = ________, so that $P(x)$ = ________. Thus, $P'(x)$ = ________.

To maximize her profit, set $P'(x)$ = ____, so that x = ____. Thus, the manufacturer should sell 2500 thousands toys to maximize her profit.

$R(x) - C(x)$

$x - 0.0002x^2 - 600$, $1 - 0.0004x$

0, 2500

175 An airline estimates that at \$30 per person all 400 seats in a 747 jet will be taken, but for each increase of \$1 per person, 10 seats

will be vacant. What price should the airline charge to maximize its income?

Let \$$x$ be the increase in the price, so that the price per person is \$________. At this price ____________ will go. Hence, the total income y from this trip is given by

$y =$ ____________________

$=$ ____________________,

so that $\frac{dy}{dx} =$ __________ $=$ ____ when $x =$ ____. Hence, an additional charge of \$5 per person will give the maximum income, and the price should be \$35 per person.

$30 + x$, $400 - 10x$

$(30 + x)(400 - 10x)$

$12{,}000 + 100x - 10x^2$

$100 - 20x$, 0, 5

8 Implicit Functions and Implicit Differentiation

8.1 Implicit Differentiation

176 Given an equation that determines y implicitly as a differentiable function of x, calculate $\frac{dy}{dx}$ as follows:

(1) Differentiate both sides of the equation with respect to ____ to obtain an equation involving x, y, and $\frac{dy}{dx}$. When necessary use the __________.

(2) Solve the resulting equation for ____ in terms of x and y.

The procedure is called ____________________.

x

chain rule

$\frac{dy}{dx}$

implicit differentiation

In Problems 177 through 187, find $\frac{dy}{dx}$ using implicit differentiation.

177 $4x^2 - 3y^2 = 12$.

$8x - 6y\left(\_\_\_\_\right) =$ ____, so that $\frac{dy}{dx} = \frac{\_\_\_}{6y} = \frac{\_\_\_}{3y}$.

$\frac{dy}{dx}$, 0, $8x$, $4x$

178 $3xy^2 + 4x^2y = 7$.

$3x($______$) + 3y^2 + 4x^2 D_x y +$ _____ $= 0$, so that

$D_x y($__________$) = -3y^2 - 8xy$ or

$D_x y = \frac{-(3y^2 + 8xy)}{\_\_\_\_\_\_}$.

$2yD_x y$, $8xy$

$6xy + 4x^2$

$6xy + 4x^2$

179 $x^3 + y^3 - 4xy = 0$.

$3x^2 +$ ______ $- 4x\frac{dy}{dx} - 4($____$) = 0$, so that $(3y^2 - 4x)\frac{dy}{dx} =$ __________ or $\frac{dy}{dx} = \frac{\_\_\_\_\_\_}{3y^2 - 4x}$.

$3y^2\frac{dy}{dx}$, y

$-3x^2 + 4y$, $3x^2 + 4y$

180 $x^{1/2} + y^{1/2} = 1$.

$\frac{1}{2}x^{-1/2} + \frac{1}{2}y^{-1/2}\left(\rule{1.5cm}{0.4pt}\right) = \rule{1cm}{0.4pt}$, so that $\frac{dy}{dx} = \rule{1.5cm}{0.4pt}$.

$\frac{dy}{dx}$, 0, $-\sqrt{\frac{y}{x}}$

181 $xy = (x - y)^2$.

$x\frac{dy}{dx} + y = 2(x - y)\left(\rule{2cm}{0.4pt}\right)$, so that $x\frac{dy}{dx} + y =$

$1 - \frac{dy}{dx}$

$2\left(x - x\frac{dy}{dx} - y + \rule{1.5cm}{0.4pt}\right)$, or $\frac{dy}{dx}(\rule{3cm}{0.4pt}) =$

$y\frac{dy}{dx}$, $x + 2x - 2y$

$2x - y - 2y$. Thus, $\frac{dy}{dx}(\rule{2.5cm}{0.4pt}) = 2x - 3y$. Therefore,

$3x - 2y$

$\frac{dy}{dx} = \rule{2.5cm}{0.4pt}$.

$\frac{2x - 3y}{3x - 2y}$

182 $(x + y)^{1/2} + (x - y)^{1/2} = 1$.

$\frac{1}{2}(x + y)^{-1/2}\left(\rule{2.5cm}{0.4pt}\right) + \frac{1}{2}(x - y)^{-1/2}\left(\rule{2.5cm}{0.4pt}\right) = 0$,

$1 + \frac{dy}{dx}$, $1 - \frac{dy}{dx}$

so that $\frac{1 + \frac{dy}{dx}}{\sqrt{x + y}} + \frac{\rule{2cm}{0.4pt}}{\sqrt{x - y}} = \rule{1cm}{0.4pt}$ or $\frac{1 + \frac{dy}{dx}}{\sqrt{x + y}} =$

$1 - \frac{dy}{dx}$, 0

$\frac{\rule{2cm}{0.4pt}}{\sqrt{x - y}}$, so that $\left(\frac{dy}{dx} - 1\right)\sqrt{x + y} = \left(\frac{dy}{dx} + 1\right)\sqrt{\rule{2cm}{0.4pt}}$. Thus,

$\frac{dy}{dx} - 1$, $x - y$

$\frac{dy}{dx}[\sqrt{x + y} - \sqrt{x - y}] = \sqrt{x - y} + \rule{2.5cm}{0.4pt}$ or

$\sqrt{x + y}$

$\frac{dy}{dx} = \frac{\rule{5cm}{0.4pt}}{\sqrt{x + y} - \sqrt{x - y}}$.

$\sqrt{x + y} + \sqrt{x - y}$

183 $x^{2/3} + y^{2/3} = 1$.

$\frac{2}{3}x^{-1/3} + \frac{2}{3}y^{-1/3}\left(\rule{1.5cm}{0.4pt}\right) = \rule{1cm}{0.4pt}$, so that $\frac{dy}{dx} = -\frac{\rule{1.5cm}{0.4pt}}{y^{-1/3}} =$

$-\sqrt[3]{\frac{y}{x}}$.

$\frac{dy}{dx}$, 0, $x^{-1/3}$

184 $y^3 + y = x^4$.

$3y^2\frac{dy}{dx} + \rule{1.2cm}{0.4pt} = \rule{1cm}{0.4pt}$, so that $(\rule{2.5cm}{0.4pt})\frac{dy}{dx} = 4x^3$ or

$\frac{dy}{dx}$, $4x^3$, $3y^2 + 1$

$\frac{dy}{dx} = \frac{\rule{1.5cm}{0.4pt}}{3y^2 + 1}$.

$4x^3$

185 $\frac{x}{x - y} + \frac{y}{x} = 1$.

$x^2 + y(\rule{2cm}{0.4pt}) = x(\rule{2cm}{0.4pt})$, so that $2xy - y^2 = \rule{1cm}{0.4pt}$. Then

$x - y$, $x - y$, 0

$2xD_x y + \rule{1cm}{0.4pt} - 2yD_x y = 0$ or $D_x y(2x - 2y) = \rule{1cm}{0.4pt}$, so that

$2y$, $-2y$

$D_x y = \frac{\rule{1cm}{0.4pt}}{2x - 2y} = \frac{\rule{1cm}{0.4pt}}{y - x}$.

$-2y$, y

186 $\sqrt{y} + \sqrt[3]{y} = 7x$.

$\frac{1}{2}y^{-1/2}\frac{dy}{dx} + \frac{1}{3}y^{-2/3}\frac{dy}{dx} = \rule{1cm}{0.4pt}$, so that

7

7

$$\frac{dy}{dx}\left(\frac{1}{2}y^{-1/2} + \frac{1}{3}y^{-2/3}\right) = \underline{\qquad} \text{ or}$$

$$\frac{dy}{dx} = \frac{7}{\frac{1}{2}y^{-1/2} + \frac{1}{3}y^{-2/3}} = \frac{7}{\frac{1}{2\sqrt{y}} + \frac{1}{3\sqrt[3]{y^2}}}$$

$\sqrt{y}\,\sqrt[3]{y^2},\ y^{7/6}$

$$= \frac{42(\underline{\qquad\qquad})}{3\sqrt[3]{y^2} + 2\sqrt{y}} = \frac{42(\underline{\qquad})}{3\sqrt[3]{y^2} + 2\sqrt{y}}.$$

187 $x^2 y = (x - y)^{10}$.

$2xy,\ 1 - \frac{dy}{dx}$

$$x^2\frac{dy}{dx} + \underline{\qquad} = 10(x - y)^9\left(\underline{\qquad\qquad}\right)$$

$x - y$

$$x^2\frac{dy}{dx} + 2xy = 10(x - y)^9 - 10\frac{dy}{dx}(\underline{\qquad})^9, \text{ so that}$$

$10(x - y)^9 - 2xy$

$$[x^2 + 10(x - y)^9]\frac{dy}{dx} = \underline{\qquad\qquad\qquad\qquad} \text{ or}$$

$10(x - y)^9 - 2xy$

$$\frac{dy}{dx} = \frac{\underline{\qquad\qquad\qquad}}{x^2 + 10(x - y)^9}.$$

In Problems 188 through 190, find the equations of the tangent and normal lines to the graph at the point indicated.

188 $x^3 + y^3 = 9$ at $(1, 2)$.

$3y^2\frac{dy}{dx},\ -\frac{3x^2}{3y^2},\ \frac{x^2}{y^2}$

$3x^2 + \underline{\qquad} = 0$, so that $\frac{dy}{dx} = \underline{\qquad} = -\underline{\qquad}$. Then

$-\frac{1}{4}$

$\frac{dy}{dx}$ at the point $(1, 2)$ is $\underline{\quad}$. The equation of the tangent line is

$x - 1,\ x + 4y$

$y - 2 = -\frac{1}{4}(\underline{\qquad})$ or $\underline{\qquad} = 9$. The equation of the normal line is $y - 2 = \underline{\quad}(x - 1)$ or $4x - y = \underline{\quad}$.

$4,\ 2$

189 $y + \sqrt{x + y} = x$ at $(3, 1)$.

1

$$\frac{dy}{dx} + \frac{1}{2}(x + y)^{-1/2}\left(1 + \frac{dy}{dx}\right) = \underline{\quad}, \text{ so that } \frac{dy}{dx} + \frac{1 + \frac{dy}{dx}}{2\sqrt{x + y}} = 1.$$

$2\sqrt{x + y} - 1$

Then $\frac{dy}{dx} = \frac{\underline{\qquad\qquad}}{2\sqrt{x + y} + 1}$. Therefore, $\frac{dy}{dx}$ at the point $(3, 1)$

$\frac{3}{5},\ x - 3$

is $\underline{\quad}$. The equation of the tangent line is $y - 1 = \frac{3}{5}(\underline{\qquad})$ or $3x - 5y = 4$. The equation of the normal line is $y - 1 =$

$-\frac{5}{3},\ 18$

$\underline{\quad}(x - 3)$ or $5x + 3y = \underline{\quad}$.

190 $x^2 - 2xy - 2y^2 = 1$ at $(-3, -1)$.

$4y\frac{dy}{dx},\ x - y$

$2x - 2x\frac{dy}{dx} - 2y - \underline{\qquad} = 0$, so that $\frac{dy}{dx} = \frac{\underline{\qquad}}{x + 2y}$. Then

$\frac{2}{5}$

$\frac{dy}{dx}$ at the point $(-3, -1)$ is $\underline{\quad}$. The equation of the tangent

$\frac{2}{5},\ -1$

line is $y + 1 = \underline{\quad}(x + 3)$ or $2x - 5y = \underline{\quad}$. The equation of

$-\frac{5}{2}$, $5x + 2y$

the normal line is $y + 1 =$ ____ $(x + 3)$ or ______ $= -17$.

In Problems 191 through 193, find $\frac{d^2y}{dx^2}$.

191 $x^2 + y^2 = 81$.

$2y$, $-\frac{x}{y}$

$2x +$ ____ $\frac{dy}{dx} = 0$, so that $\frac{dy}{dx} =$ ____. Then

$$\frac{d^2y}{dx^2} = -\frac{y - x\frac{dy}{dx}}{y^2} = -\frac{y + \frac{x^2}{y}}{y^2}$$

$x^2 + y^2$, -81

$$= -\frac{\overline{}}{y^3} = \frac{\overline{}}{y^3}.$$

192 $xy = 36$.

y, $-\frac{y}{x}$

$x\frac{dy}{dx} +$ ____ $= 0$, so that $\frac{dy}{dx} =$ ____. Then

$$\frac{d^2y}{dx^2} = -\frac{x\frac{dy}{dx} - y}{x^2} = -\frac{x\left(\frac{-y}{x}\right) - y}{x^2}$$

$2xy$, 72

$$= \frac{\overline{}}{x^3} = \frac{\overline{}}{x^3}.$$

193 $x^2 - y^2 = a^2$.

0, $\frac{x}{y}$

$2x - 2y\frac{dy}{dx} =$ ____, so that $\frac{dy}{dx} =$ ____. Then

$\frac{x^2}{y}$

$$\frac{d^2y}{dx^2} = \frac{y - x\frac{dy}{dx}}{y^2} = \frac{y - \overline{}}{y^2}$$

$y^2 - x^2$, $-a^2$

$$= \frac{\overline{}}{y^3} = \frac{\overline{}}{y^3}.$$

In Problems 194 through 196, find $\frac{dx}{dy}$.

194 $5x^2 + 2y^2 = 10$.

$4y$, 0, $4y$, $2y$

$10x\frac{dx}{dy} +$ ____ $=$ ____, so that $\frac{dx}{dy} = -\frac{\overline{}}{10x} = -\frac{\overline{}}{5x}$.

195 $x^3 - x^2y + 2y^3 = 6$.

$6y^2$

$3x^2\frac{dx}{dy} - x^2 - 2xy\frac{dx}{dy} +$ ____ $= 0$, so that $\frac{dx}{dy}(3x^2 - 2xy) =$

$x^2 - 6y^2$, $x^2 - 6y^2$

________ or $\frac{dx}{dy} = \frac{\overline{}}{3x^2 - 2xy}$.

196 $\sqrt{x} + xy^4 = 15$.

$\frac{dx}{dy}$, 0

$\frac{1}{2}x^{-1/2}\frac{dx}{dy} + 4xy^3 + y^4$ (____) $=$ ____, so that

$-4xy^3$, $-4xy^3$

$-8x^{3/2}y^3$

$\frac{dx}{dy} \cdot \left(\frac{1}{2}x^{-1/2} + y^4\right) = -$ ______ or $\frac{dx}{dy} = \frac{\text{\_\_\_\_\_\_}}{\frac{1}{2}x^{-1/2} + y^4} =$

$\frac{\text{\_\_\_\_\_\_}}{1 + 2y^4\sqrt{x}}$.

197 Find the dimensions of the rectangle of maximum perimeter that can be inscribed in the circle whose equation is $x^2 + y^2 = 36$.

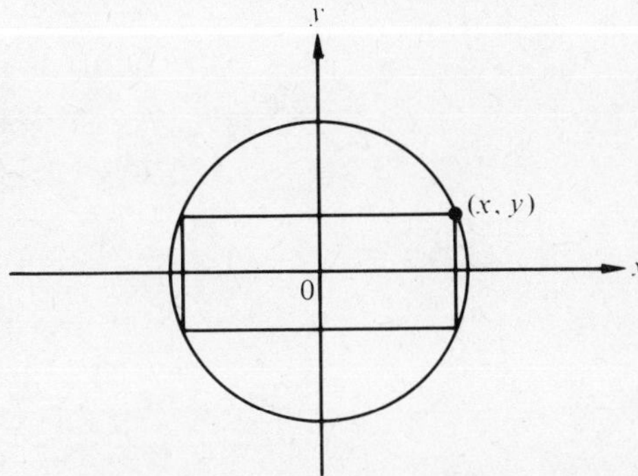

$4x + 4y$, $4\frac{dy}{dx}$

$\frac{dy}{dx}$, 0, $-\frac{x}{y}$

0, y

36, 36, 18

Let $2x$ and $2y$ represent the dimensions of the rectangle, so that the perimeter $P =$ ______, and $\frac{dP}{dx} = 4 + \text{\_\_\_\_\_\_}$. Since $x^2 + y^2 = 36$, then $2x + 2y(\text{\_\_\_\_}) =$ ____ and $\frac{dy}{dx} =$ ____. Therefore, $\frac{dP}{dx} = 4 - 4\frac{x}{y} =$ ____ or $x =$ ____. Then $y^2 + y^2 =$ ____, so that $2y^2 =$ ____ or $y^2 =$ ____; that is, $y = 3\sqrt{2}$ and $x = 3\sqrt{2}$.

8.2 Remarks Concerning Implicit Functions

explicitly

$f(x)$

x, domain

$\sqrt{4 - x^2}$

$4 - x^2$, 4

198 The equation $y = f(x)$ gives y ______ as a function of x.

199 A continuous function f, defined at least on an open interval, is said to be implicit in an equation involving x and y, provided that, when y is replaced by ____ in this equation, the resulting equation is true for all values of ____ in the ______ of f.

200 The function defined by $f(x) = \sqrt{4 - x^2}$ is implicit in the equation $x^2 + y^2 = 4$, since if y is replaced by ______, then $y^2 =$ ______ and so $x^2 + y^2 =$ ____.

In Problems 201 through 205, find functions implicit in each equation by solving for y in terms of x.

$7x^2 - 1$, $\frac{7x^2 - 1}{9}$, $\frac{\sqrt{7x^2 - 1}}{3}$

201 $7x^2 - 9y^2 = 1$.

$9y^2 =$ ______, so that $y^2 =$ ______ and $y =$ ______

$-\dfrac{\sqrt{7x^2 - 1}}{3}$

or $y =$ ________.

$\sqrt[3]{8x},\ 2\sqrt[3]{x}$

202 $8x = y^3$.

$y^3 = 8x$, so that $y =$ _____ = ______.

203 $\dfrac{x}{y} + \dfrac{y}{x} = 2$.

$xy,\ 2xy,\ 0$

$\dfrac{x^2 + y^2}{\_\_\_\_} = 2$, so that $x^2 + y^2 =$ ____ or $(x - y)^2 =$ ____.

$0,\ x$

Then $x - y =$ ____ or $y =$ ____.

204 $y^3 - 3y^2 + 3y = 3(x + 1)$.

$3x + 3$

$y^3 - 3y^2 + 3y =$ ________, so that $y^3 - 3y^2 + 3y - 1 =$

$3x + 2,\ 3x + 2$

________. Then $(y - 1)^3 =$ ________ or $y - 1 =$

$3x + 2,\ 3x + 2$

$\sqrt[3]{\_\_\_\_\_\_\_\_}$. Thus, $y = 1 + \sqrt[3]{\_\_\_\_\_\_\_\_}$.

205 $x = \dfrac{y - 1}{y + 2}$.

$y - 1,\ -2x - 1$

$x = \dfrac{y - 1}{y + 2}$ or $xy + 2x =$ ______ or $xy - y =$ _________, so that

$x - 1,\ -2x - 1$

$y($______$) = -2x - 1$ or $y = \dfrac{\_\_\_\_\_\_\_\_}{x - 1}$. Therefore,

$\dfrac{2x + 1}{1 - x}$

$y =$ ________.

9 Related Rates of Change

206 If x and y are related according to the equation $y = f(x)$, then implicit differentiation is used to find the relationship between

$\dfrac{dx}{dt}$

their rates of change, ____ and $\dfrac{dy}{dt}$.

207 If $x^3 + y^3 = 15$, find a relationship between $\dfrac{dx}{dt}$ and $\dfrac{dy}{dt}$.

$\dfrac{dy}{dt},\ 0$

We have $3x^2 \dfrac{dx}{dt} + 3y^2 \left(\_\_\_\_\right) =$ ____, so that

$-y^2 \dfrac{dy}{dt}$

$x^2 \dfrac{dx}{dt} =$ ________.

208 At a certain instant the dimensions of a rectangle are 5 and 6 inches and they are increasing at the rates of 2 and 3 inches per second, respectively. At what rate is the area increasing?

Let x and y represent the length and the width of the rectangle, so

2 in./sec

that when $x = 5$ and $y = 6$, then $\dfrac{dx}{dt} =$ ____________ and

3 in./sec, xy, $\dfrac{dx}{dt}$

$\dfrac{dy}{dt} =$ ____________. $A =$ ____. Thus, $\dfrac{dA}{dt} = x \dfrac{dy}{dt} + y \left(\_\_\_\_\right)$,

27

so that $\dfrac{dA}{dt} = 5(3) + 6(2) =$ ____ in.^{2}/sec.

209 A kite, at a height 20 meters, is moving horizontally at a rate of 2 meters per second away from the boy who flies it. How fast is the chord being released when 30 meters is out?

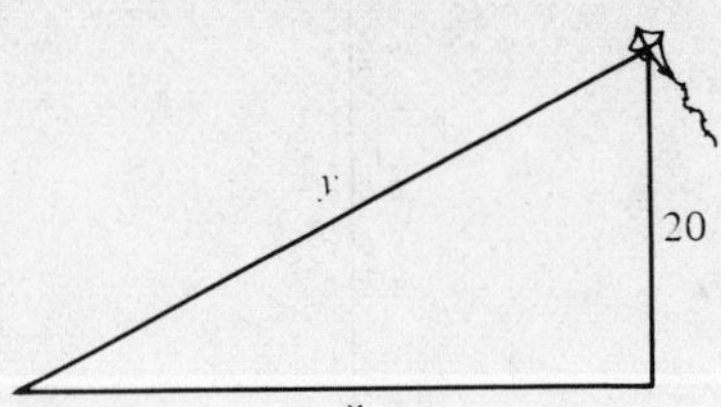

Let x meters denote the horizontal displacement of the kite and let the length of the chord be y meters. Then $x^2 + (20)^2 =$ ____, so that $2x \dfrac{dx}{dt} = (\_\_\_\_) \dfrac{dy}{dt}$, $\dfrac{dx}{dt} = 2$, and $x = \sqrt{y^2 - (20)^2}$. When $y = 30$, then $x = \sqrt{\_\_\_\_} =$ ______. Therefore, $(10\sqrt{5})(2) = (30)\dfrac{dy}{dt}$ or $\dfrac{dy}{dt} = \dfrac{\_\_\_\_}{3}$ meters per second.

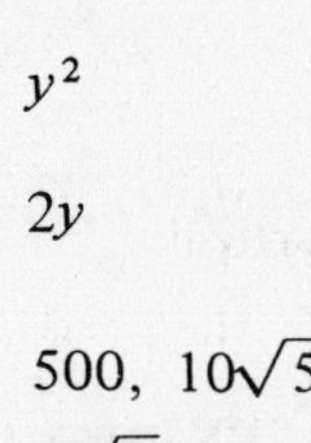

y^2

$2y$

$500,\ 10\sqrt{5}$

$2\sqrt{5}$

210 If the radius of a circle increases at a rate of 0.01 inch per second, find the rate of change of the area when the radius is 3 inches long.

Let r and A represent the radius and the area of the circle at time t. $A =$ ____, so that $\dfrac{dA}{dt} = 2\pi r$ ____. Therefore, $\dfrac{dA}{dt} =$ $2\pi(3)(\_\_\_\_) \approx 2(3.14)(3)(\_\_\_\_) \approx$ __________.

$\pi r^2,\ \dfrac{dr}{dt}$

0.01, 0.01, 0.19 in.^{2}/sec

211 One leg of a right triangle is always 6 centimeters long, and the other leg is increasing at a rate of 2 centimeters per second. Find the rate of change of the hypotenuse when it is 10 centimeters long.

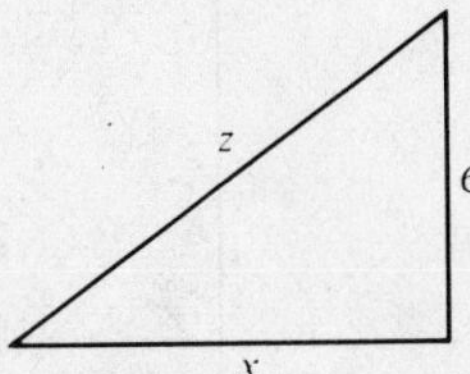

Let x and z represent the lengths of the other leg and hypotenuse at the time t seconds. $z^2 = x^2 +$ ____, so that $2z \dfrac{dz}{dt} = 2x\left(\_\_\_\_\right)$. When $z = 10$, $2(10)\dfrac{dz}{dt} = 2x(\_\_\_\_)$. Also, when $z = 10$, then $x =$ ____. Therefore, $\dfrac{dz}{dt} = \dfrac{2(8)(\_\_\_\_)}{20} =$ __________.

36

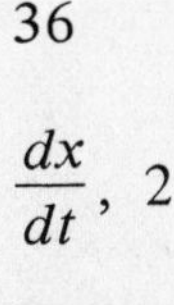

$\dfrac{dx}{dt},\ 2$

8, 2

1.6 cm/sec

212 One end of a ladder 13 meters long slides down a vertical wall while the other end moves away from the wall horizontally at a rate of 1 meter per minute. How fast is the top of the ladder descending when its foot is 5 meters from the wall?

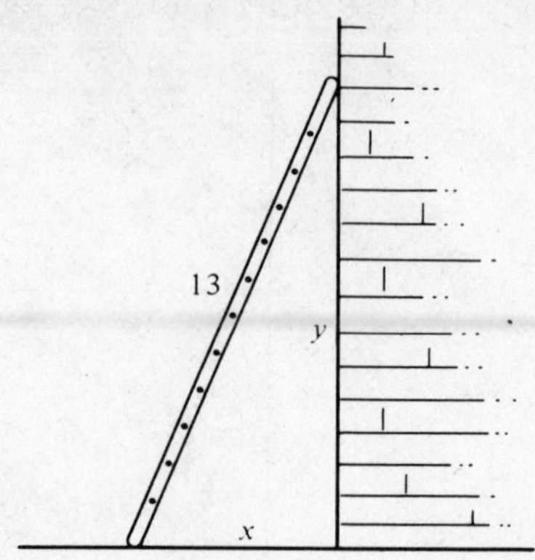

Let x and y represent the horizontal distance and the vertical

169 — distance at time t. Then $x^2 + y^2 =$ ____. When $x = 5$, then

12, y, 1 — $y =$ ____. Also, $2x \dfrac{dx}{dt} + 2(\_\_\_\_) \dfrac{dy}{dt} = 0$. But $\dfrac{dx}{dt} =$ ____, so that

0, $\dfrac{-5}{12}$ — $2(5)(1) + 2(12)\dfrac{dy}{dt} =$ ____, so that $\dfrac{dy}{dt} =$ ____ meter per minute.

213 Two motorcyclists start from the same point. One goes north at a rate of 40 miles per hour and the other goes east at a rate of 50 miles per hour. How fast is the distance between them changing after 2 hours?

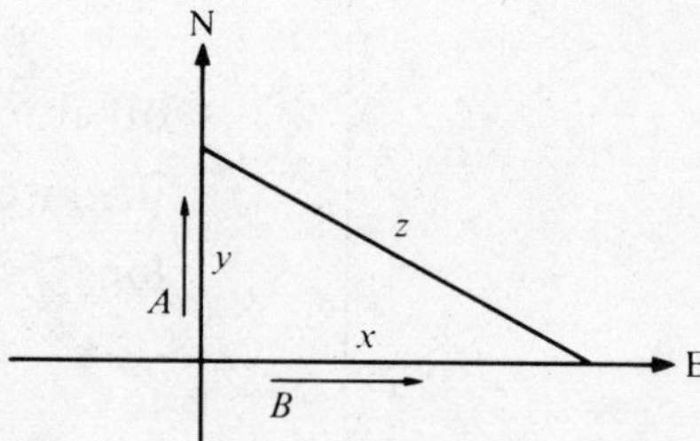

Let A travel y miles at 40 miles per hour and B travel x miles at 50 miles per hour. Let z miles be the distance between them, so

z^2 — that $x^2 + y^2 =$ ____. After 2 hours, $x = 100$, $y = 80$, so that

128.1, $\dfrac{dz}{dt}$ — $z \approx$ ____. Here, $2x \dfrac{dx}{dt} + 2y \dfrac{dy}{dt} = 2z\left(\_\_\_\_\right)$, so that

128.1 — $(100)(50) + (80)(40) \approx (\_\_\_\_)\dfrac{dz}{dt}$. Therefore, $\dfrac{dz}{dt} \approx$

128.1, 64.01 — $\dfrac{5000 + 3200}{\_\_\_\_} \approx$ ____. The distance is changing approximately at the rate of 64.01 miles per hour.

214 A commercial jet is flying due north at a constant altitude at the rate of 600 miles per hour and is directly over a car traveling due west at the rate of 60 miles per hour. How fast is the horizontal distance between the plane and the car changing 3 minutes later?

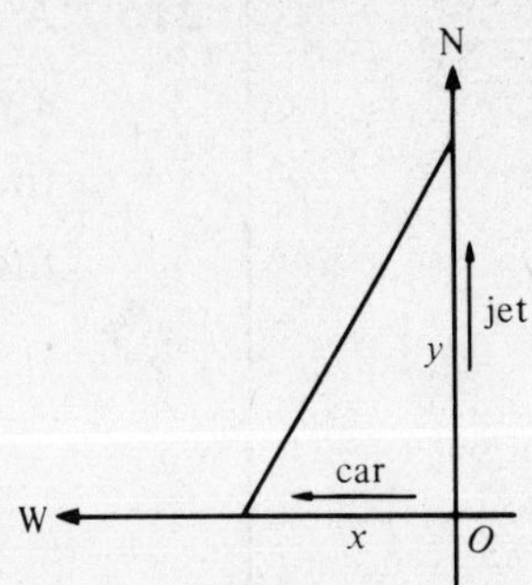

Let x and y miles represent the horizontal and vertical distances of the car and the jet, respectively, at t hours and let z miles be the

30 | distance between them. After 3 minutes the jet is ____ miles

$3,\ z^2$ | from 0 and the car is ____ miles from 0. Also, $x^2 + y^2 =$ ____.

30.2 | When $x = 30$ and $y = 3$, then $z \approx$ ____. Since $2x\dfrac{dx}{dt} + 2y\dfrac{dy}{dt} =$

$2z,\ z$ | ____ $\dfrac{dz}{dt}$, then $x\dfrac{dx}{dt} + y\dfrac{dy}{dt} =$ ____ $\dfrac{dz}{dt}$. When $x = 3$, $\dfrac{dx}{dt} =$

60, 600, 30.2 | ____, $y = 30$, $\dfrac{dy}{dt} =$ ____, and $z \approx$ ____, we have

30, 18,180 | $(3)(60) + (\text{\_\_\_\_})(600) \approx (30.2)\dfrac{dz}{dt}$, so that $\dfrac{dz}{dt} \approx \dfrac{\text{\_\_\_\_}}{30.2} \approx$

602 | ____. Therefore, they are separating at the approximate rate of

602 | ____ miles per hour.

215 Jamal is walking 5 ft/sec on a bridge above a river directly over a boat traveling 10 ft/sec and at right angles to the bridge. How fast is the distance between Jamal and the boat increasing 8 seconds later?

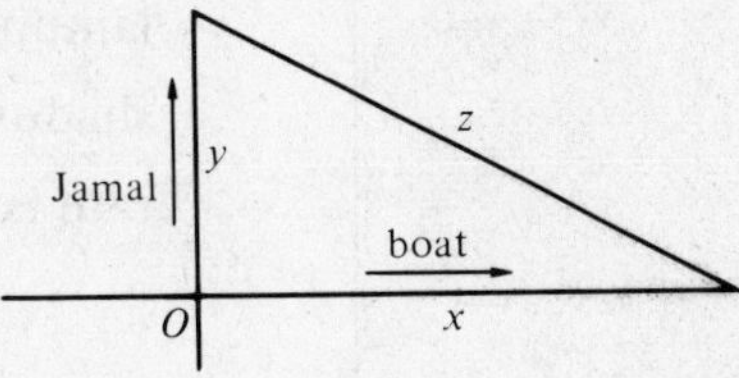

Let x and y feet represent the distances traveled by the boat and Jamal during t seconds, and let z be the distance between them.

z^2, 80 | Thus, $x^2 + y^2 =$ ____. After 8 seconds the boat is ____ feet from

40, 89.44 | 0 and Jamal is ____ feet from 0, so that $z \approx$ ____ feet. Here,

$x^2 + y^2,\ 2y\dfrac{dy}{dt}$ | $z^2 =$ ________, so that $2z\dfrac{dz}{dt} = 2x\dfrac{dx}{dt} +$ ____. Therefore,

2(40) (5), 89.44 | $2(89.44)\dfrac{dz}{dt} \approx 2(80)(10) +$ ________ or $\dfrac{dz}{dt} \approx \dfrac{1000}{\text{\_\_\_\_}} \approx$

11.2 | ____. Hence, they are separating at the approximate rate of 11.2 ft/sec.

216 A light is 20 feet above the ground and a man 6 feet tall is walking away from it on a level road at the rate of 4 feet per second. Find the rate at which the shadow is lengthening and the rate at which the tip of the shadow is moving.

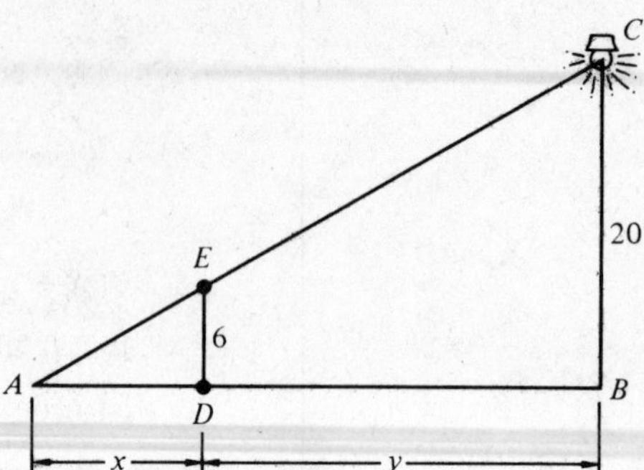

Let C and E represent the light and the man, respectively. Let x feet be the length of the shadow of the man and y feet be the distance of the man from the light. Since ΔABC is similar to

x, 0 | ΔADE, we have $\dfrac{x+y}{20} = \dfrac{\_\_\_\_}{6}$, so that $7x - 3y = \_\_\_\_$. Then

$\dfrac{dy}{dt}$, $\dfrac{12}{7}$ | $7\dfrac{dx}{dt} = 3\left(\_\_\_\_\right)$. Since $\dfrac{dy}{dt} = 4$, then $\dfrac{dx}{dt} = \_\_\_\_$, so that the rate

$\dfrac{12}{7}$ | at which his shadow is lengthening is ____ feet per second. The rate at which the tip of his shadow is moving is $\dfrac{d(y+x)}{dt} =$

$\dfrac{dx}{dt}$, $\dfrac{12}{7}$, $\dfrac{40}{7}$ | $\dfrac{dy}{dt} + \_\_\_\_ = 4 + \_\_\_\_ = \_\_\_\_$ feet per second.

217 Joe, who is 5 feet tall, walks away from the base of a 16-foot-high lamppost at the rate of 4 ft/sec. At what rate is the tip of Joe's shadow moving when he is 18 feet from the pole, and at what rate is the length of his shadow increasing?

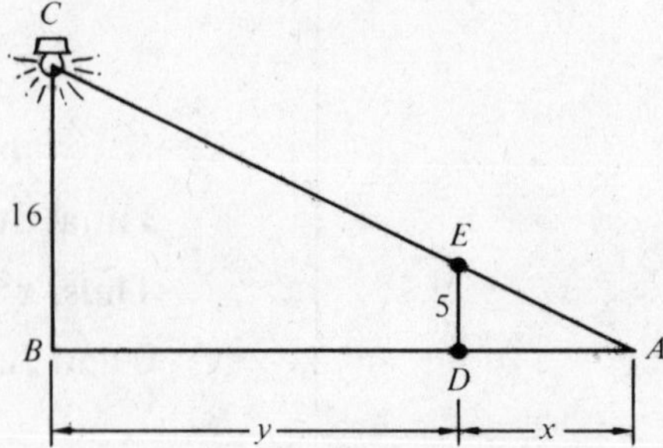

Let C represent the position of the lamp. DE is Joe's height and DA is his shadow. Let x ft be the length of his shadow and let y

ft be the distance of Joe from the lamppost. Then since ΔABC is

ΔADE, $x + y$

similar to _____, we have $\dfrac{\_\_\_\_}{16} = \dfrac{x}{5}$, so that $11x - 5y =$

0, $\dfrac{dy}{dt}$, 4

____. Then $11\dfrac{dx}{dt} = 5\left(\_\_\_\_\right)$, so that $\dfrac{dx}{dt} = \dfrac{5(\_\_\_\_)}{11} =$

20, 1.82

$\dfrac{\_\_\_\_}{11} \approx$ ____. Thus, Joe's shadow is lengthening at the approxi-

1.82

mate rate of ____ feet per second. The tip of Joe's shadow is

$\dfrac{dy}{dt}$, 4

moving at the rate of $\dfrac{d(x+y)}{dt} = \dfrac{dx}{dt} + \left(\_\_\_\_\right) \approx 1.82 +$ ____ $=$

5.82

____ ft/sec.

218 The volume of a cube is decreasing at the rate of 2 cubic centimeters per hour. Find the rate at which the surface area is decreasing at the time when the volume is 343 cubic centimeters. Let x cm represent the edge of the cube, V cubic centimeters represent its volume, and S square centimeters represent its surface area.

x^3, V, 7

$V =$ ____, so that $x = \sqrt[3]{\_\_\_\_}$. When $V = 343$, then $x =$ ____.

$6x^2$, $12x$

Also, $S =$ ____, so that $\dfrac{dS}{dt} = \_\_\_\_ \dfrac{dx}{dt}$, and when $x = 7$, then

84

$\dfrac{dS}{dt} = \_\_\_\_ \dfrac{dx}{dt}$. But $\dfrac{dV}{dt} = 3x^2 \dfrac{dx}{dt}$, so that when $x = 7$,

147, -0.014, -0.014

$\dfrac{dx}{dt} = \dfrac{-2}{\_\_\_\_} \approx$ ______. Then $\dfrac{dS}{dt} \approx (84)(\_\_\_\_\_\_) =$

-1.176

______, so that the area is decreasing at the approximate rate of 1.176 square centimeters per hour.

219 A triangular trough is 12 feet long, 3 feet wide at the top, and 3 feet deep. If water is poured into the trough at a rate of 10 cubic feet per minute, how fast is the surface rising when the depth is 2 feet?

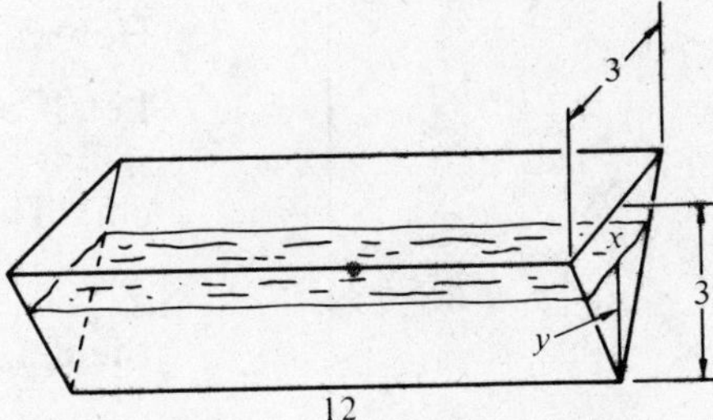

Let x feet and y feet represent the base and the depth of the

$\dfrac{1}{2}xy(12)$

water. The volume of water in the trough $V =$ ______ $=$

$6xy$, $\dfrac{dy}{dt}$, 2

____. We wish to find $\dfrac{\_\_\_\_}{}$ when $y =$ ____. From similar tri-

$3,\ y,\ 6y^2$

angles, we have $\dfrac{x}{y} = \dfrac{\_\_\_\_}{3}$, so that $x =$ ____. Then $V =$ ____.

$\dfrac{dy}{dt}$

$\dfrac{dV}{dt} = 12y$ ____, so that $10 = 12y\,\dfrac{dy}{dt}$. When $y = 2$, then

$\dfrac{5}{12}$ foot per minute

$\dfrac{dy}{dt} =$ ____. Therefore, the surface is

$\dfrac{5}{12}$

rising at the rate of ____ foot per minute.

220 A barge whose deck is 5 feet below the level of a dock is drawn up to it by means of a cable running over a pulley at the edge of the dock. When the barge is 12 feet away from the dock, the cable is being hauled in at the rate of 4 feet per minute. At what rate is the barge moving at this time?

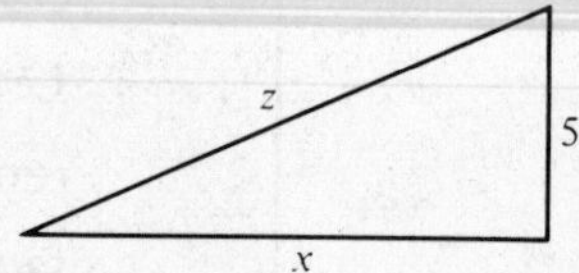

Let z feet represent the length of the cable from the pulley to the barge and let x represent the distance of the barge from the dock

$-4,\ 12$

at t minutes. $\dfrac{dz}{dt} =$ ____, when $x =$ ____. To find $\dfrac{dx}{dt}$ notice

$x^2 + 25,\ 2x$

that $z^2 =$ ____, so that $2z\,\dfrac{dz}{dt} = (\_\_\_\_)\,\dfrac{dx}{dt}$ or $\dfrac{dx}{dt} =$

$2x,\ 13$

$\dfrac{2z\,\dfrac{dz}{dt}}{\_\_\_\_}$. Therefore, when $x = 12$, $\dfrac{dx}{dt} = \dfrac{z(-4)}{12}$. But $z =$ ____

$-\dfrac{13}{3}$

when $x = 12$, so that $\dfrac{dx}{dt} = \dfrac{(13)(-4)}{12} =$ ____. The barge is

$\dfrac{13}{3}$

moving toward the dock at the rate of ____ ft/min.

221 The volume of a sphere is increasing at the rate of 10 cm³/sec. How fast is the radius increasing when it is 5 cm?

Let V cm³ represent the volume of the sphere whose radius is r

$\dfrac{4}{3}\pi r^3,\ 4\pi r^2$

cm. Then $V =$ ____, so that $\dfrac{dV}{dt} =$ ____ $\dfrac{dr}{dt}$. When $r = 5$,

$10,\ \dfrac{1}{10\pi}$

then ____ $= 4\pi(25)\,\dfrac{dr}{dt}$, so that $\dfrac{dr}{dt} =$ ____. The radius is increasing at the rate of $\dfrac{1}{10\pi}$ cm/sec.

222 A stone is thrown into a calm body of water, sending out a series of concentric ripples. If the radius of the outer ripple increases uniformly at the rate of 5 cm/sec, find how rapidly the disturbed area is increasing at the end of 2 seconds.

Let r cm be the radius of the outer ripple and let A cm^2 be the area enclosed by the outer ripple.

πr^2, $2\pi r$, 10

$A =$ ____, $\dfrac{dA}{dt} =$ ____ $\dfrac{dr}{dt}$, so that at $t = 2$, $r =$ ____, and

5, 314

$\dfrac{dA}{dt} \approx 2(3.14)(10)($____$) \approx$ ____.

314

The area is increasing at the approximate rate of ____ cm^2/sec.

223 The radius of the base of a right circular cone is decreasing at the rate 3 cm/sec, and the height is increasing at the rate of 5 cm/sec. At what rate is the volume changing when the height is 10 cm and the radius is 6 cm?

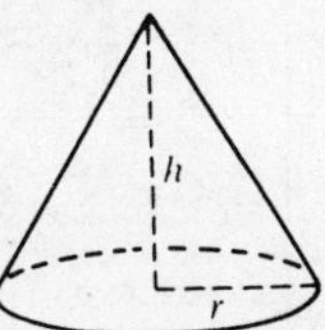

$\frac{1}{3}\pi r^2 h$

Let r cm and h cm represent the radius and the height of the cone whose volume is V cm^3. $V =$ ________, so that $\dfrac{dV}{dt} =$

h

$\dfrac{1}{3}\pi r^2 \dfrac{dh}{dt} + \dfrac{2}{3}\pi r($____$)\dfrac{dr}{dt}$. Then, when $h = 10$, $r = 6$, $\dfrac{dh}{dt} = 5$,

10, -3

and $\dfrac{dr}{dt} = -3$, $\dfrac{dV}{dt} = \dfrac{1}{3}\pi(36)(5) + \dfrac{2}{3}\pi(6)($____$)($____$) =$

60π, -60π, -188.5

____ $- 120\pi =$ ____ $\approx$ ______. The volume is changing at

-188.5

the approximate rate of _______ cm^3/sec.

224 Sand is being poured on the ground, forming a conical pile with its altitude equal to $\frac{2}{3}$ of the radius of the base. If the sand is falling at the rate of 10 cubic feet per second, find how fast the altitude is increasing when the pile is 5 feet high.

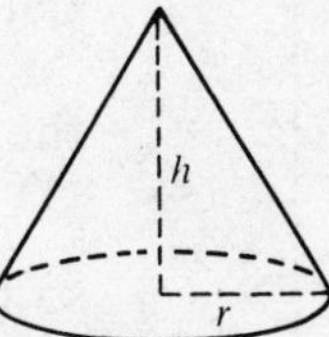

$\frac{1}{3}\pi r^2 h$, 2, $\frac{3}{2}h$

Let r and h be the radius and the height of the cone whose volume is V, so that $V =$ ________. $\dfrac{h}{r} = \dfrac{\_\_}{3}$ or $r =$ ______. Then

$\frac{3}{2}h$, $\frac{3}{4}\pi h^3$, $\frac{dh}{dt}$

$V = \dfrac{1}{3}\pi\left(\_\_\_\_\right)^2 h =$ ________, so that $\dfrac{dV}{dt} = \dfrac{9}{4}\pi h^2\left(\_\_\_\_\right)$.

225π

0.057

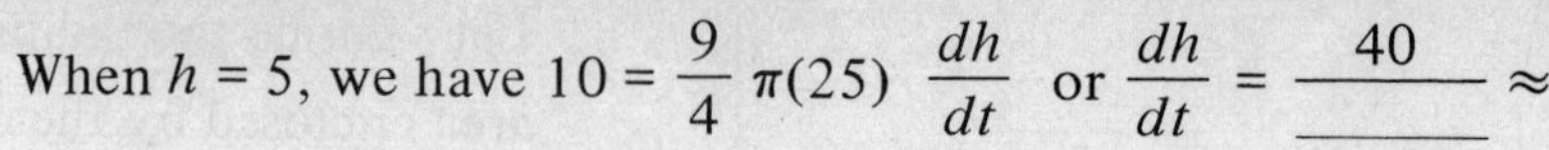

When $h = 5$, we have $10 = \frac{9}{4}\pi(25)\frac{dh}{dt}$ or $\frac{dh}{dt} = \frac{40}{\_\_\_\_} \approx$ ______, so that the altitude is increasing at the approximate rate of 0.057 ft/sec.

225 A tank has the form of an inverted regular square pyramid with vertical axis; its altitude is 6 feet and its base edge is 3 feet. If water flows into the tank at the rate of 5 cubic feet per minute, how fast is the surface rising when the tank is about to overflow?

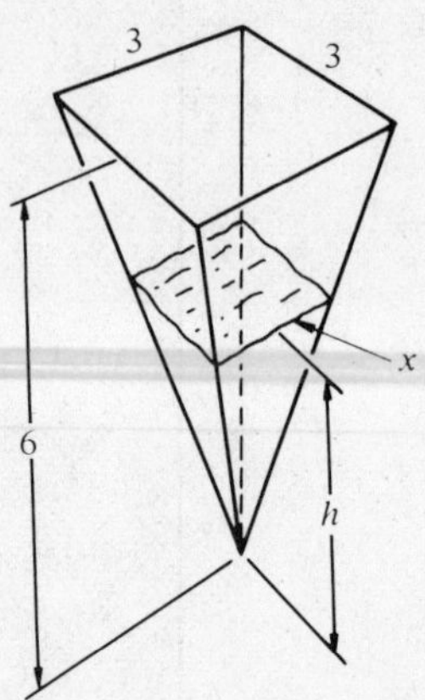

h

$6, \frac{h}{2}, \frac{h}{2}$

$6, \frac{dh}{dt}$

$36, \frac{20}{36}, 0.556$

Let x feet represent the base edge and h feet represent the height of the water, whose volume is V cubic feet. $V = \frac{1}{3}x^2(\_\_\_\_)$; but $\frac{x}{h} = \frac{3}{\_\_\_\_}$, so that $x = \_\_\_\_$ and $V = \frac{1}{3}\left(\_\_\_\_\right)^2 h = \frac{h^3}{12}$. To find $\frac{dh}{dt}$ when $h = \_\_\_\_$, we calculate $\frac{dV}{dt} = \frac{1}{4}h^2\left(\_\_\_\_\right)$, so that $5 = \frac{1}{4}(\_\_\_\_)\frac{dh}{dt}$ or $\frac{dh}{dt} = \frac{\_\_\_\_}{\_\_\_\_} \approx$ ______. The surface is rising at the approximate rate of 0.556 ft/min.

6

$r, \frac{1}{3}\pi r^2 h$

$h^2, 16\pi, -0.02$

226 A conical funnel of height and radius each 6 cm contains a liquid which escapes at the rate of 1 cubic cm/min. How fast is the surface falling when it is 4 centimeters from the top of the funnel? Let r and h represent the base radius and height in centimeters of the liquid. Let V represent the volume of the liquid. $\frac{h}{r} = \frac{\_\_\_\_}{6}$, so that $h = \_\_\_\_$. $V = \_\_\_\_\_\_ = \frac{1}{3}\pi h^3$. $\frac{dV}{dt} = \frac{3\pi}{3}\_\_\_\_\frac{dh}{dt}$, so that $\frac{dh}{dt} = \frac{-1}{\_\_\_\_} \approx$ ______; and so the surface is falling at the approximate rate of 0.02 cm/min.

227 The adiabatic law of the expansion of air is $PV^{1.4} = C$, where C is a constant. If at a given time the volume V is 12 cubic inches and the pressure P is 70 lb/in.^{2}, find the rate at which the pressure is changing if the volume is decreasing 1 cubic in./min.

$\frac{dP}{dt}$ — $PV^{1.4} = C$, so that $P(1.4)V^{0.4}\frac{dV}{dt} + V^{1.4}\left(\underline{\qquad}\right) = 0.$

12 — $(70)(1.4)(12)^{0.4}(-1) + (\underline{\quad})^{1.4}\frac{dP}{dt} = 0$, so that $\frac{dP}{dt} =$

$12,\ 98,\ \frac{49}{6}$ — $\frac{(70)(1.4)(\underline{\quad})^{0.4}}{(12)^{1.4}} = \frac{\underline{\quad}}{12} = \underline{\quad}$. The pressure is increasing at

$\frac{49}{6}$ — the rate of $\underline{\quad}$ lb/in.²/min.

228 If p is the distance between a convex lens and some object, while q is the distance between the lens and the image, and f is the (constant) focal length of the lens, then it is known that $\frac{1}{q} + \frac{1}{p} = \frac{1}{f}$. For a lens whose focal length is 16 inches, is the image distance increasing or decreasing when the object distance is 10 feet and is increasing at the rate of 30 ft/sec? What is the rate?

$-\frac{1}{p^2}$ — $\frac{1}{q} + \frac{1}{p} = \frac{1}{f}$, so that $-\frac{1}{q^2}\frac{dq}{dt} + \left(\underline{\qquad}\right)\frac{dp}{dt} = 0.$ If $p = 10$

$\frac{4}{3}$ ft, $\frac{20}{13}$ ft — and $f = \underline{\qquad}$, then $q = \underline{\qquad}$; so that

$100,\ \frac{120}{169}$ — $-\frac{169}{400}\frac{dq}{dt} - \frac{1}{\underline{\quad}}(30) = 0$ or $\frac{dq}{dt} = -\underline{\qquad}$, so that the

0.71 — image distance is decreasing at the approximate rate of $\underline{\qquad}$ ft/sec.

Chapter Test

1 Let f be a polynomial function defined by the equation $f(x) = x^3 - 2x - 3$. Use the intermediate value theorem to show that there is a root of f between 1 and 2.

2 Let f be the function defined by the equation $f(x) = x^2 - 6x + 8$. Find a number c between 2 and 6 such that the tangent to the graph of f at the point $(c, f(c))$ is parallel to the secant line containing the points $(2, f(2))$ and $(6, f(6))$.

3 Let f be defined by the equation $f(x) = |x| - \frac{1}{2}$. Determine whether any of the hypotheses of Rolle's theorem fail to hold in the interval $\left[-\frac{1}{2}, \frac{1}{2}\right]$.

4 Let f be defined by $f(x) = x^3 + x^2 - 5x - 5$. Find:

(a) The intervals on which f is increasing and decreasing.

(b) The intervals on which f is concave upward and downward.

(c) The relative maximum and the relative minimum values of f.

(d) The point of inflection.

(e) Sketch the graph of f.

5 Let $f(x) = \dfrac{3}{x+7}$. Find:

(a) $f'(x)$ (b) $f''(x)$ (c) $f'''(x)$

6 Let f and g be twice-differentiable functions so that $f(2) = -1, f'(2) = 3, f''(2) = 5, g(2) = 2, g'(2) = -4$, and $g''(2) = 1$. Find:

(a) $(f+g)''(2)$ (b) $(fg)''(2)$

7 Let $f(x) = \dfrac{6x}{x^2+1}$. Find the relative maximum and relative minimum values of f and sketch the graph of f.

8 Find the absolute maximum and absolute minimum of $f(x) = \sqrt{x} - 2\sqrt[4]{x}$ for x in the interval $[0, 100]$.

9 It is desired to make an open box of maximum possible volume from a square piece of tin whose side is 12 inches by cutting squares out of the corners and then folding up the tin to form the sides. What should be the length of a side of the cut out squares?

10 Find the altitude of the right circular cylinder of maximum volume that can be inscribed in a given right circular cone of altitude h units.

11 A circular plate of metal expands by heat so that its radius increases at the rate of 0.01 inch per second. At what rate is the surface area increasing when the radius is 2 inches?

12 A spherical weather balloon is expanding due to the heating of its gas by solar radiation. When the balloon's radius r reaches 6 meters, its surface area S is increasing at the rate $\dfrac{1}{9}$ square meter per second. At what rate is the volume V changing?

13 Given the equation $x^3 + y^3 = 8$. Find:

(a) $\dfrac{dy}{dx}$ (b) $\dfrac{d^2y}{dx^2}$

Answers

1 $f(1) = -4$ and $f(2) = 1$

2 $c = 4$

3 f is not differentiable at 0.

4 (a) Increasing on $\left(-\infty, -\frac{5}{3}\right]$ and $[1, \infty)$; decreasing on $\left[-\frac{5}{3}, 1\right]$.

(b) Concave upward on $\left(-\frac{1}{3}, \infty\right)$; concave downward on $\left(-\infty, -\frac{1}{3}\right)$.

(c) Relative maximum at $\left(-\frac{5}{3}, \frac{40}{27}\right)$; relative minimum at $(1, -8)$.

(d) Point of inflection is $\left(-\frac{1}{3}, -\frac{88}{27}\right)$.

(e)

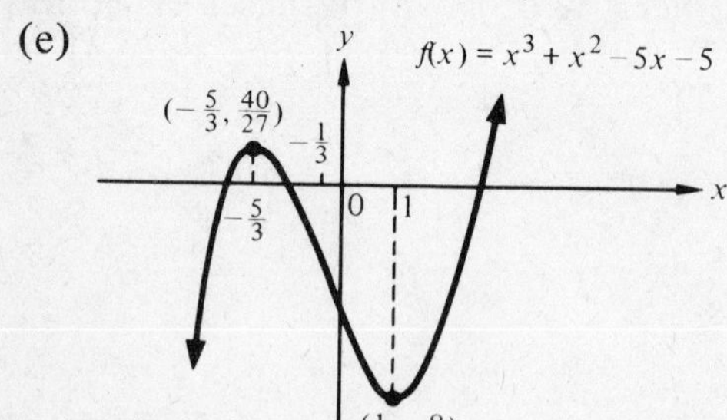

5 (a) $f'(x) = \dfrac{-3}{(x+7)^2}$ (b) $f''(x) = \dfrac{6}{(x+7)^3}$ (c) $f'''(x) = \dfrac{-18}{(x+7)^4}$

6 (a) 6 (b) -15

7 Relative minimum is -3 at -1 and relative maximum is 3 at 1. The graph is

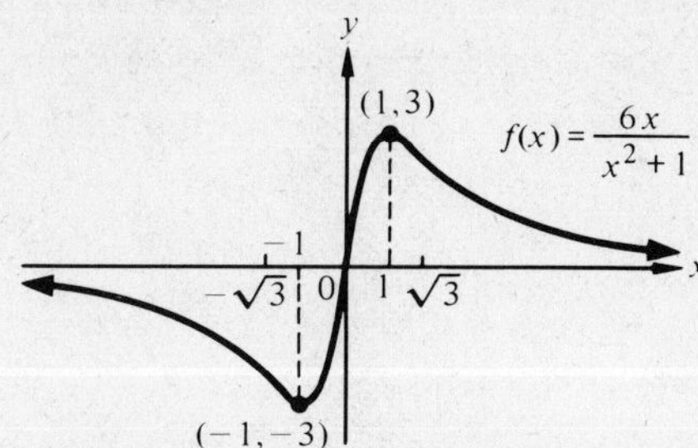

8 Absolute maximum of f is $10 - 2\sqrt{10}$. Absolute minimum of f is -1.

9 2 inches

10 Altitude of the cylinder is $\frac{1}{3}h$.

11 0.04π square inch per second

12 $\frac{1}{3}$ cubic meter per second

13 (a) $\frac{dy}{dx} = -\frac{x^2}{y^2}$ (b) $\frac{d^2y}{dx^2} = -\frac{16x}{y^5}$

Chapter 4 ANALYTIC GEOMETRY AND THE CONICS

In this chapter we shall investigate special curves obtained by sectioning or cutting circular cones by planes. These curves are called conic sections. We shall use translation and rotation to study the conics—the circle, the ellipse, the parabola, and the hyperbola. In solving the problems of this chapter, the student is able to:

1 Find the equation of a circle satisfying certain conditions, and, conversely, to find the center and radius of a circle from a given equation.

2 Apply translation equations to change the equations of the conics from one form to another.

3 Find the center, vertices, and foci of an ellipse from a given equation, and sketch its graph.

4 Find the vertex, focus, directrix, and latus rectum of a parabola, and sketch its graph.

5 Find the center, vertices, foci, and equations of the asymptotes of a hyperbola, and sketch its graph.

6 Find and identify the conics directly from the general definition in terms of the directrix, eccentricity, and focus.

1 The Circle and Translation of Axes

1.1 The Circle

r

diameter

radius

1 A circle of radius r with center at the point C is defined to be the set of all points P in the plane whose distance from C is ____. Any line segment containing C between two points on this circle is called a ________ of the circle. Any line segment between a point P on this circle and the center C is called a ______ of the circle.

$(y-k)^2$, circle

(h, k)

2 Let r be a positive number and C the point with coordinates (h, k) in the xy plane. Then the graph of the equation $(x-h)^2 +$ ________ $= r^2$ is a ______ of radius r whose center is $C =$ ______.

In Problems 3 through 11, find the equation of the circle with the given information.

1, 4, 5

1, 4, $(x-1)^2 + (y-4)^2 = 25$

3 Radius 5 and center at (1, 4).
Here $h =$ ___, $k =$ ___, and $r =$ ___, so that
$(x -$ ___$)^2 + (y -$ ___$)^2 = 25$ or ________________.

$(-1, 2)$, 3

-1, 2, $(x+1)^2 + (y-2)^2 = 9$

4 Radius 3 and center at $(-1, 2)$.
$(h, k) =$ ________ and $r =$ ___, so that
$(x -$ ___$)^2 + (y -$ ___$)^2 = 9$ or ________________.

$(2, -3)$, $\sqrt{7}$

2, -3, $(x-2)^2 + (y+3)^2 = 7$

5 Radius $\sqrt{7}$ and center at $(2, -3)$.
$(h, k) =$ ________ and $r =$ ___, so that
$(x -$ ___$)^2 + (y -$ ___$)^2 = 7$ or ________________.

$(0, -2)$, 6

0, -2, $x^2 + (y+2)^2 = 36$

6 Radius 6 and center at $(0, -2)$.
$(h, k) =$ ________ and $r =$ ___, so that
$(x -$ ___$)^2 + (y -$ ___$)^2 = 36$ or ________________.

$(0, 0)$, 5, $x^2 + y^2 = 25$

7 Radius 5 and center at (0, 0).
$(h, k) =$ ________ and $r =$ ___, so that ____________.

$5 - 2$, 5

$(y-2)^2$, $(x+1)^2 + (y-2)^2 = 25$

8 Center at $(-1, 2)$ and containing the point (3, 5).
The radius $r = \sqrt{(3+1)^2 + (\_\_\_\_)^2} =$ ___, so that
$(x+1)^2 + ($________$)^2 = 25$ or ________________.

6

diameter, 4, 1

0, 1, 0

$(x-1)^2 + y^2 = 9$

9 Radius 3 and containing the points $(-2, 0)$ and $(4, 0)$.
The distance between these points is ___. Hence, they are the endpoints of a ________. Therefore, $h = \frac{-2 + \_\_\_}{2} =$ ___
and $k =$ ___, so that $(x -$ ___$)^2 + (y -$ ___$)^2 = 9$ or
________________.

-5, 0, -2, 3

10 The points (5, 8) and $(-5, -2)$ are the endpoints of a diameter.
$h = \frac{5 + \_\_\_}{2} =$ ___, $k = \frac{8 + \_\_\_}{2} =$ ___, and

$3,\ 5\sqrt{2}$

$0,\ x^2 + (y-3)^2 = 50$

$r = \sqrt{(5-0)^2 + (8-\_\_\_\_)^2} = \_\_\_\_\_\_$, so that

$(x - \_\_\_\_)^2 + (y-3)^2 = 50$ or $\_\_\_\_\_\_\_\_\_\_\_\_$.

11 The circle contains the points (2, 4), (0, 0), and (3, 3).
Let (h, k) be the coordinates of the center and let r be the radius, so that

$4-k$

$0-k$

$3-h$

r^2

$r^2,\ h^2+k^2,\ r^2$

5

$3,\ 1,\ 2,\ \sqrt{5}$

$y-2$

$(x-1)^2 + (y-2)^2 = 5$

$(2-h)^2 + (\_\_\_\_\_\_)^2 = r^2$,

$(0-h)^2 + (\_\_\_\_\_\_)^2 = r^2$,

$(\_\_\_\_\_\_)^2 + (3-k)^2 = r^2$.

Expanding these, we obtain $h^2 + k^2 - 4h - 8k + 20 = \_\_\_\_$, $h^2 + k^2 = \_\_\_\_$, and $\_\_\_\_\_\_\_\_ - 6h - 6k + 18 = \_\_\_\_$.

Simplifying and combining, we have $h + 2k = \_\_\_\_$ and $h + k = \_\_\_\_$, so that $h = \_\_\_\_$, $k = \_\_\_\_$, and $r = \_\_\_\_$. Then the equation of the circle is $(x-1)^2 + (\_\_\_\_\_\_)^2 = (\sqrt{5})^2$ or $\_\_\_\_\_\_\_\_\_\_\_\_$.

In Problems 12 through 16, find the radius r and the coordinates (h, k) of the center of the circle with each equation. Sketch the graph.

$-2,\ 1,\ -2$

3

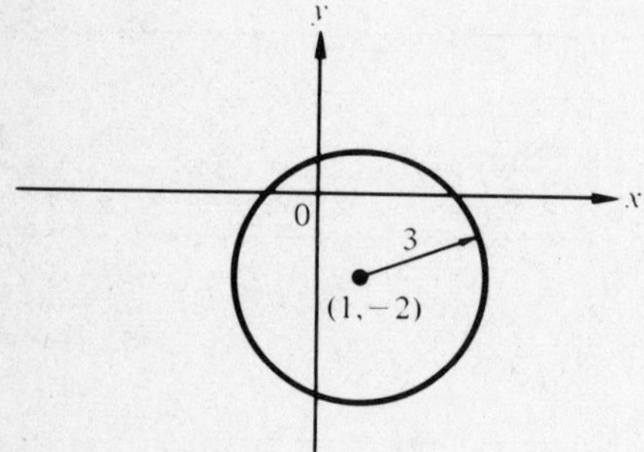

12 $(x-1)^2 + (y+2)^2 = 9$.

We have $(x-1)^2 + (y - \_\_\_\_)^2 = 9$, so that $h = \_\_\_\_$, $k = \_\_\_\_$, and $r = \_\_\_\_$. The graph is

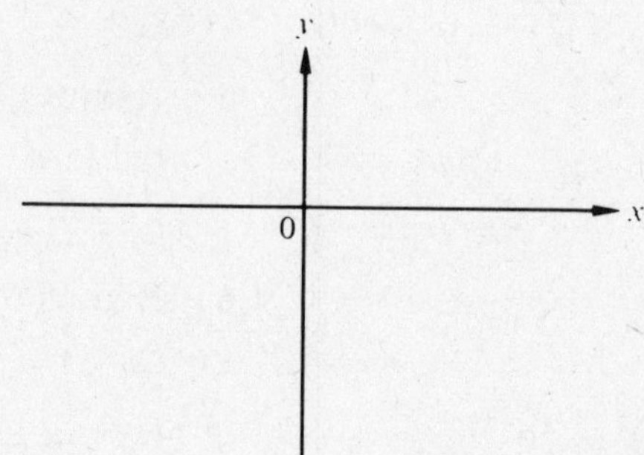

$1,\ 4,\ 1,\ 4$

$y-2,\ (1, 2)$

5

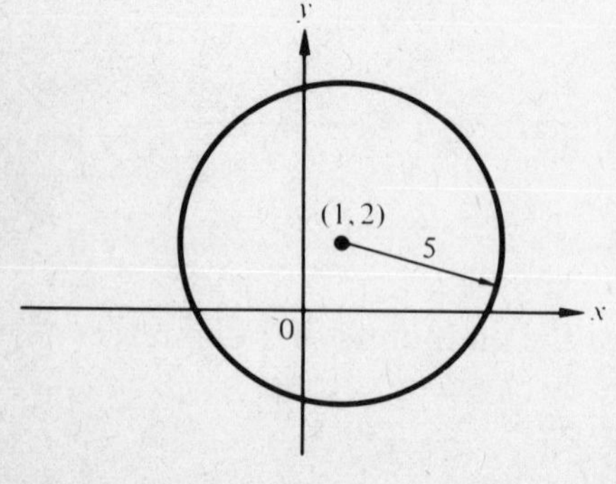

13 $x^2 + y^2 - 2x - 4y - 20 = 0$.

Rewrite the equation in standard form by completing the square, so that $(x^2 - 2x + \_\_\_\_) + (y^2 - 4y + \_\_\_\_) = 20 + \_\_\_\_ + \_\_\_\_$ or $(x-1)^2 + (\_\_\_\_\_\_)^2 = 25$. Then $(h, k) = \_\_\_\_\_\_$ and $r = \_\_\_\_$. The graph is

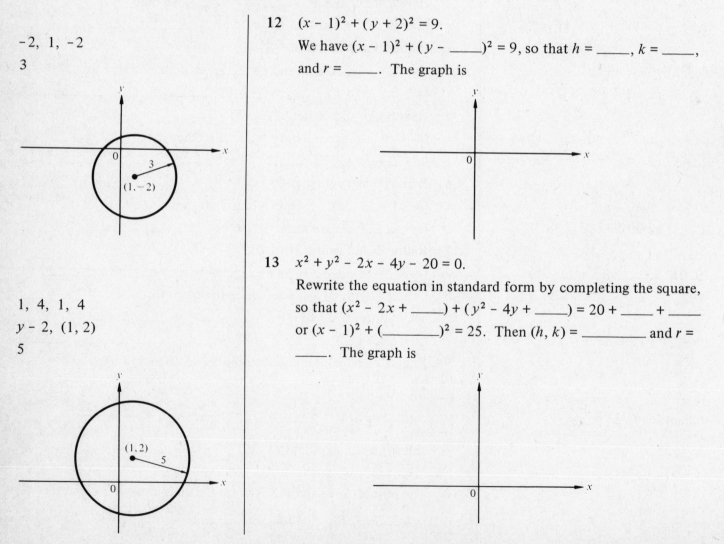

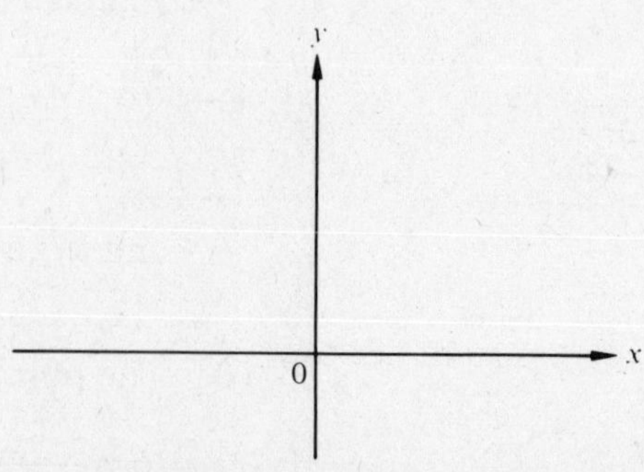

14 $x^2 + y^2 + 6x = 0$.

Rewrite the equation in standard form by completing the square, so that $(x^2 + 6x + \_\_\_\_) + y^2 = \_\_\_\_$ or $(x + 3)^2 + \_\_\_\_ = 9$. Then $(h, k) = \_\_\_\_\_\_\_\_$ and $r = \_\_\_\_$. The graph is

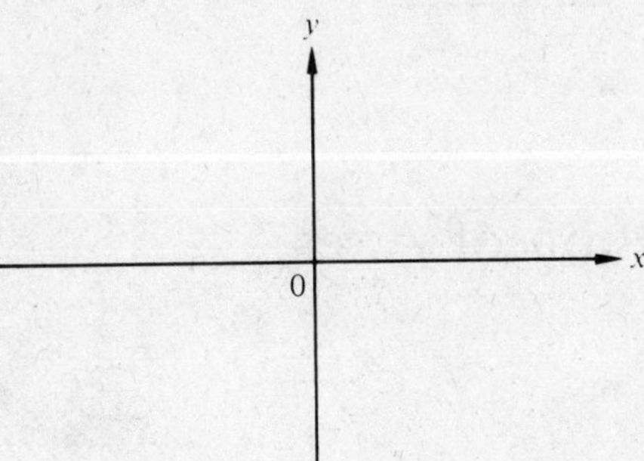

9, 9, y^2

$(-3, 0)$, 3

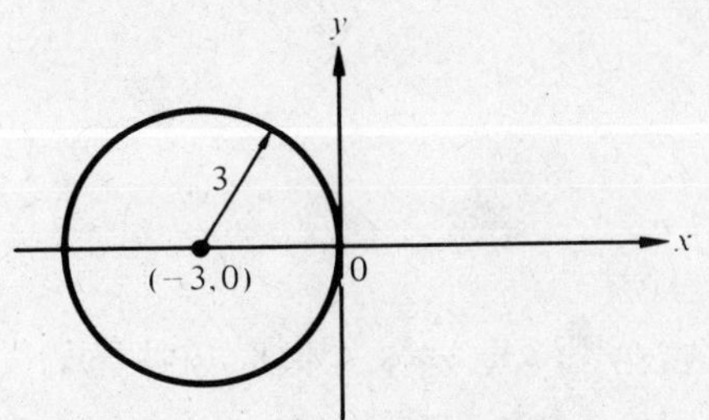

15 $3x^2 + 3y^2 + 6x - 12y - 10 = 0$.

First, we divide both sides of the equation by ____, so that $x^2 + y^2 + 2x - 4y = \_\_\_\_$; then we write it in standard form by completing the square, so that

$(x^2 + 2x + \_\_\_\_) + (y^2 - 4y + \_\_\_\_) = \frac{10}{3} + \_\_\_\_ + \_\_\_\_$ or

$(x + 1)^2 + (\_\_\_\_\_\_)^2 = \frac{25}{3}$. Then $(h, k) = \_\_\_\_\_\_$ and $r =$ ______. The graph is

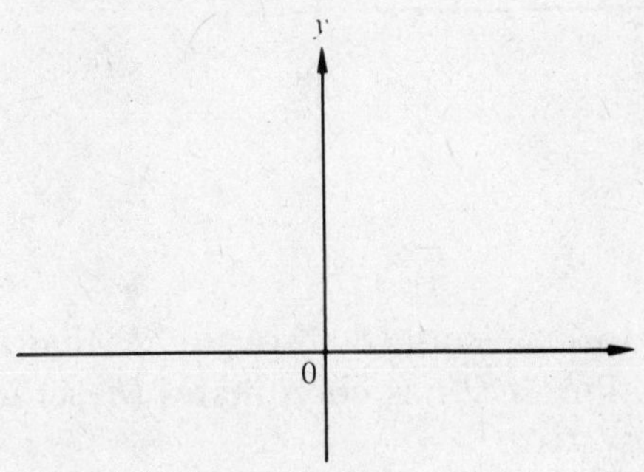

3

$\frac{10}{3}$

1, 4, 1, 4

$y - 2$, $(-1, 2)$

$\frac{5}{\sqrt{3}}$

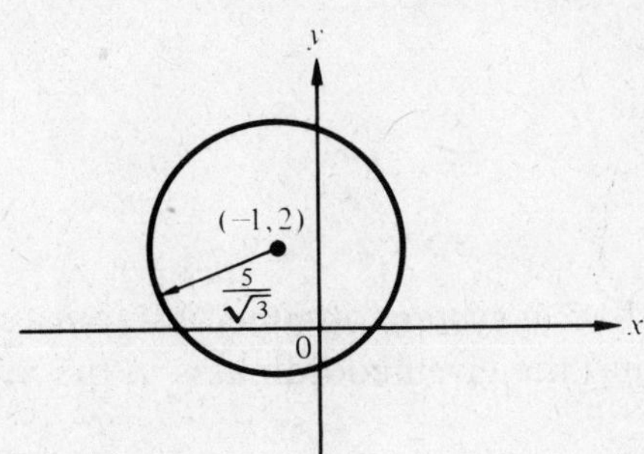

16 $4x^2 + 4y^2 = 25y$.

Divide both sides of the equation by ____, so that $x^2 + y^2 =$ ______ y; then $x^2 + y^2 - \frac{25y}{4} + \_\_\_\_\_\_ = \_\_\_\_\_\_$ or

$x^2 + \left(y - \frac{25}{8}\right)^2 = \_\_\_\_\_\_$. Therefore, $(h, k) = \_\_\_\_\_\_\_\_$ and $r = \_\_\_\_\_\_$. The graph is

4

$\frac{25}{4}$, $\frac{625}{64}$, $\frac{625}{64}$

$\frac{625}{64}$, $\left(0, \frac{25}{8}\right)$

$\frac{25}{8}$

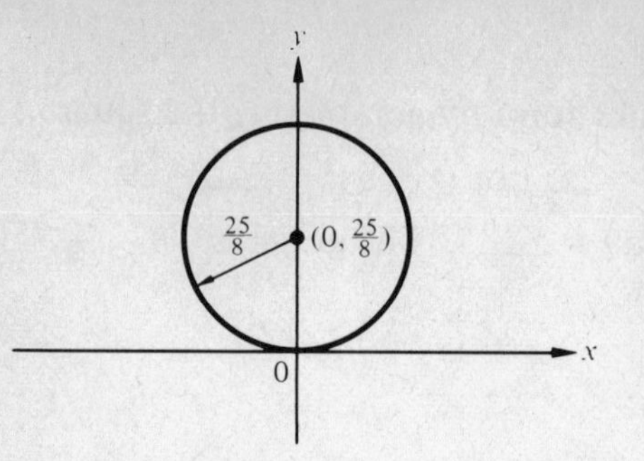

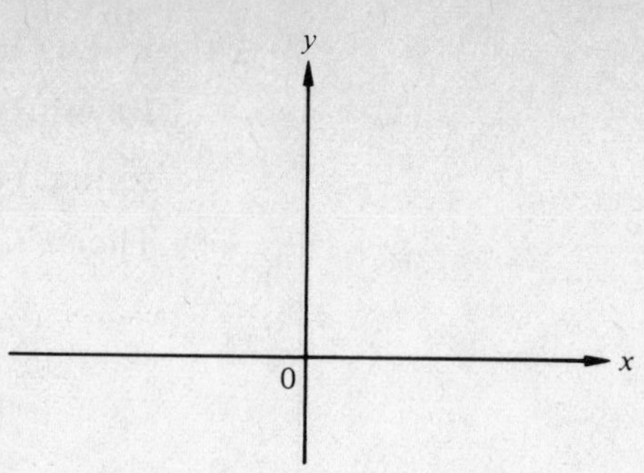

1.2 Translation of Axes

17 **Suppose that a pair of cartesian coordinate axes, designated the $\overline{x}\,\overline{y}$ axes and called the "new" coordinate system, is obtained by translation from a pair of cartesian coordinate axes, designated the xy axes and called the "old" coordinate system. Furthermore, suppose that the origin $\overline{O}$ of the $\overline{x}\,\overline{y}$ system has coordinates (h, k) with respect to the xy system. Then, if a point P has coordinates (x, y) in the xy system and $(\overline{x}, \overline{y})$ in the $\overline{x}\,\overline{y}$ system, the following translation equations must hold: $\overline{x}$ = ______ and $\overline{y}$ = ______ or, equivalently, x = ______ and y = ______. The graph is**

$x - h, \quad y - k$

$\overline{x} + h, \quad \overline{y} + k$

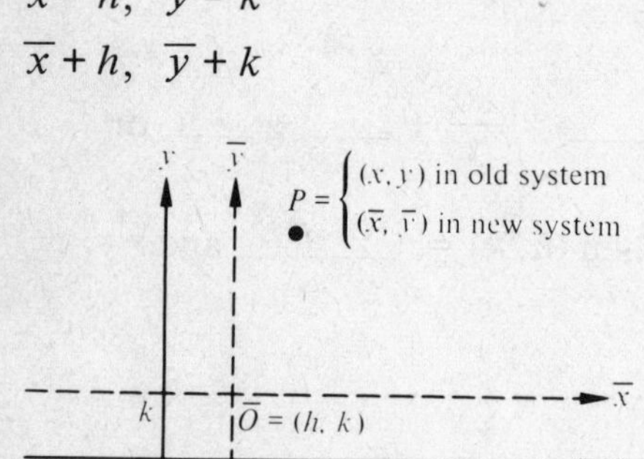

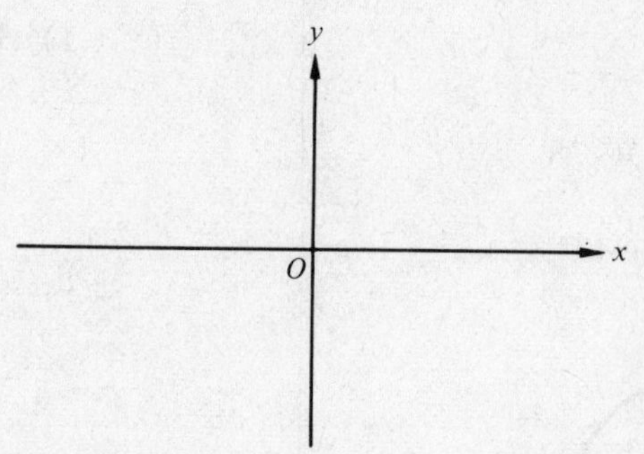

In Problems 18 through 22, find the coordinates of P in the $\overline{x}\,\overline{y}$ system obtained by translation, if $\overline{O}$ has coordinates (h, k) in the xy system and P has the given coordinates in the xy system.

18 $P = (2, -1)$ and $(h, k) = (3, -2)$.
$\overline{x} = x - h$ and $\overline{y} = y - k$, so that $\overline{x} = 2 -$ ____ = ____ and $\overline{y} =$ $-1 - ($____$) =$ ____. Therefore, $(\overline{x}, \overline{y}) =$ ________.

$3, \ -1$

$-2, \ 1, \ (-1, 1)$

19 $P = (4, 3)$ and $(h, k) = (-1, 2)$.
$\overline{x} = x - h$ and $\overline{y} = y - k$, so that $\overline{x} = 4 -$ ____ = ____ and $\overline{y} =$ $3 -$ ____ = ____. Therefore, $(\overline{x}, \overline{y}) =$ ________.

$-1, \ 5$

$2, \ 1, \ (5, 1)$

20 $P = (-5, -3)$ and $(h, k) = (-2, 1)$.
$\overline{x} = x - h$ and $\overline{y} = y - k$, so that $\overline{x} = -5 + 2 =$ ____ and $\overline{y} =$ $-3 - 1 =$ ____. Therefore, $(\overline{x}, \overline{y}) =$ ________.

-3

$-4, \ (-3, -4)$

Answers	Problems
	21 $P = (0, 3)$ and $(h, k) = (-1, -3)$.
1, 6	$\overline{x} = x - h$ and $\overline{y} = y - k$, so that $\overline{x} =$ ____ and $\overline{y} =$ ____. There-
(1, 6)	fore, $(\overline{x}, \overline{y}) =$ ________.
	22 $P = (-3, 0)$ and $(h, k) = (1, 7)$.
-4	$\overline{x} = x - h$ and $\overline{y} = y - k$, so that $\overline{x} = -3 - 1 =$ ____ and $\overline{y} =$
-7, (-4, -7)	$0 - 7 =$ ____. Therefore, $(\overline{x}, \overline{y}) =$ ________.

In Problems 23 through 25, find the coordinates of $\overline{P}$ in the xy system whose coordinates are given in the $\overline{x}\,\overline{y}$ system if $\overline{O} = (h, k)$.

Answers	Problems
	23 $\overline{P} = (5, 3)$ and $(h, k) = (2, 1)$.
2, 7	$x = \overline{x} + h$ and $y = \overline{y} + k$, so that $x = 5 +$ ____ $=$ ____ and $y =$
1, 4, (7, 4)	$3 +$ ____ $=$ ____. Therefore, $(x, y) =$ ________.
	24 $\overline{P} = (6, -1)$ and $(h, k) = (-3, -2)$.
3, -3	$x = \overline{x} + h$ and $y = \overline{y} + k$. $x =$ ____ and $y =$ ____. Therefore,
(3, -3)	$(x, y) =$ ________.
	25 $\overline{P} = (-2, -7)$ and $(h, k) = (1, 0)$.
-1, -7	$x = \overline{x} + h$ and $y = \overline{y} + k$. $x =$ ____ and $y =$ ____. Therefore,
(-1, -7)	$(x, y) =$ ________.

In Problems 26 through 28, convert each equation to the form $\overline{x}^2 + \overline{y}^2 = r^2$, using $x = \overline{x} + h$ and $y = \overline{y} + k$.

Answers	Problems
	26 $x^2 + y^2 - 4x + 8y + 8 = 0$.
$\overline{y} + k$	$(\overline{x} + h)^2 + (\overline{y} + k)^2 - 4(\overline{x} + h) + 8($________$) + 8 = 0$.
	Simplifying, we get
$2h - 4$, $2k + 8$	$\overline{x}^2 + \overline{y}^2 + ($________$)\overline{x} + ($________$)\overline{y} + h^2 + k^2 - 4h + 8k + 8 = 0$.
	The terms involving $\overline{x}$ and $\overline{y}$ to the first power will drop out if we
2, -4, 12	put $h =$ ____ and $k =$ ____. Hence, $\overline{x}^2 + \overline{y}^2 =$ ____.
	27 $x^2 + y^2 - 8x - 4y + 11 = 0$.
	$(\overline{x} + h)^2 + (\overline{y} + k)^2 - 8(\overline{x} + h) - 4(\overline{y} + k) + 11 = 0$. Simplifying,
	we get
$2h - 8$, $2k - 4$	$\overline{x}^2 + \overline{y}^2 + ($________$)\overline{x} + ($________$)\overline{y} + h^2 + k^2 - 8h - 4k + 11 = 0$.
4, 2	If $2h - 8 = 0$ and $2k - 4 = 0$, then $h =$ ____ and $k =$ ____. Hence,
7	$\overline{x}^2 + \overline{y}^2 =$ ____.
	28 $x^2 + y^2 + 6x - 4y - 3 = 0$.
	The new equation can also be obtained by completing the square
9, 4, 16	so that $(x^2 + 6x +$ ____$) + (y^2 - 4y +$ ____$) =$ ____ or
$y - 2$	$(x + 3)^2 + ($________$)^2 = 16$. Let $\overline{x} = x + 3$ and $\overline{y} = y - 2$, so that
16	$\overline{x}^2 + \overline{y}^2 =$ ____.

In Problems 29 and 30, find the equations of the tangent and normal lines to each circle at the given point.

29 $x^2 + y^2 + 5x - 6y + 2 = 0$ at (1, 2).

The slope of the tangent line is obtained by finding $\frac{dy}{dx}$ at ______, so that $2x + 2y\frac{dy}{dx} + 5 - 6\left(\_\_\_\right) = 0$ or $\frac{dy}{dx} = \frac{\_\_\_\_}{2y - 6} = \frac{-2(\_\_\_) - 5}{2(2) - 6} = \_\_\_$. The equation of the tangent line is $y - 2 = \_\_\_ (x - 1)$. The equation of the normal line is $y - 2 = \_\_\_ (x - 1)$.

(1, 2), $\frac{dy}{dx}$

$-2x - 5$, 1, $\frac{7}{2}$

$\frac{7}{2}$

$-\frac{2}{7}$

30 $3x^2 + 3y^2 + 6x - 12y = 0$ at (1, 1).

The slope $\frac{dy}{dx}$ of the tangent line is given by

$6x + 6y\frac{dy}{dx} + 6 - 12\left(\_\_\_\right) = 0$, so that $\frac{dy}{dx} = \frac{\_\_\_\_}{6y - 12} = \_\_\_$. The equation of the tangent line is $y - 1 = 2(\_\_\_\_)$.

The equation of the normal line is $y - 1 = \_\_\_ (x - 1)$.

$\frac{dy}{dx}$, $-6x - 6$

2, $x - 1$

$-\frac{1}{2}$

2 The Ellipse

In Problems 31 through 35, use the adjacent figure.

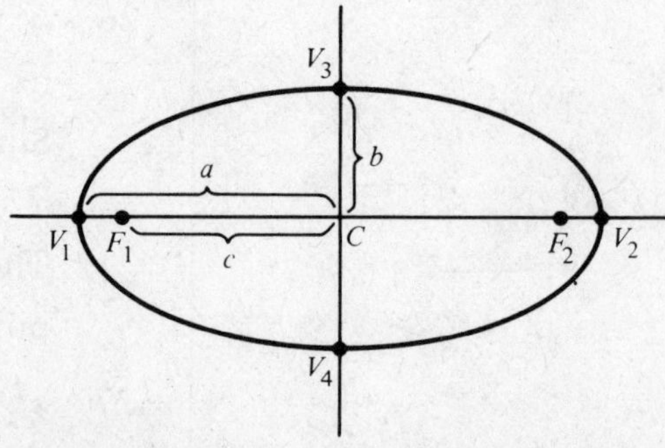

31 An ellipse is defined to be the set of all points P in the plane such that the sum, $|\overline{PF_1}| + |\overline{PF_2}|$, of the distances from P to ______ points F_1 and F_2 is ______.

32 F_1 and F_2 are points in the plane called the ______ or the ______ of the ellipse. The midpoint C of the line segment $\overline{F_1F_2}$ between the foci is called the ______ of the ellipse.

33 The four points V_1, V_2, V_3, and V_4 where the axes of symmetry intersect the ellipse are called the ______ of the ellipse.

two fixed

constant

focal points

foci

center

vertices

major axis

minor

semimajor, semiminor

34 The line segment $\overline{V_1V_2}$ between the two vertices that contains the two foci F_1 and F_2 is called the ______________ of the ellipse, while the line segment $\overline{V_3V_4}$ between the remaining vertices is called the ________ axis. The numbers a and b are called the ______________ axis and the ______________ axis of the ellipse, respectively.

$\frac{x^2}{a^2} + \frac{y^2}{b^2} = 1$, semimajor

semiminor, $b^2 + c^2$

standard form

35 The equation of the ellipse with foci $F_1 = (-c, 0)$ and $F_2 = (c, 0)$ is ______________, where a is the ______________ axis, b is the ______________ axis, and $a^2 =$ ____________. The equation $\frac{x^2}{a^2} + \frac{y^2}{b^2} = 1$ is called the ______________ for the equation of the ellipse.

In Problems 36 through 39, find the coordinates of the vertices and foci of the ellipse. Also, sketch the graph.

16

1, 4, 2

(0, −2)

$a^2 - b^2$, 12

$2\sqrt{3}$, $(-2\sqrt{3}, 0)$, $(2\sqrt{3}, 0)$

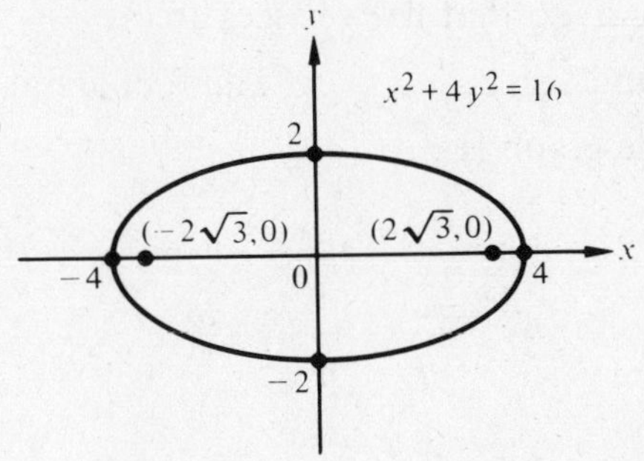

36 $x^2 + 4y^2 = 16$.

Divide both sides of the equation by _____, so that $\frac{x^2}{16} + \frac{y^2}{4} =$ _____. Then $a =$ _____ and $b =$ _____; therefore, the vertices are at (4, 0), (−4, 0), (0, 2), and __________. The foci are at $(-c, 0)$ and $(c, 0)$, where $c^2 =$ __________ $= 16 - 4 =$ _____ and $c = \sqrt{12} =$ ______. Thus, $F_1 =$ ______________ and $F_2 =$ ______________. The graph is

32

1, $4\sqrt{2}$, 4

(0, −4)

$a^2 - b^2$, 16, 4, (−4, 0)

(4, 0)

37 $x^2 + 2y^2 = 32$.

Divide both sides of the equation by _____, so that $\frac{x^2}{32} + \frac{y^2}{16} =$ _____. $a = \sqrt{32} =$ _______ $b =$ _____. Therefore, the vertices are at $(4\sqrt{2}, 0)$, $(-4\sqrt{2}, 0)$, (0, 4), and __________. $c^2 =$ __________ $=$ _____, so that $c =$ _____. The foci are at __________ and __________. The graph is

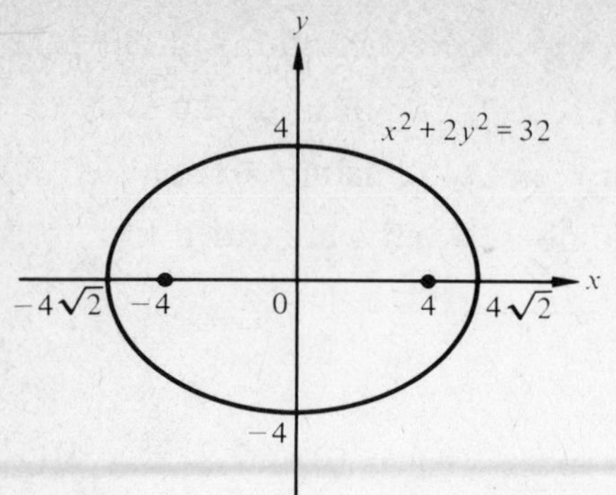

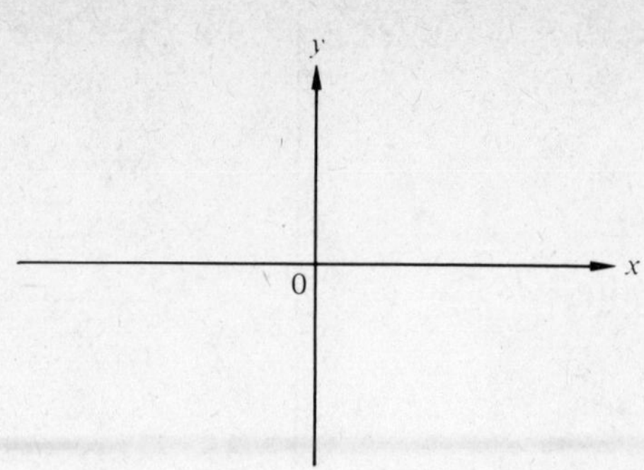

38 $4x^2 + 9y^2 = 144$.

Divide both sides of the equation by _____, so that $\frac{x^2}{36} + \frac{y^2}{16} =$ _____. $a =$ _____ and $b =$ _____. Therefore, the vertices are at (6, 0), (-6, 0), __________, and __________. $c^2 =$ __________ = _____, so that $c =$ _____. The foci are at $(2\sqrt{5}, 0)$ and ______________. The graph is

144

1, 6, 4

(0, 4), (0, -4), $a^2 - b^2$

20, $2\sqrt{5}$

$(-2\sqrt{5}, 0)$

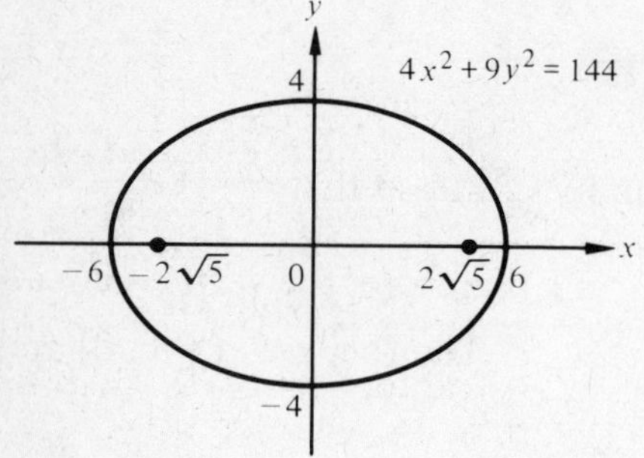

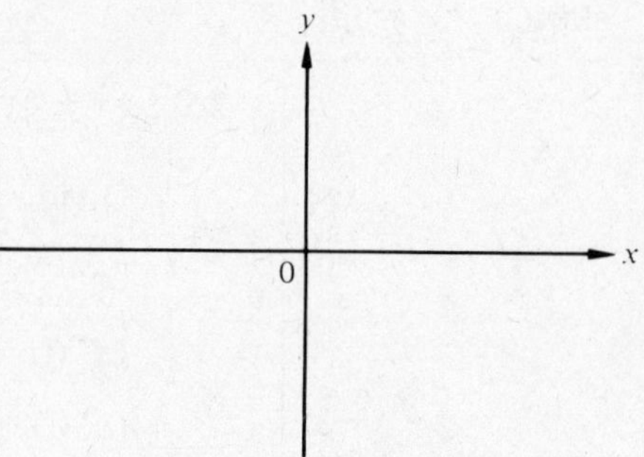

39 $\frac{x^2}{169} + \frac{y^2}{144} = 1.$

$a =$ _____, $b =$ _____, and $c =$ _____, so that the vertices are at (13, 0), (-13, 0), __________, and ____________. The foci are at __________ and __________. The graph is

13, 12, 5

(0, 12), (0, -12)

(5, 0), (-5, 0)

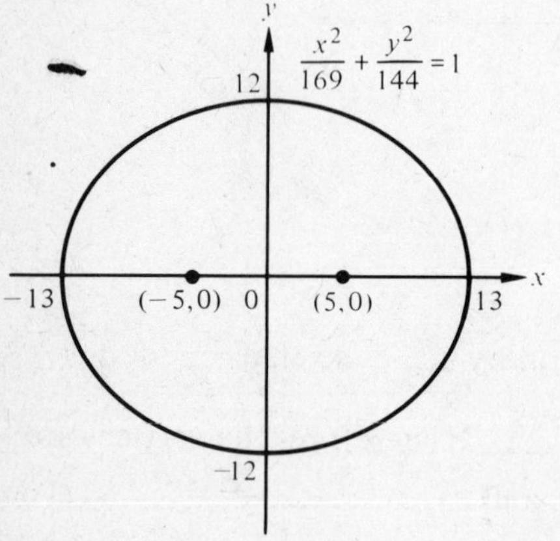

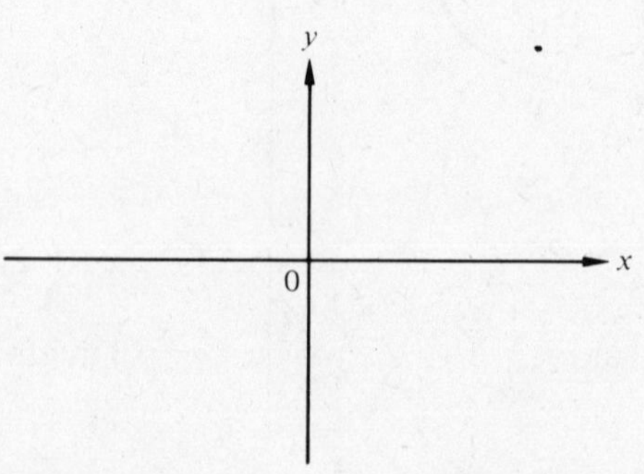

40 If $F_1 = (0, -c)$, $F_2 = (0, c)$, $V_1 = (0, -a)$, $V_2 = (0, a)$, $V_3 = (-b, 0)$, and $V_4 = (b, 0)$, then the equation of the ellipse is

$\frac{x^2}{b^2} + \frac{y^2}{a^2} = 1$, semimajor

semiminor

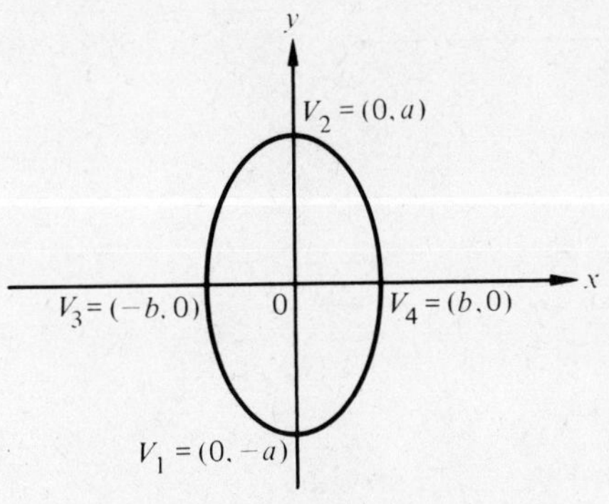

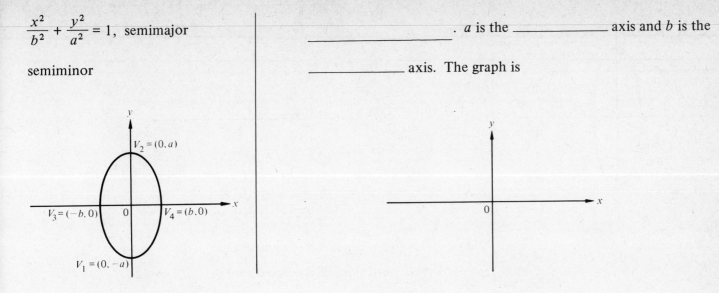

____________________. a is the _______________ axis and b is the _______________ axis. The graph is

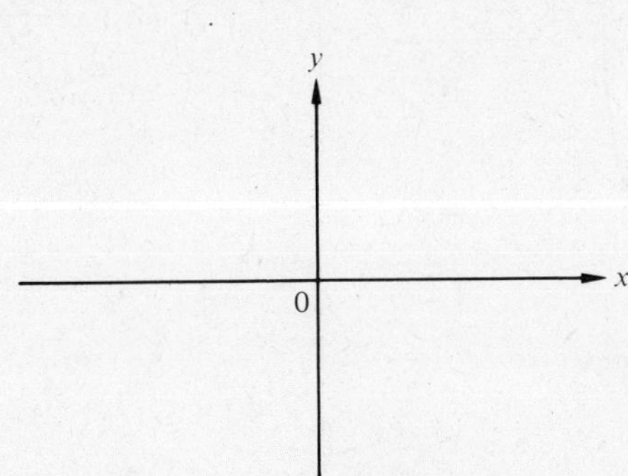

In Problems 41 through 43, find the coordinates of the vertices and foci of the ellipse. Also, sketch the graph.

225

$\frac{x^2}{9} + \frac{y^2}{25} = 1$, 5, 3

$a^2 - b^2$, 16, 4

(−3, 0)

(0, −4)

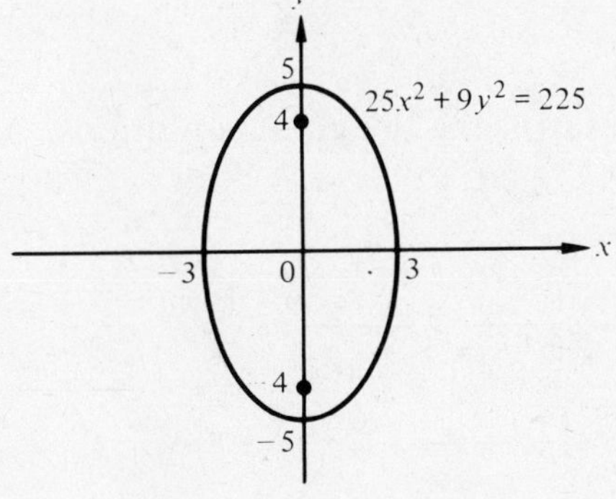

41 $25x^2 + 9y^2 = 225$.

Divide both sides of the equation by _____, so that ____________________. a = _____ and b = _____. c^2 = __________ = 25 − 9 = _____, then c = _____. The vertices are at (0, 5), (0, −5), (3, 0), and __________. The foci are at (0, 4) and __________. The graph is

36

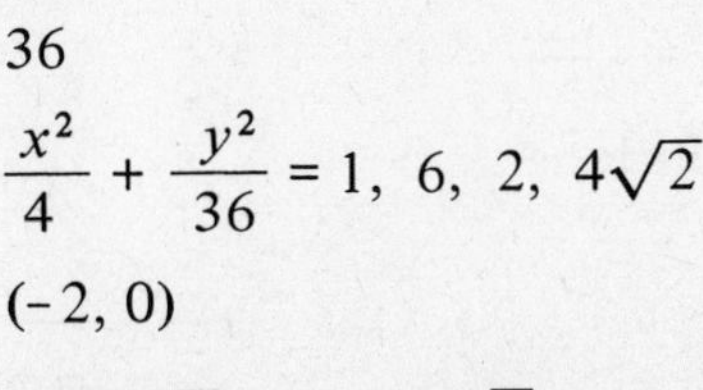

$\frac{x^2}{4} + \frac{y^2}{36} = 1$, 6, 2, $4\sqrt{2}$

(−2, 0)

$(0, 4\sqrt{2})$, $(0, -4\sqrt{2})$

42 $9x^2 + y^2 = 36$.

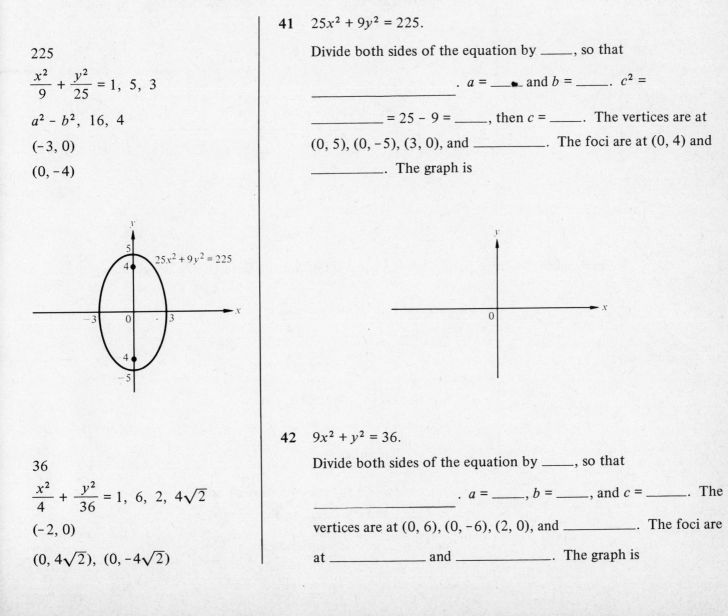

Divide both sides of the equation by _____, so that ____________________. a = _____, b = _____, and c = _____. The vertices are at (0, 6), (0, −6), (2, 0), and __________. The foci are at __________ and __________. The graph is

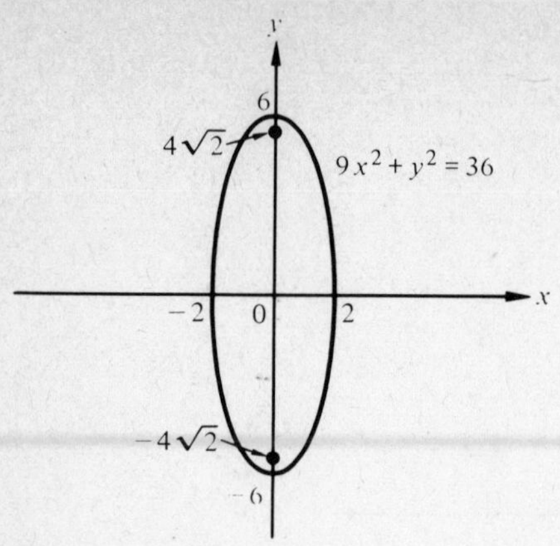

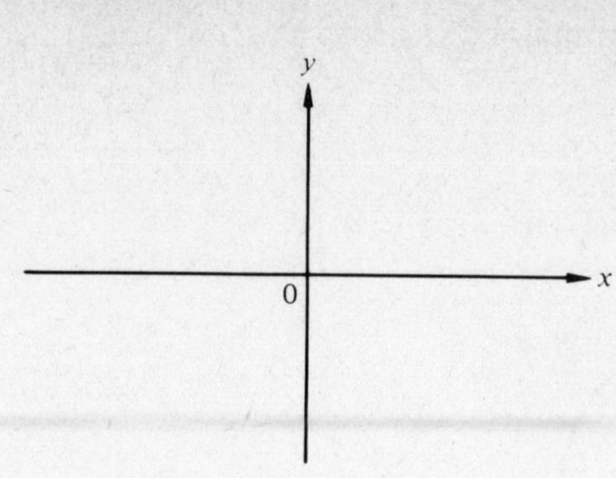

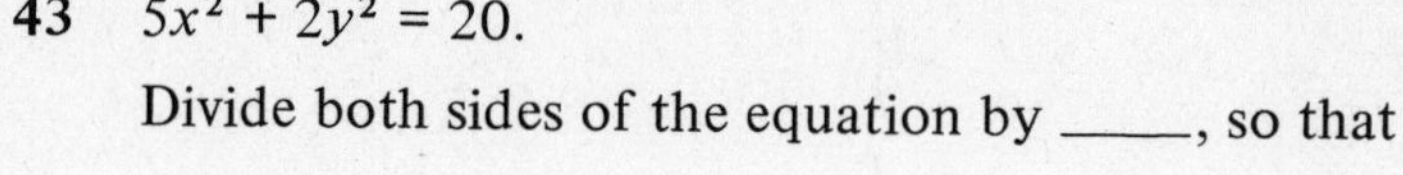

43 $5x^2 + 2y^2 = 20$.

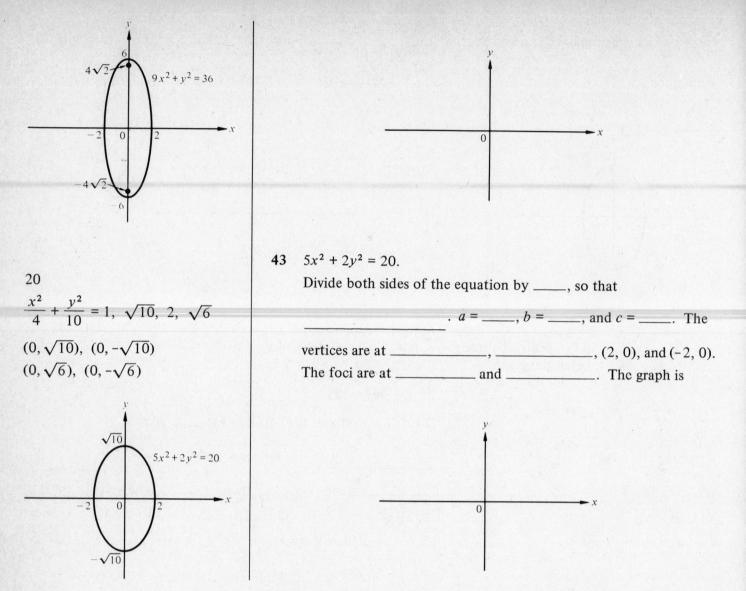

Divide both sides of the equation by ____, so that ________________. $a =$ ____, $b =$ ____, and $c =$ ____. The vertices are at ____________, ____________, (2, 0), and (−2, 0). The foci are at __________ and ____________. The graph is

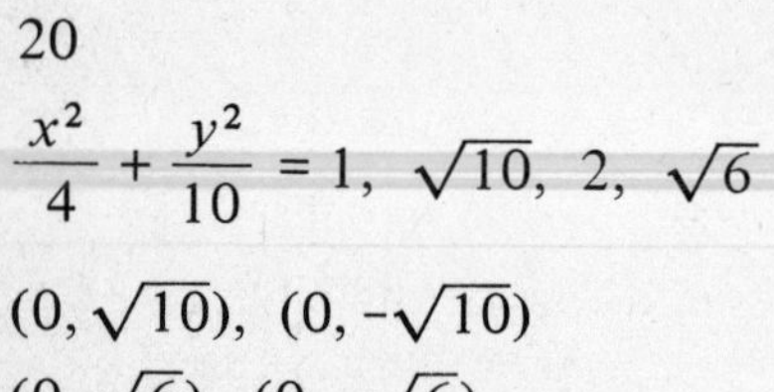

20

$\frac{x^2}{4} + \frac{y^2}{10} = 1$, $\sqrt{10}$, 2, $\sqrt{6}$

$(0, \sqrt{10})$, $(0, -\sqrt{10})$

$(0, \sqrt{6})$, $(0, -\sqrt{6})$

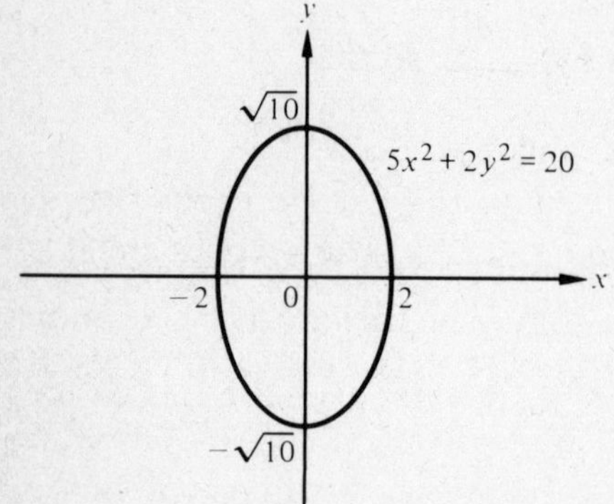

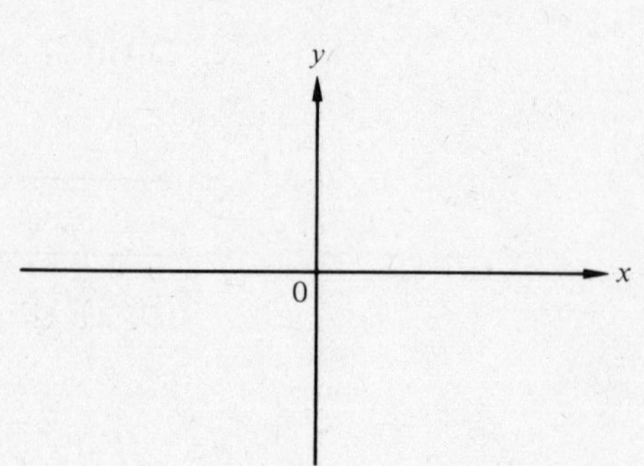

In Problems 44 through 49, find the equation of the ellipse satisfying the given conditions.

44 Foci at (±3, 0) and vertices at (±5, 0).

$a =$ ____, $c =$ ____, and $b^2 =$ __________ $=$ ____, so that $b =$ ____. The equation is ____________________.

5, 3, $a^2 - c^2$, 16

4, $\frac{x^2}{25} + \frac{y^2}{16} = 1$

45 Vertices at (±2, 0) and (0, ±1).

$a =$ ____, $b =$ ____, and $c^2 =$ __________ $=$ ____, so that $c =$ ____. The equation is ____________________.

2, 1, $a^2 - b^2$, 3

$\sqrt{3}$, $\frac{x^2}{4} + \frac{y^2}{1} = 1$

46 Foci at (±5, 0) and minor axis is 24 units long.

$c =$ ____ and $b =$ ____, so that $a^2 = b^2 + c^2 = 5^2 + 12^2 =$ ____ and $a =$ ____. The equation is ____________________.

5, 12, 169

13, $\frac{x^2}{169} + \frac{y^2}{144} = 1$

47 Vertices at $(0, \pm 5)$ and $(\pm 4, 0)$.
$a = 5$, $b =$ ____, and $c =$ ____. The major axis is on the ________ axis, so that the equation is written in the form __________. Thus, __________.

4, 3

vertical

$\frac{x^2}{b^2} + \frac{y^2}{a^2} = 1$, $\frac{x^2}{16} + \frac{y^2}{25} = 1$

48 Vertices at $(\pm 4, 0)$ and containing the point $\left(3, \frac{3\sqrt{7}}{4}\right)$.

Let $a =$ ____ and the standard equation is __________.

Since the point $\left(3, \frac{3\sqrt{7}}{4}\right)$ is on the ellipse, then $\frac{9}{16} + \frac{63}{16b^2} = 1$ or $7b^2 =$ ____ or $b^2 =$ ____. The equation is __________.

4, $\frac{x^2}{a^2} + \frac{y^2}{b^2} = 1$

63, 9

$\frac{x^2}{16} + \frac{y^2}{9} = 1$

49 Foci on the y axis, the distance between the foci is half the length of the major axis; center at the origin. Contains the point $\left(\frac{9}{2}, 3\right)$.

$c = \frac{1}{2}$ (____) and $a^2 =$ ________, so that $a^2 =$ ________ or $b^2 =$ ____. The standard equation is of the form $\frac{x^2}{b^2} + \frac{y^2}{a^2} =$ 1 or __________. Since the point $\left(\frac{9}{2}, 3\right)$ is on the ellipse, then $\frac{\frac{81}{4}}{\frac{3a^2}{4}} + \frac{9}{a^2} =$ ____ or $a^2 =$ ____ and $b^2 =$ ____. The equation is __________.

a, $b^2 + c^2$, $b^2 + \frac{a^2}{4}$

$\frac{3a^2}{4}$

$\frac{x^2}{\frac{3a^2}{4}} + \frac{y^2}{a^2} = 1$

1, 36, 27

$\frac{x^2}{27} + \frac{y^2}{36} = 1$

In Problems 50 and 51, use the translation $\overline{x} = x - h$, $\overline{y} = y - k$ to find the equation of an ellipse whose axes of symmetry are parallel to the coordinate axes and whose center is at (h, k), if the ellipse has:

50 A horizontal major axis.
Here a is the semimajor axis, $b < a$, and $c = \sqrt{\rule{2cm}{0.4pt}}$. The center is at ______. The equation is __________. The graph is __________

$a^2 - b^2$

(h, k)

$\frac{(x - h)^2}{a^2} + \frac{(y - k)^2}{b^2} = 1$

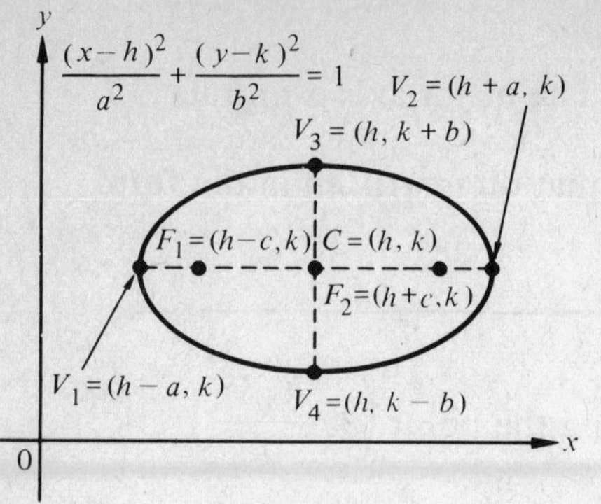

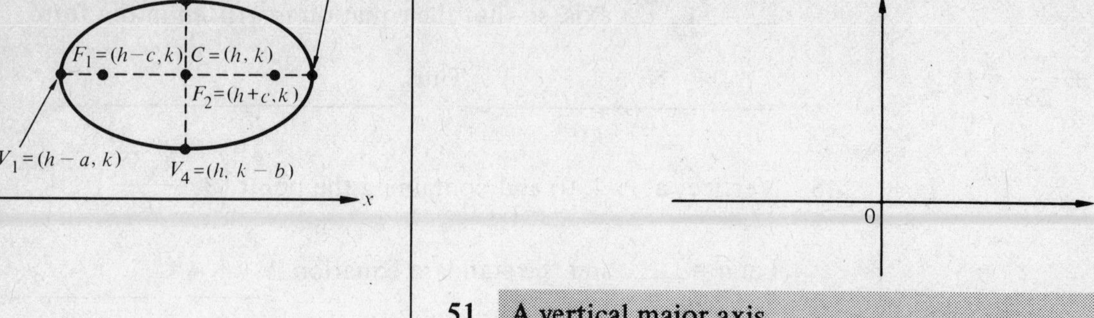

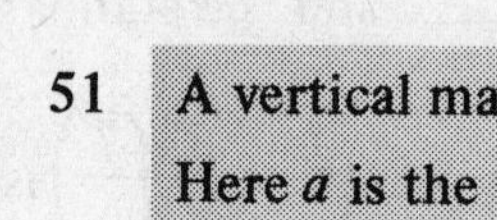

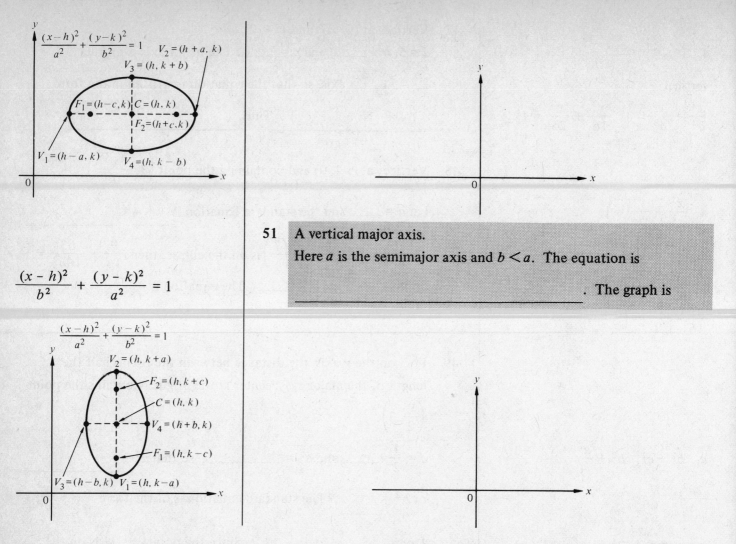

51 A vertical major axis.
Here a is the semimajor axis and $b < a$. The equation is

____. The graph is

$$\frac{(x-h)^2}{b^2} + \frac{(y-k)^2}{a^2} = 1$$

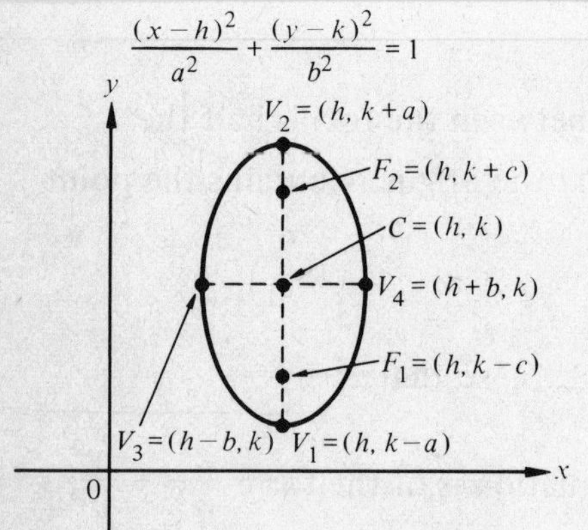

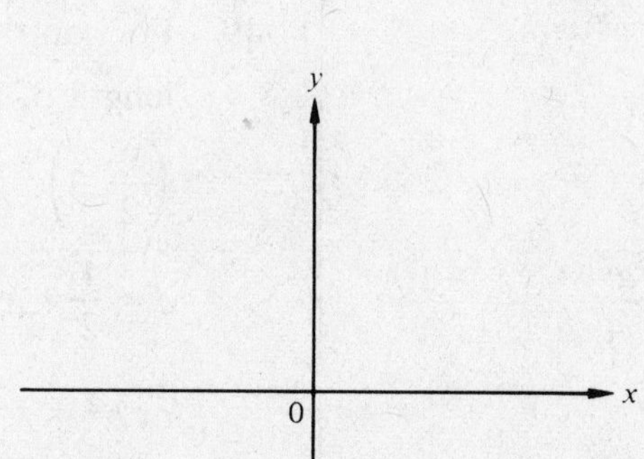

In Problems 52 through 56, find the coordinates of the center, the vertices, and the foci of the ellipse. Also, sketch the graph.

52 $\frac{(x-8)^2}{4} + \frac{(y+2)^2}{1} = 1.$

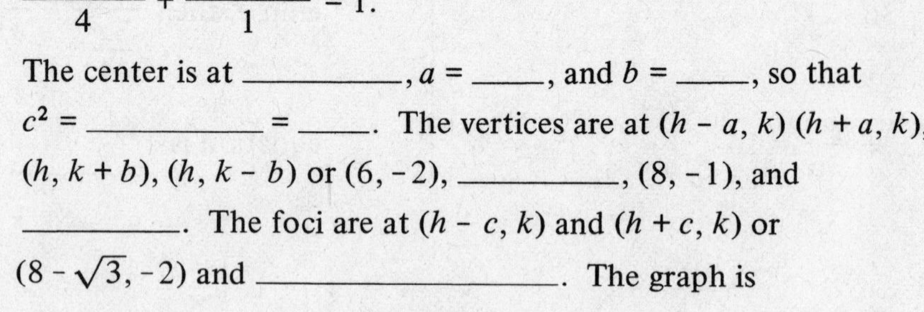

The center is at ________, $a =$ ____, and $b =$ ____, so that $c^2 =$ __________ $=$ ____. The vertices are at $(h - a, k)$ $(h + a, k)$, $(h, k + b)$, $(h, k - b)$ or $(6, -2)$, __________, $(8, -1)$, and __________. The foci are at $(h - c, k)$ and $(h + c, k)$ or $(8 - \sqrt{3}, -2)$ and ______________. The graph is

$(8, -2)$, 2, 1
$a^2 - b^2$, 3
$(10, -2)$
$(8, -3)$
$(8 + \sqrt{3}, -2)$

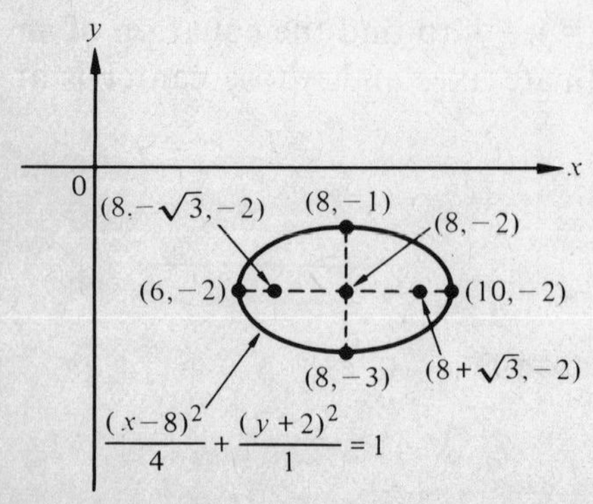

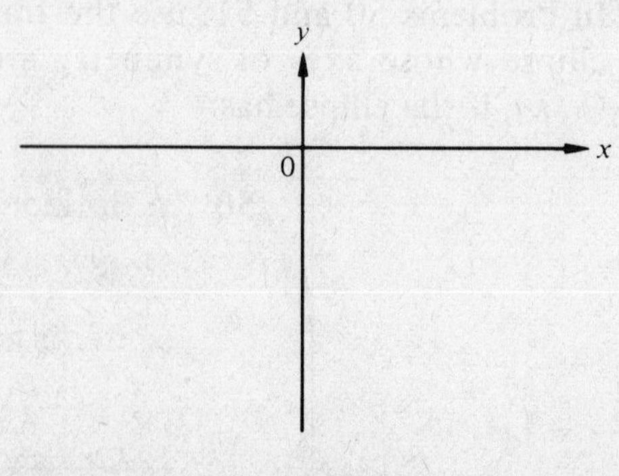

$(-2, 3)$, 2, 1
$\sqrt{3}$
$(-3, 3)$
$(-1, 3)$
$(-2, 3 + \sqrt{3})$

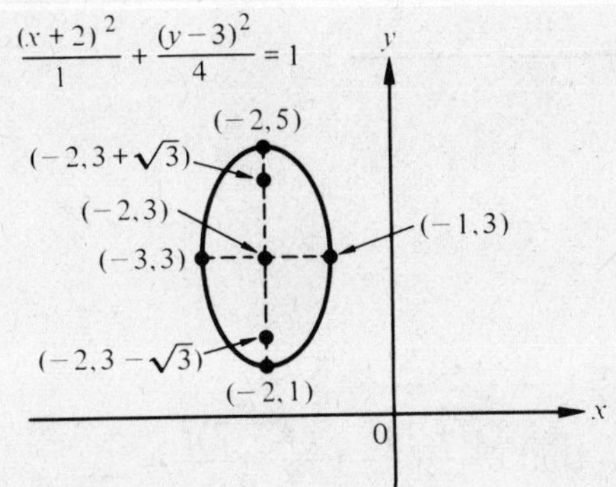

53 $\dfrac{(x+2)^2}{1} + \dfrac{(y-3)^2}{4} = 1.$

The center is at ________, $a =$ ____, and $b =$ ____, so that $c = \sqrt{a^2 - b^2} =$ _____. The vertices are at $(h, k - a)$, $(h, k + a)$, $(h - b, k)$, and $(h + b, k)$ or $(-2, 1)$, $(-2, 5)$, ________, and ________. The foci are at $(h, k - c)$ and $(h, k + c)$ or $(-2, 3 - \sqrt{3})$ and ____________. The graph is

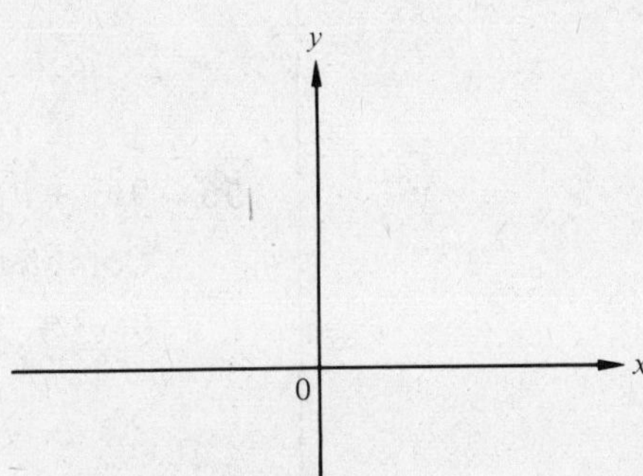

1, 1, 16
$y + 1$, 144
1, $(1, -1)$, 4
3, $\sqrt{7}$
$(1, 2)$
$(1 - \sqrt{7}, -1)$

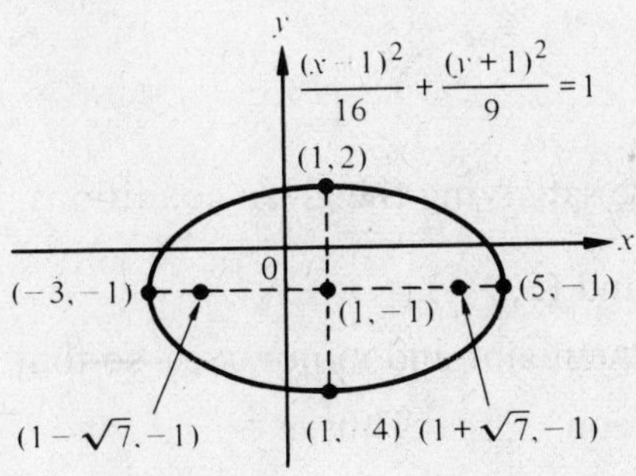

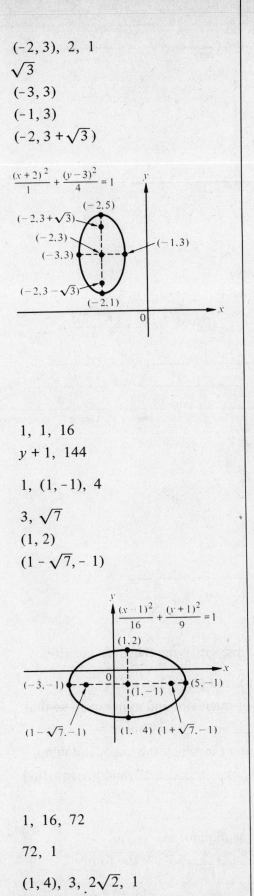

54 $9x^2 + 16y^2 - 18x + 32y - 119 = 0.$

Completing the square, we get

$9(x^2 - 2x +$ ____$) + 16(y^2 + 2y +$ ____$) = 119 + 9 +$ ____,

so that $9(x-1)^2 + 16($________$)^2 =$ _____ or

$\dfrac{(x-1)^2}{16} + \dfrac{(y+1)^2}{9} =$ ____. $(h, k) =$ __________, $a =$ _____,

and $b =$ ____, so that $c = \sqrt{16 - 9} =$ _____. Vertices are at $(-3, -1)$, $(5, -1)$, $(1, -4)$, and ________. Foci at $(1 + \sqrt{7}, -1)$ and ____________. The graph is

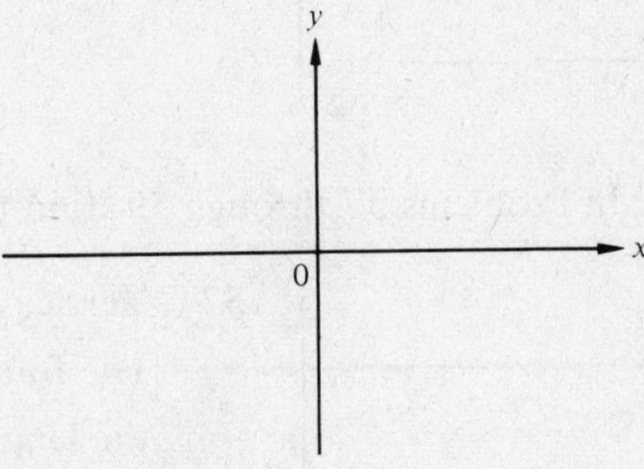

1, 16, 72

72, 1

$(1, 4)$, 3, $2\sqrt{2}$, 1
$(1, 4 + 2\sqrt{2})$

55 $8x^2 + 9y^2 - 16x - 72y + 80 = 0.$

Completing the square, we have

$8(x^2 - 2x +$ ____$) + 9(y^2 - 8y +$ ____$) =$ ____ or

$8(x-1)^2 + 9(y-4)^2 =$ ____ or $\dfrac{(x-1)^2}{9} + \dfrac{(y-4)^2}{8} =$ ____.

$(h, k) =$ ________. $a =$ ____, $b =$ ______, and $c =$ ____.

Vertices at $(-2, 4)$, $(4, 4)$, ____________, and

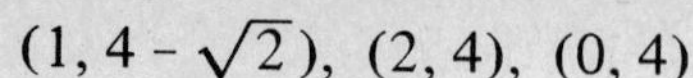

$(1, 4 - \sqrt{2})$, $(2, 4)$, $(0, 4)$

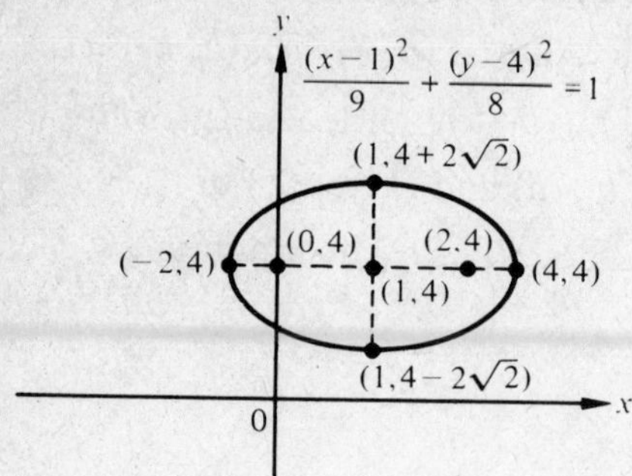

________________. Foci at ________ and ________.

The graph is

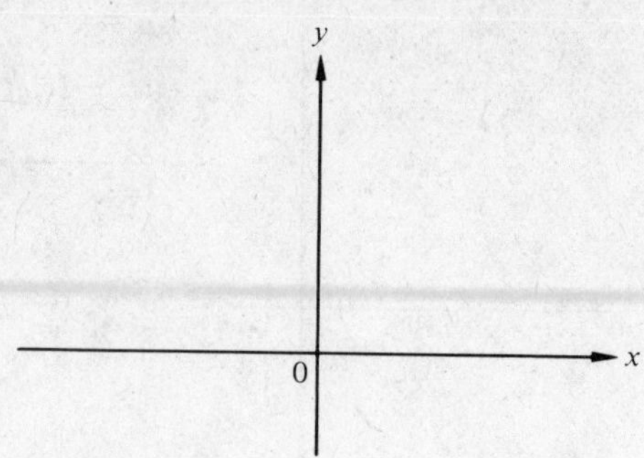

1, 4, 36

36, 1

$(-1, 2)$, 3, 2, $\sqrt{5}$

$(-1, -1)$, $(-1, 5)$

$(-1, 2 - \sqrt{5})$, $(-1, 2 + \sqrt{5})$

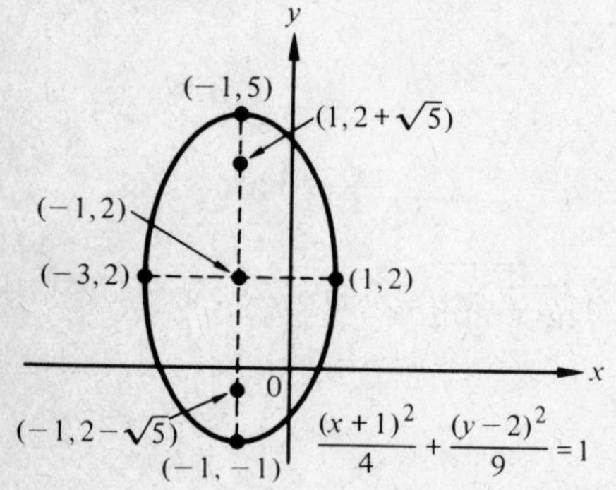

56 $9x^2 + 4y^2 + 18x - 16y - 11 = 0.$

Completing the square, we have

$9(x^2 + 2x + \_\_\_\_) + 4(y^2 - 4y + \_\_\_\_) = \_\_\_\_$ or

$9(x + 1)^2 + 4(y - 2)^2 = \_\_\_\_$, so that $\frac{(x+1)^2}{4} + \frac{(y-2)^2}{9} = \_\_\_\_$.

$(h, k) = \_\_\_\_\_\_\_\_$, $a = \_\_\_\_$, $b = \_\_\_\_$, and $c = \_\_\_\_$.

Vertices are at $(-3, 2)$, $(1, 2)$, ________, and ________.

Foci are at ____________ and ____________.

The graph is

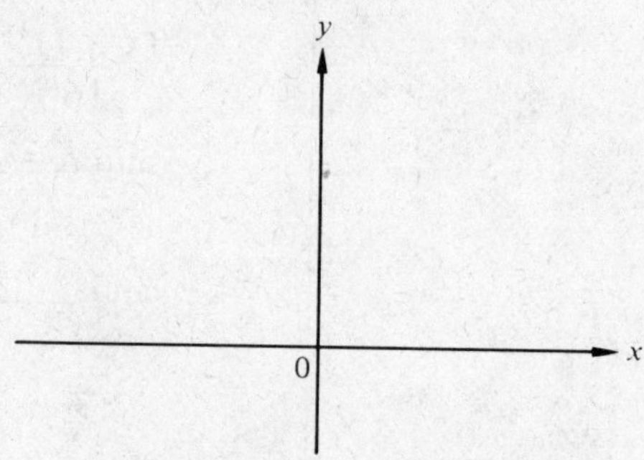

In Problems 57 through 59, find the equation of the ellipse satisfying the given conditions.

10, 4, 5

2

$(3, 1)$

$\frac{(x - 3)^2}{4} + \frac{(y - 1)^2}{25} = 1$

vertical

$\frac{(x - h)^2}{b^2} + \frac{(y - k)^2}{a^2} = 1$, $(2, 3)$

57 Vertices at $(1, 1)$, $(5, 1)$, $(3, 6)$, and $(3, -4)$.

The four points are the ends of the major and minor axes, so that the lengths of the axes are ____ and ____. Thus, $a = \_\_\_\_$ and $b = \_\_\_\_$. The center is the point in which the major and minor axes intersect. In this case $(h, k) = \_\_\_\_\_\_\_\_$ and the equation is ________________.

58 Foci at $(2, 9)$ and $(2, -3)$ and semiminor axis is 5.

Since the major axis is ________, we use the form ________________. $(h, k) = \_\_\_\_\_\_\_\_$,

6, 5, 61

$$\frac{(x-2)^2}{25}+\frac{(y-3)^2}{61}=1$$

$c =$ ____, and $b =$ ____, so that $a^2 = b^2 + c^2 =$ ____. The equation is ________________.

horizontal

$$\frac{(x-h)^2}{a^2}+\frac{(y-k)^2}{b^2}=1,\ (4, 4)$$

4, 6

$$\frac{(x-4)^2}{36}+\frac{(y-4)^2}{16}=1$$

59 Vertices at $(-2, 4)$ and $(10, 4)$ and semiminor axis is 4. Since the major axis is ________________, we use the form ________________. $(h, k) =$ ________, $b =$ ____, and $a = \frac{10+2}{2} =$ ____. The equation is ________________.

$-\frac{18x}{32y}$

$-\frac{3}{4}$

$y - 3 = -\frac{3}{4}(x - 4)$

$y - 3 = \frac{4}{3}(x - 4)$

60 Find the equations of the tangent and normal lines to the ellipse $9x^2 + 16y^2 = 288$ at the point $(4, 3)$.
The slope of the tangent line is obtained by differentiating both sides of the equation, so that $18x + 32y\frac{dy}{dx} = 0$ or $\frac{dy}{dx} =$ ________, so at $(4, 3)$, $\frac{dy}{dx} =$ ____. The equation of the tangent line is ________________, and the equation of the normal line is ________________.

3 The Parabola

focus

directrix

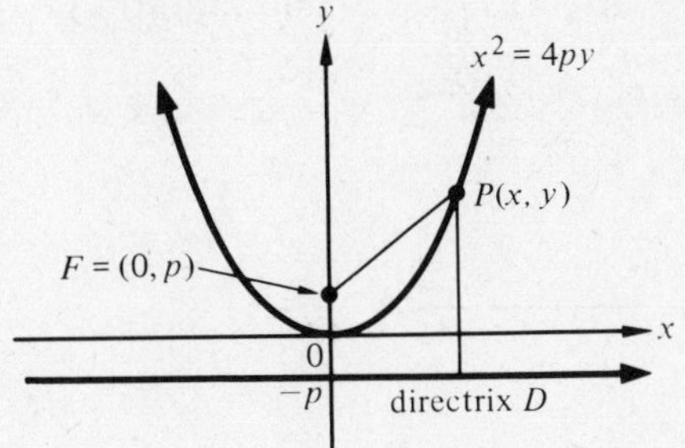

$x^2 = 4py$

symmetric

61 A parabola is defined to be the set of all points P in the plane such that the distance $|\overline{PF}|$ from P to a fixed point F, called the ________, is equal to the distance from P to a fixed straight line D, called the ____________. The graph is

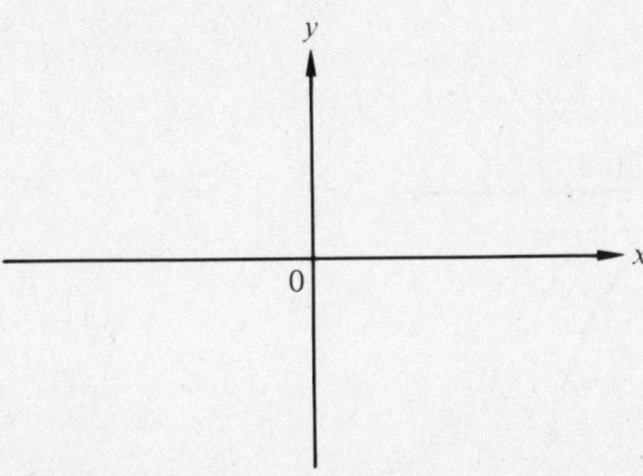

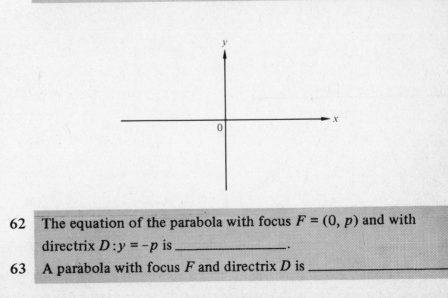

62 The equation of the parabola with focus $F = (0, p)$ and with directrix $D: y = -p$ is ____________.

63 A parabola with focus F and directrix D is ________________

perpendicular

symmetry

the vertex

latus rectum

focal chord

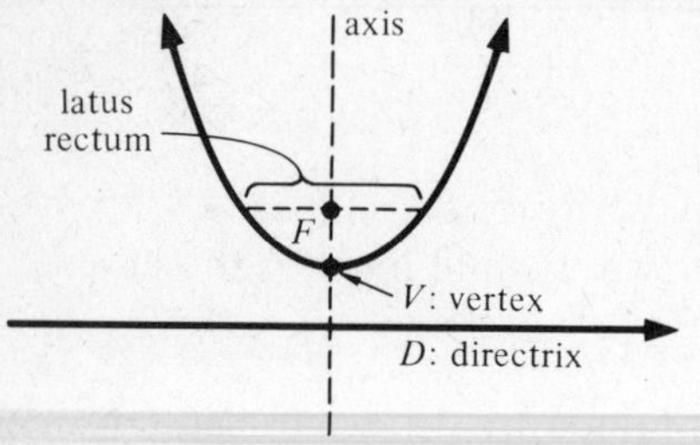

about the straight line through F __________ to D. The straight line is called the axis of __________ or the axis of the parabola. The point V where the parabola intersects its axis is called __________ of the parabola. The segment cut by the parabola on the straight line through the focus and perpendicular to the axis is called the __________ or the __________ of the parabola. The graph is

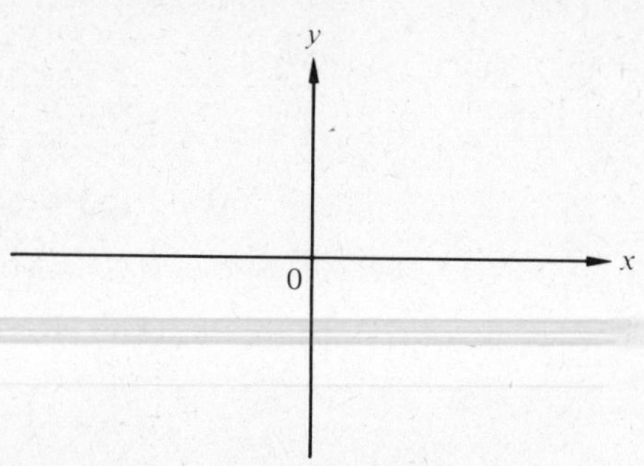

midway

64 The vertex of the parabola is located on its axis __________ between the focus F and the directrix D.

upward

65 A parabola with focus $F = (0, p)$, directrix $y = -p$, and equation $x^2 = 4py$ opens __________. The graph is

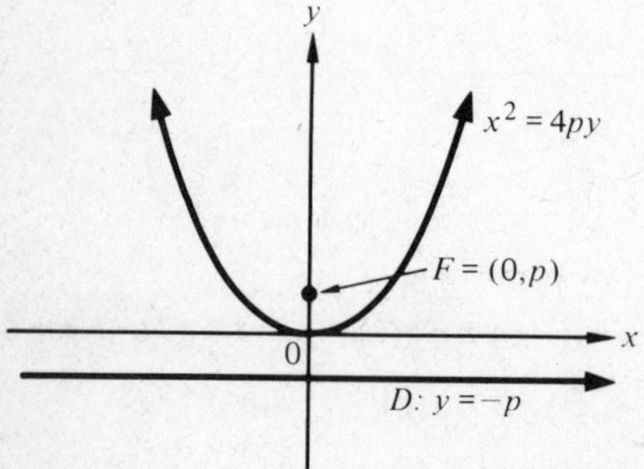

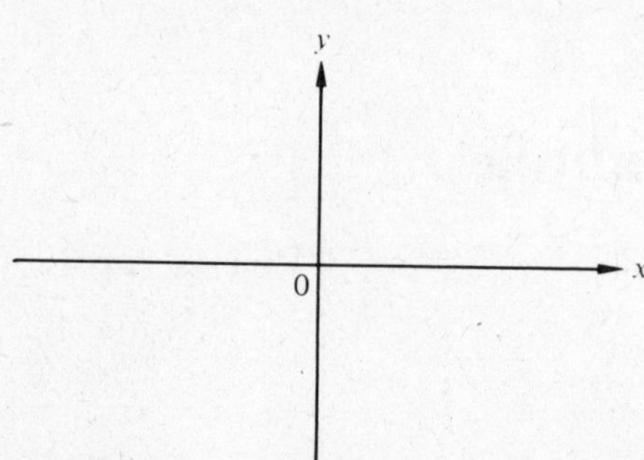

downward

66 A parabola with focus $F = (0, -p)$, directrix $y = p$, and equation $x^2 = -4py$ opens __________. The graph is

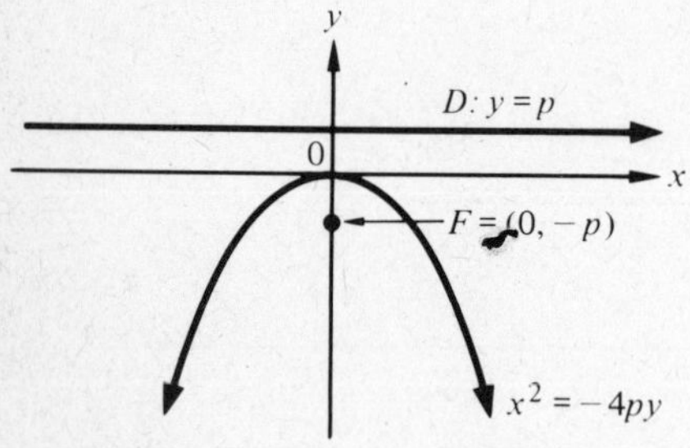

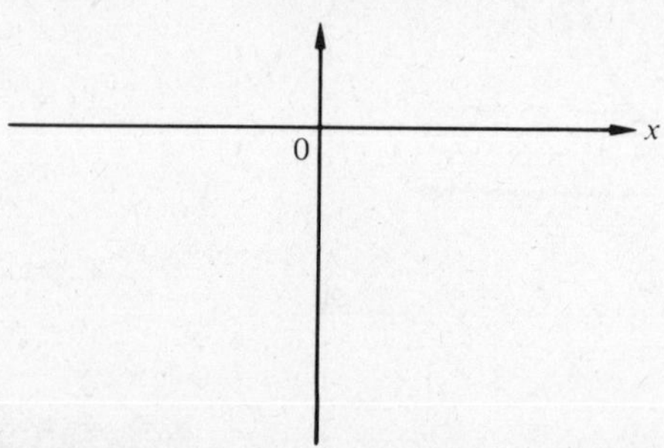

right

67 A parabola with focus $F = (p, 0)$, directrix $x = -p$, and equation $y^2 = 4px$ opens to the __________. The graph is

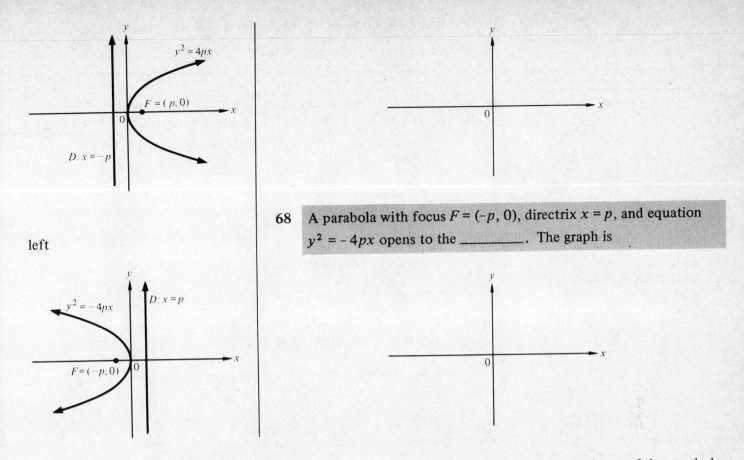

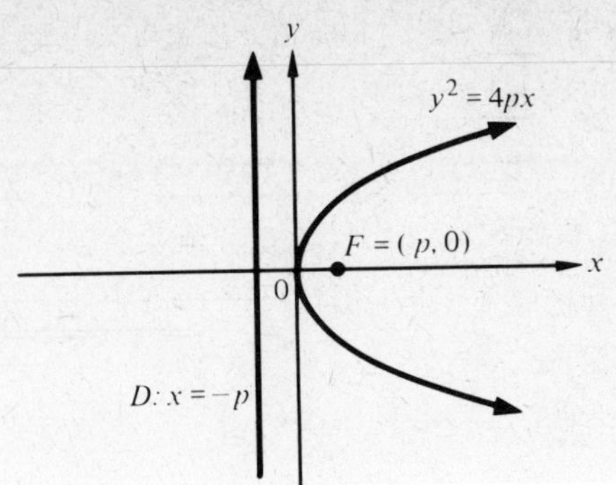

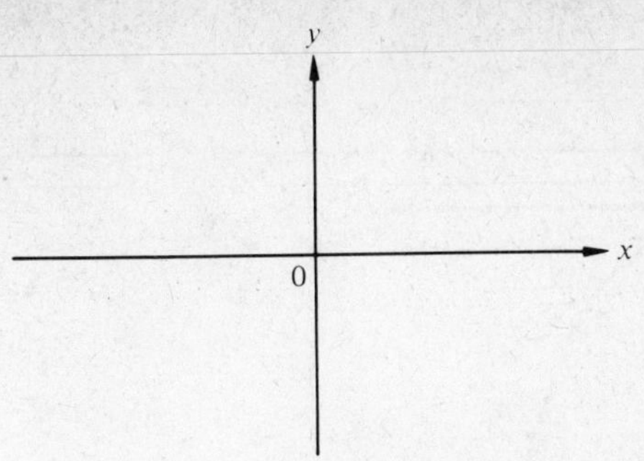

68 **A parabola with focus $F = (-p, 0)$, directrix $x = p$, and equation $y^2 = -4px$ opens to the ______. The graph is**

left

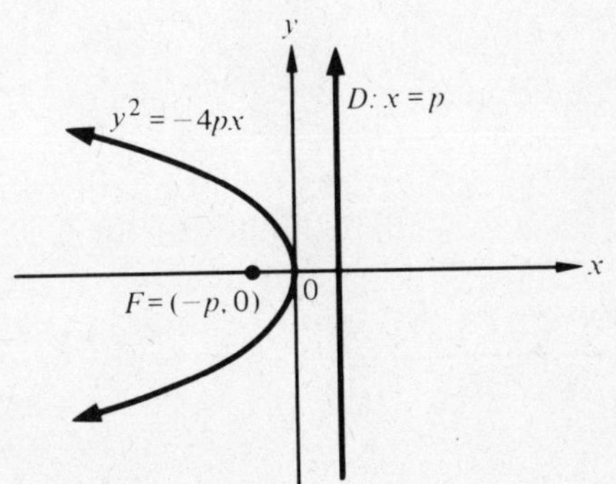

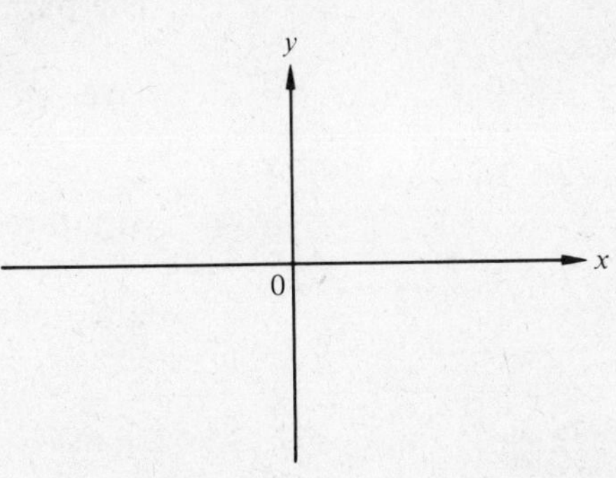

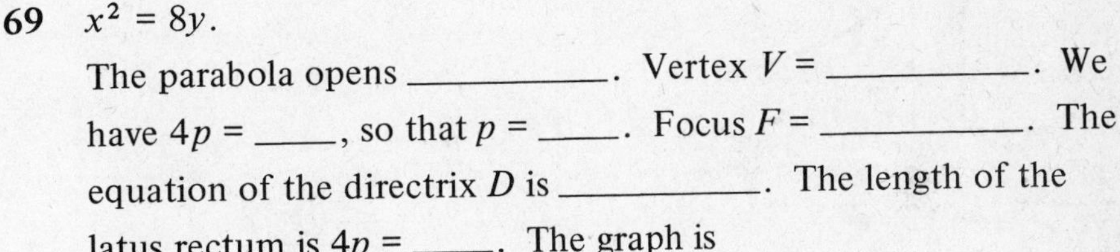
In Problems 69 through 72, find the coordinates of the vertex and the focus of the parabola. Also, find the equation of the directrix and the length of the latus rectum. Sketch the graph.

upward, (0, 0)
8, 2, (0, 2)
$y = -2$
8

69 $x^2 = 8y$.

The parabola opens ______. Vertex $V =$ ______. We have $4p =$ ____, so that $p =$ ____. Focus $F =$ ______. The equation of the directrix D is ______. The length of the latus rectum is $4p =$ ____. The graph is

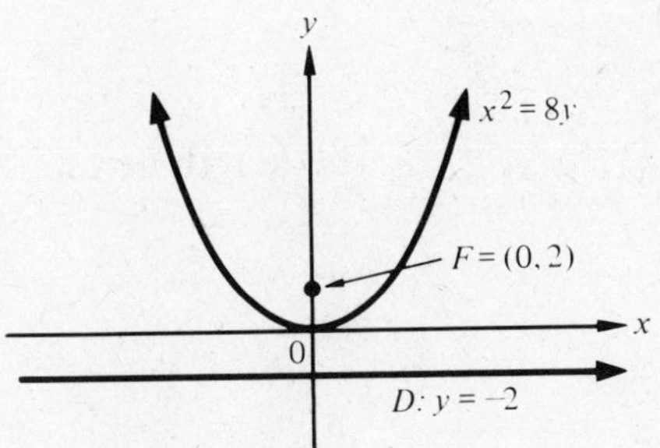

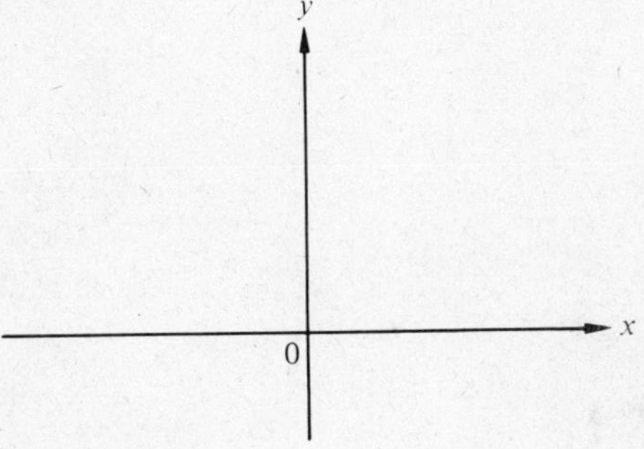

downward, (0, 0)
25, $\frac{25}{4}$, $\left(0, -\frac{25}{4}\right)$
$y = \frac{25}{4}$

70 $x^2 = -25y$.

The parabola opens ______. Vertex $V =$ ______.

We have $4p =$ ____, so that $p =$ ____. Focus $F =$ ______.

The equation of the directrix D is ______. The length of the

25

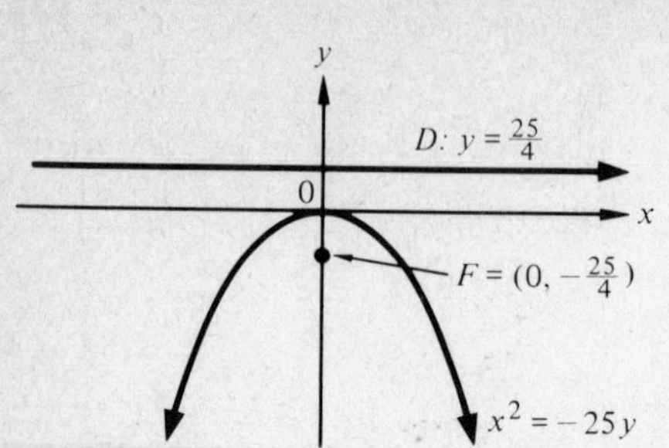

latus rectum is _____. The graph is

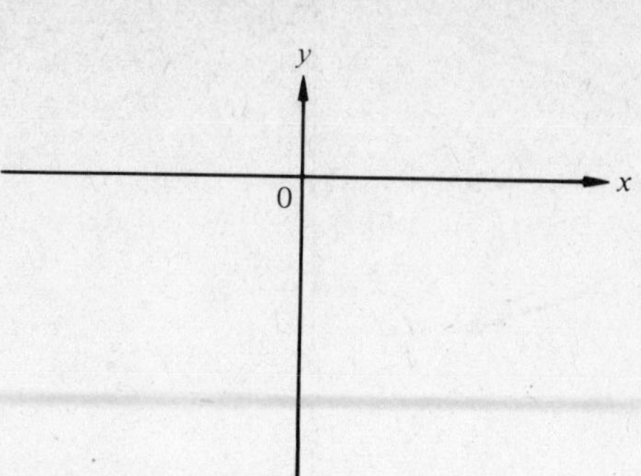

right, (0, 0)

$6, \frac{3}{2}, \left(\frac{3}{2}, 0\right)$

$x = -\frac{3}{2}$

6

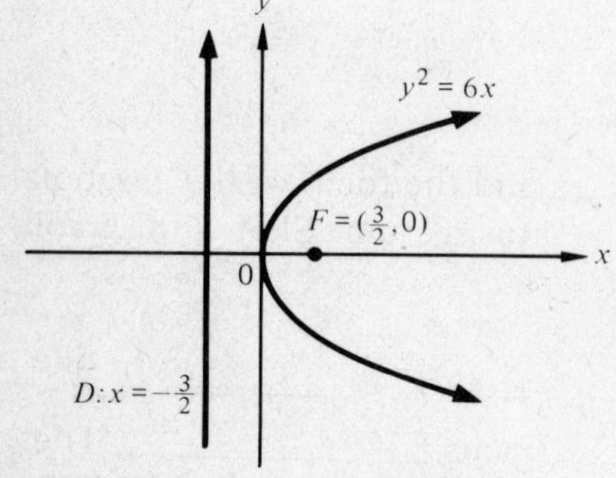

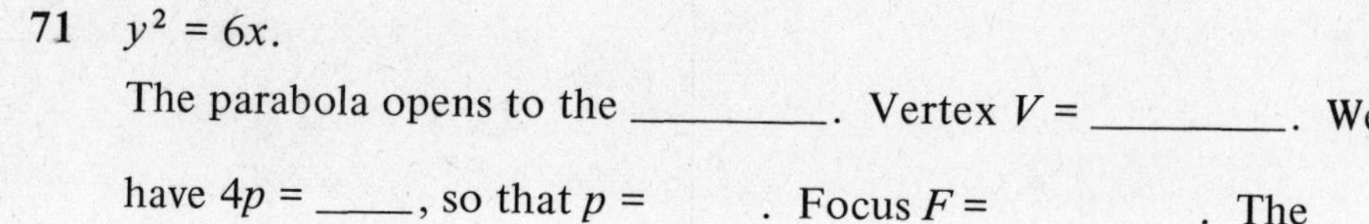

71 $y^2 = 6x$.

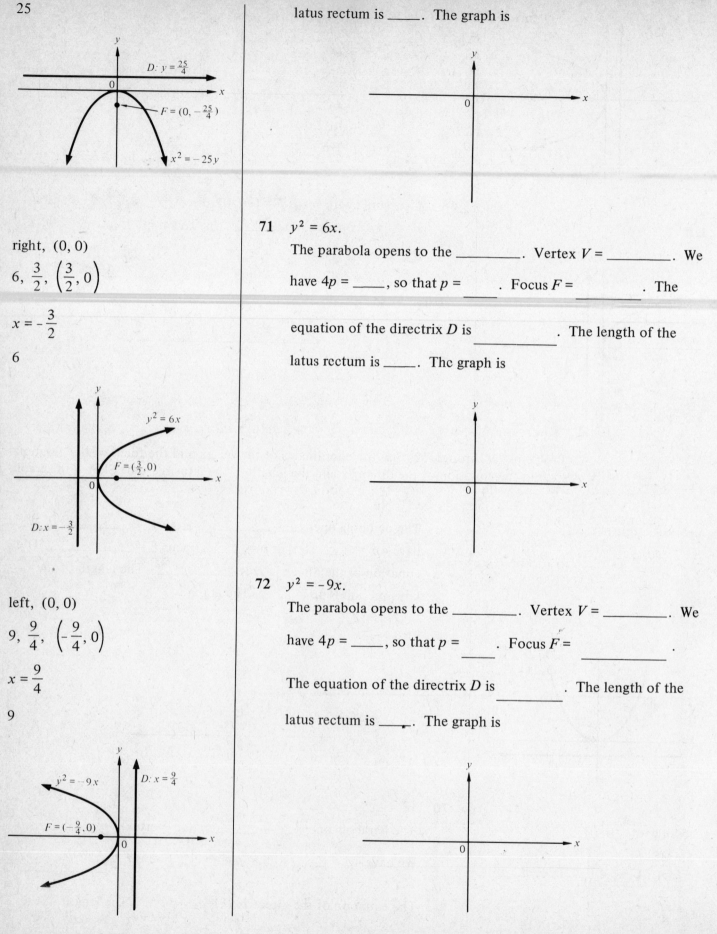

The parabola opens to the _________. Vertex $V =$ _________. We have $4p =$ _____, so that $p =$ _____. Focus $F =$ _________. The equation of the directrix D is _________. The length of the latus rectum is _____. The graph is

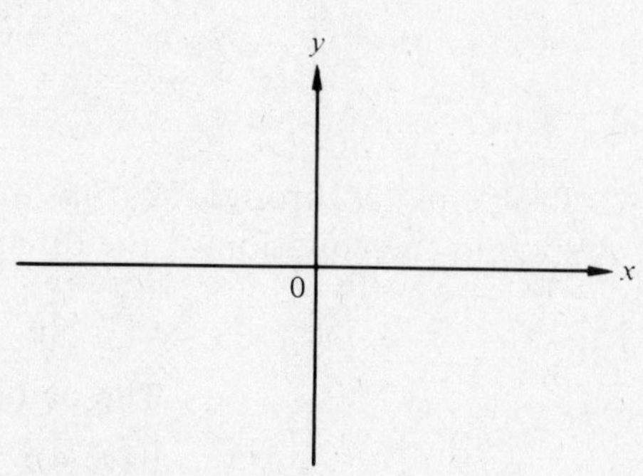

left, (0, 0)

$9, \frac{9}{4}, \left(-\frac{9}{4}, 0\right)$

$x = \frac{9}{4}$

9

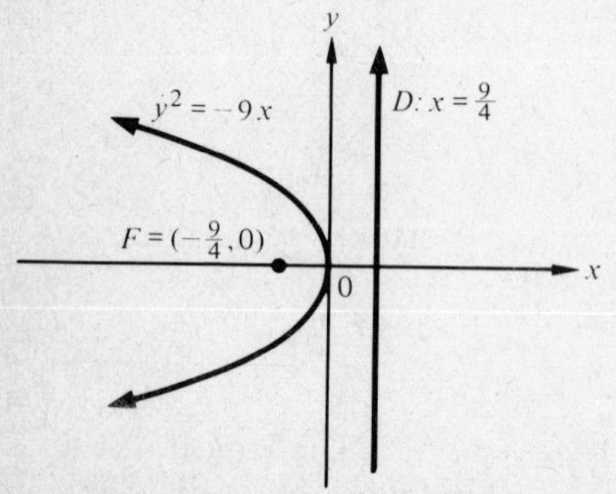

72 $y^2 = -9x$.

The parabola opens to the _________. Vertex $V =$ _________. We have $4p =$ _____, so that $p =$ _____. Focus $F =$ _________.

The equation of the directrix D is _________. The length of the latus rectum is _____. The graph is

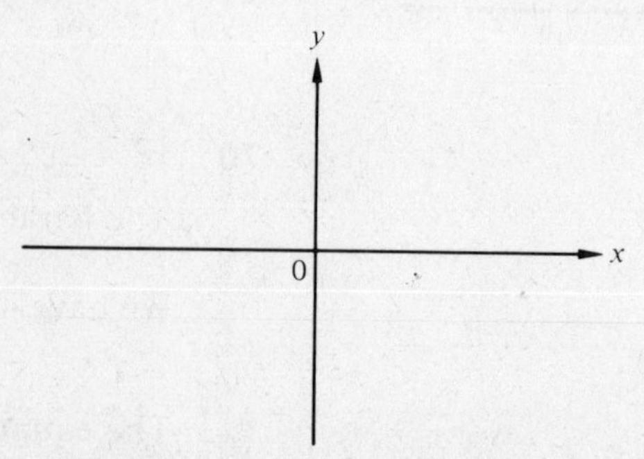

In Problems 73 through 77, find the equation of the parabola satisfying the given conditions.

73 Vertex at origin, focus at (3, 0).

The parabola opens to the ________, so its equation has the form ____________. Since $p =$ ____, the equation is ____________.

right

$y^2 = 4px$, 3, $y^2 = 12x$

74 Vertex at origin, focus at $\left(-\frac{5}{4}, 0\right)$.

The parabola opens to the ________, so its equation has the form ____________. Since $p =$ ____, the equation is ____________.

left

$y^2 = -4px$, $\frac{5}{4}$

$y^2 = -5x$

75 Vertex at origin, focus at $\left(0, \frac{3}{2}\right)$.

The parabola opens ________, so its equation has the form ____________. Since $p =$ ____, the equation is ____________.

upward

$x^2 = 4py$, $\frac{3}{2}$

$x^2 = 6y$

76 Vertex at origin, directrix $2y - 3 = 0$.

The parabola opens ____________, so its equation has the form ____________. Since $p =$ ____, the equation is ____________.

downward

$x^2 = -4py$, $\frac{3}{2}$

$x^2 = -6y$

77 Vertex at origin, focus on the y axis, and it contains the point (−3, 2).

The parabola opens ________, so its equation has the form ____________. Since it contains the point (−3, 2), 9 = ______ or $p =$ ____. The equation is ____________.

upward

$x^2 = 4py$, $8p$

$\frac{9}{8}$, $x^2 = \frac{9}{2}y$

In Problems 78 through 81, assume that the axis of symmetry of the parabola is parallel to one of the coordinate axes and its vertex V is at the point (h, k). If $p > 0$, then the graph of the equation

78 $(x - h)^2 = 4p(y - k)$ opens ________, with focus F at the point ____________ and the equation of the directrix is $y =$ ________. The graph is

upward

$(h, k + p)$

$k - p$

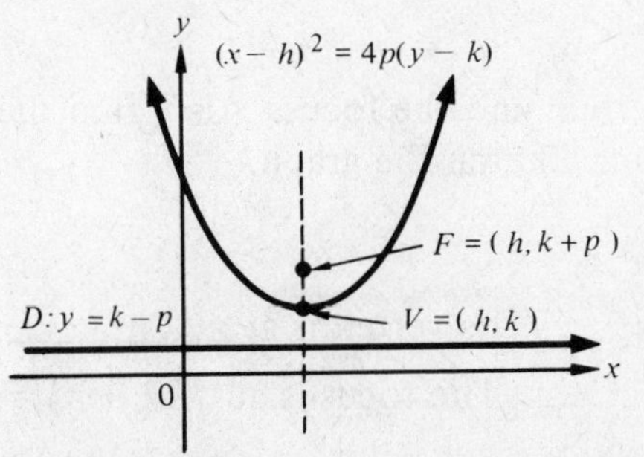

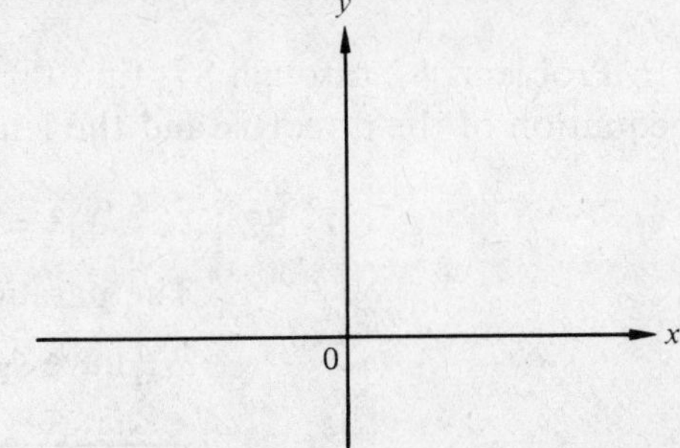

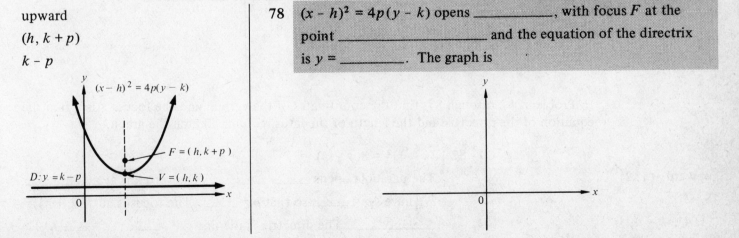

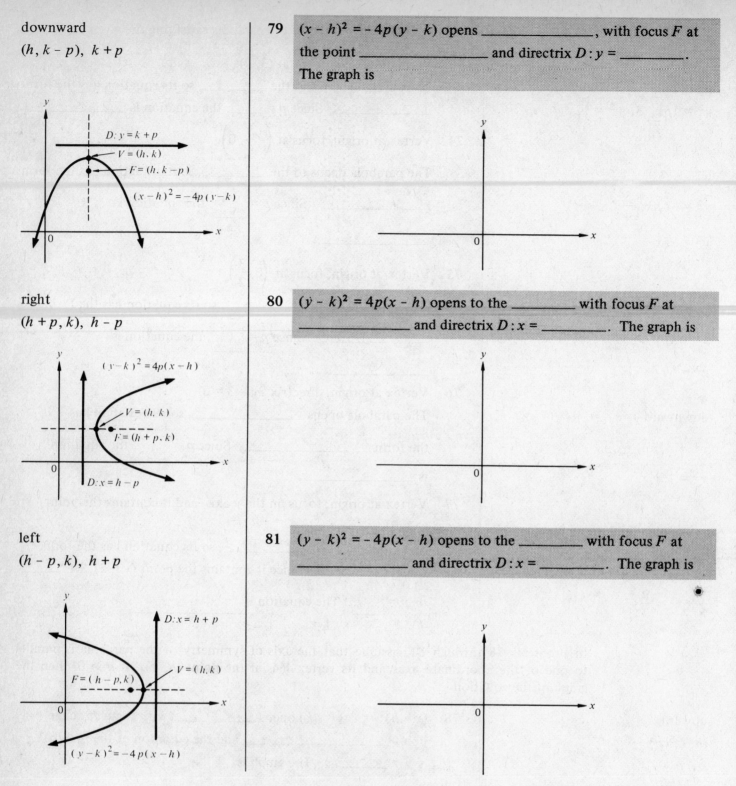

downward

$(h, k - p)$, $k + p$

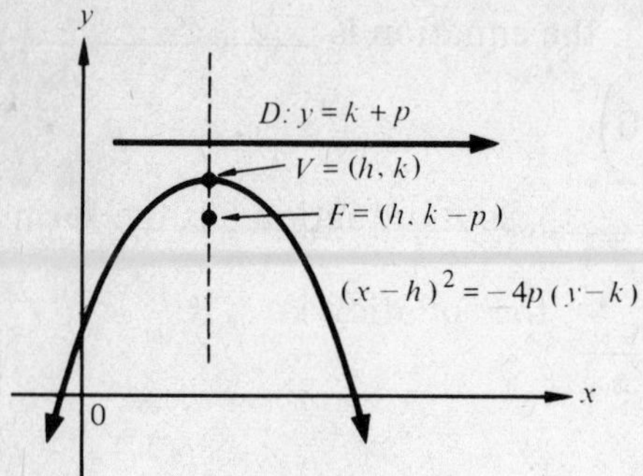

79 $(x - h)^2 = -4p(y - k)$ opens ______________, with focus F at the point ______________ and directrix $D : y =$ __________. The graph is

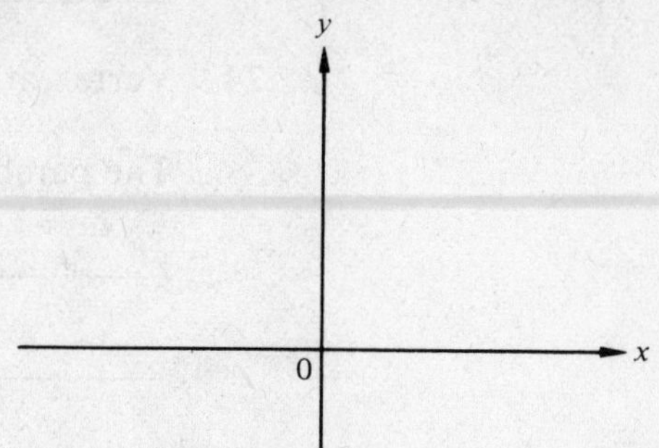

right

$(h + p, k)$, $h - p$

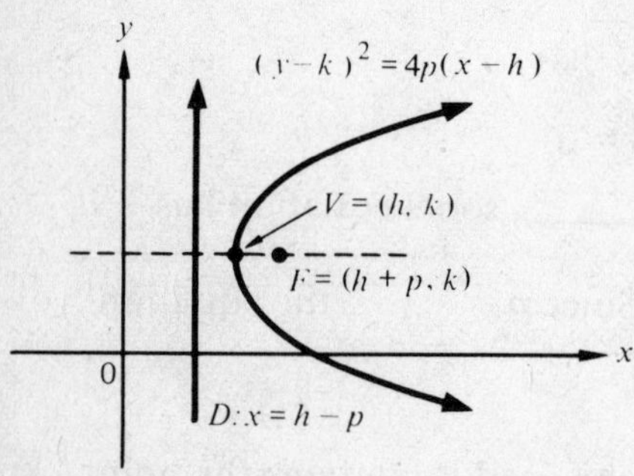

80 $(y - k)^2 = 4p(x - h)$ opens to the __________ with focus F at ______________ and directrix $D : x =$ __________. The graph is

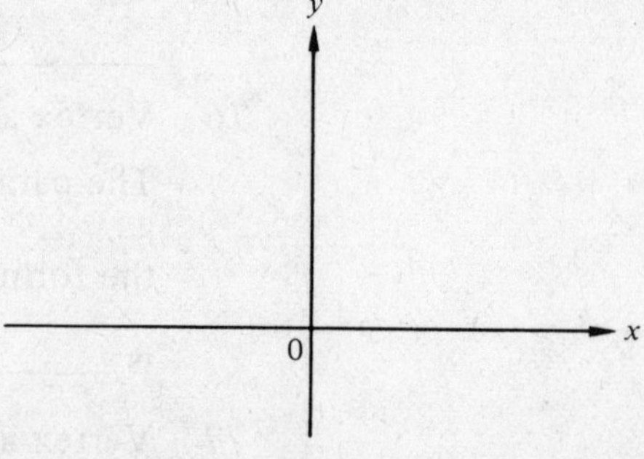

left

$(h - p, k)$, $h + p$

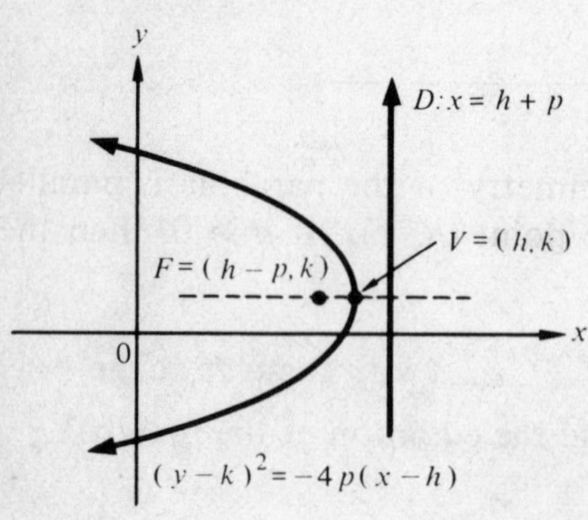

81 $(y - k)^2 = -4p(x - h)$ opens to the __________ with focus F at ______________ and directrix $D : x =$ __________. The graph is

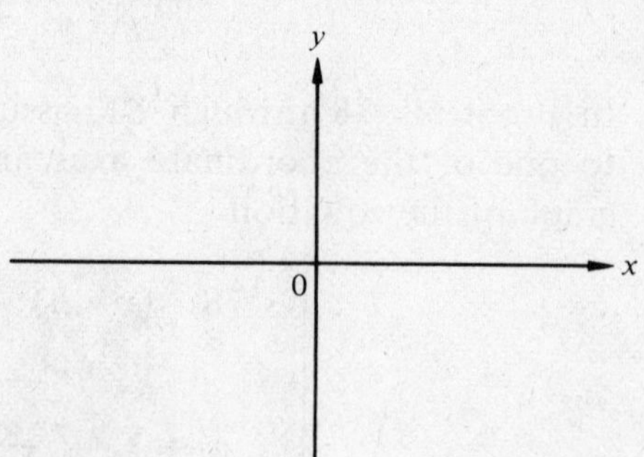

In Problems 82 through 87, find the coordinates of the vertex, and the focus. Also, find the equation of the directrix and the length of the latus rectum. Sketch the graph.

upward, $(1, 2)$

8, 2

$(1, 4)$, $k - p$, 0

82 $(x - 1)^2 = 8(y - 2)$.

The parabola opens ______________. The vertex is at __________.

We have $4p =$ _____, so that $p =$ _____. The focus is at $(h, k + p) =$ __________. The directrix is the line $y =$ __________ $=$ _____. The

8

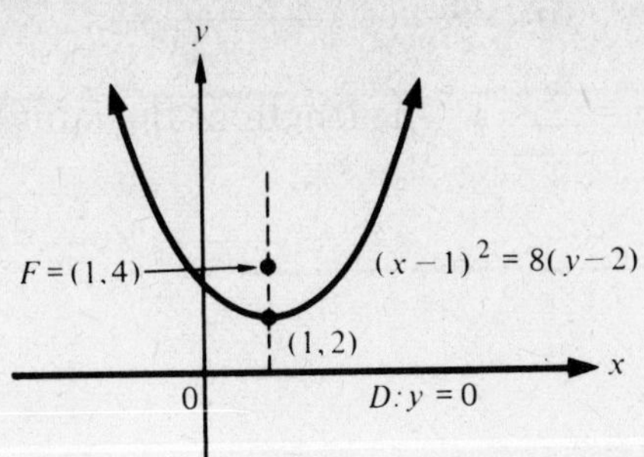

length of the latus rectum is ____. The graph is

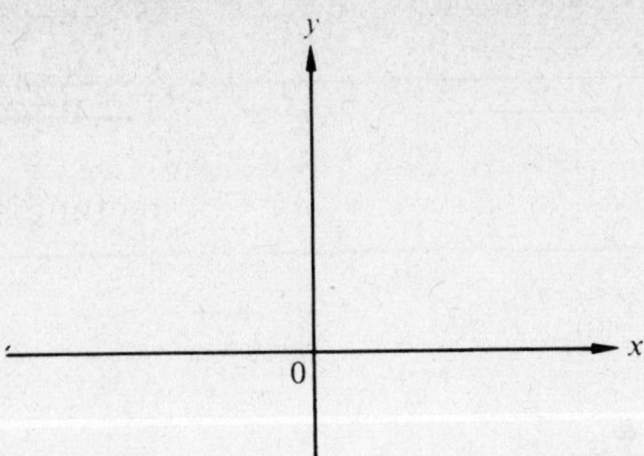

downward

(2, 3), 4, 1

(2, 2)

4, 4

83 $(x - 2)^2 = -4(y - 3)$.

The parabola opens ____________. The vertex is at ________. We have $4p =$ ____, so that $p =$ ____. The focus is at $(h, k - p) =$ ________. The directrix is the line $y = k + p =$ ____. The length of the latus rectum is ____. The graph is

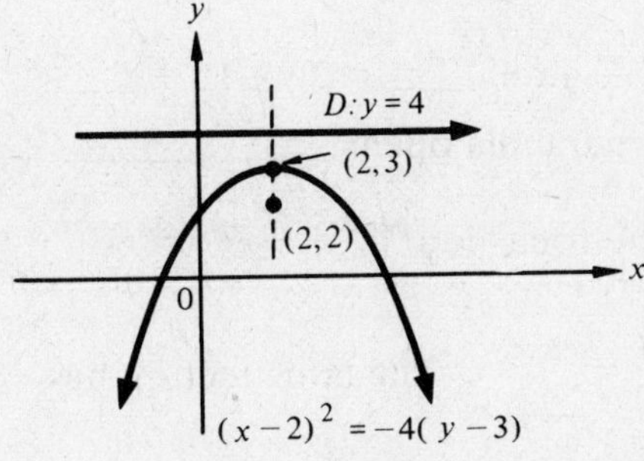

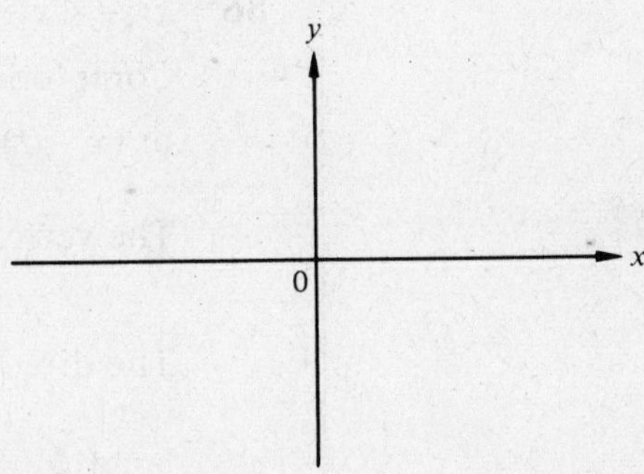

right, (3, 2)

2, (5, 2)

1

8

84 $(y - 2)^2 = 8(x - 3)$.

The parabola opens to the ________. The vertex is at ________. Since $4p = 8$, then $p =$ ____, so that the focus F is at ________. The directrix is the line $x =$ ____. The latus rectum has length ____. The graph is

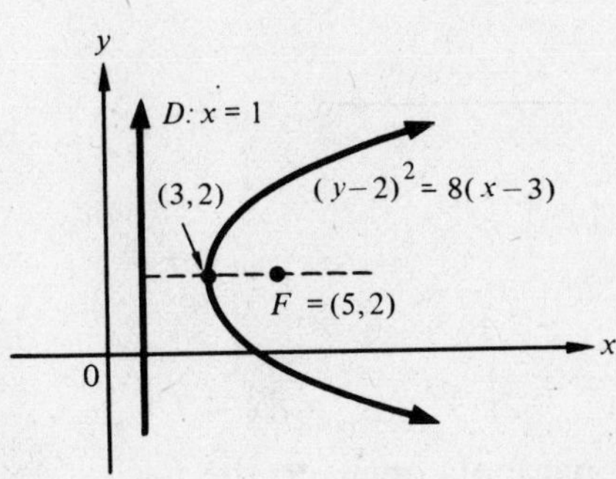

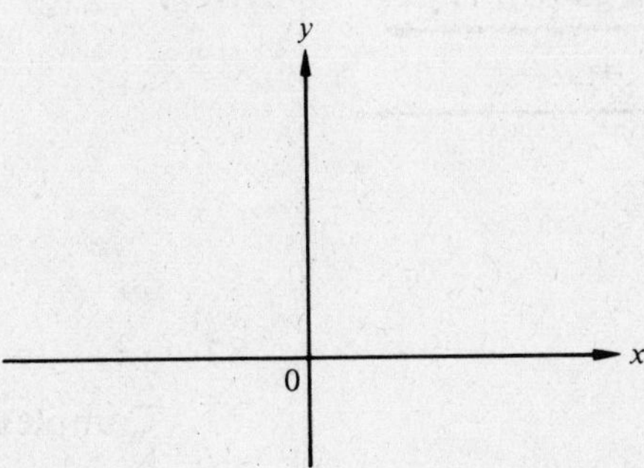

9, 9

$-6(x + 1)$, left

$(-1, 3)$, 6

85 $y^2 - 6y + 6x + 15 = 0$.

Completing the square, we have $y^2 - 6y +$ ____ $= -6x - 15 +$ ____ or $(y - 3)^2 =$ ____________. The parabola opens to the ____. The vertex is at ____________. Since $4p =$ ____, so that

$\frac{3}{2}$, $\left(-\frac{5}{2}, 3\right)$

$\frac{1}{2}$

6

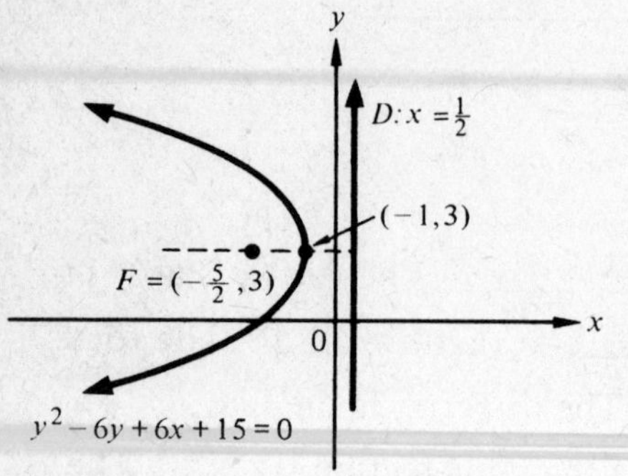

$p =$ ____, then the focus is at $(h - p, k) =$ ____________.

The directrix is the line $x = h + p =$ ____. The length of the latus rectum is ____. The graph is

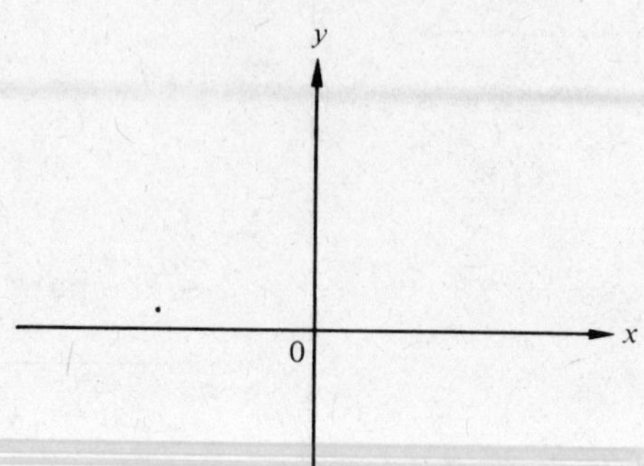

4, 4

$y - 3$, upward

(2, 3), $\left(2, \frac{7}{2}\right)$

$\frac{5}{2}$

2

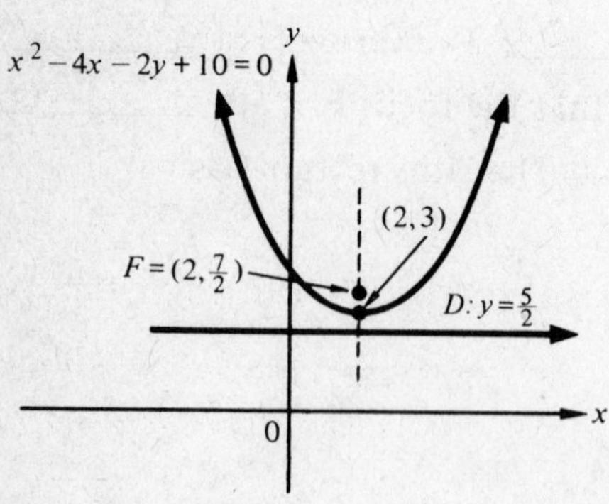

86 $x^2 - 4x - 2y + 10 = 0$.

Completing the square, we get $x^2 - 4x +$ ____ $= 2y - 10 +$ ____ or $(x - 2)^2 = 2($________$)$. The parabola opens ____________.

The vertex is at __________. The focus is at $\left(2, 3 + \frac{1}{2}\right) =$ _____.

The directrix is the line $y = 3 - \frac{1}{2} =$ ____. The latus rectum has length ____. The graph is

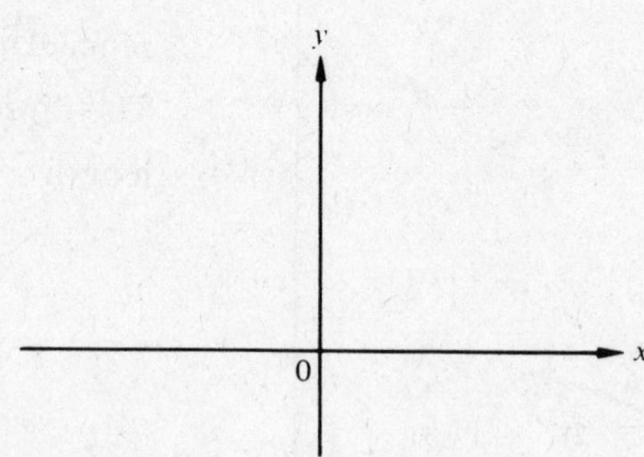

1, 1

$x - 1$, right

8, 2, (1, 1)

(3, 1), −1

8

87 $y^2 - 2y - 8x + 9 = 0$.

Completing the square, we have $y^2 - 2y +$ ____ $= 8x - 9 +$ ____ or $(y - 1)^2 = 8($________$)$. The parabola opens to the ________. Since $4p =$ ____, $p =$ ____. The vertex is at __________. The focus is at __________. The directrix is the line $x =$ ____. The length of the latus rectum is ____. The graph is

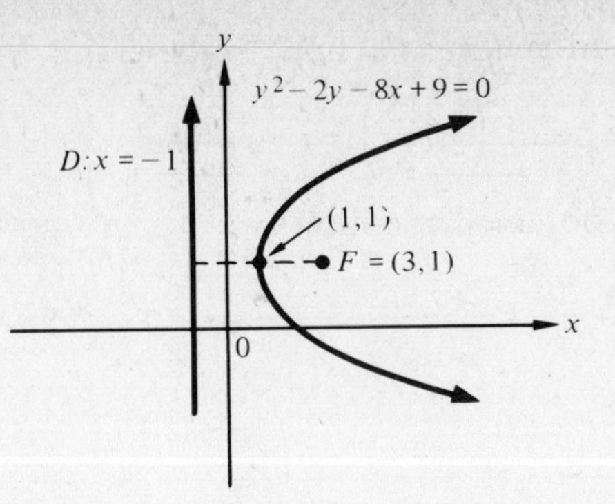

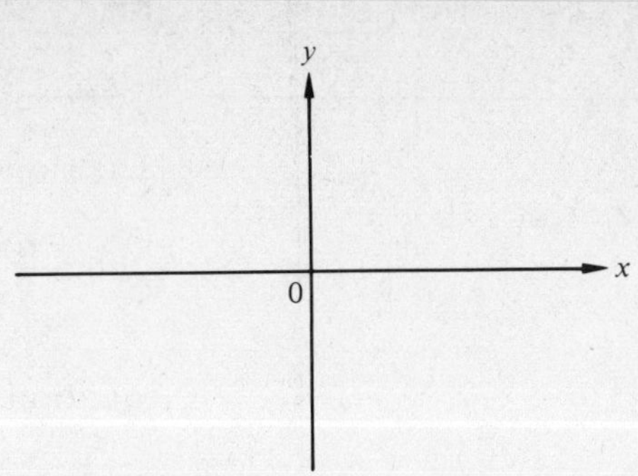

In Problems 88 through 92, find the equation of the parabola satisfying the given conditions.

88 Focus at $(1, 4)$ and directrix $y = 0$.
The parabola opens __________. The vertex is midway between the directrix and the ________, so that $(h, k) =$ ________. Since $p =$ ____, then $4p =$ ____. The equation is ____________________.

upward
focus, $(1, 2)$
$2,\ 8$
$(x - 1)^2 = 8(y - 2)$

89 Vertex at $\left(-\frac{1}{2}, \frac{1}{6}\right)$ and focus at $\left(-\frac{1}{2}, -\frac{5}{24}\right)$.
The parabola opens ____________. The distance between the vertex and the focus is p, so that $p =$ ____ and $4p =$ ____.
The equation is ____________________.

downward
$\frac{3}{8},\ \frac{3}{2}$
$\left(x + \frac{1}{2}\right)^2 = -\frac{3}{2}\left(y - \frac{1}{6}\right)$

90 Vertex at $(2, -3)$ and directrix $y = -7$.
The parabola opens __________. The distance between the vertex and the directrix is p, so that $p =$ ____ and $4p =$ ____. The equation is ____________________.

upward
$4,\ 16$
$(x - 2)^2 = 16(y + 3)$

91 Vertex at $(2, 4)$ and directrix $x = -3$.
The parabola opens to the ________. $p = 2 - (-3) =$ ____ and $4p =$ ____. The equation is ____________________.

right, 5
$20,\ (y - 4)^2 = 20(x - 2)$

92 The endpoints of the latus rectum are $(-4, 1)$ and $(2, 1)$ and the parabola opens upward.
The focus is at $\left(\frac{-4 + 2}{2}, \frac{1 + 1}{2}\right) =$ ____________. $4p =$ ____, the length of the ____________. Thus, $p =$ ____ and the vertex is at $\left(-1, 1 - \frac{3}{2}\right) =$ ____________. The equation is ____________________.

$(-1, 1),\ 6$
latus rectum, $\frac{3}{2}$
$\left(-1, -\frac{1}{2}\right)$
$(x + 1)^2 = 6\left(y + \frac{1}{2}\right)$

93 Find the equations of the tangent and normal lines to the graph of $x^2 + 8y + 4x - 20 = 0$ at the point $\left(1, \frac{15}{8}\right)$.

The slope of the tangent line is obtained as follows:

$2x + 8\frac{dy}{dx} + 4 = 0$, so at $\left(1, \frac{15}{8}\right)$, $\frac{dy}{dx} = \frac{-(\underline{\qquad})}{8} = \underline{\quad}$. The equation of the tangent line is $\underline{\qquad\qquad}$, and the equation of the normal is $\underline{\qquad\qquad}$.

$2x + 4,\ -\frac{3}{4}$

$y - \frac{15}{8} = -\frac{3}{4}(x - 1)$

$y - \frac{15}{8} = \frac{4}{3}(x - 1)$

94 Find the dimensions of the rectangle of largest area with vertices at the origin, on the x axis, the y axis, and on the parabola $y = 4 - x^2$ as shown.

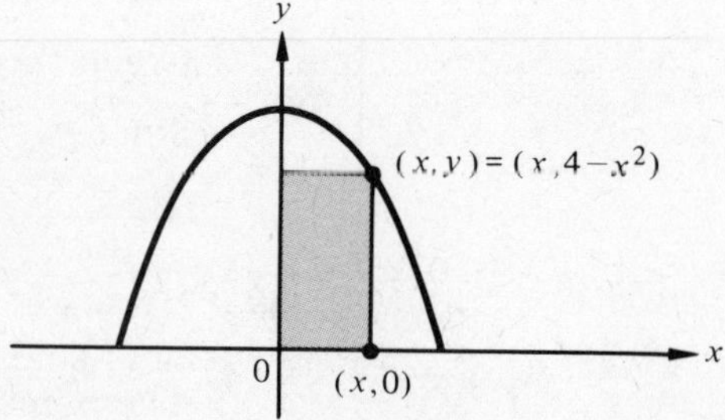

Let (x, y) be the corner of the rectangle that lies on the parabola. Then the dimensions of the rectangle are x and y units. The area is $A = x \cdot y = x(\underline{\qquad}) = \underline{\qquad}$. To find the maximum area, we find $\frac{dA}{dx} = \underline{\qquad} = 0$, so that $x = \underline{\quad}$ and $y = 4 - \underline{\quad} = \underline{\quad}$. The dimensions of the rectangle are $\underline{\quad}$ and $\underline{\quad}$.

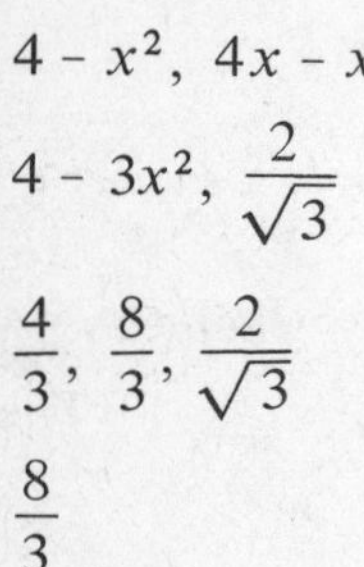

$4 - x^2,\ 4x - x^3$

$4 - 3x^2,\ \frac{2}{\sqrt{3}}$

$\frac{4}{3},\ \frac{8}{3},\ \frac{2}{\sqrt{3}}$

$\frac{8}{3}$

4 The Hyperbola

95 A hyperbola is defined to be the set of all points P in the plane such that the distance $|\overline{PF_1}| - |\overline{PF_2}|$ from P to two $\underline{\qquad}$ points F_1 and F_2 in the plane is a $\underline{\qquad\qquad}$.

fixed

constant positive number

In Problems 96 through 98, refer to the following figure.

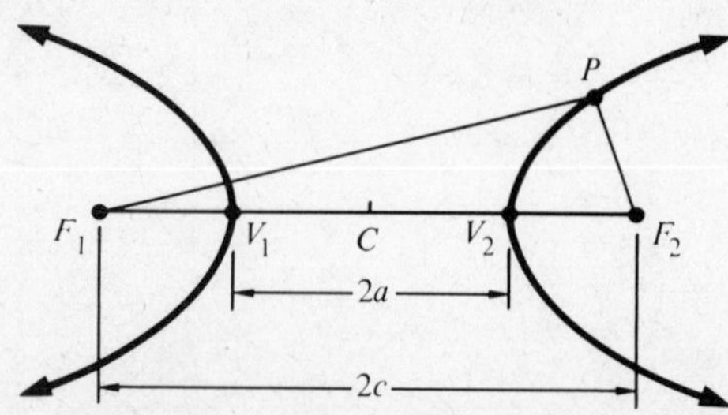

96 The points F_1 and F_2 are called the ________ of the hyperbola. The midpoint C of F_1F_2 is called the ________ of the hyperbola.

foci
center

97 The two points V_1 and V_2 where the two branches of the hyperbola intersect the axis of symmetry passing through the foci are called the ________ of the hyperbola.

vertices

98 The line segment $\overline{V_1V_2}$ between the two vertices is called the ________ axis.

transverse

99 The equation of the hyperbola with foci $F_1 = (-c, 0)$, $F_2 = (c, 0)$ and vertices $V_1 = (-a, 0)$, $V_2 = (a, 0)$ is ________, where $b = \sqrt{}$. The equation $\frac{x^2}{a^2} - \frac{y^2}{b^2} = 1$ is called the ________ of the equation of a hyperbola.

$\frac{x^2}{a^2} - \frac{y^2}{b^2} = 1$
$c^2 - a^2$
standard form

In Problems 100 through 104, consider the hyperbola whose equation is $\frac{x^2}{a^2} - \frac{y^2}{b^2} = 1$ and whose graph is the figure below. Then:

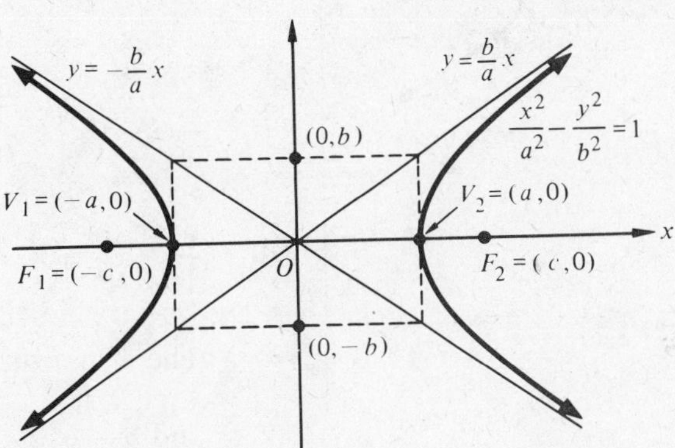

100 The x and y axes are axes of ________ and the origin is the ________.

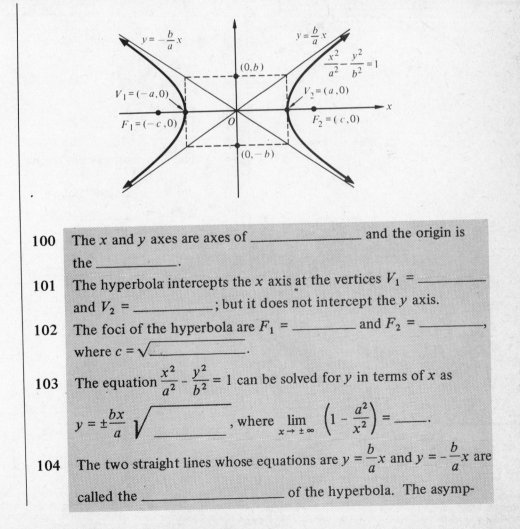

symmetry
center

101 The hyperbola intercepts the x axis at the vertices $V_1 =$ ________ and $V_2 =$ ________; but it does not intercept the y axis.

$(-a, 0)$
$(a, 0)$

102 The foci of the hyperbola are $F_1 =$ ________ and $F_2 =$ ________, where $c = \sqrt{}$.

$(-c, 0)$, $(c, 0)$
$a^2 + b^2$

103 The equation $\frac{x^2}{a^2} - \frac{y^2}{b^2} = 1$ can be solved for y in terms of x as $y = \pm\frac{bx}{a}\sqrt{}$, where $\lim_{x \to \pm\infty}\left(1 - \frac{a^2}{x^2}\right) =$ ________.

$1 - \frac{a^2}{x^2}$, 1

104 The two straight lines whose equations are $y = \frac{b}{a}x$ and $y = -\frac{b}{a}x$ are called the ________ of the hyperbola. The asymp-

asymptotes

$2a$, $2b$

totes are the extensions of the diagonals of the rectangle whose dimensions are ____ and ____.

In Problems 105 through 108, find the coordinates of the vertices and the foci of the hyperbola. Also, find the equations of the asymptotes and sketch the graph.

$\frac{x^2}{a^2} - \frac{y^2}{b^2} = 1$, 3

1, $\sqrt{10}$

$(-3, 0)$, $(3, 0)$

$(-\sqrt{10}, 0)$, $(\sqrt{10}, 0)$

$\frac{1}{3}x$, $-\frac{1}{3}x$

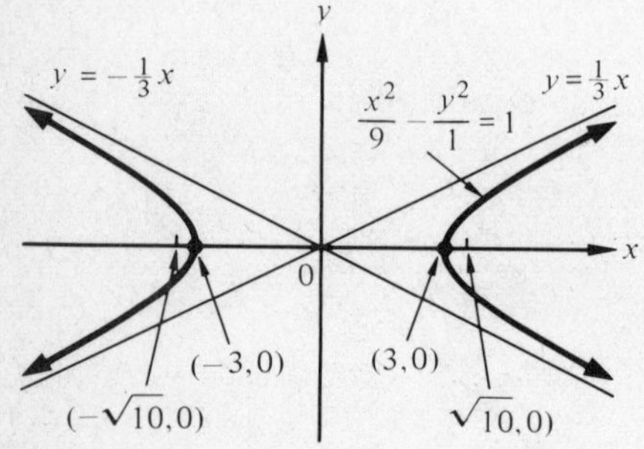

105 $\frac{x^2}{9} - \frac{y^2}{1} = 1.$

The equation has the form ______________ with a = ____ and b = ____. Hence, $c = \sqrt{a^2 + b^2}$ = ____. Thus, the vertices are at V_1 = __________ and V_2 = __________, and the foci are at F_1 = ______________ and F_2 = ______________. The equations of the asymptotes are $y = \frac{b}{a}x$ = ____ and $y = -\frac{b}{a}x$ = ____. The graph is

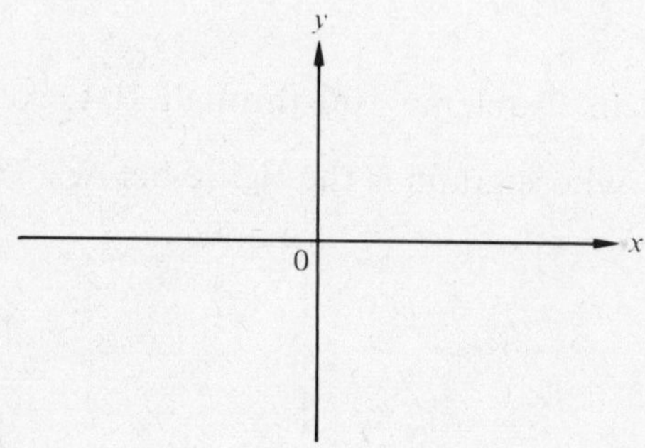

$\frac{x^2}{a^2} - \frac{y^2}{b^2} = 1$, 4

3, 5

$(-4, 0)$, $(4, 0)$

$(-5, 0)$, $(5, 0)$

$\frac{3}{4}x$, $-\frac{3}{4}x$

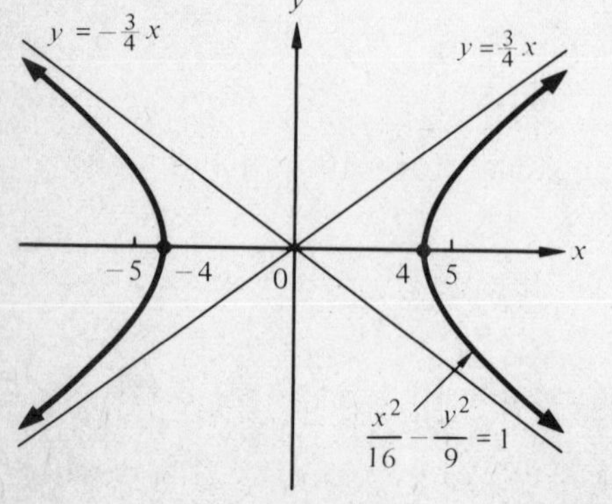

106 $\frac{x^2}{16} - \frac{y^2}{9} = 1.$

The equation has the form ______________ with a = ____ and b = ____, so that $c = \sqrt{a^2 + b^2}$ = ____. Thus, the vertices are V_1 = __________ and V_2 = __________ and the foci are F_1 = __________ and F_2 = __________. The equations of the asymptotes are $y = \frac{b}{a}x$ = ____ and $y = -\frac{b}{a}x$ = ____. The graph is

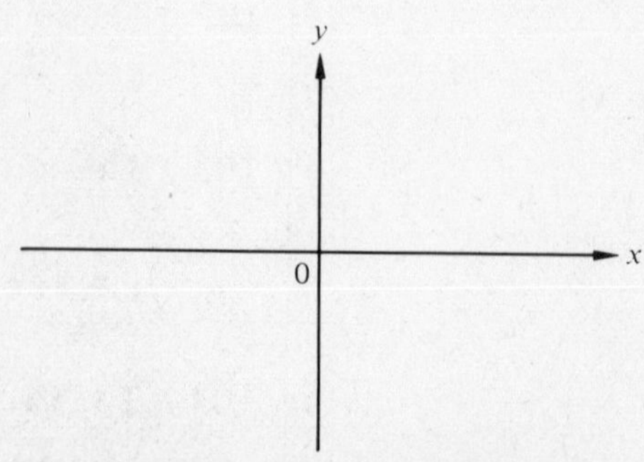

$16x^2 - y^2 = 16$

1, 1, 4

$\sqrt{17}$, (-1, 0), (1, 0)

$(-\sqrt{17}, 0)$, $(\sqrt{17}, 0)$

$4x$, $-4x$

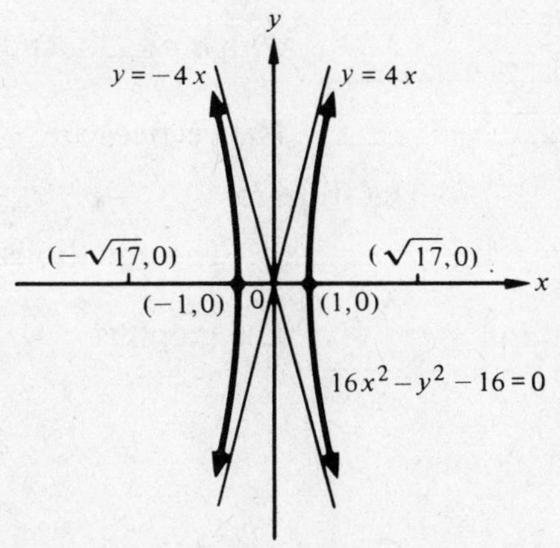

107 $16x^2 - y^2 - 16 = 0$.

The equation is written as ________________, so that $\frac{x^2}{1} - \frac{y^2}{16} =$ ____. $a =$ ____ and $b =$ ____, so that $c = \sqrt{a^2 + b^2} =$ ____. The vertices are $V_1 =$ ________ and $V_2 =$ ________. The foci are $F_1 =$ ____________ and $F_2 =$ ____________. The equations of the asymptotes are $y =$ ____ and $y =$ ____. The graph is

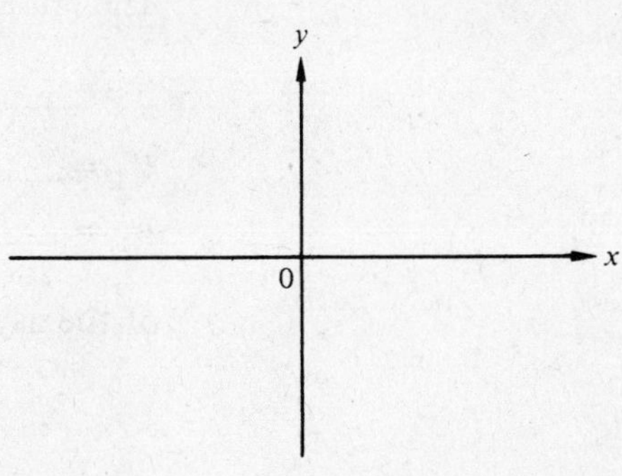

$\frac{x^2}{\frac{1}{4}} - \frac{y^2}{\frac{1}{2}} = 1$, $\frac{1}{2}$, $\frac{1}{\sqrt{2}}$, $\frac{\sqrt{3}}{2}$

$\left(-\frac{1}{2}, 0\right)$, $\left(\frac{1}{2}, 0\right)$

$\left(-\frac{\sqrt{3}}{2}, 0\right)$, $\left(\frac{\sqrt{3}}{2}, 0\right)$

$\sqrt{2}x$, $-\sqrt{2}x$

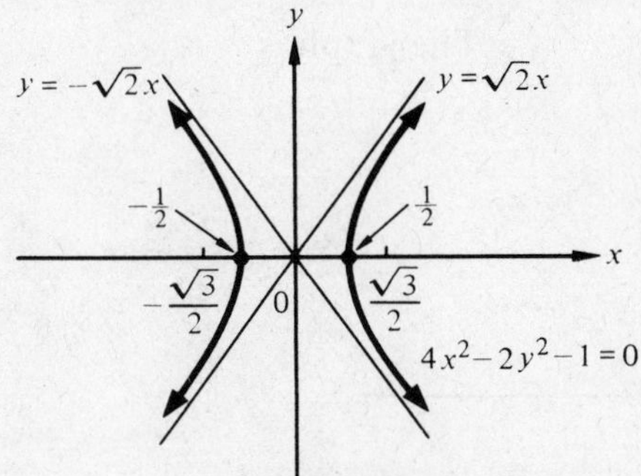

108 $4x^2 - 2y^2 - 1 = 0$.

The equation can be written in the form $\frac{x^2}{a^2} - \frac{y^2}{b^2} = 1$ as

with $a =$ ____, $b =$ ______, and $c =$ ______.

____

The vertices are $V_1 =$ ____________ and $V_2 =$ __________.

The foci are $F_1 =$ ____________ and $F_2 =$ ____________. The equations of the asymptotes are $y =$ ______ and $y =$ ______. The graph is

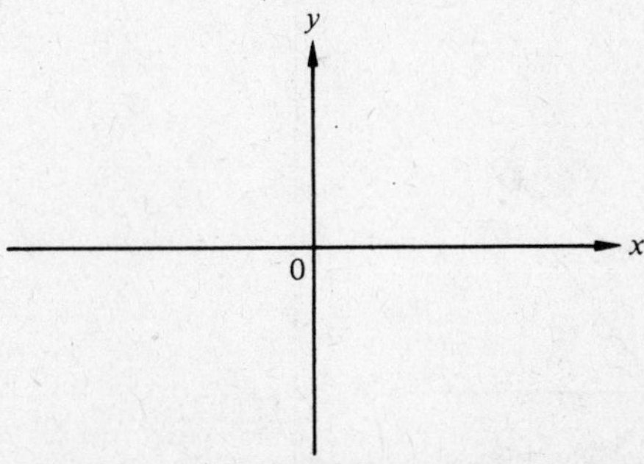

109 The equation of a hyperbola which has a vertical transverse axis, center at the origin, vertices $V_1 = (0, -b)$ and $V_2 = (0, b)$, and foci

$\frac{y^2}{b^2} - \frac{x^2}{a^2} = 1$

$\frac{b}{a}x, \frac{-b}{a}x$

$F_1 = (0, -c)$ and $F_2 = (0, c)$ is given by ______________. The equations of the asymptotes are given by $y =$ ____ and $y =$ ____.

In Problems 110 through 112, find the coordinates of the vertices and the foci of the hyperbola. Also, find the equations of the asymptotes and sketch the graph.

$\frac{y^2}{b^2} - \frac{x^2}{a^2} = 1$, 3

2, $a^2 + b^2$, $\sqrt{13}$

$(0, -2)$, $(0, 2)$

$(0, -\sqrt{13})$, $(0, \sqrt{13})$

$-\frac{2}{3}x$, $\frac{2}{3}x$

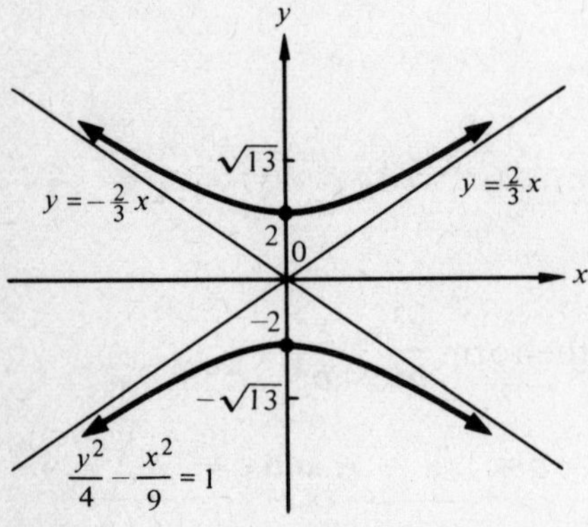

110 $\frac{y^2}{4} - \frac{x^2}{9} = 1.$

The equation has the form ______________ with $a =$ ____ and $b =$ ____, so that $c = \sqrt{(\_\_\_\_\_\_\_\_)} =$ ____. The vertices are $V_1 =$ ________ and $V_2 =$ ________. The foci are $F_1 =$ ____________ and $F_2 =$ ____________. The equations of the asymptotes are $y =$ ____ and $y =$ ____. The graph is

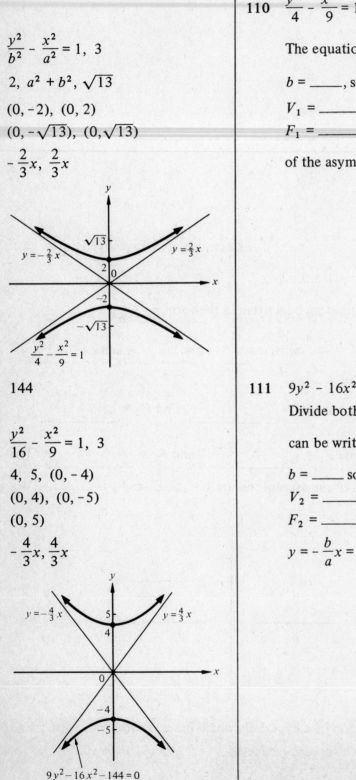

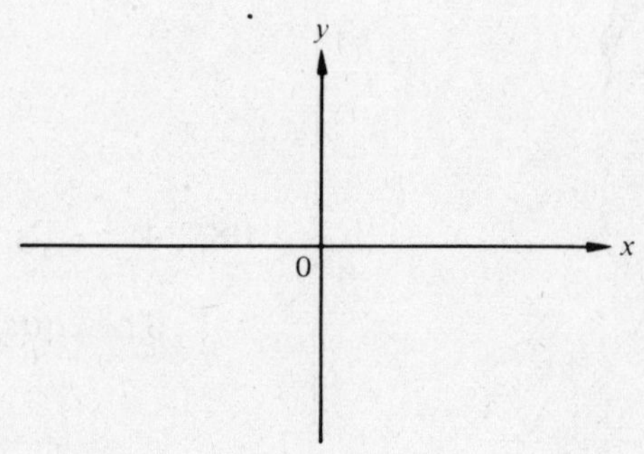

144

$\frac{y^2}{16} - \frac{x^2}{9} = 1$, 3

4, 5, $(0, -4)$

$(0, 4)$, $(0, -5)$

$(0, 5)$

$-\frac{4}{3}x$, $\frac{4}{3}x$

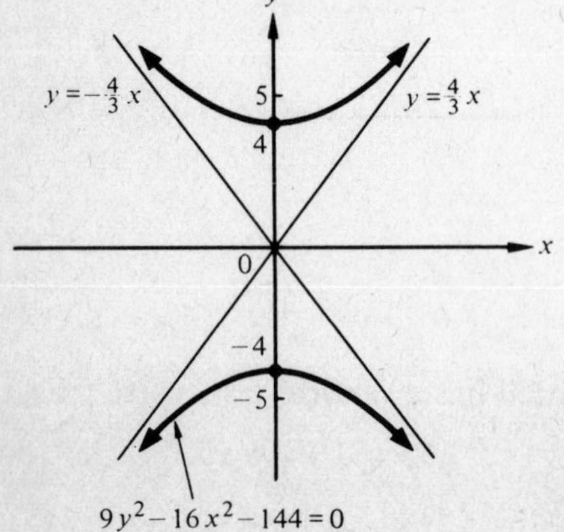

111 $9y^2 - 16x^2 - 144 = 0.$

Divide both sides of the equation by ____, so that the equation can be written as ______________. Here $a =$ ____ and $b =$ ____ so that $c =$ ____. The vertices are $V_1 =$ ________ and $V_2 =$ ________. The foci are $F_1 =$ ________ and $F_2 =$ ________. The equations of the asymptotes are $y = -\frac{b}{a}x =$ ____ and $y = \frac{b}{a}x =$ ____. The graph is

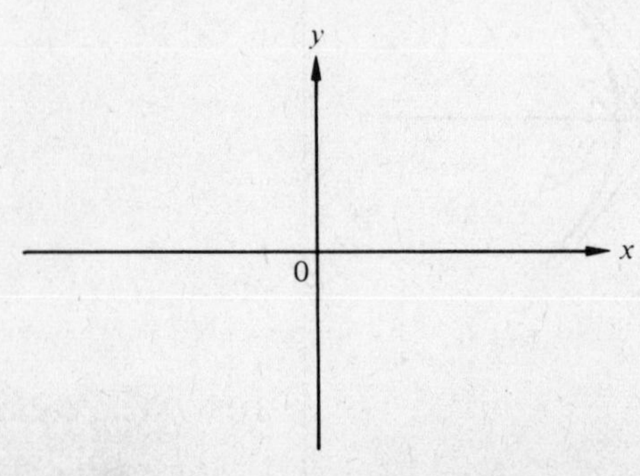

112 $5y^2 - 4x^2 = 20$.

Divide both sides of the equation by ____ so that ____________.

Here a = ____ and b = ____, so that c = ____. The vertices are V_1 = __________ and V_2__________. The foci are F_1 = __________ and F_2 = __________. The equations of the asymptotes are

y = ________ and y = ______. The graph is

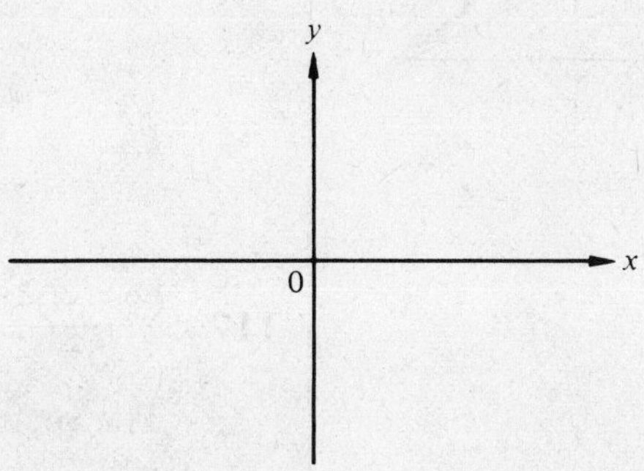

Answers: $20, \frac{y^2}{4} - \frac{x^2}{5} = 1$

$\sqrt{5}$, 2, 3

(0, −2), (0, 2), (0, −3)

(0, 3)

$-\frac{2}{\sqrt{5}}x, \frac{2}{\sqrt{5}}x$

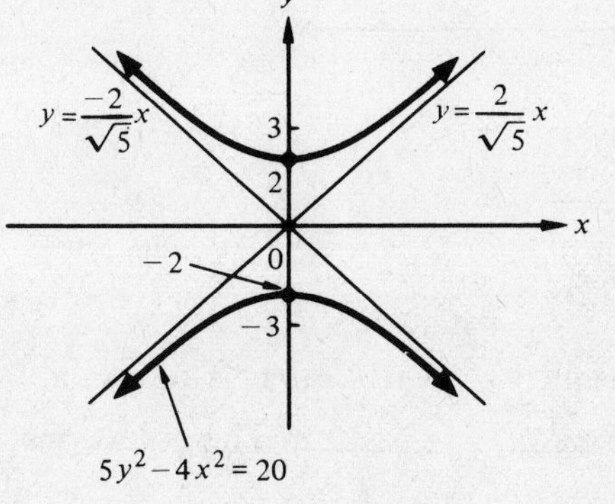

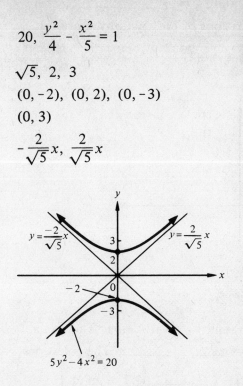

In Problems 113 through 115, find the equation of the hyperbola that satisfies the given conditions.

113 Vertices at (−5, 0) and (5, 0). Foci at $(-\sqrt{41}, 0)$ and $(\sqrt{41}, 0)$. The transverse axis is on the ______________ axis. $a = 5$ and $c = \sqrt{41}$, so that $b^2 = c^2 - a^2 = 41$ − ____ = ____. The equation is of the form $\frac{x^2}{a^2} - \frac{y^2}{b^2} = 1$, so that ______________.

Answers: horizontal

25, 16

$\frac{x^2}{25} - \frac{y^2}{16} = 1$

114 Vertices at (−2, 0) and (2, 0). Equations of the asymptotes are $y = -2x$ and $y = 2x$.

The transverse axis is on the ______________ axis. a = ____ and $\frac{b}{a}$ = ____, so that b = ____. The equation is $\frac{x^2}{4} - \frac{y^2}{16}$ = ____.

Answers: horizontal, 2

2, 4, 1

115 Vertices at (0, −6) and (0, 6). Equations of the asymptotes $y = -2x$ and $y = 2x$.

The transverse axis is on the ______________. b = ____ and $\frac{b}{a}$ = ____, so that a = ____. The equation is of the form $\frac{y^2}{b^2} - \frac{x^2}{a^2} = 1$, so that __________.

Answers: vertical, 6

2, 3

$\frac{y^2}{36} - \frac{x^2}{9} = 1$

In Problems 116 and 117, the hyperbola has its center at (h, k). Use the translation $\bar{x} = x - h$ and $\bar{y} = y - k$ to describe the hyperbola.

116 $\frac{(x-h)^2}{a^2} - \frac{(y-k)^2}{b^2} = 1$.

The vertices are $V_1 = (h - a, k)$ and V_2 = ______________. The

Answer: $(h + a, k)$

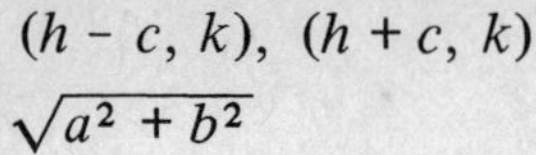

$(h - c, k)$, $(h + c, k)$

$\sqrt{a^2 + b^2}$

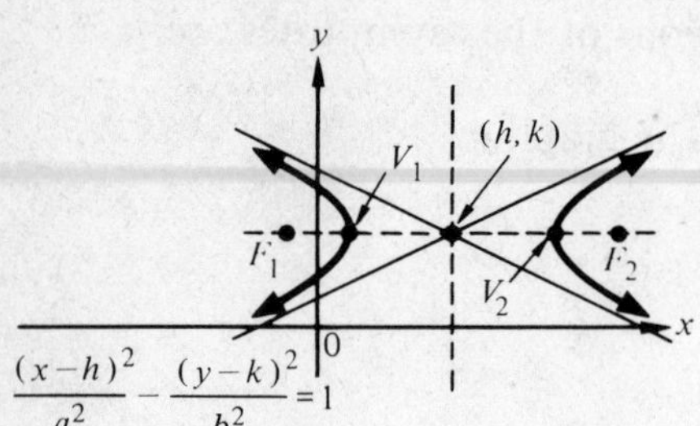

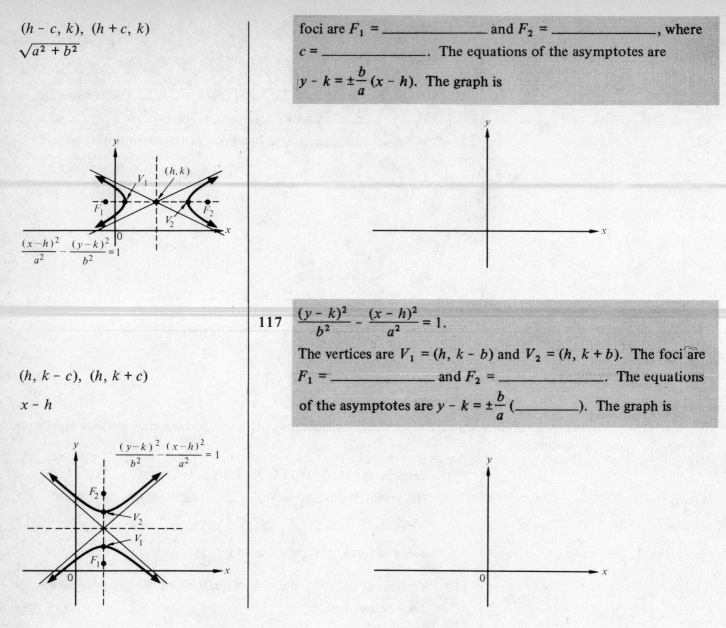

foci are F_1 = ______________ and F_2 = ______________, where c = ______________. The equations of the asymptotes are $y - k = \pm\frac{b}{a}(x - h)$. The graph is

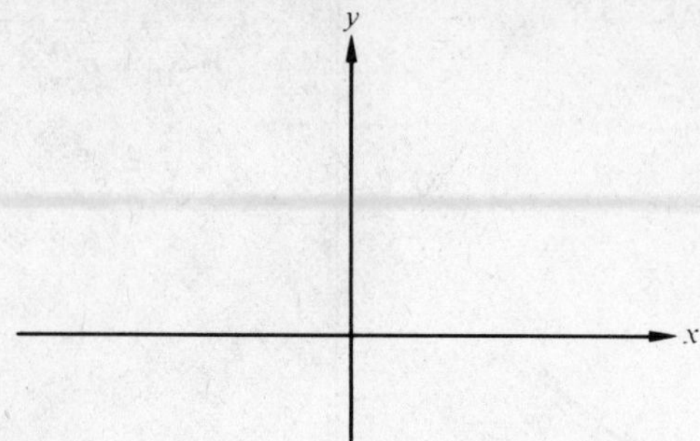

$(h, k - c)$, $(h, k + c)$

$x - h$

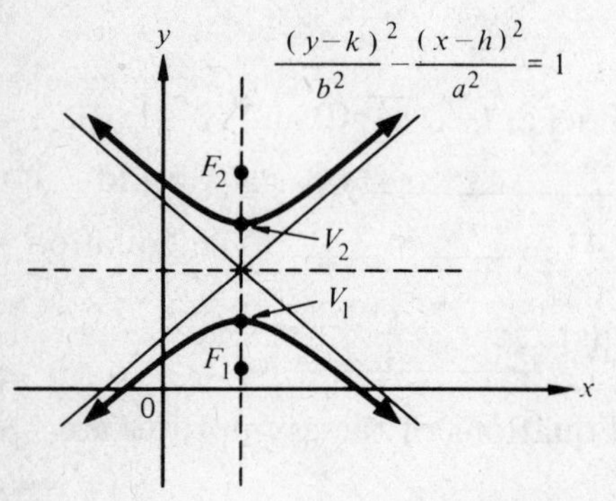

117 $\frac{(y - k)^2}{b^2} - \frac{(x - h)^2}{a^2} = 1.$

The vertices are $V_1 = (h, k - b)$ and $V_2 = (h, k + b)$. The foci are F_1 = ______________ and F_2 = ______________. The equations of the asymptotes are $y - k = \pm\frac{b}{a}$(__________). The graph is

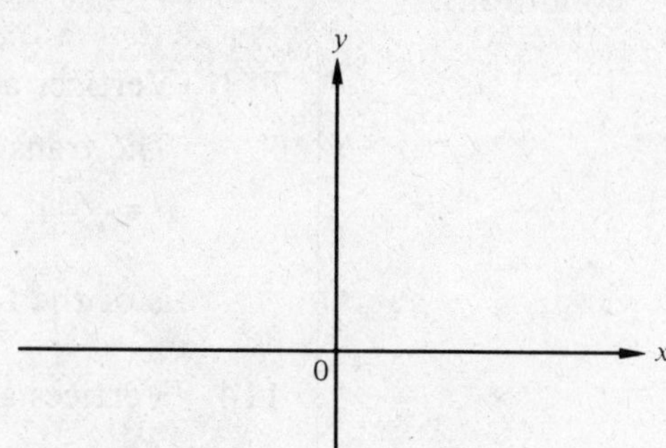

In Problems 118 through 121, find the coordinates of the center, the vertices, and the foci. Also, find the equations of the asymptotes and sketch the graph.

horizontal

$(1, -2)$, 4, 3, 5

$(h - a, k)$, $(-3, -2)$

$(h + a, k)$, $(5, -2)$

$(h - c, k)$, $(-4, -2)$

$(h + c, k)$, $(6, -2)$

$x - 1$, $y + 2 = \frac{3}{4}(x - 1)$

118 $\frac{(x - 1)^2}{16} - \frac{(y + 2)^2}{9} = 1.$

The transverse axis is on the ______________ axis. (h, k) = __________. a = _____, b = _____, and c = _____. The vertices are V_1 = ______________ = ______________ and V_2 = ______________ = __________. The foci are F_1 = ______________ = ______________ and F_2 = ______________ = __________. The asymptotes are the lines $y + 2 = -\frac{3}{4}$(_______) and ______________________. The graph is

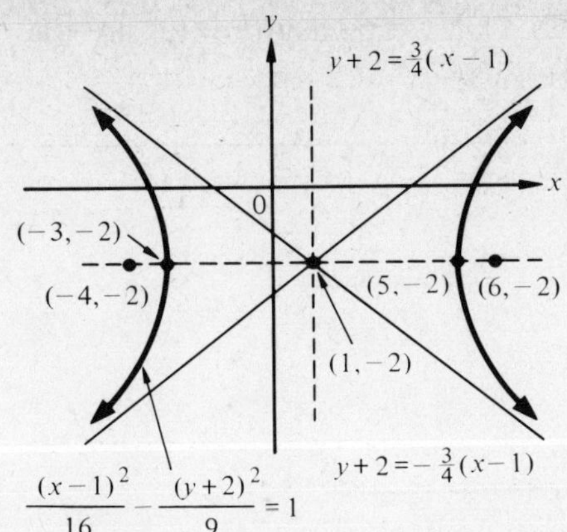

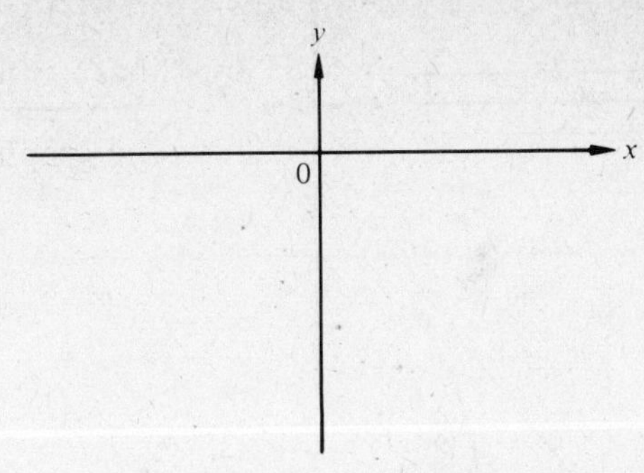

119 $\dfrac{(y-1)^2}{9} - \dfrac{(x-1)^2}{7} = 1.$

The transverse axis is on the ____________ axis. $(h, k) =$ ________. $a =$ ____ and $b =$ ____, so that $c =$ ____. The vertices are $V_1 =$ ____________ = ________ and $V_2 =$ ____________ = ________. The foci are $F_1 =$ ____________ = ________ and and $F_2 =$ ____________ = ________. The asymptotes are the lines ____________________ and ____________________.

The graph is

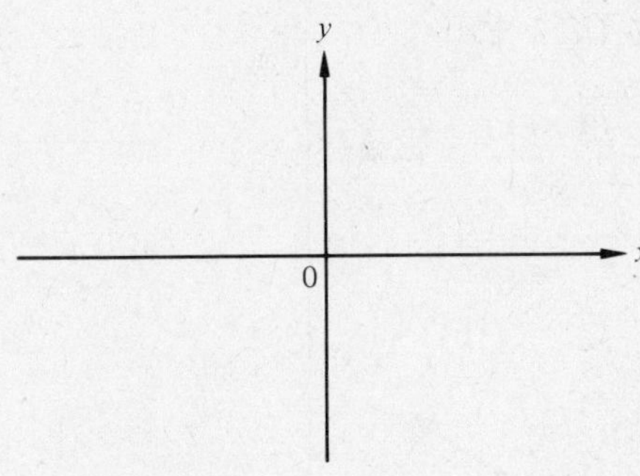

vertical
(1, 1), $\sqrt{7}$, 3, 4
$(h, k - b)$, (1, −2), $(h, k + b)$
(1, 4), $(h, k - c)$, (1, −3)
$(h, k + c)$, (1, 5)
$y - 1 = \dfrac{3}{\sqrt{7}}(x - 1)$, $y - 1 = \dfrac{-3}{\sqrt{7}}(x - 1)$

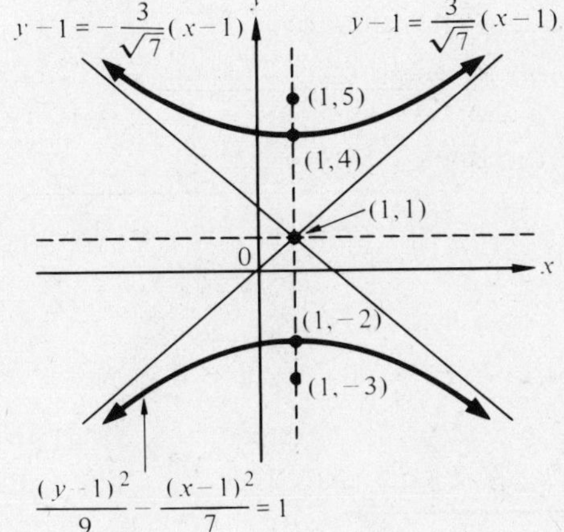

120 $9x^2 - 16y^2 - 36x - 32y - 124 = 0.$

Completing the square, we have

$9(x^2 - 4x +$ ____$) - 16(y^2 + 2y +$ ____$) = 124 +$ ____ $+$ ____,

so that $9(x-2)^2 - 16(y+1)^2 =$ ____, or

$\dfrac{(x-2)^2}{16} - \dfrac{(y+1)^2}{9} =$ ____. The transverse axis is on the ____________ axis. $a =$ ____, $b =$ ____, and $c =$ ____.

$(h, k) =$ ________. The vertices are $V_1 =$ ____________ = ________ and $V_2 =$ ____________ = ____________. The foci are $F_1 =$ ____________ = ________ and

4, 1, 36, −16
144
1
horizontal, 4, 3, 5
(2, −1), $(h + a, k)$
(6, −1), $(h - a, k)$, (−2, −1)
$(h + c, k)$, (7, −1)

$(h - c, k)$, $(-3, -1)$

$x - 2$, $y + 1 = \frac{3}{4}(x - 2)$

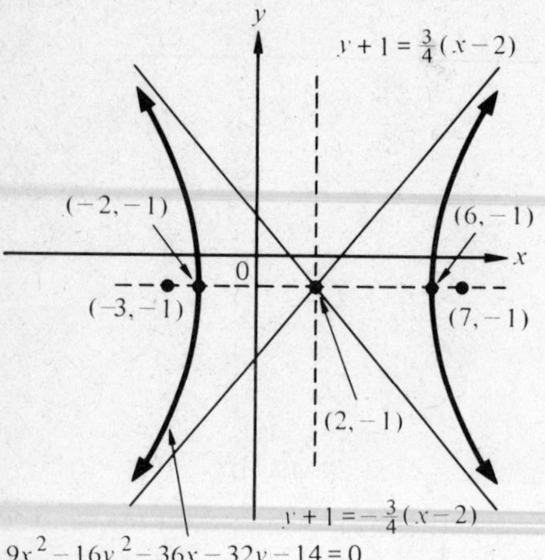

F_2 = ______________ = ______________. The asymptotes are the lines $y + 1 = -\frac{3}{4}$(________) and ______________________.

The graph is

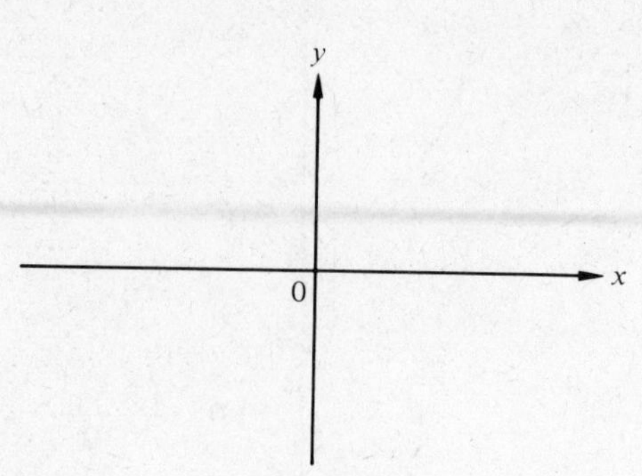

121 $16y^2 - 9x^2 + 64y + 18x - 89 = 0$.

Completing the square, we have

$16(y^2 + 4y +$ ____$) - 9(x^2 - 2x +$ ____$) = 89 +$ ____ $+$ ____ or

$16(y + 2)^2 - 9(x - 1)^2 =$ ____ or $\frac{(y + 2)^2}{9} - \frac{(x - 1)^2}{16} =$ ____.

$a =$ ____, $b =$ ____, and $c =$ ____. $(h, k) =$ __________. The vertices are V_1 = ______________ = __________, and V_2 = ______________ = __________. The foci are F_1 = ______________ = __________ and F_2 = ______________ = __________. The asymptotes are the lines ______________________ and $y + 2 = \frac{3}{4}(x - 1)$. The graph is

4, 1, 64, −9

144, 1

4, 3, 5, (1, −2)

$(h, k - b)$, $(1, -5)$

$(h, k + b)$, $(1, 1)$

$(h, k - c)$, $(1, -7)$, $(h, k + c)$

$(1, 3)$, $y + 2 = -\frac{3}{4}(x - 1)$

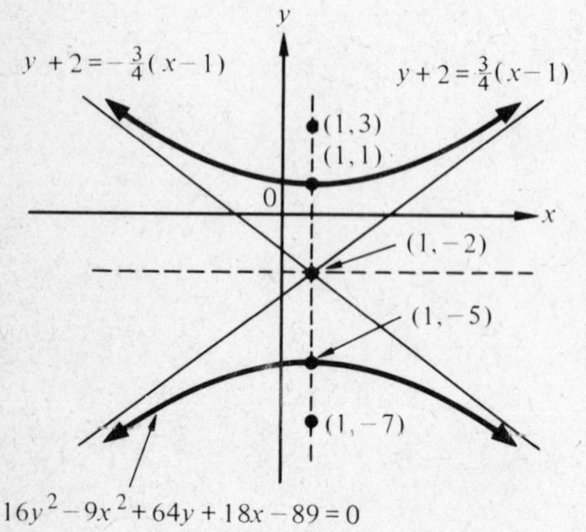

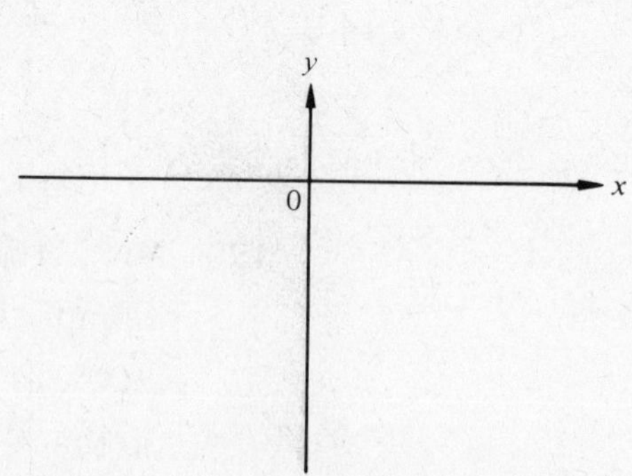

In Problems 122 through 124, find the equation of the hyperbola that satisfies the given conditions.

122 Vertices at (1, 1) and (5, 1). Foci at (0, 1) and (6, 1).

The transverse axis is on the ______________ axis.

horizontal

(3, 1), 1
2, 0, 3
5
$(y - 1)^2$

$(h, k) = \left(\frac{1+5}{2}, \frac{1+1}{2}\right) =$ ________. $h - a =$ _____ or $3 - a = 1$, so that $a =$ _____. Also, $h - c = 0$ or $3 - c =$ _____, so that $c =$ _____. Then $b^2 = c^2 - a^2 =$ _____. The equation is

$$\frac{(x-3)^2}{4} - \frac{\qquad}{5} = 1.$$

vertical
(4, 3), 5
2, 4
12
$\frac{(y-3)^2}{4} - \frac{(x-4)^2}{12} = 1$

123 Vertices at (4, 1) and (4, 5). Foci at (4, −1) and (4, 7). The transverse axis is on the __________ axis.

$(h, k) = \left(\frac{4+4}{2}, \frac{1+5}{2}\right) =$ ________, $k + b =$ _____ or $3 + b = 5$, so that $b =$ _____. Also, $k + c = 7$, so that $c =$ _____. $a^2 = c^2 - b^2 = 16 - 4 =$ _____. Thus, the equation is

____________________.

vertical
$x - h$
constant (say M)
3
3
0, $\frac{(y-2)^2}{3} - \frac{\left(x - \frac{1}{2}\right)^2}{\frac{3}{4}} = 1$

124 Contains the point (1, 4), and the equations of its asymptotes are $y = -2x + 3$ and $y = 2x + 1$. The transverse axis is on the __________ axis. The asymptotes of the hyperbola $\frac{(y-k)^2}{b^2} - \frac{(x-h)^2}{a^2} = 1$ are given by $y - k = \pm\frac{b}{a}$ (________). The equation of the hyperbola has the form

$$\left[(y-k) - \frac{b}{a}(x-h)\right]\left[(y-k) + \frac{b}{a}(x-h)\right] = \text{constant},$$

so that $(y - 2x - 1)(y + 2x - 3) =$ ________________.
Substituting the coordinates of point (1, 4), we have $M =$ _____.
Hence, the required equation is $(y - 2x - 1)(y + 2x - 3) =$ _____,

or $y^2 - 4x^2 - 4y + 4x =$ _____, or

____

5 The Conic Sections

fixed point
fixed line
focus
directrix
eccentricity

125 A conic is the set of all points P in the plane such that the distance from P to a ____________ F in the plane bears a constant ratio e to the distance from P to a ____________ D in the plane. The fixed point F is called the ________, the fixed straight line D is called the __________, and the constant number e is called the ____________ of the conic.

In Problems 126 and 127, use the definition in Problem 125 to find the equation of the conic with the given conditions.

126 Focus $F = (1, 0)$, directrix D: $x = -1$, and eccentricity $e = \frac{1}{2}$.

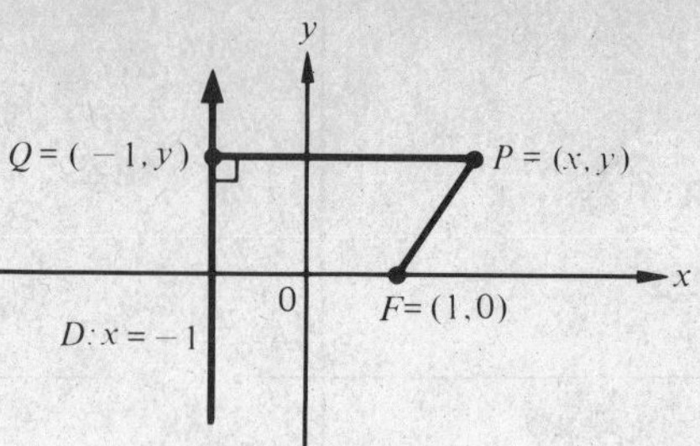

$Q =$ ________, $|\overline{PF}| = \sqrt{(x-1)^2 + \_\_\_\_}$, and $|\overline{PQ}| =$ ________.
Also, $\dfrac{|\overline{PF}|}{|\overline{PQ}|} = e =$ ____, so that $\dfrac{\sqrt{(x-1)^2+y^2}}{|x+1|} =$ ____ or
$2\sqrt{(x-1)^2+y^2} =$ ________. Squaring both sides of the equation, we have $4[(x-1)^2+y^2] =$ __________; that is,
________________ $= 0$. Completing the square, we have $3\left(x^2 - \dfrac{10}{3}x + \dfrac{25}{9}\right) + 4y^2 = -3 +$ ____ or $3\left(x - \dfrac{5}{3}\right)^2 + 4y^2 =$
____. Dividing both sides of the equation by $\dfrac{16}{3}$, we have

$$\frac{\left(x-\frac{5}{3}\right)^2}{\frac{16}{9}} + \frac{y^2}{\frac{4}{3}} = \_\_\_\_.$$

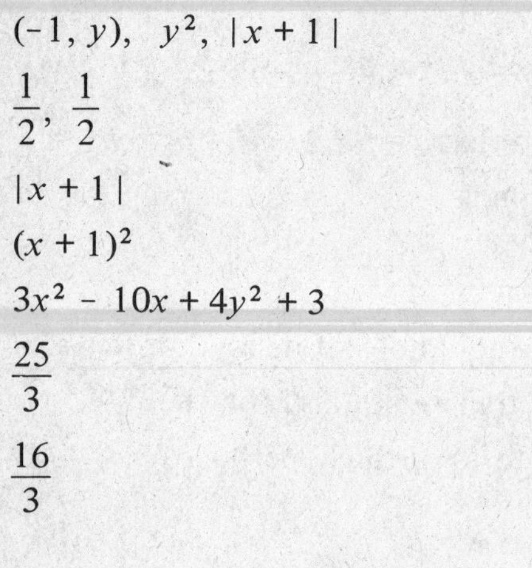

$(-1, y)$, y^2, $|x+1|$

$\dfrac{1}{2}$, $\dfrac{1}{2}$

$|x+1|$

$(x+1)^2$

$3x^2 - 10x + 4y^2 + 3$

$\dfrac{25}{3}$

$\dfrac{16}{3}$

1

127 Focus $F = (3, 0)$, directrix is the y axis, and eccentricity $e = 2$.

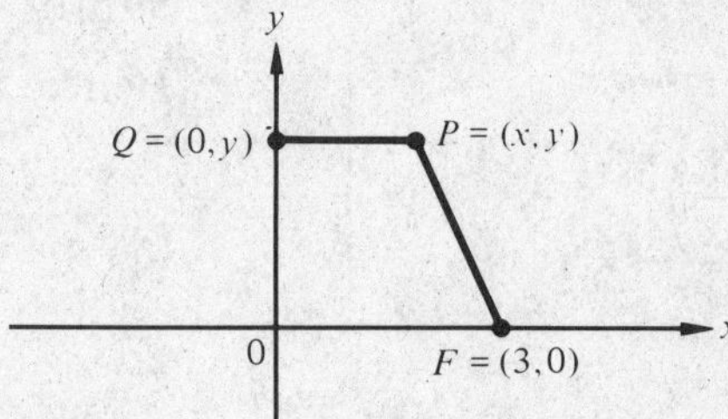

$Q =$ ________. $|\overline{PF}| = \sqrt{(x-3)^2 + \_\_\_\_}$ $|\overline{PQ}| =$ _____. Also,
$\dfrac{|\overline{PF}|}{|\overline{PQ}|} = e =$ ____, so that $\dfrac{\sqrt{(x-3)^2+y^2}}{|x|} =$ ____ or
$\sqrt{(x-3)^2+y^2} =$ ____. Squaring both sides of the equation, we have $(x-3)^2 + y^2 =$ ____ or ________________ $=$
0. Completing the square, we have $3(x^2 + 2x +$ ____$) - y^2 =$
$9 +$ ____ or $3(x+1)^2 - y^2 =$ ____. Dividing both sides by 12,
we obtain ____________ $= 1$.

$(0, y)$, y^2, $|x|$

2, 2

$2|x|$

$4x^2$, $3x^2 - y^2 + 6x - 9 = 0$

1

3, 12

$\dfrac{(x+1)^2}{4} - \dfrac{y^2}{12}$

128 Let the focus F of a conic with eccentricity $e > 0$ be at the origin and suppose that the directrix is D: $x = -d$, where $d > 0$. Then the equation of the conic is

$(1 - e^2)x^2 - 2e^2dx + y^2 = e^2d^2$

$y^2 - 2dx = d^2$

______________________. If $e = 1$, the equation reduces to ______________.

129 Suppose that a conic with focus F at the origin and directrix D: $x = -d$ has eccentricity e, where e and d are positive. Then exactly one of the following holds:

(i) $e < 1$ and the conic is an __________ with the equation ______________________, where a = __________,

$$b = \frac{ed}{\sqrt{1 - e^2}}, \text{ and } c = \sqrt{a^2 - b^2} = \_\_\_\_.$$

(ii) $e = 1$ and the conic is a __________ with the equation ______________________, where p = ____.

(iii) $e > 1$ and the conic is a __________ with the equation ______________________, where $b = \dfrac{ed}{\sqrt{e^2 - 1}}$,

$$a = \_\_\_\_\_\_, \text{ and } c = \sqrt{a^2 + b^2} = \_\_\_\_.$$

ellipse

$\frac{(x - c)^2}{a^2} + \frac{y^2}{b^2} = 1, \quad \frac{ed}{1 - e^2}$

ae

parabola

$4p(x + p) = y^2, \quad \frac{d}{2}$

hyperbola

$\frac{(x + c)^2}{a^2} - \frac{y^2}{b^2} = 1$

$\frac{ed}{e^2 - 1}, \quad ae$

In Problems 130 through 134, use Problems 128 and 129 to find the equation of the conic.

130 Focus $F = (0, 0)$, directrix $x = -2$, and eccentricity $e = \frac{1}{2}$. Using Problem 128, $(1 - e^2)x^2 - 2e^2\,dx + y^2 = e^2d^2$,

$\left(1 - \frac{1}{4}\right)x^2 - 2\left(\frac{1}{4}\right)(2)x + y^2 = \left(\frac{1}{4}\right)(\_\_\_\_)$ or $\frac{3}{4}x^2 - x + y^2 =$ ____, so that $\frac{3}{4}\left(x^2 - \frac{4}{3}x + \frac{4}{9}\right) + y^2 = 1 +$ ____ or

$\frac{3}{4}\left(x - \frac{2}{3}\right)^2 + y^2 =$ ____. Thus, $\dfrac{\left(x - \frac{2}{3}\right)^2}{\frac{16}{9}} + \dfrac{y^2}{\frac{4}{3}} =$ ____.

131 Focus $F = (0, 0)$, directrix $x = -2$, and eccentricity $e = 1$. Using Problem 129 with $e = 1$ and $p = \frac{d}{2} =$ ____, we have

$4(1)(x +$ ____$) = y^2$ or ________ $= y^2$.

132 Focus $F = (0, 0)$, directrix $x = -1$, and eccentricity $e = 2$. Using Problem 129 with $e = 2$ and $d = 1$, we have

$a = \dfrac{ed}{e^2 - 1} = \dfrac{(2)(1)}{\_\_\_\_} =$ ____,

$b = \dfrac{ed}{\sqrt{e^2 - 1}} = \dfrac{(2)(\_\_\_\_)}{\sqrt{3}} =$ ____,

$c = ae = \left(\frac{2}{3}\right)(\_\_\_\_) =$ ____.

4

$1, \frac{1}{3}$

$\frac{4}{3}, 1$

1

$1, 4x + 4$

$3, \frac{2}{3}$

$1, \frac{2}{\sqrt{3}}$

$2, \frac{4}{3}$

$\dfrac{\left(x+\dfrac{4}{3}\right)^2}{\dfrac{4}{9}} - \dfrac{y^2}{\dfrac{4}{3}} = 1$

The equation is

$$\frac{(x+c)^2}{a^2} - \frac{y^2}{b^2} = 1 \text{ or } \underline{\qquad\qquad\qquad}.$$

4

2

$\sqrt{5}, \dfrac{5}{4}$

$\dfrac{\left(x+\dfrac{5}{4}\right)^2}{\dfrac{5}{16}} - \dfrac{y^2}{\dfrac{5}{4}} = 1$

133 $e = \sqrt{5}$, directrix $x = -1$, and $F = (0, 0)$.

Using Problem 129, we have

$$a = \frac{ed}{e^2 - 1} = \frac{\sqrt{5}}{\underline{\quad}},$$

$$b = \frac{ed}{\sqrt{e^2 - 1}} = \frac{\sqrt{5}}{\underline{\quad}},$$

$$c = ae = \frac{\sqrt{5}}{4}(\underline{\quad}) = \underline{\quad}.$$

The equation is

$$\frac{(x+c)^2}{a^2} - \frac{y^2}{b^2} = 1 \text{ or } \underline{\qquad\qquad\qquad}.$$

$\dfrac{16}{25}, \dfrac{25}{27}$

$\dfrac{1}{3}, \dfrac{5}{9}$

$\dfrac{4}{5}, \dfrac{20}{27}$

$\dfrac{\left(x-\dfrac{20}{27}\right)^2}{\dfrac{625}{729}} + \dfrac{y^2}{\dfrac{25}{81}} = 1$

major axis, $\dfrac{a^2}{c}$

transverse axis

$\dfrac{a^2}{c}$

134 $e = \dfrac{4}{5}$, directrix $x = -\dfrac{5}{12}$, and $F = (0, 0)$.

Using Problem 129, we have

$$a = \frac{ed}{1 - e^2} = \frac{\left(\frac{4}{5}\right)\left(\frac{5}{12}\right)}{1 - \underline{\quad}} = \underline{\quad},$$

$$b = \frac{ed}{\sqrt{1 - e^2}} = \frac{\underline{\quad}}{\frac{3}{5}} = \underline{\quad},$$

$$c = ae = \frac{25}{27}\left(\underline{\quad}\right) = \underline{\quad}.$$

The equation is

$$\frac{(x-c)^2}{a^2} + \frac{y^2}{b^2} = 1 \text{ or } \underline{\qquad\qquad\qquad}.$$

135 The directrices of an ellipse are two lines perpendicular to the ______________ of the ellipse at a distance ____ units from the center. The directrices of a hyperbola are two lines perpendicular to the ____________________ of the hyperbola at a distance ____ from the center.

In Problems 136 through 142, find the eccentricity and the equations of the directrices for each conic.

136 $\dfrac{x^2}{16} + \dfrac{y^2}{9} = 1.$

Answers	Frames
4, 3	The semimajor axis is $a =$ ____ and the semiminor axis is $b =$ ____. The distance from the center $C = (0, 0)$ to either focus is
$7, \frac{\sqrt{7}}{4}$	$c = \sqrt{\quad}$, so that the eccentricity $e = \frac{c}{a} =$ ____. The distance
16	from the center to either directrix is $\frac{a^2}{c} = \frac{\quad}{\sqrt{7}}$ units, so that the equations of the directrices are:
$\frac{16}{\sqrt{7}}, \frac{16}{\sqrt{7}}$	D_1: $x = -$ ____ and D_2: $x =$ ____.
	137 $\frac{x^2}{9} + \frac{y^2}{25} = 1.$
vertical, 5	The major axis is ____. The semimajor axis is $a =$ ____
3	and the semiminor axis is $b =$ ____. The distance from the center
4	to either focus is $\sqrt{25 - 9} =$ ____. Hence, the eccentricity is
$\frac{4}{5}, \frac{25}{4}$	$e =$ ____ and the directrices are $\frac{a^2}{c} =$ ____ units from the center.
$\frac{25}{4}$	The equations of the directrices are D_1: $y = -$ ____ and
$\frac{25}{4}$	D_2: $y =$ ____.
	138 $\frac{x^2}{4} - \frac{y^2}{1} = 1.$
(0, 0)	The center is $C =$ ____, the distance from the center to either
2	vertex is $a =$ ____, and the distance from the center to either focus
5	is $c = \sqrt{\quad}$. Hence, the eccentricity is $e = \frac{c}{a} =$ ____.
$\frac{\sqrt{5}}{2}$.	The distance from the center to either directrix is $\frac{a^2}{c} =$ ____.
$\frac{4}{\sqrt{5}}$	Thus, the equations of the directrices are $D_1 : x = -$ ____ and
$\frac{4}{\sqrt{5}}, \frac{4}{\sqrt{5}}$	$D_2 : x =$ ____.
	139 $\frac{y^2}{9} - \frac{x^2}{1} = 1.$
vertical, 3, 1	The transverse axis is ____. $b =$ ____ and $a =$ ____, so
10, 1, $\sqrt{10}$	that $c = \sqrt{\quad}$. Therefore, $e = \frac{c}{a} = \frac{\sqrt{10}}{\quad} =$ ____. The center is
(0, 0)	$C =$ ____. The distance from the center to either directrix is
$\frac{1}{\sqrt{10}}$	$\frac{a^2}{c} =$ ____. Thus, the equations of the directrices are D_1: $y =$
$-\frac{1}{\sqrt{10}}, \frac{1}{\sqrt{10}}$	____ and D_2: $y =$ ____.
	140 $\frac{(x-3)^2}{4} + \frac{(y-1)^2}{25} = 1.$
vertical, (3, 1)	The major axis is ____ and the center is ____. The

Answers	
5, 2	semimajor axis is $a =$ ____ and the semiminor axis is $b =$ ____.
21	The distance from the center to either focus is $\sqrt{25-4} = \sqrt{}$ ____.
$\frac{\sqrt{21}}{5}$	Hence, the eccentricity is $e = \frac{c}{a} =$ ____ and the directrices are
$\frac{25}{\sqrt{21}}$	$\frac{a^2}{c} =$ ____ units from the center. The equation of the di-
$1 - \frac{25}{\sqrt{21}}$, $1 + \frac{25}{\sqrt{21}}$	rectrices are D_1: $y =$ ____ and D_2: $y =$ ____.
	141 $\frac{(x+3)^2}{9} - \frac{(y+2)^2}{7} = 1.$
horizontal	The transverse axis is ____ and the center is
(−3, −2), 3, $\sqrt{7}$, 4	____. $a =$ ____, $b =$ ____, and $c = \sqrt{9+7} =$ ____.
$\frac{4}{3}$	Hence, the eccentricity $e = \frac{c}{a} =$ ____. The distance from the
$\frac{9}{4}$	center to either directrix is $\frac{a^2}{c} =$ ____. The equations of the di-
$-\frac{21}{4}$, $-\frac{3}{4}$	rectrices are D_1: $x = -3 - \frac{9}{4} =$ ____ and D_2: $x = -3 + \frac{9}{4} =$ ____.
	142 $\frac{(y+7)^2}{6} - \frac{x^2}{10} = 1.$
vertical, (0, −7)	The transverse axis is ____. The center is ____.
$\sqrt{10}$, $\sqrt{6}$, 4	$a =$ ____, $b =$ ____, and $c = \sqrt{10+6} =$ ____. Hence, the eccen-
$\frac{4}{\sqrt{10}}$	tricity $e = \frac{c}{a} =$ ____. The distance from the center to either
$\frac{5}{2}$	directrix is $\frac{a^2}{c} =$ ____. Thus, the equations of the directrices are
$-\frac{19}{2}$, $-\frac{9}{2}$	D_1: $y =$ ____ and D_2: $y =$ ____.

Chapter Test

1 Find the equation of the circle satisfying the given conditions.
(a) Center at (2, −1) and radius 5 units. (b) Center at (−2, 4) and containing the point (−5, 0).

2 Find the center and radius of the circle whose equation is $36x^2 + 36y^2 - 24x + 36y - 131 = 0$.

3 Find a translation of axes that simplifies the equation $(x+7)^2 + (y-3)^2 = 81$.

4 Find the equation of the ellipse with foci at the points (0, 4) and (0, −4) and with vertices at (0, 5) and (0, −5).

5 Find the coordinates of the center, the foci, and the vertices of the ellipse whose equation is $9x^2 + 4y^2 + 36x - 8y + 4 = 0$. Also, sketch the graph.

6 Find the coordinates of the center, the vertices, and the foci of the following hyperbolas. Also, find the equations of the asymptotes.

(a) $\frac{y^2}{9} - \frac{x^2}{16} = 1$ (b) $16x^2 - 9y^2 - 32x - 36y - 164 = 0$

7 Find the coordinates of the vertex and the focus of the parabola. Also, find the equation of the directrix and the length of the latus rectum.

(a) $x^2 = -4y$ (b) $y^2 + 4y - 12x + 16 = 0$

8 Find the equation of the conic satisfying the given conditions:

(a) $e = \frac{4}{5}$, focus at (0, 4), and directrix $y = \frac{25}{4}$ (b) $e = \frac{5}{4}$, focus at (5, 0), and directrix $x = \frac{16}{5}$

9 Given the equation of the conic $\frac{(x-1)^2}{4} + \frac{(y+2)^2}{3} = 1$. Find the eccentricity and the equations of the directrices.

Answers

1 (a) $(x-2)^2 + (y+1)^2 = 25$ (b) $(x+2)^2 + (y-4)^2 = 25$

2 $C = \left(\frac{1}{3}, -\frac{1}{2}\right)$, radius = 2

3 $\bar{x} = x + 7$ and $\bar{y} = y - 3$

4 $\frac{x^2}{9} + \frac{y^2}{25} = 1$

5 $C = (-2, 1), F_1 = (-2, 1 - \sqrt{5}), F_2 = (-2, 1 + \sqrt{5})$. $V_1 = (-2, -2), V_2 = (-2, 4), V_3 = (-4, 1), V_4 = (0, 1)$.
The graph is

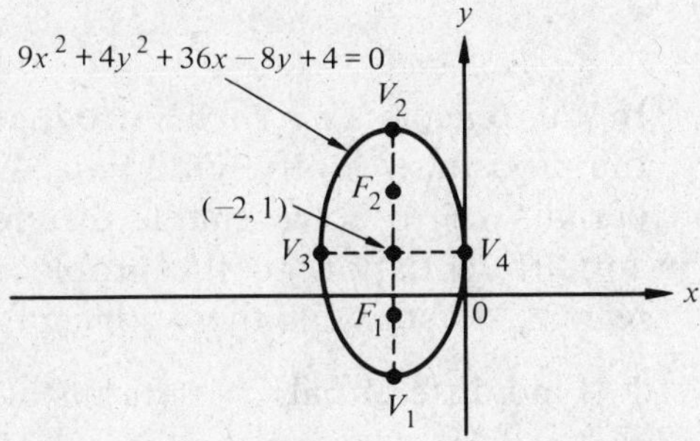

6 (a) $C = (0, 0)$
$V_1 = (0, 3), V_2 = (0, -3)$
$F_1 = (0, 5), F_2 = (0, -5)$
$y = \pm\frac{3}{4}x$

(b) $C = (1, -2)$
$V_1 = (-2, -2), V_2 = (4, -2)$
$F_1 = (-4, -2), F_2 = (6, -2)$
$y + 2 = \pm\frac{4}{3}(x - 1)$

7 (a) $V = (0, 0)$
$F = (0, -1)$
$y = 1$
Length of latus rectum is 4.

(b) $V = (1, -2)$
$F = (4, -2)$
$x = -2$
Length of latus rectum is 12.

8 (a) $\frac{x^2}{9} + \frac{y^2}{25} = 1$ (b) $\frac{x^2}{16} - \frac{y^2}{9} = 1$.

9 $e = \frac{1}{2}$, $x = 5$, $x = -3$

Chapter 5 ANTIDIFFERENTIATION, DIFFERENTIAL EQUATIONS, AND AREA

In this chapter we shall introduce antidifferentiation, which reverses the procedure of differentiation. We shall show how antidifferentiation permits us to solve simple differential equations and we shall relate antidifferentiation to the problem of calculating the areas of certain regions. In studying these concepts the student will be able to:

1 Find differentials dy in terms of x and dx.

2 Find the antiderivative directly and by change of variable (substitution).

3 Apply the method of substitution to solve simple differential equations by separation of variables.

4 Solve second-order differential equations and discuss linear motion.

5 Apply differential equations in the areas of work done by a variable force, economics, Newton's law of motion, momentum, and energy.

6 Set up differential equations using "infinitesimals."

7 Find areas of regions in the plane by slicing.

8 Find the area under the graph of a function using the definite integral.

1 Differentials

1 **Let f be a function and let x and y be variables so related that $y = f(x)$. Then the differential dx is a variable that can take on any value in R. If x is any number in the domain of f for which $f'(x)$ exists, then the differential dy is defined by the equation $dy =$ ________.**

$f'(x)\,dx$

In Problems 2 through 8, find dy in terms of x and dx.

2 $y = 3x^2 + 7$.

$dy = (\_\_\_\_)\,dx$.

$6x$

3 $y = \dfrac{x^3}{3} + \dfrac{x^2}{2} + 7$.

$dy = (\_\_\_\_\_\_)\,dx$.

$x^2 + x$

4 $y = \sqrt{1 - x^2}$.

$dy = \dfrac{1}{2}(1 - x^2)^{-1/2}\,(\_\_\_\_)\,dx$

$-2x$

$= \_\_\_\_\_\_\_\_.$

$-\dfrac{x}{\sqrt{1 - x^2}}\,dx$

5 $y = x\sqrt{4 - 2x}$.

$dy = \left[(x)\left(\dfrac{1}{2}\right)(4 - 2x)^{-1/2}\,(\_\_\_\_) + \sqrt{4 - 2x}\right]dx$

-2

$= \_\_\_\_\_\_\_\_\;dx.$

$-\dfrac{3x + 4}{\sqrt{4 - 2x}}$

6 $y = (3x - 6x^2)^{2/3}$.

$dy = \dfrac{2}{3}(3x - 6x^2)^{-1/3}\,(\_\_\_\_\_\_)\,dx$

$3 - 12x$

$= \_\_\_\_\_\_\_\_\;dx.$

$\dfrac{2(1 - 4x)}{(3x - 6x^2)^{1/3}}$

7 $y = \dfrac{x^3}{2x^3 + 3}$.

$dy = \dfrac{(2x^3 + 3)\,(\_\_\_\_) - x^3\,(\_\_\_\_)\,dx}{(2x^3 + 3)^2}$

$3x^2,\ 6x^2$

$= \_\_\_\_\_\_\_\_\;dx.$

$\dfrac{9x^2}{(2x^2 + 3)^2}$

8 $y = (1 + x^3)^{4/5}$.

$dy = \dfrac{4}{5}(1 + x^3)^{-1/5}\,(\_\_\_\_)\,dx$

$3x^2$

$= \_\_\_\_\_\_\_\_\;dx.$

$\dfrac{12x^2}{5\sqrt[5]{1 + x^3}}$

In Problems 9 through 12, set $\Delta y = f(x_1 + \Delta x) - f(x_1)$ and $dx = \Delta x$. Find Δy, dy, and $\Delta y - dy$ for the given values of x_1 and Δx.

Answers	Problems
	9 f is defined by $y = x^3$, $x_1 = 1, \Delta x = 0.01$.
	$\Delta y = f(x_1 + \Delta x) - f(x_1)$
x_1^3	$= (x_1 + \Delta x)^3 - \_\_\_\_,$
	so that Δy at $x_1 = 1$ and $\Delta x = 0.01$ is
1.0303	$\Delta y = (1.01)^3 - 1^3 = \_\_\_\_\_\_\_\_ - 1$
0.0303	$= \_\_\_\_\_\_\_\_.$
$3x_1^2$	$dy = \_\_\_\_ dx$, so that
0.03	$dy = 3(1^2)(0.01) = \_\_\_\_,$
0.0003	$\Delta y - dy = 0.0303 - 0.03 = \_\_\_\_\_\_\_\_.$
	10 f is defined by $y = x(x^2 + 1)$, $x_1 = 2$, and $\Delta x = 0.1$.
	$\Delta y = f(x_1 + \Delta x) - f(x_1) = f(2.1) - f(2)$
11.361, 1.361	$= \_\_\_\_\_\_\_\_ - 10 = \_\_\_\_\_\_,$
13	$dy = (3x_1^2 + 1)\,dx = (\_\_\_\_)(0.1)$
	$= 1.3,$
0.061	$\Delta y - dy = 1.361 - 1.300 = \_\_\_\_\_\_.$
	11 f is defined by $y = \sqrt{x}$, $x_1 = 4$, and $\Delta x = -1$.
$-0.2679\ldots$	$\Delta y = f(x_1 + \Delta x) - f(x_1) = \sqrt{3} - \sqrt{4} = \_\_\_\_\_\_\_\_\_\_\_\_.$
-0.25	$dy = \dfrac{1}{2\sqrt{x_1}}\,dx = \dfrac{1}{2\sqrt{4}}(-1) = \_\_\_\_\_\_.$
-0.0179	$\Delta y - dy \approx -0.2679 + 0.25 = \_\_\_\_\_\_\_\_.$
	12 f is defined by $y = \dfrac{1}{\sqrt[3]{x}}$, $x_1 = 8$, and $\Delta x = 0.05$.
$\sqrt[3]{8.05}$, 0.5	$\Delta y = f(x_1 + \Delta x) - f(x_1) = \dfrac{1}{\_\_\_\_\_\_} - \_\_\_\_$
-0.001037	$\approx \_\_\_\_\_\_\_\_\_\_,$
	$dy = -\dfrac{1}{3}x_1^{-4/3}\,dx = -\dfrac{1}{3\sqrt[3]{(8)^4}}(0.05)$
16, 48, 960	$= \dfrac{-0.05}{3(\_\_\_\_)} = \dfrac{-0.05}{\_\_\_\_} = \dfrac{-1}{\_\_\_\_},$
$\dfrac{1}{960}$, 0.00004	$\Delta y - dy = \dfrac{1}{\sqrt[3]{8.05}} - 0.5 + \dfrac{}{\_\_\_\_} \approx \_\_\_\_\_\_\_\_.$

In Problems 13 through 16, use differentials to estimate the following expressions.

Answers	Problems
	13 $\sqrt{17}$.
	Let $y = \sqrt{x}$ and let Δy be the change in y as x increases from
16, Δy, 1	$\_\_\_\_$ to 17. Thus, $\sqrt{17} = \sqrt{16} + \_\_\_\_$. Put $dx = \Delta x = \_\_\_\_$.
$\dfrac{1}{2\sqrt{x}}$, 0.125	$\dfrac{dy}{dx} = \_\_\_\_$. $\dfrac{dy}{dx}$ at $x = 16$ is $\_\_\_\_$. Therefore,

0.125, 0.125

4.125

27, 1, Δy

$\dfrac{1}{27}$

$\dfrac{1}{27}, \dfrac{1}{27}$

$\dfrac{82}{27}$, 3.0370

81, 1

$\dfrac{1}{108}, \dfrac{1}{108}$

$\dfrac{1}{108}$, 3.0093

0.04

$\sqrt{0.04}$, 2.5

2.5, 0.0075

0.0075, 0.2075

x^3, $3x^2\,dx$

$3x^2$, $0.06x^3$

πr^2

$2 + \Delta r$

0.01

$2 + \Delta r$

$\pi(2^2)$

$\Delta y \approx dy = \dfrac{dy}{dx}\,dx =$ (______) (1), so that $\sqrt{17} \approx 4 +$ ______ =

______.

14 $\sqrt[3]{28}$.

Let $y = \sqrt[3]{x}$ and let Δy be the change in y as x increases from ____ to 28. Put $dx = \Delta x =$ ____. $\sqrt[3]{28} = \sqrt[3]{27} +$ ____. $\dfrac{dy}{dx} = \dfrac{1}{3x^{2/3}}$. $\dfrac{dy}{dx}$ at $x = 27$ is $\dfrac{\quad}{\_\_\_\_}$. Therefore,

$\Delta y \approx dy = \dfrac{dy}{dx}\,dx = \dfrac{\quad}{\_\_\_\_}$ (1), so that $\sqrt[3]{28} \approx 3 + \dfrac{\quad}{\_\_\_\_} =$

$\dfrac{\quad}{\_\_\_\_} \approx$ ________.

15 $\sqrt[4]{82}$.

Let $y = \sqrt[4]{x}$ and let Δy be the change in y as x increases from ____ to 82. Put $dx = \Delta x =$ ____. $\dfrac{dy}{dx} = \dfrac{1}{4x^{3/4}}$. $\dfrac{dy}{dx}$ at $x = 81$ is $\dfrac{\quad}{\_\_\_\_}$. Therefore, $\Delta y \approx dy = \dfrac{dy}{dx}\,dx = \left(\dfrac{\quad}{\_\_\_\_}\right)(1)$, so that

$\sqrt[4]{82} \approx 3 + \dfrac{\quad}{\_\_\_\_} = \dfrac{325}{108} \approx$ ________.

16 $\sqrt{0.043}$.

Let $y = \sqrt{x}$ and let Δy be the change in y as x increases from ______ to 0.043. Put $dx = \Delta x = 0.003$. Then $\sqrt{0.043} =$ ______ $+ \Delta y$, $\dfrac{dy}{dx} = \dfrac{1}{2\sqrt{x}}$. $\dfrac{dy}{dx}$ at $x = 0.04$ is ____.

$\Delta y \approx dy = \dfrac{dy}{dx}\,dx =$ (____) (0.003) = ________, so that

$\sqrt{0.043} \approx 0.2 +$ ________ = ________.

17 Use differentials to find the approximate change in the volume V of a cube whose side is x cm caused by increasing the sides by 2 percent.

$V =$ ____, so that $dV =$ ________. $\Delta x = dx = (0.02)x$. Then $\Delta V \approx dV =$ (____) $(0.02x) =$ ________ cubic centimeters.

18 The radius r of a circle is measured to be 2 in. and it is known that the error involved in this measurement does not exceed 0.01 in. Use differentials to estimate the possible error in the calculated area A.

The area A of a circle of radius r is $A =$ ____. The true value of the radius is of the form ________, where the measurement error Δr satisfies $|\Delta r| \leqslant$ _____. The true value of the area is π(________)^2, while the value of the area calculated from the measurement of the radius is _______. The difference $\Delta A =$

4π	$\pi(2 + \Delta r)^2$ - _____ gives the error in the calculated area. Put
2	$dr = \Delta r$; then, $\Delta A \approx dA = \frac{dA}{dr}\,dr = 2\pi r\,\Delta r = 2\pi(\_\_\_\_)\,\Delta r$.
$4\pi(0.01)$, 0.04π	Therefore, $\lvert \Delta A \rvert \approx 4\pi\,\lvert \Delta r \rvert \leqslant$ ____________ = ________ square inch.
	19 The altitude h of a certain right circular cone is constant and equal to 3 in. If the radius r of its base is increased from 10 in. to 10.5 in., find the approximate increase in the volume V of the cone.
$\frac{1}{3}\pi r^2 h$, $\frac{2\pi r}{3} h$	$V =$ ________, so that $\frac{dV}{dr} =$ ________. Since $h = 3$ in.,
$2\pi r$	$\frac{dV}{dr} =$ _____. The approximate increase in the volume is given by
$\frac{dV}{dr} \cdot \Delta r$, Δr	$dV =$ ________, so that $dV = (2\pi r)$ (_____). Therefore, dV at
10, 0.5	$r = 10$ in. and $\Delta r = 0.5$ is given by $dV = 2\pi$(_____) (_____) =
10π, 31.4	_____ $\approx$ _____ cubic inches.
	20 A particle moves along a straight line in accordance with the equation $s = 2t^3 - 4t^2 + 2t - 1$, where t is the elapsed time in seconds and s is its directed distance, measured in feet, from the origin to the particle. Find the approximate distance covered by the particle in the interval from $t = 3$ to $t = 3.2$ seconds.
	The approximate distance covered by the particle is given by
$6t^2 - 8t + 2$	$ds =$ (________________) dt. Set $dt = \Delta t$. We have $ds =$
$6t^2 - 8t + 2$	(________________) Δt, so that ds at $t = 3$ and $\Delta t = 0.2$ is
32, 6.4	$ds =$ (_____) (0.2) = _____ feet.
	21 The attractive force between unlike electrically charged particles is expressed by $F = \frac{k}{x^2}$, where x is the distance between the particles and k is a certain positive constant. If x is increased by 3 percent, find the approximate percent decrease in F.
$-\frac{2}{x^3}$, $-\frac{2k}{x^3}$, $-\frac{2}{x}$	$F = \frac{k}{x^2}$, so that $dF = k\left(\_\_\_\_\right) dx$. $\frac{dF}{F} = \frac{\_\_\_\_}{\frac{k}{x^2}}\,dx = \left(\_\_\_\_\right) dx$.
	Set $dx = \Delta x$, so that
Δx, $\frac{\Delta x}{x}$	$\frac{\Delta F}{F} \approx \frac{-2}{x}(\_\_\_\_) = -2\left(\_\_\_\_\right)$
0.03	$\frac{\Delta F}{F} \approx -2\left(\frac{\Delta x}{x}\right) = -2(\_\_\_\_)$
0.06, 6%	$= -$ _____ $= -$ _____.
6%	Therefore, the approximate decrease is _____.

1.1 Formulas Involving Differentials

In Problems 22 through 29, let u and v be variables which are differentiable functions of x, let c represent a constant function or a constant number, and let n be a rational exponent. Find the differential in each case.

0	22 $dc =$ ____.
du	23 $d(cu) = c($____$)$.
dv	24 $d(u + v) = du +$ ____.
$v\,du$	25 $d(uv) = u\,dv +$ ____.
$v\,du - u\,dv$	26 $d\left(\frac{u}{v}\right) = \frac{\rule{3cm}{0.4pt}}{v^2}$.
nu^{n-1}	27 $d(u^n) = ($____$)\,du$.
ncu^{n-1}	28 $d(cu^n) = ($____$)\,du$.
ncx^{n-1}	29 $d(cx^n) = ($____$)\,dx$.

In Problems 30 through 33, use Problems 22 through 29 to find dy.

	30 $y = x^{-2/3} + 7$.
$-\frac{2}{3}x^{-5/3}$	$dy =$ ____ dx.
	31 $y = (x^2 + 4)^2 (2x^3 - 1)^3$.
$d[(2x^3 - 1)^3],\ d[(x^2 + 4)^2]$	$dy = (x^2 + 4)^2($____$) + ($____$)(2x^3 - 1)^3$
$d(2x^3 - 1)$	$= (x^2 + 4)^2 \cdot 3(2x^3 - 1)^2($____$) +$
$d(x^2 + 4)$	$2(x^2 + 4)($____$)(2x^3 - 1)^3$
$6x^2\,dx$	$= (x^2 + 4)\,3(2x^3 - 1)^2($____$) +$
$2x\,dx$	$(2x^3 - 1)^3 \cdot 2(x^2 + 4)($____$)$
$2x(x^2 + 4)(2x^3 - 1)^2(4x^3 + 9x - 2)$	$= ($____$)\,dx$.
	32 $y = \frac{5 - x}{5 + x}$.
$d(5 - x),\ d(5 + x)$	$dy = \frac{(5 + x)(\rule{2cm}{0.4pt}) - (5 - x)(\rule{2cm}{0.4pt})}{(5 + x)^2}$
$-dx,\ dx$	$= \frac{(5 + x)(\rule{1cm}{0.4pt}) - (5 - x)(\rule{1cm}{0.4pt})}{(5 + x)^2}$
-10	$= \frac{\rule{1cm}{0.4pt}}{(5 + x)^2}\,dx$.
	33 $xy + x - 3y = 13$.
$3y,\ d(13),\ x\,dy + y\,dx$	$d(xy) + dx - d($____$) =$ ____ or $($____$) + dx - 3dy =$
$0,\ x - 3,\ y + 1$	____ or ____ $dy + (y + 1)\,dx = 0$, so that $dy = -\frac{\rule{1cm}{0.4pt}}{x - 3}\,dx$.

2 Antiderivatives

$g'(x) = f(x)$	34 A function g is called an antiderivative of a function f on a set of numbers I if ____ holds for every value of x in I.

antidifferentiation

The procedure of finding antiderivatives is called ____________.

In Problems 35 through 38, verify that each function g is an antiderivative of the function f on the set I.

$g'(x) = f(x)$

$8,\ f(x)$

35 $f(x) = 28x^3 - 6x^2 + 8$ and $g(x) = 7x^4 - 2x^3 + 8x + 5$.
We must show that ____________.
$g'(x) = 28x^3 - 6x^2 +$ ____ = ____.

$g'(x) = f(x)$

$6x^2 - 10x,\ f(x)$

36 $f(x) = 5(2x^3 - 5x^2 + 4)^4(6x^2 - 10x)$ and $g(x) = (2x^3 - 5x^2 + 4)^5$.
We must show that ____________. Thus,
$g'(x) = 5(2x^3 - 5x^2 + 4)^4$ (________) = ____.

3

$1,\ -1$

$5,\ f(x)$

37 $f(x) = \dfrac{5}{(3-x)^2}$ and $g(x) = \dfrac{2+x}{3-x}$.
The domain I of f consists of all real numbers except ____.

$$g'(x) = \frac{(3-x)(\_\_\_) - (2+x)(\_\_\_)}{(3-x)^2}$$

$$= \frac{\_\_\_}{(3-x)^2} = \_\_\_.$$

-1

$1,\ 1$

$2(x+1) - (x-1),\ x+3$

38 $f(x) = \dfrac{x+3}{2(x+1)^{3/2}}$ and $g(x) = \dfrac{x-1}{\sqrt{x+1}}$.

The domain I of f consists of all real numbers greater than ____.

$$g'(x) = \frac{(\sqrt{x+1})(\_\_\_) - (x-1)\frac{1}{2}(x+1)^{-1/2}(\_\_\_)}{x+1}$$

$$= \frac{\_\_\_\_\_\_}{2(x+1)^{3/2}} = \frac{\_\_\_}{2(x+1)^{3/2}} = f(x).$$

$g(x) = C$

39 Let I be an open interval and suppose that f is the function with domain I defined by the equation $f(x) = 0$ for all values of x in I. Then a function g with domain I is an antiderivative of f on I if and only if there is a constant number C such that ________ for every value of x in I.

$g + C$

40 Suppose that g is an antiderivative of the function f on the open interval I. Then, a function h with domain I is an antiderivative of f on I if and only if there is a constant C such that $h =$ ____.

$2(x-3),\ 2x-6$

$g'(x) = h'(x),\ 9$

41 Given $g(x) = (x-3)^2$ and $h(x) = x(x-6)$. Then $g'(x) =$ ________ and $h'(x) =$ ______, so that ________. Here, $g(x) = h(x) +$ ____.

42 Let $g(x) = \dfrac{x}{1+x}$ and $h(x) = \dfrac{-1}{1+x}$. Then,

1, 1, 1

1, $h'(x)$, 1

$g'(x) = \dfrac{(1+x)(\_\_\_) - x(\_\_\_)}{(1+x)^2} = \dfrac{\_\_\_}{(1+x)^2}$ and

$h'(x) = \dfrac{\_\_\_}{(1+x)^2}$, so that $g'(x) = $ _____. Here, $g(x) = h(x) +$ _____.

2.1 Notation for Antiderivatives

antiderivative, $f(x)$

antiderivative

constant of integration

integral sign

integrand

indefinite integration

$g'(x)$

$3x^2$

$\dfrac{1}{x^3}$

$\cos x$

43 The notation $\int f(x)\,dx = g(x) + C$, where C denotes an arbitrary constant, will mean, by definition, that the function g is an __________ of the function f, so that $g'(x) = $ _____ holds for all values of x in the domain of f. If I is a set of numbers, the assertion that $\int f(x)\,dx = g(x) + C$ on I will likewise mean that g is an __________ of f on I. The constant C is called the __________, the symbol $\int$ is called the __________, and the function f is called the __________ of the expression $\int f(x)\,dx$. One also says that $f(x)$ is under the integral sign in the expression $\int f(x)\,dx$. The procedure of evaluating $\int f(x)\,dx$, that is, of finding $g(x) + C$ is called __________.

44 To verify that $\int f(x)\,dx = g(x) + C$, it is only necessary to show that $f(x) = $ _____ for all values of x in the domain of f.

45 $\int 3x^2\,dx = x^3 + C$, since $D_x(x^3) = $ _____.

46 $\int \dfrac{dx}{x^3} = -\dfrac{1}{2x^2} + C$, since $D_x\left(-\dfrac{1}{2x^2}\right) = $ _____.

47 $\int \cos x\,dx = \sin x + C$, since $D_x(\sin x) = $ _____.

2.2 Basic Rules for Antidifferentiation

$f(x)$

$f(x) + C$

$x + C$

$\dfrac{x^{n+1}}{n+1} + C$

$\int f(x)\,dx$

$\int g(x)\,dx$

$\int f_2(x)\,dx$

48 $D_x \int f(x)\,dx = $ _____.

49 $\int f'(x)\,dx = $ __________.

50 $\int dx = $ _____.

51 If n is a rational number other than -1, then

$$\int x^n\,dx = \_\_\_\_\_\_\_\_\_\_.$$

52 If a is a constant, then $\int af(x)\,dx = a($__________$)$.

53 $\int [f(x) + g(x)]\,dx = \int f(x)\,dx +$ __________.

54 $\int [a_1 f_1(x) + a_2 f_2(x) + \cdots + a_n f_n(x)]\,dx$
$= a_1 \int f_1(x)\,dx + a_2($__________$) + \cdots + a_n \int f_n(x)\,dx$,
where $a_1, a_2, \ldots, a_n$ are constants.

In Problems 55 through 63, evaluate each antiderivative.

55 $\int (4x + 5)\,dx.$

$\int (4x+5)\,dx = \int 4x\,dx + $ ________

$= 4\int x\,dx + 5($_____$)$

$= 4\,\dfrac{___}{___} + 5($_____$) + C$

$=$ ________________.

$\int 5\,dx$

$\int dx$

$\dfrac{x^2}{2},\ x$

$2x^2 + 5x + C$

56 $\int (9x^2 - 4x + 7)\,dx.$

$\int (9x^2 - 4x + 7)\,dx = \int 9x^2\,dx - \int 4x\,dx + ($________$)$

$= 9\int x^2\,dx - 4($________$) + 7\int dx$

$= 9\left(\dfrac{x^3}{3}\right) - 4\left(____\right) + 7($_____$) + C$

$=$ ________________________.

$\int 7\,dx$

$\int x\,dx$

$\dfrac{x^2}{2},\ x$

$3x^3 - 2x^2 + 7x + C$

57 $\int (6t^2 - 12t + 11)\,dt.$

$\int (6t^2 - 12t + 11)\,dt = \int 6t^2\,dt - \int 12t\,dt + ($________$)$

$= 6($________$) - 12\int t\,dt + 11($_____$)$

$= 6\left(\_\_\_\_\right) - 12\left(\_\_\_\_\right) + 11t + C$

$=$ ________________________.

$\int 11\,dt$

$\int t^2\,dt,\ \int dt$

$\dfrac{t^3}{3},\ \dfrac{t^2}{2}$

$2t^3 - 6t^2 + 11t + C$

58 $\int (5 - 3x^2 - 4x^{-3})\,dx.$

$\int (5 - 3x^2 - 4x^{-3})\,dx = \int 5\,dx - ($__________$) - \int 4x^{-3}\,dx$

$= 5($_____$) - 3($________$) - 4\int x^{-3}\,dx$

$= 5x - 3\left(\dfrac{x^3}{3}\right) - 4\left(\_\_\_\_\right) + C$

$=$ ____________________.

$\int 3x^2\,dx$

$dx,\ \int x^2\,dx$

$\dfrac{x^{-2}}{-2}$

$5x - x^3 + \dfrac{2}{x^2} + C$

59 $\displaystyle\int \frac{2y^2 - 6y + 1}{\sqrt{y}}\,dy.$

$\displaystyle\int \frac{2y^2 - 6y + 1}{\sqrt{y}}\,dy = \int\left(\frac{2y^2}{y^{1/2}} - \frac{6y}{y^{1/2}} + \_\_\_\_\right)dy$

$= \int (2y^{3/2} - 6y^{1/2} + $______$)\,dy$

$= 2($________$) - 6($________$) + \int y^{-1/2}\,dy$

$= 2\left(\dfrac{y^{5/2}}{\frac{5}{2}}\right) - 6\left(\dfrac{y^{3/2}}{\frac{3}{2}}\right) + \_\_\_\_ + C$

$=$ ______________________.

$\dfrac{1}{y^{1/2}}$

$y^{-1/2}$

$\int y^{3/2}\,dy,\ \int y^{1/2}\,dy$

$\dfrac{y^{1/2}}{\frac{1}{2}}$

$\dfrac{4}{5}y^{5/2} - 4y^{3/2} + 2y^{1/2} + C$

60 $\displaystyle\int \frac{5x^3 - 4x^2 + x}{\sqrt[3]{x}}\,dx.$

$\displaystyle\int \frac{5x^3 - 4x^2 + x}{\sqrt[3]{x}}\,dx = \int\left(\frac{5x^3}{x^{1/3}} - \frac{4x^2}{x^{1/3}} + \_\_\_\_\right)dx$

$= \int (5x^{8/3} - 4x^{5/3} + $______$)\,dx$

$\dfrac{x}{x^{1/3}}$

$x^{2/3}$

$\int x^{8/3}\,dx$, $\int x^{5/3}\,dx$

$\dfrac{x^{11/3}}{\frac{11}{3}}$, $\dfrac{x^{8/3}}{\frac{8}{3}}$

$\frac{15}{11}x^{11/3} - \frac{3}{2}x^{8/3} + \frac{3}{5}x^{5/3} + C$

$= 5(\_\_\_\_\_) - 4(\_\_\_\_\_) + \int x^{2/3}\,dx$

$= 5\left(\dfrac{}{\_\_\_\_\_}\right) - 4\left(\dfrac{}{\_\_\_\_\_}\right) + \dfrac{x^{5/3}}{\frac{5}{3}} + C$

$= \_\_\_\_\_\_\_\_\_\_.$

$x^{-2/3}$

$\int x^{-2/3}\,dx$

$\dfrac{x^{7/3}}{\frac{7}{3}}$

$\frac{3}{7}x^{7/3} + 21x^{1/3} + C$

61 $\int x^{-2/3}(x^2 + 7)\,dx.$

$\int x^{-2/3}(x^2 + 7)\,dx = \int [x^{4/3} + 7(\_\_\_\_\_)]\,dx$

$= \int x^{4/3}\,dx + 7(\_\_\_\_\_)$

$= \_\_\_\_\_ + 7\left(\dfrac{x^{1/3}}{\frac{1}{3}}\right) + C$

$= \_\_\_\_\_\_\_\_\_\_.$

4

$\int dt$

$\dfrac{t^{5/3}}{\frac{5}{3}}$

$\frac{12}{5}t^{5/3} - 4t + 3t^{1/3} + C$

62 $\int \left(2\sqrt[3]{t} - \dfrac{1}{\sqrt[3]{t}}\right)^2 dt.$

$\int \left(2\sqrt[3]{t} - \dfrac{1}{\sqrt[3]{t}}\right)^2 dt = \int (4t^{2/3} - \_\_\_\_ + t^{-2/3})\,dt$

$= 4\int t^{2/3}\,dt - 4(\_\_\_\_\_) + \int t^{-2/3}\,dt$

$= 4\left(\dfrac{}{\_\_\_\_\_}\right) - 4t + \dfrac{t^{1/3}}{\frac{1}{3}} + C$

$= \_\_\_\_\_\_\_\_\_\_.$

$6x^2 - 3$

$\dfrac{x^3}{3}$

x

$2x^7 - \frac{7}{5}x^5 + 2x^3 - 3x + C$

63 $\int (7x^4 + 3)(2x^2 - 1)\,dx.$

$\int (7x^4 + 3)(2x^2 - 1)\,dx = \int (14x^6 - 7x^4 + \_\_\_\_\_)\,dx$

$= 14\left(\dfrac{x^7}{7}\right) - 7\left(\dfrac{x^5}{5}\right) + 6\left(\dfrac{}{\_\_\_\_}\right) - 3(\_\_\_\_) + C$

$= \_\_\_\_\_\_\_\_\_\_.$

2.3 Change of Variable (Substitution)

$\int f(u)\,du$, $h[g(x)]$

64 **Let g be a differentiable function of x, and let the range of g be an interval I. Let f be a continuous function defined on I and h an antiderivative of f on I. Then, if $u = g(x)$ and g' is continuous, $\int f[g(x)]\,g'(x)\,dx = \_\_\_\_\_\_\_\_ = h(u) + C = \_\_\_\_\_\_\_\_ + C.$**

In Problems 65 through 80, find the antiderivative by using the indicated change of variable.

$3x - 7$, $\frac{1}{3}\,du$

65 $\int (3x - 7)^8\,dx.$

Let $u = \_\_\_\_\_\_$. Then $du = 3\,dx$, so that $dx = \_\_\_\_$ and

$u^8, \ \frac{u^9}{9}$

$(3x-7)^9$

$1-2x^2, \ -4x\,dx, \ -\frac{1}{2}\,du$

$\frac{u^{3/2}}{\frac{3}{2}}$

$1-2x^2$

$\frac{-(1-2x^2)^{3/2}}{3}+C$

$2x\,dx, \ \frac{1}{2}\,du$

$\frac{u^{-3}}{-3}$

$u^3, \ 7+x^2$

$3x^2+5, \ 6x\,dx, \ \frac{1}{6}\,du$

$u^{11}, \ \frac{u^{12}}{12}$

$3x^2+5$

$3x^2\,dx, \ \frac{1}{3}\,du$

$\frac{u^{4/3}}{\frac{4}{3}}, \ x^3+7$

$-\frac{1}{15}\,du$

$$\int (3x-7)^8\,dx = \frac{1}{3}\int (\_\_\_)\,du = \frac{1}{3}\left(\frac{}{\_\_\_}\right)+C$$

$$= \frac{1}{27}(\_\_\_\_\_\_\_\_)+C.$$

66 $\int 2x\sqrt{1-2x^2}\,dx.$

Let $u =$ ______. Then $du =$ ______ and $2x\,dx =$ ______.

$$\int 2x\sqrt{1-2x^2}\,dx = -\frac{1}{2}\int u^{1/2}\,du$$

$$= -\frac{1}{2}\left(\frac{}{\_\_\_}\right)+C$$

$$= -\frac{1}{3}(\_\_\_\_\_\_)^{3/2}+C$$

$$= \_\_\_\_\_\_\_\_.$$

67 $\int \frac{x\,dx}{(7+x^2)^4}.$

Let $u = 7+x^2$. Then $du =$ ______, so that $x\,dx =$ ______ and

$$\int \frac{x\,dx}{(7+x^2)^4} = \frac{1}{2}\int \frac{du}{u^4} = \frac{1}{2}\left(\frac{}{\_\_\_}\right)$$

$$= -\frac{1}{6(\_\_\_)}+C = -\frac{1}{6(\_\_\_\_\_\_)^3}+C.$$

68 $\int x(3x^2+5)^{11}\,dx.$

Let $u =$ ______. Then $du =$ ______, so that $x\,dx =$ ______,

and

$$\int x(3x^2+5)^{11}\,dx = \frac{1}{6}\int (\_\_\_)\,du = \frac{1}{6}\left(\frac{}{\_\_\_}\right)+C$$

$$= \frac{1}{72}(\_\_\_\_\_\_)^{12}+C.$$

69 $\int x^2\sqrt[3]{x^3+7}\,dx.$

Let $u = x^3+7$. Then $du =$ ______, so that $x^2\,dx =$ ______, and

$$\int x^2\sqrt[3]{x^3+7}\,dx = \frac{1}{3}\int \sqrt[3]{u}\,du$$

$$= \frac{1}{3}\left(\frac{}{\_\_\_}\right)+C = \frac{1}{4}(\_\_\_\_\_\_)^{4/3}+C.$$

70 $\int \frac{y^2\,dy}{(1-5y^3)^{2/3}}.$

Let $u = 1-5y^3$. Then, $du = -15y^2\,dy$, so that $y^2\,dy =$ ______, and

$$\int \frac{y^2\,dy}{(1-5y^3)^{2/3}} = -\frac{1}{15}\int \frac{du}{u^{2/3}} = -\frac{1}{15}\int u^{-2/3}\,du$$

$$= -\frac{1}{15}\left(\frac{\quad}{\quad}\right) + C = -\frac{1}{5}(\underline{\qquad})^{1/3} + C.$$

$\dfrac{u^{1/3}}{\frac{1}{3}}$, $1 - 5y^3$

71 $\displaystyle\int \sqrt[5]{1+\frac{1}{t}} \cdot \frac{1}{t^2}\,dt.$

Let $u = 1 + \dfrac{1}{t}$. Then, $du = \underline{\qquad}$, so that $\dfrac{1}{t^2}\,dt = \underline{\qquad}$, and

$-\dfrac{1}{t^2}\,dt$, $-du$

$$\int \sqrt[5]{1+\frac{1}{t}} \cdot \frac{1}{t^2}\,dt = -\int \sqrt[5]{u}\,du$$

$$= -\left(\frac{\quad}{\quad}\right) + C = -\frac{5}{6}\left(\underline{\qquad}\right)^{6/5} + C.$$

$\dfrac{u^{6/5}}{\frac{6}{5}}$, $1 + \dfrac{1}{t}$

72 $\displaystyle\int \frac{\sqrt[3]{1+\sqrt{x}}}{x}\,dx.$

Let $u = 1 + \sqrt{x}$. Then, $du = \dfrac{1}{2\sqrt{x}}\,dx$, so that $\dfrac{dx}{\sqrt{x}} = \underline{\qquad}$ and

$2\,du$

$$\int \frac{\sqrt[3]{1+\sqrt{x}}}{x}\,dx = 2\int \sqrt[3]{u}\,du = 2\left(\frac{\quad}{\quad}\right) + C$$

$\dfrac{u^{4/3}}{\frac{4}{3}}$

$$= \frac{3}{2}(\underline{\qquad})^{4/3} + C.$$

$1 + \sqrt{x}$

73 $\displaystyle\int \left(x+\frac{1}{x}\right)^6 \left(\frac{x^2-1}{x^2}\right)dx.$

Let $u = x + \dfrac{1}{x}$; then $du = \left(1 - \dfrac{1}{x^2}\right)dx = \underline{\qquad\qquad}$ and

$\dfrac{x^2-1}{x^2}\,dx$

$$\int \left(x+\frac{1}{x}\right)^6 \left(\frac{x^2-1}{x^2}\right)dx = \int u^6\,du = \frac{\underline{\qquad}}{7} + C$$

u^7

$$= \frac{1}{7}\left(\underline{\qquad}\right)^7 + C.$$

$x + \dfrac{1}{x}$

74 $\int (x^2 - 6x + 9)^{1/3}\,dx.$

$$\int (x^2-6x+9)^{1/3}\,dx = \int [(x-3)^2]^{1/3}\,dx$$

$$= \int (\underline{\qquad})^{2/3}\,dx.$$

$x - 3$

Let $u = x - 3$. Then $du = \underline{\qquad}$ and

dx

$$\int (x-3)^{2/3}\,dx = \int u^{2/3}\,du = \frac{\quad}{\underline{\qquad}} + C$$

$\dfrac{u^{5/3}}{\frac{5}{3}}$

$$= \frac{3}{5}(\underline{\qquad})^{5/3} + C.$$

$x - 3$

75 $\int x\sqrt{1-x}\,dx.$

Let $u = \sqrt{1-x}$, so that $u^2 = \underline{\qquad}$ and $x = \underline{\qquad}$. Then $dx = \underline{\qquad}$ and

$1 - x$, $1 - u^2$

$-2u\,du$

$-2u\,du$

$2u^4$

$\frac{2}{5}u^5$

$1 - x$

$$\int x\sqrt{1-x}\,dx = \int (1-u^2)\cdot u(\_\_\_\_\_\_\_\_)$$
$$= \int (-2u^2 + \_\_\_\_)\,du$$
$$= -\frac{2}{3}u^3 + \_\_\_\_\_\_\_\_ + C$$
$$= -\frac{2}{3}(1-x)^{3/2} + \frac{2}{5}(\_\_\_\_\_\_\_\_)^{5/2} + C.$$

$2\,dx,\ \frac{1}{2}du$

$u^2 + 2u + 1$

$\sqrt{u}$

$\frac{u^{5/2}}{\frac{5}{2}},\ \frac{u^{3/2}}{\frac{3}{2}}$

$2x - 1$

$2x - 1$

$\frac{(2x-1)^{7/2}}{7} + \frac{2}{5}(2x-1)^{5/2} + \frac{(2x-1)^{3/2}}{3}$

76 $\int x^2\sqrt{2x-1}\,dx.$

Let $u = 2x - 1$, so that $du = \_\_\_\_\_\_$, $dx = \_\_\_\_\_\_\_\_$, and $x = \frac{1}{2}(u+1)$. Then $x^2 = \frac{1}{4}(\_\_\_\_\_\_\_\_\_\_\_\_\_\_)$.

$$\int x^2\sqrt{2x-1}\,dx = \frac{1}{4}\int (u^2 + 2u + 1)(\_\_\_\_)\,\frac{1}{2}du$$
$$= \frac{1}{8}\int (u^{5/2} + 2u^{3/2} + u^{1/2})\,du$$
$$= \frac{1}{8}\left[\frac{u^{7/2}}{\frac{7}{2}} + 2\left(\_\_\_\_\_\_\_\_\right) + \left(\_\_\_\_\_\_\_\_\right)\right] + C$$
$$= \frac{1}{8}\left[\frac{2}{7}(2x-1)^{7/2} + \frac{4}{5}(\_\_\_\_\_\_\_\_)^{5/2} + \frac{2}{3}(\_\_\_\_\_\_\_\_)^{3/2}\right] + C$$
$$= \frac{1}{4}\left(\_\_\_\_\_\_\_\_\_\_\_\_\_\_\_\_\_\_\_\_\_\_\_\_\right) + C.$$

$y + 1,\ u^2 - 1$

$2u\,du$

$u^2 - 1$

u

$(y+1)^{3/2}$

77 $\int \frac{y}{\sqrt{y+1}}\,dy.$

Let $u = \sqrt{y+1}$. Then $u^2 = \_\_\_\_\_\_\_\_$, so that $y = \_\_\_\_\_\_\_\_$ and $dy = \_\_\_\_\_\_\_\_$.

$$\int \frac{y}{\sqrt{y+1}}\,dy = \int \frac{\_\_\_\_\_\_\_\_}{u}\cdot 2u\,du$$
$$= 2\int (u^2 - 1)\,du = 2\left(\frac{u^3}{3} - \_\_\_\_\right) + C$$
$$= 2\left[\frac{\_\_\_\_\_\_\_\_}{3} - (y+1)^{1/2}\right] + C.$$

$2t\,dt$

$\frac{1}{2}du$

$u - 1$

$u^{-1/2}$

78 $\int \frac{t^3}{\sqrt{1+t^2}}\,dt.$

Let $u = 1 + t^2$. Then $du = \_\_\_\_\_\_$, so that $t^3\,dt = t^2 \cdot t\,dt = (u-1)\left(\_\_\_\_\_\_\_\_\right) = \frac{1}{2}(u-1)\,du.$

$$\int \frac{t^3}{\sqrt{1+t^2}}\,dt = \frac{1}{2}\int \frac{\_\_\_\_\_\_\_\_}{\sqrt{u}}\,du$$
$$= \frac{1}{2}\int (u^{1/2} - \_\_\_\_)\,du$$

$u^{1/2}$

$1 + t^2$

$\dfrac{(1 + t^2)^{3/2}}{3} - (1 + t^2)^{1/2}$

$$= \frac{1}{2}\left(\frac{u^{3/2}}{\frac{3}{2}} - \frac{\rule{1.5em}{0.4pt}}{\frac{1}{2}}\right) + C$$

$$= \frac{1}{2}\left[\frac{2}{3}(\rule{4em}{0.4pt})^{3/2} - 2(1 + t^2)^{1/2}\right] + C$$

$$= \rule{10em}{0.4pt} + C.$$

$x + 2,\ \ u^2 - 1$

$2u\,du$

$2u$

$2u^2$

$u^5,\ \ u^3$

$x + 2$

$\dfrac{2}{3}$

79 $\int (x + 1)^2 \sqrt{x + 2}\, dx.$

Let $u = \sqrt{x + 2}$, so that $u^2 =$ ________, $x + 1 =$ ________, and $dx =$ ________.

$$\int (x + 1)^2 \sqrt{x + 2}\, dx = \int (u^2 - 1)^2 \cdot u(\rule{2em}{0.4pt})\, du$$

$$= \int (2u^6 - 4u^4 + \rule{2em}{0.4pt})\, du$$

$$= \frac{2}{7}u^7 - \frac{4}{5}(\rule{2em}{0.4pt}) + \frac{2}{3}(\rule{2em}{0.4pt}) + C$$

$$= \frac{2}{7}(x + 2)^{7/2} - \frac{4}{5}(\rule{4em}{0.4pt})^{5/2} + C$$

$$= \left(\rule{2em}{0.4pt}\right)(x + 2)^{3/2} + C.$$

$4 - u,\ -du$

$4 - u$

$u^{-2/3}$

$u^{1/3}$

$2 - t$

80 $\displaystyle\int \frac{t + 2}{(2 - t)^{2/3}}\, dt.$

Let $u = 2 - t$, so that $t + 2 =$ ________ and $dt =$ _____.

$$\int \frac{t + 2}{(2 - t)^{2/3}}\, dt = \int \frac{(\rule{4em}{0.4pt})(-du)}{u^{2/3}}$$

$$= \int [u^{1/3} - 4(\rule{3em}{0.4pt})]\, du$$

$$= \left[\frac{u^{4/3}}{\frac{4}{3}} - \frac{4}{\frac{1}{3}}(\rule{3em}{0.4pt})\right] + C$$

$$= \frac{3}{4}(\rule{4em}{0.4pt})^{4/3} - 12(2 - t)^{1/3} + C.$$

3 Simple Differential Equations and Their Solution

81 An equation such as $\dfrac{dy}{dx} = f(x)$ which involves derivatives or differentials is called a ____________________.

differential equation

82 Any one solution $y = g(x)$ of $\dfrac{dy}{dx} = f(x)$ is called a ____________ solution.

particular

83 A general or ____________ solution to the differential equation $\dfrac{dy}{dx} = f(x)$ on an open interval I will have the form ________________.

complete

$y = g(x) + C$

84 The physical or geometrical circumstances that give rise to differential equations often enforce additional conditions called

constraints, initial conditions	________, side conditions, ________, or boundary conditions.

In Problems 85 through 90, find the complete solution or the particular solution of each differential equation.

	85 $\frac{dy}{dx} = 5x.$
$\int x\, dx,\ \frac{x^2}{2}$	$y = \int 5x\, dx = 5(\_\_\_\_) = 5(\_\_) + C.$
	86 $y' = 3x^2 - 5.$
$\int x^2\, dx$	$y = \int (3x^2 - 5)\, dx = 3(\_\_\_\_\_\_) - 5\int dx$
$\frac{x^3}{3},\ x^3 - 5x + C$	$= 3(\_\_) - 5x + C = \_\_\_\_\_\_.$
	87 $\frac{dy}{dx} = \left(x^2 - \frac{1}{x^2}\right)^2.$
$\frac{1}{x^4}$	$y = \int \left(x^2 - \frac{1}{x^2}\right)^2 dx = \int \left(x^4 - 2 + \_\_\_\right) dx$
$\int \frac{1}{x^4}\, dx$	$= \int x^4\, dx - 2\int dx + (\_\_\_\_\_\_)$
$\frac{x^5}{5}$	$= (\_\_\_) - 2x - \frac{1}{3x^3} + C.$
	88 $\frac{dy}{dx} = x - 7$ with side condition $y = 3$ when $x = 2$.
$\int dx$	$\frac{dy}{dx} = x - 7$, so that $y = \int (x - 7)\, dx = \int x\, dx - 7(\_\_\_\_) =$
$\frac{x^2}{2} - 7x + C,\ 14$	________. When $x = 2$, $y = 3$, so that $3 = 2 - \_\_\_ + C$
$15,\ \frac{x^2}{2} - 7x + 15$	or $C = \_\_\_$. Thus, $y = \_\_\_\_\_\_$.
	89 $y' = 9x^8 - 4x^3 - x$, with side condition $y = 3$ when $x = 1$.
$x^9 - x^4 - \frac{1}{2}x^2 + C$	$y' = 9x^8 - 4x^3 - x$, so that $y = \_\_\_\_\_\_$.
$-\frac{1}{2},\ \frac{7}{2}$	$y = 3$ when $x = 1$, so that $3 = 1 - 1 - (\_\_\_) + C$ or $C = \_\_\_$.
$x^9 - x^4 - \frac{1}{2}x^2 + \frac{7}{2}$	Then $y = \_\_\_\_\_\_$.
	90 $\frac{dy}{dx} = \sqrt{x} + \frac{1}{\sqrt{x}}$ with side condition $y = \frac{5}{2}$ when $x = 4$.
	$\frac{dy}{dx} = \sqrt{x} + \frac{1}{\sqrt{x}}$, so that
$x^{-1/2}$	$y = \int \left(\sqrt{x} + \frac{1}{\sqrt{x}}\right) dx = \int (x^{1/2} + \_\_\_\_)\, dx$

$2x^{1/2}$

$= \frac{2}{3}x^{3/2} +$ _______ $+ C$.

8

$y = \frac{5}{2}$ when $x = 4$, so that $\frac{5}{2} = \frac{2}{3}$(____) $+ 2(2) + C$ or

$-\frac{41}{6}$, $\frac{2}{3}x^{3/2} + 2\sqrt{x} - \frac{41}{6}$

$C =$ ______. Then $y =$ ________________.

separable

91 The differential equation $\frac{dy}{dx} = f(x)$ is said to be ____________, since it can be rewritten in the form $dy =$ __________, in which the variables x and y are separated so that all terms involving x are on the right and all terms involving y are on the left.

$f(x)\,dx$

92 An equation that can be rewritten in the form $G(y)\,dy = F(x)\,dx$, where F and G are functions, is called ____________.

separable

93 To find the "general solution" of a separable differential equation, simply separate the variables, so that the equation takes the form ____________________. Then take the ____________________ of both sides. The two constants of integration corresponding to $\int G(y)\,dy$ and $\int F(x)\,dx$ can be combined into a single constant.

$G(y)\,dy = F(x)\,dx$

indefinite integral

94 A solution that cannot be obtained directly from a general solution by assigning a value to the constant of integration is called a ____________ solution.

singular

In Problems 95 through 98, separate the variables and solve the differential equation.

95 $\frac{dy}{dx} = x(x^2 + 1)^5$.

$dy =$ ________________, and so $y = \int x(x^2 + 1)^5\,dx$. Let $u = x^2 + 1$, so that $du =$ ______. $y = \frac{1}{2}\int u^5\,du =$ ________ $+ C$ or

$x(x^2 + 1)^5\,dx$

$2x\,dx$, $\frac{1}{12}u^6$

$y =$ ____________________.

$\frac{1}{12}(x^2 + 1)^6 + C$

96 $y\frac{dy}{dx} = x^3\sqrt{4y^2 + 3}$.

Separating the variables, we have $\frac{y\,dy}{\sqrt{4y^2 + 3}} =$ ________, so that

$x^3\,dx$

$\int \frac{y\,dy}{\sqrt{4y^2 + 3}} = \int x^3\,dx$. Let $u = 4y^2 + 3$, so that $du =$ ________.

$8y\,dy$

Then $\frac{1}{4}(4y^2 + 3)^{1/2} =$ ______ $+ C$.

$\frac{x^4}{4}$

97 $y' = \frac{x\sqrt[3]{y^4 + 5}}{2y^3}$.

Using the fact that $y' = \frac{dy}{dx}$ and separating the variables, we have

$\frac{2y^3}{\sqrt[3]{y^4+5}}$, $\frac{x^2}{2}+C$ | ______ $dy = x\,dx$, so that $\int \frac{2y^3}{\sqrt[3]{y^4+5}}\,dy =$ ______.

$4y^3\,dy$ | Let $u = y^4 + 5$. Then $du =$ ______. Therefore, $\frac{3}{4}(y^4+5)^{2/3} =$

$\frac{x^2}{2}$ | ______ $+ C$.

98 $y' = \frac{x^3 + \sqrt[3]{x}}{y^3 - \sqrt[3]{y}}$.

$\frac{dy}{dx}$ | $y' =$ ______, so that, separating the variables, we have

$x^3 + \sqrt[3]{x}$ | $(y^3 - \sqrt[3]{y})\,dy = ($______$)\,dx$, so that $\frac{y^4}{4} - \frac{3}{4}y^{4/3} =$

$\frac{x^4}{4} + \frac{3}{4}x^{4/3}$ | ______ $+ C$.

99 Find the equation of the particular curve satisfying the condition that the slope of the tangent line is 4 times the abscissa of the point (x, y), if the curve contains the point $(3, 5)$. The differential equation expressing the condition on the slope of the tangent

$4x$ | is $\frac{dy}{dx} =$ ______. The complete solution of this differential equation

$2x^2 + C$ | is $y =$ ______. If we impose the side condition $y = 5$ when

$18, -13$ | $x = 3$, then $5 =$ ______ $+ C$ or $C =$ ______. The desired equation is $y =$

$2x^2 - 13$ | ______.

100 At each point (x, y) on a certain curve the tangent line has slope $4x^3 + 2$. Find the equation of the curve if it contains the point $(1, 7)$.

The differential equation expressing the condition above is $\frac{dy}{dx} =$

$4x^3 + 2$, $x^4 + 2x + C$ | ______. The complete solution is $y =$ ______.

Imposing the side condition $y = 7$ when $x = 1$, we have $7 =$

$1, 2, 4$ | ______ $+$ ______ $+ C$, so that $C =$ ______. The desired equation is $y =$

$x^4 + 2x + 4$ | ______.

101 At each point (x, y) on a certain curve, the tangent line has slope $\left(\frac{y}{x}\right)^3$. Find the equation of the curve if it contains the point $(1, 2)$.

The differential equation expressing the above condition is $\frac{dy}{dx} =$

$\frac{y^3}{x^3}$, $\frac{dx}{x^3}$, $-\frac{1}{2x^2}$ | ______, so that $\frac{dy}{y^3} =$ ______ or $-\frac{1}{2y^2} =$ ______ $+ C$. When $y = 2$,

$1, -\frac{1}{2}, \frac{3}{8}$ | $x =$ ______, so that $-\frac{1}{8} =$ ______ $+ C$ or $C =$ ______. The desired

equation is $-\frac{1}{2y^2} = -\frac{1}{2x^2} + \frac{3}{8}$.

Answers	Frames
	102 The slope of the tangent line at each point (x, y) on a curve is $5\sqrt{x}$. Find an equation of the curve if it contains the point (9, 4).
$5\sqrt{x}$	The differential equation is $\frac{dy}{dx} =$ ______. Solving the equation,
$\frac{10}{3}x^{3/2} + C$	we have $y = 5\int\sqrt{x}\,dx =$ __________. When $y = 4$, $x =$
9, 27, −86	____, so that $4 = \frac{10}{3}$ (____) $+ C$ or $C =$ ____. The desired equa-
$\frac{10}{3}x^{3/2} - 86$	tion is $y =$ __________.

3.1 Second-Order Differential Equations

Answers	Frames
order	**103** The highest order of all derivatives involved in a differential equation is called the ______ of the equation.
two successive	**104** The general solution of a second-order differential equation can be obtained by __________ antidifferentiations and the
two	resulting solution will involve ____ arbitrary constants which cannot be combined into one constant.

In Problems 105 through 112, find the solution of the differential equation.

Answers	Frames
	105 $\frac{d^2y}{dx^2} = 3x^2 + 5.$
$\int(3x^2 + 5)\,dx$	Since $\frac{d^2y}{dx^2} = \frac{d}{dx}\left(\frac{dy}{dx}\right)$, we have $\frac{dy}{dx} =$ __________, so
$x^3 + 5x + C_1$	that $\frac{dy}{dx} =$ __________. Now, since
$\int(x^3 + 5x + C_1)\,dx$	$y =$ __________, we have
$\frac{x^4}{4} + \frac{5x^2}{2} + C_1x + C_2$	$y =$ __________.
	106 $y'' = 3x^2 + 13x.$
$\frac{d}{dx}\left(\frac{dy}{dx}\right)$, $\int(3x^2 + 13x)\,dx$	Since $y'' = \frac{d^2y}{dx^2} =$ ______, we have $y' =$ ________,
$x^3 + \frac{13x^2}{2} + C_1$	so that $y' =$ __________. Now,
$\int\left(x^3 + \frac{13x^2}{2} + C_1\right)dx$	$y =$ __________. Therefore,
$\frac{x^4}{4} + \frac{13x^3}{6} + C_1x + C_2$	$y =$ __________.
	107 $\frac{d^2y}{dx^2} = 24.$
$\frac{d}{dx}\left(\frac{dy}{dx}\right)$, $\int 24\,dx$	Since $\frac{d^2y}{dx^2} =$ ______, we have $\frac{dy}{dx} =$ ______, so that

$24x + C_1$

$12x^2 + C_1x + C_2$

$\frac{dy}{dx} =$ __________. Now $y = \int (24x + C_1)\,dx$, so that $y =$ __________.

108 $\frac{d^2y}{dx^2} = (3x+1)^4$.

$\frac{1}{15}(3x+1)^5 + C_1$

$\frac{1}{270}(3x+1)^6 + C_1x + C_2$

Since $\frac{d^2y}{dx^2} = \frac{d}{dx}\left(\frac{dy}{dx}\right) = (3x+1)^4$, we have $\frac{dy}{dx} = \int (3x+1)^4\,dx =$ __________. Therefore, $y = \int \left[\frac{1}{15}(3x+1)^5 + C_1\right] dx =$ __________.

109 $s'' = (t+2)^2$.

$\frac{(t+2)^3}{3} + C_1$

$\frac{(t+2)^4}{12} + C_1t + C_2$

Since $s'' = \frac{d}{dt}(s') = (t+2)^2$, we have $s' =$ __________. But $s' = \frac{ds}{dt}$, so $s =$ __________.

110 $\frac{d^2y}{dx^2} = 6x^2 + 1$, $y' = 5$ and $y = 3$ when $x = 0$.

$6x^2 + 1$

$2x^3 + x + C_1$

$0,\ 5$

$\frac{x^2}{2}$

0

$3,\ \frac{x^4}{2} + \frac{x^2}{2} + 5x + 3$

Since $\frac{d^2y}{dx^2} = \frac{d}{dx}(y') =$ __________, we have $y' =$ __________. Substitute $x = 0$ and $y' = 5$ into the latter equation to obtain $5 =$ ____ $+ 0 + C_1$, so that $C_1 =$ ____. Now, $y = \int (2x^3 + x + 5)\,dx$, so $y = \frac{x^4}{2} +$ ____ $+ 5x + C_2$. Substituting $y = 3$ when $x = 0$, we have $3 = 0 + 0 +$ ____ $+ C_2$, so that $C_2 =$ ____. Therefore, $y =$ __________.

111 $\frac{d^2y}{dx^2} = \sqrt{x+1}$, $\frac{dy}{dx} = 5$, and $y = 2$ when $x = 3$.

$\sqrt{x+1}$

$\frac{2}{3}(x+1)^{3/2} + C_1$

$\frac{16}{3},\ -\frac{1}{3}$

$\frac{2}{3}(x+1)^{3/2} - \frac{1}{3}$

$\frac{1}{3}x$

$32,\ 1$

$-\frac{83}{15},\ \frac{4}{15}(x+1)^{5/2} - \frac{x}{3} - \frac{83}{15}$

Since $\frac{d^2y}{dx^2} = \frac{d}{dx}\left(\frac{dy}{dx}\right) =$ __________, we have $\frac{dy}{dx} = \int \sqrt{x+1}\,dx =$ __________. Substituting $\frac{dy}{dx} = 5$ and $x = 3$, we have $5 =$ ____ $+ C_1$, so that $C_1 =$ ____. Now, $\frac{dy}{dx} =$ __________, so that $y = \int \left[\frac{2}{3}(x+1)^{3/2} - \frac{1}{3}\right] dx = \frac{4}{15}(x+1)^{5/2} -$ ____ $+ C_2$. $y = 2$ when $x = 3$, so that $2 = \frac{4}{15}$(____) $-$ ____ $+ C_2$. Then $C_2 =$ ____. Therefore, $y =$ __________.

112 $\frac{d^2s}{dt^2} = t^{-3/2}$, $s = 2$ when $t = 1$, and $s = -4$ when $t = 9$.

$t^{-3/2}$, $-2t^{-1/2} + C_1$

$-4\sqrt{t} + C_1 t + C_2$

-12, $\frac{1}{4}$

$\frac{23}{4}$, $-4\sqrt{t} + \frac{1}{4}t + \frac{23}{4}$

$\frac{d^2s}{dt^2} = \frac{d}{dt}\left(\frac{ds}{dt}\right) =$ ______, so that $\frac{ds}{dt} =$ __________, and we have $s =$ __________. $s = 2$ when $t = 1$, so that $2 = -4 + C_1 + C_2$, and $s = -4$ when $t = 9$, so that $-4 =$ ____ $+ 9C_1 + C_2$. Solving for C_1 and C_2, we obtain $C_1 =$ ____ and $C_2 =$ ____. Therefore, $y =$ __________.

3.2 Linear Motion

$\frac{ds}{dt}$, $f'(t)$

$\frac{dv}{dt}$, $\frac{d^2s}{dt^2}$, $f''(t)$

$s = f(t)$

113 In considering the motion of a particle along a linear scale when an equation of motion $s = f(t)$ is given, the instantaneous velocity is $v =$ ____ $=$ ______ and the instantaneous acceleration is $a =$ ____ $=$ ____ $=$ ______.

114 If the instantaneous velocity v or the instantaneous acceleration a is given as a known function of t, subject to suitable initial conditions, it may be possible to solve the resulting differential equation to obtain the law of motion ________.

In Problems 115 through 117, a particle is moving along a linear scale according to the equation $s = f(t)$, where s is in feet and t is in seconds. Suppose that v, in feet per second, is the velocity at t and a, in feet per square second, is the acceleration at t. Find each of the following as a function of t.

$2t + 1$, $t^2 + t + C_1$

0, 0, $t^2 + t$

$t^2 + t$

$\frac{t^3}{3} + \frac{t^2}{2} + C_2$

0

$\frac{t^3}{3} + \frac{t^2}{2}$

115 v and s if $a = 2t + 1$, $v = 0$ when $t = 0$ and $s = 0$ when $t = 0$.

$v = \int a\, dt = \int($______$)\, dt =$ __________. $v = 0$ when $t = 0$, so that $0 = 0 +$ ____ $+ C_1$ or $C_1 =$ ____. Thus, $v =$ ______.

$s = \int v\, dt = \int($______$)\, dt$

$=$ __________.

$s = 0$ when $t = 0$, so that $0 = 0 + 0 + C_2$ or $C_2 =$ ____. Thus, $s =$ ________.

$2 - (5t - 1)^{-3/2}$

$5t - 1$

$\frac{1}{2}$, $-\frac{1}{5}$

$2t + \frac{2}{5}(5t - 1)^{-1/2} - \frac{1}{5}$

116 $a = 2 - (5t - 1)^{-3/2}$, $v = 2$ when $t = 1$ and $s = 0$ when $t = 1$.

$v = \int a\, dt = \int($__________$)\, dt$

$= 2t + \frac{2}{5}($______$)^{-1/2} + C_1$.

$v = 2$ when $t = 1$, so that $2 = 2 + \frac{2}{5}($____$) + C_1$ or $C_1 =$ ____.

Thus, $v =$ __________.

$s = \int v\,dt = \int \left(\rule{6cm}{0.4pt} \right) dt$

$= t^2 + \frac{4}{25}(\rule{2cm}{0.4pt}) - \frac{1}{5}t + C_2.$

$s = 0$ when $t = 1$, so that $0 = 1 + \frac{4}{25}(\rule{1cm}{0.4pt}) - \frac{1}{5} + C_2$ or $C_2 =$ ______. Therefore, $s =$ ______________________.

$2t + \frac{2}{5}(5t-1)^{-1/2} - \frac{1}{5}$

$\sqrt{5t-1}$

2

$-\frac{28}{25}$, $t^2 + \frac{4}{25}(5t-1)^{1/2} - \frac{t}{5} - \frac{28}{25}$

117 $a = t(t+1)^2$, $v = 3$ when $t = 1$ and $s = 1$ when $t = 1$.

$v = \int a\,dt = \int(\rule{3cm}{0.4pt})\,dt$

$= \int (t^3 + 2t^2 + t)\,dt =$ ______________________.

$v = 3$ when $t = 1$, so that $3 = \frac{1}{4} + \frac{2}{3} + \frac{1}{2} + C_1$ or $C_1 =$ ______.

Thus, $v =$ ______________________.

$s = \int v\,dt = \int \left(\rule{6cm}{0.4pt} \right) dt$

$= \frac{t^5}{20} + \frac{1}{6}(\rule{1cm}{0.4pt}) + \frac{1}{6}(\rule{1cm}{0.4pt}) + \frac{19}{12}t + C_2.$

$s = 1$ when $t = 1$, so that $1 = \frac{1}{20} + \frac{1}{6} + \frac{1}{6} + \frac{19}{12} + C_2$ or $C_2 =$ ______. Therefore, $s =$ ______________________.

$t(t+1)^2$

$\frac{t^4}{4} + \frac{2}{3}t^3 + \frac{1}{2}t^2 + C_1$

$\frac{19}{12}$

$\frac{t^4}{4} + \frac{2}{3}t^3 + \frac{1}{2}t^2 + \frac{19}{12}$

$\frac{t^4}{4} + \frac{2}{3}t^3 + \frac{1}{2}t^2 + \frac{19}{12}$

t^4, t^3

$-\frac{29}{30}$, $\frac{1}{20}t^5 + \frac{1}{6}t^4 + \frac{1}{6}t^3 + \frac{19}{12}t - \frac{29}{30}$

118 A car is braked to a stop with constant acceleration. The car stops 10 seconds after the brakes are applied and travels 250 feet during this time. Find the acceleration and the speed of the car when the brakes were applied.

Here the acceleration a is a constant, so that $\frac{d^2s}{dt^2} =$ ______ and $s =$ ______ when $t = 0$. Also, $v =$ ______ when $t = 10$. Then $v = \int a\,dt =$ __________. $v = 0$ when $t = 10$, so that $0 = 10a + C_1$ or $C_1 =$ ______. Therefore, $v =$ __________. $s = \int v\,dt = \int(\rule{3cm}{0.4pt})\,dt =$ ______________ $+ C_2$. $s = 0$ when $t = 0$, so that $C_2 =$ ______. $s = 250$ when $t = 10$, so that $250 = \frac{a}{2}(\rule{1cm}{0.4pt}) - 100a$. Hence, $a =$ ______. Therefore, the deceleration is __________. The speed of the car at the instant when the

a

0, 0

$at + C_1$

$-10a$, $at - 10a$

$at - 10a$, $\frac{at^2}{2} - 10at$

0

100, -5

5 ft/sec^2

0, −5
0, 50
50 ft/sec

$-a,\ -at + C_1$

$4a$

$4a,\ -\dfrac{at^2}{2} + 4at + C_2$

0
$16a$, 60, 7.5
7.5, 30
30, 30
30

2

45, 2

32 ft/sec²
96 ft/sec, 0, 96

$-32t + 96,\ -16t^2 + 96t + C_2$

144 ft, 0, 0
144, $-16t^2 + 96t + 144$
$-16t^2 + 96t + 144$
0
0, $t^2 - 6t - 9$

brakes were applied is obtained by putting t = ____ and a = ____ in the equation $v = at - 10a$, so that v = ____ $- 10(-5)$ = ____. Therefore, the speed of the car was ________ at the moment the brakes were applied.

119 Assume that a hockey puck sliding across the ice decelerates at a constant rate. When it stops after 4 seconds, it has moved 60 feet from the position where it was first observed. Find its velocity when it was first observed and find how far it has moved after 2 seconds.

Let s ft, v ft/sec, and a ft/sec² represent the distance traveled by the puck, its velocity, and its deceleration at time t seconds. $\dfrac{dv}{dt}$ = ____, so that v = ________. $v = 0$ when $t = 4$, so that $0 = -4a + C_1$ or C_1 = ____. $v = \dfrac{ds}{dt} = -at + C_1$, so that v = $-at$ + ____ and s = ________. $s = 0$ when $t = 0$, so that $0 = 0 + 0 + C_2$ or C_2 = ____. Also, $s = 60$ when $t = 4$, so that $60 = -8a$ + ____ or $8a$ = ____, so that a = ____. Therefore, the acceleration is −7.5 ft/sec². $C_1 = 4a = 4($____$)$ = ____. Thus, $v = -at + C_1 = -7.5t$ + ____. At $t = 0$, v = ____. Therefore, the velocity of the puck when it was first observed was ____ ft/sec. Also, $s = -\dfrac{15t^2}{4} + 30t$. When $t = 2$, $s = -\dfrac{15}{4}(4) + 30($____$)$ = ____. Therefore, the puck traveled 45 feet after ____ seconds.

120 From the top edge of a building 144 feet high, a stone is thrown vertically upward with an initial velocity of 96 ft/sec. Find its distance above the ground after t seconds. In how many seconds will the stone strike the ground?

Let s ft represent the distance above the ground at t seconds. Let v ft/sec represent the velocity and a ft/sec² represent the acceleration of the stone. The acceleration due to gravity is a = −________, so that $v = -32t + C_1$. When $t = 0$, v = ________, so that 96 = ____ + C_1, so that C_1 = ____. Then v = ________. Since $\dfrac{ds}{dt} = v$, s = ________.

When $t = 0$, s = ________, so that 144 = ____ + ____ + C_2 or C_2 = ____. Then s = ________. Hence, its distance above the ground is s = ________.

The stone will strike the ground when s = ____ so that $-16t^2 + 96t + 144$ = ____ or ________ = 0. By using

$3 + 3\sqrt{2} \approx 7.2$

$-9.8t + C_1$

500 meters/sec, 0

500

0

$\frac{500}{4.9}$

$-9.8t$

$\frac{500}{4.9}$, 500

-32 ft/sec^2

$\frac{dv}{dt}$, 15

15, C_2, 150

150, 150

0

$16t^2 - 15t - 150 = 0$

$\frac{15 \pm \sqrt{9825}}{32} \approx 3.566$ sec

$-32t + 15$, $\sqrt{9825}$, -99.1

99.1

the quadratic formula, we find that $t =$ ______________ seconds.

121 A projectile is fired vertically upward by a cannon with an initial velocity of 500 meters/sec. At what speed will the projectile be moving when it returns and strikes the hapless canoneer? (Neglect air resistance.)

Let s meters represent the distance at t seconds. Let v meters/sec represent the velocity and a meters/sec^2 represent the acceleration (due to gravity). $a = -9.8$ meters/sec.^{2} $v =$ ______________. At $t = 0$, $v =$ ______________, so that $500 =$ ____ $+ C_1$ or $C_1 =$ ____. $v = \frac{ds}{dt}$, so that $s = -4.9t^2 + 500t + C_2$. $s = 0$ when $t = 0$, so that $C_2 =$ ____. Thus, $s = -4.9t^2 + 500t$. When $s = 0$, $t = \frac{\quad}{\quad}$. To find the speed at which the canoneer is struck, v_0, we find that $\frac{ds}{dt}$ at $t = \frac{500}{4.9} \cdot \frac{ds}{dt} =$ ________ $+ 500$, so that $v_0 = -9.8\left(\frac{\quad}{\quad}\right) + 500 = -$____. The projectile strikes the canoneer at a velocity of 500 meters/sec.

122 A balloon is rising vertically at the constant rate of 15 ft/sec and is 150 ft from the ground at the instant when the aeronaut drops her binoculars. How long will it take the binoculars to strike the ground? With what speed will the binoculars strike the ground?

Let s ft be the distance above ground of the binoculars at t seconds. Let v ft/sec represent the velocity and a ft/sec^2 represent the acceleration of the binoculars. $a =$ ______________ and $a = \frac{\quad}{\quad}$, so that $v = -32t + C_1$. At $t = 0$, $v =$ ____, so that $C_1 =$ ____. $v = \frac{ds}{dt}$, so that $s = -16t^2 + 15t +$ ____. At $t = 0$, $s =$ ____, so that $C_2 =$ ____. Therefore, $s = -16t^2 + 15t +$ ____. When the binoculars strike the ground $s =$ ____, so that $-16t^2 + 15t + 150 = 0$ or ______________________, and

$$t = \frac{15 \pm \sqrt{225 + 4(16)(150)}}{32} = \text{\_\_\_\_\_\_\_\_\_\_\_\_\_\_}.$$

The speed with which the binoculars strike the ground is found as follows:

$$\frac{ds}{dt} = \text{\_\_\_\_\_\_\_\_\_\_} = -(\text{\_\_\_\_\_\_}) \approx \text{\_\_\_\_\_\_} \text{ ft/sec}$$

so that the speed is approximately ________ ft/sec.

4 Applications of Differential Equations

4.1 Work Done by a Variable Force

123 If a constant force F having a direction parallel to a number line, called the s axis, acts on a particle P which moves along the ______ from an original position with s coordinate a to a final position with s coordinate b, then the force does an amount of work equal to ______.

s axis

$F \cdot (b - a)$

124 The work W done when the applied force F is variable and depends on the time t, but still acts at all times in the direction parallel to the s axis, is given by ______, with initial condition $W = 0$ when $t =$ ____ or when $s = s_0$.

$dW = F\,ds$

t_0

In Problems 125 through 127, the force F acting on a particle P moving along the s axis is given in terms of the coordinate s of P. Find the work W done by F in moving P from $s = a$ to $s = b$.

125 $F = 4s$, s in feet, F in pounds, $a = 1$, and $b = 5$.
$dW =$ ______ so that $W = \int 4s\,ds$ and $W =$ ______. When $s = 1$, $W =$ ____, so that $C =$ ____. Thus, $W =$ ______.
When $s = 5$, $W = 2($____$) - 2 = 48$, so that the work done is ______.

$4s\,ds,\ 2s^2 + C$

$0,\ -2,\ 2s^2 - 2$

25

48 foot-pounds

126 $F = \sqrt{s}$, s in centimeters, F in dynes, $a = 0$, and $b = 9$.
$dW =$ ______, so that $W = \int s^{1/2}\,ds =$ ______. When $s = 0$, $W =$ ____, so that $C =$ ____. When $s = 9$, $W = \frac{2}{3}(9^{3/2}) =$ ____, so that the work done is ______.

$\sqrt{s}\,ds,\ \frac{2}{3}s^{3/2} + C$

$0,\ 0$

18, 18 ergs

127 $F = (3 + s)^{2/3}$, s in feet, F in pounds, $a = -5$, and $b = 5$.
$dW =$ ______, so that

$$W = \int (3 + s)^{2/3}\,ds$$

$$= \frac{3}{5}(\text{\_\_\_\_\_\_}) + C.$$

When $s = -5$, $W =$ ____, so that $0 = \frac{3}{5}(-2)^{5/3} + C$ or $C =$ ______. When $s = 5$,

$$W = \frac{3}{5}(\text{\_\_\_\_})^{5/3} + \frac{6}{5}\sqrt[3]{4}$$

$$= \frac{3}{5}(\text{\_\_\_\_}) + \frac{6}{5}\sqrt[3]{4}$$

$$= \text{\_\_\_\_\_\_}.$$

$(3 + s)^{2/3}\,ds$

$(3 + s)^{5/3}$

0

$\frac{6}{5}\sqrt[3]{4}$

8

32

$\frac{96 + 6\sqrt[3]{4}}{5}$ foot-pounds

In Problems 128 and 129, a perfectly elastic spring is stretched from its relaxed position through b units. When extended these b units the stretching force on the spring is F_b. The spring constant is k and $F = ks$. Use the given information to find the total work done.

128 $b = 20$ feet and $F_b = 120$ pounds.

$dW =$ _____ $=$ _____, so that $W =$ _____. When $s = 0$, $W =$ _____, so that $C =$ _____. Thus, $W =$ _____. When $s = 20$, $F = 20k =$ _____, so that $k = \dfrac{120}{20} =$ _____. Therefore, the work done is $W = \dfrac{6(20)^2}{2} =$ _____.

Answers: $F\,ds,\ ks\,ds,\ \dfrac{ks^2}{2} + C$; $0,\ 0,\ \dfrac{ks^2}{2}$; $120,\ 6$; 1200 foot-pounds

129 $k = 2000$ dynes per centimeter, $F_b = 8000$ dynes.

$dW = F\,ds =$ _____, so that $W =$ _____. When $s = 0$, $W =$ _____, so that $C =$ _____. Thus, $W =$ _____. $F = 8000 = ks = 2000s$, so that $s =$ _____. Therefore, the work done is $W = \dfrac{2000(16)}{2} =$ _____.

Answers: $ks\,ds,\ \dfrac{ks^2}{2} + C$; $0,\ 0,\ \dfrac{ks^2}{2}$; 4 centimeters; 16,000 ergs

130 A bucket containing sand is lifted, starting at ground level, at a constant speed of 3 ft/sec. The bucket weighs 4 pounds and, at the start, is filled with 80 pounds of sand. As the bucket is lifted, sand runs out of a hole in the bottom at the constant rate of 1 pound per second. How much work is done in lifting the bucket up to the height at which the last of the sand runs out?

Let s feet represent the height of the bucket. At height s feet, the weight of the bucket and the sand is $\left(80 + 4 - \dfrac{s}{3}\right)$ pounds, so that the work $W = \int\left(\text{\_\_\_\_\_}\right) ds$ or $W = 84s -$ _____ $+ C$. When $s = 0$, $W =$ _____, so that $C =$ _____. Thus, $W =$ _____. When all the sand runs out, $s = \dfrac{(3\text{ ft/sec})(80\text{ pounds})}{1\text{ pound/sec}} =$ _____.

The work done is $W = 84(240) - \dfrac{1}{6}(240)^2 =$ _____.

Answers: $84 - \dfrac{s}{3},\ \dfrac{s^2}{6}$; $0,\ 0,\ 84s - \dfrac{s^2}{6}$; 240 feet; 10,560 foot-pounds

4.2 Remarks on Setting Up Differential Equations

131 In thinking of dx as an "infinitesimal" bit of x, we regard $\int dx$ as a summation of all "infinitesimal" bits of x to give the quantity x; that is, apart from an additive constant of integration, $x =$ _____.

Answer: $\int dx$

4.3 Application to Economics

132 The marginal cost $\frac{dC}{dx}$ for producing x items is given by $\frac{dC}{dx} = 0.05 + \frac{5000}{x^2}$. Find the total cost C as a function of x, given that $C = \$5500$ when $x = 1000$.

$\frac{dC}{dx} = 0.05 + \frac{5000}{x^2}$, so that

$$C = \int \left(0.05 + \frac{5000}{x^2}\right) dx = 0.05 \int dx + 5000 \left(\_\_\_\_\_\right)$$

$$= 0.05x - \frac{5000}{x} + k.$$

Put $C = 5500$ and $x = 1000$; then $k =$ _____. Therefore, $C =$ __________.

Answers: $\int \frac{dx}{x^2}$; 5455; $0.05x - \frac{5000}{x} + 5455$

133 A company knows that its marginal cost of production of a certain item is given by $\frac{dC}{dx} = 3.2x - 9.2$, where x is the number of production units and $\frac{dC}{dx}$ is in dollars per unit. Find the total cost C if the fixed cost is \$10.

$\frac{dC}{dx} = 3.2x - 9.2$, so that $C = \int (3.2x - 9.2)\, dx =$ __________.

$C = 10$ when $x =$ _____, so that $k =$ _____. Therefore, $C =$ __________.

Answers: $1.6x^2 - 9.2x + k$; 0, 10; $1.6x^2 - 9.2x + 10$

134 The marginal revenue $\frac{dR}{dx}$ for selling x items is given by $\frac{dR}{dx} = 4x + 3x^2$. If the total revenue received from the sale of 10 items is \$1500, find the total revenue R in terms of x.

$\frac{dR}{dx} = 4x + 3x^2$, so that $R = \int (4x + 3x^2)\, dx =$ __________.

$R = 1500$ when $x = 10$, so that $1500 = 2(100) +$ _____ $+ k$, or $k =$ _____. Therefore, $R =$ __________.

Answers: $2x^2 + x^3 + k$; 1000; 300, $2x^2 + x^3 + 300$

135 A company's marginal revenue function is given by $\frac{dR}{dx} = \frac{1000}{\sqrt{x-1}}$, where $x > 1$ is the number of items manufactured. Find the total revenue R in terms of x, if $R = \$13{,}000$ when $x = 26$.

$\frac{dR}{dx} = \frac{1000}{\sqrt{x-1}}$, so that $R = \int ($__________$)\, dx$. Let $u = x - 1$. Then $du =$ _____, so that $R = \int 1000u^{-1/2}\, du =$ _____ $+ k$. $R = 13{,}000$ when $x = 26$, so that $k =$ _____.

Therefore, $R = 2000\sqrt{x-1} +$ _____.

Answers: $1000(x-1)^{-1/2}$; dx; $2000u^{1/2}$, 3000; 3000

136 A marginal revenue function for selling a certain product is given by $\frac{dR}{dx} = 8\sqrt[3]{x+1}$, where x is the amount sold. Find the total revenue function R if $R = \$100$ when $x = 7$.

$\frac{dR}{dx} = 8\sqrt[3]{x+1}$, so that $R = \int 8\sqrt[3]{x+1}\,dx$. Let $u =$ ________.
Then $du =$ _____. Thus, $R = 8\int u^{1/3}\,du = 6($_____$)^{4/3} + k$, so that $R =$ ____________________. When $R = 100$ and $x = 7$, $100 = 6(8)^{4/3} + k$, so that $k =$ _____. Then, $R =$ ____________________.

Answers: $x + 1$; dx, u; $6(x+1)^{4/3} + k$; 4, $6(x+1)^{4/3} + 4$

4.4 Newton's Law of Motion—Momentum and Energy

137 If a particle P of mass m moves on a linear scale due to an unopposed force F, then Newton's law of motion can be written $F = ma$ or $F = m$ _____ or $F = m$ _____, where v is the velocity, a is the acceleration, and s is the distance of the particle.

Answers: $\frac{dv}{dt}$, $\frac{d^2s}{dt^2}$

138 The quantity mv is called the ____________________ of the particle P of Problem 131. Also, the quantity $\frac{1}{2}mv^2$ is called the ____________________ of the particle P. The quantity $V = -\int F\,ds$ is called the ____________________ of the particle P. The sum of the kinetic and potential energy of a particle P is called the ____________________ E, so that $E =$ ____________________.

Answers: linear momentum; kinetic energy; potential energy; total energy; $\frac{1}{2}mv^2 + V$

139 What is the potential energy of a stone of weight 6 pounds at a height of 30 feet?
The potential energy $V =$ ________________ $= -\int(-6)\,ds =$ _____ $+ C$.
Putting $C = 0$, we have $V = 6s = 6($_____$) =$ ____________________.

Answers: $-\int F\,ds$, $6s$; 30, 180 foot-pounds

140 What velocity must a 1-pound snowball have in order to have the same kinetic energy as a 4000-pound car traveling at 10 miles/hour?

Let v ft/sec be the velocity of the snowball. 10 miles/hour $= \frac{(10)(5280)}{3600} =$ ________________. The kinetic energy of the snowball is $\frac{1}{2}mv^2$. Therefore, $\frac{1}{2}(1)v^2 = \frac{1}{2}(4000)\left(\_\_\_\_\right)^2$, so that $v = \frac{44}{3}\sqrt{\_\_\_\_} =$ ____________________.

Answers: $\frac{44}{3}$ ft/sec; $\frac{44}{3}$; 4000, $\frac{880}{3}\sqrt{10} \approx 927.6$ ft/sec

141 A particle P is moving along the positive s axis under the influence of a force F. The total energy E of P is zero and the potential energy is given by $V = -\dfrac{2}{s^3}$. Find the velocity $v = \dfrac{ds}{dt}$ of P in terms of s.

V

The total energy $E = \dfrac{1}{2}mv^2 + $ ____. Since $E = 0$, then

$0, \dfrac{2}{s^3}, \dfrac{4}{ms^3}$

$\dfrac{1}{2}mv^2 - \dfrac{2}{s^3} = $ ____, so that $\dfrac{1}{2}mv^2 = $ ____ or $v^2 = $ ____ and

ms

$v = \dfrac{2}{s\sqrt{\_\_\_\_}}$.

142 A particle P is moving along the positive s axis under the influence of a force F. The total energy E of P is zero and the potential energy is given by $V = -\dfrac{2}{s^4}$. If the particle is at $s = 4$ feet when $t = 0$, find the equation of motion of P.

$0, \dfrac{\pm 2}{\sqrt{m}\, s^2}$

$E = \dfrac{1}{2}mv^2 + V, \dfrac{1}{2}mv^2 - \dfrac{2}{s^4} = $ ____, so that $v = $ ________.

$\pm\dfrac{2}{\sqrt{m}}\, dt$

But $v = \dfrac{ds}{dt}$, so that $\dfrac{ds}{dt} = \pm\dfrac{2}{\sqrt{m}\, s^2}$ or $s^2\, ds = $ ________.

C

Thus, $\dfrac{s^3}{3} = \pm\dfrac{2}{\sqrt{m}}t + $ ____. When $t = 0$, $s = 4$, so that

$0, \dfrac{64}{3}, \dfrac{64}{3}$

$\dfrac{64}{3} = \pm\dfrac{2}{\sqrt{m}}(\_\_\_\_) + C$; therefore, $C = $ ____. $\dfrac{s^3}{3} = \pm\dfrac{2}{\sqrt{m}}t + $ ____

$1/3$, positive

or $s = \left(\pm\dfrac{6}{\sqrt{m}}t + 64\right)^{(\_\_\_\_)}$. Since $v = \dfrac{ds}{dt}$ and v is ________ when $t = 0$, we must use the plus sign. Therefore,

$\dfrac{6}{\sqrt{m}}t + 64$

$s = \sqrt[3]{\_\_\_\_\_\_\_\_}$.

143 The force F in dynes acting on an electrically charged particle of mass m grams due to an electrostatic field is given by $F = s^{-2}$, where s is measured in centimeters. The potential energy V of the particle satisfies $V = 4$ ergs when $s = 4$ centimeters. The particle starts at $s = 4$ with an initial velocity $v_0 = 0$. Find both V as a function of s and $\lim\limits_{s\to+\infty} V$. Also, find the velocity of the particle when it reaches the position with coordinate s.

$F\, ds, -F\, ds$

$V = -W$ and $dW = $ ____, so that $dV = $ ______. Therefore,

$\dfrac{1}{s} + C, 4$

$V = \displaystyle\int (-F)\, ds = \int -\frac{1}{s^2}\, ds = $ ______. When $s = 4$, $V = $ ____, so

$0, \dfrac{1}{s}, 0$

that $C = $ ____ and therefore $V = $ ____. $\lim\limits_{s\to+\infty} V = \lim\limits_{s\to+\infty} \dfrac{1}{s} = $ ____.

$\dfrac{1}{2}mv^2 + V, \dfrac{1}{s}$

The total energy E, given by $E = $ ________ $= \dfrac{1}{2}mv^2 + $ ____,

$0, \frac{1}{4}$

$\frac{1}{4}, \frac{1}{s}$

$s - 4,\ s - 4,\ \frac{s-4}{2ms}$

is constant. When $s = 4$, $v =$ ____, so that $E = \frac{1}{2}m(0)^2 +$ ____ $=$ ____. Therefore, $\frac{1}{4} = \frac{1}{2}mv^2 +$ ____, so that $\frac{1}{2}mv^2 = \frac{1}{4} - \frac{1}{s} = \frac{\rule{2cm}{0.4pt}}{4s}$. It follows that $v = \sqrt{\frac{2(\rule{2cm}{0.4pt})}{4ms}} = \sqrt{\rule{2cm}{0.4pt}}$.

5 Areas of Regions in the Plane by the Method of Slicing

5.1 Areas by Slicing

$dA = l\,ds$

s

0

b

144 The area A of the portion of a region between the perpendicular at $s = a$ and the perpendicular at $s = b$ can be obtained by solving the differential equation ____________, where l is the length of a perpendicular line segment intersecting the region. At each point s, l is a function of ____. The differential equation is subject to the side condition $A =$ ____ when $s = a$ and A is the desired area when $s =$ ____. The graph is

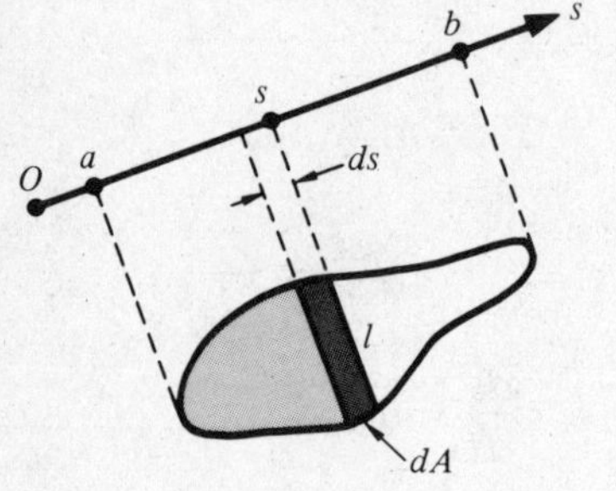

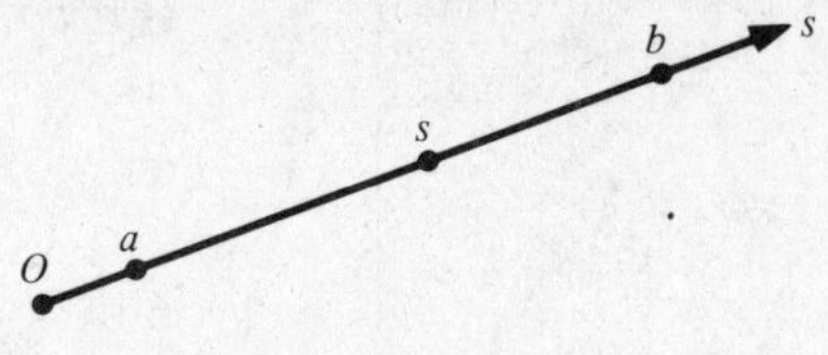

In Problems 145 through 153, find the area of each region by the method of slicing.

$l\,ds$

145 A triangle with base 4 inches and height 7 inches.
Take the reference s axis perpendicular to the base of the triangle with the origin at the level of the base, so that $dA =$ ____.

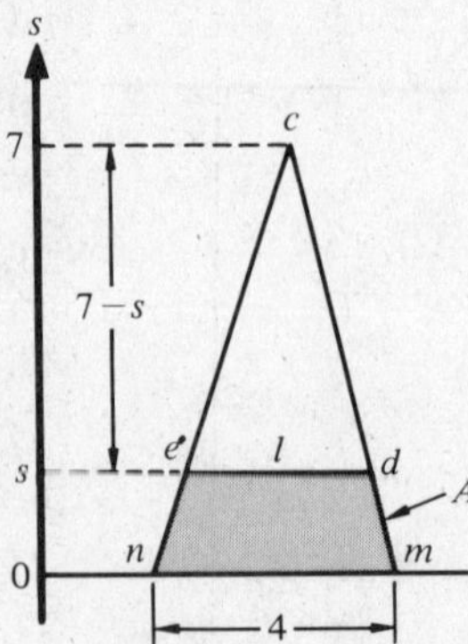

cmn

$\frac{4}{7}, \frac{4}{7}(7 - s), \frac{4}{7}(7 - s)\,ds$

The triangle cde is similar to the triangle ____ so that $\frac{l}{7 - s} =$ ____. Then $l =$ ____________ and so $dA =$ ____________.

$\frac{4}{7}(7 - s)$, $4s - \frac{2s^2}{7} + C$	Therefore, $A = \int l\,ds = \int (\_\_\_\_\_\_\_\_\_\_\_\_)\,ds =$ __________.
0, 0	Since $A = 0$ at $s =$ ____, we know that $C =$ ____; hence,
$4s - \frac{2s^2}{7}$	$A =$ __________. The area of the entire triangle is obtained by
7, 14	putting $s = 7$, so that $A = 4(7) - \frac{2}{7}(\_\_\_\_)^2 =$ ____ square inches.

146 The triangle in the figure.

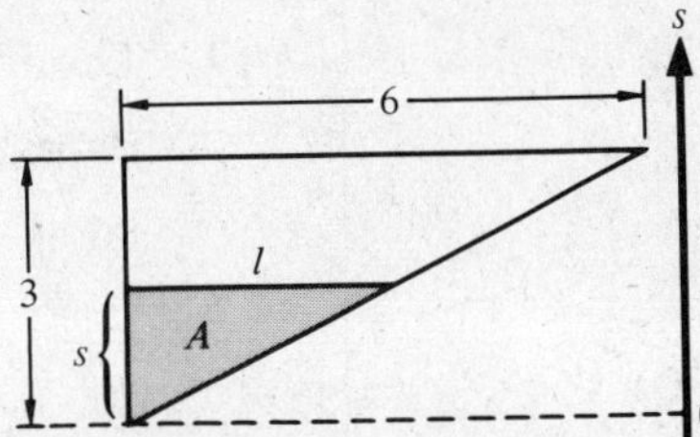

$\frac{6}{3}$, 2	Using similar triangles, we have $\frac{l}{s} = \frac{\_\_\_\_}{\_\_\_\_} =$ ____, so that
$2s$, $2s\,ds$, s^2	$l =$ ____. $dA = l\,ds =$ ________, so that $A =$ ____ $+ C$.
0, 0	When $s = 0$, $A =$ ____ and $C =$ ____, so that $A = s^2$. Therefore,
9	when $s = 3$, $A = 3^2 =$ ____ square units.

147 The rectangle in the figure.

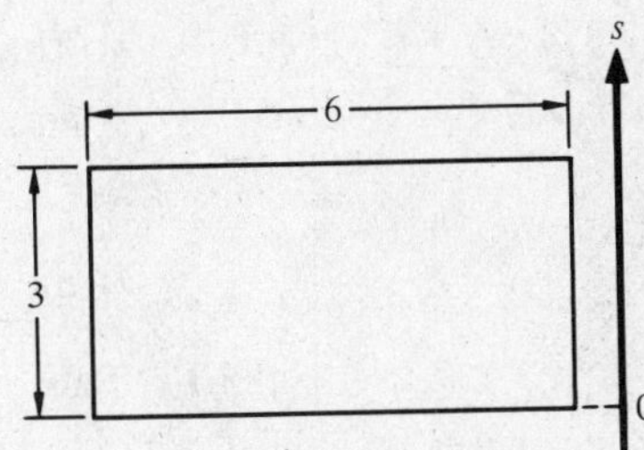

6, $l\,ds$, $6\,ds$, $6s + C$	$l =$ ____ and $dA =$ _____, so that $dA =$ _____ and $A =$ _______.
0, 0, $6s$	Then $s = 0$, $A =$ ____, and $C =$ ____, so that $A =$ ____. When
18	$s = 3$, $A = 6(3) =$ ____ square units.

148 The region in the xy plane bounded below by $y = x^3$, above by the line $y = 8$, and on the left by the line $x = 0$.

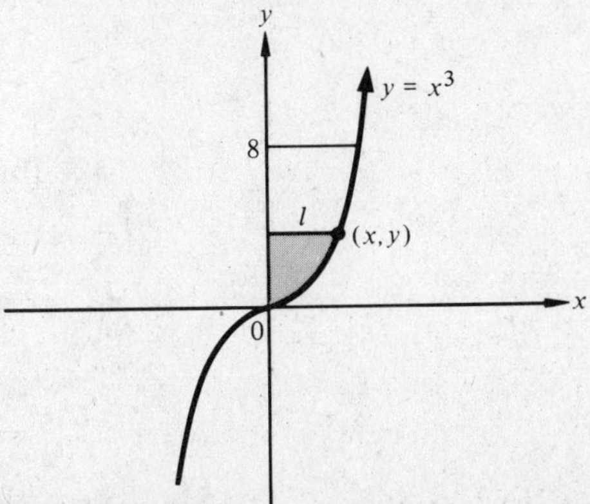

$l\,dy$, x
$y = x^3$, $\sqrt[3]{y}$
$\sqrt[3]{y}$, $\frac{3}{4}y^{4/3} + C$, 0
0, 0, $\frac{3}{4}y^{4/3}$
8, 12

Take the y axis as the reference axis for the method of slicing, so that $dA =$ ______, $l =$ _____ for $x > 0$ and consider (x, y) on the graph __________. Then $x =$ ______ and $dA = \sqrt[3]{y}\,dy$. Therefore, $A = \int($______$)\,dy =$ ________________. $A =$ _____ when $y =$ _____, so we have $C =$ _____ and $A =$ __________. When $y = 8$, then $A = \frac{3}{4}y^{4/3} = \frac{3}{4}($_____$)^{4/3} =$ _____ square units.

149 The region in the first quadrant of the xy plane bounded above by $y = x^3$ below by the x axis and on the right by the line $x = 2$.

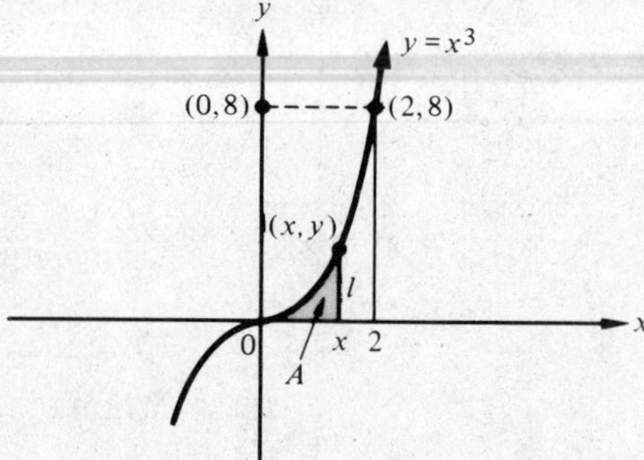

$l\,dx$, x^3
$x^3\,dx$, $\frac{x^4}{4} + C$
0, 0, $\frac{x^4}{4}$
4

Take the x axis as the axis of reference. $dA =$ ______. $l = y =$ _____ so that $dA =$ ________ and $A =$ __________. Since $A = 0$ when $x =$ _____, we know that $C =$ _____, so that $A =$ _____. For $x = 2$, $A =$ _____ square units.

150 The region in the xy plane between the line $y = x$ and the curve $y = x^2$.

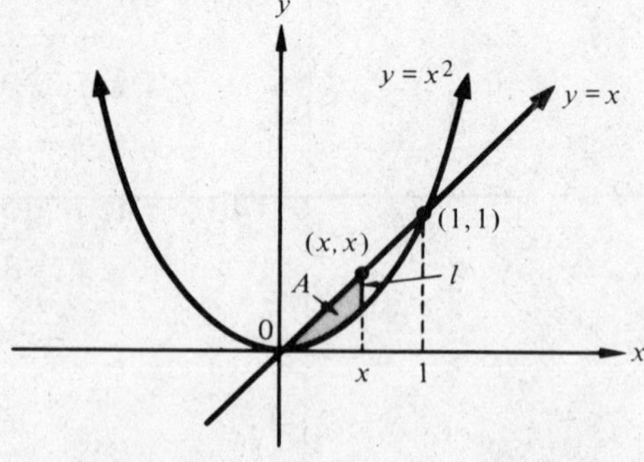

$l\,dx$
$x - x^2$, $x - x^2$
$\frac{x^2}{2} - \frac{x^3}{3} + C$, 0
0, $\frac{1}{3}$, $\frac{1}{6}$

Take the x axis as the axis of reference. Then $dA =$ ______. Since $l =$ __________, $dA = ($__________$)\,dx$, so that $A = \int l\,dx =$ $\int(x - x^2)\,dx =$ ________________. When $x = 0$, $A =$ _____, so $C =$ _____. When $x = 1$, then $A = \frac{1}{2} -$ _____ $=$ _____ square unit.

151 The region between $y = x^2$ and $x = y^2$.

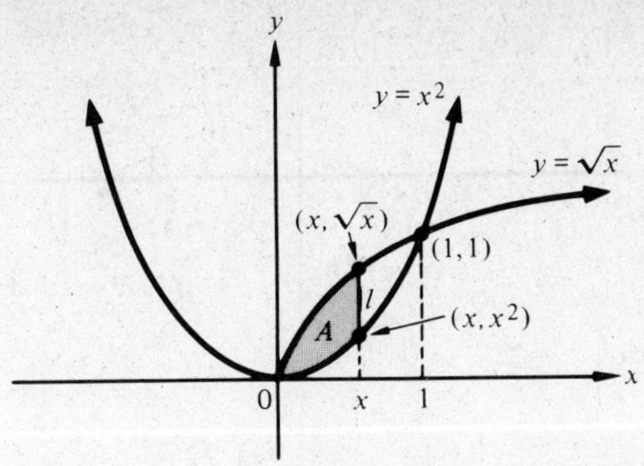

Take the x axis as the axis of reference. Then $dA = l\,dx$, where $l =$ _______________, so that $A = \int(\sqrt{x} - x^2)\,dx =$ ___________________________. When $x = 0$, $A =$ ____ and $C =$ ____. When $x = 1$, $A = \frac{2}{3}(1) -$ ____ $=$ ____ square unit.

$\sqrt{x} - x^2$

$\frac{2}{3}x^{3/2} - \frac{x^3}{3} + C,\ 0$

$0,\ \frac{1}{3},\ \frac{1}{3}$

152 The region under the curve $y = x\sqrt{1 - x^2}$ between $x = -1$ and $x = 0$.

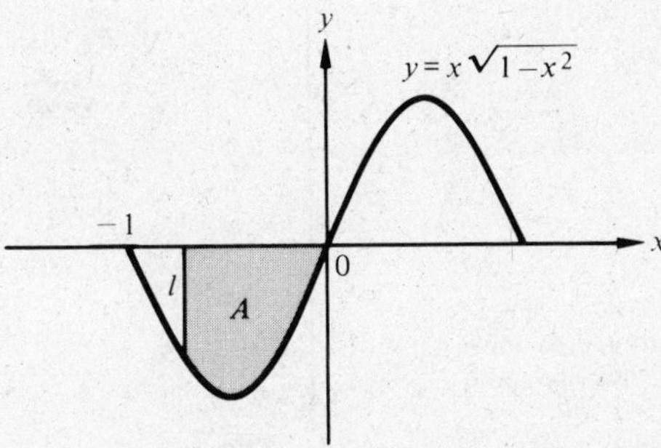

Take the x axis as the axis of reference, so that $dA = l\,dx$, where $l = -$ _______________ and $A = -\int x\sqrt{1 - x^2}\,dx$. Let $u = 1 - x^2$ and $du =$ ________, so that $x\,dx =$ ________.

$A = \frac{1}{2}\int u^{1/2}\,du =$ ________ $+ C =$ ________________________.

When $x = -1$, $A =$ ____ and $C =$ ____. When $x = 0$,

$A = \frac{1}{3}(1 - 0^2)^{3/2} =$ ____ square unit.

$x\sqrt{1 - x^2}$

$-2x\,dx,\ -\frac{1}{2}\,du$

$\frac{1}{3}u^{3/2},\ \frac{1}{3}(1 - x^2)^{3/2} + C$

$0,\ 0$

$\frac{1}{3}$

153 The region bounded by the curves $y = 1 - x^2$ and $y = x^2 - 1$.

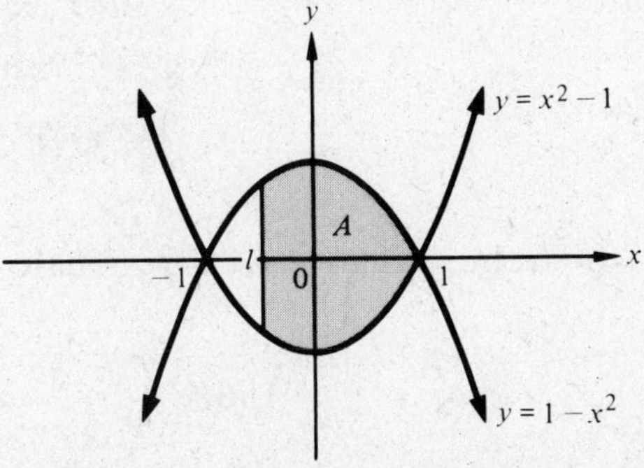

$l\,dx$

$x^2 - 1,\ 2 - 2x^2$

$2 - 2x^2,\ 2x - \frac{2}{3}x^3 + C$

$0,\ \frac{4}{3}$

$\frac{4}{3},\ \frac{8}{3}$

Take the x axis as the axis of reference so that $dA =$ ______, where $l = (1 - x^2) -$ ________ $=$ __________.

Then $A = \int($____________$)\,dx =$ ____________________. When $x = -1$, $A =$ ____, and so $C =$ ____. When $x = 1$, $A = 2(1) - \frac{2}{3}(1) +$ ____ $=$ ____ square units.

6 Area Under the Graph of a Function—The Definite Integral

$A_1 - A_2$

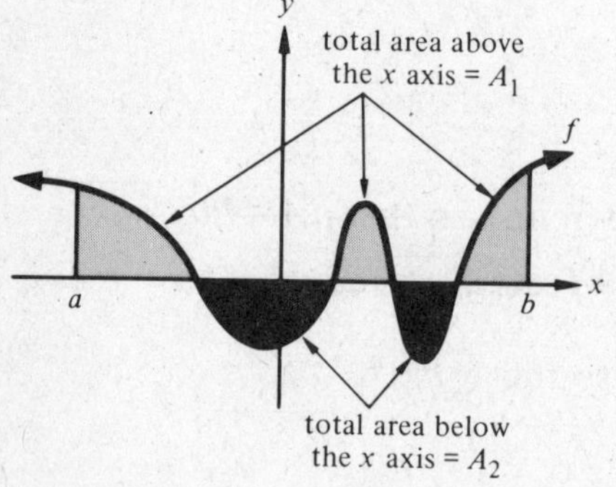

154 Let f be a function defined at least on the closed interval $[a, b]$ and denote the region between the graph of f and the x axis and between the lines $x = a$ and $x = b$ by R. Suppose that A_1 is the area of the portion of R that lies above the x axis and that A_2 is the area of the portion of R below the x axis. Then the definite integral of the function f between a and b, in symbols, $\int_a^b f(x)\,dx$, is defined by $\int_a^b f(x)\,dx =$ __________. The graph is

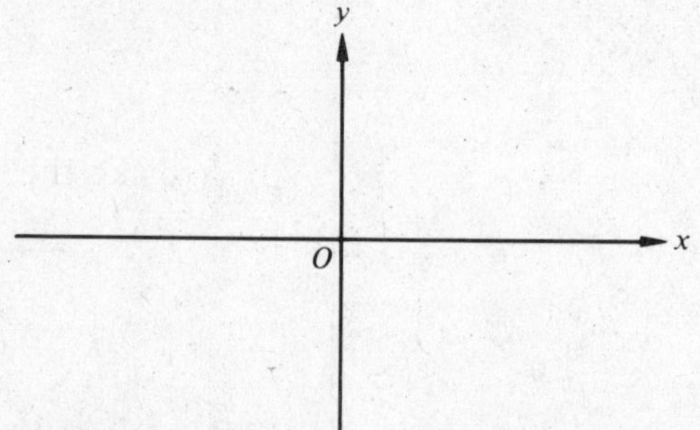

integral

$f(x)\,dx$

integrand

interval of integration

lower, limits

155 The expression $\int_a^b f(x)\,dx$ is called the _______________ from a to b of ______________. The function f [or the expression $f(x)$] is called the ________________ and the interval $[a, b]$ is called the ________________________________. The numbers a and b are called the __________ and the upper ______________ of integration, respectively.

In Problems 156 and 157, evaluate each definite integral by using elementary geometry.

156 $\int_{-2}^{3} (2 - x)\,dx.$

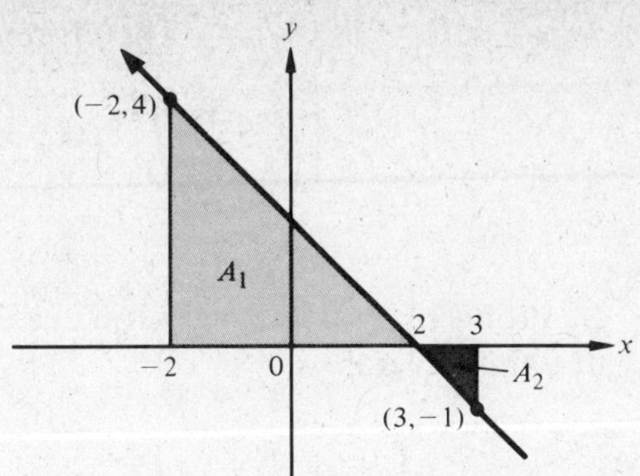

4, 4

8, 1, $\frac{1}{2}$

$\frac{1}{2}$, $\frac{15}{2}$

In the figure, the area A_1 above the x axis is $\frac{1}{2}$(____)(____) = ____, and the area A_2 below the x axis is $\frac{1}{2}(1)$(____) = ____. Therefore,

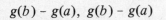

$$\int_{-2}^{3} (2 - x)\,dx = A_1 - A_2 = 8 - \_\_\_\_ = \_\_\_\_ \text{ square units.}$$

157 $\int_{-1}^{2} |x - 1|\,dx$.

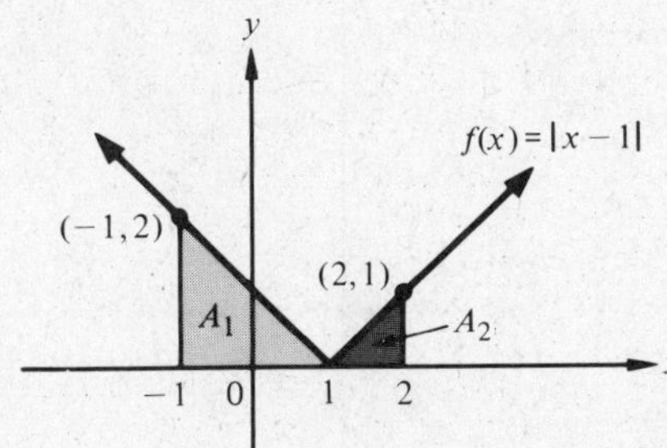

2, 2

1, $\frac{1}{2}$

$\frac{1}{2}$, $\frac{5}{2}$

The area $A_1 = \frac{1}{2}(2)$(____) = ____, and the area $A_2 = \frac{1}{2}(1)$(____) = ____, so that

$$\int_{-1}^{2} |x - 1|\,dx = A_1 + A_2 = 2 + \_\_\_\_ = \_\_\_\_ \text{ square units.}$$

g(b) − g(a)

158 The preliminary version of the fundamental theorem of calculus states: If $\int f(x)\,dx = g(x) + C$ and f is a continuous function on the closed interval $[a, b]$, then

$$\int_a^b f(x)\,dx = \_\_\_\_\_\_\_\_.$$

g(b) − g(a), g(b) − g(a)

159 If g is any function and if the numbers a and b belong to the domain of g, then the notation $g(x)\Big|_a^b$ will by definition represent the number ________; that is, $g(x)\Big|_a^b$ = ________.

	Therefore, if f and g are as in Problem 158, then
$g(b) - g(a)$	$\int_a^b f(x)\,dx = g(x)\Big\vert_a^b =$ _____________.

In Problems 160 through 165, use the fundamental theorem of calculus to evaluate each definite integral.

	160 $\int_1^3 x^3\,dx.$
$\frac{x^4}{4}$, 1	$\int_1^3 x^3\,dx = \underline{\quad\quad}\Big\vert_1^3 = \frac{1}{4}(3^4 - \underline{\quad\quad})$
81, 20	$= \frac{1}{4}(\underline{\quad\quad} - 1) = \underline{\quad\quad}.$
	161 $\int_1^{27} \frac{1}{\sqrt[3]{x}}\,dx.$
$x^{2/3}$	$\int_1^{27} \frac{1}{\sqrt[3]{x}}\,dx = \int_1^{27} x^{-1/3}\,dx = \frac{3}{2}(\underline{\quad\quad})\Big\vert_1^{27}$
1	$= \frac{3}{2}(27^{2/3} - \underline{\quad\quad})$
12	$= \frac{3}{2}(9 - 1) = \underline{\quad\quad}.$
	162 $\int_0^2 (y - 1)^2\,dy.$
	$\int_0^2 (y - 1)^2\,dy = \int_0^2 (y^2 - 2y + 1)\,dy$
$\frac{y^3}{3} - y^2 + y$	$= \underline{\quad\quad\quad\quad}\Big\vert_0^2$
0	$= \frac{8}{3} - 4 + 2 - (\underline{\quad\quad})$
$\frac{2}{3}$	$= \underline{\quad\quad}.$
	163 $\int_{-2}^0 x^2(3 - x)\,dx.$
	$\int_{-2}^0 x^2(3 - x)\,dx = \int_{-2}^0 (3x^2 - x^3)\,dx$
$x^3 - \frac{x^4}{4}$	$= \underline{\quad\quad\quad\quad}\Big\vert_{-2}^0$
0	$= \underline{\quad\quad} - \left(-8 - \frac{16}{4}\right)$
12	$= \underline{\quad\quad}.$

164 $\int_1^2 \frac{dx}{(3-x)^3}$.

$-dx$

$\int_1^2 \frac{dx}{(3-x)^3} = \int_1^2 (3-x)^{-3}\,dx$. Let $u = 3 - x$. Then $du =$ _____,

so that

$\frac{1}{2u^2}$

$\int (3-x)^{-3}\,dx = -\int u^{-3}\,du =$ _____

$\frac{1}{2(3-x)^2} + C$

$=$ ____________________.

Therefore,

$\frac{1}{(3-x)^2}$

$\int_1^2 \frac{dx}{(3-x)^2} = \frac{1}{2}\Big[\_\_\_\_\_\_\_\_\Big]\Big|_1^2$

$\frac{1}{4}$

$= \frac{1}{2}\left[\frac{1}{(3-2)^2} - \_\_\_\_\right]$

$\frac{3}{8}$

$=$ ____.

165 $\int_0^1 x\sqrt{9-5x^2}\,dx$. Let $u = 9 - 5x^2$.

$-10x\,dx$

Then $du =$ ____________, so that

$\int x\sqrt{9-5x^2}\,dx = -\frac{1}{10}\int u^{1/2}\,du$

$u^{3/2}$

$= -\frac{1}{15}(\_\_\_\_\_) + C$

$= -\frac{1}{15}(9-5x^2)^{3/2} + C.$

Therefore,

$9 - 5x^2$

$\int_0^1 x\sqrt{9-5x^2}\,dx = -\frac{1}{15}(\_\_\_\_\_\_\_\_\_\_)^{3/2}\Big|_0^1$

$4^{3/2}$

$= -\frac{1}{15}(\_\_\_\_ - 9^{3/2})$

8

$= -\frac{1}{15}(\_\_\_\_ - 27)$

$\frac{19}{15}$

$=$ ____.

In Problems 166 through 170, use the fundamental theorem of calculus to find the area of each region.

166 The region below the graph of $f(x) = -x^2 + 12x - 20$, above the x axis and between $x = 2$ and $x = 10$.

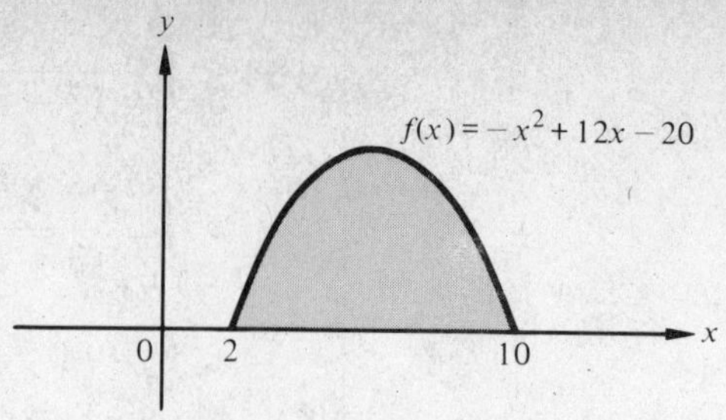

The area is given by

$$\int_2^{10} (\underline{\qquad\qquad}) \, dx = \left[\int (-x^2 + 12x - 20) \, dx\right]\Bigg|_2^{10}$$

$$= \underline{\qquad\qquad}\Bigg|_2^{10}$$

$$= \left(-\frac{1000}{3} + 600 - 200\right) - \underline{\qquad\qquad}$$

$$= \underline{\qquad} \text{ square units.}$$

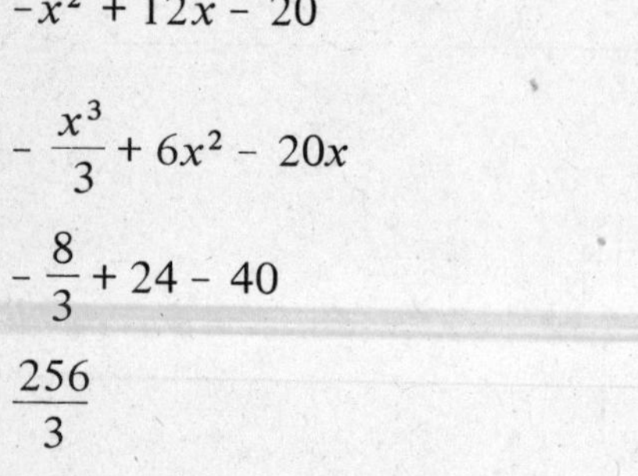

$-x^2 + 12x - 20$

$-\dfrac{x^3}{3} + 6x^2 - 20x$

$-\dfrac{8}{3} + 24 - 40$

$\dfrac{256}{3}$

167 The region below the graph of $f(x) = x^3 + 1$, above the x axis and between $x = -1$ and $x = 0$.

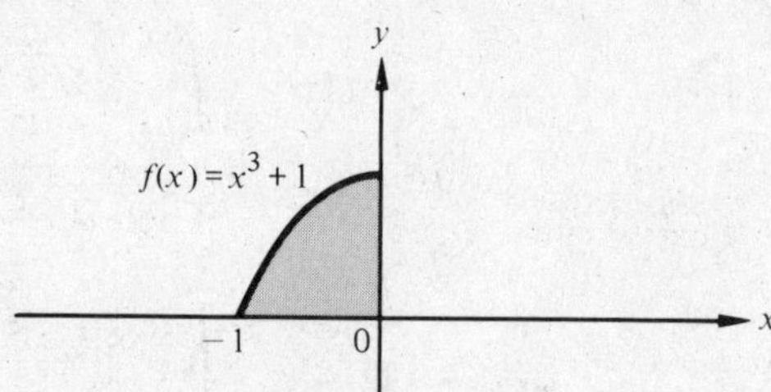

The area is given by

$$\int_{-1}^{0} (\underline{\qquad\qquad}) \, dx = \left[\int (x^3 + 1) \, dx\right]\Bigg|_{-1}^{0}$$

$$= \underline{\qquad\qquad}\Bigg|_{-1}^{0}$$

$$= \underline{\qquad} - \underline{\qquad}$$

$$= \underline{\qquad} \text{ square unit.}$$

$x^3 + 1$

$\dfrac{x^4}{4} + x$

$0, -\dfrac{3}{4}$

$\dfrac{3}{4}$

168 The region below the graph of $f(x) = x^3 - 2x^2 - 5x + 6$, above the x axis and between $x = -1$ and $x = 1$.

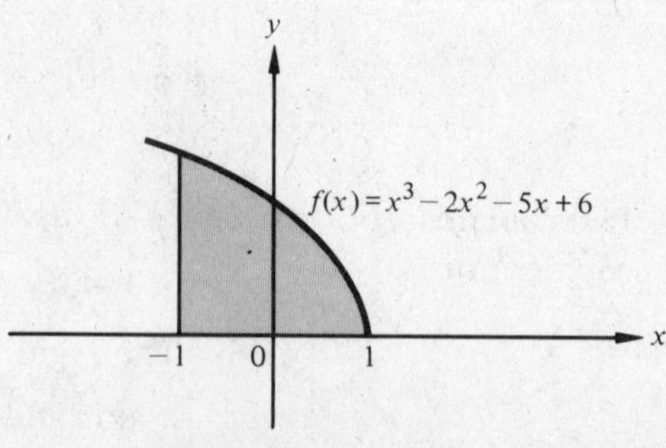

The area is given by

$$\int_{-1}^{1} (\underline{\qquad\qquad}) \, dx = \left[\int (x^3 - 2x^2 - 5x + 6) \, dx\right]\Bigg|_{-1}^{1}$$

$x^3 - 2x^2 - 5x + 6$

$$= \underline{\qquad\qquad\qquad}\Bigg|_{-1}^{1}$$

$\frac{1}{4}x^4 - \frac{2}{3}x^3 - \frac{5}{2}x^2 + 6x$

$$= \left(\frac{1}{4} - \frac{2}{3} - \frac{5}{2} + 6\right) - \underline{\qquad\qquad}$$

$\frac{1}{4} + \frac{2}{3} - \frac{5}{2} - 6$

$$= \underline{\quad} \text{ square units.}$$

$\frac{32}{3}$

169 The region below the graph of $f(x) = -x^2 + 4x$, above the x axis and between $x = 0$ and $x = 2$.

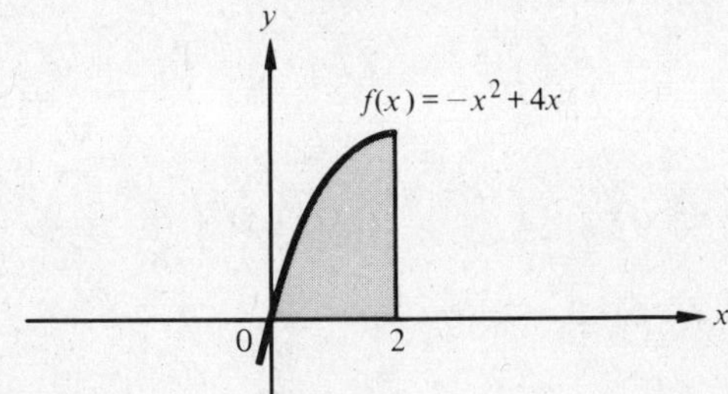

The area is given by

$$\int_{0}^{2} (\underline{\qquad\qquad}) \, dx = \left[\int (-x^2 + 4x) \, dx\right]\Bigg|_{0}^{2}$$

$-x^2 + 4x$

$$= \underline{\qquad\qquad\qquad}\Bigg|_{0}^{2}$$

$-\frac{x^3}{3} + 2x^2$

$$= -\frac{8}{3} + 8 - \underline{\qquad}$$

0

$$= \underline{\quad} \text{ square units.}$$

$\frac{16}{3}$

170 The region below the graph of $f(x) = x\sqrt{x^2 + 1}$, above the x axis and between $x = 0$ and $x = 3$.

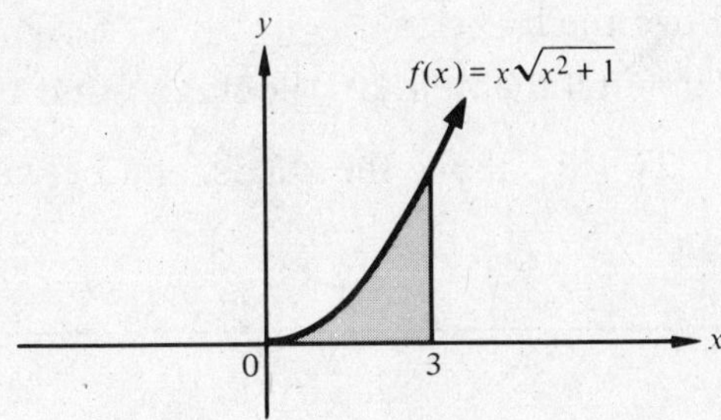

The area is given by

$$\int_{0}^{3} (\underline{\qquad\qquad}) \, dx = \left(\int x\sqrt{x^2 + 1} \, dx\right)\Bigg|_{0}^{3}$$

$x\sqrt{x^2 + 1}$

$$= \frac{1}{3}(\underline{\qquad\qquad\qquad})\Bigg|_{0}^{3}$$

$(x^2 + 1)^{3/2}$

1

$\frac{1}{3}(10\sqrt{10} - 1)$

$= \frac{1}{3}(10^{3/2} - \_\_\_\_)$

$= \_\_\_\_\_\_\_\_$ square units.

Chapter Test

1 Let $y = x^4 - 2x^3 + 9x + 7$.

(a) Find dy.

(b) Approximate the value of y when $x = 1.997$ and $\Delta x = dx = -0.003$.

2 Use differentials to approximate $\sqrt{27}$.

3 Evaluate the following antiderivatives.

(a) $\int (x + 1)(2 - x)\,dx$

(b) $\int \frac{(1 + \sqrt{3x})^2}{\sqrt{x}}\,dx$

(c) $\int x\sqrt{1 - x^2}\,dx$

(d) $\int \frac{x\,dx}{\sqrt{x + 1}}$

4 Solve the differential equations.

(a) $\frac{dy}{dx} = 2x + 1$

(b) $\frac{dy}{dx} = (3x - 1)^4$, $y = -4$ when $x = 0$.

5 Find v and s as functions of t if $a = 36t^2$, $v = -6$ when $t = 0$, and $s = 5$ when $t = 0$.

6 Let the acceleration of a given particle be given by $a = 24\sqrt[3]{s}$. Find the equation of motion if the initial velocity $v_0 = 6$ ft/sec at the initial position $s_0 = 1$ foot. $\left(\textit{Hint: } a\,ds = \frac{dv}{dt}\,ds = \frac{ds}{dt}\,dv = v\,dv.\right)$

7 A force F acting on a particle P moving along the s axis is given by $F = 400s\sqrt{1 + s^2}$, where s is in centimeters and F is in dynes. Find the work done by F in moving P from $s = 0$ to $s = 3$.

8 A particle P is moving along the positive s axis under the influence of a force F. The total energy E of P is zero and the potential energy is given by $V = -\frac{1}{s}$. Find the velocity $v = \frac{ds}{dt}$ of P in terms of s.

9 The marginal cost function $\frac{dC}{dx}$ is given by $\frac{dC}{dx} = 4x - 8$. If the cost of producing 5 units is \$20, find the total cost of producing x units.

10 Use the method of slicing to find the area of the region in the xy plane bounded below by the parabola $y = x^2$ and above the line $y = 4$.

11 Use the fundamental theorem of calculus to find the area of the region below the graph of $f(x) = \frac{1}{8}(x^2 - 2x + 8)$, above the x axis, and between $x = -2$ and $x = 4$.

Answers

1 (a) $dy = (4x^3 - 6x^2 + 9)\,dx$

(b) 24.949

2 5.2

3 (a) $-\frac{x^3}{3} + \frac{x^2}{2} + 2x + C$

(b) $2x^{1/2} + \frac{4\sqrt{3}}{3}x^{3/2} + \frac{6}{5}x^{5/2} + C$

(c) $-\frac{1}{3}(1 - x^2)^{3/2} + C$

(d) $\frac{2}{3}(x + 1)^{3/2} - 2(x + 1)^{1/2} + C$

4 (a) $y = x^2 + x + C$ (b) $y = \frac{(3x - 1)^5}{15} - \frac{59}{15}$

5 $v = 12t^3 - 6; s = 3t^4 - 6t + 5$

6 $s = (2t + 1)^3$

7 $\frac{400}{3}[(10)^{3/2} - 1]$ ergs

8 $v = \pm \sqrt{\frac{2}{ms}}$

9 $C = 2x^2 - 8x + 10$

10 $A = \frac{32}{3}$ square units

11 $\frac{15}{2}$ square units

Chapter 6 THE DEFINITE OR RIEMANN INTEGRAL

In this chapter we shall give a brief indication of the manner in which the definite integral can be defined and evaluated. We also develop methods for approximating values of definite integrals. In working the problems in this chapter, the student is able to:

1 Find sums and apply the basic properties of summation.
2 Find Riemann sums and make use of the formal definition of the definite integral.
3 Understand the basic properties of the definite integral, and find the mean value of a function on an interval.
4 Evaluate the definite integral using the fundamental theorem of calculus. Understand that the fundamental theorem of calculus connects differentiation and integration.
5 Approximate the value of the definite integral by using the trapezoidal rule and Simpson's rule.
6 Find the area of the region bounded by two graphs.

1 The Sigma Notation for Sums

1 The Greek letter sigma, which is written Σ, stands for the words, "the ____ of all terms of the form . . . ."

sum

2 The notation $\sum_{k=1}^{n} k^3$ means: The sum of all terms of the form k^3 as k runs through the integers from 1 to ____.

n

In Problems 3 through 9, evaluate the numerical value of each sum.

3 $\sum_{k=1}^{5} 3k$.

$$\sum_{k=1}^{5} 3k = 3(1) + 3(2) + 3(\_\_\_\_) + 3(\_\_\_\_) + 3(5)$$

3, 4

$$= 3 + 6 + 9 + 12 + 15 = \_\_\_\_.$$

45

4 $\sum_{k=1}^{3} (4k^2 - 3k)$.

$$\sum_{k=1}^{3} (4k^2 - 3k) = [4(1^2) - 3(1)] + [\_\_\_\_\_\_\_\_\_\_\_\_] + [4(3^2) - 3(3)]$$

$4(2^2) - 3(2)$

$$= 1 + \_\_\_\_ + \_\_\_\_ = \_\_\_\_.$$

10, 27, 38

5 $\sum_{i=3}^{6} i(i-2)$.

$$\sum_{i=3}^{6} i(i-2) = [3(3-2)] + [\_\_\_\_\_\_\_\_] + [5(5-2)] + [\_\_\_\_\_\_\_\_]$$

$4(4-2)$

$6(6-2)$

$$= 3 + \_\_\_\_ + 15 + \_\_\_\_ = \_\_\_\_.$$

8, 24, 50

6 $\sum_{k=3}^{5} \log_{10} k$.

$$\sum_{k=3}^{5} \log_{10} k = \log_{10} 3 + \_\_\_\_\_\_\_\_ + \_\_\_\_\_\_\_\_$$

$\log_{10} 4$, $\log_{10} 5$

$$= \log_{10}(\_\_\_\_\_\_\_\_) = \_\_\_\_\_\_\_\_.$$

$3 \times 4 \times 5$, $\log_{10} 60$

7 $\sum_{k=1}^{4} \sin \frac{\pi}{k}$.

$$\sum_{k=1}^{4} \sin \frac{\pi}{k} = \sin \frac{\pi}{1} + \sin \frac{\pi}{2} + \_\_\_\_\_\_\_\_ + \_\_\_\_\_\_\_\_$$

$\sin \frac{\pi}{3}$, $\sin \frac{\pi}{4}$

$$= \_\_\_\_ + \_\_\_\_ + \frac{\sqrt{3}}{2} + \frac{\sqrt{2}}{2}$$

0, 1

$$= \_\_\_\_\_\_\_\_\_\_\_\_.$$

$\frac{1}{2}(2 + \sqrt{3} + \sqrt{2})$

8 $\sum_{k=2}^{5} \frac{k-1}{k+1}$.

$\frac{3-1}{3+1}, \frac{5-1}{5+1}$

$\sum_{k=2}^{5} \frac{k-1}{k+1} = \frac{2-1}{2+1} + \_\_\_\_ + \frac{4-1}{4+1} + \_\_\_\_$

$\frac{3}{5}, \frac{4}{6}, \frac{21}{10}$

$= \frac{1}{3} + \frac{2}{4} + \_\_\_\_ + \_\_\_\_ = \_\_\_\_$.

9 $\sum_{k=1}^{4} 3^k$.

$3^3, 3^4$

$\sum_{k=1}^{4} 3^k = 3^1 + 3^2 + \_\_\_\_ + \_\_\_\_$

$27, 81$

$= 3 + 9 + \_\_\_\_ + \_\_\_\_$

120

$= \_\_\_\_$.

a_n

10 $\sum_{k=1}^{n} a_k = a_1 + a_2 + \cdots + \_\_\_\_$.

The variable k which runs from 1 to n is called the ________.

summation index

Basic Properties of Summation

In Problems 11 through 21, let $a_1, a_2, a_3, \ldots$ and $b_1, b_2, b_3, \ldots$ denote sequences of numbers and let A, B, and C be constant numbers. Then:

nC

11 *Constant Property:* $\sum_{k=1}^{n} C = \_\_\_\_$.

$\sum_{k=1}^{n} a_k$

12 *Homogeneous Property:* $\sum_{k=1}^{n} Ca_k = C(\_\_\_\_)$.

$\sum_{k=1}^{n} b_k$

13 *Additive Property:* $\sum_{k=1}^{n} (a_k + b_k) = \sum_{k=1}^{n} a_k + (\_\_\_\_)$.

$\sum_{k=1}^{n} b_k$

14 *Linear Property:* $\sum_{k=1}^{n} (Aa_k + Bb_k) = A\left(\sum_{k=1}^{n} a_k\right) + B(\_\_\_\_)$.

$\sum_{k=1}^{n} |a_k|$

15 *Generalized Triangle Inequality:* $\left|\sum_{k=1}^{n} a_k\right| \leq \_\_\_\_$.

16 *Sum of Arithmetic Sequence:*

$\frac{n}{2}[2A + (n-1)C]$

$\sum_{k=1}^{n} [A + (k-1)C] = \_\_\_\_$.

$1 - C^{n+1}$

17 *Sum of Geometric Sequence:* $\sum_{k=1}^{n} AC^k = A\frac{\_\_\_\_}{1-C}$, if $C \neq 1$.

$\frac{n(n+1)}{2}$

18 *Sum of Successive Integers:* $\sum_{k=1}^{n} k = \_\_\_\_$.

$\frac{n(n+1)(2n+1)}{6}$

19 *Sum of Successive Squares:* $\sum_{k=1}^{n} k^2 = \_\_\_\_$.

$\frac{n^2(n+1)^2}{4}$

$b_n - b_0$

20 *Sum of Successive Cubes:* $\sum_{k=1}^{n} k^3 =$ __________.

21 *Telescoping Property:* $\sum_{k=1}^{n} (b_k - b_{k-1}) =$ __________.

In Problems 22 through 28, use the properties in Problems 11 through 21 to evaluate the given sum.

22 $\sum_{k=1}^{10} (3k^2 - 5k + 7)$.

$\sum_{k=1}^{10} 7$

$\sum_{k=1}^{10} k^2, \quad \sum_{k=1}^{10} k$

21, 11, 70

1155, 275

950

$$\sum_{k=1}^{10} (3k^2 - 5k + 7) = \sum_{k=1}^{10} 3k^2 - \sum_{k=1}^{10} 5k + \left(\underline{\qquad}\right)$$

$$= 3\left(\underline{\qquad}\right) - 5\left(\underline{\qquad}\right) + 7\sum_{k=1}^{10} 1$$

$$= 3\left[\frac{(10)(11)(\underline{\quad})}{6}\right] - 5\left[\frac{10(\underline{\quad})}{2}\right] + \underline{\qquad}$$

$$= \underline{\qquad} - \underline{\qquad} + 70$$

$$= \underline{\qquad}.$$

23 $\sum_{k=1}^{20} k^2(3k + 2)$.

$3k^3 + 2k^2$

$\sum_{k=1}^{20} 2k^2$

$\sum_{k=1}^{20} k^2$

21, 41

132,300, 5740

138,040

$$\sum_{k=1}^{20} k^2(3k + 2) = \sum_{k=1}^{20} (\underline{\qquad\qquad})$$

$$= \sum_{k=1}^{20} 3k^3 + \underline{\qquad\qquad}$$

$$= 3\sum_{k=1}^{20} k^3 + 2\left(\underline{\qquad}\right)$$

$$= 3\left[\frac{(20)^2(\underline{\quad})^2}{4}\right] + 2\left[\frac{(20)(21)(\underline{\quad})}{6}\right]$$

$$= \underline{\qquad} + \underline{\qquad}$$

$$= \underline{\qquad}.$$

24 $\sum_{k=1}^{n} (3k^2 + 2k)$.

$\sum_{k=1}^{n} k^2, \quad \sum_{k=1}^{n} k$

$n(n+1)(2n+1), \quad n(n+1)$

2

$\frac{n(n+1)(2n+3)}{2}$

$$\sum_{k=1}^{n} (3k^2 + 2k) = 3\left(\underline{\qquad}\right) + 2\left(\underline{\qquad}\right)$$

$$= \frac{3}{6}(\underline{\qquad\qquad\qquad}) + \frac{2}{2}(\underline{\qquad\qquad})$$

$$= \frac{n(n+1)[(2n+1) + \underline{\quad}]}{2}$$

$$= \underline{\qquad\qquad\qquad}.$$

25 $\sum_{k=0}^{n} \frac{1}{3^k}$.

$1-\left(\frac{1}{3}\right)^{n+1}$

$$\sum_{k=0}^{n} \frac{1}{3^k} = \sum_{k=0}^{n}\left(\frac{1}{3}\right)^k = \frac{\underline{\qquad\qquad}}{1-\frac{1}{3}}$$

$1-\left(\frac{1}{3}\right)^{n+1}$, $\frac{1}{2}(3-3^{-n})$

$$= \frac{3}{2}\left(\underline{\qquad\qquad}\right) = \underline{\qquad\qquad}.$$

26 $\sum_{k=1}^{n}(5^k - 5^{k-1})$.

Using Problem 21, we have

5^n

$$\sum_{k=1}^{n}(5^k - 5^{k-1}) = (\underline{\qquad}) - 5^0$$

$5^n - 1$

$$= \underline{\qquad\qquad}.$$

27 $\sum_{k=1}^{1000}\left(\frac{1}{k} - \frac{1}{k+1}\right)$.

Using Problem 21, we have

$$\sum_{k=1}^{1000}\left(\frac{1}{k} - \frac{1}{k+1}\right) = -\sum_{k=1}^{1000}\left(\frac{1}{k+1} - \frac{1}{k}\right)$$

$$= -\sum_{k=1}^{1000}\left[\frac{1}{k+1} - \frac{1}{(k-1)+1}\right]$$

1001, 1

$$= -\left(\frac{1}{\underline{\qquad}} - \frac{1}{\underline{\qquad}}\right)$$

$\frac{1000}{1001}$

$$= 1 - \frac{1}{1001} = \underline{\qquad}.$$

28 $\sum_{k=1}^{40}(\sqrt{2k+1} - \sqrt{2k-1})$.

Using Problem 21, we have

$2(0) + 1$

$$\sum_{k=1}^{40}(\sqrt{2k+1} - \sqrt{2k-1}) = \sqrt{2(40)+1} - \sqrt{\underline{\qquad\qquad}}$$

1

$$= \sqrt{81} - \sqrt{\underline{\qquad}}$$

1

$$= 9 - \underline{\qquad}$$

8

$$= \underline{\qquad}.$$

1.1 The Area Under a Parabola

29 Using the figure on page 267, we can obtain an estimate of the area A under the parabola $y = x^2$ between $x = 0$ and $x = 1$ by

circumscribed

summing the areas of the n _______________ rectangles; that is,

$\frac{k}{n}$

$A \approx \sum_{k=1}^{n} \frac{1}{n}\left(\underline{\qquad}\right)^2$. Now,

$\frac{1}{n^3}$, k^2

$$\sum_{k=1}^{n}\left(\frac{1}{n}\right)\left(\frac{k}{n}\right)^2 = \sum_{k=1}^{n}\left(\underline{\qquad}\right)^3 k^2 = \left(\frac{1}{n}\right)^3 \sum_{k=1}^{n}(\underline{\qquad})$$

$\frac{n(n+1)(2n+1)}{6}$, $2n+1$

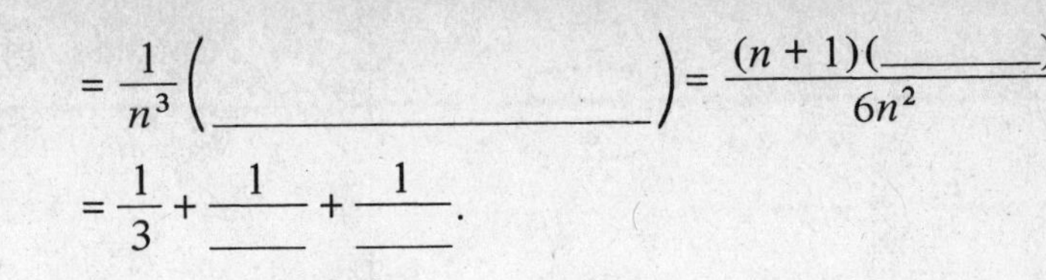

$= \frac{1}{n^3}\left(\rule{5em}{0.4pt}\right) = \frac{(n+1)(\rule{3em}{0.4pt})}{6n^2}$

$2n$, $6n^2$

$= \frac{1}{3} + \frac{1}{\rule{2em}{0.4pt}} + \frac{1}{\rule{2em}{0.4pt}}.$

500^2

When $n = 500$, $A \approx \sum_{k=1}^{500} \frac{1}{n}\left(\frac{k}{n}\right)^2 = \frac{1}{3} - \frac{1}{1000} + \frac{1}{6(\rule{3em}{0.4pt})} =$

0.33233400

__________ square unit.

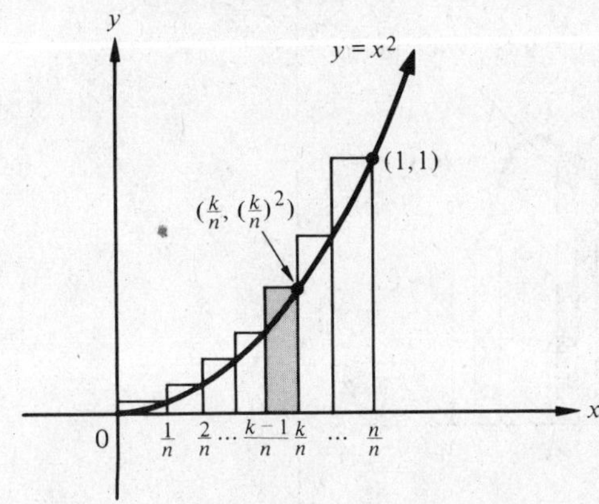

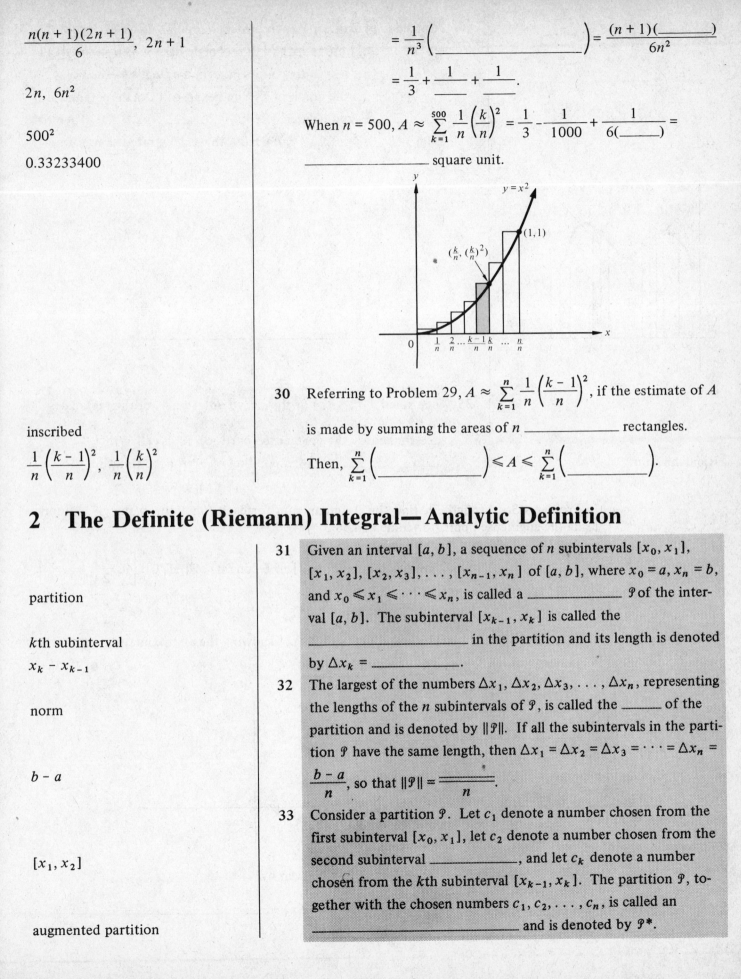

30 Referring to Problem 29, $A \approx \sum_{k=1}^{n} \frac{1}{n}\left(\frac{k-1}{n}\right)^2$, if the estimate of A

inscribed

is made by summing the areas of n __________ rectangles.

$\frac{1}{n}\left(\frac{k-1}{n}\right)^2$, $\frac{1}{n}\left(\frac{k}{n}\right)^2$

Then, $\sum_{k=1}^{n}\left(\rule{5em}{0.4pt}\right) \leqslant A \leqslant \sum_{k=1}^{n}\left(\rule{5em}{0.4pt}\right)$.

2 The Definite (Riemann) Integral—Analytic Definition

31 Given an interval $[a, b]$, a sequence of n subintervals $[x_0, x_1]$, $[x_1, x_2]$, $[x_2, x_3]$, . . . , $[x_{n-1}, x_n]$ of $[a, b]$, where $x_0 = a$, $x_n = b$, and $x_0 \leqslant x_1 \leqslant \cdots \leqslant x_n$, is called a __________ $\mathcal{P}$ of the interval $[a, b]$. The subinterval $[x_{k-1}, x_k]$ is called the __________ in the partition and its length is denoted by $\Delta x_k =$ __________.

partition

kth subinterval

$x_k - x_{k-1}$

32 The largest of the numbers $\Delta x_1, \Delta x_2, \Delta x_3, \ldots, \Delta x_n$, representing the lengths of the n subintervals of $\mathcal{P}$, is called the ______ of the partition and is denoted by $\|\mathcal{P}\|$. If all the subintervals in the partition $\mathcal{P}$ have the same length, then $\Delta x_1 = \Delta x_2 = \Delta x_3 = \cdots = \Delta x_n = \frac{b-a}{n}$, so that $\|\mathcal{P}\| = \frac{\rule{3em}{0.4pt}}{n}$.

norm

$b - a$

33 Consider a partition $\mathcal{P}$. Let c_1 denote a number chosen from the first subinterval $[x_0, x_1]$, let c_2 denote a number chosen from the second subinterval __________, and let c_k denote a number chosen from the kth subinterval $[x_{k-1}, x_k]$. The partition $\mathcal{P}$, together with the chosen numbers $c_1, c_2, \ldots, c_n$, is called an __________ and is denoted by $\mathcal{P}^*$.

$[x_1, x_2]$

augmented partition

34 Consider a partition $\mathcal{P}$ and construct a rectangle on each subinterval of $\mathcal{P}$. Suppose that the kth rectangle has the kth subinterval $[x_{k-1}, x_k]$ of length Δx_k as its base, and extends to the point $(c_k, f(c_k))$ on the graph of f. The height of the kth rectangle is ________ and its area is $\Delta A_k =$ ________. Therefore, $f(c_k)\,\Delta x_k = \pm$____. Graphically, this is illustrated as follows:

$|f(c_k)|$, $|f(c_k)|\,\Delta x_k$

ΔA_k

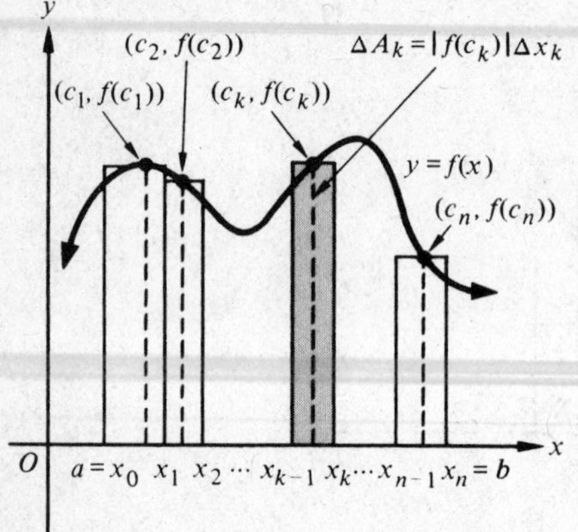

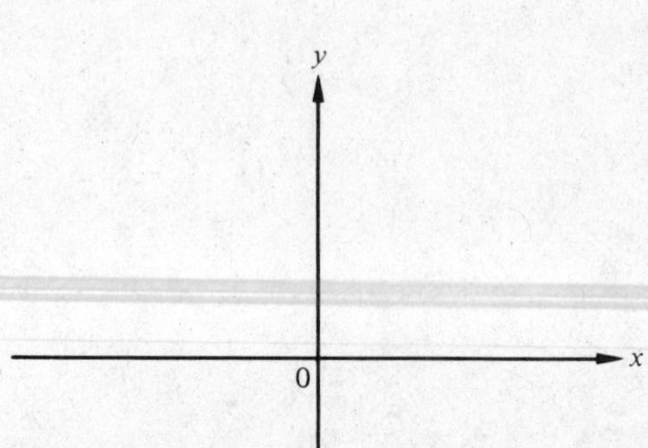

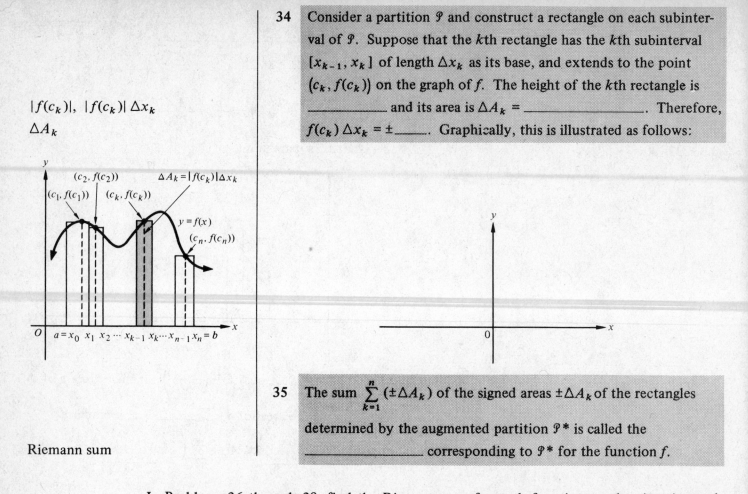

35 The sum $\sum_{k=1}^{n} (\pm \Delta A_k)$ of the signed areas $\pm\Delta A_k$ of the rectangles determined by the augmented partition $\mathcal{P}^*$ is called the ________ corresponding to $\mathcal{P}^*$ for the function f.

Riemann sum

In Problems 36 through 39, find the Riemann sum for each function on the given interval using the augmented partition indicated.

36 $f(x) = 4 - x^2$ on $[-1, 2]$. $\mathcal{P}^*$ consists of $[-1, 0]$, $\left[0, \frac{1}{2}\right]$, $\left[\frac{1}{2}, \frac{5}{4}\right]$, and $\left[\frac{5}{4}, 2\right]$, with $c_1 = -\frac{3}{4}$, $c_2 = \frac{1}{4}$, $c_3 = \frac{3}{4}$, and $c_4 = \frac{3}{2}$.

The graph of $f(x) = 4 - x^2$ showing the approximating rectangles is

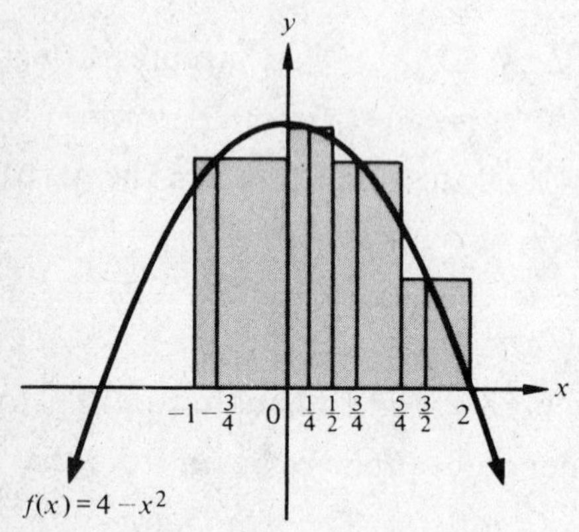

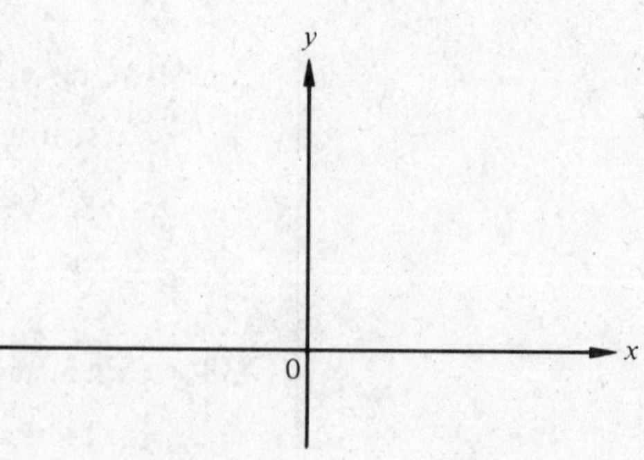

The Riemann sum is given by

$\sum_{k=1}^{4} f(c_k)\,\Delta x_k = f(c_1)\,\Delta x_1 +$ ________________

$f(c_2)\,\Delta x_2 + f(c_3)\,\Delta x_3 + f(c_4)\,\Delta x_4$

$$= f\left(-\frac{3}{4}\right)\cdot 1 + \left(\_\_\_\_\_\_\right)\cdot\frac{1}{2} + \left(\_\_\_\_\_\_\right)\cdot\frac{3}{4}$$

$$+ f\left(\frac{3}{2}\right)\cdot\left(\_\_\_\right)$$

$$= \left(\frac{55}{16}\right)\cdot 1 + \left(\frac{63}{16}\right)\cdot\frac{1}{2} + \left(\_\_\_\right)\cdot\frac{3}{4} + \left(\_\_\_\right)\cdot\frac{3}{4}$$

$$= \frac{220}{64} + \left(\_\_\_\right) + \frac{165}{64} + \frac{84}{64}$$

$$= \_\_\_\_ .$$

$f\left(\frac{1}{4}\right)$, $f\left(\frac{3}{4}\right)$

$\frac{3}{4}$

$\frac{55}{16}$, $\frac{7}{4}$

$\frac{126}{64}$

$\frac{595}{64}$

37 $f(x) = 3x + 2$ on $[0, 3]$. $\mathcal{P}^*$ consists of $\left[0, \frac{1}{2}\right]$, $\left[\frac{1}{2}, 1\right]$, $\left[1, \frac{3}{2}\right]$, $\left[\frac{3}{2}, 2\right]$, $\left[2, \frac{5}{2}\right]$, and $\left[\frac{5}{2}, 3\right]$, with $c_1 = \frac{1}{2}$, $c_2 = 1$, $c_3 = \frac{3}{2}$, $c_4 = 2$, $c_5 = \frac{5}{2}$, and $c_6 = 3$. The graph of $f(x) = 3x + 2$ showing the approximating rectangles is

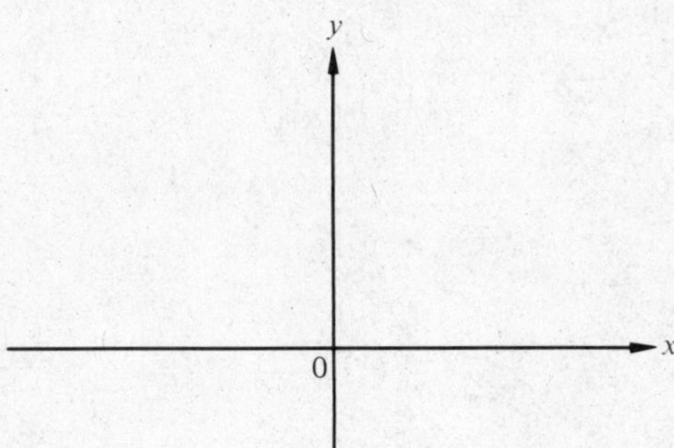

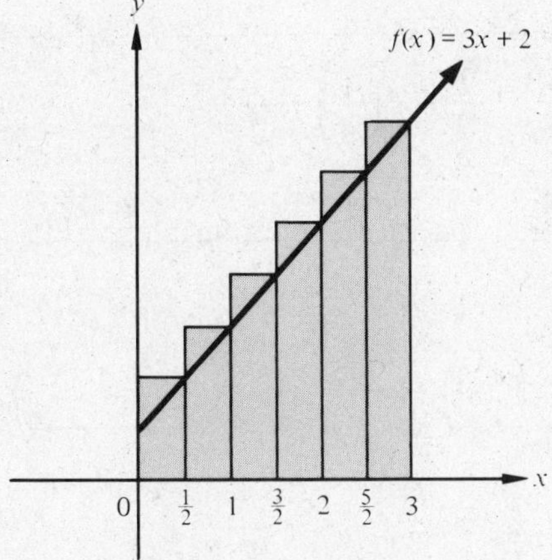

$$\sum_{k=1}^{6} f(c_k)\,\Delta x_k = f(c_1)\,\Delta x_1 + f(c_2)\,\Delta x_2 + f(c_3)\,\Delta x_3$$

$$+ \_\_\_\_\_\_\_\_\_\_\_\_\_\_\_\_\_\_\_\_\_\_\_\_$$

$f(c_4)\,\Delta x_4 + f(c_5)\,\Delta x_5 + f(c_6)\,\Delta x_6$

$$= f\left(\frac{1}{2}\right)\cdot\frac{1}{2} + f(1)\cdot\frac{1}{2} + f\left(\frac{3}{2}\right)\cdot\frac{1}{2}$$

$$+ \_\_\_\_\_\_\_\_\_\_\_\_\_\_\_\_\_\_\_\_\_\_\_\_$$

$f(2)\cdot\frac{1}{2} + f\ \frac{5}{2}\ \cdot\frac{1}{2} + f(3)\cdot\frac{1}{2}$

$$= \frac{7}{2}\cdot\frac{1}{2} + 5\cdot\frac{1}{2} + \frac{13}{2}\cdot\frac{1}{2} + \_\_\_\_\_\_\_\_\_\_\_\_$$

$8\cdot\frac{1}{2} + \frac{19}{2}\cdot\frac{1}{2} + 11\cdot\frac{1}{2}$

$$= \frac{7}{4} + \left(\_\_\_\right) + \frac{13}{4} + \left(\_\_\_\right) + \frac{19}{4} + \left(\_\_\_\right)$$

$\frac{5}{2}$, $\frac{8}{2}$, $\frac{11}{2}$

$$= \_\_\_\_ .$$

$\frac{87}{4}$

38 $f(x) = -4x^2$ on $[0, 3]$. $\mathcal{P}^*$ consists of $\left[0, \frac{1}{2}\right]$, $\left[\frac{1}{2}, 1\right]$, $\left[1, \frac{3}{2}\right]$,

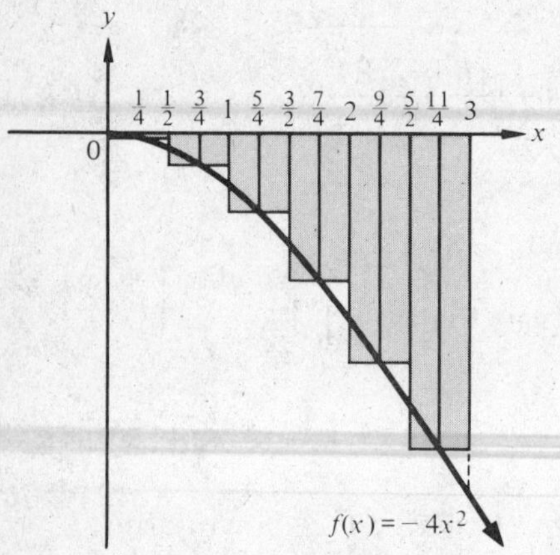

$f(c_4)\,\Delta x_4 + f(c_5)\,\Delta x_5 + f(c_6)\,\Delta x_6$

$f\left(\frac{7}{4}\right)\cdot\frac{1}{2} + f\left(\frac{9}{4}\right)\cdot\frac{1}{2} +$
$f\left(\frac{11}{4}\right)\cdot\frac{1}{2}$

$\frac{9}{4}\cdot\frac{1}{2}, \frac{49}{4}\cdot\frac{1}{2}$

$\frac{121}{4}\cdot\frac{1}{2}$

$-\frac{286}{8}, -\frac{143}{4}$

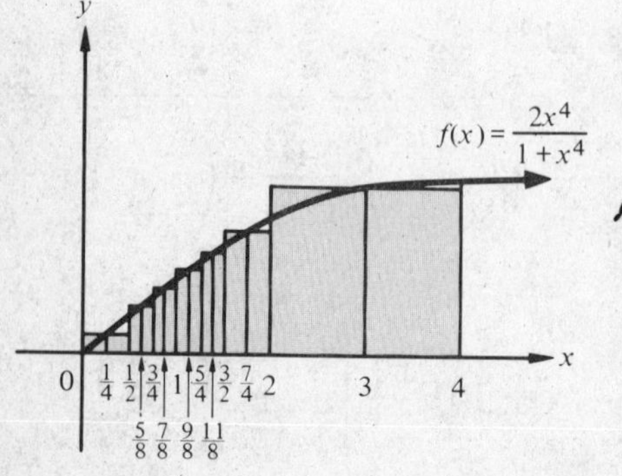

$\left[\frac{3}{2}, 2\right], \left[2, \frac{5}{2}\right]$, and $\left[\frac{5}{2}, 3\right]$, with $c_1 = \frac{1}{4}, c_2 = \frac{3}{4}, c_3 = \frac{5}{4}, c_4 = \frac{7}{4}, c_5 = \frac{9}{4}$, and $c_6 = \frac{11}{4}$. The graph of $f(x) = -4x^2$ showing the approximating rectangles is

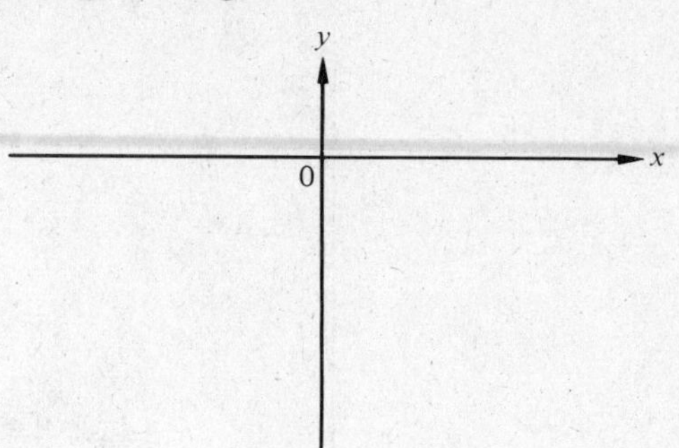

$$\sum_{k=1}^{6} f(c_k)\,\Delta x_k = f(c_1)\,\Delta x_1 + f(c_2)\,\Delta x_2 + f(c_3)\,\Delta x_3$$

$$+ \underline{\hspace{6cm}}$$

$$= f\left(\frac{1}{4}\right)\cdot\frac{1}{2} + f\left(\frac{3}{4}\right)\cdot\frac{1}{2} + f\left(\frac{5}{4}\right)\cdot\frac{1}{2}$$

$$+ \underline{\hspace{6cm}}$$

$$= -\frac{1}{4}\cdot\frac{1}{2} - \left(\underline{\hspace{2cm}}\right) - \frac{25}{4}\cdot\frac{1}{2} - \left(\underline{\hspace{2cm}}\right)$$

$$-\frac{81}{4}\cdot\frac{1}{2} - \left(\underline{\hspace{2cm}}\right)$$

$$= \underline{\hspace{1.5cm}} = \underline{\hspace{1.5cm}}.$$

39 $f(x) = \frac{2x^4}{1 + x^4}$ on $[0, 4]$. $\mathcal{P}^*$ consists of $\left[0, \frac{1}{2}\right], \left[\frac{1}{2}, \frac{3}{4}\right], \left[\frac{3}{4}, 1\right], \left[1, \frac{5}{4}\right], \left[\frac{5}{4}, \frac{3}{2}\right], \left[\frac{3}{2}, 2\right]$, and $[2, 4]$, with $c_1 = \frac{1}{4}, c_2 = \frac{5}{8}, c_3 = \frac{7}{8}, c_4 = \frac{9}{8}, c_5 = \frac{11}{8}, c_6 = \frac{7}{4}$, and $c_7 = 3$. The graph of $f(x) = \frac{2x^4}{1 + x^4}$ showing the approximating rectangles is

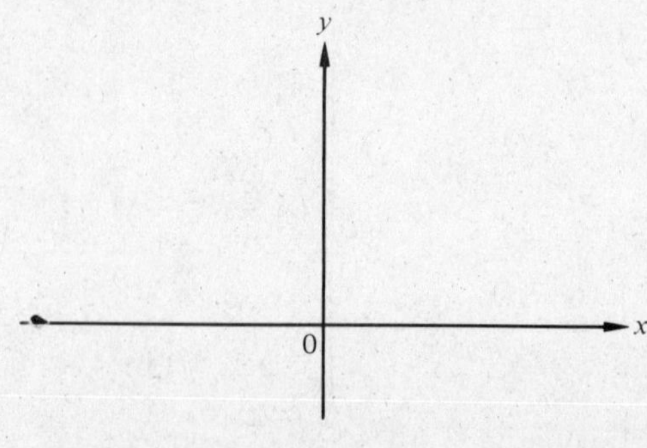

$$\sum_{k=1}^{7} f(c_k)\,\Delta x_k = f(c_1)\,\Delta x_1 + f(c_2)\,\Delta x_2 + f(c_3)\,\Delta x_3$$

$+$ ____

$f(c_4)\,\Delta x_4 + f(c_5)\,\Delta x_5 + f(c_6)\,\Delta x_6 + f(c_7)\,\Delta x_7$

$= f\left(\frac{1}{4}\right)\cdot\frac{1}{2} +$ __________________________ $\cdot\frac{1}{4} +$

$f\left(\frac{5}{8}\right)+f\left(\frac{7}{8}\right)+f\left(\frac{9}{8}\right)+f\left(\frac{11}{8}\right)$

$f\left(\frac{7}{4}\right)\cdot\frac{1}{2} + f(3)\cdot 2$

$$= \frac{2}{257}\cdot\frac{1}{2} + \left(\frac{1250}{4721} + \frac{4802}{6407} + \frac{12{,}122}{10{,}657} + \frac{29{,}282}{18{,}737}\right)\cdot\frac{1}{4} + \frac{4802}{2657}\cdot\frac{1}{2} + \frac{162}{82}\cdot 2$$

$$\approx 0.0039 + (0.2648 + 0.7495 + 1.1375 + 1.5628)\frac{1}{4} + 0.9036 + 3.9512$$

$$= 0.0039 + 0.9287 + 0.9036 + 3.9512$$

$\approx$ ______.

5.79

40 Let f be a function defined at least on the closed interval $[a, b]$. For each augmented partition $\mathcal{P}^*$ of $[a, b]$ consisting of n subintervals, $[x_0, x_1]$, $[x_1, x_2]$, . . . , $[x_{n-1}, x_n]$ and the chosen numbers $c_1, c_2, c_3, \ldots, c_n$ belonging to the respective subintervals, form the corresponding Riemann sum $\sum_{k=1}^{n} f(c_k)\,\Delta x_k$. Here $\Delta x_k =$ __________ is the length of the kth subinterval for $k = 1, 2, 3, \ldots, n$. The norm $\|\mathcal{P}\|$ of the partition $\mathcal{P}$ is defined to be the largest of the numbers $\Delta x_1, \Delta x_2, \ldots, \Delta x_n$. If the limit __________________ exists, then the function f is said to be ______________ on $[a, b]$ in the sense of Riemann. If f is integrable, then the definite (Riemann) integral of f on the interval $[a, b]$ is defined by $\int_a^b f(x)\,dx =$ __________________________.

$x_k - x_{k-1}$

$\lim_{\|\mathcal{P}\|\to 0} \sum_{k=1}^{n} f(c_k)\,\Delta x_k$

integrable

$\lim_{\|\mathcal{P}\|\to 0} \sum_{k=1}^{n} f(c_k)\,\Delta x_k$

41 The limit in the definition in Problem 40 is to be understood as follows: To assert that a number I is the limit of the Riemann sum $\sum_{k=1}^{n} f(c_k)\,\Delta x_k$ as the norm $\|\mathcal{P}\|$ approaches zero means that for every positive number ϵ there exists a positive number δ such that ________________________________ holds for every augmented partition $\mathcal{P}^*$ such that __________.

$\left|\sum_{k=1}^{n} f(c_k)\,\Delta x_k - I\right| < \epsilon$

$\|\mathcal{P}\| < \delta$

2.1 Existence of Riemann Integrals

Answers	Problems
continuous Riemann-integrable	**42** If f is a ______ function on the closed interval $[a, b]$, then f is ______ on $[a, b]$.
continuous Riemann-integrable	**43** If f is a bounded and piecewise ______ function on the closed interval $[a, b]$, then f is ______ on $[a, b]$.
bounded	**44** If f is defined and Riemann-integrable on $[a, b]$, then f is ______ on $[a, b]$.
finite Riemann-integrable $\int_a^b f(x)\,dx$	**45** If f is defined and Riemann-integrable on $[a, b]$, and if h is also defined on $[a, b]$ and satisfies $h(x) = f(x)$ for all but a ______ number of values of x in $[a, b]$, then h is also ______ on $[a, b]$ and $\int_a^b h(x)\,dx =$ ______.

In Problems 46 through 48, indicate whether each Riemann integral exists.

Answers	Problems
$[1, 300]$ exists	**46** $\int_1^{300} \frac{dx}{x^2}$. Since $f(x) = \frac{1}{x^2}$ is continuous on ______, the integral $\int_1^{300} \frac{dx}{x^2}$ ______.
$[-2, 5]$ does not exist	**47** $\int_{-2}^{5} \frac{dx}{x+1}$. Since $f(x) = \frac{1}{x+1}$ is not continuous on ______, $\int_{-2}^{5} \frac{dx}{x+1}$ ______.
not continuous, $\left[0, \frac{3\pi}{4}\right]$ does not exist	**48** $\int_0^{3\pi/4} \tan x\,dx$. Since $f(x) = \tan x$ is ______ on ______, $\int_0^{3\pi/4} \tan x\,dx$ ______.

2.2 Calculation of Riemann Integrals by Direct Use of the Definition

49 The definite integral $\int_a^b f(x)\,dx$ of any Riemann-integrable function f can be calculated by selecting any convenient sequence of

$\lim_{n\to+\infty} \|\mathcal{P}\| = 0$

augmented partitions $\mathcal{P}_1^*, \mathcal{P}_2^*, \mathcal{P}_3^*, \ldots$ such that ________, calculating the Riemann sums corresponding to each augmented partition, and finding the limit of the resulting sequence of Riemann sums. In doing this, it is often convenient to consider only partitions of $[a, b]$ consisting of n equal subintervals each of which has length $\Delta x = \dfrac{\_\_\_\_}{n}$.

$b - a$

In Problems 50 through 53, evaluate each Riemann integral directly by calculating the limit of Riemann sums. Use partitions consisting of subintervals of equal lengths and use the indicated rectangles.

50 $\int_0^3 4x\,dx$ (circumscribed rectangles).

Let $\mathcal{P}$ denote the partition of $[0, 3]$ consisting of n equal subintervals of length $\Delta x = \dfrac{\_\_\_\_}{n}$. Since f is an increasing function, the circumscribed rectangles are obtained by forming an augmented partition $\mathcal{P}^*$ with $c_1, c_2, c_3, \ldots, c_n$ as the right-hand endpoints of the corresponding subintervals. Thus, $\mathcal{P}^*$ consists of the intervals $[0, \Delta x], [\Delta x, 2\Delta x], \ldots, [3 - \Delta x, 3]$ with $c_k =$ ____, such that $c_1 = \dfrac{3}{n}$, $c_2 =$ ____, $\ldots$, $c_n =$ ____. Therefore,

$3 - 0$

$\dfrac{3k}{n}$

$\dfrac{6}{n}, \dfrac{3n}{n}$

$\dfrac{3}{n}$

$\dfrac{36}{n^2}\sum_{k=1}^{n} k$

$\dfrac{n(n+1)}{2}$

$n^2 + n$

$\dfrac{1}{n}$

18

$$\int_0^3 4x\,dx = \lim_{n\to+\infty} \sum_{k=1}^{n} f(c_k)\,\Delta x = \lim_{n\to+\infty} \sum_{k=1}^{n} 4\left(\frac{3k}{n}\right)\cdot\left(\_\_\_\_\right)$$

$$= \lim_{n\to+\infty} \left(\_\_\_\_\_\_\_\_\right)$$

$$= \lim_{n\to+\infty} \left(\frac{36}{n^2}\,\_\_\_\_\_\_\_\_\right)$$

$$= \lim_{n\to+\infty} 18\,\frac{\_\_\_\_}{n^2}$$

$$= 18 \lim_{n\to+\infty} \left(1 + \_\_\_\_\right)$$

$$= \_\_\_\_.$$

51 $\int_0^4 x^3\,dx$ (inscribed rectangles).

Let $\mathcal{P}_n$ denote the partition of $[0, 4]$ consisting on n equal subintervals of length $\Delta x = \dfrac{4-0}{n} =$ ____. Choose c_k as the left endpoints of each subinterval so that $c_1 = 0$, $c_2 = \dfrac{4}{n}, \ldots,$

$\dfrac{4}{n}$

$c_k =$ ________. Because $f(x) = x^3$, we have

$(k-1)\left(\dfrac{4}{n}\right)$

$\dfrac{4(k-1)}{n}$	$f(c_k) = \left(\underline{\qquad\qquad}\right)^3$. Therefore,
	$\displaystyle\int_0^4 x^3\,dx = \lim_{n\to+\infty} \sum_{k=1}^{n} f(c_k)\,\Delta x_k$
$\dfrac{4}{n}$	$\displaystyle= \lim_{n\to+\infty} \sum_{k=1}^{n} \frac{4(k-1)^3}{n}\ \underline{\qquad}$
	$\displaystyle= 256 \lim_{n\to+\infty} \sum_{k=1}^{n} \frac{(k-1)^3}{n^4}$
$(k-1)^3$	$\displaystyle= 256 \lim_{n\to+\infty} \frac{1}{n^4} \sum_{k=1}^{n} (\underline{\qquad\qquad})$
k^3	$\displaystyle= 256 \lim_{n\to+\infty} \frac{1}{n^4} \sum_{k=1}^{n-1} (\underline{\qquad})$
$\dfrac{(n-1)^2 n^2}{4}$	$\displaystyle= 256 \lim_{n\to+\infty} \frac{\underline{\qquad\qquad}}{n^4}$
$n^4 - 2n^3 + n^2$	$\displaystyle= 256 \lim_{n\to+\infty} \frac{\underline{\qquad\qquad\qquad}}{4n^4}$
	$\displaystyle= \frac{256}{4} \lim_{n\to+\infty} 1 - \frac{2}{n} + \frac{1}{n^2}$
0, 0	$= 64\,(1 + \underline{\qquad} + \underline{\qquad})$
64	$= \underline{\qquad}$.
	52 $\displaystyle\int_1^3 x^2\,dx$ (circumscribed rectangles).
	Let $\mathcal{P}_n$ denote the partition of $[1, 3]$ consisting of n equal sub-
$\dfrac{2}{n}$	intervals of length $\Delta x = \underline{\qquad}$. Choose c_k as the right endpoint of
	each subinterval so that $c_1 = 1 + \dfrac{2}{n}$, $c_2 = 1 + 2\left(\dfrac{2}{n}\right), \ldots,$
$1 + k\left(\dfrac{2}{n}\right)$, $1 + \dfrac{2k}{n}$	$c_k = \underline{\qquad\qquad}$. Since $f(x) = x^2$, then $f(c_k) = \left(\underline{\qquad}\right)^2$.
	Therefore,
	$\displaystyle\int_1^3 x^2\,dx = \lim_{n\to+\infty} \sum_{k=1}^{n} \left(1 + \frac{2k}{n}\right)^2 \cdot \frac{2}{n}$
$\dfrac{n+2k}{n}$	$\displaystyle= \lim_{n\to+\infty} \sum_{k=1}^{n} \left(\underline{\qquad\qquad}\right)^2 \cdot \frac{2}{n}$
k^2	$\displaystyle= \lim_{n\to+\infty} \frac{2}{n^3} \left(n^2 \sum_{k=1}^{n} 1 + 4n \sum_{k=1}^{n} k + 4 \sum_{k=1}^{n} \underline{\qquad}\right)$
n, $\dfrac{n(n+1)}{2}$	$\displaystyle= \lim_{n\to+\infty} \frac{2}{n^3} \left[n^2\,(\underline{\qquad}) + 4n\left(\underline{\qquad\qquad}\right) + \frac{4n(n+1)(2n+1)}{6} \right]$
$n^3 + 2n^3 + 2n^2$	$\displaystyle= \lim_{n\to+\infty} \frac{2}{n^3} \left[\underline{\qquad\qquad\qquad} + \frac{2n(n+1)(2n+1)}{3} \right]$

$6 + \frac{4}{n}$ $\qquad = \lim_{n \to +\infty} \left(\_\_\_\_\_ + \frac{8n^2 + 12n + 4}{3n^2} \right)$

$\frac{4}{3n^2}$ $\qquad = \lim_{n \to +\infty} \left(6 + \frac{4}{n} + \frac{8}{3} + \frac{4}{n} + \_\_\_\_ \right)$

0, 0 $\qquad = 6 + \_\_\_\_ + \frac{8}{3} + 0 + \_\_\_\_$

$\frac{26}{3}$ $\qquad = \_\_\_\_.$

53 $\int_{-1}^{2} (8 - 2x)\, dx$ (inscribed rectangles).

Let $\mathcal{P}_n$ denote the partition of $[-1, 2]$ consisting of n equal sub-

$\frac{3}{n}$ $\qquad$ intervals of length $\Delta x = \frac{2 - (-1)}{n} = \_\_\_\_$. Choose c_k as the left

$-1 + \frac{3}{n}$ $\qquad$ endpoint of each subinterval, so that $c_1 = -1$, $c_2 = \_\_\_\_\_\_$,

$-1 + \frac{3(k-1)}{n}$ $\qquad$ $\ldots, c_k = \_\_\_\_\_\_\_\_\_\_$. Since $f(x) = 8 - 2x$, then

$8 - 2\left[-1 + \frac{3(k-1)}{n}\right]$ $\qquad$ $f(c_k) = \_\_\_\_\_\_\_\_\_\_\_\_\_\_\_\_ =$

$10 - \frac{6(k-1)}{n}$ $\qquad$ $\_\_\_\_\_\_\_\_\_\_$. Therefore,

$$\int_{-1}^{2} (8 - 2x)\, dx = \lim_{n \to +\infty} \sum_{k=1}^{n} \left[10 - \frac{6(k-1)}{n}\right] \frac{3}{n}$$

$$= \lim_{n \to +\infty} \sum_{k=1}^{n} \left(\frac{30}{n} - \frac{18k}{n^2} + \frac{18}{n^2} \right)$$

$\sum_{k=1}^{n} 1$ $\qquad = \lim_{n \to +\infty} \left[\frac{30}{n} \left(\_\_\_\_\_ \right) - \frac{18}{n^2} \sum_{k=1}^{n} k + \frac{18}{n^2} \sum_{k=1}^{n} 1 \right]$

$n, \quad \frac{n(n+1)}{2}$ $\qquad = \lim_{n \to +\infty} \left[\frac{30}{n} (\_\_\_) - \frac{18}{n^2} \left(\_\_\_\_\_\_\_\_ \right) + \frac{18}{n^2} (n) \right]$

$\frac{1}{n}$ $\qquad = \lim_{n \to +\infty} \left[30 - 9 \left(1 + \_\_\_\_ \right) + \frac{18}{n} \right]$

9 $\qquad = 30 - (\_\_\_)$

21 $\qquad = \_\_\_\_.$

3 Basic Properties of the Definite Integral

54 The integral of a constant function property states: If f is a constant function defined by the equation $f(x) = k$, where k is a constant number, then $\int_a^b f(x)\, dx = \int_a^b (\_\_\_)\, dx = k(\_\_\_\_\_)$.

$k, \quad b - a$

Riemann-integrable

$\int_a^b f(x)\,dx$

55 The homogeneous property states: If f is a Riemann-integrable function on the interval $[a, b]$ and k is a constant number, then kf is also ________________ on $[a, b]$ and $\int_a^b kf(x)\,dx = k(\_\_\_\_\_\_\_\_)$.

integrable

$\int_a^b g(x)\,dx$

56 The additive property states: If f and g are Riemann-integrable functions on the interval $[a, b]$, then $f+g$ is also Riemann-________ on $[a, b]$, and $\int_a^b [f(x)+g(x)]\,dx =$ $\int_a^b f(x)\,dx + (\_\_\_\_\_\_\_\_)$.

integrable

$\int_a^b f(x)\,dx$

57 The linear property states: If f and g are Riemann-integrable functions on the interval $[a, b]$, and if A and B are constant numbers, then $Af+Bg$ is also Riemann-________ on $[a, b]$, and $\int_a^b [Af(x)+Bg(x)]\,dx = A(\_\_\_\_\_\_\_\_) + B\int_a^b g(x)\,dx$.

$A_1 f_1 \pm A_2 f_2 \pm \cdots \pm A_n f_n$

$\int_a^b f_1(x)\,dx,\ \int_a^b f_n(x)\,dx$

58 The general linear property states: If $f_1, f_2, \ldots, f_n$ are Riemann-integrable functions on the interval $[a, b]$, and if $A_1, A_2, \ldots, A_n$ are constants, then ________________ is also Riemann-integrable, and

$$\int_a^b [A_1 f_1(x) \pm A_2 f_2(x) \pm \cdots \pm A_n f_n(x)]\,dx$$

$$= A_1(\_\_\_\_\_\_\_\_) \pm A_2 \int_a^b f_2(x)\,dx \pm \cdots \pm A_n(\_\_\_\_\_\_\_\_).$$

$\int_a^b f(x)\,dx \geqslant 0$

59 The positivity property states: If f is a Riemann-integrable function on the interval $[a, b]$, and if $f(x) \geqslant 0$ for all values of x in $[a, b]$, then ________________.

$\int_a^b g(x)\,dx$

60 The comparison property states: If f and g are Riemann-integrable functions on the interval $[a, b]$, and if $f(x) \leqslant g(x)$ holds for all values of x in $[a, b]$, then $\int_a^b f(x)\,dx \leqslant$ ________________.

$|f|,\ |f(x)|$

61 The absolute value property states: If f is Riemann-integrable on the interval $[a, b]$, then so is ____ and $\left|\int_a^b f(x)\,dx\right| \leqslant \int_a^b \_\_\_\_\ dx$.

integrable

$\int_b^c f(x)\,dx$

62 The additivity with respect to the interval of integration property states: If f is a Riemann-integrable function on the interval $[a, b]$ as well as on the interval $[b, c]$, where $a < b < c$, then f is also Riemann-________ on the interval $[a, c]$, and $\int_a^c f(x)\,dx = \int_a^b f(x)\,dx +$ ________________.

In Problems 63 through 66, assume that $\int_a^b dx = b - a$, $\int_a^b x\,dx = \frac{1}{2}(b^2 - a^2)$, $\int_a^b x^2\,dx = \frac{1}{3}(b^3 - a^3)$, and $\int_a^b x^3\,dx = \frac{1}{4}(b^4 - a^4)$. Use the basic properties in Problems 54 through 62 to evaluate each expression:

63 $\int_1^2 (3x^2 + 5)\,dx$.

5

$$\int_1^2 (3x^2 + 5)\,dx = \int_1^2 3x^2\,dx + \int_1^2 \underline{\qquad}\,dx$$

$\int_1^2 x^2\,dx$, 5

$$= 3\left(\underline{\qquad\qquad}\right) + \underline{\quad}\int_1^2 dx$$

8

$$= \frac{3}{3}(\underline{\quad} - 1) + 5(2 - 1)$$

5

$$= 7 + \underline{\quad}$$

12

$$= \underline{\quad}.$$

64 $\int_{-1}^3 (4x^3 - 3x^2 + 2x + 7)\,dx$.

$$\int_{-1}^3 (4x^3 - 3x^2 + 2x + 7)\,dx$$

$\int_{-1}^3 2x\,dx$, $\int_{-1}^3 7\,dx$

$$= \int_{-1}^3 4x^3\,dx - \int_{-1}^3 3x^2\,dx + \underline{\qquad\qquad} + \underline{\qquad\qquad}$$

$\int_{-1}^3 x\,dx$, $\int_{-1}^3 dx$

$$= 4\int_{-1}^3 x^3\,dx - 3\int_{-1}^3 x^2\,dx + 2\left(\underline{\qquad\qquad}\right) + 7\left(\underline{\qquad}\right)$$

1, 9

$$= \frac{4}{4}(81 - \underline{\quad}) - \frac{3}{3}(27 + 1) + \frac{2}{2}(\underline{\quad} - 1) + 7(4)$$

28, 8, 88

$$= 80 - \underline{\quad} + \underline{\quad} + 28 = \underline{\quad}.$$

65 $\int_0^\pi (2x + 11)\,dx + \int_\pi^4 (2x + 11)\,dx$.

$\int_0^4 (2x + 11)\,dx$

$$\int_0^\pi 2(x + 11)\,dx + \int_\pi^4 (2x + 11)\,dx = \underline{\qquad\qquad\qquad}$$

$\int_0^4 dx$

$$= 2\int_0^4 x\,dx + 11\left(\underline{\qquad}\right)$$

4 - 0

$$= \frac{2}{2}(16 - 0) + 11\,(\underline{\qquad})$$

44, 60

$$= 16 + \underline{\quad} = \underline{\quad}.$$

66 $\int_0^5 f(x)\,dx$, where $f(x) = \begin{cases} 9x^2 & \text{for } 0 \leqslant x < 3 \\ 8x & \text{for } 3 \leqslant x \leqslant 5. \end{cases}$

$9x^2$

$$\int_0^5 f(x)\,dx = \int_0^3 (\underline{\quad})\,dx + \int_3^5 8x\,dx$$

$\int_0^3 x^2\,dx, \int_3^5 x\,dx$

9

64, 145

$= 9\left(\underline{\qquad\qquad}\right) + 8\left(\underline{\qquad\qquad}\right)$

$= 3(27 - 0) + 4(25 - \underline{\quad})$

$= 81 + \underline{\quad} = \underline{\quad}.$

In Problems 67 through 70, state the property of the definite integral that justifies each assertion.

positivity property

$\int_0^2 f(x)\,dx \geqslant 0$

67 $\int_0^2 (9 + 7x - x^2)\,dx \geqslant 0.$

Since $f(x) = 9 + 7x - x^2 \geqslant 0$ in $[0, 2]$, we have, by the __________________ in Problem 59, that

__________________.

comparison property, $\int_0^1 x\,dx$

68 $\int_0^1 x^6\,dx \leqslant \int_0^1 x\,dx.$

Since $f(x) = x^6 \leqslant x = g(x)$ in $[0, 1]$, we have, by the ______________ in Problem 60, that $\int_0^1 x^6\,dx \leqslant$ ________.

5, 4, 9

additivity with respect to the interval of integration property

69 Given that $\int_{-1}^3 f(x)\,dx = 5$ and that $\int_3^8 f(x)\,dx = 4$, then

$\int_{-1}^8 f(x)\,dx = 9.$

$$\int_{-1}^8 f(x)\,dx = \int_{-1}^3 f(x)\,dx + \int_3^8 f(x)\,dx$$

$= \underline{\quad} + \underline{\quad} = \underline{\quad},$

by using the ____ in Problem 62.

absolute value property

70 $\left|\int_{-1}^7 x^5\,dx\right| \leqslant \int_{-1}^7 |x^5|\,dx.$

Use the __________________________ in Problem 61.

$f(c) \cdot (b - a)$

mean value

71 The mean value theorem for integrals states: If f is a continuous function on the interval $[a, b]$, then there exists a number c in $[a, b]$ such that $\int_a^b f(x)\,dx =$ __________________.

$f(c) = \dfrac{1}{b-a}\int_a^b f(x)\,dx$ is called the average or ____________ of f on the interval $[a, b]$.

In Problems 72 through 76, find the average (or mean) value M of each function on the interval indicated. Also, find a value of c on this interval such that $f(c)$ is this mean value. Assume that $\int_a^b dx = b - a$, $\int_a^b x\, dx = \frac{1}{2}(b^2 - a^2)$, and $\int_a^b x^2\, dx = \frac{1}{3}(b^3 - a^3)$.

Answers	Problems
	72 $f(x) = 3x + 7$ on $[1, 3]$.
$f(c)$	The mean value $M =$ ____, where
$3x + 7$	$f(c) = \frac{1}{b - a}\int_a^b f(x)\,dx = \frac{1}{3 - 1}\int_1^3 (\_\_\_\_)\,dx$
9	$= \frac{1}{2}\int_1^3 (3x + 7)\,dx = \frac{3}{4}(\_\_\_\_ - 1^2) + \frac{7}{2}(3 - 1)$
$8,\ 2,\ 13$	$= \frac{3}{4}(\_\_\_\_) + \frac{7}{2}(\_\_\_\_) = \_\_\_\_$.
$3c + 7,\ 13$	To find c, we have $f(c) =$ ____, so that $3c + 7 =$ ____ or
$6,\ 2$	$3c =$ ____. Thus, $c =$ ____.
	73 $f(x) = 4 - x^2$ on $[-3, 2]$.
$f(c)$	The mean value $M =$ ____, where
$4 - x^2$	$f(c) = \frac{1}{1 - (-3)}\int_{-3}^1 (\_\_\_\_)\,dx = \frac{1}{4}\int_{-3}^1 (4 - x^2)\,dx$
-27	$= \frac{1}{4}[4(1 + 3)] - \frac{1}{4}\left(\frac{1}{3}\right)[1 - (\_\_\_\_)]$
$\frac{5}{3}$	$= \_\_\_\_$.
$4 - c^2,\ \frac{7}{3},\ -\sqrt{\frac{7}{3}}$	But $f(c) =$ ____, so that $4 - c^2 = \frac{5}{3}$ or $c^2 =$ ____; so $c =$ ____, since only $-\sqrt{\frac{7}{3}}$ belongs to the interval $[-3, 1]$.
	74 $f(x) = x^2$ on $[-3, 0]$.
$\frac{1}{3}\int_{-3}^0 x^2\,dx$	The mean value $M = f(c)$, where $f(c) =$ ____ $=$
$27,\ 3,\ 3,\ c^2$	$\frac{1}{9}(0 + \_\_\_\_) = \_\_\_\_$, so that $f(c) =$ ____. But $f(c) =$ ____ so that
$3,\ -\sqrt{3}$	$c^2 =$ ____ and $c =$ ____ on $[-3, 0]$.
	75 $f(x) = 3 - 2x + x^2$ on $[1, 3]$.
$f(c)$	The mean value $M =$ ____, where
	$f(c) = \frac{1}{2}\int_1^3 (3 - 2x + x^2)\,dx$
$9,\ 1$	$= \frac{1}{2}\left[3(3 - 1) - \frac{2}{2}(\_\_\_\_ - 1) + \frac{1}{3}(27 - \_\_\_\_)\right]$
$\frac{10}{3}$	$= \_\_\_\_$.

$3 - 2c + c^2$, $\frac{10}{3}$

$\frac{3+2\sqrt{3}}{3}$

But $f(c) =$ ________, so that $3 - 2c + c^2 =$ ____ or $3c^2 - 6c - 1 = 0$. Solving for c, we obtain $c =$ ________.

$f(c)$

$-x^2 + 3x + 4$, $\frac{19}{6}$

$-c^2 + 3c + 4$

19

$\frac{9+\sqrt{111}}{6}$

76 $f(x) = -x^2 + 3x + 4$ on $[0, 5]$.

The mean value $M =$ ______, where

$$f(c) = \frac{1}{5-0}\int_0^5 (-x^2 + 3x + 4)\, dx$$

$$= \frac{1}{5}\int_0^5 (\underline{\qquad\qquad})\, dx = \underline{\quad}.$$

But $f(c) =$ ________, so that $-c^2 + 3c + 4 = \frac{19}{6}$ or $-6c^2 + 18c + 24 =$ ____ or $-6c^2 + 18c + 5 = 0$. Solving for c, we obtain $c =$ ________.

0

$\int_b^a f(x)\, dx$

77 If f is any function and a is a number in the domain of f, we define $\int_a^a f(x)\, dx =$ ____. If $a > b$ and f is Riemann-integrable on $[b, a]$, then $\int_a^b f(x)\, dx = -\left(\underline{\qquad\qquad}\right)$.

In Problems 78 through 82, use the basic properties of integrals and Problem 77 to simplify each integral.

$\int_{-2}^5 f(x)\, dx$

0

$\int_4^2 f(x)\, dx$

78 $\int_5^{-2} f(x)\, dx + \int_{-2}^5 f(x)\, dx$.

Using Problem 77, we have

$$\int_5^{-2} f(x)\, dx + \int_{-2}^5 f(x)\, dx = -\left(\underline{\qquad\qquad}\right) + \int_{-2}^5 f(x)\, dx = \underline{\quad}.$$

79 $\int_4^{-1} f(x)\, dx - \int_4^2 f(x)\, dx + \int_{-1}^2 f(x)\, dx$.

Using Problem 62, we have $\int_4^{-1} f(x)\, dx + \int_{-1}^2 f(x)\, dx =$ ________, so that

$$\int_4^{-1} f(x)\,dx + \int_{-1}^{2} f(x)\,dx - \int_4^2 f(x)\,dx$$

$\int_4^2 f(x)\,dx,\ 0$

$$= \underline{\hspace{3cm}} - \int_4^2 f(x)\,dx = \underline{\hspace{1cm}}.$$

80 $\int_3^5 f(x)\,dx - \int_3^{-2} f(x)\,dx.$

$\int_{-2}^{3} f(x)\,dx$

$$\int_3^5 f(x)\,dx - \int_3^{-2} f(x)\,dx = \int_3^5 f(x)\,dx + \underline{\hspace{3cm}}$$

$\int_{-2}^{5} f(x)\,dx$

$$= \underline{\hspace{3cm}}$$

81 $\int_2^b f(x)\,dx + \int_b^{a+7} f(x)\,dx - \int_2^{a+7} f(x)\,dx.$

Using Problem 62, we have $\int_2^b f(x)\,dx + \int_b^{a+7} f(x)\,dx =$

$\int_2^{a+7} f(x)\,dx$

$\underline{\hspace{3cm}}$, so that

$$\int_2^b f(x)\,dx + \int_b^{a+7} f(x)\,dx - \int_2^{a+7} f(x)\,dx$$

$\int_2^{a+7} f(x)\,dx,\ 0$

$$= \int_2^{a+7} f(x)\,dx - \left(\underline{\hspace{3cm}}\right) = \underline{\hspace{1cm}}.$$

82 $\int_1^2 f(x)\,dx + \int_2^3 f(t)\,dt + \int_3^1 f(u)\,du.$

Recall that $\int_a^b f(x)\,dx = \int_a^b f(t)\,dt = \int_a^b f(u)\,du$. Using Problem 62, we have $\int_1^2 f(x)\,dx + \int_2^3 f(x)\,dx =$

$\int_1^3 f(x)\,dx,\ \int_1^3 f(x)\,dx$

$\underline{\hspace{3cm}}$; also, $\int_3^1 f(x)\,dx = -\underline{\hspace{3cm}}$,

so that $\int_1^2 f(x)\,dx + \int_2^3 f(x)\,dx + \int_3^1 f(x)\,dx =$

$\int_1^3 f(x)\,dx,\ 0$

$$\int_1^3 f(x)\,dx - \left(\underline{\hspace{3cm}}\right) = \underline{\hspace{1cm}}.$$

4 The Fundamental Theorem of Calculus

83 The fundamental theorem of calculus can be stated informally as follows: Let f be a continuous function on an interval I, suppose that a and b are fixed numbers in I, and let x denote a variable number in I. Then:

(First part)

$$\frac{d}{dx}\int_a^x f(t)\,dt = \_\_\_\_\_.$$

$f(x)$

(Second part)

$$\int_a^b f(x)\,dx = \left(\_\_\_\_\_\_\_\_\_\_\right)\Big|_a^b$$

$\int f(x)\,dx$

$$= g(x)\big|_a^b = \_\_\_\_\_\_\_\_\_\_,$$ where g is an antiderivative of f on I.

$g(b) - g(a)$

In Problems 84 through 94, use the first part of the fundamental theorem of calculus to find $\frac{dy}{dx}$.

84 $y = \int_5^x (t^4 + 7)^{10}\,dt.$

$\frac{dy}{dx}$ results from replacing t by x in the integrand, so that

$\frac{dy}{dx} = \_\_\_\_\_\_\_\_\_\_.$

$(x^4 + 7)^{10}$

85 $y = \int_1^x (w^3 - 3w^2 + 13)^{15}\,dw.$

$\frac{dy}{dx}$ results from replacing w by x in the integrand, so that

$\frac{dy}{dx} = \_\_\_\_\_\_\_\_\_\_.$

$(x^3 - 3x^2 + 13)^{15}$

86 $y = \int_{-2}^x \frac{ds}{4 + s^2}.$

Using Problem 83, we have $\frac{dy}{dx} = \_\_\_\_\_\_\_\_.$

$\frac{1}{4 + x^2}$

87 $y = \int_1^x (t^{13} + 7)^{45}\,dt.$

Using Problem 83, we have $\frac{dy}{dx} = \_\_\_\_\_\_\_\_\_\_.$

$(x^{13} + 7)^{45}$

88 $y = \int_x^1 (t^4 + 1)^{100}\,dt.$

$y = \int_x^1 (t^4 + 1)^{100}\,dt = -\left(\_\_\_\_\_\_\_\_\_\_\right)$, so that

$\int_1^x (t^4 + 1)^{100}\,dt$

$\frac{dy}{dx} = \_\_\_\_\_\_\_\_\_\_.$

$-(x^4 + 1)^{100}$

89 $y = \int_x^4 \sqrt[5]{3s^7 + 11}\, ds.$

$y = \int_x^4 \sqrt[5]{3s^7 + 11}\, ds = -\left(\underline{\qquad\qquad}\right)$, so that $\dfrac{dy}{dx} = -\left(\underline{\qquad}\right)$.

$\int_4^x \sqrt[5]{3s^7 + 11}\, ds$

$\sqrt[5]{3x^7 + 11}$

90 $y = \int_x^0 \sqrt[3]{t^4 + 5}\, dt + \int_0^x \sqrt[3]{t^4 + 5}\, dt.$

$y = \int_x^0 \sqrt[3]{t^4 + 5}\, dt + \int_0^x \sqrt[3]{t^4 + 5}\, dt = \underline{\qquad\qquad}$.

But $\int_x^x \sqrt[3]{t^4 + 5}\, dt = \underline{\quad}$. Therefore, $\dfrac{dy}{dx} = \underline{\quad}$.

$\int_x^x \sqrt[3]{t^4 + 5}\, dt$

0, 0

91 $y = \dfrac{d}{dx} \int_1^x \dfrac{dt}{4 + t}.$

$y = \dfrac{d}{dx} \int_1^x \dfrac{dt}{4 + t} = \underline{\qquad}$, so that $\dfrac{dy}{dx} = \dfrac{d}{dx}\left(\underline{\qquad}\right)$

$= -\left(\underline{\qquad}\right)$.

$\dfrac{1}{4 + x}, \dfrac{1}{4 + x}$

$\dfrac{1}{(4 + x)^2}$

92 $y = \dfrac{d}{dx} \int_x^0 \dfrac{ds}{1 + s^2}.$

$y = \dfrac{d}{dx} \int_x^0 \dfrac{ds}{1 + s^2} = -\dfrac{d}{dx} \int_0^x \dfrac{ds}{1 + s^2} = -\left(\underline{\qquad}\right)$, so that

$y = -\left(\underline{\qquad}\right)$. Therefore, $\dfrac{dy}{dx} = \underline{\qquad}$.

$\dfrac{1}{1 + x^2}$

$\dfrac{1}{1 + x^2}, \dfrac{2x}{(1 + x^2)^2}$

93 $y = \int_{-2}^x f(t)\, dt$, where $f(t) = \begin{cases} t + 2 & \text{for } t \leqslant -2 \\ (t + 2)^3 & \text{for } t > -2. \end{cases}$

Since f is continuous, $\dfrac{dy}{dx} = f(x) = \begin{cases} \underline{\qquad} & \text{for } x \leqslant -2 \\ \underline{\qquad} & \text{for } x > -2. \end{cases}$

$x + 2$

$(x + 2)^3$

94 Let $y = \int_a^u f(t)\, dt$, where $u = g(x)$.

If we incorporate the fundamental theorem of calculus into the chain rule, then we obtain $\dfrac{dy}{dx} = \dfrac{dy}{du}\dfrac{du}{dx} = \underline{\qquad}\dfrac{du}{dx}$.

$f(u)$

In Problems 95 through 98, use the first part of the fundamental theorem of calculus together with the chain rule to find $\dfrac{dy}{dx}$.

95 $y = \int_1^{5x+1} (3t^4 + 5)^{11}\, dt.$

Using Problem 94, we have

$5x + 1$

$$\frac{dy}{dx} = [3(5x+1)^4 + 5]^{11} \frac{d}{dx}(\underline{\hspace{2cm}})$$

$5,\ 5[3(5x+1)^4 + 5]^{11}$

$$= [3(5x+1)^4 + 5]^{11}(\underline{\hspace{1cm}}) = \underline{\hspace{4cm}}.$$

96 $y = \displaystyle\int_1^{3x^2} \sqrt{s^5 + 7}\, ds.$

Using Problem 94, we have

$3x^2$

$$\frac{dy}{dx} = (\sqrt{(3x^2)^5 + 7})\frac{d}{dx}(\underline{\hspace{1cm}})$$

$6x,\ 6x\sqrt{243x^{10} + 7}$

$$= (\sqrt{(3x^2)^5 + 7})(\underline{\hspace{1cm}}) = \underline{\hspace{4cm}}.$$

97 $y = \displaystyle\int_{x^2+1}^{3} \sqrt[3]{u - 1}\, du.$

$\displaystyle\int_3^{x^2+1} \sqrt[3]{u-1}\, du$

$$y = \int_{x^2+1}^{3} \sqrt[3]{u-1}\, du = -\left(\underline{\hspace{4cm}}\right), \text{ so that}$$

$x^2 + 1$

$$\frac{dy}{dx} = -(\sqrt[3]{x^2+1-1})\frac{d}{dx}(\underline{\hspace{2cm}})$$

$2x,\ -2x\sqrt[3]{x^2}$

$$= -(\sqrt[3]{x^2+1-1})(\underline{\hspace{1cm}}) = \underline{\hspace{2.5cm}}.$$

98 $y = \displaystyle\int_{3x}^{5x^2} \sqrt[4]{t^2 + 7}\, dt.$

$\displaystyle\int_0^{5x^2} \sqrt[4]{t^2+7}\, dt$

$$y = \int_{3x}^{5x^2} \sqrt[4]{t^2+7}\, dt = \int_{3x}^{0} \sqrt[4]{t^2+7}\, dt + \left(\underline{\hspace{4cm}}\right)$$

$\displaystyle\int_0^{3x} \sqrt[4]{t^2+7}\, dt$

$$= -\left(\underline{\hspace{4cm}}\right) + \int_0^{5x^2} \sqrt[4]{t^2+7}\, dt,$$

so that

$3\sqrt[4]{9x^2+7},\ 10x$

$$\frac{dy}{dx} = -(\underline{\hspace{3cm}}) + [\sqrt[4]{(5x^2)^2 + 7}](\underline{\hspace{1cm}})$$

$-3\sqrt[4]{9x^2+7} + 10x\sqrt[4]{25x^4+7}$

$$= \underline{\hspace{6cm}}.$$

In Problems 99 through 111, use the second part of the fundamental theorem of calculus to evaluate each integral.

99 $\displaystyle\int_1^2 5x^4\, dx.$

$x^5,\ 2^5$

$$\int_1^2 5x^4\, dx = \underline{\hspace{1.5cm}}\Big|_1^2 = \underline{\hspace{1.5cm}} - 1^5$$

31

$$= 32 - 1 = \underline{\hspace{1cm}}.$$

100 $\displaystyle\int_0^1 (2 - x^2)\, dx.$

$2x - \dfrac{x^3}{3}$

$$\int_0^1 (2 - x^2)\, dx = \left[\int (2 - x^2)\, dx\right]\Big|_0^1 = \left(\underline{\hspace{3cm}}\right)\Big|_0^1$$

$$= \left(2 - \frac{1}{3}\right) - \underline{\qquad}$$

0

$$= \underline{\qquad}.$$

$\frac{5}{3}$

101 $\int_1^2 (120 + 144x^2)\,dx.$

$$\int_1^2 (120 + 144x^2)\,dx = \left[\int (120 + 144x^2)\,dx\right]\Bigg|_1^2$$

$$= (\underline{\qquad\qquad})\Big|_1^2$$

$120x + 48x^3$

$$= [(120)(2) + (48)(8)] - \underline{\qquad}$$

168

$$= \underline{\qquad}.$$

456

102 $\int_{5.5}^{231/18} (231 - 18t)\,dt.$

$$\int_{5.5}^{231/18} (231 - 18t)\,dt = \left[\int (231 - 18t)\,dt\right]\Bigg|_{5.5}^{231/18}$$

$$= (\underline{\qquad\qquad})\Big|_{5.5}^{231/18}$$

$231t - 9t^2$

$$= \left[231\left(\frac{231}{18}\right) - 9\left(\frac{231}{18}\right)^2\right] - (\underline{\qquad\qquad\qquad})$$

$231(5.5) - 9(5.5)^2$

$$= \underline{\qquad}.$$

484

103 $\int_1^2 (x^{-2} + x^{-3})\,dx.$

$$\int_1^2 (x^{-2} + x^{-3})\,dx = \left[\int (x^{-2} + x^{-3})\,dx\right]\Bigg|_1^2$$

$$(\underline{\qquad\qquad})\Big|_1^2 = \left[\left(-\frac{1}{2} - \frac{1}{8}\right) + (\underline{\qquad\qquad})\right]$$

$-\frac{1}{x} - \frac{1}{2x^2},\ 1 + \frac{1}{2}$

$$= -\frac{5}{8} + \underline{\qquad} = \underline{\qquad}.$$

$\frac{12}{8},\ \frac{7}{8}$

104 $\int_0^8 x^{4/3}\,dx.$

$$\int_0^8 x^{4/3}\,dx = \left(\int x^{4/3}\,dx\right)\Bigg|_0^8 = (\underline{\qquad})\Big|_0^8$$

$\frac{3}{7}x^{7/3}$

$$= \frac{3}{7}(8^{7/3} - \underline{\qquad})$$

0

$$= \frac{3}{7}(\underline{\qquad}) = \underline{\qquad}.$$

$128,\ \frac{384}{7}$

105 $\int_1^8 (x^{-2/3} + x^{1/3})\,dx.$

$$\int_1^8 (x^{-2/3} + x^{1/3})\,dx = \left[\int (x^{-2/3} + x^{1/3})\,dx\right]\Bigg|_1^8$$

$3x^{1/3} + \frac{3}{4}x^{4/3}$

$$= \left(\underline{\qquad\qquad}\right)\Bigg|_1^8$$

$3 + \frac{3}{4}$

$$= \left[3(8)^{1/3} + \frac{3}{4}(8)^{4/3}\right] - \left(\underline{\qquad}\right)$$

$\frac{15}{4}$, $\frac{57}{4}$

$$= 18 - \underline{\quad} = \underline{\quad}.$$

106 $\displaystyle\int_1^8 \frac{1+x^2}{x^{2/3}}\,dx.$

$$\int_1^8 \frac{1+x^2}{x^{2/3}}\,dx = \int_1^8 \left(\frac{1}{x^{2/3}} + x^{4/3}\right) dx$$

$$= \int_1^8 (x^{-2/3} + x^{4/3})\,dx = \left[\int (x^{-2/3} + x^{4/3})\,dx\right]\Bigg|_1^8$$

$3x^{1/3} + \frac{3}{7}x^{7/3}$

$$= \left(\underline{\qquad\qquad}\right)\Bigg|_1^8$$

$3 + \frac{3}{7}$

$$= \left[3(8)^{1/3} + \frac{3}{7}(8)^{7/3}\right] - \left(\underline{\qquad}\right)$$

$\frac{24}{7}$, $\frac{402}{7}$

$$= \frac{426}{7} - \underline{\quad} = \underline{\quad}.$$

107 $\displaystyle\int_{-3}^3 |x|\,dx.$

$x \geqslant 0$

$$|x| = \begin{cases} x & \text{if } \underline{\qquad} \\ -x & \text{if } x < 0, \end{cases} \quad \text{so that}$$

$\displaystyle\int_0^3 x\,dx$

$$\int_{-3}^3 |x|\,dx = \int_{-3}^0 -x\,dx + \left(\underline{\qquad\qquad}\right)$$

$\frac{x^2}{2}$

$$= \left(-\frac{x^2}{2}\right)\Bigg|_{-3}^0 + \underline{\qquad}\Bigg|_0^3$$

$\frac{9}{2}$

$$= -\left(0 - \underline{\quad}\right) + \left(\frac{9}{2} - 0\right)$$

9

$$= \underline{\quad}.$$

108 $\displaystyle\int_0^2 x|1-x|\,dx.$

$x \leqslant 1$

$$|1-x| = \begin{cases} 1-x & \text{if } \underline{\qquad} \\ x-1 & \text{if } x > 1, \end{cases}$$

so that

$x^2 - x$

$$\int_0^2 x|1-x|\,dx = \int_0^1 (x - x^2)\,dx + \int_1^2 \left(\underline{\qquad\qquad}\right) dx$$

$\frac{x^3}{3} - \frac{x^2}{2}$

$$= \left(\frac{x^2}{2} - \frac{x^3}{3}\right)\Bigg|_0^1 + \underline{\qquad\qquad}\Bigg|_1^2$$

$\frac{8}{3} - 2$

1

$$= \left(\frac{1}{2} - \frac{1}{3}\right) + \left[\left(\_\_\_\_\_\_\right) - \left(\frac{1}{3} - \frac{1}{2}\right)\right]$$

$$= \_\_\_\_.$$

109 $\int_0^2 \frac{dx}{(3+2x)^2}$.

$$\int_0^2 \frac{dx}{(3+2x)^2} = \left[\int \frac{dx}{(3+2x)^2}\right]\Bigg|_0^2.$$

$3 + 2x$

To evaluate $\int \frac{dx}{(3+2x)^2}$, use the substitution $u =$ ________, so

$2\,dx$

that $du =$ ______. Then

$\frac{1}{u^2}, \; -\frac{1}{2u}$

$$\int \frac{dx}{(3+2x)^2} = \frac{1}{2}\int \_\_\_\_\, du = \_\_\_\_\_\_ + C$$

$-\frac{1}{2(3+2x)}$

$$= \_\_\_\_\_\_\_\_\_\_\_\_ + C.$$

Therefore,

$\frac{1}{7}, \; \frac{2}{21}$

$$\int_0^2 \frac{dx}{(3+2x)^2} = \frac{1}{2}\left(\frac{-1}{3+2x}\right)\Bigg|_0^2 = -\frac{1}{2}\left(\_\_\_\_ - \frac{1}{3}\right) = \_\_\_\_.$$

110 $\int_0^1 x(x^2 - 1)^4\, dx$.

$$\int_0^1 x(x^2-1)^4\, dx = \left[\int x(x^2-1)^4\, dx\right]\Bigg|_0^1.$$

$x^2 - 1$

$2x\,dx$

To evaluate $\int x(x^2-1)^4\, dx$, use the substitution $u =$ ________, so that $du =$ _______ and

u^4

$$\int x(x^2-1)^4\, dx = \frac{1}{2}\int (\_\_\_\_)\, du$$

$\frac{u^5}{5}$

$$= \frac{1}{2}\left(\_\_\_\_\right) + C$$

$(x^2 - 1)^5$

$$= \frac{1}{10}(\_\_\_\_\_\_\_\_\_\_) + C.$$

Therefore,

$x^2 - 1$

$$\int_0^1 x(x^2-1)^4\, dx = \frac{1}{10}(\_\_\_\_\_\_\_\_)^5\Big|_0^1$$

$0, \; -1, \; \frac{1}{10}$

$$= \frac{1}{10}[\_\_\_\_ - \_\_\_\_] = \_\_\_\_.$$

111 $\int_0^2 \frac{x}{\sqrt{4x^2+9}}\, dx$.

$$\int_0^2 \frac{x}{\sqrt{4x^2+9}}\, dx = \left(\int \frac{x}{\sqrt{4x^2+9}}\, dx\right)\Bigg|_0^2.$$

$4x^2 + 9$

To evaluate $\int \frac{x\,dx}{\sqrt{4x^2+9}}$, let $u =$ ________, so that

$8x\,dx$

$u^{-1/2},\ u^{1/2}$

$4x^2+9$

$\sqrt{4x^2+9},\ 5,\ \frac{1}{2}$

$f(u)$

$du =$ ______. Then

$$\int \frac{x}{4x^2+9}\,dx = \frac{1}{8}\int (\_\_\_\_\_\_)\,du = \frac{1}{4}(\_\_\_\_\_\_) + C$$

$$= \frac{1}{4}\sqrt{(\_\_\_\_\_\_)} + C.$$

Therefore,

$$\int_0^2 \frac{x}{\sqrt{4x^2+9}}\,dx = \frac{1}{4}(\_\_\_\_\_\_)\Big|_0^2 = \frac{1}{4}(\_\_\_) - \frac{3}{4} = \_\_\_\_ .$$

112 Suppose that f and g' are continuous functions in the interval $[a, b]$. Then

$$\int_a^b f(g(x))\,g'(x)\,dx = \int_{g(a)}^{g(b)} (\_\_\_\_\_\_)\,du.$$

In Problems 113 through 117, use the type of substitution in Problem 112 to evaluate each integral.

$5x+1$

$5x+1,\ 5\,dx$

$1,\ 6$

u

$1,\ \frac{3}{10}(\sqrt[3]{36}-1)$

$x^2+1,\ 2x\,dx,\ 2$

3

$u^{-3},\ \frac{1}{u^2}$

$\frac{1}{4},\ \frac{5}{144}$

113 $\displaystyle\int_0^1 \frac{dx}{\sqrt[3]{5x+1}}.$

$$\int_0^1 \frac{dx}{\sqrt[3]{5x+1}} = \int_0^1 (\_\_\_\_\_\_)^{-1/3}\,dx.$$

Use the substitution $u =$ ______. Then $du =$ ______. $u =$ ____ when $x = 0$, and $u =$ ____ when $x = 1$. Therefore,

$$\int_0^1 \frac{dx}{\sqrt[3]{5x+1}} = \frac{1}{5}\int_1^6 (\_\_\_)^{-1/3}\,du$$

$$= \frac{1}{5}\left(\frac{3}{2}\right)(u^{2/3})\Big|_1^6$$

$$= \frac{3}{10}\left(6^{2/3} - \_\_\_\right) = \_\_\_\_\_\_\_\_ .$$

114 $\displaystyle\int_{-1}^{\sqrt{2}} \frac{x\,dx}{(x^2+1)^3}.$

$$\int_{-1}^{\sqrt{2}} \frac{x\,dx}{(x^2+1)^3} = \int_{-1}^{\sqrt{2}} x(x^2+1)^{-3}\,dx.$$

Let $u =$ ______. Then $du =$ ______. $u =$ ____ when $x = -1$, and $u =$ ____ when $x = \sqrt{2}$, so that

$$\int_{-1}^{\sqrt{2}} x(x^2+1)^{-3}\,dx = \frac{1}{2}\int_2^3 (\_\_\_)\,du = -\frac{1}{4}(\_\_\_)\Big|_2^3$$

$$= -\frac{1}{4}\left(\frac{1}{9} - \_\_\_\right) = \_\_\_\_ .$$

115 $\displaystyle\int_0^1 \sqrt{x}\,(x^{3/2}+1)^{1/2}\,dx.$

$\frac{3}{2}x^{1/2}\,dx,\ 1$

2

$u^{1/2}$

$u,\ 1$

$\frac{4}{9}(2\sqrt{2}-1)$

Let $u = x^{3/2} + 1$. Then $du =$ ______ . $u =$ ____ when $x = 0$ and $u =$ ____ when $x = 1$. Therefore,

$$\int_0^1 x^{1/2}(x^{3/2}+1)^{1/2}\,dx = \frac{2}{3}\int_1^2 \underline{\quad}\,du$$

$$= \left[\frac{4}{9}(\underline{\quad})^{3/2}\right]\Bigg|_1^2 = \frac{4}{9}(2^{3/2} - \underline{\quad})$$

$$= \underline{\qquad\qquad}.$$

$2x+2,\ 2\,dx,\ 2$

16

$2u^{1/2}$

$\sqrt{2}$

116 $\displaystyle\int_0^7 \frac{dx}{\sqrt{2x+2}}$.

Let $u =$ ______. Then $du =$ ______. $u =$ ____ when $x = 0$, and $u =$ ____ when $x = 7$, so that

$$\int_0^7 (2x+2)^{-1/2}\,dx = \frac{1}{2}\int_2^{16} u^{-1/2}\,du = \frac{1}{2}(\underline{\quad})\Big|_2^{16}$$

$$= 4 - \underline{\quad}.$$

$x^4 + 2x,\ (4x^3+2)\,dx$

$20,\ 264$

u

$\frac{1}{u^2},\ \frac{1}{20}$

$\frac{4331}{6{,}969{,}600}$

117 $\displaystyle\int_2^4 \frac{2x^3+1}{(x^4+2x)^3}\,dx$.

Let $u =$ ______. Then $du =$ ______.
$u =$ ____ when $x = 2$ and $u =$ ____ when $x = 4$, so that

$$\int_2^4 \frac{2x^3+1}{(x^4+2x)^3}\,dx = \frac{1}{2}\int_{20}^{264} (\underline{\quad})^{-3}\,du$$

$$= -\frac{1}{4}\left(\underline{\quad}\right)\Big|_{20}^{264} = -\frac{1}{4}\left[\left(\frac{1}{264}\right)^2 - \left(\underline{\quad}\right)^2\right]$$

$$= \underline{\qquad\qquad}.$$

4.1 Proof of the Fundamental Theorem of Calculus

$\int_a^x f(t)\,dt$

differentiable

$f(x),\ (b,c)$

$f(c)$

118 The first part of the fundamental theorem of calculus states: Let f be a continuous function on the closed interval $[b, c]$, and let a be a fixed number in this interval. Let the function g with domain $[b, c]$ be defined by $g(x) =$ ______ for x in $[b, c]$. Then g is ______ on the open interval (b, c) and $g'(x) =$ ______ for all x in ______.

119 In Problem 118, $f(b)$ is the derivative $g'_+(b)$ of g at b from the right and ______ is the derivative $g'_-(c)$ of g at c from the left.

120 The proof of the first part of the fundamental theorem makes

mean value theorem for integrals

use of the general additivity with respect to the interval of integration, and the theorem called the ______________________ ______________.

$g'(x) = f(x)$

$g(b) - g(a)$

121 The second part of the fundamental theorem of calculus states: If f is a continuous function on the closed interval $[a, b]$, and if g is a continuous function on $[a, b]$ such that ______________ holds for all values of x in the open interval (a, b), then

$$\int_a^b f(x)\,dx = \underline{\qquad\qquad}.$$

constant

122 The proof of the second part of the fundamental theorem of calculus makes use of the first part of the fundamental theorem of calculus, and the fact that if a function has a derivative zero on an open interval, then it is a ______________ on that interval.

differentiation
integration

123 The fundamental theorem of calculus establishes a profound connection between ______________ and ______________.

$\int_a^x f(t)\,dt$, continuous

124 Integration is a "smoothing" process. Even if the integrable function f is not continuous, the function g defined by $g(x) =$ ______________ will be ______________.

5 Approximation of Definite Integrals—Simpson and Trapezoidal Rules

5.1 Direct Use of the Analytic Definition

$\int_a^b f(x)\,dx$

125 The definite integral $\int_a^b f(x)\,dx$ can be approximated by selecting a partition $\mathcal{P}$ with a small norm, then augmenting $\mathcal{P}$ in some convenient way to obtain $\mathcal{P}^*$, evaluating the corresponding Riemann sum $\sum_{k=1}^{n} f(c_k)\,\Delta x_k$ and observing that

$$\sum_{k=1}^{n} f(c_k)\,\Delta x_k \approx \underline{\qquad\qquad}.$$

126 Obtain an approximation to $\int_0^1 (1 + x^3)^{1/2}\,dx$ by using a partition with four equal subintervals, and augment the partition by choosing c_1, c_2, c_3, and c_4 to be the left endpoints of the subintervals. The partition $\mathcal{P}$ consists of the subintervals

$\left[\frac{1}{2}, \frac{3}{4}\right], \left[\frac{3}{4}, 1\right]$

$\frac{1}{4}, \frac{1}{4}, \frac{1}{2}, \frac{3}{4}$

$\left[0, \frac{1}{4}\right], \left[\frac{1}{4}, \frac{1}{2}\right],$ ________, and ________, each having length $\Delta x =$ ____. Here, $c_1 = 0, c_2 =$ ____, $c_3 =$ ____, and $c_4 =$ ____.

The graph is

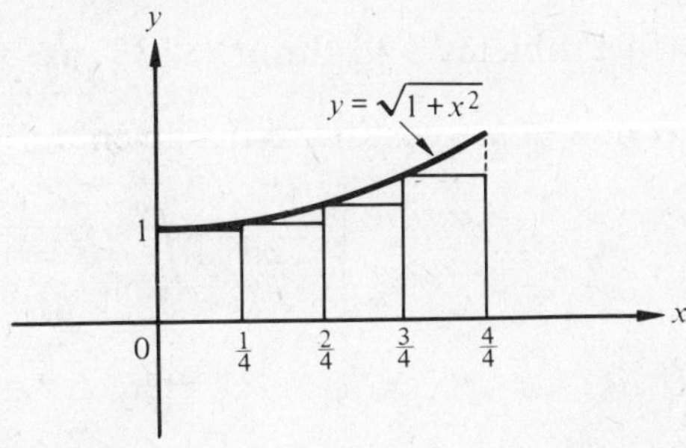

$f(x) = \sqrt{1+x^3}$, so that

1 — $f(0) = \sqrt{1+0^3} =$ ____,

1.008 — $f\left(\frac{1}{4}\right) = \sqrt{1+\frac{1}{64}} = \sqrt{\frac{65}{64}} =$ ____,

1.061 — $f\left(\frac{1}{2}\right) = \sqrt{1+\frac{1}{8}} = \sqrt{\frac{9}{8}} =$ ____,

1.192 — $f\left(\frac{3}{4}\right) = \sqrt{1+\frac{27}{64}} = \sqrt{\frac{91}{64}} =$ ____.

Using the corresponding Riemann sum as an approximation to $\int_0^1 \sqrt{1+x^3}\,dx$, we have

$\frac{\sqrt{65}}{8}, \frac{3\sqrt{2}}{16}$

$$\sum_{k=1}^{4} f(c_k)\,\Delta x_k = (1)\left(\frac{1}{4}\right) + (\quad\quad)^{1/4} + (\quad\quad) + \frac{\sqrt{91}}{8}\left(\frac{1}{4}\right)$$

$\frac{8+\sqrt{65}+6\sqrt{2}+\sqrt{91}}{8}$

$$= \frac{1}{4}(\qquad\qquad\qquad)$$

1.0652

$$\approx \text{\_\_\_\_\_\_}.$$

Therefore,

1.0652

$$\int_0^1 \sqrt{1+x^3}\,dx \approx \text{\_\_\_\_\_\_}.$$

5.2 The Trapezoidal Rule

$\frac{y_0}{2} + y_1 + y_2 + \cdots + y_{n-1} + \frac{y_n}{2}$

$\frac{b-a}{n}$

127 The trapezoidal rule states: If the function f is defined and Riemann-integrable on the closed interval $[a, b]$ and if for each positive integer n, T_n is defined by

$$T_n = (\qquad\qquad\qquad)\Delta x,$$

where $\Delta x =$ ______ and $y_k = f(a + k\,\Delta x)$ for $k = 0, 1, 2, \ldots, n$,

$\int_a^b f(x)\,dx$	then $T_n \approx$ ________, the approximation becoming better
$\int_a^b f(x)\,dx$	as n increases in the sense that $\lim_{n\to+\infty} T_n =$ ________.

In Problems 128 through 133, use the trapezoidal rule $T_n \approx \int_a^b f(x)\,dx$, with the indicated value of n to estimate the given definite integral.

	128 $\int_0^1 \sqrt{1+x^3}\,dx$, $n = 4$.
$\frac{1}{4}$	Here $\Delta x = \frac{1-0}{4} =$ ____, $y = \sqrt{1+x^3}$,
1	$y_0 = \sqrt{1+0} =$ ____,
1.008	$y_1 = \sqrt{1+\left(\frac{1}{4}\right)^3} = \frac{\sqrt{65}}{8} \approx$ ______.
$\sqrt{\frac{9}{8}}$, 1.061	$y_2 = \sqrt{1+\left(\frac{1}{2}\right)^3} =$ ______ $\approx$ ______,
1.192	$y_3 = \sqrt{1+\left(\frac{3}{4}\right)^3} = \frac{\sqrt{91}}{8} =$ ______,
1.414	$y_4 = \sqrt{1+(1)^3} = \sqrt{2} =$ ______.
	Therefore,
	$T_4 = \left(\frac{y_0}{2} + y_1 + y_2 + y_3 + \frac{y_4}{2}\right)\Delta x$
0.500, 0.707	$\approx$ (______ + 1.008 + 1.061 + 1.192 + ______) $\cdot \frac{1}{4}$
4.468, 1.117	= (______) $\cdot \frac{1}{4} \approx$ ______.
1.117	Hence, $\int_0^1 \sqrt{1+x^3}\,dx \approx$ ______. (Compare with Problem 126.)
	129 $\int_1^3 \frac{dx}{x}$, $n = 3$.
$\frac{2}{3}$	Here $\Delta x = \frac{3-1}{3} =$ ____, $y = \frac{1}{x}$.
1	$y_0 = \frac{1}{1} =$ ____,
$\frac{3}{5}$, 0.6	$y_1 = \frac{1}{\frac{5}{3}} =$ ____ = ____,
$\frac{3}{7}$, 0.4286	$y_2 = \frac{1}{\frac{7}{3}} =$ ____ $\approx$ ______,

Answers	
0.3333	$y_3 = \frac{1}{3} \approx$ ______.
Δx	$T_3 = \left(\frac{y_0}{2} + y_1 + y_2 + \frac{y_3}{2}\right)$ (____)
0.6, 0.4286, 0.1667	$\approx (0.5 +$ ____ $+$ ______ $+$ ______$)\left(\frac{2}{3}\right)$
1.6953, 1.130	$= ($______$)\left(\frac{2}{3}\right) \approx$ ______,
1.130	so that $\int_1^3 \frac{dx}{x} \approx$ ______.
	130 $\int_0^3 \sqrt{9 - x^2}\, dx; n = 6.$
$\frac{1}{2}$	Here $\Delta x = \frac{3 - 0}{6} =$ ____, $y = \sqrt{9 - x^2}$.
3	$y_0 = \sqrt{9 - 0} =$ ____,
$\frac{\sqrt{35}}{2}$, 2.9580	$y_1 = \sqrt{9 - \left(\frac{1}{2}\right)^2} =$ ____ $\approx$ ______,
$\sqrt{8}$, 2.8284	$y_2 = \sqrt{9 - 1^2} =$ ____ $\approx$ ______,
$\frac{\sqrt{27}}{2}$, 2.5981	$y_3 = \sqrt{9 - \left(\frac{3}{2}\right)^2} = \sqrt{\frac{27}{4}} =$ ____ $\approx$ ______,
2.2361	$y_4 = \sqrt{9 - (2)^2} = \sqrt{5} \approx$ ______,
$\frac{\sqrt{11}}{2}$, 1.6583	$y_5 = \sqrt{9 - \left(\frac{5}{2}\right)^2} = \sqrt{\frac{11}{4}} =$ ____ $\approx$ ______,
0	$y_6 = \sqrt{9 - 3^2} =$ ____.
Δx	$T_6 = \left(\frac{y_0}{2} + y_1 + y_2 + y_3 + y_4 + y_5 + \frac{y_6}{2}\right)$ (____)
2.5981, 2.2361	$\approx (1.5 + 2.9580 + 2.8284 +$ ______ $+$ ______
1.6583	$+$ ______$)(0.5)$
13.7789, 6.8895	$= ($______$)(0.5) \approx$ ______,
6.8895	so that $\int_0^3 \sqrt{9 - x^2}\, dx \approx$ ______.
	131 $\int_0^1 \sqrt{0.2 + x^3}\, dx; n = 4.$
$\frac{1}{4}$	Here $\Delta x = \frac{1 - 0}{4} =$ ____, $y = \sqrt{0.2 + x^3}$,
$\sqrt{0.2}$, 0.4472	$y_0 = \sqrt{0.2 + 0^3} =$ ____ $\approx$ ______,
0.4643	$y_1 = \sqrt{0.2 + \left(\frac{1}{4}\right)^3} = \sqrt{0.2 + \frac{1}{64}} \approx$ ______,
0.5701	$y_2 = \sqrt{0.2 + \left(\frac{1}{2}\right)^3} = \sqrt{0.2 + 0.125} =$ ______,

0.7886

1.0954

Δx

0.2236, 0.5477

2.5943, 0.6486

0.6486

0.1

-1

-0.6667

-0.4286

-0.25

-0.1111

0

0.0909

0.1667

0.2308

Δx

-0.5000

0.1154

-1.5834, -0.15834

-0.1583

1

$$y_3 = \sqrt{0.2 + \left(\frac{3}{4}\right)^3} \approx \sqrt{0.2 + 0.4219} \approx \_\_\_\_\_\_\_\_,$$

$$y_4 = \sqrt{0.2 + 1^3} = \sqrt{1.2} \approx \_\_\_\_\_\_\_\_.$$

$$T_4 = \left(\frac{y_0}{2} + y_1 + y_2 + y_3 + \frac{y_4}{2}\right)(\_\_\_\_)$$

$$\approx (\_\_\_\_\_\_\_\_ + 0.4643 + 0.5701 + 0.7886 + \_\_\_\_\_\_\_\_) \cdot \frac{1}{4}$$

$$= (\_\_\_\_\_\_\_\_)(0.25) \approx \_\_\_\_\_\_\_\_,$$

so that $\int_0^1 \sqrt{0.2 + x^3}\, dx \approx \_\_\_\_\_\_\_\_$.

132 $\int_{-0.5}^{0.3} \frac{x}{1+x}\, dx;\ n = 8.$

Here $\Delta x = \frac{0.3 + 0.5}{8} = \_\_\_\_$, $y = \frac{x}{1+x}$,

$$y_0 = \frac{-0.5}{1 - 0.5} = \_\_\_\_,$$

$$y_1 = \frac{-0.4}{1 - 0.4} \approx \_\_\_\_\_\_\_\_,$$

$$y_2 = \frac{-0.3}{1 - 0.3} \approx \_\_\_\_\_\_\_\_,$$

$$y_3 = \frac{-0.2}{1 - 0.2} = \_\_\_\_\_\_\_\_,$$

$$y_4 = \frac{-0.1}{1 - 0.1} \approx \_\_\_\_\_\_\_\_,$$

$$y_5 = \frac{0}{1 + 0} = \_\_\_\_,$$

$$y_6 = \frac{0.1}{1 + 0.1} \approx \_\_\_\_\_\_\_\_,$$

$$y_7 = \frac{0.2}{1 + 0.2} \approx \_\_\_\_\_\_\_\_,$$

$$y_8 = \frac{0.3}{1 + 0.3} \approx \_\_\_\_\_\_\_\_.$$

$$T_8 = \left(\frac{y_0}{2} + y_1 + y_2 + y_3 + y_4 + y_5 + y_6 + y_7 + \frac{y_8}{2}\right)(\_\_\_\_)$$

$$\approx (\_\_\_\_\_\_\_\_ - 0.6667 - 0.4286 - 0.25 - 0.1111 + 0$$
$$+ 0.0909 + 0.1667 + \_\_\_\_\_\_\_\_)(0.1)$$

$$= (\_\_\_\_\_\_\_\_)(0.1) \approx \_\_\_\_\_\_\_\_,$$

so that $\int_{-0.5}^{0.3} \frac{x}{1+x}\, dx \approx \_\_\_\_\_\_\_\_$.

133 $\int_0^3 x\sqrt{16 + x^2}\, dx;\ n = 3.$

Here $\Delta x = \frac{3 - 0}{3} = \_\_\_\_$, $y = x\sqrt{16 + x^2}$,

0

4.1231

8.9443

15

Δx

0, 7.5000

20.5674

20.567

$$y_0 = 0\sqrt{16+0} = \underline{\quad},$$
$$y_1 = 1\sqrt{16+1^2} \approx \underline{\qquad},$$
$$y_2 = 2\sqrt{16+2^2} \approx \underline{\qquad},$$
$$y_3 = 3\sqrt{16+3^2} = \underline{\quad}.$$

Then

$$T_3 = \left(\frac{y_0}{2} + y_1 + y_2 + \frac{y_3}{2}\right)(\underline{\quad})$$
$$\approx (\underline{\quad} + 4.1231 + 8.9443 + \underline{\qquad})(1)$$
$$= \underline{\qquad}.$$

Therefore, $\int_0^3 x\sqrt{16+x^2}\,dx \approx \underline{\qquad}$.

$\dfrac{M(b-a)^3}{12n^2}$

134 **Suppose that f'' is defined and continuous on $[a, b]$ and that M is the maximum value of $|f''(x)|$ for x in $[a, b]$. Then, if T_n is the approximation to $\int_a^b f(x)\,dx$ given by the trapezoidal rule, $\left|T_n - \int_a^b f(x)\,dx\right| \leq \underline{\qquad}$.**

$-\dfrac{1}{x^2}$

$\dfrac{2}{x^3}$, 1.5, 3, 6

$\dfrac{2}{(1.5)^3}$

1.5, $\dfrac{1}{6^3}$

0.00463

135 Find a bound on the error E in the approximation of $\int_{1.5}^3 \frac{dx}{x}$ with $n = 6$ when using the trapezoidal rule. $f(x) = \frac{1}{x}$, $f'(x) = \underline{\quad}$, and $f''(x) = \underline{\quad}$. $a = \underline{\quad}$, $b = \underline{\quad}$, and $n = \underline{\quad}$; also, $|f''(x)| \leq |f''(1.5)| = \underline{\qquad}$ for $1.5 \leq x \leq 3$. Using Problem 134, we have

$$|E| = \left|T_n - \int_a^b f(x)\,dx\right| \leq \frac{M(b-a)^3}{12n^2}$$
$$\leq \frac{\left[\frac{2}{(1.5)^3}\right](\underline{\quad})^3}{12(6)^2} = \underline{\quad}$$
$$\approx \underline{\qquad}.$$

5.3 Simpson's Rule

$b - a$

136 **Simpson's parabolic rule states: Let the function f be defined and Riemann-integrable on the closed interval $[a, b]$. For each positive integer n, define**

$$S_{2n} = \frac{\Delta x}{3}(y_0 + 4y_1 + 2y_2 + 4y_3 + 2y_4 + \cdots + 2y_{2n-2} + 4y_{2n-1} + y_{2n}),$$ **where $\Delta x = \frac{\qquad}{2n}$ and $y_k = f(a + k\,\Delta x)$ for**

$\int_a^b f(x)\,dx$

$\int_a^b f(x)\,dx$

$k = 0, 1, 2, \ldots, 2n$. Then $S_{2n} \approx$ ________, the approximation becoming better and better as n increases in the sense that $\lim_{n \to +\infty} S_n =$ ________.

In Problems 137 through 140, use Simpson's parabolic rule, $S_{2n} \approx \int_a^b f(x)\,dx$, to estimate each definite integral using the indicated value of n.

137 $\int_0^1 \frac{dx}{1+x^2}$; $n = 2$.

even

The interval must be subdivided into an ____ number of parts for Simpson's rule. Here $\Delta x = \frac{1-0}{2(2)} =$ ____, $y = \frac{1}{1+x^2}$,

$\frac{1}{4}$

$y_0 = \frac{1}{1+0^2} =$ ____,

1

$y_1 = \frac{1}{1+\left(\frac{1}{4}\right)^2} = \underline{\qquad} \approx$ ________,

$\frac{16}{17}$, 0.9412

$y_2 = \frac{1}{1+\left(\frac{1}{2}\right)^2} = \underline{\qquad} =$ ____,

$\frac{4}{5}$, 0.8

$y_3 = \frac{1}{1+\left(\frac{3}{4}\right)^2} = \underline{\qquad} =$ ____,

$\frac{16}{25}$, 0.64

$y_4 = \frac{1}{1+1^2} = \underline{\qquad} =$ ____.

$\frac{1}{2}$, 0.5

Therefore,

$$S_{2n} = S_4 = \frac{\Delta x}{3}(y_0 + 4y_1 + 2y_2 + 4y_3 + y_4)$$

$$= \left(\underline{\qquad}\right)\left(1 + \frac{64}{17} + \frac{8}{5} + \frac{64}{25} + \frac{1}{2}\right)$$

$\frac{1}{12}$

$$\approx \frac{1}{12}(1 + 3.7645 + \underline{\qquad} + \underline{\qquad} + 0.5)$$

1.6, 2.56

$$= \frac{1}{12}(\underline{\qquad\quad}) \approx \underline{\qquad\quad},$$

9.4247, 0.7854

so that $\int_0^1 \frac{dx}{1+x^2} \approx$ ________.

0.7854

138 $\int_1^2 \frac{dx}{x}$; $n = 2$.

Here $\Delta x = \frac{2-1}{2(2)} = \underline{\qquad}$, $y = \frac{1}{x}$, $y_0 = \frac{1}{1} =$ ____,

$\frac{1}{4}$, 1

$\frac{4}{5}$, 0.8, $\frac{2}{3}$, 0.6667

$\frac{4}{7}$, 0.5714, 0.5

$\frac{\Delta x}{3}$

$\frac{4}{3}$, $\frac{1}{2}$

1.3333, 2.2857

8.3190, 0.6933

0.6933

$\frac{1}{5}$

1

1.0198

1.0770

1.1662

1.2806

1.4142

1.5621

1.7205

1.8868

$2y_4 + 4y_5 + 2y_6 + 4y_7 + y_8$

4.0792, 2.5612

3.1242, 6.8820

32.009, 2.1339

2.1339

$\frac{1}{2}$

$y_1 = \frac{1}{\frac{5}{4}} = \underline{\quad} = \underline{\quad}$, $y_2 = \frac{1}{\frac{3}{2}} = \underline{\quad} \approx \underline{\qquad}$,

$y_3 = \frac{1}{\frac{7}{4}} = \underline{\quad} \approx \underline{\qquad}$, and $y_4 = \frac{1}{2} = \underline{\quad}$. Therefore,

$$S_{2n} = S_4 = \left(\underline{\quad}\right)(y_0 + 4y_1 + 2y_2 + 4y_3 + y_4)$$

$$= \frac{1}{12}\left(1 + \frac{16}{5} + \underline{\quad} + \frac{16}{7} + \underline{\quad}\right)$$

$$\approx \frac{1}{12}(1 + 3.2 + \underline{\qquad} + \underline{\qquad} + 0.5)$$

$$= \frac{1}{12}(\underline{\qquad}) \approx \underline{\qquad},$$

so that $\int_1^2 \frac{dx}{x} \approx \underline{\qquad}$.

139 $\int_0^{1.6} \sqrt{1 + x^2}\, dx;\ n = 4.$

Here $\Delta x = \frac{1.6 - 0}{2(4)} = \underline{\quad}$, $y = \sqrt{1 + x^2}$,

$y_0 = \sqrt{1 + 0^2} = \underline{\quad}$,

$y_1 = \sqrt{1 + (0.2)^2} \approx \underline{\qquad}$,

$y_2 = \sqrt{1 + (0.4)^2} \approx \underline{\qquad}$,

$y_3 = \sqrt{1 + (0.6)^2} \approx \underline{\qquad}$,

$y_4 = \sqrt{1 + (0.8)^2} \approx \underline{\qquad}$,

$y_5 = \sqrt{1 + (1)^2} \approx \underline{\qquad}$,

$y_6 = \sqrt{1 + (1.2)^2} \approx \underline{\qquad}$,

$y_7 = \sqrt{1 + (1.4)^2} \approx \underline{\qquad}$,

$y_8 = \sqrt{1 + (1.6)^2} \approx \underline{\qquad}$.

$$S_{2n} = S_8 = \frac{\Delta x}{3}(y_0 + 4y_1 + 2y_2 + 4y_3 + \underline{\qquad\qquad})$$

$$\approx \frac{1}{15}(1 + \underline{\qquad} + 2.154 + 4.6648 + \underline{\qquad}$$

$$+ 5.6568 + \underline{\qquad} + \underline{\qquad} + 1.8868)$$

$$= \frac{1}{15}(\underline{\qquad}) = \underline{\qquad},$$

so that $\int_0^{1.6} \sqrt{1 + x^2}\, dx \approx \underline{\qquad}$.

140 $\int_0^2 \frac{dx}{\sqrt{1 + x^2}};\ n = 2.$

Here $\Delta x = \frac{2 - 0}{2(2)} = \underline{\quad}$, $y = \frac{1}{\sqrt{1 + x^2}}$,

$$y_0 = \frac{1}{\sqrt{1+0^2}} = \_\_\_\_,$$

$$y_1 = \frac{1}{\sqrt{1+\left(\frac{1}{2}\right)^2}} = \_\_\_\_ \approx \_\_\_\_,$$

$$y_2 = \frac{1}{\sqrt{1+1^2}} = \_\_\_\_ \approx \_\_\_\_,$$

$$y_3 = \frac{1}{\sqrt{1+\left(\frac{3}{2}\right)^2}} = \_\_\_\_ \approx \_\_\_\_,$$

$$y_4 = \frac{1}{\sqrt{1+2^2}} = \_\_\_\_ \approx \_\_\_\_.$$

$$S_{2n} = S_4 = \frac{1}{6}(y_0 + 4y_1 + 2y_2 + 4y_3 + y_4)$$

$$= \frac{1}{6}\left(1 + \_\_\_\_ + \_\_\_\_ + \_\_\_\_ + \frac{1}{\sqrt{5}}\right)$$

$$\approx \frac{1}{6}(1 + 3.5776 + \_\_\_\_ + 2.2188 + \_\_\_\_)$$

$$= \frac{1}{6}(\_\_\_\_) = \_\_\_\_,$$

so that $\int_0^2 \frac{dx}{1+x^2} \approx$ ____.

Answers: 1; $\frac{2}{\sqrt{5}}$, 0.8944; $\frac{1}{\sqrt{2}}$, 0.7071; $\frac{2}{\sqrt{13}}$, 0.5547; $\frac{1}{\sqrt{5}}$, 0.4472; $\frac{8}{\sqrt{5}}$, $\sqrt{2}$, $\frac{8}{\sqrt{13}}$; 1.4142, 0.4472; 8.6578, 1.443; 1.443

141 Suppose that $f^{(4)}(x)$ is defined and continuous on $[a, b]$ and that N is the maximum value of $f^{(4)}(x)$ for x in $[a, b]$. Then, if S_{2n} is the approximation to $\int_a^b f(x)\,dx$ given by Simpson's rule,

$$\left|S_{2n} - \int_a^b f(x)\,dx\right| \leqslant \_\_\_\_.$$

Answer: $N\frac{(b-a)^5}{2880n^4}$

142 Find a bound on the error E in the approximation of $\int_1^2 \frac{dx}{x}$ given by Simpson's rule for $n = 2$.

$f(x) = \frac{1}{x}$, $f'(x) =$ ____, $f''(x) =$ ____, $f'''(x) =$ ____, and

$f^{(4)}(x) =$ ____, so that $|f^{(4)}(x)| = \left|\frac{24}{x^5}\right| \leqslant$ ____ for $1 \leqslant x \leqslant 2$.

Take $N = 24$ and apply Problem 142 with $a = 1$, $b = 2$, and $2n =$ ____. Thus,

$$|E| \leqslant \frac{24(2-1)^5}{2880(2^4)} = \frac{24}{2880(\_\_\_\_)}$$

$$\approx \_\_\_\_.$$

Answers: $-\frac{1}{x^2}$, $\frac{2}{x^3}$, $-\frac{6}{x^4}$; $\frac{24}{x^5}$, 24; 4; 16; 0.00052

6 Areas of Planar Regions

signed

143 Let f be a piecewise continuous and bounded function on the closed interval $[a, b]$. Then the definite (Riemann) integral is numerically equal to the ________ area under the graph of f between $x = a$ and $x = b$.

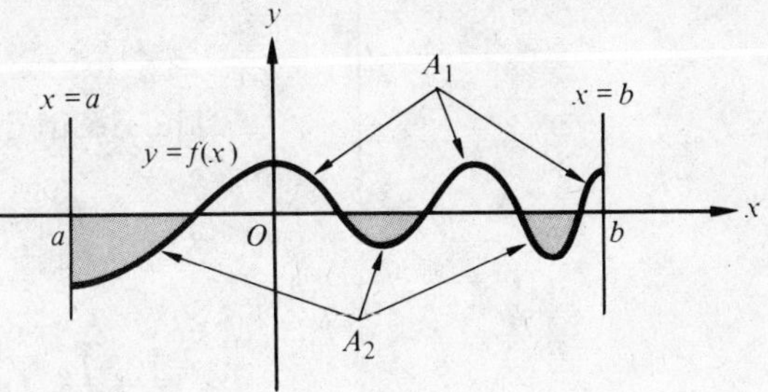

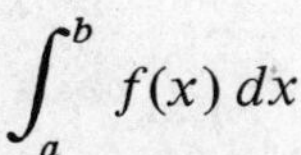

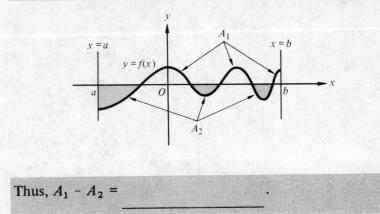

$\int_a^b f(x)\,dx$

Thus, $A_1 - A_2 =$ ________.

In Problems 144 through 148, find the area under the graph of the function f in each interval.

144 $f(x) = x^2 - 2x$ between $x = 0$ and $x = 3$.

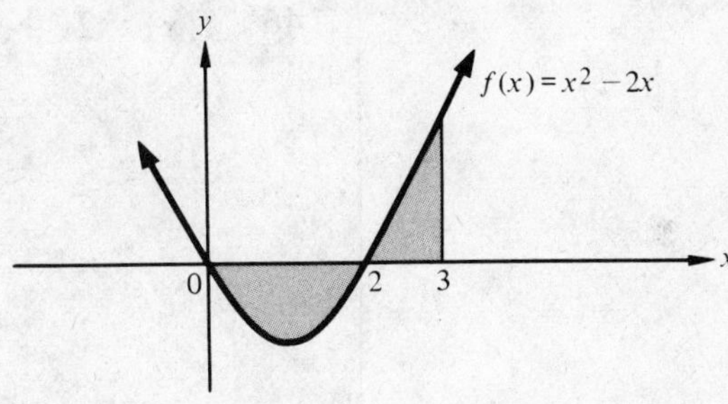

The desired area is given by

$\int_2^3 f(x)\,dx$

$$A = -\int_0^2 f(x)\,dx + \underline{\qquad\qquad}$$

$\int_2^3 (x^2 - 2x)\,dx$

$$= -\int_0^2 (x^2 - 2x)\,dx + \left(\underline{\qquad\qquad}\right)$$

$x^2 - \dfrac{x^3}{3}, \ \dfrac{x^3}{3} - x^2$

$$= \left(\underline{\qquad\qquad}\right)\Big|_0^2 + \left(\underline{\qquad\qquad}\right)\Big|_2^3$$

$0, \ -\dfrac{4}{3}$

$$= \left[\left(4 - \frac{8}{3}\right) - \underline{\quad}\right] + \left[\left(\frac{27}{3} - 9\right) - \underline{\quad}\right]$$

$\dfrac{8}{3}$

$$= \underline{\quad} \text{ square units.}$$

145 $f(x) = x^4 + x^3 - 2$ between $x = 0$ and $x = 2$.

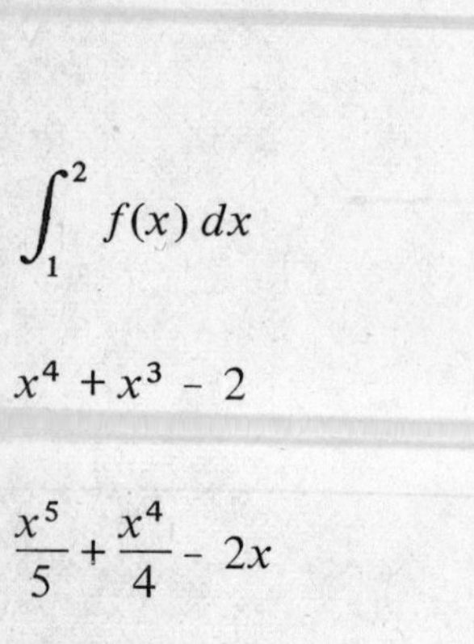

The desired area is given by

$$A = -\int_0^1 f(x)\,dx + \underline{\qquad\qquad}$$

$$= -\int_0^1 (x^4 + x^3 - 2)\,dx + \int_1^2 (\underline{\qquad\qquad})\,dx$$

$$= -\left(\frac{x^5}{5} + \frac{x^4}{4} - 2x\right)\Bigg|_0^1 + \left(\underline{\qquad\qquad}\right)\Bigg|_1^2$$

$$= -\left(\underline{\quad} - 0\right) + \left[\underline{\quad} - \left(-\frac{31}{20}\right)\right]$$

$$= \underline{\quad} \text{ square units.}$$

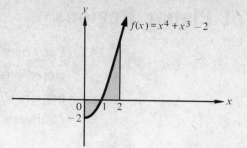

Answers: $\int_1^2 f(x)\,dx$; $x^4 + x^3 - 2$; $\frac{x^5}{5} + \frac{x^4}{4} - 2x$; $-\frac{31}{20}, \frac{32}{5}$; $\frac{19}{2}$

146 $f(x) = 2x - x^2$ between $x = 1$ and $x = 3$.

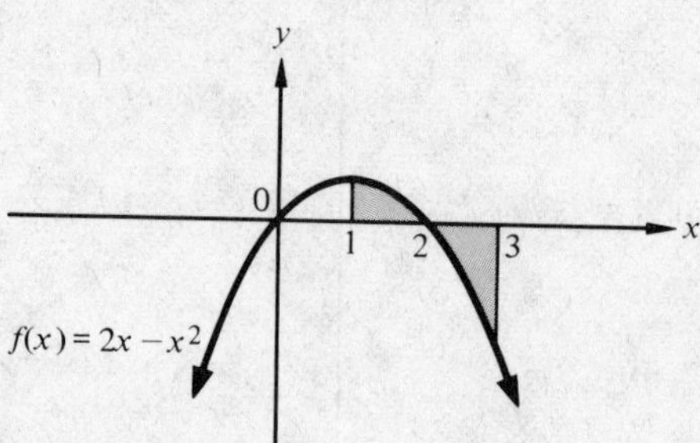

The desired area is given by

$$A = \int_1^2 (2x - x^2)\,dx - \int_2^3 (\underline{\qquad\qquad})\,dx$$

$$= \left(x^2 - \frac{x^3}{3}\right)\Bigg|_1^2 - \left(\underline{\qquad\qquad}\right)\Bigg|_2^3$$

$$= \left(\frac{4}{3} - \underline{\quad}\right) - \left(\underline{\quad} - \underline{\quad}\right)$$

$$= \underline{\quad} \text{ square units.}$$

Answers: $2x - x^2$; $x^2 - \frac{x^3}{3}$; $\frac{2}{3}, 0, \frac{4}{3}$; 2

147 $f(x) = x^2 - x - 2$ between $x = 0$ and $x = 4$.

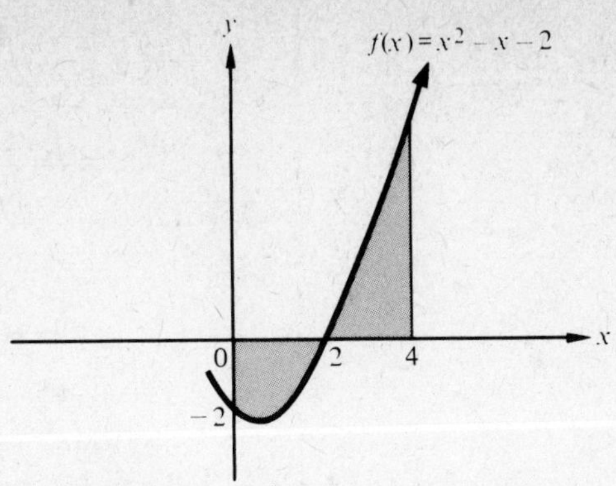

The desired area is given by

$x^2 - x - 2$

$$A = -\int_0^2 (x^2 - x - 2)\,dx + \int_2^4 (\underline{\qquad\qquad})\,dx$$

$\frac{x^3}{3} - \frac{x^2}{2} - 2x$

$$= -\left(\frac{x^3}{3} - \frac{x^2}{2} - 2x\right)\Big|_0^2 + \left(\underline{\qquad\qquad}\right)\Big|_2^4$$

$0, \frac{16}{3}$

$$= -\left(-\frac{10}{3} - \underline{\quad}\right) + \left(\underline{\quad} + \frac{10}{3}\right)$$

$\frac{16}{3}$

$$= \frac{10}{3} + 0 + \underline{\quad} + \frac{10}{3}$$

12

$$= \underline{\quad} \text{ square units.}$$

148 $f(x) = 4 + 3x - x^2$ and the x axis.

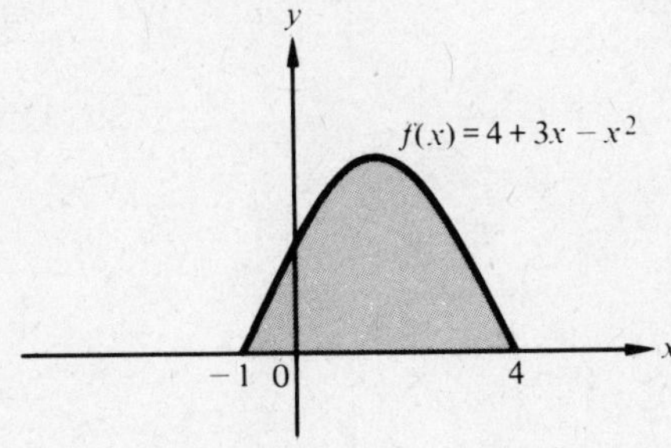

−1, 4

The x intercepts of the graph are $x = \underline{\quad}$ and $x = \underline{\quad}$. The desired area is given by

$4 + 3x - x^2$

$$A = \int_{-1}^4 f(x)\,dx = \int_{-1}^4 (\underline{\qquad\qquad})\,dx$$

$4x + \frac{3x^2}{2} - \frac{x^3}{3}, \frac{56}{3}$

$$= \left(\underline{\qquad\qquad}\right)\Big|_{-1}^4 = \underline{\quad} - \left(-\frac{13}{6}\right)$$

$\frac{125}{6}$

$$= \underline{\quad} \text{ square units.}$$

6.1 Areas by Slicing

149 Suppose that R is a region whose boundary consists of a finite number of straight line segments or smooth areas that meet in

bounded
distances
admissible

a finite number of "corners" or "vertices"; and that R is ________________ in the sense that there is an upper bound to the ________________ between the points of R. Then R is called an ________________ region.

$\int_a^b l(s)\,ds$

150 Let R be an admissible region and let a convenient coordinate axis be chosen as a "reference axis." At each point along this reference axis, construct a perpendicular line and suppose that the region R is entirely contained between the two perpendiculars at the points with coordinates a and b, respectively. Let the perpendicular at the point with coordinate s intersect the region R in one or more line segments of total length $l(s)$. Then the area of the region R is given by $A(R) =$ ________.

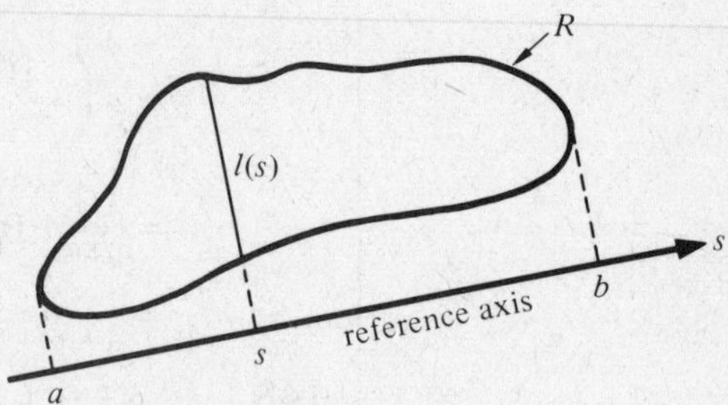

151 Use Problem 150 to find the area of the triangle whose vertices are (0, 0), (4, 2), and (6, 0).

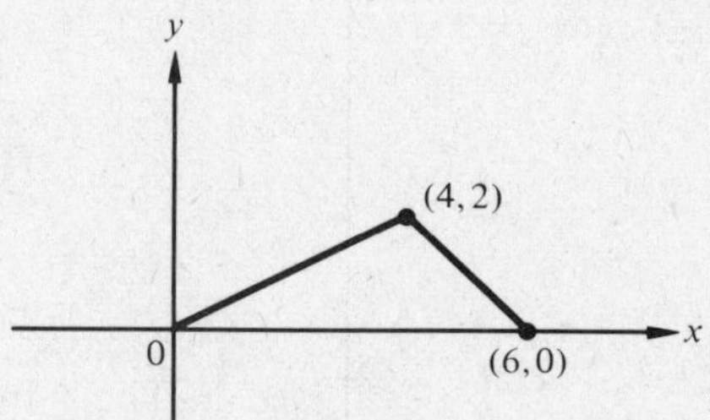

$\frac{1}{2}x,\ \frac{1}{2}x$

$-x+6$

$-x+6$

$\int_4^6 l(x)\,dx$

$-x+6$

$-\frac{x^2}{2}+6x$

0, 18, 6

The straight line through (0, 0) and (4, 2) has the equation $y =$ ____; hence, $l(x) =$ ____ for $0 \leqslant x \leqslant 4$. The straight line through (4, 2) and (6, 0) has the equation $y =$ __________; hence, $l(x) =$ __________ for $4 \leqslant x \leqslant 6$. The desired area is

$$A = \int_0^6 l(x)\,dx = \int_0^4 l(x)\,dx + \left(\rule{4cm}{0.4pt}\right)$$

$$= \int_0^4 \frac{1}{2}x\,dx + \int_4^6 (\rule{2cm}{0.4pt})\,dx$$

$$= \frac{x^2}{4}\Big|_0^4 + \left(\rule{3cm}{0.4pt}\right)\Big|_4^6$$

$$= (4 - \rule{1cm}{0.4pt}) + (\rule{1cm}{0.4pt} - 16) = \rule{1cm}{0.4pt} \text{ square units.}$$

6.2 Area Between Two Graphs

152 Let f and g be continuous functions on the closed interval $[a, b]$. Then the area of the region R between the graph of f and the graph of g to the right of $x = a$ and to the left of $x = b$ is given by

$$A(R) = \int_a^b (\_\_\_\_\_\_\_\_\_\_\_\_\_\_\_)\,dx.$$

$|f(x) - g(x)|$

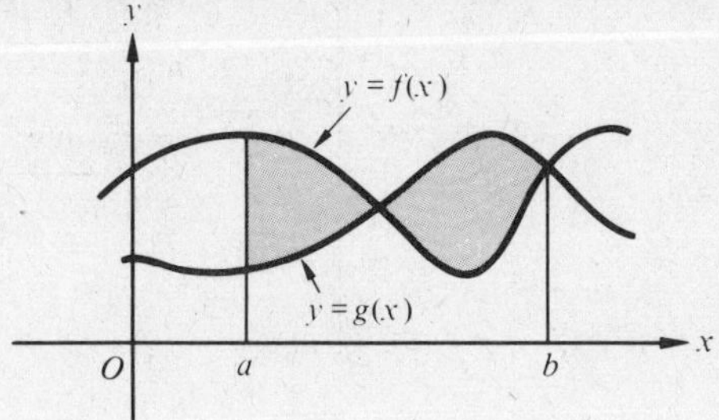

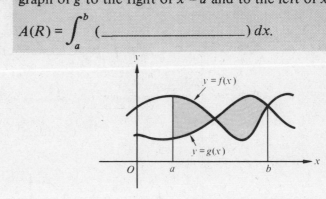

In Problems 153 through 161, find the area of the region bounded by the graphs of the two given equations.

153 $f(x) = x$ and $g(x) = x^3$.

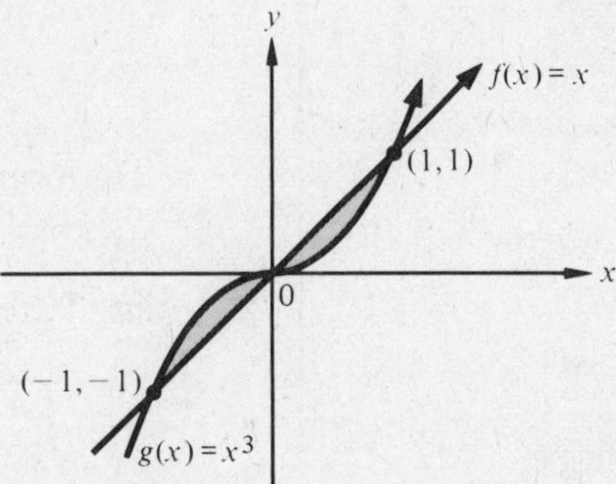

The points of intersection of the two graphs are $(-1, -1)$ and ________. The desired area is given by

$$A = \int_{-1}^{0} [g(x) - f(x)]\,dx + \int_0^1 (\_\_\_\_\_\_\_\_\_\_\_\_\_\_\_)\,dx$$

$$= \int_{-1}^{0} (x^3 - x)\,dx + \int_0^1 (\_\_\_\_\_\_\_\_\_\_)\,dx$$

$$= \left(\frac{x^4}{4} - \frac{x^2}{2}\right)\Big|_{-1}^{0} + \left(\_\_\_\_\_\_\_\_\_\_\_\_\right)\Big|_0^1$$

$$= \_\_\_\_ + \_\_\_\_ = \_\_\_\_ \text{ square unit.}$$

$(1, 1)$

$f(x) - g(x)$

$x - x^3$

$\frac{x^2}{2} - \frac{x^4}{4}$

$\frac{1}{4}, \frac{1}{4}, \frac{1}{2}$

154 $f(x) = 6x - x^2$ and $g(x) = x$.

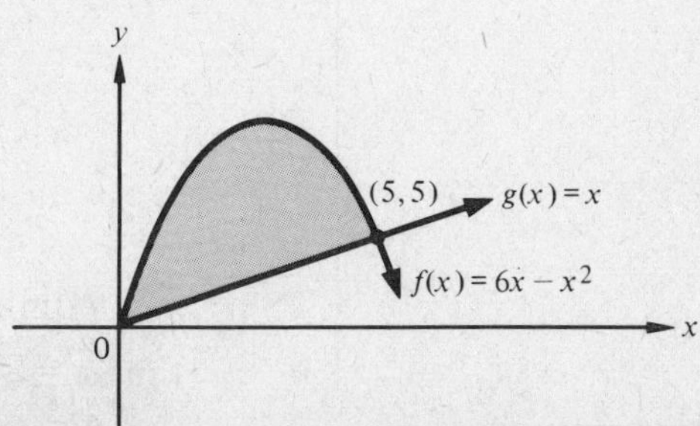

x, 0, 5

The points of intersection are obtained by solving $f(x) = g(x)$. For $6x - x^2 =$ ____, $x =$ ____ or $x =$ ____. The desired area is given by

$6x - x^2 - x$

$$A = \int_0^5 [f(x) - g(x)]\, dx = \int_0^5 (\_\_\_\_\_\_\_\_)\, dx$$

$5x - x^2$

$$= \int_0^5 (\_\_\_\_\_\_)\, dx = \left(\frac{5x^2}{2} - \frac{x^3}{3}\right)\Bigg|_0^5$$

$\frac{125}{6}$, $\frac{125}{6}$

$$= \_\_\_\_ - 0 = \_\_\_\_ \text{ square units.}$$

155 $f(x) = 6x - x^2$ and $g(x) = x^2 - 2x$.

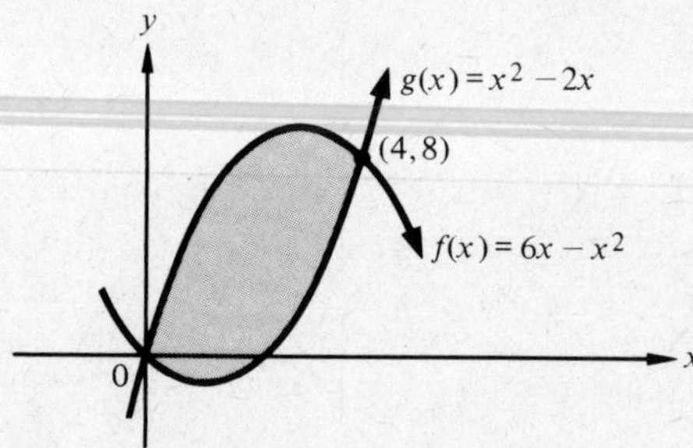

$x^2 - 2x$, 0, 4

The points of intersection are obtained by solving $f(x) = g(x)$. For $6x - x^2 =$ ______ we have $x =$ ____ or $x =$ ____. The desired area is

$x^2 - 2x$

$$A = \int_0^4 [f(x) - g(x)]\, dx = \int_0^4 [(6x - x^2) - (\_\_\_\_\_\_)]\, dx$$

$8x - 2x^2$

$$= \int_0^4 (\_\_\_\_\_\_)\, dx = \left(4x^2 - \frac{2}{3}x^3\right)\Bigg|_0^4$$

$\frac{64}{3}$, $\frac{64}{3}$

$$= \_\_\_\_ - 0 = \_\_\_\_ \text{ square units.}$$

156 $y = x^3$ and $y = 4x^2$.

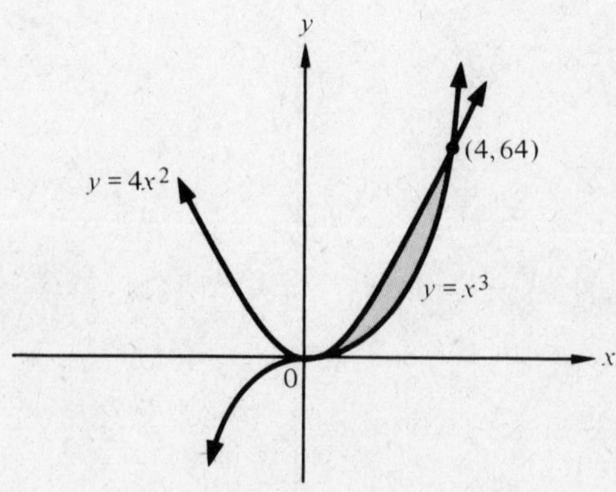

0, 4

The points of intersection are obtained by solving $x^3 = 4x^2$, so that $x =$ ____ or $x =$ ____. The desired area is given by

$\frac{4}{3}x^3 - \frac{x^4}{4}$

$$A = \int_0^4 (4x^2 - x^3)\,dx = \left(\underline{\qquad\qquad}\right)\Big|_0^4$$

$0, \ \frac{64}{3}$

$$= \left(\frac{4^4}{3} - \frac{4^4}{4}\right) - (\underline{\quad}) = \underline{\quad} \text{ square units.}$$

157 $y = -x^2 - 1$ and $y = -2x - 1$.

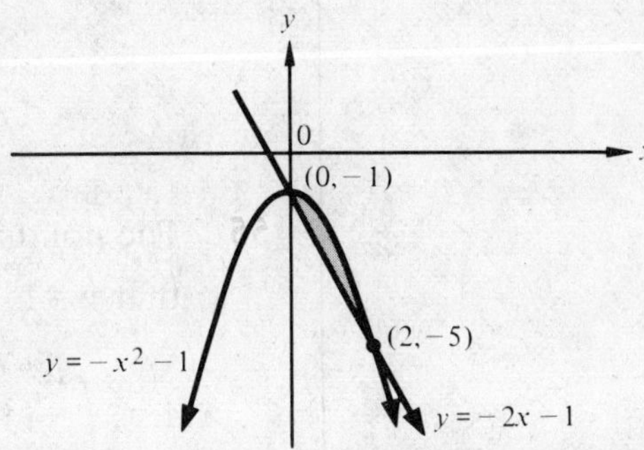

$0, \ 2$

The points of intersection are obtained by solving $-x^2 - 1 = -2x - 1$, so that $x = \underline{\quad}$ or $x = \underline{\quad}$. The desired area is

$-2x - 1, \ 2x - x^2$

$$A = \int_0^2 [(-x^2 - 1) - (\underline{\qquad\qquad})]\,dx = \int_0^2 (\underline{\qquad\qquad})\,dx$$

$\frac{4}{3}$

$$= \left(x^2 - \frac{x^3}{3}\right)\Big|_0^2 = \underline{\quad} - 0$$

$\frac{4}{3}$

$$= \underline{\quad} \text{ square units.}$$

158 $y = 4 - 4x^2$ and $y = 1 - x^2$.

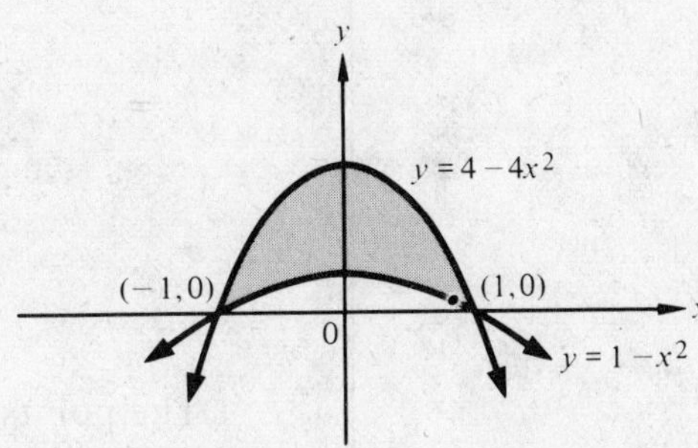

$-1, \ 1$

The points of intersection are obtained by solving $4 - 4x^2 = 1 - x^2$, so that $x = \underline{\quad}$ or $x = \underline{\quad}$. The desired area is given by

$1 - x^2, \ 3 - 3x^2$

$$A = \int_{-1}^{1} [(4 - 4x^2) - (\underline{\qquad\qquad})]\,dx = \int_{-1}^{1} (\underline{\qquad\qquad})\,dx$$

$-2, \ 4$

$$= (3x - x^3)\Big|_{-1}^{1} = 2 - \underline{\quad} = \underline{\quad} \text{ square units.}$$

159 $y^2 = x + 2$ and $y = x$.

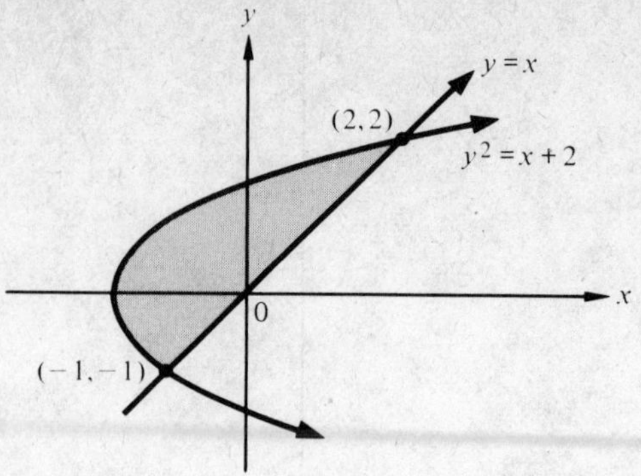

$-1,\ 2$

The points of intersection are obtained by solving $y^2 - 2 = y$, so that $y =$ ____ or $y =$ ____. The desired area is given by

$y^2 - 2,\ y - y^2 + 2$

$$A = \int_{-1}^{2} [y - (\_\_\_\_\_\_\_\_)]\, dy = \int_{-1}^{2} (\_\_\_\_\_\_\_\_\_\_\_\_)\, dy$$

$$= \left(\frac{y^2}{2} - \frac{y^3}{3} - 2y\right)\Bigg|_{-1}^{2}$$

$-\frac{7}{6},\ \frac{27}{6},\ \frac{9}{2}$

$$= \frac{10}{3} - \_\_\_\_ = \_\_\_\_ = \_\_\_\_ \text{ square units.}$$

160 $y = 2x - 4$ and $y^2 = 4x$.

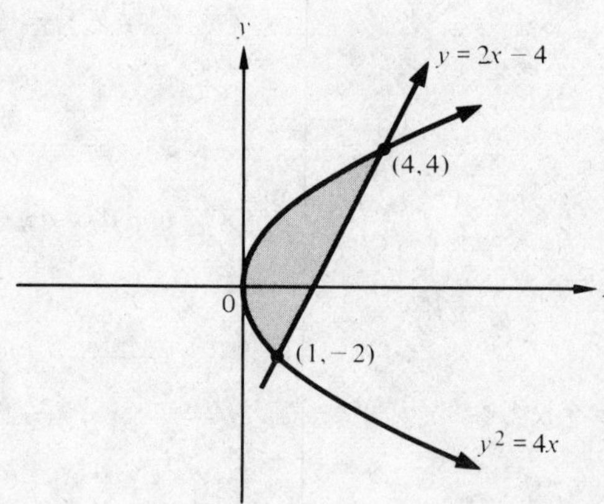

$(1, -2),\ (4, 4)$

The points of intersection are ________ and ________. The desired area is given by

$\frac{y^2}{4},\ 2 + \frac{1}{2}y - \frac{1}{4}y^2$

$$A = \int_{-2}^{4} \left[\left(2 + \frac{1}{2}y\right) - \_\_\_\_\right] dy = \int_{-2}^{4} \left(\_\_\_\_\_\_\_\_\_\_\_\_\right) dy$$

$2y + \frac{y^2}{4} - \frac{y^3}{12},\ \frac{20}{3},\ -\frac{7}{3},\ 9$

$$= \left(\_\_\_\_\_\_\_\_\_\_\_\_\right)\Bigg|_{-2}^{4} = \_\_\_\_ - \_\_\_\_ = \_\_\_\_ \text{ square units.}$$

161 $y^2 = -4(x - 1)$ and $y^2 = -2(x - 2)$.

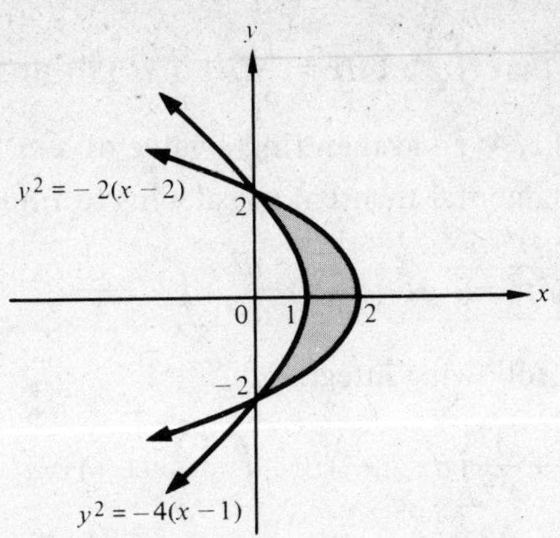

$(0, -2)$, $(0, 2)$

$\frac{1}{4}(4 - y^2)$

$\frac{1}{4}(4 - y^2)$

$\frac{16}{3}$, $\frac{8}{3}$

The points of intersection are ________ and ________. The desired area is given by

$$A = \int_{-2}^{2} \left[\frac{1}{2}(4 - y^2) - (\_\_\_\_\_\_\_\_)\right] dy$$

$$= \int_{-2}^{2} (\_\_\_\_\_\_\_\_)\, dy = \frac{1}{4}\left(4y - \frac{y^3}{3}\right)\Big|_{-2}^{2}$$

$$= \frac{1}{4}\left[(\_\_\_\_) - \left(-\frac{16}{3}\right)\right] = \_\_\_\_ \text{ square units.}$$

Chapter Test

1 Evaluate the following sums.

(a) $\sum_{k=1}^{5} (k + 10)$ (b) $\sum_{i=10}^{13} (2i + 1)$

2 Let $f(x) = 2x - 1$ be defined on the interval $[1, 4]$.

(a) Find the Riemann sum if $\mathcal{P}^*$ consists of $\left[1, \frac{7}{4}\right]$, $\left[\frac{7}{4}, \frac{10}{4}\right]$, $\left[\frac{10}{4}, \frac{13}{4}\right]$, and $\left[\frac{13}{4}, 4\right]$ with $c_1 = \frac{7}{4}$, $c_2 = \frac{10}{4}$, $c_3 = \frac{13}{4}$, and $c_4 = 4$.

(b) Use a partition consisting of subintervals of equal length, circumscribed rectangles, and the fact that $\int_1^4 f(x)\, dx = \lim_{n \to +\infty} \sum_{k=1}^{n} f(c_k)\, \Delta x_k$ to evaluate $\int_1^4 (2x - 1)\, dx$.

3 Given that $\int_{-7}^{13} f(x)\, dx = 2.77$, $\int_{13}^{17} f(x)\, dx = 2.23$, and $\int_{-7}^{13} g(x) = -1.32$, find

(a) $\int_{-7}^{13} [f(x) + g(x)]\, dx$ (b) $\int_{-7}^{13} 4f(x)\, dx$

(c) $\int_{-7}^{13} [2f(x) + 3g(x)]\, dx$ (d) $\int_{-7}^{17} f(x)\, dx$

4 Use the fact that $\int_a^b x^2\,dx = \frac{1}{3}(b^3 - a^3)$ to find the mean value of the function f defined by $f(x) = x^2$ on the interval $[1, 4]$, and then find a value of c in this interval such that $f(c)$ gives this mean value.

5 Use the fundamental theorem of calculus to find the derivative of each expression.

(a) $\frac{d}{dx}\int_1^x \sqrt{t^3 + 7}\,dt$ (b) $\frac{d^2}{dt^2}\int_x^0 \frac{ds}{1 + s^2}$ (c) $\frac{d}{dx}\int_0^{-2x^2} \sqrt{1 + w^2}\,dw$

6 Evaluate the following integrals.

(a) $\int_2^5 \left(x^2 + \frac{1}{x^2}\right) dx$ (b) $\int_{-3}^{-2} (x + 1)^2 x\,dx$

(c) $\int_1^2 \frac{t^2\,dt}{(t^3 + 2)^2}$ (d) $\int_0^3 |x - 1|\,(x - 1)\,dx$

7 Use the trapezoidal rule to approximate $\int_0^1 \sqrt{1 + x^3}\,dx$ with $n = 4$.

8 Use Simpson's rule to approximate $\int_1^2 \frac{dx}{x}$ with $n = 2$.

9 Find the area of the region between the x axis and the graph of $f(x) = \frac{1}{4}(x^3 + x - 2)$ in the interval $[0, 2]$.

10 Find the area bounded by the curves $y = 9$ and $y = x^2$.

Answers

1 (a) 65 (b) 96

2 (a) $\frac{57}{4}$ (b) 12

3 (a) 1.45 (b) 11.08 (c) 1.58 (d) 5

4 $7, \sqrt{7}$

5 (a) $\sqrt{x^3 + 7}$ (b) $\frac{2x}{(1 + x^2)^2}$ (c) $-4x\sqrt{1 + 4x^4}$

6 (a) 39.3 (b) $-\frac{73}{12}$ (c) $\frac{7}{90}$ (d) $\frac{7}{3}$

7 1.117

8 0.6933

9 $\frac{1}{2}$ square unit

10 36 square units

Chapter 7 APPLICATIONS OF THE DEFINITE INTEGRAL

In this chapter we apply the definite integral to problems arising in engineering, physics, economics, and the life sciences. After working the problems in this chapter, the student should be able to:

1 Find the volumes of solids of revolution by the circular disk method.
2 Find the volumes of solids of revolution by the cylindrical shell method.
3 Find the volumes of solids by the slicing method.
4 Find the arc lengths of curves and areas of surfaces of revolution.
5 Apply the definite integral to problems in economics and the life sciences.
6 Calculate work and energy.
7 Find force caused by fluid pressure.

1 Volumes of Solids of Revolution

1 Suppose that a three-dimensional region S has the following two properties:

(i) The boundary of S consists of a finite number of smooth surfaces which can intersect in a finite number of edges. These edges, in turn, can intersect in a finite number of vertices.

(ii) S is bounded in the sense that there is an upper bound to the distances between the points of S.

Then S is called a ______.

solid

2 Two solids S_1 and S_2 are said to be congruent if it is possible to establish a ______ between the points in S_1 and the points in S_2 in such a way that the distance between any two points in S_1 is the same as the distance between the two corresponding points in S_2.

one-to-one correspondence

3 If two solids share a common point other than a common boundary point, we say that they ______.

overlap

4 A solid consisting of all points lying between an admissible planar region B_1 and a second admissible planar region B_2 obtained by parallel translation of B_1 is called a ______. All line segments joining points on the base B_1 to corresponding points on the base B_2 are ______ to one another.

solid cylinder with bases B_1 and B_2

parallel

5 If all the line segments (in Problem 4) are perpendicular to the bases, then the solid cylinder is called ______. The distance measured perpendicularly between the two bases of a solid cylinder is called its ______.

a solid right cylinder

height

In Problems 6 through 10, state the basic properties of volume.

6 Every solid S has a definite volume $V(S)$ cubic units and $V(S)$ ______.

$\geqslant 0$

7 If a solid S is decomposed into a finite number of nonoverlapping pieces $S_1, S_2, \ldots, S_n$, each of which is a solid, then $V(S) =$ ______.

$V(S_1) + V(S_2) + \cdots + V(S_n)$

8 If a solid S_1 is contained in a solid S_2, then ______.

$V(S_1) \leqslant V(S_2)$

9 Two congruent solids have the ______.

same volume

10 The volume of a solid right cylinder is its ______ times the ______ of one of its bases.

height

area

1.1 Solids of Revolution—Method of Circular Disks

11 Let R be an admissible planar region and let l be a straight line lying in the same plane as R, but touching R, if at all, only at boundary points of R. The solid S swept out or generated when R is revolved about the line l as an axis is called a ____________________.

solid of revolution

12 Let f be a nonnegative continuous function on the interval $[a, b]$, and let R be the planar region under the graph of f, above the x axis, and between $x = a$ and $x = b$. The volume V of the solid S generated by revolving the planar region R about the x axis is, according to the method of circular disks, given by $V = \int_a^b (\_\_\_\_\_\_\_\_)\, dx$. The solid is illustrated by the figure

$\pi[f(x)]^2$

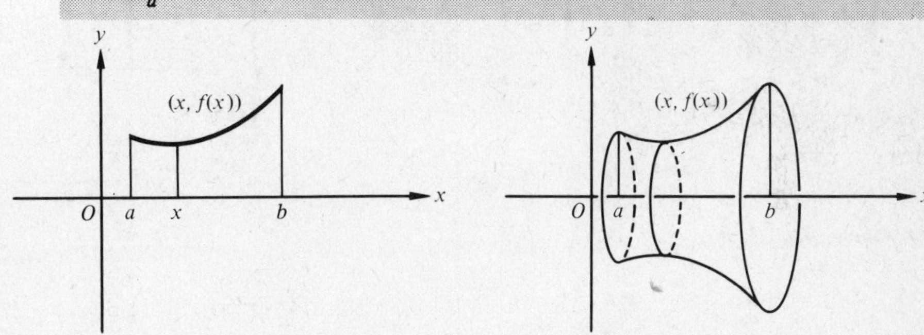

In Problems 13 through 21, use the method of circular disks to find the volume V of the solid S generated by revolving the region R under the graph of each function f about the x axis on the indicated interval.

13 $f(x) = x^2$ on $[-1, 2]$.

The graph is

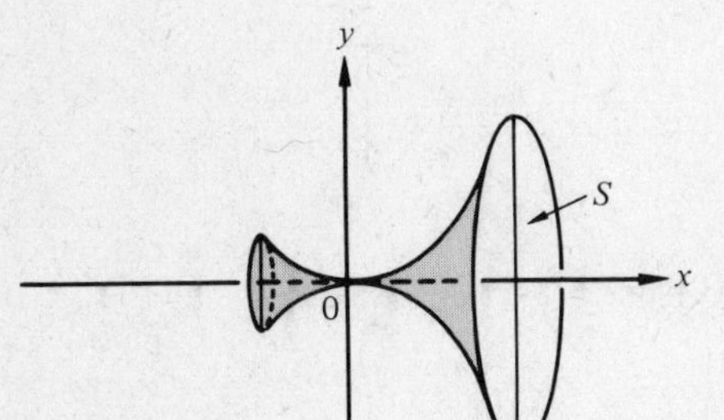

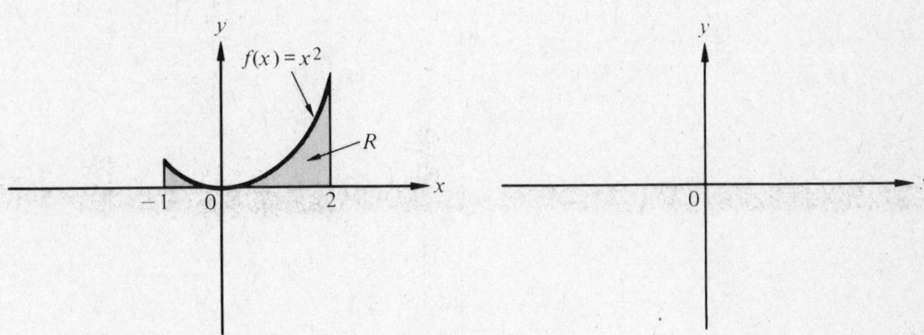

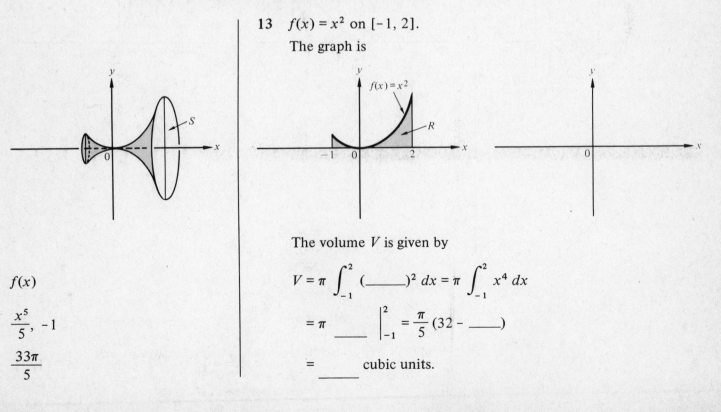

The volume V is given by

$$V = \pi \int_{-1}^{2} (\_\_\_\_\_\_)^2\, dx = \pi \int_{-1}^{2} x^4\, dx$$

$$= \pi \; \_\_\_\_ \Big|_{-1}^{2} = \frac{\pi}{5}(32 - \_\_\_\_)$$

$$= \_\_\_\_ \text{ cubic units.}$$

$f(x)$

$\frac{x^5}{5}$, -1

$\frac{33\pi}{5}$

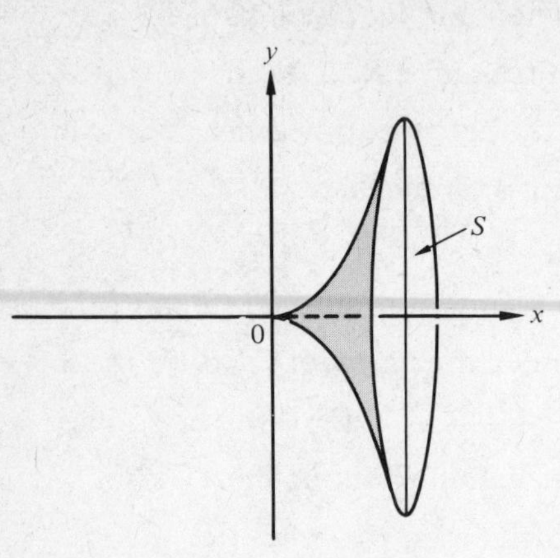

14 $f(x) = x^3$ on $[0, 2]$.

The graph is

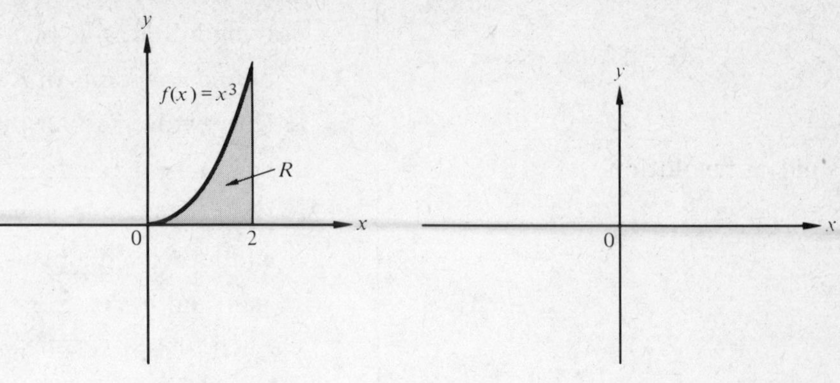

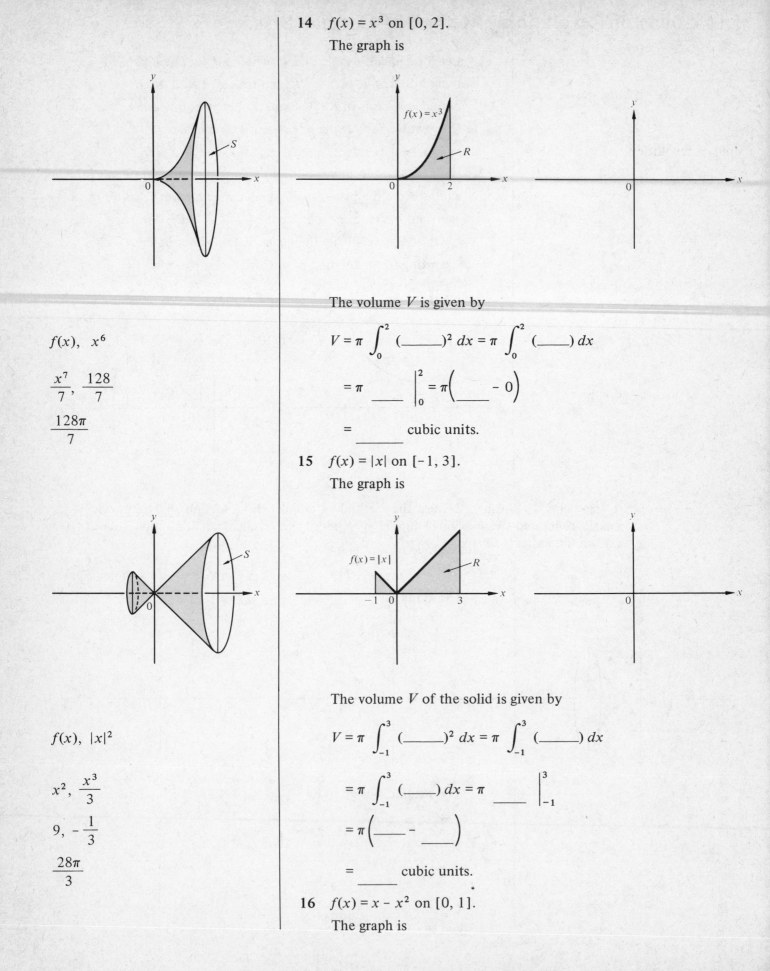

The volume V is given by

$$V = \pi \int_0^2 (\underline{\qquad})^2\, dx = \pi \int_0^2 (\underline{\qquad})\, dx$$

$$= \pi \; \underline{\qquad} \Big|_0^2 = \pi\left(\underline{\qquad} - 0\right)$$

$$= \underline{\qquad} \text{ cubic units.}$$

$f(x)$, x^6

$\dfrac{x^7}{7}$, $\dfrac{128}{7}$

$\dfrac{128\pi}{7}$

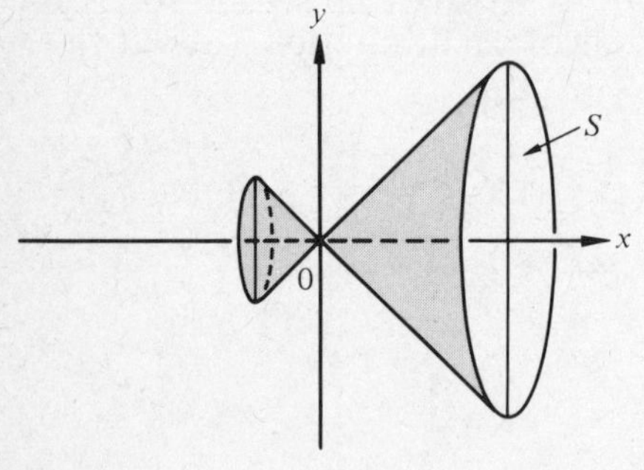

15 $f(x) = |x|$ on $[-1, 3]$.

The graph is

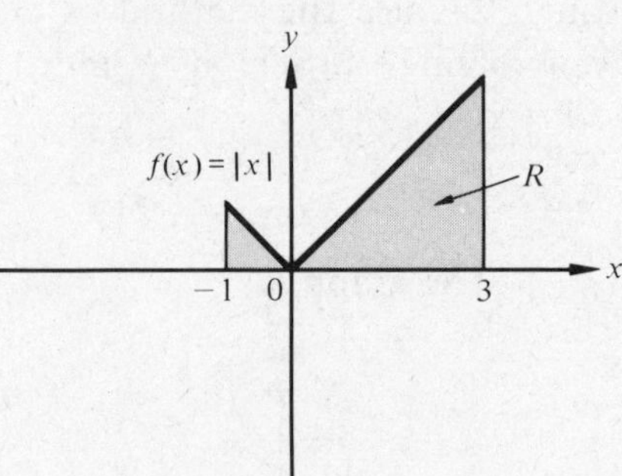

The volume V of the solid is given by

$$V = \pi \int_{-1}^{3} (\underline{\qquad})^2\, dx = \pi \int_{-1}^{3} (\underline{\qquad})\, dx$$

$$= \pi \int_{-1}^{3} (\underline{\qquad})\, dx = \pi \; \underline{\qquad} \Big|_{-1}^{3}$$

$$= \pi\left(\underline{\qquad} - \underline{\qquad}\right)$$

$$= \underline{\qquad} \text{ cubic units.}$$

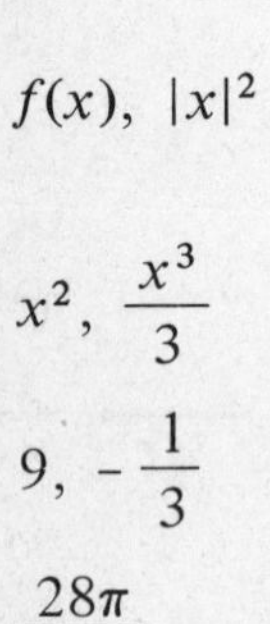

$f(x)$, $|x|^2$

x^2, $\dfrac{x^3}{3}$

9, $-\dfrac{1}{3}$

$\dfrac{28\pi}{3}$

16 $f(x) = x - x^2$ on $[0, 1]$.

The graph is

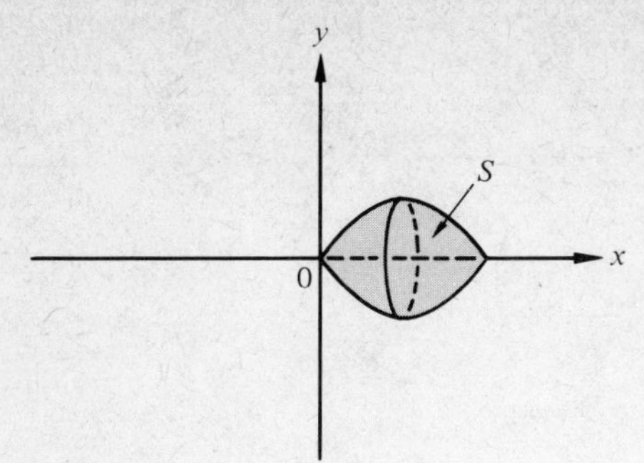

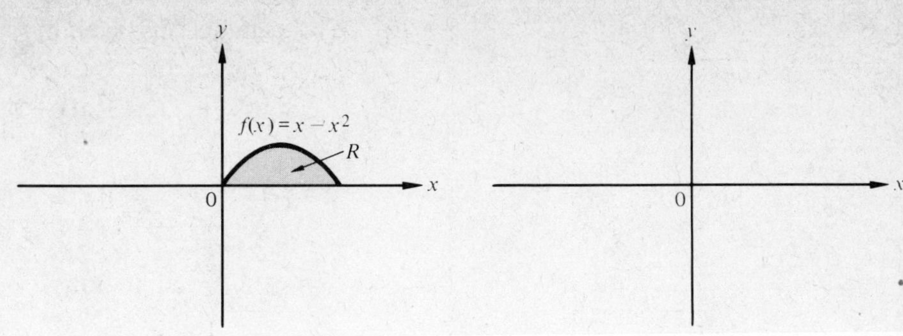

The volume V is given by

$f(x)$, $x - x^2$

$$V = \pi \int_0^1 (\_\_\_\_\_)^2\, dx = \pi \int_0^1 (\_\_\_\_\_\_)^2\, dx$$

$x^2 - 2x^3 + x^4$, $\dfrac{x^3}{3} - \dfrac{x^4}{2} + \dfrac{x^5}{5}$

$$= \pi \int_0^1 (\_\_\_\_\_\_\_\_\_\_\_\_\_\_)\, dx = \pi\left(\_\_\_\_\_\_\_\_\_\_\_\_\_\_\right)\Big|_0^1$$

$\dfrac{1}{30}$, 0, $\dfrac{\pi}{30}$

$$= \pi\left(\_\_\_\_ - \_\_\_\_\right) = \_\_\_\_ \text{ cubic unit.}$$

17 $f(x) = \dfrac{1}{x}$ on [1, 2].

The graph is

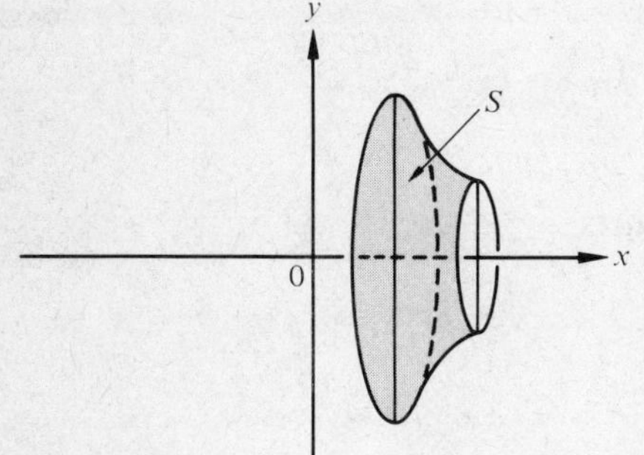

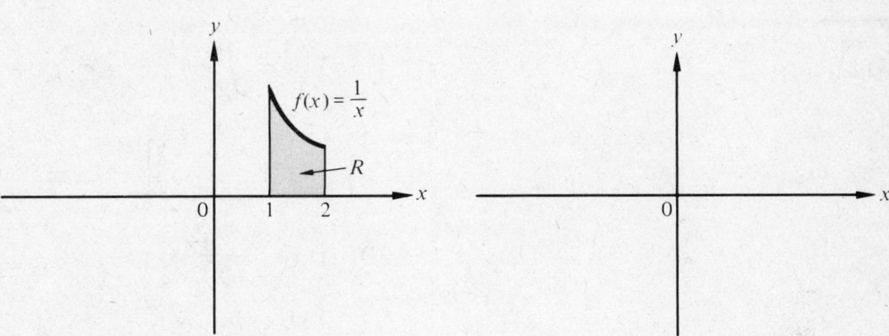

The volume V is given by

$f(x)$, $\dfrac{1}{x^2}$

$$V = \pi \int_1^2 (\_\_\_\_\_)^2\, dx = \pi \int_1^2 \left(\_\_\_\_\right) dx$$

$-\dfrac{1}{x}$, $\dfrac{\pi}{2}$

$$= \pi\left(\_\_\_\_\_\_\right)\Big|_1^2 = \_\_\_\_ \text{ cubic units.}$$

18 $f(x) = \sqrt{1 - x^2}$ on [−1, 1].

The graph is

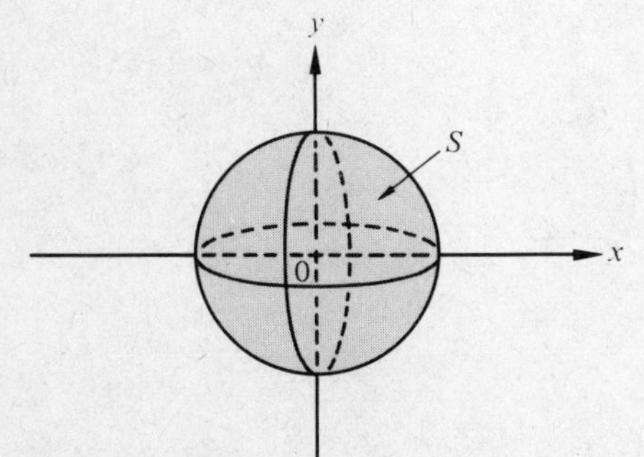

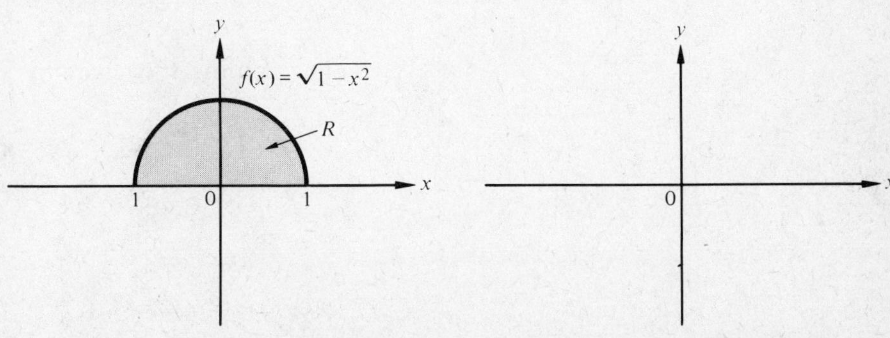

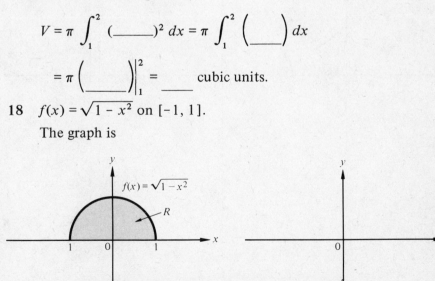

The volume V is given by

$f(x)$, $1 - x^2$

$$V = \pi \int_{-1}^{1} (\_\_\_\_\_\_)^2\, dx = \pi \int_{-1}^{1} (\_\_\_\_\_\_\_\_)\, dx$$

$x - \dfrac{x^3}{3}$, $-\dfrac{2}{3}$

$$= \pi \left(\_\_\_\_\_\_\_\_\right)\Big|_{-1}^{1} = \pi\left(\frac{2}{3} - \_\_\_\_\_\_\right)$$

$\dfrac{4}{3}\pi$

$$= \_\_\_\_\_\_ \text{ cubic units.}$$

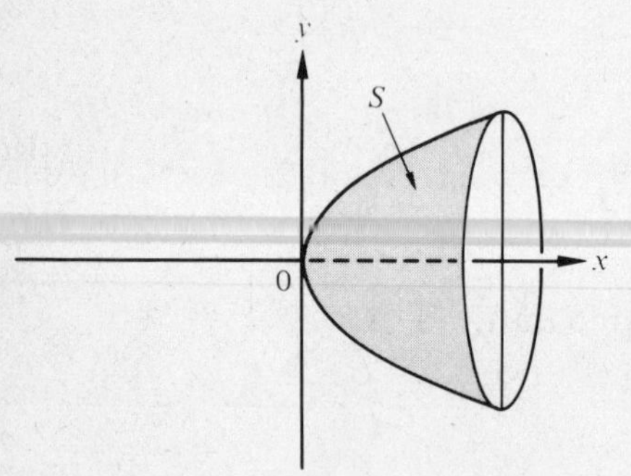

19 $f(x) = \sqrt{x}$ on $[0, 4]$.

The graph is

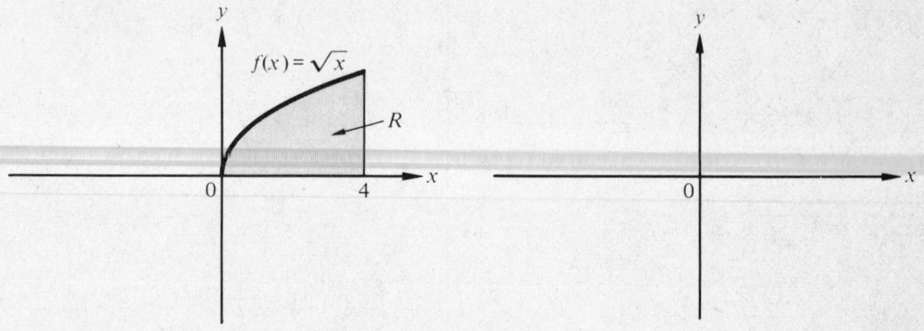

The volume V is given by

$f(x)$, x

$$V = \pi \int_{0}^{4} (\_\_\_\_\_\_)^2\, dx = \pi \int_{0}^{4} (\_\_\_\_\_)\, dx$$

$\dfrac{x^2}{2}$, 8π

$$= \pi \left(\_\_\_\_\_\right)\Big|_{0}^{4} = \_\_\_\_\_ \text{ cubic units.}$$

20 $f(x) = \dfrac{1}{\sqrt[3]{x}}$ on $[1, 8]$.

The graph is

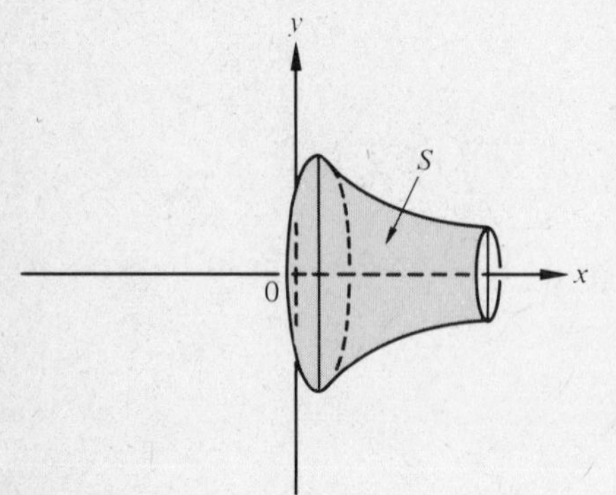

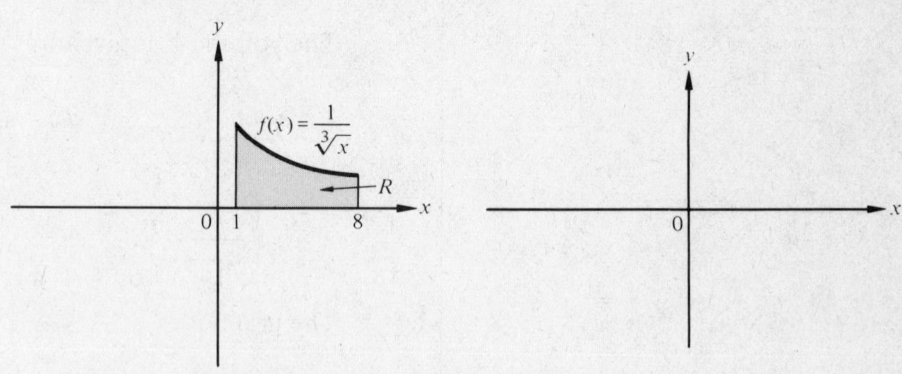

The volume V is given by

$f(x)$, $\dfrac{1}{\sqrt[3]{x^2}}$

$$V = \pi \int_{1}^{8} (\_\_\_\_\_\_)^2\, dx = \pi \int_{1}^{8} \left(\_\_\_\_\_\_\_\right) dx$$

$3\sqrt[3]{x}$, 3

$$= \pi(\_\_\_\_\_\_\_\_)\Big|_{1}^{8} = \pi(6 - \_\_\_\_\_)$$

3π

$$= \_\_\_\_\_ \text{ cubic units.}$$

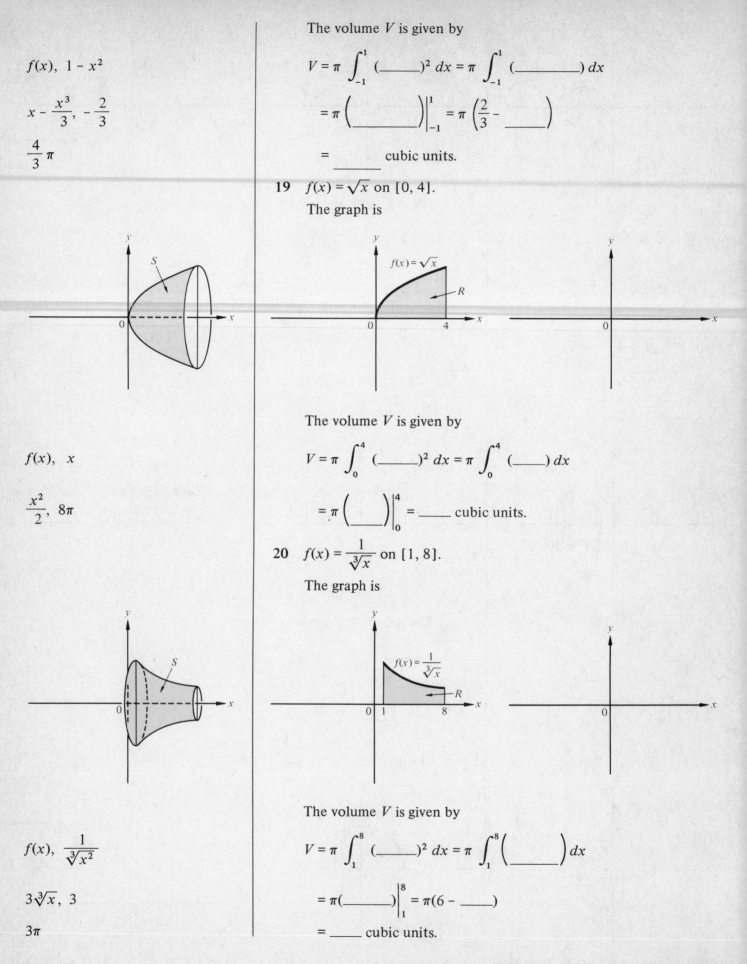

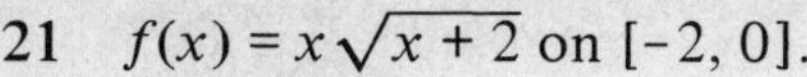

21 $f(x) = x\sqrt{x+2}$ on $[-2, 0]$.

The graph is

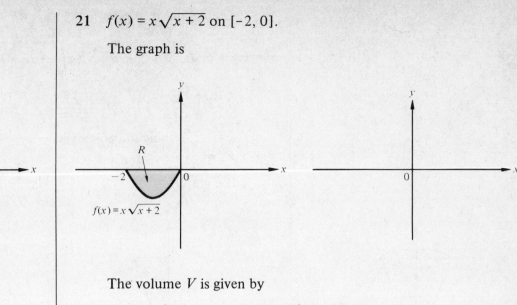

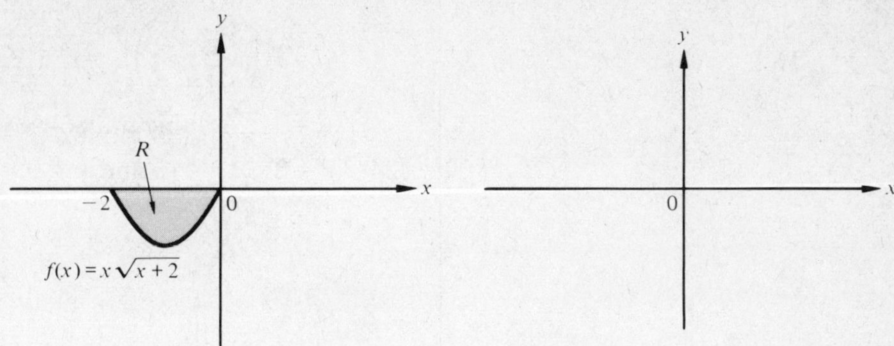

The volume V is given by

$$V = \pi \int_{-2}^{0} (\underline{\qquad})^2\,dx = \pi \int_{-2}^{0} (\underline{\qquad\qquad})\,dx$$

$$= \pi \int_{-2}^{0} (\underline{\qquad\qquad})\,dx = \pi \left(\underline{\qquad\qquad}\right)\Big|_{-2}^{0}$$

$$= \pi\left(\underline{\quad}\right) = \underline{\quad} \text{ cubic units.}$$

$f(x)$, $x^2(x + 2)$

$x^3 + 2x^2$, $\dfrac{x^4}{4} + \dfrac{2}{3}x^3$

$\dfrac{4}{3}$, $\dfrac{4\pi}{3}$

22 Let g be a continuous nonnegative function defined on the interval $[a, b]$, where $a < b$. Let R be the planar region bounded by the y axis, the horizontal lines $y = a$ and $y = b$, and by the graph of $x = g(y)$. If R is revolved about the y axis, a solid of revolution S is generated. The volume V of S is, according to the method of circular disks, given by $V = \pi \int_a^b (\underline{\qquad})^2\,dy$.

$g(y)$

In Problems 23 through 27, use the method of circular disks to find the volume of the solid generated by revolving the finite region enclosed by the graphs of the given equations about the y axis.

23 $y = x^2$, $y = 4$, and $x = 0$.

The graph is

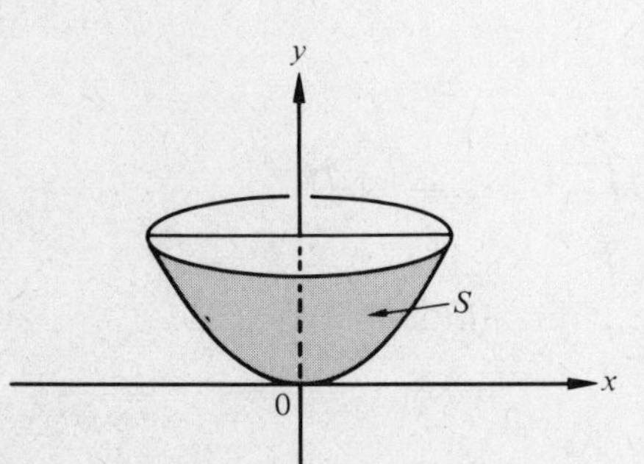

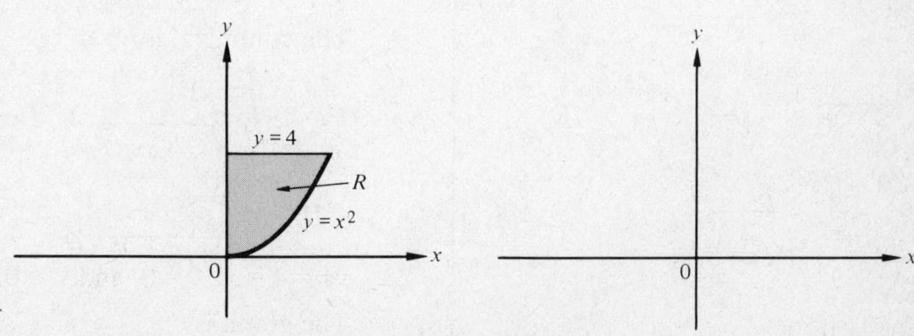

The volume V is given by

$$V = \pi \int_0^4 (\_\_\_\_)^2\, dy = \pi \int_0^4 (\_\_\_\_)\, dy$$

$$= \pi \left(\_\_\_\_\right)\Big|_0^4 = \frac{\pi}{2}(\_\_\_\_)$$

$= \_\_\_\_$ cubic units.

$\sqrt{y},\ y$

$\frac{1}{2}y^2,\ 16$

8π

24 $y = 2x^3$, $y = 4$, and $x = 0$.

The graph is

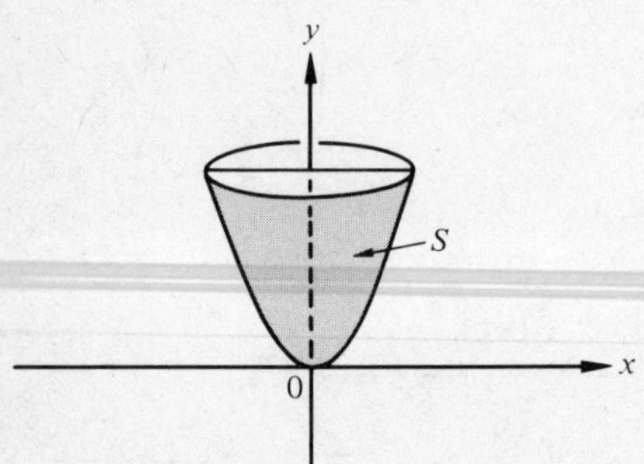

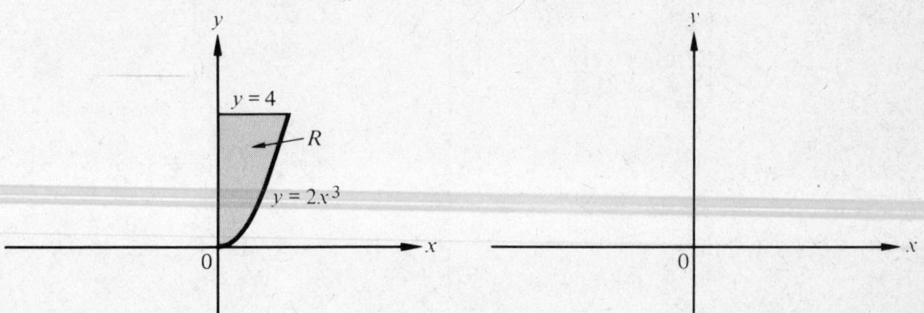

The volume V is given by

$$V = \pi \int_0^4 \left(\_\_\_\_\right)^2 dy = \pi \int_0^4 \left(\_\_\_\_\right) dy$$

$$= \frac{\pi}{\sqrt[3]{4}} \int_0^4 (\_\_\_\_)\, dy = \frac{\pi}{\sqrt[3]{4}} \left(\_\_\_\_\right)\Big|_0^4$$

$$= \frac{3\pi}{5\sqrt[3]{4}}(\_\_\_\_) = \_\_\_\_ \text{ cubic units.}$$

$\left(\frac{y}{2}\right)^{1/3},\ \left(\frac{y}{2}\right)^{2/3}$

$y^{2/3},\ \frac{3}{5}y^{5/3}$

$4^{5/3},\ \frac{12\sqrt[3]{4}}{5}$

25 $y = 9 - x^2$, $x = 0$, and $y = 0$.

The graph is

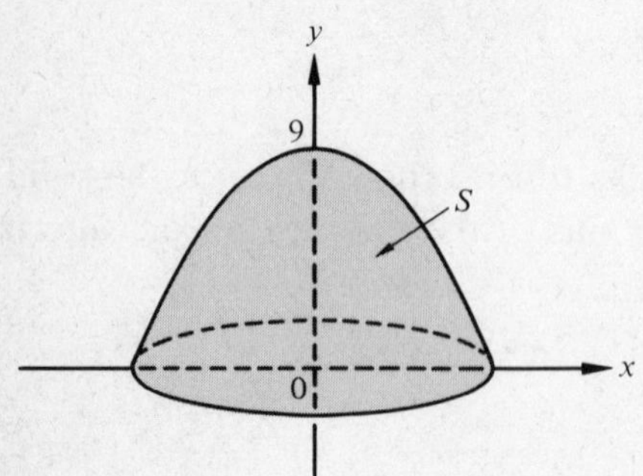

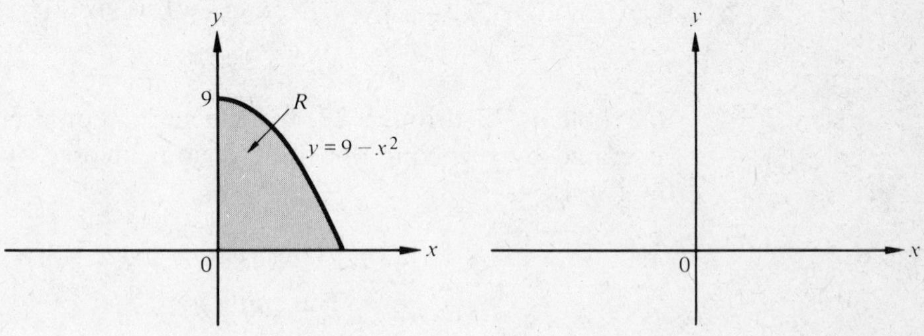

The volume V is given by

$$V = \pi \int_0^9 (\_\_\_\_)^2\, dy = \pi \int_0^9 (\_\_\_\_)\, dy$$

$$= \pi\left(\_\_\_\_\right)\Big|_0^9 = \_\_\_\_ \text{ cubic units.}$$

$\sqrt{9-y},\ 9-y$

$9y - \frac{y^2}{2},\ \frac{81\pi}{2}$

26 $y^2 = 4 - x$, $x = 0$, and $y = 0$.

The graph is

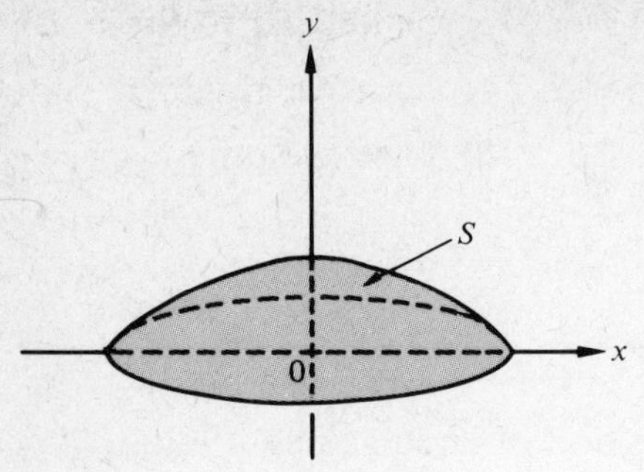

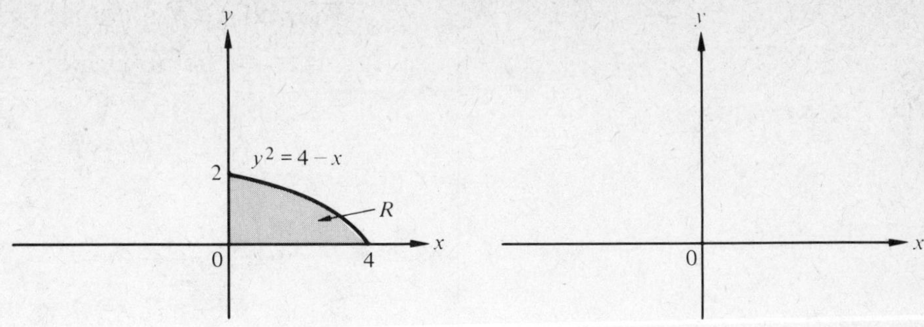

The volume V is given by

$$V = \pi \int_0^2 (\underline{\qquad})^2\, dy$$

$$= \pi \int_0^2 (\underline{\qquad\qquad})\, dy = \pi \left(\underline{\qquad\qquad}\right)\Big|_0^2$$

$$= \pi \left(\underline{\quad}\right) = \underline{\quad} \text{ cubic units.}$$

$4 - y^2$

$16 - 8y^2 + y^4,\ 16y - \frac{8y^3}{3} + \frac{y^5}{5}$

$\frac{256}{15},\ \frac{256\pi}{15}$

27 $y = 3x + 5$, $x = 0$, and $y = 0$.
The graph is

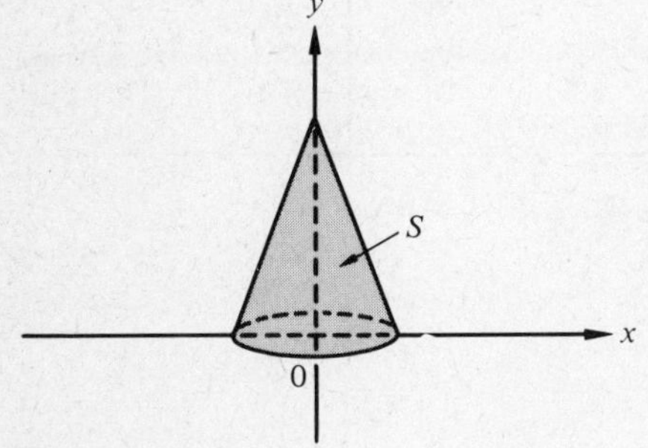

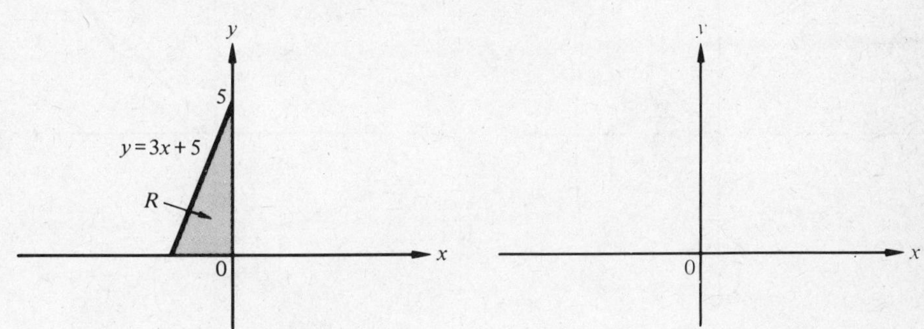

The volume V is given by

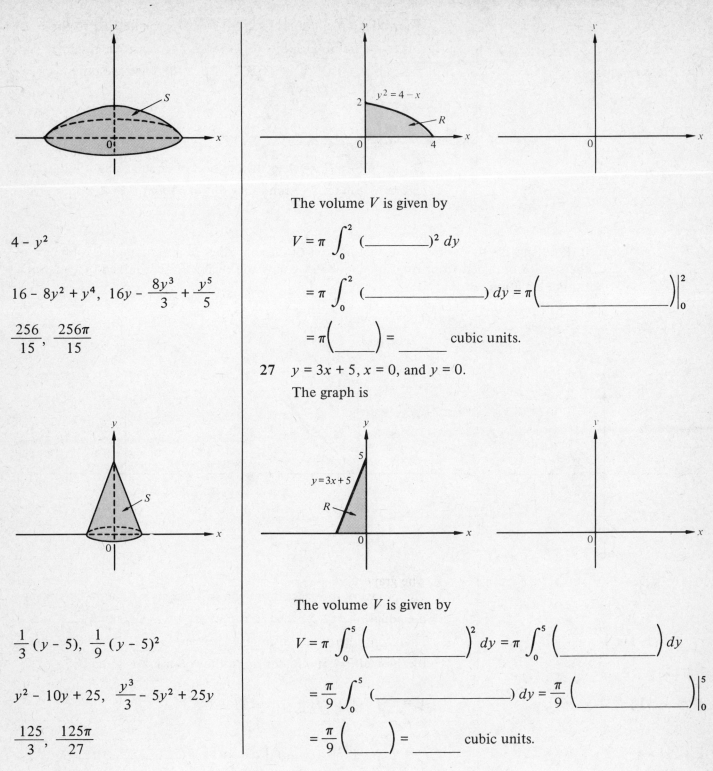

$$V = \pi \int_0^5 \left(\underline{\qquad}\right)^2 dy = \pi \int_0^5 \left(\underline{\qquad}\right) dy$$

$$= \frac{\pi}{9} \int_0^5 (\underline{\qquad\qquad})\, dy = \frac{\pi}{9} \left(\underline{\qquad\qquad}\right)\Big|_0^5$$

$$= \frac{\pi}{9} \left(\underline{\quad}\right) = \underline{\quad} \text{ cubic units.}$$

$\frac{1}{3}(y - 5),\ \frac{1}{9}(y - 5)^2$

$y^2 - 10y + 25,\ \frac{y^3}{3} - 5y^2 + 25y$

$\frac{125}{3},\ \frac{125\pi}{27}$

1.2 The Method of Circular Rings

28 Suppose that f and g are nonnegative continuous functions on the interval $[a, b]$ such that $f(x) \geqslant g(x)$ holds for all x in $[a, b]$. Let R be the planar region bounded by the graphs of f and g between $x = a$ and $x = b$. Let S be the solid generated by revolving R about

the x axis. Then, the cross section of S perpendicular to the x axis at the point with coordinate x will be a circular ring with out-

$f(x)$, $g(x)$

side radius _____, inside radius _____, and with volume V given by

$[f(x)]^2 - [g(x)]^2$

$$V = \pi \int_a^b (\_\_\_\_\_\_\_\_\_\_\_\_\_\_\_\_)\, dx$$

The same formula, called the method of circular rings, holds if f and g are nonpositive functions on $[a, b]$ and $f(x) \leqslant g(x)$.

In Problems 29 through 37, use the method of circular rings to determine the volume V of the solid S obtained from rotating about the x axis the region R determined by each function or equation.

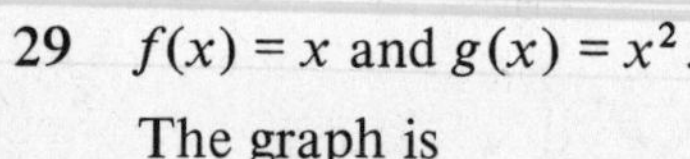

29 $f(x) = x$ and $g(x) = x^2$.

The graph is

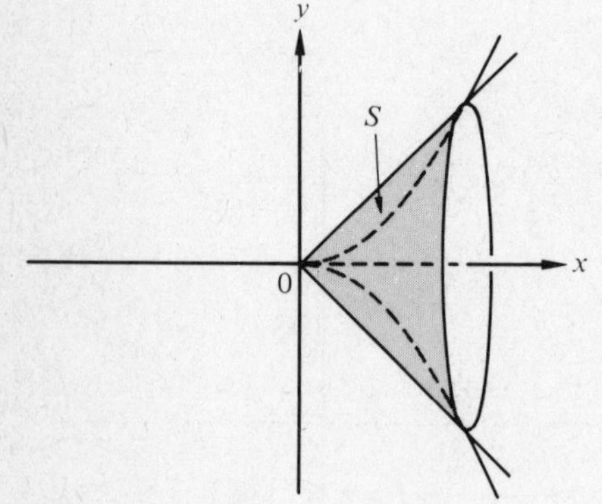

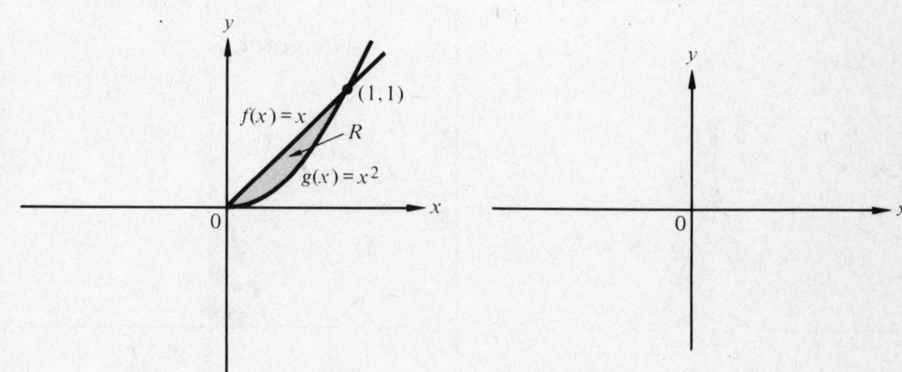

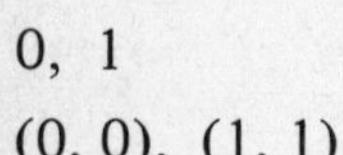

0, 1

(0, 0), (1, 1)

The points of intersection of the two graphs are found by solving the equation $x^2 = x$. The solutions are $x =$ _____ and $x =$ _____. Therefore, the points of intersection are ________ and ________. By the method of circular rings, the volume V is given by

$g(x)$

$$V = \pi \int_0^1 \{[f(x)]^2 - [\_\_\_\_\_]^2\}\, dx$$

x^2, $x^2 - x^4$

$$= \pi \int_0^1 [x^2 - (\_\_\_\_)^2]\, dx = \pi \int_0^1 (\_\_\_\_\_\_\_\_)\, dx$$

$\frac{x^3}{3} - \frac{x^5}{5}$, $\frac{2}{15}$

$$= \pi \left(\_\_\_\_\_\_\_\_\_\_\right)\Big|_0^1 = \pi\left(\_\_\_\_ - 0\right)$$

$\frac{2\pi}{15}$

$$= \_\_\_\_ \text{ cubic unit.}$$

30 $f(x) = \sqrt{x}$ and $g(x) = x$.

The graph is

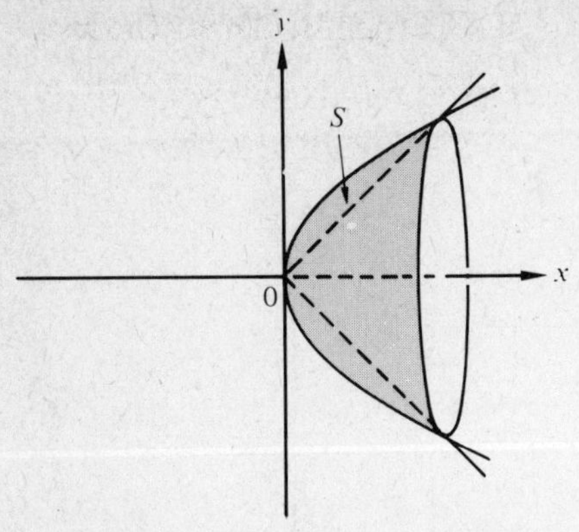

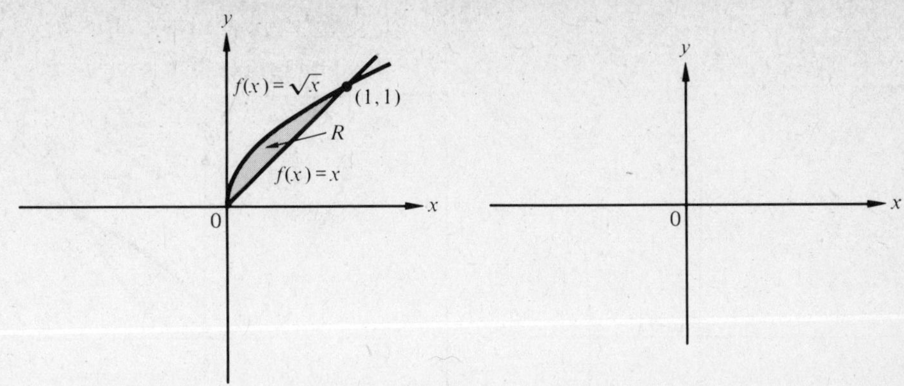

The points of intersection are ________ and ________. The volume V of the solid is given by

$$V = \pi \int_0^1 \{[f(x)]^2 - [\_\_\_\_]^2\}\, dx$$

$$= \pi \int_0^1 [(\sqrt{x})^2 - \_\_\_\_]\, dx = \pi \int_0^1 (\_\_\_\_\_\_)\, dx$$

$$= \pi \left(\_\_\_\_\_\_\_\_ \right)\Big|_0^1 = \pi\left(\_\_\_\_ - 0\right)$$

$$= \_\_\_\_ \text{ cubic unit.}$$

(0, 0), (1, 1)

$g(x)$

x^2, $x - x^2$

$\frac{x^2}{2} - \frac{x^3}{3}$, $\frac{1}{6}$

$\frac{\pi}{6}$

31 $f(x) = x$ and $g(x) = x^3$.
The graph is

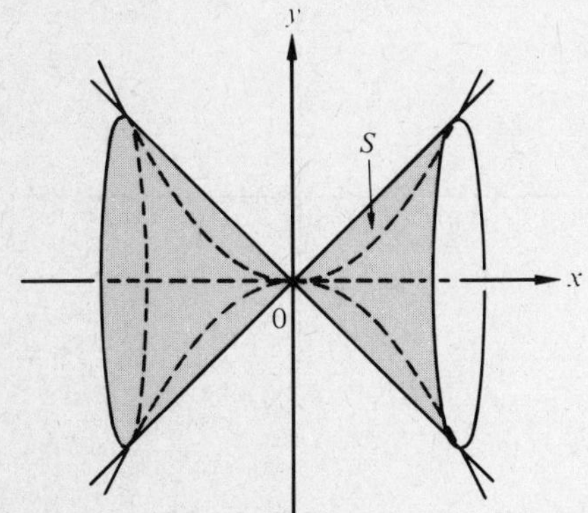

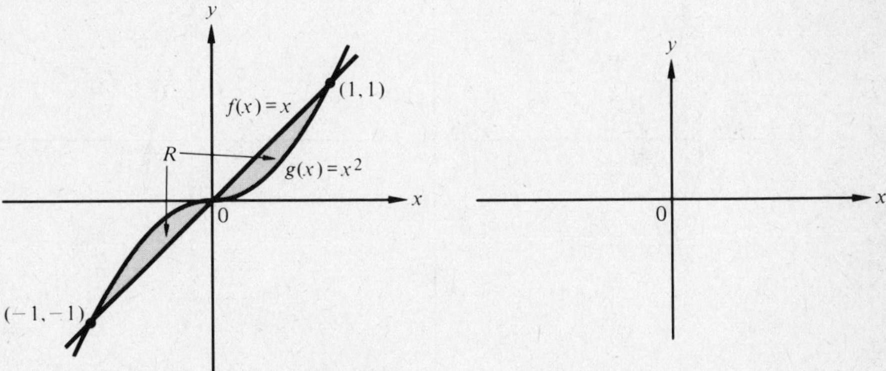

The points of intersection are ________, ________, and ________. The volume V of the solid is given by

$$V = \pi \int_{-1}^1 \{[f(x)]^2 - [g(x)]^2\}\, dx$$

$$= \pi \int_{-1}^1 (\_\_\_\_\_\_)\, dx = \pi \left(\_\_\_\_\_\_\_\_ \right)\Big|_{-1}^1$$

$$= \pi\left(\_\_\_\_ - \_\_\_\_\right) = \_\_\_\_ \text{ cubic unit.}$$

(−1, −1), (0, 0)
(1, 1)

$x^2 - x^6$, $\frac{x^3}{3} - \frac{x^7}{7}$

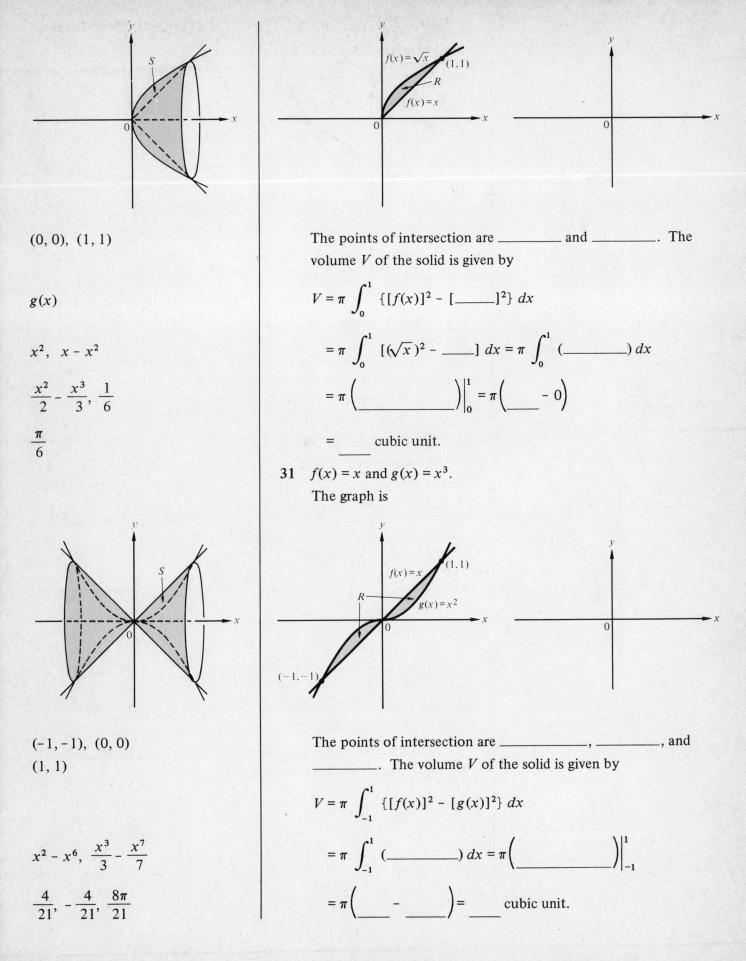

$\frac{4}{21}$, $-\frac{4}{21}$, $\frac{8\pi}{21}$

32 $f(x) = \sqrt{x}$ and $g(x) = x^2$.

The graph is

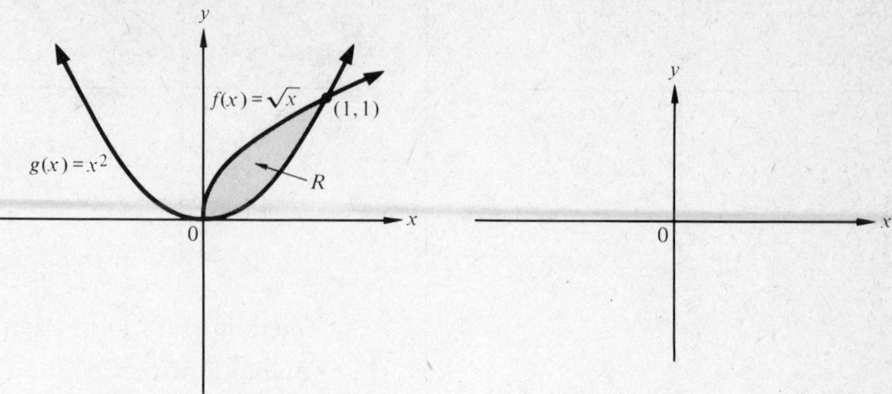

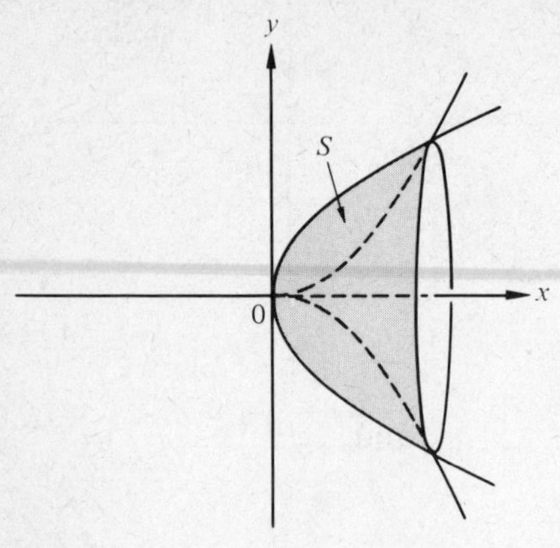

(0, 0), (1, 1)

The points of intersection are ________ and ________. The volume V of the solid is given by

x^4, $x - x^4$

$$V = \pi \int_0^1 [(\sqrt{x})^2 - \_\_\_\_]\, dx = \pi \int_0^1 (\_\_\_\_\_\_\_\_)\, dx$$

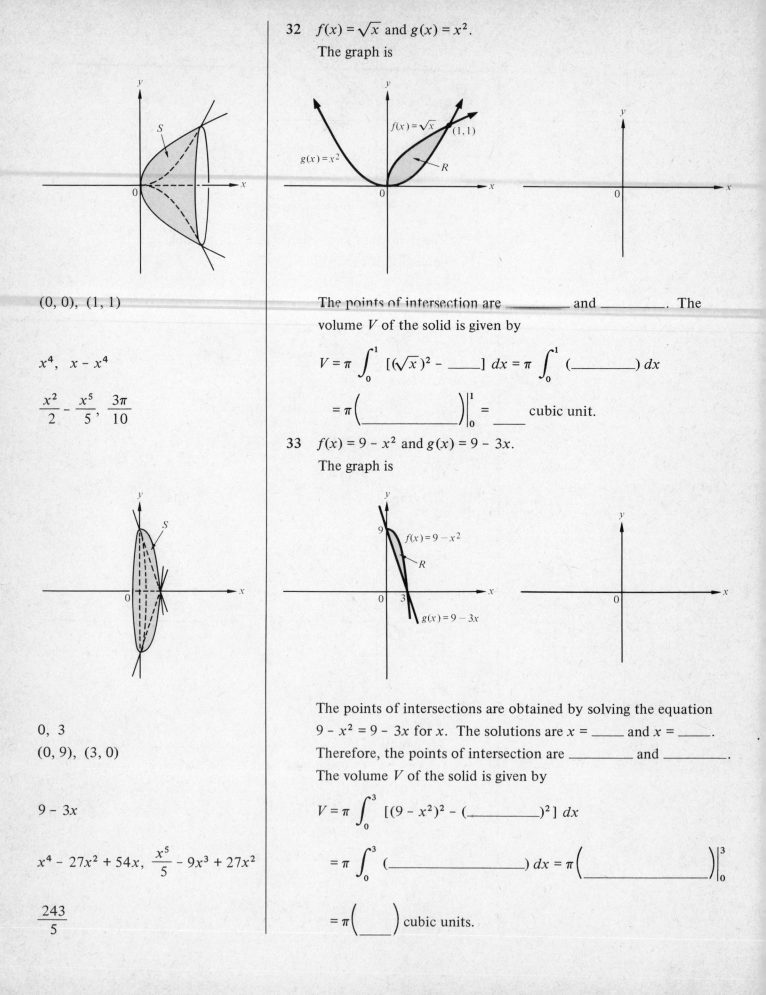

$\dfrac{x^2}{2} - \dfrac{x^5}{5}$, $\dfrac{3\pi}{10}$

$$= \pi\left(\_\_\_\_\_\_\_\_\_\_\_\_\right)\Big|_0^1 = \_\_\_\_ \text{ cubic unit.}$$

33 $f(x) = 9 - x^2$ and $g(x) = 9 - 3x$.

The graph is

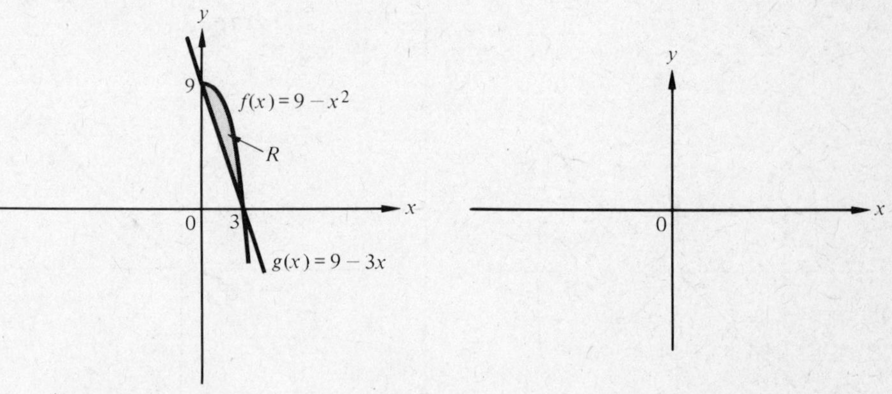

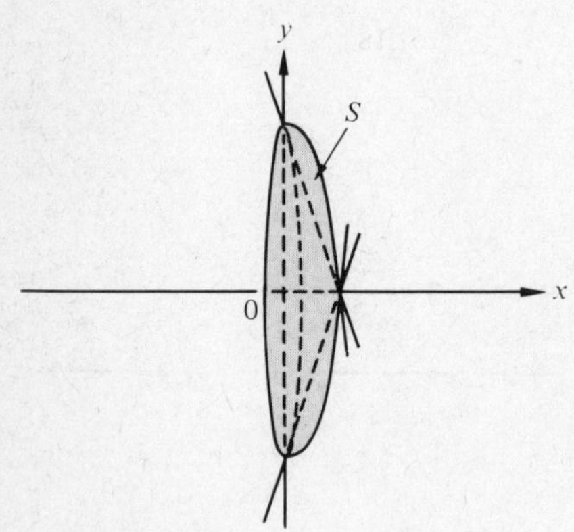

The points of intersections are obtained by solving the equation

0, 3

$9 - x^2 = 9 - 3x$ for x. The solutions are $x =$ ____ and $x =$ ____.

(0, 9), (3, 0)

Therefore, the points of intersection are ________ and ________. The volume V of the solid is given by

$9 - 3x$

$$V = \pi \int_0^3 [(9 - x^2)^2 - (\_\_\_\_\_\_\_\_)^2]\, dx$$

$x^4 - 27x^2 + 54x$, $\dfrac{x^5}{5} - 9x^3 + 27x^2$

$$= \pi \int_0^3 (\_\_\_\_\_\_\_\_\_\_\_\_)\, dx = \pi\left(\_\_\_\_\_\_\_\_\_\_\_\_\right)\Big|_0^3$$

$\dfrac{243}{5}$

$$= \pi\left(\_\_\_\_\right) \text{ cubic units.}$$

34 $f(x) = \sqrt{1 - x^2}$ and $g(x) = |x|$.

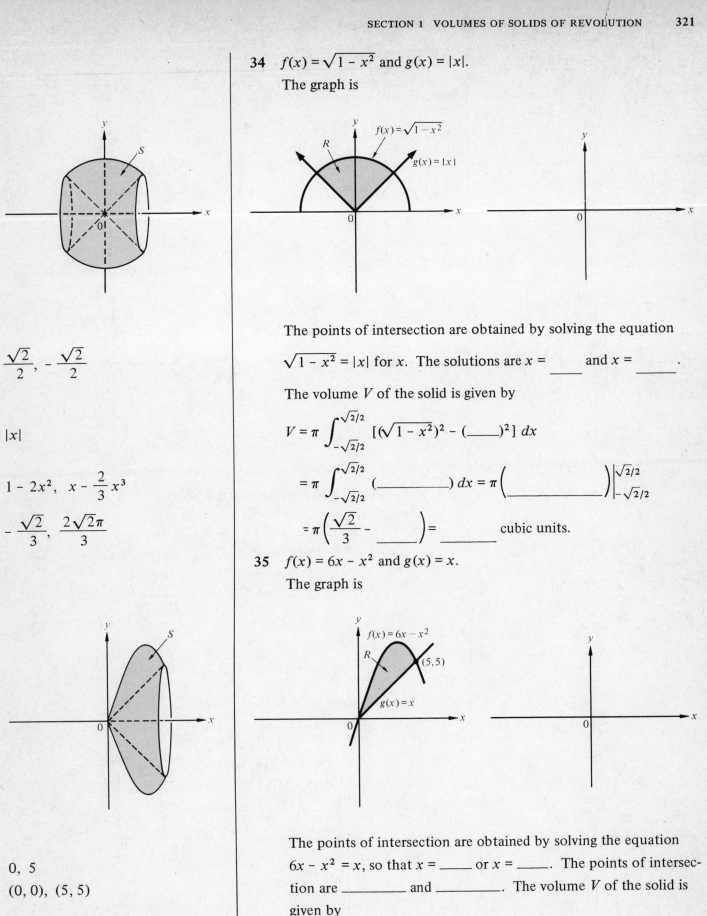

The graph is

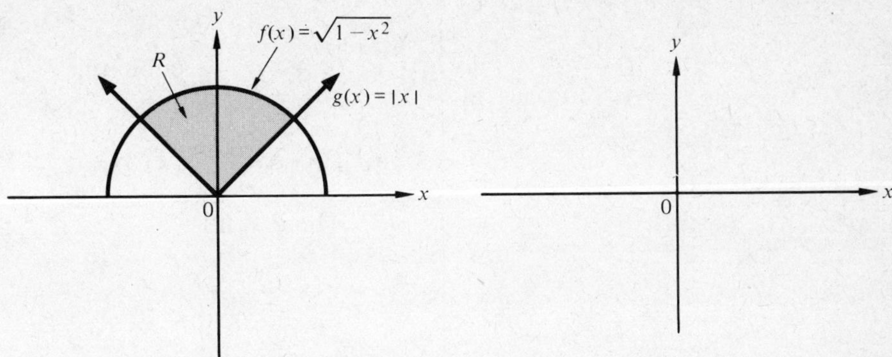

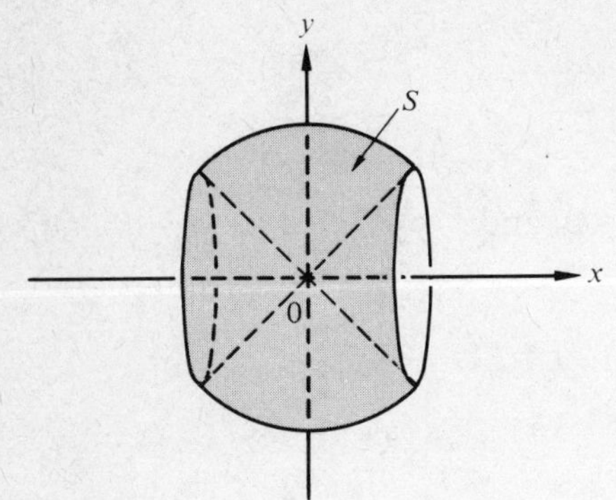

The points of intersection are obtained by solving the equation $\sqrt{1 - x^2} = |x|$ for x. The solutions are $x =$ ____ and $x =$ ____.

$\frac{\sqrt{2}}{2}, -\frac{\sqrt{2}}{2}$

The volume V of the solid is given by

$$V = \pi \int_{-\sqrt{2}/2}^{\sqrt{2}/2} [(\sqrt{1 - x^2})^2 - (\_\_\_\_)^2]\, dx$$

$|x|$

$$= \pi \int_{-\sqrt{2}/2}^{\sqrt{2}/2} (\_\_\_\_\_\_)\, dx = \pi \left(\_\_\_\_\_\_\_\_\right)\Big|_{-\sqrt{2}/2}^{\sqrt{2}/2}$$

$1 - 2x^2, \quad x - \frac{2}{3}x^3$

$$= \pi\left(\frac{\sqrt{2}}{3} - \_\_\_\_\right) = \_\_\_\_\_ \text{ cubic units.}$$

$-\frac{\sqrt{2}}{3}, \quad \frac{2\sqrt{2}\pi}{3}$

35 $f(x) = 6x - x^2$ and $g(x) = x$.

The graph is

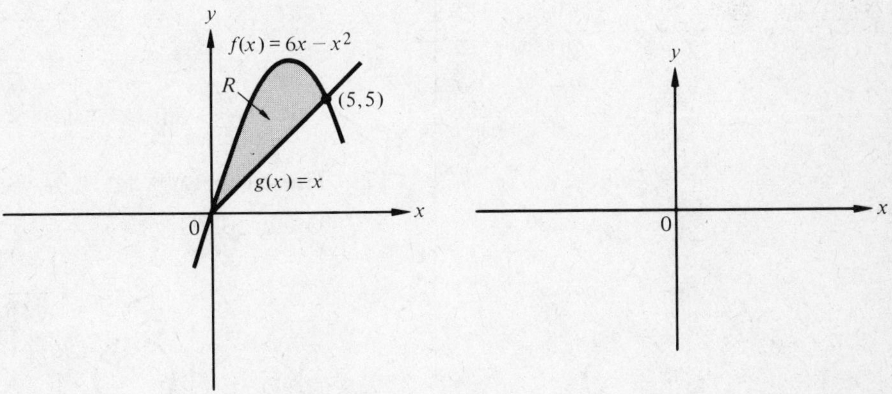

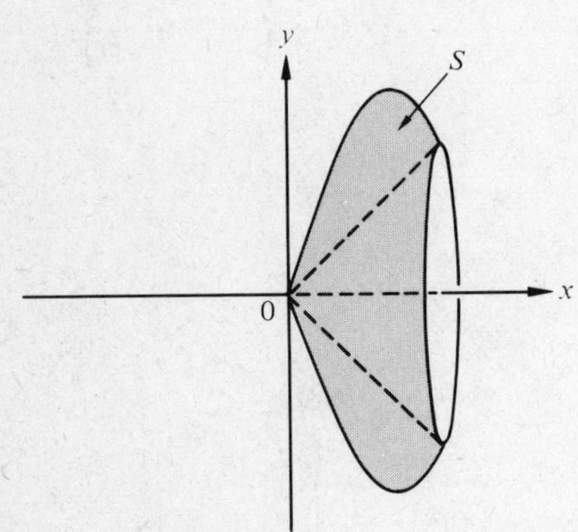

The points of intersection are obtained by solving the equation $6x - x^2 = x$, so that $x =$ ____ or $x =$ ____. The points of intersection are ________ and ________. The volume V of the solid is given by

0, 5

(0, 0), (5, 5)

$$V = \pi \int_0^5 [(\_\_\_\_\_\_)^2 - x^2]\, dx$$

$6x - x^2$

$x^4 - 12x^3 + 35x^2$, $\frac{x^5}{5} - 3x^4 + \frac{35x^3}{3}$

$\frac{625}{3}\pi$

$= \pi \int_0^5$ (________________) $dx = \pi\Big($________________$\Big)\Big|_0^5$

= ______ cubic units.

36 $f(x) = x^2 + 2$, $g(x) = \frac{1}{2}x + 1$, $x = 0$, and $x = 1$.

The graph is

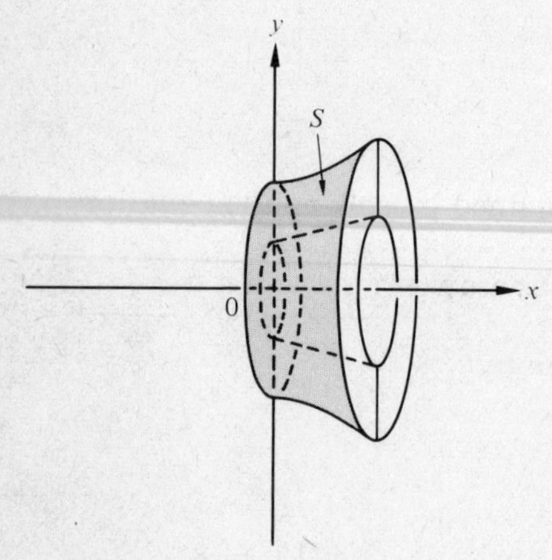

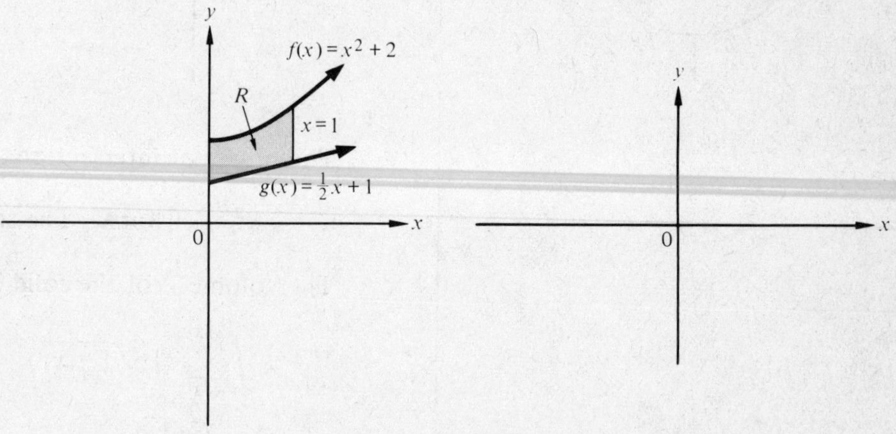

The volume V of the solid is given by

$x^2 + 2$

$V = \pi \int_0^1 \left[(\underline{\qquad\qquad})^2 - \left(\frac{1}{2}x + 1\right)^2\right] dx$

$x^4 + \frac{15}{4}x^2 - x + 3$

$= \pi \int_0^1 \Big($________________________$\Big)\, dx$

$\frac{x^5}{5} + \frac{5}{4}x^3 - \frac{x^2}{2} + 3x$, $\frac{79}{20}$

$= \pi\Big($________________________$\Big)\Big|_0^1 = \pi\,($____ $- 0)$

$\frac{79\pi}{20}$

= ______ cubic units.

37 The circle $(x - 3)^2 + (y - 4)^2 = 4$.

The graph is

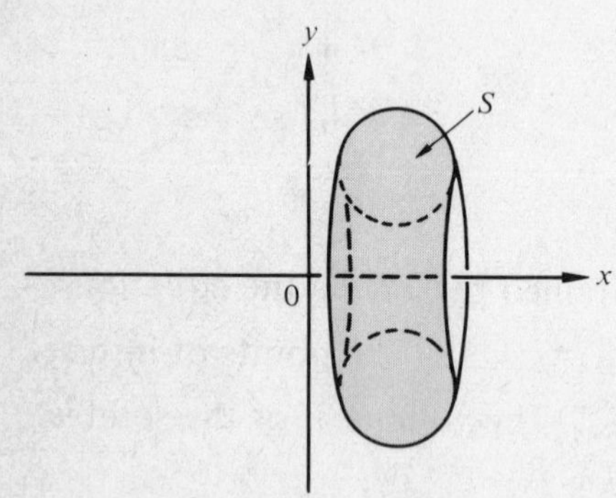

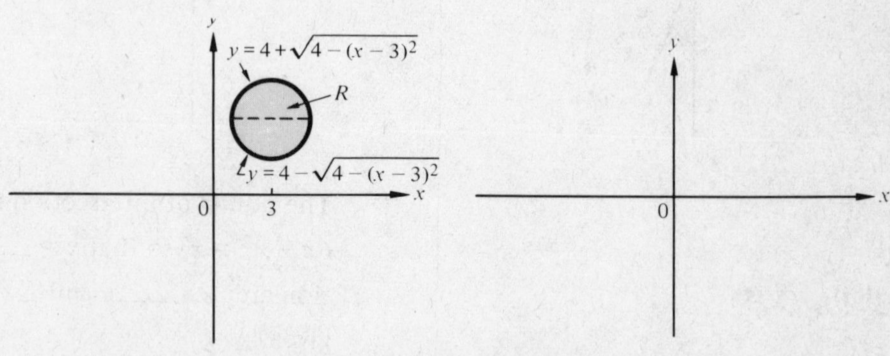

$4 - \sqrt{4 - (x - 3)^2}$

Solving the equation for y in terms of x, we find that $y = 4 + \sqrt{4 - (x - 3)^2}$ or $y =$ ________________________. The volume V of the solid is given by

$4 - \sqrt{4-(x-3)^2}$

$$V = \pi \int_1^5 \{[4 + \sqrt{4-(x-3)^2}]^2 - [\_\_\_\_\_\_\_\_\_\_]^2\}\, dx$$

$16\sqrt{4-(x-3)^2}$

$$= \pi \int_1^5 (\_\_\_\_\_\_\_\_\_\_)\, dx = 16\pi \int_1^5 \sqrt{4-(x-3)^2}\, dx$$

$2\pi,\ 32\pi^2$

$$= 16\pi(\_\_\_\_) = \_\_\_\_ \text{ cubic units.}$$

38 Let the planar region R be bounded on the right by the graph of the equation $x = F(y)$ and on the left by the graph of the equation $x = G(y)$. If R is revolved about the y axis, then the volume V of the solid S generated over the interval $[a, b]$ is, according to the method of circular rings, given by

$F(y)$

$$V = \pi \int_a^b \{[\_\_\_\_]^2 - [G(y)]^2\}\, dy$$

The graph is

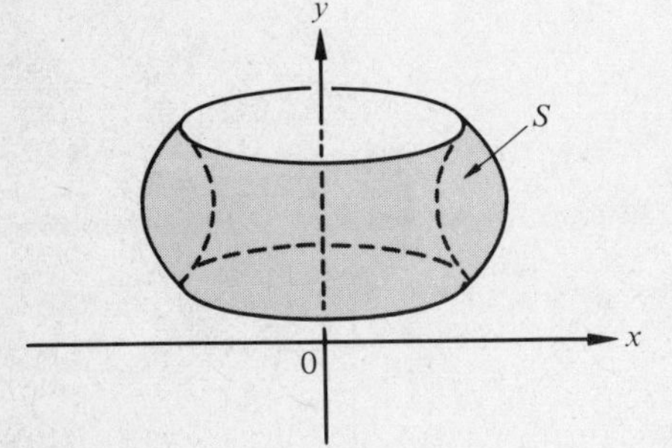

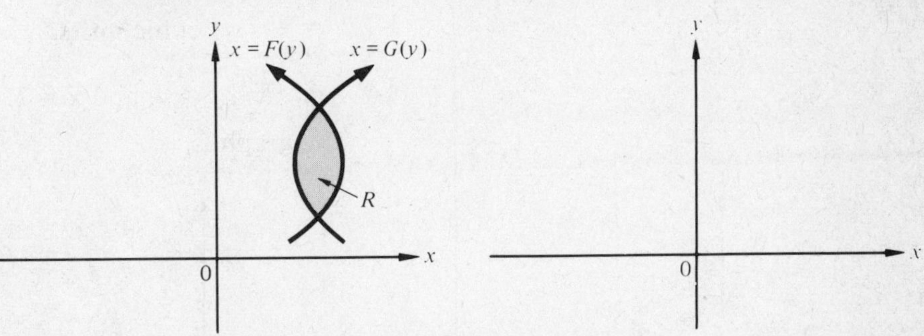

In Problems 39 through 43, use the method of circular rings to find the volume V of the solid S generated by rotating the region bounded by each curve about the y axis.

39 $y = x^2$, $x = 2$, and $y = 0$.

The graph is

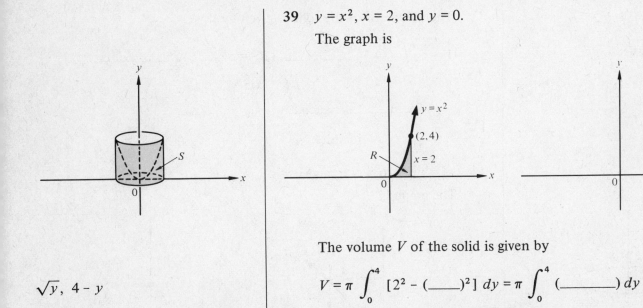

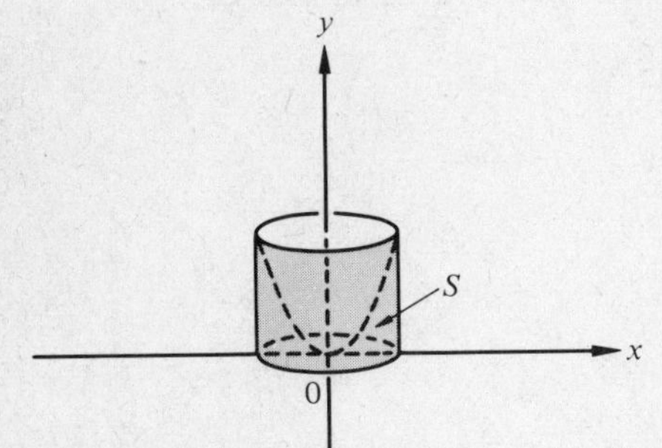

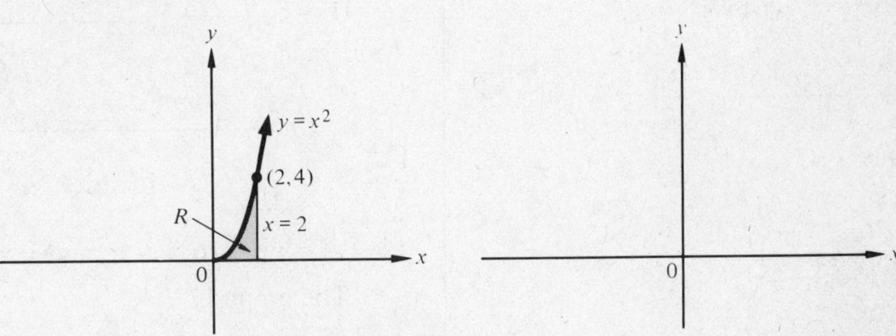

The volume V of the solid is given by

$\sqrt{y},\ 4 - y$

$$V = \pi \int_0^4 [2^2 - (\_\_\_\_)^2]\, dy = \pi \int_0^4 (\_\_\_\_\_)\, dy$$

$4y - \dfrac{y^2}{2},\ 8\pi$

$$= \left(\_\_\_\_\_\_\_\_\right)\Big|_0^4 = \_\_\_\_ \text{ cubic units.}$$

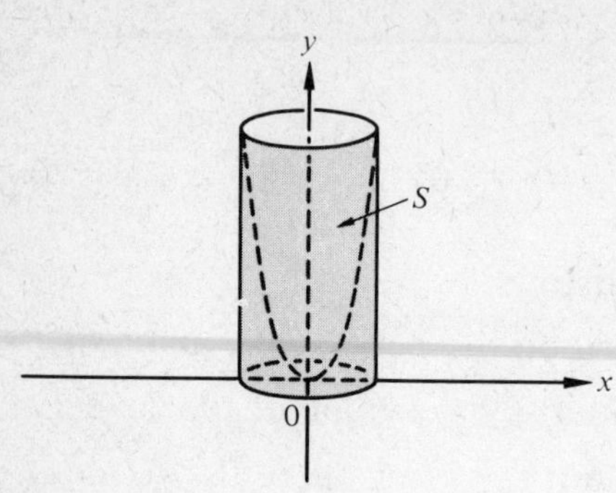

$(\sqrt[3]{y})^2$, $y^{2/3}$

$4y - \frac{3}{5}y^{5/3}$, $\frac{64}{5}$

$\frac{64\pi}{5}$

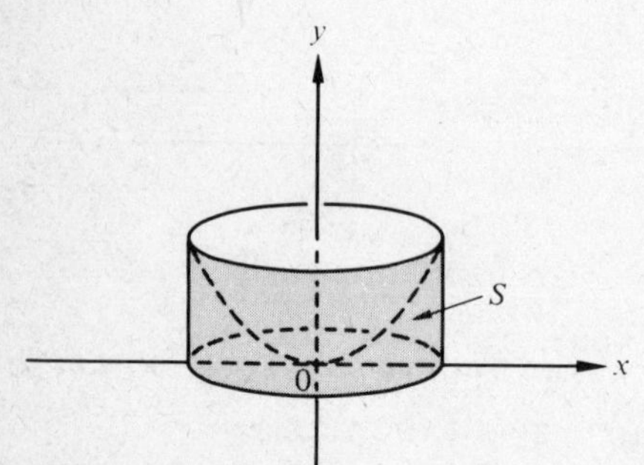

$(2y^2)^{2/3}$, $\sqrt[3]{4}\,y^{4/3}$

$4y - \frac{3\sqrt[3]{4}\,y^{7/3}}{7}$, $\frac{32}{7}$

$\frac{32\pi}{7}$

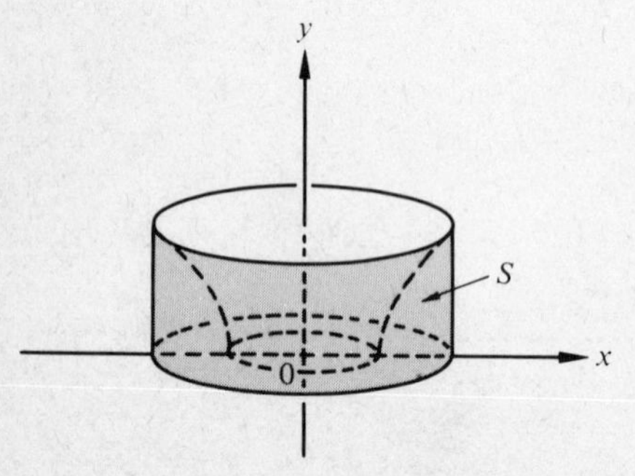

40 $y = x^3$, $y = 0$, and $x = 2$.

The graph is

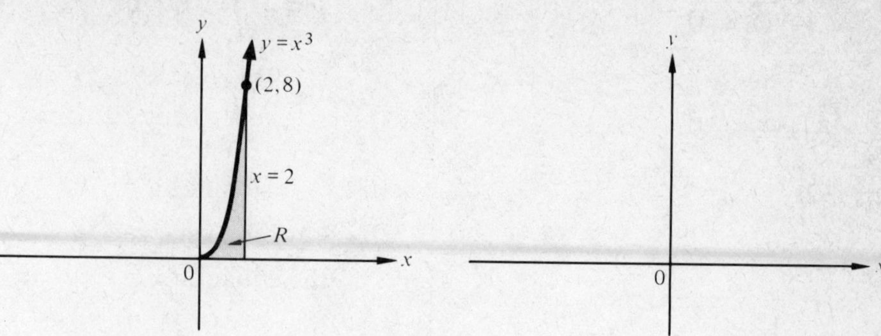

The volume V of the solid is given by

$$V = \pi \int_0^8 (2^2 - \_\_\_\_\_\_)\,dy = \pi \int_0^8 (4 - \_\_\_\_\_\_)\,dy$$

$$= \pi \left(\_\_\_\_\_\_\_\_\_\_ \right)\Big|_0^8 = \pi \left(\_\_\_\_ - 0 \right)$$

$= \_\_\_\_\_\_$ cubic units.

41 $2y^2 = x^3$, $y = 0$, and $x = 2$.

The graph is

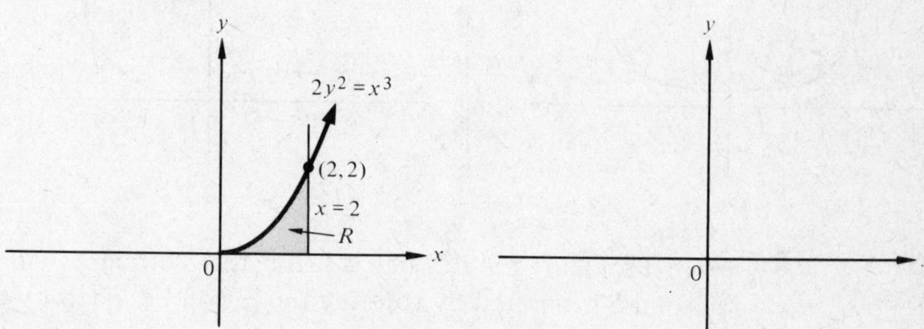

The volume V of the solid is given by

$$V = \pi \int_0^2 (4 - \_\_\_\_\_\_\_\_)\,dy = \pi \int_0^2 (4 - \_\_\_\_\_\_\_\_)\,dy$$

$$= \pi \left(\_\_\_\_\_\_\_\_\_\_ \right)\Big|_0^2 = \pi \left(\_\_\_\_ - 0 \right)$$

$= \_\_\_\_\_\_$ cubic units.

42 $y^2 = x^2 - 16$, $y = 0$, and $x = 8$.

The graph is

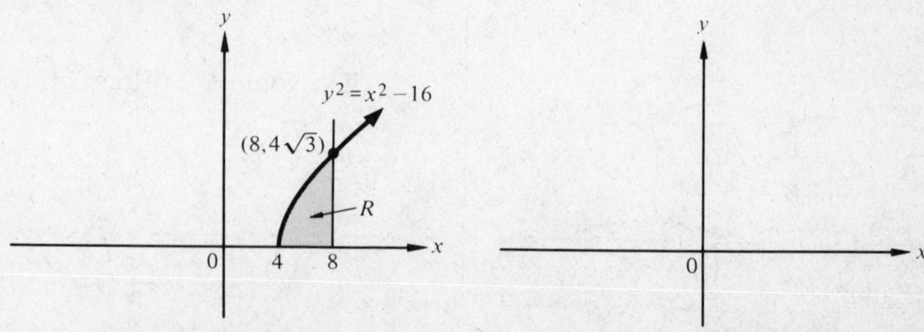

$\sqrt{y^2 + 16}$

$y^2 + 16$

$48 - y^2, \ 48y - \dfrac{y^3}{3}$

$128\sqrt{3}, \ 128\sqrt{3}\pi$

The volume V of the solid is given by

$$V = \pi \int_0^{4\sqrt{3}} [8^2 - (\_\_\_\_\_\_)^2]\, dy$$

$$= \pi \int_0^{4\sqrt{3}} [64 - (\_\_\_\_\_\_)]\, dy$$

$$= \pi \int_0^{4\sqrt{3}} (\_\_\_\_\_\_)\, dy = \pi\left(\_\_\_\_\_\_\right)\Big|_0^{4\sqrt{3}}$$

$$= \pi(\_\_\_\_\_ - 0) = \_\_\_\_\_\_ \text{ cubic units.}$$

43 $y = x^2 + 4$ and $y = 2x^2$.

The graph is

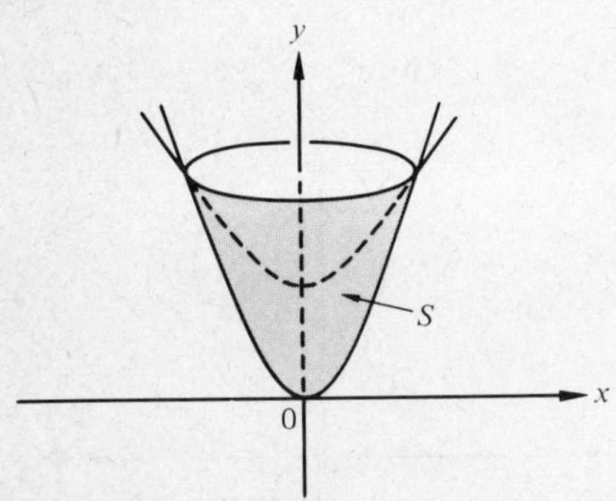

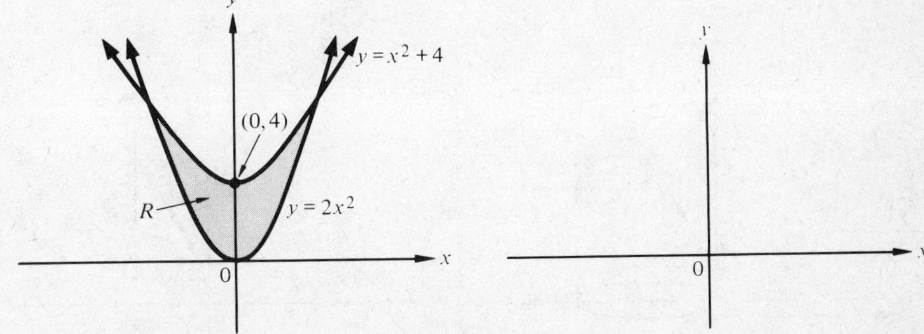

$(2, 8), \ (-2, 8)$

$(0, 0), \ (0, 4)$

$0, \ \ y - 4$

$4 - \dfrac{y}{2}$

$4y - \dfrac{y^2}{4}$

$4, \ 16, \ 12$

8π

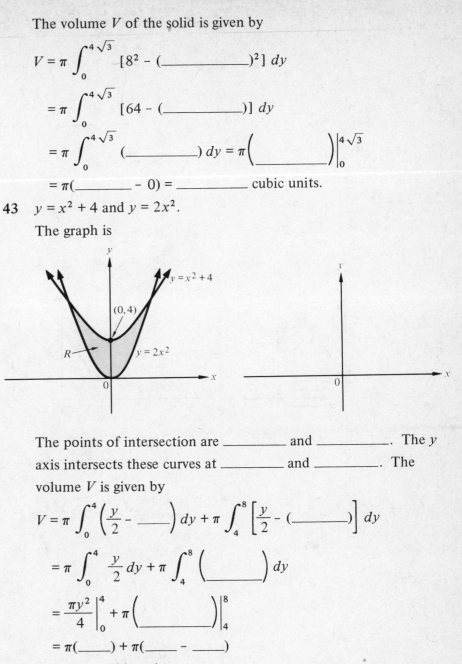

The points of intersection are ______ and ______. The y axis intersects these curves at ______ and ______. The volume V is given by

$$V = \pi \int_0^4 \left(\frac{y}{2} - \_\_\_\_\right) dy + \pi \int_4^8 \left[\frac{y}{2} - (\_\_\_\_\_\_)\right] dy$$

$$= \pi \int_0^4 \frac{y}{2}\, dy + \pi \int_4^8 \left(\_\_\_\_\_\_\right) dy$$

$$= \frac{\pi y^2}{4}\Big|_0^4 + \pi\left(\_\_\_\_\_\_\right)\Big|_4^8$$

$$= \pi(\_\_\_\_) + \pi(\_\_\_\_ - \_\_\_\_)$$

$$= \_\_\_\_ \text{ cubic units.}$$

In Problems 44 through 49, find the volume of the solid generated by rotating the region bounded by the given curves about the axis indicated.

44 $y^2 = 8x$, $y = 0$, and $x = 2$ about the line $x = 2$.

The graph is

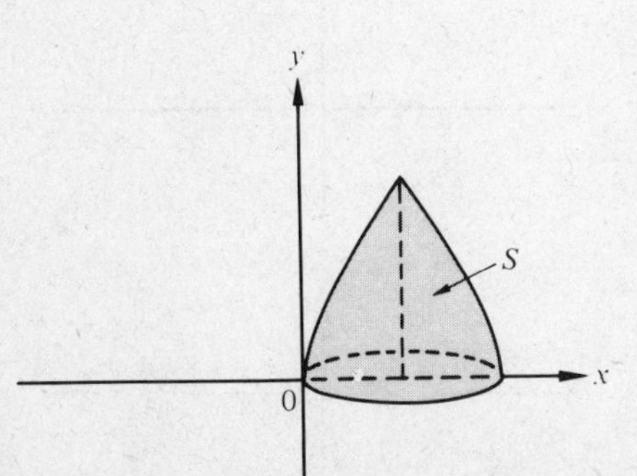

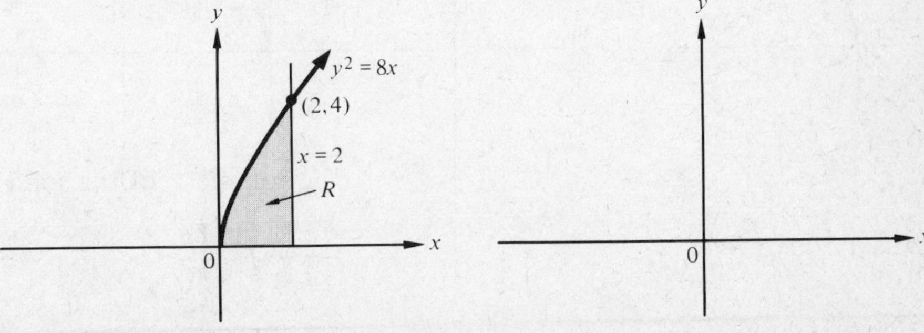

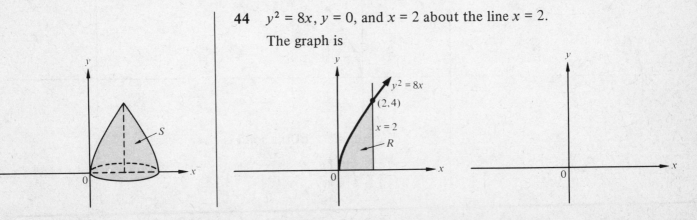

$2 - \frac{y^2}{8}$

$4 - \frac{y^2}{2} + \frac{y^4}{64}$

$4y - \frac{y^3}{6} + \frac{y^5}{320}$

$\frac{128}{15}, \frac{128\pi}{15}$

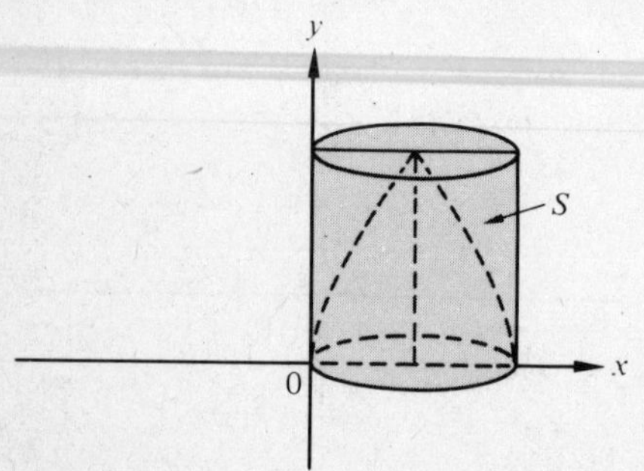

$2 - \frac{y^2}{8}$

$4y$

$16, \frac{128\pi}{15}$

$\frac{112\pi}{15}$

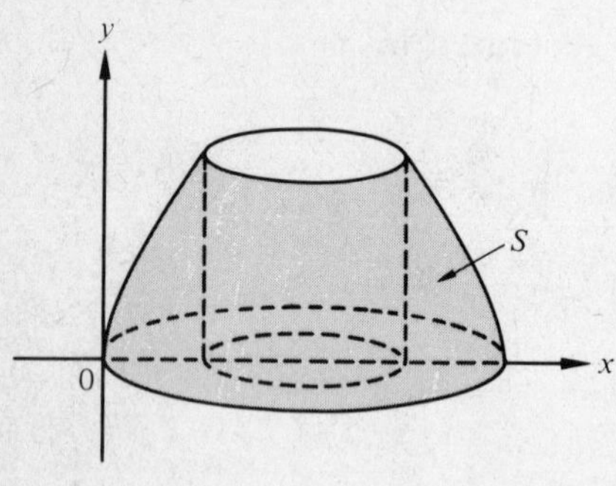

2^2

The volume V of the solid is given by

$$V = \pi \int_0^4 \left(\_\_\_\_\_\_\right)^2 dy$$

$$= \pi \int_0^4 \left(\_\_\_\_\_\_\_\_\_\_\right) dy$$

$$= \pi\left(\_\_\_\_\_\_\_\_\_\_\right)\Big|_0^4$$

$$= \pi\left(\_\_\_\_ - 0\right) = \_\_\_\_ \text{ cubic units.}$$

45 $y^2 = 8x$, $y = 4$, and $x = 0$ about the line $x = 2$.

The graph is

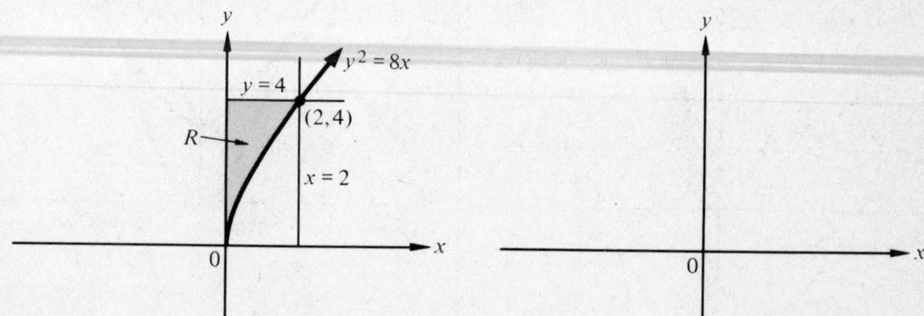

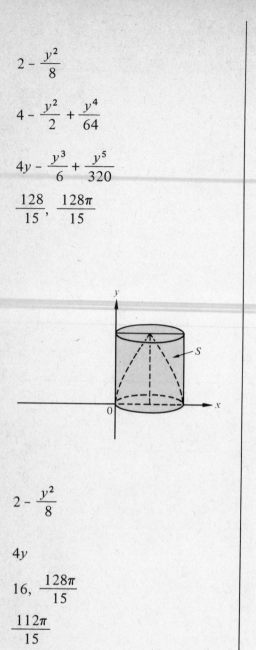

The volume V of the solid is given by

$$V = \pi \int_0^4 \left[2^2 - \left(\_\_\_\_\_\_\right)^2\right] dy$$

$$= \pi(\_\_\_\_)\Big|_0^4 - \pi \int_0^4 \left(4 - \frac{y^2}{2} + \frac{y^4}{64}\right) dy$$

$$= \pi(\_\_\_\_) - \left(\_\_\_\_\right)$$

$$= \_\_\_\_ \text{ cubic units.}$$

46 $y^2 = 8x$, $x = 2$, and $y = 0$ about the line $x = 4$.

The graph is

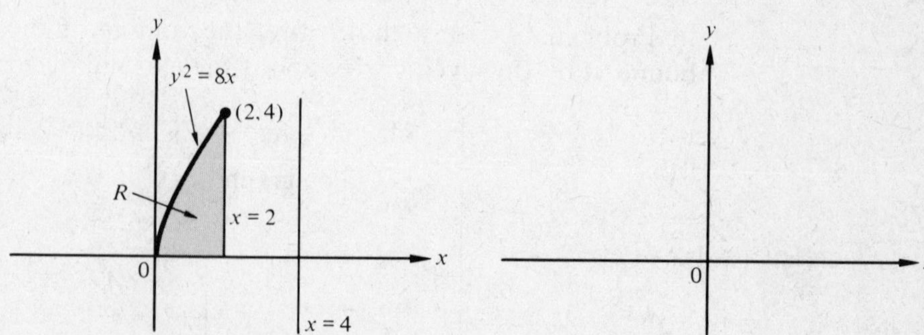

The volume V of the solid is given by

$$V = \pi \int_0^4 \left[\left(4 - \frac{y}{8}\right)^2 - \_\_\_\_\right] dy$$

$16 - y^2 + \frac{y^4}{64}$

$12 - y^2 + \frac{y^4}{64}$, $12y - \frac{y^3}{3} + \frac{y^5}{320}$

$\frac{448}{15}$

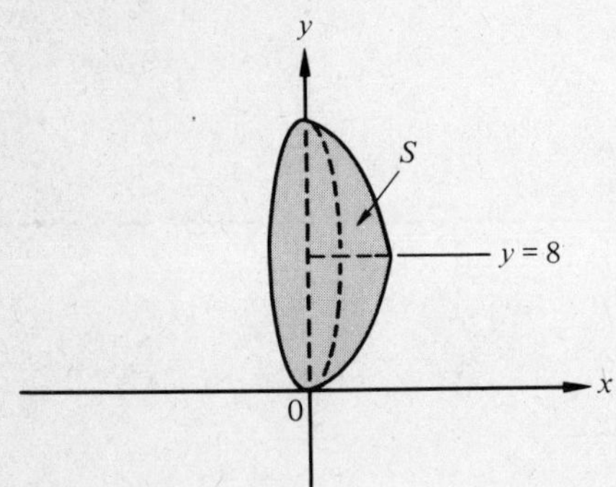

$8 - x^{3/2}$

$64 - 16x^{3/2} + x^3$

$64x - \frac{32}{5}x^{5/2} + \frac{x^4}{4}$

$\frac{576\pi}{5}$

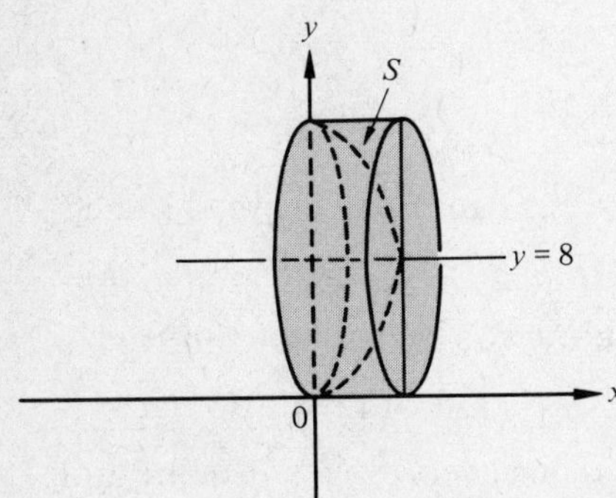

$8 - x^{3/2}$

$$= \pi \int_0^4 \left[\left(\underline{\qquad\qquad}\right) - 4\right] dy$$

$$= \pi \int_0^4 \left(\underline{\qquad\qquad}\right) dy = \pi \left(\underline{\qquad\qquad}\right)\Big|_0^4$$

$$= \pi \left(\underline{\qquad}\right) \text{ cubic units.}$$

47 $y^2 = x^3$, $y = 8$, and $x = 0$ about the line $y = 8$.

The graph is

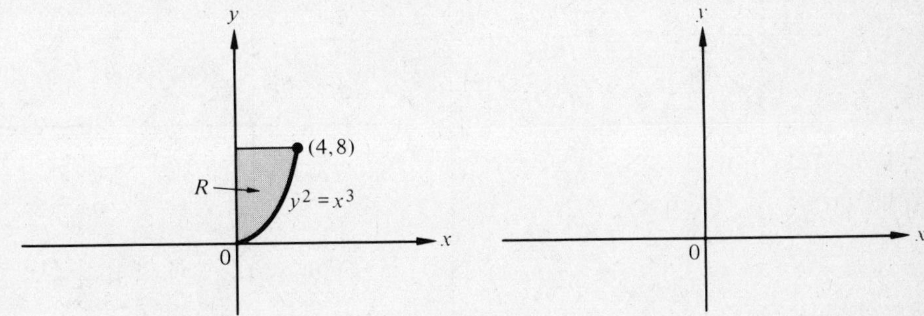

The volume V of the solid is given by

$$V = \pi \int_0^4 (\underline{\qquad\qquad})^2\, dx$$

$$= \pi \int_0^4 (\underline{\qquad\qquad})\, dx$$

$$= \pi \left(\underline{\qquad\qquad}\right)\Big|_0^4$$

$$= \underline{\qquad} \text{ cubic units.}$$

48 $y^2 = x^3$, $x = 4$, and $y = 0$ about the line $y = 8$.

The graph is

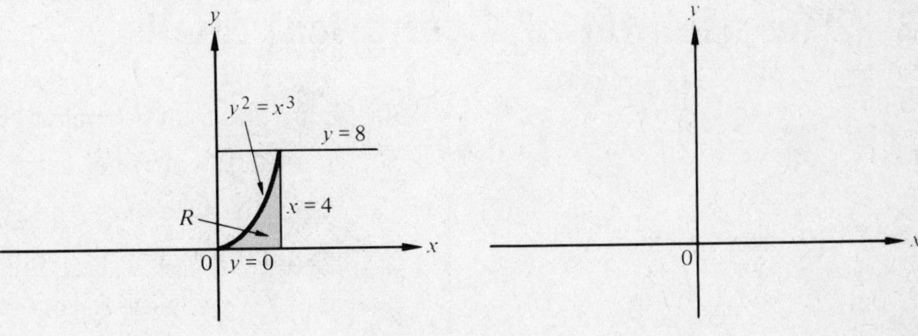

The volume V of the solid is given by

$$V = \pi \int_0^4 [8^2 - (\underline{\qquad\qquad})^2]\, dx$$

$64y$

$$= \pi(\_\_\_\_)\Big|_0^4 - \pi \int_0^4 (8 - x^{3/2})^2\, dx$$

$256, \dfrac{704\pi}{5}$

$$= \pi(\_\_\_\_) - \frac{576\pi}{5} = \_\_\_\_ \text{ cubic units.}$$

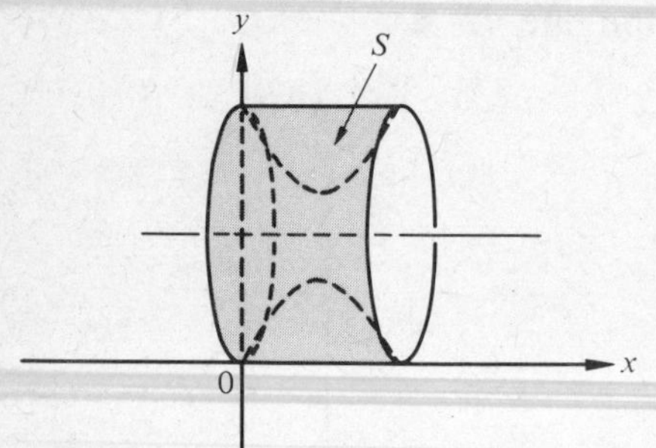

49 $y = 3x - x^2$, $y = 0$ about the line $y = 5$.
The graph is

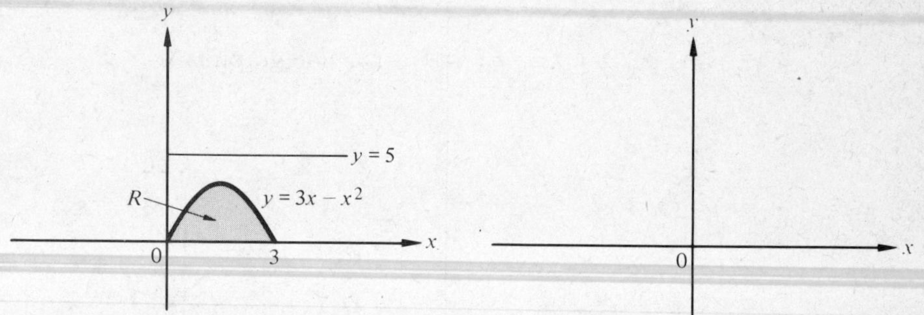

The volume V of the solid is given by

$5 - (3x - x^2)$

$$V = \pi \int_0^3 [5^2 - (\_\_\_\_\_\_\_\_)^2]\, dx$$

$10(3x - x^2) - (3x - x^2)^2$

$$= \pi \int_0^3 (\_\_\_\_\_\_\_\_\_\_\_\_)\, dx$$

$30x - 19x^2 + 6x^3 - x^4$

$$= \pi \int_0^3 (\_\_\_\_\_\_\_\_\_\_\_\_)\, dx$$

$15x^2 - \dfrac{19}{3}x^3 + \dfrac{3}{2}x^4 - \dfrac{x^5}{5}$

$$= \pi \left(\_\_\_\_\_\_\_\_\_\_\_\_\right)\Big|_0^3$$

$\dfrac{369\pi}{10}$

$$= \_\_\_\_ \text{ cubic units.}$$

2 The Method of Cylindrical Shells

$x[f(x) - g(x)]$

50 Let f and g be nonnegative continuous functions on $[a, b]$ such that $f(x) \geqslant g(x)$ holds for all x in $[a, b]$. Let R be the planar region between $x = a$ and $x = b$, below the graph f and above the graph of g. Then the volume V of the solid of revolution generated by revolving R about the y axis is, according to the method of cylindrical shells, given by $V = 2\pi \int_a^b (\_\_\_\_\_\_\_\_)\, dx$.

The graph is

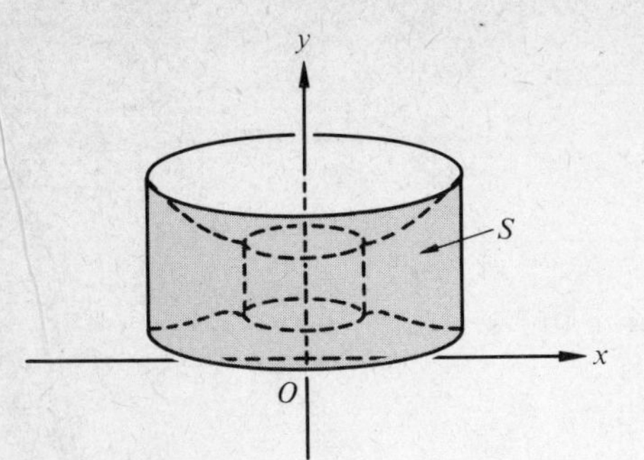

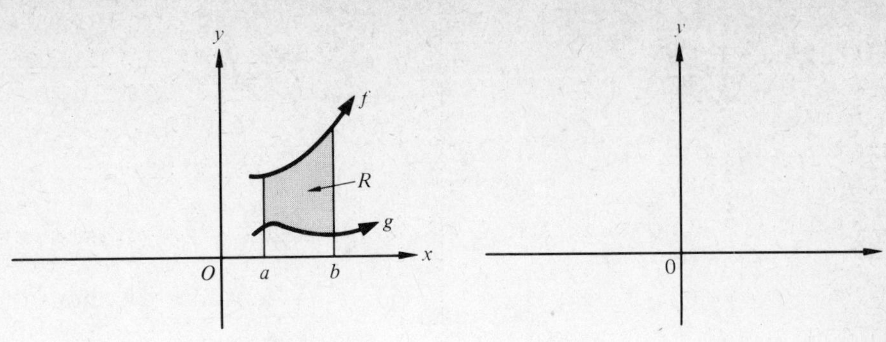

In Problems 51 through 61, use the cylindrical shell method to find the volume of the solid if the planar region bounded by the given curves is revolved about the indicated axis.

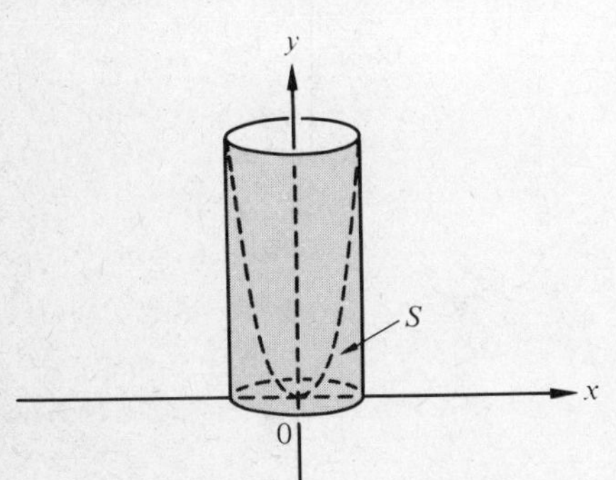

51 $y = x^2$, $y = 0$, and $x = 4$ about the y axis.

The graph is

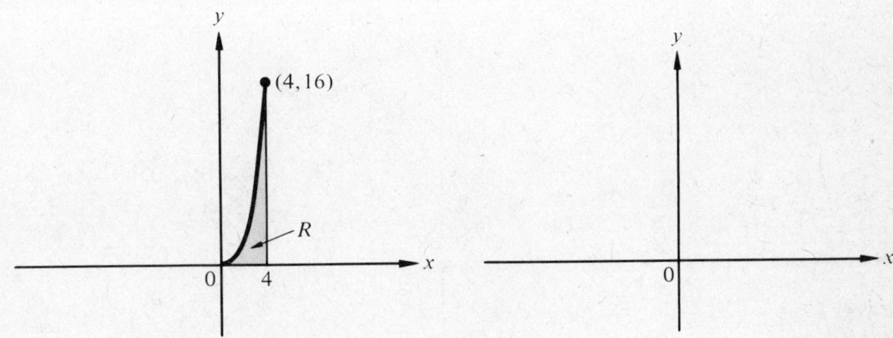

The volume V of the solid is given by

$x^2, \; x^3$

$$V = 2\pi \int_0^4 x(\_\_\_\_)\,dx = 2\pi \int_0^4 (\_\_\_\_)\,dx$$

$\frac{x^4}{4}, \; 128\pi$

$$= 2\pi\left(\_\_\_\_\right)\Big|_0^4 = \_\_\_\_ \text{ cubic units.}$$

52 $y = x^2 + 1$, $y = 0$, $x = 0$, and $x = 2$ about the line $x = -2$.

The graph is

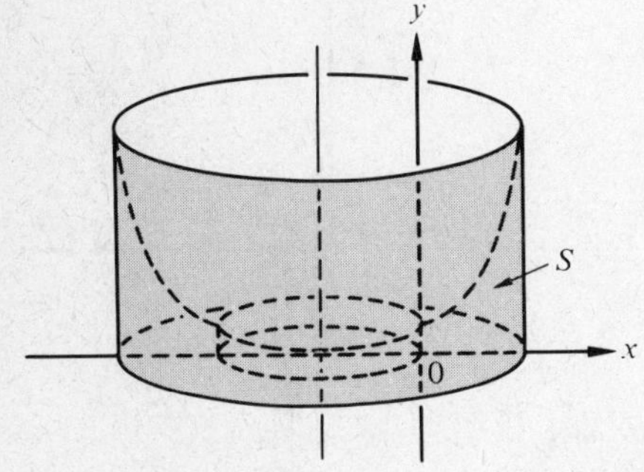

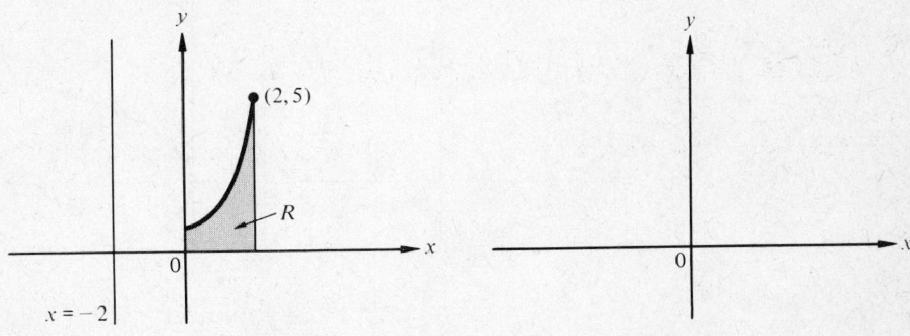

The volume V of the solid is given by

$x + 2, \; x^2 + 1$

$$V = 2\pi \int_0^2 (\_\_\_\_\_\_)(\_\_\_\_\_\_)\,dx$$

$x^3 + 2x^2 + x + 2$

$$= 2\pi \int_0^2 (\_\_\_\_\_\_\_\_\_\_\_\_) \, dx$$

$\frac{x^4}{4} + \frac{2x^3}{3} + \frac{x^2}{2} + 2x$

$$= 2\pi \left(\_\_\_\_\_\_\_\_\_\_\_\_ \right) \Big|_0^2$$

$\frac{46}{3}, \frac{92\pi}{3}$

$$= 2\pi \left(\_\_\_\_ - 0 \right) = \_\_\_\_ \text{ cubic units.}$$

53 $y^2 = 4x$ and $x = 4$ about the line $x = 4$.

The graph is

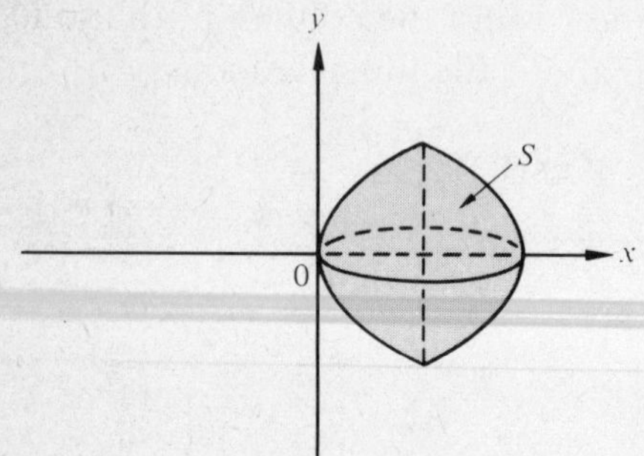

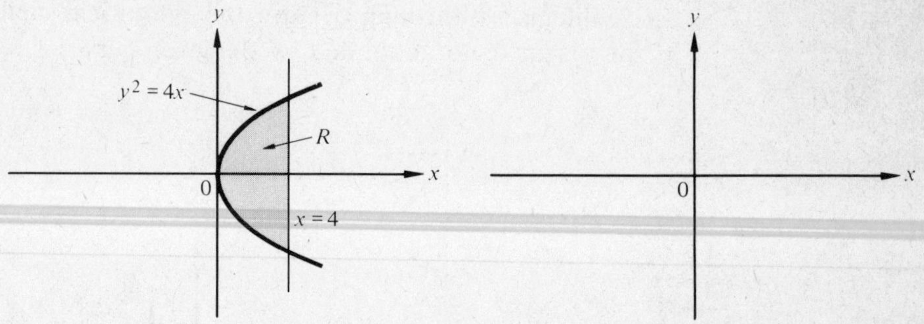

The volume V of the solid is given by

$4 - x$

$$V = 2\pi \int_0^4 (\_\_\_\_\_\_)(2\sqrt{4x}) \, dx$$

$4\sqrt{x} - x\sqrt{x}$

$$= 8\pi \int_0^4 (\_\_\_\_\_\_\_\_\_\_) \, dx$$

$\frac{8}{3}x^{3/2} - \frac{2}{5}x^{5/2}$

$$= 8\pi \left(\_\_\_\_\_\_\_\_\_\_\_\_ \right) \Big|_0^4$$

$\frac{128}{15}, \frac{1024\pi}{15}$

$$= 8\pi \left(\_\_\_\_ - 0 \right) = \_\_\_\_ \text{ cubic units.}$$

54 $y = x^2$ and $y^2 = x$ about the line $x = -1$.

The graph is

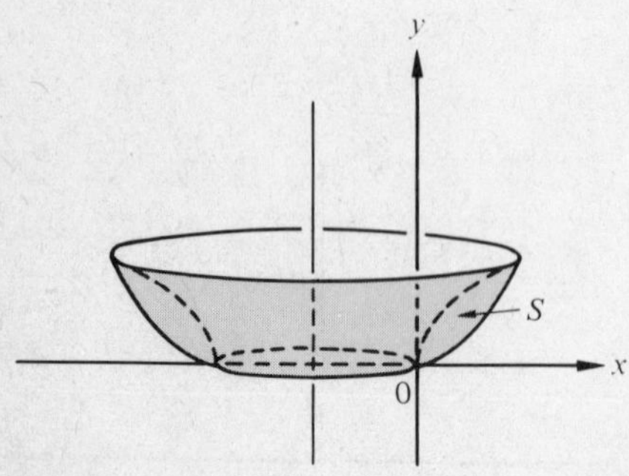

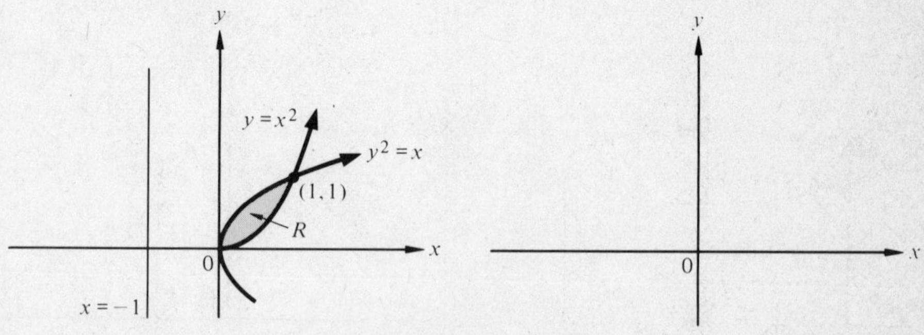

The volume V of the solid is given by

$\sqrt{x} - x^2$

$$V = 2\pi \int_0^1 (x + 1)(\_\_\_\_\_\_\_\_) \, dx$$

$-x^3 - x^2 + x\sqrt{x} + \sqrt{x}$

$$= 2\pi \int_0^1 (\_\_\_\_\_\_\_\_\_\_\_\_\_\_) \, dx$$

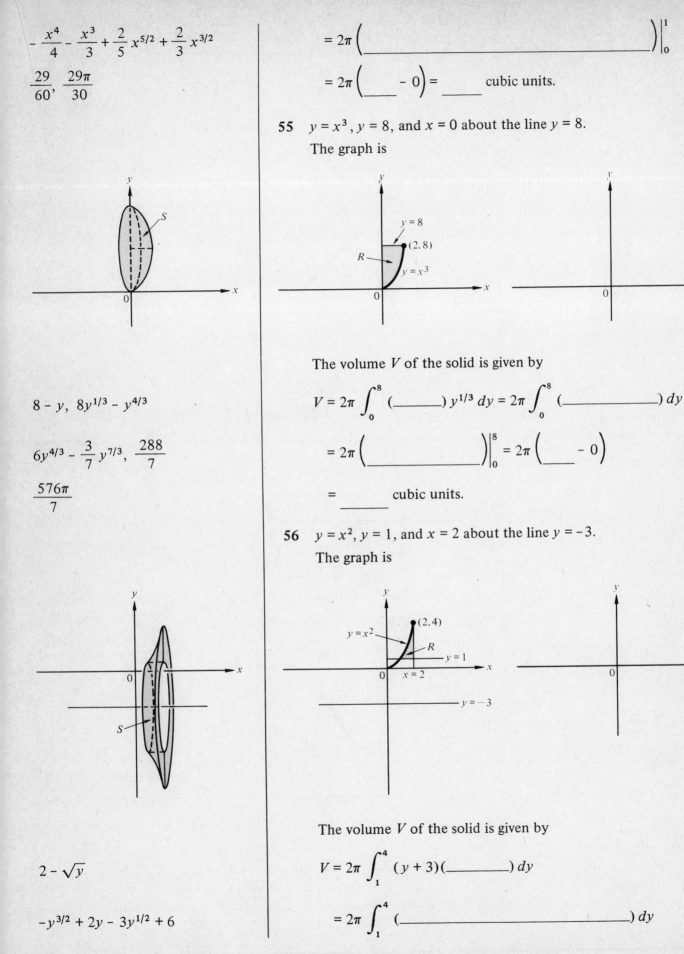

$-\frac{x^4}{4} - \frac{x^3}{3} + \frac{2}{5}x^{5/2} + \frac{2}{3}x^{3/2}$

$\frac{29}{60}, \frac{29\pi}{30}$

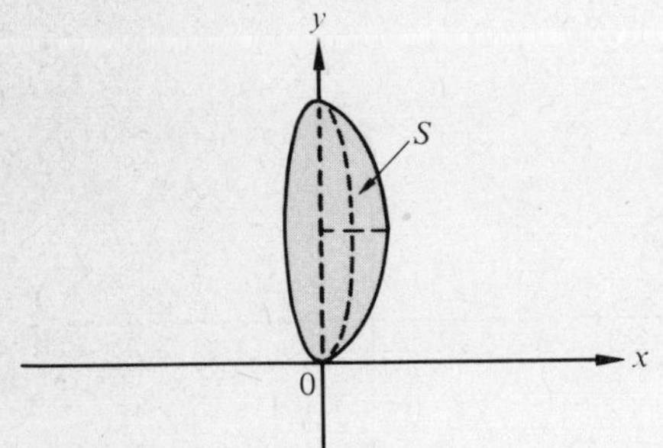

$8 - y,\ 8y^{1/3} - y^{4/3}$

$6y^{4/3} - \frac{3}{7}y^{7/3},\ \frac{288}{7}$

$\frac{576\pi}{7}$

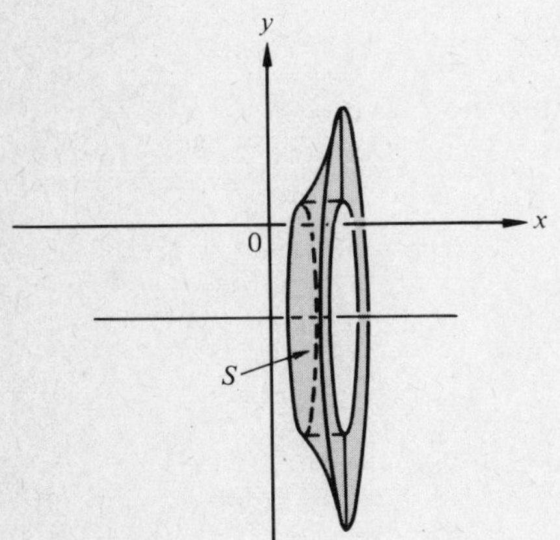

$2 - \sqrt{y}$

$-y^{3/2} + 2y - 3y^{1/2} + 6$

$= 2\pi\left(\underline{\hspace{6cm}} \right)\Big|_0^1$

$= 2\pi\left(\underline{\hspace{1cm}} - 0\right) = \underline{\hspace{1cm}}$ cubic units.

55 $y = x^3$, $y = 8$, and $x = 0$ about the line $y = 8$.

The graph is

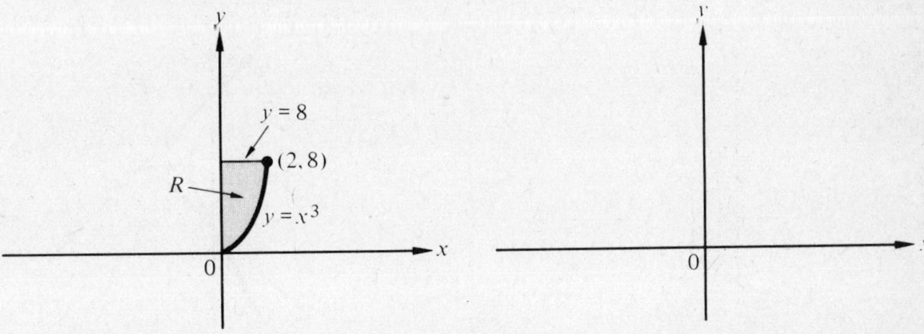

The volume V of the solid is given by

$V = 2\pi \int_0^8 (\underline{\hspace{2cm}})\, y^{1/3}\, dy = 2\pi \int_0^8 (\underline{\hspace{3cm}})\, dy$

$= 2\pi\left(\underline{\hspace{4cm}} \right)\Big|_0^8 = 2\pi\left(\underline{\hspace{1cm}} - 0\right)$

$= \underline{\hspace{1.5cm}}$ cubic units.

56 $y = x^2$, $y = 1$, and $x = 2$ about the line $y = -3$.

The graph is

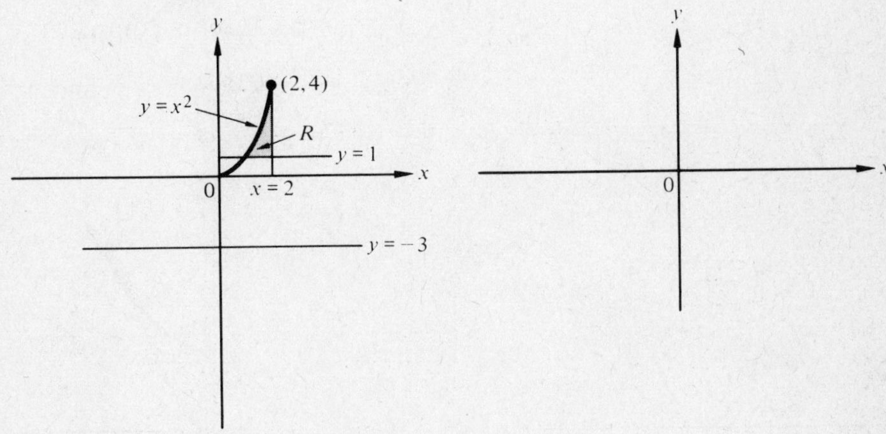

The volume V of the solid is given by

$V = 2\pi \int_1^4 (y + 3)(\underline{\hspace{2cm}})\, dy$

$= 2\pi \int_1^4 (\underline{\hspace{6cm}})\, dy$

$-\frac{2}{5}y^{5/2} + y^2 - 2y^{3/2} + 6y$

$\frac{23}{5}, \frac{66\pi}{5}$

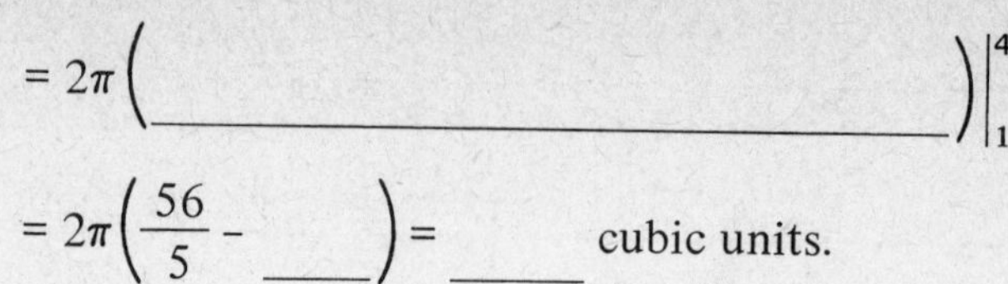

$= 2\pi\Big(\underline{\hspace{4cm}}\Big)\Big|_1^4$

$= 2\pi\left(\frac{56}{5} - \underline{\quad}\right) = \underline{\quad}$ cubic units.

57 $y = 2x$, $y = x$, and $x + y = 6$ about the x axis.

The graph is

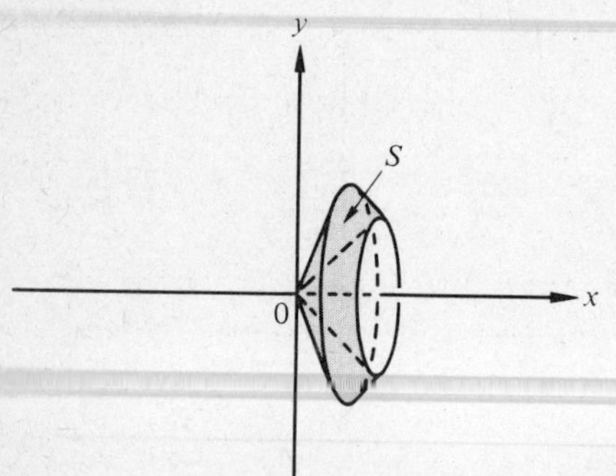

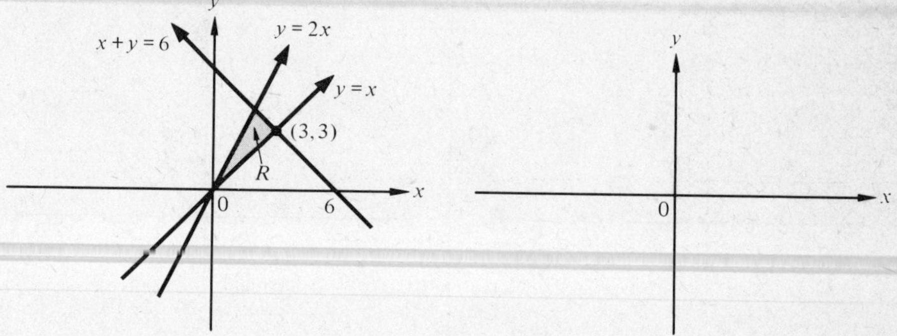

The volume V of the solid is given by

$V = 2\pi \int_0^3 y\left(y - \frac{y}{2}\right) dy + 2\pi \int_3^4 \Big(\underline{\hspace{3cm}}\Big) dy$

$= 2\pi \int_0^3 \Big(\underline{\quad}\Big) dy + 2\pi \int_3^4 \Big(\underline{\hspace{2cm}}\Big) dy$

$= 2\pi\Big(\underline{\quad}\Big)\Big|_0^3 + 2\pi\Big(\underline{\hspace{2cm}}\Big)\Big|_3^4$

$= 2\pi\Big(\underline{\quad}\Big) + 2\pi\Big(\underline{\quad}\Big)$

$= \underline{\quad}$ cubic units.

$y\left(6 - y - \frac{y}{2}\right)$

$\frac{y^2}{2}, 6y - \frac{3y^2}{2}$

$\frac{y^3}{6}, 3y^2 - \frac{y^3}{2}$

$\frac{9}{2}, \frac{5}{2}$

14π

58 $y = x^2 + 4$, $y = 2x^2$, and $x = 0$ about the y axis.

The graph is

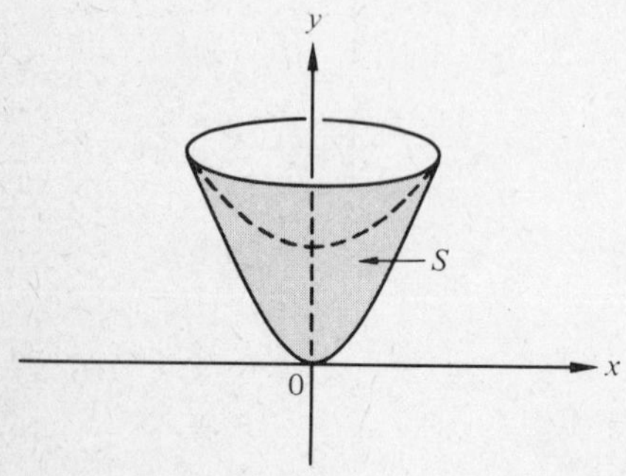

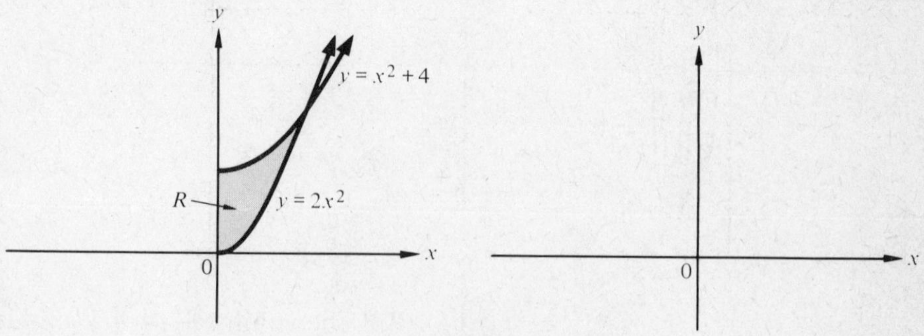

The volume V of the solid is given by

$V = 2\pi \int_0^2 x(\underline{\hspace{2cm}} - 2x^2)\, dx$

$x^2 + 4$

$4x - x^3,\ 2x^2 - \dfrac{x^4}{4}$

8π

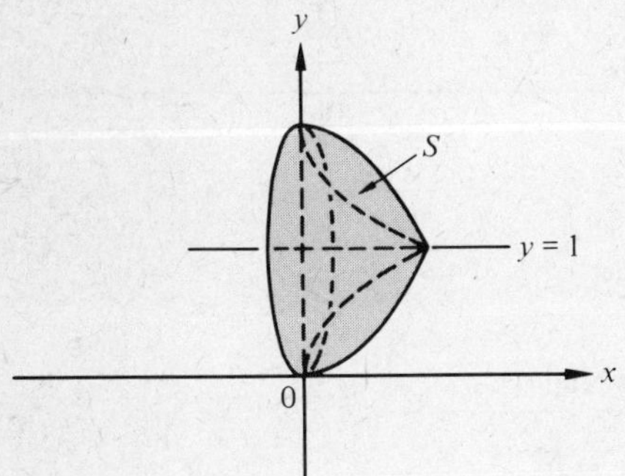

$\sqrt{y} - y^2$

$y^3 - y^2 - y^{3/2} + y^{1/2}$

$\dfrac{y^4}{4} - \dfrac{y^3}{3} - \dfrac{2}{5}y^{5/2} + \dfrac{2}{3}y^{3/2},\ \dfrac{11}{60}$

$\dfrac{11\pi}{30}$

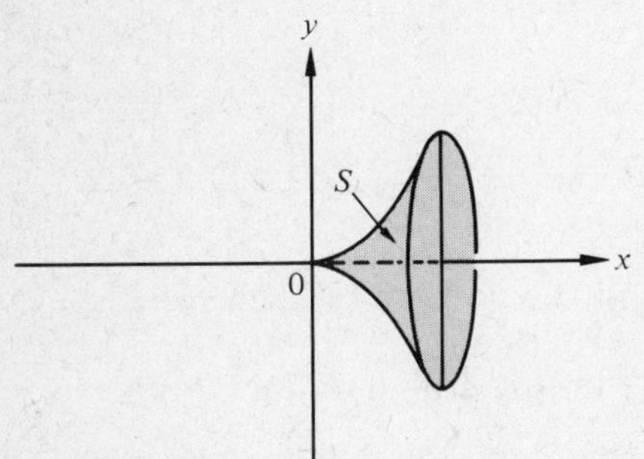

$1 - \sqrt{y},\ \ y - y^{3/2}$

$\dfrac{y^2}{2} - \dfrac{2y^{5/2}}{5},\ \dfrac{\pi}{5}$

$$= 2\pi \int_0^2 (\_\_\_\_\_\_)\,dx = 2\pi\left(\_\_\_\_\_\_\right)\Big|_0^2$$

$= \_\_\_\_$ cubic units.

59 $y = x^2$ and $y = \sqrt{x}$ about the line $y = 1$.

The graph is

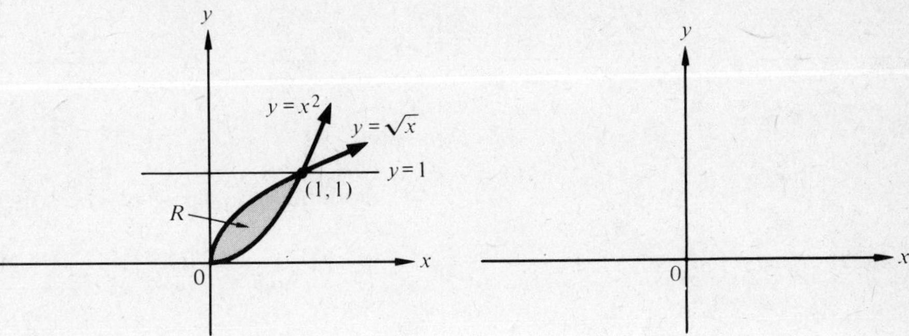

The volume V of the solid is given by

$$V = 2\pi \int_0^1 (1 - y)(\_\_\_\_\_\_)\,dy$$

$$= 2\pi \int_0^1 (\_\_\_\_\_\_\_\_\_\_\_\_)\,dy$$

$$= 2\pi\left(\_\_\_\_\_\_\_\_\_\_\_\_\right)\Big|_0^1 = 2\pi\left(\_\_\_\right)$$

$= \_\_\_\_$ cubic unit.

60 $y = x^2$, $y = 0$, and $x = 1$ about the x axis.

The graph is

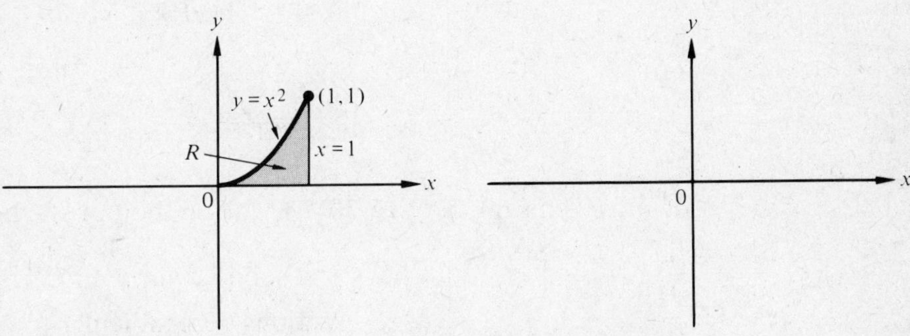

The volume V of the solid is given by

$$V = 2\pi \int_0^1 y(\_\_\_\_\_\_)\,dy = 2\pi \int_0^1 (\_\_\_\_\_\_)\,dy$$

$$= 2\pi\left(\_\_\_\_\_\_\right)\Big|_0^1 = \_\_\_\_$$ cubic unit.

61 $y = x^2 + 2$, $y = 3x$, and $x = 0$ about the line $x = 2$.

The graph is

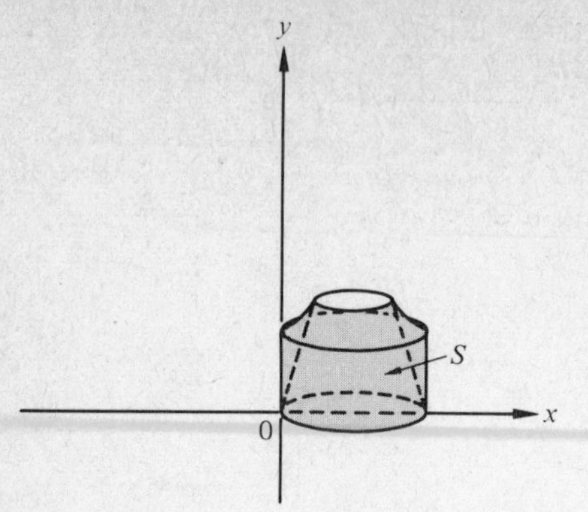

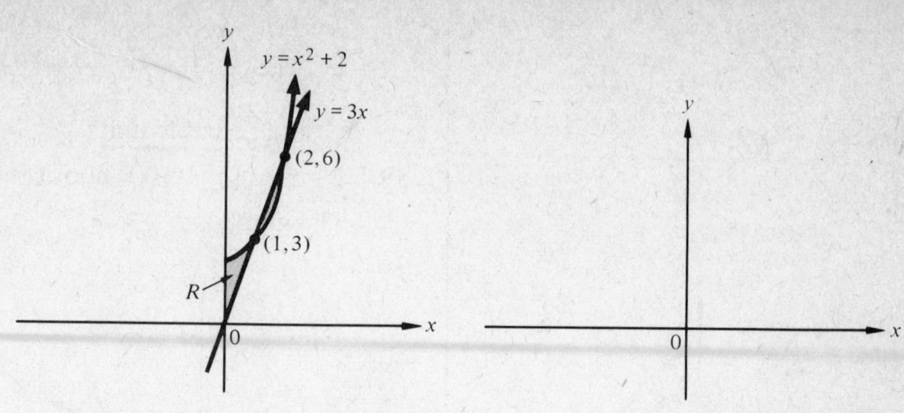

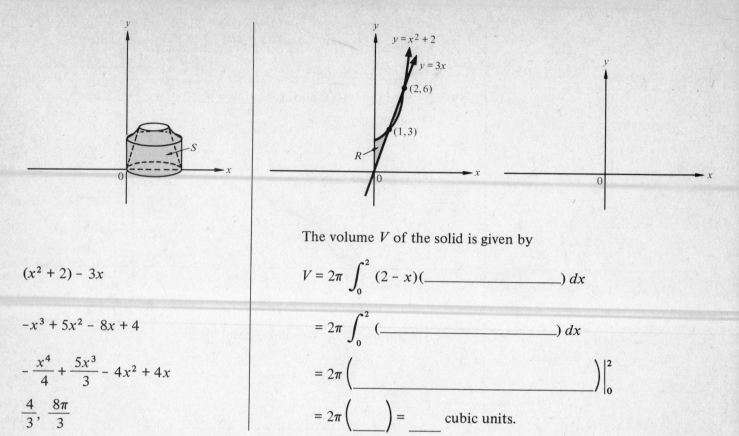

The volume V of the solid is given by

$(x^2 + 2) - 3x$

$$V = 2\pi \int_0^2 (2 - x)(\_\_\_\_\_\_\_\_)\, dx$$

$-x^3 + 5x^2 - 8x + 4$

$$= 2\pi \int_0^2 (\_\_\_\_\_\_\_\_)\, dx$$

$-\dfrac{x^4}{4} + \dfrac{5x^3}{3} - 4x^2 + 4x$

$$= 2\pi \left(\_\_\_\_\_\_\_\_\right)\Big|_0^2$$

$\dfrac{4}{3}, \dfrac{8\pi}{3}$

$$= 2\pi \left(\_\_\_\right) = \_\_\_ \text{ cubic units.}$$

3 Volumes by the Method of Slicing

62 Let S be a solid lying between two planes that are perpendicular to a reference axis, denoted as the s axis, at $s = a$ and $s = b$. If the area of the plane section of S perpendicular to the s axis at point s is given by $A(s)$, then the volume V of S is given by

$V =$ ________.

$\int_a^b A(s)\, ds$

In Problems 63 through 70, use the method of slicing to find the volume of each solid.

63 The solid having a circular base of radius 4 inches and whose cross sections perpendicular to one of the diameters of the base are squares.

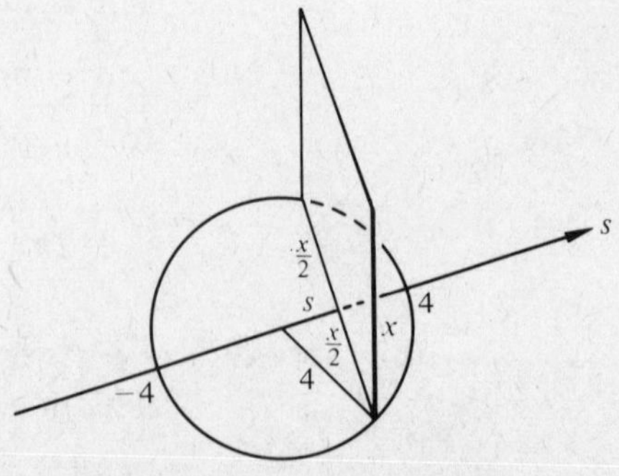

A typical cross section of the solid is illustrated in the figure. Let x inches be the length of the side of the square. The area of the square $A(s) =$ ____. But $\frac{x^2}{4} + s^2 =$ ____, so that $\frac{x^2}{4} =$ ________

x^2, 16, $16 - s^2$

or $x^2 =$ __________. Therefore, the volume V of the solid is given by

$64 - 4s^2$

$$V = \int_{-4}^{4} (\_\_\_\_\_\_)\, ds = \left(\_\_\_\_\_\_\right)\Big|_{-4}^{4}$$

$64 - 4s^2$, $64s - \frac{4s^3}{3}$

$$= \left(256 - \frac{256}{3}\right) - \left(\_\_\_\_\_\_\right)$$

$-256 + \frac{256}{3}$

$=$ ____ cubic inches.

$\frac{1024}{3}$

64 The solid having a circular base of radius 4 centimeters and whose cross sections perpendicular to a fixed diameter of the base are equilateral triangles.

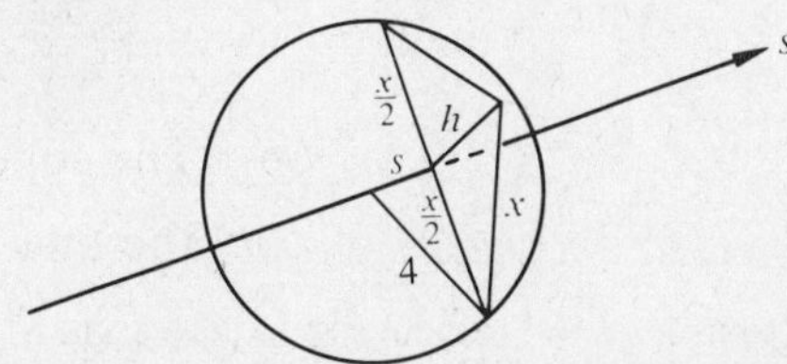

A typical cross section of the solid is illustrated in the figure. Let x centimeters be the length of the side of the triangle and h meters be its height. Then its area A is given by $A(s) =$ ____.

$\frac{1}{2}xh$

But $\left(\frac{x}{2}\right)^2 + s^2 = 16$, so that $x^2 =$ ________. Also,

$64 - 4s^2$

$\left(\frac{x}{2}\right)^2 + h^2 = x^2$, so that $h^2 =$ ____ or $h =$ ____. Thus, $A(s) =$

$\frac{3x^2}{4}$, $\frac{\sqrt{3}x}{2}$

$\frac{1}{2}x \cdot \frac{\sqrt{3}x}{2} =$ ____ $= \frac{\sqrt{3}}{4}$(______) $=$ ________.

$\frac{\sqrt{3}x^2}{4}$, $64 - 4s^2$, $\sqrt{3}(16 - s^2)$

Therefore, the volume V of the solid is given by

$$V = \int_{-4}^{4} (\_\_\_\_\_\_)\, ds = \sqrt{3}\left(\_\_\_\_\_\_\right)\Big|_{-4}^{4}$$

$\sqrt{3}(16 - s^2)$, $16s - \frac{s^3}{3}$

$$= \sqrt{3}\left(\frac{128}{3} - \_\_\_\_\right) = \_\_\_\_ \text{ cubic centimeters.}$$

$-\frac{128}{3}$, $\frac{256\sqrt{3}}{3}$

65 The solid having an elliptical base with semimajor axis 5 inches and semiminor axis 4 inches and every cross section perpendicular to the major axis is an isosceles triangle with altitude 7 inches.

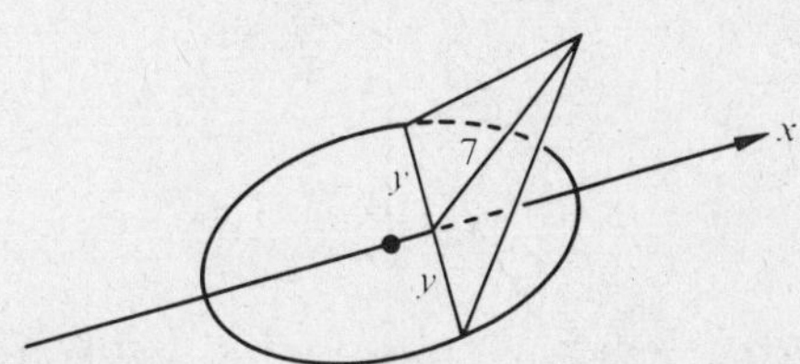

A typical cross section of the solid is illustrated in the figure. Take the x axis as a reference axis with origin at the center of the

$\frac{x^2}{25} + \frac{y^2}{16} = 1$

ellipse. The equation of the ellipse is ________________. Let the base of the isosceles triangle by $2y$ inches. Its altitude is 7

$7y, \; \frac{4}{5}\sqrt{25 - x^2}$

inches. Thus, its area $A(x) =$ ____ $= 7($________$)$.

Therefore, the volume V of the solid is given by

$$V = \int_{-5}^{5} A(x)\,dx$$

$\sqrt{25 - x^2}$

$$= \frac{28}{5} \int_{-5}^{5} (\_\_\_\_\_\_\_\_)\,dx$$

$\frac{1}{2}$ area of a circle of radius 5

$$= \frac{28}{5} \left(\_\_\_\_\_\_\_\_\_\_\_\_\_\_\_\_ \right)$$

$\frac{25\pi}{2}, \; 70\pi$

$$= \frac{28}{5} \left(\_\_\_\_ \right) = \_\_\_\_ \text{ cubic inches.}$$

66 The solid whose base is a triangle cut from the first quadrant by the line $2x + 3y = 6$ and whose cross sections perpendicular to the x axis are semicircles.

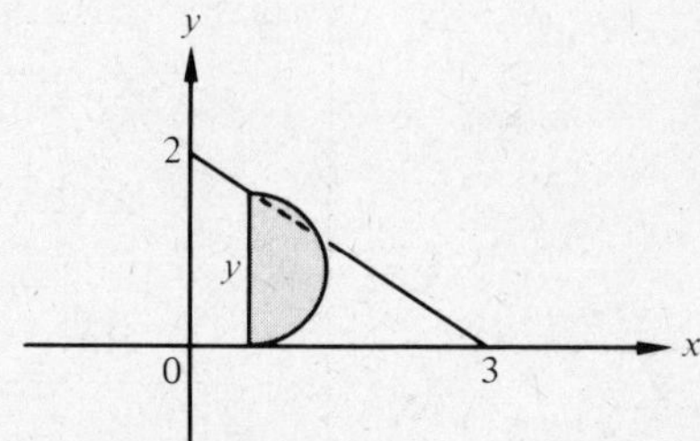

A typical cross section of the solid is illustrated in the figure.

$\frac{2}{3}(3 - x)$

Since $2x + 3y = 6$, then $y =$ ____________. y is the diameter of

$\left(\frac{y}{2}\right)^2$

the semicircle, so that its area is $A(x) = \pi($______$) = \frac{1}{4}\pi y^2 =$

$\frac{4}{9}(3 - x)^2, \; \frac{\pi}{9}(3 - x)^2$

$\frac{1}{4}\pi($____________$) =$ ____________. The volume V is given by

$\frac{\pi}{9}(3 - x)^2$

$$V = \int_0^3 A(x)\,dx = \int_0^3 \left(\_\_\_\_\_\_\_\_\_\_\_\_ \right) dx$$

$$= -\frac{\pi}{9}\left[\frac{(3 - x)^3}{3}\right]\Bigg|_0^3$$

$-\pi$

$$= (0 - \_\_\_\_)$$

π

$$= \_\_\_\_ \text{ cubic units.}$$

67 A tower 80 feet tall whose horizontal cross sections x feet from the top are squares of side $\frac{1}{400}(x+40)^2$ feet.

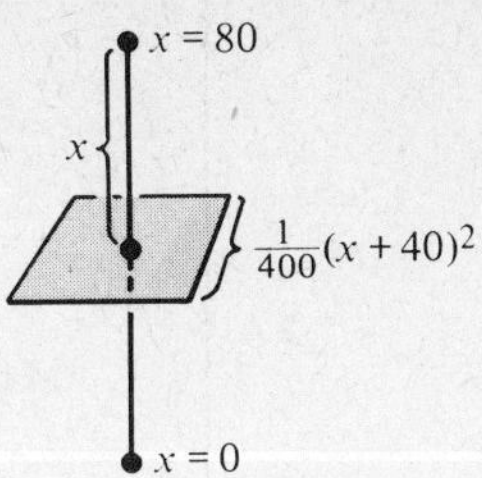

A typical cross section of the solid is illustrated in the figure. The area A of the cross section of the tower is given by $A(x) =$ ________________ . Therefore, the volume V of the tower is given by

> $\frac{1}{160{,}000}(x+40)^4$

$$V = \int_0^{80} A(x)\,dx = \int_0^{80} \left(\underline{\qquad\qquad} \right) dx$$

> $\frac{1}{160{,}000}(x+40)^4$

$$= \frac{1}{160{,}000} \left(\underline{\qquad\qquad} \right) \Big|_0^{80}$$

> $\frac{(x+40)^5}{5}$

$$= \frac{1}{800{,}000} [(120)^5 - \underline{\qquad}]$$

> $(40)^5$

$$= \underline{\qquad} \text{ cubic feet.}$$

> 30,976

68 The solid right circular cone of height 10 inches and of base radius 3 inches.

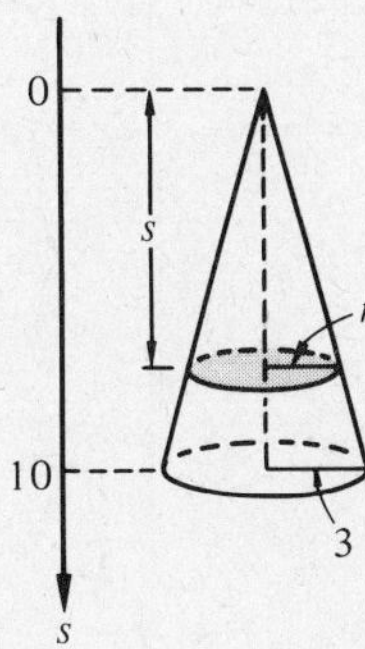

Choose the reference axis to be parallel to the axis of the cone. From the similar triangles in the figure, we have $\frac{s}{10} = \underline{\quad}$, so that $r = \underline{\quad}$. The area A of the cross section is $A(s) = \pi r^2 = \pi\left(\underline{\quad}\right)$. Therefore, the volume V of the solid right circular cone is

> $\frac{r}{3}$
>
> $\frac{3s}{10}$
>
> $\frac{9s^2}{100}$

$$V = \int_0^{10} A(s)\,ds = \int_0^{10} \left(\underline{\quad} \right) ds$$

> $\frac{9s^2}{100}$

$\frac{s^3}{3}$

$= \frac{9}{100}\left(\_\_\_\_\right)\Big|_0^{10}$

1000

$= \frac{3}{100}(\_\_\_\_ - 0)$

30

$= \_\_\_\_$ cubic inches.

69 A solid sphere of radius 3 centimeters.

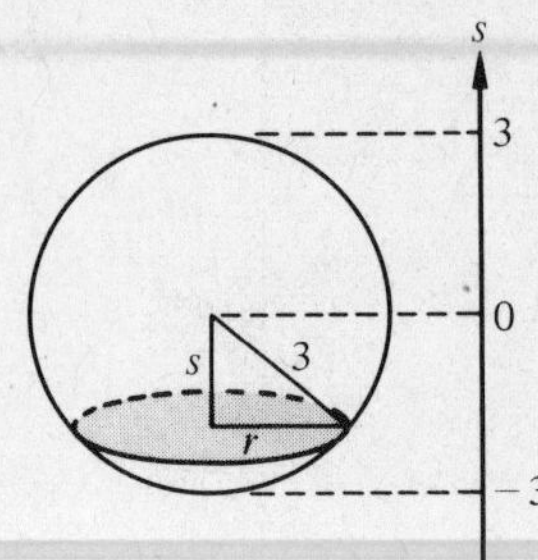

The cross section of the sphere at level s is a circle of radius r and area $A(s) = \_\_\_\_$. But $r^2 = \_\_\_\_$ so that $A(s) = \_\_\_\_$. Hence, the volume V is given by

$\pi r^2,\ 9 - s^2,\ \pi(9 - s^2)$

$$V = \int_{-3}^{3} (\_\_\_\_)\, ds = \pi\left(\_\_\_\_\right)\Big|_{-3}^{3}$$

$\pi(9 - s^2),\ 9s - \frac{s^3}{3}$

$= \pi(18 - \_\_\_\_)$

-18

$= \_\_\_\_$ cubic centimeters.

36π

70 The solid is a football which is 16 inches long and a plane section containing a seam of the cover is an ellipse 8 inches broad. The leather is so stiff that every plane section is a square.

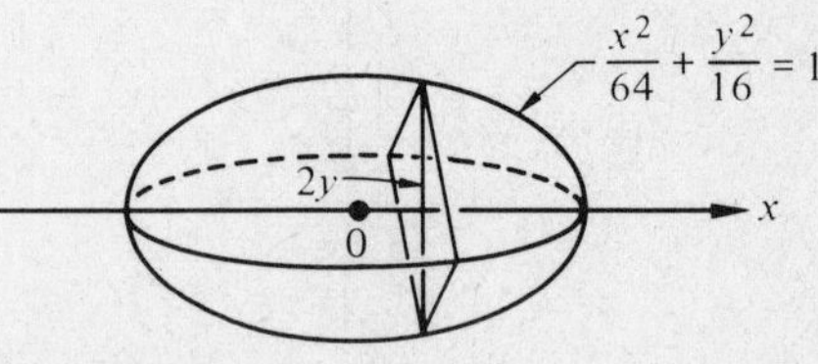

Let the center of the football coincide with the origin of the x axis. The equation of the ellipse is $\_\_\_\_$ so that

$\frac{x^2}{64} + \frac{y^2}{16} = 1$

$y^2 = \_\_\_\_$. The diameter of a square cross section lies

$16\left(1 - \frac{x^2}{64}\right)$

in the plane of the ellipse and has length $\_\_\_\_$. Therefore, the

$2y$

side of the square is $\_\_\_\_$. The area of the cross section is

$\sqrt{2}\,y$

$A(x) = \_\_\_\_ = \_\_\_\_$. The volume V of the football is

$2y^2,\ 32\left(1 - \frac{x^2}{64}\right)$

$$V = \int_{-8}^{8} A(x)\, dx = \int_{-8}^{8} \left(\_\_\_\_\right) dx$$

$32 - \frac{x^2}{2}$

$32x - \dfrac{x^3}{6}$

$-\dfrac{512}{3}$

$\dfrac{1024}{3}$

$= \left(\underline{\qquad\qquad} \right)\Big|_{-8}^{8}$

$= \dfrac{512}{3} - \left(\underline{\qquad} \right)$

$= \underline{\qquad}$ cubic inches.

a solid cone

71 **If B is an admissible planar region and p is a point not lying in the same plane as B, then the three-dimensional region R consisting of all points lying on straight line segments between p and the points of B is called a __________________ with vertex p and base B.**

$A(s)$, $\dfrac{A(h)\,h}{3}$

72 **Using the method of slicing, the volume V of a solid cone whose cross-sectional area is A and altitude h is given by**

$$V = \int_0^h (\underline{\qquad})\, ds = \underline{\qquad\qquad}.$$

5, 75

73 Find the volume of a pyramid with a square base 5 inches on each side if the perpendicular distance from the vertex to the base is 9 inches. Such a pyramid is a solid cone with a square base; hence, using Problem 71 we have $V = \dfrac{1}{3}(\underline{\quad})^2 \cdot 9 = \underline{\quad}$ cubic inches.

74 A solid right circular cylinder has a radius of 4 inches. A wedge is cut from this cylinder by a plane through a diameter of the base and inclined to the base at an angle of 30°. Find the volume of the wedge.

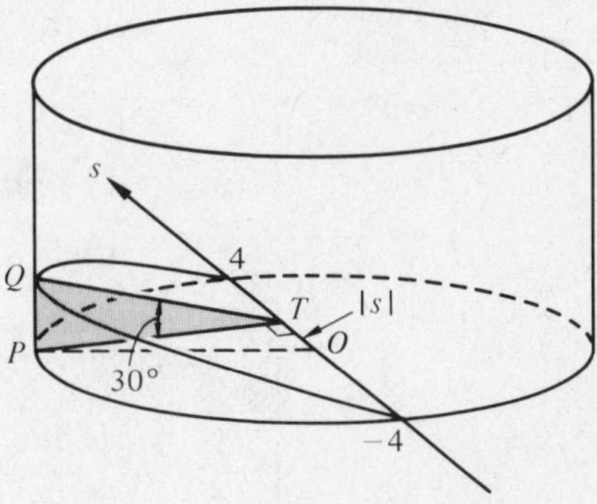

A typical wedge with a triangular cross section TPQ is illustrated in the figure. The area $A(s) = \dfrac{1}{2}\,|\overline{PQ}| \cdot |\overline{PT}|$. Since OTP is a right triangle, $|\overline{OT}| = |s|$ and $|\overline{OP}| = \underline{\quad}$, we have $|\overline{OP}|^2 =$ __________________, so that $|\overline{PT}|^2 =$ __________________ = __________. In the triangle TPQ, we have $|\overline{PQ}| =$ ____________.

Therefore, $A(s) = \dfrac{1}{2}\,|\overline{PQ}| \cdot |\overline{PT}| =$ __________________ = $\dfrac{1}{2}(\underline{\qquad\qquad})\tan 30°$.

4

$|\overline{OT}|^2 + |\overline{PT}|^2$, $|\overline{OP}|^2 - |\overline{OT}|^2$

$16 - s^2$, $|\overline{PT}| \tan 30°$

$\dfrac{1}{2}\,|\overline{PT}|^2 \tan 30°$

$16 - s^2$

$(16 - s^2) \tan 30°$

$16s - \dfrac{s^3}{3}$

$\dfrac{128}{3\sqrt{3}}$

By the method of slicing, the volume V is given by

$$V = \int_{-4}^{4} A(s)\, ds = \frac{1}{2}\int_{-4}^{4} (\underline{\qquad\qquad\qquad})\, ds$$

$$= \frac{1}{2\sqrt{3}}\left(\underline{\qquad\qquad}\right)\Big|_{-4}^{4}$$

$$= \underline{\qquad\quad} \text{ cubic inches.}$$

75 A solid right circular cylinder has a radius of 3 centimeters. A wedge is cut from this cylinder by a plane through a diameter of the base and inclined to the base at an angle of 45°. Find the volume of the wedge.

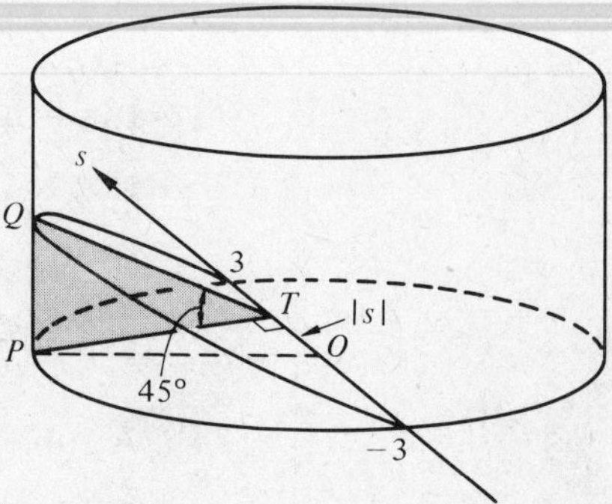

$\dfrac{1}{2}|\overline{PQ}| \cdot |\overline{PT}|$

$|s|$, $|\overline{OT}|^2 + |\overline{PT}|^2$

$|\overline{OP}|^2 - |\overline{OT}|^2$, $9 - s^2$

$|\overline{PT}| \tan 45°$, $|\overline{PT}|$

$\dfrac{1}{2}|\overline{PT}|^2$, $9 - s^2$

A typical wedge with a triangular cross section TPQ is illustrated in the figure. The area A is given by $A(s) = \underline{\qquad\qquad}$. $|\overline{OT}| = \underline{\quad}$ and $|\overline{OP}| = 3$; also, $|\overline{OP}|^2 = \underline{\qquad\qquad\qquad}$, so that $|\overline{PT}|^2 = \underline{\qquad\qquad\qquad} = \underline{\qquad\quad}$. In the triangle TPQ, we have $|\overline{PQ}| = \underline{\qquad\qquad\qquad} = \underline{\qquad\quad}$. Therefore, $A(s) = \frac{1}{2}|\overline{PQ}| \cdot |\overline{PT}| = \underline{\qquad\qquad} = \frac{1}{2}(\underline{\qquad\quad})$.

$9 - s^2$

$9s - \dfrac{s^3}{3}$

18

Hence, the volume V is given by

$$V = \int_{-3}^{3} A(s)\, ds = \frac{1}{2}\int_{-3}^{3} (\underline{\qquad\qquad})\, ds$$

$$= \frac{1}{2}\left(\underline{\qquad\qquad}\right)\Big|_{-3}^{3}$$

$$= \underline{\qquad} \text{ cubic centimeters.}$$

76 A tent is made by stretching canvas from a circle of radius 3 meters to a semicircular rib erected at right angles to the base and meeting the base at the ends of a diameter. Find the volume enclosed by the tent.

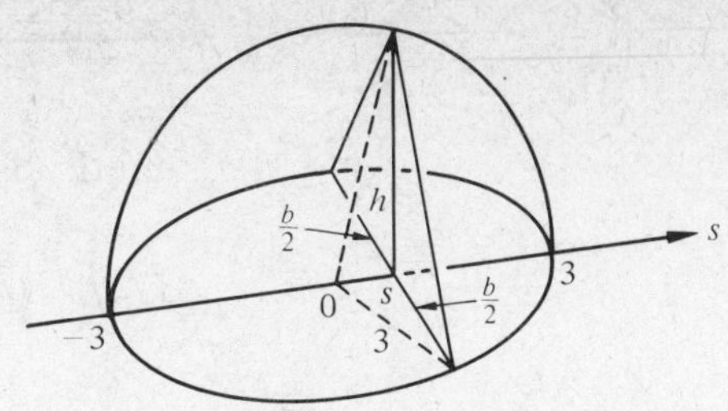

A typical cross section of the base is illustrated in the figure. The area A of the base is given by $A(s) =$ ______ . Notice that

$\frac{1}{2}bh$

$s^2 + h^2 =$ ____ and $s^2 + \left(\frac{b}{2}\right)^2 =$ ____, so that $h^2 =$ ________ and

$9,\ 9,\ 9 - s^2$

$b^2 =$ ______________. Then

$4(9 - s^2)$

$A(s) = \frac{1}{2}bh = \frac{1}{2}(2\sqrt{9 - s^2})$ (____________) = __________.

$\sqrt{9 - s^2},\ 9 - s^2$

Therefore, the volume is given by

$$V = \int_{-3}^{3} A(s)\,ds = \int_{-3}^{3} (\_\_\_\_\_\_)\,ds$$

$9 - s^2$

$$= \left(\_\_\_\_\_\_\right)\Big|_{-3}^{3}$$

$9s - \frac{s^3}{3}$

$=$ ____ cubic meters.

36

4 Arc Length and Surface Area

4.1 Arc Length of the Graph of a Function

77 Let f be a function with a continuous first derivative on some open interval I containing the closed interval $[a, b]$. Then the arc length s of the portion of the graph of f between $(a, f(a))$ and $(b, f(b))$ is given by $s = \int_a^b$ (______________________) dx.

$\sqrt{1 + [f'(x)]^2}$

78 If the equation of the curve between two points is expressed in the form $x = g(y)$, then the arc length s of the graph of g between $(g(c), c)$ and $(g(d), d)$ is given by

$$s = \int_c^d (\_\_\_\_\_\_\_\_\_\_\_\_)\,dy.$$

$\sqrt{1 + [g'(y)]^2}$

In Problems 79 through 89, find the arc length of the graph of each equation over the indicated interval.

79 $y = \frac{2}{3}x^{3/2}$ in the interval $[0, 3]$.

$x^{1/2}$, $1 + x$

$\frac{dy}{dx} =$ _____, so that $1 + \left(\frac{dy}{dx}\right)^2 =$ ________. Therefore, the arc length s is given by

$1 + \left(\frac{dy}{dx}\right)^2$, $1 + x$

$(1 + x)^{3/2}$

8

$\frac{14}{3}$

$$s = \int_0^3 \sqrt{\_\_\_\_\_\_\_\_}\, dx = \int_0^3 \sqrt{\_\_\_\_\_\_}\, dx$$
$$= \frac{2}{3}(\_\_\_\_\_\_\_\_)\Big|_0^3$$
$$= \frac{2}{3}(\_\_\_\_ - 1)$$
$$= \underline{\quad\quad} \text{ units.}$$

80 $y = \frac{2}{3}(x - 1)^{3/2}$ in the interval [1, 4].

$(x - 1)^{1/2}$, x

$\frac{dy}{dx} =$ ________, so that $1 + \left(\frac{dy}{dx}\right)^2 =$ ____.

The arc length s is given by

$1 + \left(\frac{dy}{dx}\right)^2$, x

$\frac{2}{3}x^{3/2}$

1

$\frac{14}{3}$

$$s = \int_1^4 \sqrt{\_\_\_\_\_\_\_\_}\, dx = \int_1^4 \sqrt{\_\_\_\_}\, dx$$
$$= \left(\_\_\_\_\_\_\right)\Big|_1^4$$
$$= \frac{2}{3}(8 - \_\_\_\_)$$
$$= \underline{\quad\quad} \text{ units.}$$

81 $y = \frac{2}{3}(x^2 + 1)^{3/2}$ in the interval [0, 3].

$2x(x^2 + 1)^{1/2}$

$1 + 4x^4 + 4x^2$

$\frac{dy}{dx} =$ __________, so that

$1 + \left(\frac{dy}{dx}\right)^2 =$ __________. The arc length s is given by

$1 + \left(\frac{dy}{dx}\right)^2$, $4x^4 + 4x^2 + 1$

$(2x^2 + 1)^2$

$\frac{2x^3}{3} + x$

21

$$s = \int_0^3 \sqrt{\_\_\_\_\_\_\_\_}\, dx = \int_0^3 \sqrt{\_\_\_\_\_\_\_\_\_\_}\, dx$$
$$= \int_0^3 \sqrt{\_\_\_\_\_\_\_\_\_\_}\, dx$$
$$= \int_0^3 (2x^2 + 1)\, dx$$
$$= \left(\_\_\_\_\_\_\right)\Big|_0^3$$
$$= \_\_\_\_ \text{ units.}$$

82 $y = \frac{1}{6}\left(x^3 + \frac{3}{x}\right)$ in the interval [1, 3].

$\frac{x^2}{2} - \frac{1}{2x^2}$

$\frac{x^2}{2} - \frac{1}{2x^2}, \frac{x^4}{4} + \frac{1}{2} + \frac{1}{4x^4}$

$\left(\frac{x^2}{2} + \frac{1}{2x^2}\right)^2$

$1 + \left(\frac{dy}{dx}\right)^2, \frac{x^2}{2} + \frac{1}{2x^2}$

$\frac{x^3}{6} - \frac{1}{2x}$

$\frac{13}{3}$

$\frac{14}{3}$

5, 26

$1 + \left(\frac{dy}{dx}\right)^2$, 26

$\sqrt{26}$

$3x^2, 3x^2, 1 + \frac{9x^4}{4y^2}$

$1 + \frac{9x}{4}$

$1 + \left(\frac{dy}{dx}\right)^2, 1 + \frac{9}{4}x$

$\frac{2}{3}\left(1 + \frac{9}{4}x\right)^{3/2}$

$10\sqrt{10}$

$\frac{x^3}{2} - \frac{1}{2x^3}, \frac{x^6}{4} + \frac{1}{2} + \frac{1}{4x^6}$

$\left(\frac{x^3}{2} + \frac{1}{2x^3}\right)^2$

$\frac{x^3}{2} + \frac{1}{2x^3}$

$\frac{x^3}{2} + \frac{1}{2x^3}, \frac{x^4}{8} - \frac{1}{4x^2}$

$\frac{727}{72}, -\frac{1}{8}, \frac{92}{9}$

$\frac{dy}{dx} =$ ____________, so that $1 + \left(\frac{dy}{dx}\right)^2 =$

$1 + \left(\text{\_\_\_\_\_\_\_\_\_\_\_\_}\right)^2 =$ ____________ $=$

____________. The arc length s is given by

$$s = \int_1^3 \sqrt{\text{\_\_\_\_\_\_\_\_\_\_\_\_}}\, dx = \int_1^3 \left(\text{\_\_\_\_\_\_\_\_\_\_}\right) dx$$

$$= \left(\text{\_\_\_\_\_\_\_\_}\right)\Big|_1^3$$

$$= \left[\text{\_\_\_\_} - \left(-\frac{1}{3}\right)\right]$$

$=$ ____ units.

83 $y = 5x$ in the interval $[0, 1]$.

$\frac{dy}{dx} =$ ____, so that $1 + \left(\frac{dy}{dx}\right)^2 =$ ____. The arc length s is given by

$$s = \int_0^1 \sqrt{\text{\_\_\_\_\_\_\_\_\_\_\_\_}}\, dx = \int_0^1 \sqrt{\text{\_\_\_\_}}\, dx$$

$=$ ____ units.

84 $y^2 = x^3$ in the interval $[0, 4]$.

$2y\frac{dy}{dx} =$ ____, so that $\frac{dy}{dx} = \frac{\text{\_\_\_\_}}{2y}$ and $1 + \left(\frac{dy}{dx}\right)^2 =$ ____________ $=$

________. The arc length s is given by

$$s = \int_0^4 \sqrt{\text{\_\_\_\_\_\_\_\_\_\_\_\_}}\, dx = \int_0^4 \sqrt{\text{\_\_\_\_\_\_\_\_}}\, dx$$

$$= \frac{4}{9}\left(\text{\_\_\_\_\_\_\_\_\_\_\_\_\_\_\_\_}\right)\Big|_0^4$$

$= \frac{8}{27}$(________ − 1) units.

85 $y = \frac{1}{8}\left(x^4 + \frac{2}{x^2}\right)$ in the interval $[1, 3]$.

$\frac{dy}{dx} =$ ____________, so that $1 + \left(\frac{dy}{dx}\right)^2 =$ ____________ $=$

____________. Therefore, the arc length s is given by

$$s = \int_1^3 \sqrt{1 + \left(\frac{dy}{dx}\right)^2}\, dx = \int_1^3 \sqrt{\left(\text{\_\_\_\_\_\_\_\_\_\_}\right)^2}\, dx$$

$$= \int_1^3 \left(\text{\_\_\_\_\_\_\_\_\_\_}\right) dx = \left(\text{\_\_\_\_\_\_\_\_\_\_}\right)\Big|_1^3$$

$= \left[\left(\text{\_\_\_\_} - \text{\_\_\_\_}\right)\right] =$ ____ units.

$\frac{3}{2}y^{1/2}$, $1 + \frac{9}{4}y$

$1 + \left(\frac{dx}{dy}\right)^2$, $1 + \frac{9}{4}y$

$\frac{2}{3}\left(1 + \frac{9}{4}y\right)^{3/2}$, 8

$\frac{56}{27}$

86 $x = y\sqrt{y}$ from $y = 0$ to $y = \frac{4}{3}$.

$\frac{dx}{dy} =$ ________, so that $1 + \left(\frac{dx}{dy}\right)^2 =$ ________. Therefore, the arc length s is given by

$$s = \int_0^{4/3} \sqrt{\rule{3cm}{0.4pt}}\, dy = \int_0^{4/3} \sqrt{\rule{2cm}{0.4pt}}\, dy$$

$$= \frac{4}{9}\Big(\rule{3cm}{0.4pt}\Big)\Big|_0^{4/3} = \frac{8}{27}(\rule{1cm}{0.4pt} - 1)$$

$= $ ____ units.

$\frac{3y^2}{16x}$, $1 + \frac{9}{32}y$

$1 + \left(\frac{dx}{dy}\right)^2$, $1 + \frac{9}{32}y$

$\frac{2}{3}\left(1 + \frac{9}{32}y\right)^{3/2}$

$\frac{8}{27}(13\sqrt{13} - 8)$

87 $x^2 = \frac{1}{8}y^3$ from $y = 0$ to $y = 8$.

$\frac{dx}{dy} =$ ____, so that $1 + \left(\frac{dx}{dy}\right)^2 =$ ________. The arc length s is given by

$$s = \int_0^{8} \sqrt{\rule{3cm}{0.4pt}}\, dy = \int_0^{8} \sqrt{\rule{2cm}{0.4pt}}\, dy$$

$$= \frac{32}{9}\ \rule{4cm}{0.4pt}\ \Big|_0^{8}$$

$= $ ________________ units.

$\sqrt{y} - \frac{1}{4\sqrt{y}}$, $\frac{1}{2} + y + \frac{1}{16y}$

$\sqrt{y} + \frac{1}{4\sqrt{y}}$

$1 + \left(\frac{dx}{dy}\right)^2$, $\sqrt{y} + \frac{1}{4\sqrt{y}}$

$\frac{2}{3}y^{3/2} + \frac{1}{2}y^{1/2}$

$\frac{55}{3}$

88 $x = \frac{1}{6}\sqrt{y}\,(4y - 3)$ from $y = 1$ to $y = 9$.

$\frac{dx}{dy} =$ ________, so that $1 + \left(\frac{dx}{dy}\right)^2 =$ ____________ $=$ $\Big(\rule{3cm}{0.4pt}\Big)^2$. The arc length s is given by

$$s = \int_1^{9} \sqrt{\rule{3cm}{0.4pt}}\, dy = \int_1^{9} \Big(\rule{3cm}{0.4pt}\Big)\, dy$$

$$= \Big(\rule{3cm}{0.4pt}\Big)\Big|_1^{9}$$

$= $ ____ units.

$\frac{y^2}{8} - \frac{2}{y^2}$, $\frac{y^4}{64} + \frac{1}{2} + \frac{4}{y^4}$

$\frac{y^2}{8} + \frac{2}{y^2}$

$1 + \left(\frac{dx}{dy}\right)^2$, $\frac{y^2}{8} + \frac{2}{y^2}$

89 $x = \frac{1}{24}\left(y^3 + \frac{48}{y}\right)$ from $y = 2$ to $y = 4$.

$\frac{dx}{dy} =$ ________, so that $1 + \left(\frac{dx}{dy}\right)^2 =$ ____________ $=$ $\Big(\rule{2cm}{0.4pt}\Big)^2$. The arc length s is given by

$$s = \int_2^{4} \sqrt{\rule{3cm}{0.4pt}}\, dy = \int_2^{4} \sqrt{\Big(\rule{2cm}{0.4pt}\Big)^2}\, dy$$

$\frac{y^2}{8}+\frac{2}{y^2}, \frac{y^3}{24}-\frac{2}{y}$

$\frac{17}{6}$

$$= \int_2^4 \left(\_\_\_\_\_\right) dy = \left(\_\_\_\_\_\right)\Big|_2^4$$

$= \_\_\_\_$ units.

4.2 Area of a Surface of Revolution

$f(x)\sqrt{1+[f'(x)]^2}$

90 If a function f has a continuous first derivative, then the surface area A generated by revolving the portion of the graph of f in the interval $[a, b]$ about the x axis is given by

$$A = 2\pi \int_a^b (\_\_\_\_\_\_\_\_)\, dx.$$

In Problems 91 through 98, find the (lateral) surface area of the surface of revolution obtained by revolving each curve about the x axis.

$\frac{4}{3}$

$y\sqrt{1+\left(\frac{dy}{dx}\right)^2}$

$1+\frac{16}{9}$

$\frac{x^2}{2}$

20π

91 $y = \frac{4}{3}x$ from $x = 0$ to $x = 3$.

Here $\frac{dy}{dx} = \_\_\_\_$ and the surface area A is given by

$$A = 2\pi \int_0^3 \left(\_\_\_\_\_\_\_\_\right) dx$$

$$= 2\pi \int_0^3 \frac{4}{3}x\sqrt{\_\_\_\_\_\_}\, dx$$

$$= \frac{40}{9}\pi \left(\_\_\_\right)\Big|_0^3$$

$= \_\_\_\_$ square units.

$\frac{1}{3}x^2$

$y\sqrt{1+\left(\frac{dy}{dx}\right)^2}$

$1+\frac{x^4}{9}$

$\left(1+\frac{x^4}{9}\right)^{3/2}$

$\frac{98\pi}{81}$

92 $y = \frac{1}{9}x^3$ from $x = 0$ to $x = 2$.

Here $\frac{dy}{dx} = \_\_\_\_$ and the surface area A is given by

$$A = 2\pi \int_0^2 \left(\_\_\_\_\_\_\_\_\right) dx$$

$$= 2\pi \int_0^2 \frac{1}{9}x^3\sqrt{\_\_\_\_\_\_}\, dx$$

$$= \frac{\pi}{3}\left(\_\_\_\_\_\_\_\_\right)\Big|_0^2$$

$= \_\_\_\_$ square units.

$4, \frac{2}{y}, \frac{1}{\sqrt{x}}$

93 $y^2 = 4x$ from $x = 1$ to $x = 3$.

Here $2y\frac{dy}{dx} = \_\_\_\_$, so that $\frac{dy}{dx} = \_\_\_\_ = \_\_\_\_$ and the surface area

$y\sqrt{1+\left(\frac{dy}{dx}\right)^2}$

$1+\frac{1}{x}$

$x+1$

$(x+1)^{3/2}$, 8

$\frac{16\pi}{3}(4-\sqrt{2})$

$-2x(3-x)+(3-x)^2$

$-\frac{1}{2}\left(\frac{x-1}{\sqrt{x}}\right)$

$y\sqrt{1+\left(\frac{dy}{dx}\right)^2}$

$1+\frac{1}{4}\left[\frac{(x-1)^2}{x}\right]$

$-x^2+2x+3$

$-\frac{x^3}{3}+x^2+3x$

$\frac{5\pi}{9}$

$\frac{1}{2}\left(x^2-\frac{1}{x^2}\right)$, $\frac{x^4-1}{2x^2}$

$y\sqrt{1+\left(\frac{dy}{dx}\right)^2}$

$1+\frac{(x^4-1)^2}{4x^4}$

$\frac{x^4+1}{2x^2}$

$x^5+4x+\frac{3}{x^3}$

$\frac{x^6}{6}+2x^2-\frac{3}{2x^2}$

$\frac{47\pi}{16}$

A is given by

$$A = 2\pi\int_1^3 \left(\underline{\qquad\qquad}\right)dx$$

$$= 2\pi\int_1^3 2\sqrt{x}\sqrt{\underline{\qquad}}\,dx$$

$$= 4\pi\int_1^3 \sqrt{\underline{\qquad}}\,dx$$

$$= \frac{8\pi}{3}(\underline{\qquad\qquad})\Big|_1^3 = \frac{8\pi}{3}(\underline{\quad} - 2\sqrt{2})$$

$= \underline{\qquad\qquad}$ square units.

94 $9y^2 = x(3-x)^2$ from $x = 2$ to $x = 3$.

Here $18y\frac{dy}{dx} = \underline{\qquad\qquad\qquad}$, so that

$\frac{dy}{dx} = \underline{\qquad\qquad}$, and the surface area A is given by

$$A = 2\pi\int_2^3 \left(\underline{\qquad\qquad}\right)dx$$

$$= 2\pi\int_2^3 \frac{1}{3}\sqrt{x}\,(3-x)\sqrt{\underline{\qquad\qquad}}\,dx$$

$$= \frac{2\pi}{6}\int_2^3 (\underline{\qquad\qquad})\,dx$$

$$= \frac{\pi}{3}\left(\underline{\qquad\qquad}\right)\Big|_2^3$$

$= \underline{\quad}$ square units.

95 $y = \frac{1}{6}\left(x^3+\frac{3}{x}\right)$ from $x = 1$ to $x = 2$.

Here $\frac{dy}{dx} = \underline{\qquad\qquad} = \underline{\qquad}$, so that the surface area A is given by

$$A = 2\pi\int_1^2 \left(\underline{\qquad\qquad}\right)dx$$

$$= 2\pi\int_1^2 \frac{1}{6}\left(x^3+\frac{3}{x}\right)\sqrt{\underline{\qquad\qquad}}\,dx$$

$$= 2\pi\int_1^2 \frac{x^4+3}{6x}\left(\underline{\qquad}\right)dx$$

$$= \frac{\pi}{6}\int_1^2 \left(\underline{\qquad\qquad}\right)dx$$

$$= \frac{\pi}{6}\left(\underline{\qquad\qquad}\right)\Big|_1^2$$

$= \underline{\quad}$ square units.

96 $y = \frac{1}{6}\sqrt{x}\,(4x - 3)$ from $x = 1$ to $x = 9$.

Here $\frac{dy}{dx} =$ ____________, so that the surface area A is given by

$A = 2\pi \int_1^9 \left(\_\_\_\_\_\_\_\_\_\_ \right) dx$

$= 2\pi \int_1^9 \frac{1}{6}\sqrt{x}\,(4x - 3)\sqrt{\_\_\_\_\_\_\_\_\_\_}\; dx$

$= 2\pi \int_1^9 \frac{1}{6}\sqrt{x}\,(4x - 3)\left(\_\_\_\_\_\_\_\_\_\_ \right) dx$

$= \frac{\pi}{12} \int_1^9 (\_\_\_\_\_\_\_\_\_\_)\, dx$

$= \frac{\pi}{12} \left(\_\_\_\_\_\_\_\_\_\_ \right) \Big|_1^9$

$=$ __________ square units.

$x^{1/2} - \frac{1}{4}x^{-1/2}$

$y\sqrt{1 + \left(\frac{dy}{dx}\right)^2}$

$1 + \left(\sqrt{x} - \frac{1}{4\sqrt{x}}\right)^2$

$\sqrt{x} + \frac{1}{4\sqrt{x}}$

$16x^2 - 8x - 3$

$\frac{16x^3}{3} - 4x^2 - 3x$

$\frac{2654}{9}\pi$

97 $y = \frac{x^4}{4} + \frac{1}{8x^2}$ from $x = 1$ to $x = 2$.

Here $\frac{dy}{dx} =$ ____________, so that the surface area A is given by

$A = 2\pi \int_1^2 \left(\_\_\_\_\_\_\_\_\_\_ \right) dx$

$= 2\pi \int_1^2 \left(\frac{x^4}{4} + \frac{1}{8x^2}\right)\sqrt{\_\_\_\_\_\_\_\_\_\_}\; dx$

$= 2\pi \int_1^2 \left(\frac{x^4}{4} + \frac{1}{8x^2}\right)\sqrt{\_\_\_\_\_\_\_\_\_\_}\; dx$

$= 2\pi \int_1^2 \left(\_\_\_\_\_\_\_\_\_\_ \right) dx$

$= 2\pi \left(\_\_\_\_\_\_\_\_\_\_ \right) \Big|_1^2$

$=$ __________ square units.

$x^3 - \frac{1}{4x^3}$

$y\sqrt{1 + \left(\frac{dy}{dx}\right)^2}$

$1 + \left(x^3 - \frac{1}{4x^3}\right)^2$

$\left(x^3 + \frac{1}{4x^3}\right)^2$

$\frac{x^7}{4} + \frac{3x}{8} + \frac{1}{32x^5}$

$\frac{x^8}{32} + \frac{3x^2}{16} - \frac{1}{128x^4}$

$\frac{17{,}487\pi}{1024}$

98 $8y^2 = x^2 - x^4$ from $x = \frac{1}{4}$ to $x = \frac{1}{2}$.

Here $16y\frac{dy}{dx} =$ ____________, so that $\frac{dy}{dx} =$ ____________, and the surface area A is given by

$A = 2\pi \int_{1/4}^{1/2} \left(\_\_\_\_\_\_\_\_\_\_ \right) dx$

$= 2\pi \int_{1/4}^{1/2} \frac{x\sqrt{1 - x^2}}{2\sqrt{2}}\sqrt{\_\_\_\_\_\_\_\_\_\_}\; dx$

$= 2\pi \int_{1/4}^{1/2} \frac{x\sqrt{1 - x^2}}{2\sqrt{2}}\sqrt{\_\_\_\_\_\_\_\_\_\_}\; dx$

$2x - 4x^3, \quad \frac{x - 2x^3}{8y}$

$y\sqrt{1 + \left(\frac{dy}{dx}\right)^2}$

$1 + \left(\frac{x - 2x^3}{8y}\right)^2$

$\frac{(3 - 2x^2)^2}{8(1 - x^2)}$

$3 - 2x^2$

$= 2\pi \int_{1/4}^{1/2} \frac{x\sqrt{1-x^2}}{2\sqrt{2}} \frac{\rule{4em}{0.4pt}}{2\sqrt{2}\sqrt{1-x^2}}\, dx$

$3x - 2x^3$

$= \frac{2\pi}{8} \int_{1/4}^{1/2} (\rule{6em}{0.4pt})\, dx$

$\frac{3x^2}{2} - \frac{x^4}{2}$

$= \frac{\pi}{4}\left(\rule{6em}{0.4pt}\right)\Big|_{1/4}^{1/2}$

$\frac{129\pi}{2048}$

$= \rule{4em}{0.4pt}$ square unit.

99 If the axis of revolution is the y axis, the corresponding formula for the surface area of the surface of revolution is

$A = 2\pi \int_a^b (\rule{2em}{0.4pt})\, ds$, where $ds =$ ________________ or $ds =$ ________________, depending upon whether the curve revolved about the y axis is expressed in the form $y = f(x)$ or $x = g(y)$, respectively.

x, $\sqrt{1 + [f'(x)]^2}\, dx$

$\sqrt{1 + [g'(y)]^2}\, dy$

In Problems 100 through 104, find the (lateral) surface area of the surface of revolution obtained by revolving the given curve about the y axis.

100 $12y = 5x$ from $x = 0$ to $x = 2$.

Here $\frac{dy}{dx} =$ ____, so that the surface area A is given by

$\frac{5}{12}$

$A = 2\pi \int_0^2 \left(\rule{8em}{0.4pt}\right) dx$

$x\sqrt{1 + \left(\frac{dy}{dx}\right)^2}$

$= 2\pi \int_0^2 x\sqrt{\rule{6em}{0.4pt}}\, dx$

$1 + \frac{25}{144}$

$= 2\pi \int_0^2 \left(\rule{3em}{0.4pt}\right) dx$

$\frac{13x}{12}$

$= \frac{13\pi}{12} (\rule{2em}{0.4pt})\Big|_0^2$

x^2

$= \rule{3em}{0.4pt}$ square units.

$\frac{13\pi}{3}$

101 $y = x^2$ from $x = 0$ to $x = \frac{6}{5}$.

Here $\frac{dy}{dx} =$ ____, so that the surface area A is given by

$2x$

$A = 2\pi \int_0^{6/5} \left(\rule{8em}{0.4pt}\right) dx$

$x\sqrt{1 + \left(\frac{dy}{dx}\right)^2}$

$= 2\pi \int_0^{6/5} x\sqrt{\rule{6em}{0.4pt}}\, dx$

$1 + 4x^2$

$(1+4x^2)^{3/2}$ | $= \frac{\pi}{6}(\_\_\_\_\_\_\_\_)\Big|_0^{6/5}$

$\frac{1036\pi}{375}$ | $= \_\_\_\_\_\_$ square units.

102 $x = y^3$ from $y = 0$ to $y = 1$.

$3y^2$ | Here $\frac{dx}{dy} = \_\_\_\_$, so that the surface area A is given by

$1 + \left(\frac{dx}{dy}\right)^2$ | $A = 2\pi \int_0^1 x\sqrt{\_\_\_\_\_\_\_\_}\, dy$

$1 + 9y^4$ | $= 2\pi \int_0^1 x\sqrt{\_\_\_\_\_\_}\, dy$

$= 2\pi \int_0^1 y^3\sqrt{1+9y^4}\, dy$

$\frac{1}{54}(1+9y^4)^{3/2}$ | $= 2\pi\left(\_\_\_\_\_\_\_\_\right)\Big|_0^1$

$10\sqrt{10} - 1$ | $= \frac{\pi}{27}(\_\_\_\_\_\_\_\_)$ square units.

103 $x = \frac{y^3}{24} + \frac{2}{y}$ from $y = 2$ to $y = 4$.

$\frac{y^2}{8} - \frac{2}{y^2}$, $\frac{y^2}{8} - \frac{2}{y^2}$ | Here $\frac{dx}{dy} = \_\_\_\_\_\_$, so that $1 + \left(\frac{dx}{dy}\right)^2 = 1 + \left(\_\_\_\_\_\_\right)^2$.

The surface area A is given by

$1 + \left(\frac{dx}{dy}\right)^2$ | $A = 2\pi \int_2^4 x\sqrt{\_\_\_\_\_\_\_\_}\, dy$

$\left(\frac{y^2}{8} + \frac{2}{y^2}\right)^2$ | $= 2\pi \int_2^4 \left(\frac{y^3}{24} + \frac{2}{y}\right)\sqrt{\_\_\_\_\_\_\_\_}\, dy$

$\frac{y^5}{192} + \frac{y}{3} + \frac{4}{y^3}$ | $= 2\pi \int_2^4 \left(\_\_\_\_\_\_\_\_\right) dy$

$\frac{y^6}{1152} + \frac{y^2}{6} - \frac{2}{y^2}$ | $= 2\pi\left(\_\_\_\_\_\_\_\_\right)\Big|_2^4$

$\frac{47\pi}{4}$ | $= \_\_\_\_$ square units.

104 $x^2 = 4y$ from $y = 1$ to $y = 2$.

$y^{-1/2}$ | Here $\frac{dx}{dy} = \_\_\_\_$, so that the surface area A is given by

$1 + \left(\frac{dx}{dy}\right)^2$ | $A = 2\pi \int_1^2 x\sqrt{\_\_\_\_\_\_\_\_}\, dy$

$1 + y^{-1}$ | $= 2\pi \int_1^2 2\sqrt{y}\sqrt{\_\_\_\_\_\_}\, dy$

$y + 1$	$= 4\pi \int_1^2 \sqrt{\_\_\_\_\_\_} \, dy$
$(y + 1)^{3/2}$	$= \frac{8\pi}{3} (\_\_\_\_\_\_) \Big\vert_1^2$
$\frac{8\pi}{3}(3\sqrt{3} - 2\sqrt{2})$	$=$ ______ square units.

5 Applications to Economics and the Life Sciences

5.1 Consumer's Surplus

105 Let x be the number of units of a certain commodity that are demanded when the price is $\$y$ per unit. Suppose that the demand equation $x = g(y)$ takes the form $y = f(x)$, where f is the ______ of g. If the market price $\$y_0$ is less than $\$c$ per unit, where $c = f(0)$ and $x_0 = g(y_0)$ is positive, then the gain to the customers who are willing to pay more than $\$y_0$ is measured by the integral ______. This gain is called ______.

Answers: inverse; $\int_{y_0}^{c} g(y)\, dy$; consumer's surplus

106 An alternative formula for consumer's surplus is $\int_0^{x_0} f(x)\, dx - (\_\_\_\_\_\_)$.

Answer: $x_0 y_0$

107 Consider the demand equation $x = g(y)$, where $g(y) = 50(5 - \sqrt{9y})$. Calculate the consumer's surplus when $y_0 = \$1$ per unit, using the formula in Problem 106.

Since $x = 50(5 - \sqrt{9y})$, we can solve for y to obtain $y =$ ______, so that $f(x) = \left(\frac{250 - x}{150}\right)^2$. When $y_0 = 1$, we must solve the equation $x_0 = 50[5 - \sqrt{\_\_\_\_}]$. Thus, $x_0 =$ ____. Therefore,

$$\text{consumer's surplus} = \int_0^{x_0} f(x)\, dx - \_\_\_\_$$

$$= \int_0^{100} \left(\_\_\_\_\_\_\right) dx - \_\_\_\_$$

$$= \left[-\frac{(\_\_\_\_)^3}{3(150)^2}\right]\Bigg\vert_0^{100} - 100$$

$$= -\_\_\_\_ + \frac{(250)^3}{3(150)^2} - 100$$

$$\approx \$\_\_\_\_.$$

Answers: $\frac{(250 - x)^2}{(150)^2}$; 9; 100; $x_0 y_0$; $\frac{(250 - x)^2}{(150)^2}$, 100; $250 - x$; 50; 81.48

5.2 Production over a Period of Time

108 Let x denote the number of units of a certain commodity produced in the first t units of working time by a new process, so that ____ represents the rate of production at time t. Then the total number of units produced during the interval of time from $t = a$ to $t = b$ is given by ________.

$\frac{dx}{dt}$

$\int_a^b \frac{dx}{dt}\, dt$

109 A new employee at Solar Electric Company can solder $40 - \frac{40}{\sqrt{x+1}}$ connections per hour at the end of t working hours. How many connections does the employee solder during the first 3 working hours?

Here $\frac{dx}{dt} =$ ________, so that by Problem 108, the total number of units produced is given by

$$\int_0^3 \left(\underline{\qquad\qquad}\right) dx = (40x - \underline{\qquad\qquad})\Big|_0^3$$

$$= 120 - 160 + \underline{\quad}$$

$$= \underline{\quad} \text{ connections.}$$

$40 - \frac{40}{\sqrt{x+1}}$

$40 - \frac{40}{\sqrt{x+1}}$, $80\sqrt{x+1}$

80

40

5.3 Pollution

110 If we denote by x the number of units of pollutant accumulated in a given ecosystem after t units of time, then the rate of pollution of the ecosystem is given by ____. Thus, the total number of units of pollutant accumulated in the ecosystem during the interval of time from $t = a$ to $t = b$ is given by ________.

$\frac{dx}{dt}$

$\int_a^b \frac{dx}{dt}\, dt$

111 A factory is dumping pollutants into a lake at the rate of $\frac{t^{2/3}}{540}$ tons per week, where t is the time in weeks since the factory commenced operations. After 5 years of operation, how much pollutant has the factory dumped into the lake?

By Problem 110, the amount of pollutant accumulated is given by

$$\int_0^{260} \left(\underline{\quad}\right) dt = \left[\frac{3}{5}\left(\frac{\underline{\quad}}{540}\right)\right]\Bigg|_0^{260}$$

$\frac{t^{2/3}}{540}$, $t^{5/3}$

900

11.77

$= \frac{(260)^{5/3}}{\_\_\_\_}$

$\approx$ ______ tons.

5.4 Blood Flow in the Circulatory System

continuously

maximum, width

r

$\int_0^R vr\,dr$

112 Suppose that blood flows smoothly through a cylindrical blood vessel in such a way that the velocity v of flow increases ______ from a value close to zero at the wall vessel to a ______ value at its center. Let dr be the ______ of an "infinitesimal" circular ring at a distance r from the center of the blood vessel. Assume that v depends only on ____, and let R denote the radius of the blood vessel. Then the total volume of blood passing the entire cross section in unit time is given by $V = \int_{r=0}^{r=R} dV = 2\pi\left(\_\_\_\_\_\_\right)$.

vr

$2.5 - 36r^2$

$36r^3$

r^4

0.22

113 A cylindrical blood vessel has a radius $R = 0.2$ cm and is flowing through this vessel with a velocity $v = 2.5 - 36r^2$ cm/sec at points r cm from the center. Find the rate of flow of the blood.

By Problem 112, the rate of flow is given by

$$V = 2\pi \int_0^{0.2} (\_\_\_\_)\,dr$$

$$= 2\pi \int_0^{0.2} (\_\_\_\_\_\_\_\_)\,r\,dr$$

$$= 2\pi \int_0^{0.2} (2.5r - \_\_\_\_)\,dr$$

$$= 2\pi\left[\frac{2.5}{2}r^2 - 9(\_\_\_\_)\right]\Bigg|_0^{0.2}$$

$\approx$ ______ cm^3/sec.

6 Force, Work, and Energy

6.1 Work Done by a Variable Force

$\int_a^b F\,ds$

114 If a particle P moves along the s axis under the influence of a possibly variable force F acting parallel to the s axis, then the cumulative work W done by the force F between the coordinates a and b on the s axis is given by $W =$ ______.

115 A force $F = 3s^2$ pounds acts on a particle P on the s axis and moves the particle from $s = 1$ foot to $s = 3$ feet. How much work

is done? The work W done is given by

$3s^2$, s^3

$$W = \int_1^3 (\_\_\_\_)\, ds = (\_\_\_\_)\Big|_1^3$$

26

$= \_\_\_\_$ foot-pounds.

116 How much work must be done to stretch a spring of natural length 12 inches, if the spring constant is 25 pounds per inch?

$25s$

By Hooke's law, the force $F = ks = \_\_\_\_$. The work W done is

$25s$, $\dfrac{25s^2}{2}$

$$W = \int_3^6 (\_\_\_\_)\, ds = \left(\_\_\_\_\_\_\right)\Big|_3^6$$

337.5

$= \_\_\_\_\_\_$ inch-pounds.

117 It requires 8 inch-pounds of work to stretch a certain spring 2 inches from its rest position. Assuming that the spring follows Hooke's law, what is the spring constant k?

ks

By Hooke's law, the force F is given by $F = \_\_\_\_$. The work W done is given by

ks

$$W = \int_0^2 (\_\_\_\_)\, ds = \left(\frac{ks^2}{2}\right)\Big|_0^2$$

$2k$

$= \_\_\_\_$.

8, 4

But $W = 8$ inch-pounds, so that $2k = \_\_\_\_$ or $k = \_\_\_\_$ pounds per inch.

6.2 Work Done in Pumping a Liquid

118 Suppose that all the liquid in a tank is to be pumped upward to the same level. We establish a vertical s axis with its origin at the level to which the liquid is to be pumped. (Here, we let the positive direction on the s axis be downward.) Assume that the level of the liquid at the start of the pumping is $s = a$ and its level at the end of the pumping is $s = b$. Suppose that the cross-sectional area of the surface of the liquid at level s is $A(s)$ square units. Then the total work W required to pump the liquid from level $s = a$ to level $s = b$ is given by

$swA(s)$

$$W = \int_a^b dW = \int_a^b (\_\_\_\_\_\_)\, ds$$

$$= w \int_a^b sA(s)\, ds,$$

where w is the weight density of the liquid.

119 A reservoir is in the form of a hemisphere of radius 3 meters. If it is filled with salt water weighing 1032 kilograms per cubic meter, how much work is required to pump all the salt water over the rim of the reservoir?

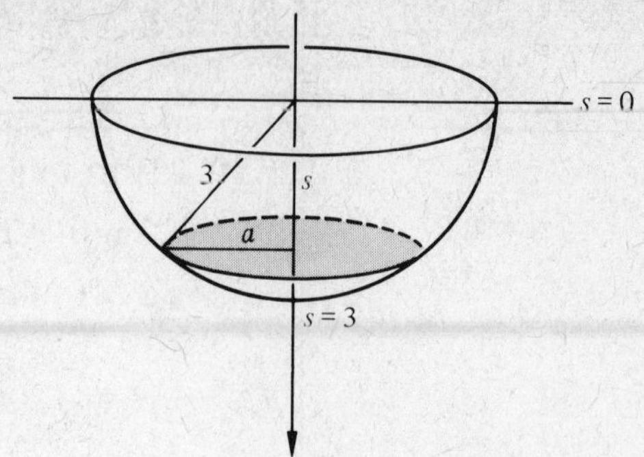

9

$9 - s^2,\ \pi(9 - s^2)$

Using the Pythagorean theorem, we have $a^2 + s^2 =$ ____, so that $a^2 =$ __________, and $A(s) = \pi a^2 =$ ______________. The work W done is

$sw\pi(9 - s^2),\ 9s - s^3$

$$W = \int_0^3 (\_\_\_\_\_\_\_\_\_\_\_\_\_\_\_\_)\,ds = 1032\pi \int_0^3 (\_\_\_\_\_\_\_\_)\,ds$$

$\dfrac{9s^2}{2} - \dfrac{s^4}{4}$

$$= 1032\pi\left(\_\_\_\_\_\_\_\_\_\_\_\_\right)\Bigg|_0^3$$

$20{,}898\pi$

$=$ ____________ kilogram-meters.

120 Water is pumped directly up from the surface of a lake into a water tower. The tank for this water is a vertical right circular cylinder of height 20 feet, with radius 5 feet, whose bottom is 60 feet above the surface of the lake. The pump is driven by a 1.5-horsepower motor. Neglecting friction, how long will it take to fill the tank with water?

Assume that water weighs 62.4 pounds per cubic foot and that 1 horsepower is 33,000 foot-pounds per minute.

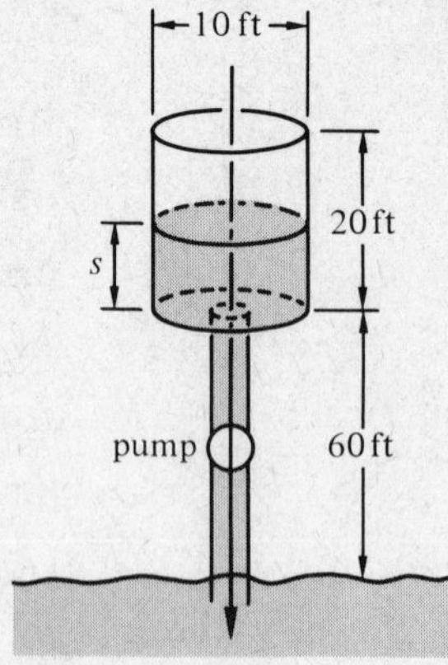

Let the distance from the bottom of the tank to the surface of the water be s feet. The work W required to lift the water from the surface of the lake to fill the tank is

$1560\pi(s + 60),\ \dfrac{s^2}{2} + 60s$

$$W = \int_0^{20} (\_\_\_\_\_\_\_\_\_\_\_\_\_\_\_\_)\,ds = 1560\pi\left(\_\_\_\_\_\_\_\_\right)\Bigg|_0^{20}$$

$2{,}184{,}000\pi$

$=$ ________________ foot-pounds.

$\frac{728\pi}{11}$

This will require $\frac{2,184,000\pi}{33,000} =$ ______ horsepower-minutes of work. Since we have a 1.5-horsepower motor, it will require $\frac{728\pi}{(11)(1.5)} \approx$ ________ minutes to fill the tank.

138.6

6.3 Compression or Expansion of a Gas

121 Consider a quantity of gas in a cylinder of radius r closed by a movable piston. Choose the reference axis parallel to the central axis of the cylinder. Let the coordinate of the position of the piston be s and let c be the coordinate of the end of the cylinder. Denote the pressure of the gas by P units of force per unit area and let V represent the volume of the gas in cubic units. The force F in the piston is given by $F =$ ________, and the volume V is given by $V =$ __________. The total work W done compressing the gas by F from an initial volume V_0 to a final volume V_1 is given by $W = \int_{V_0}^{V_1} \pi r^2 P \left(\underline{\hspace{2cm}} \right) dV = \int_{V_0}^{V_1} (\underline{\hspace{1cm}})\, dV.$

$\pi r^2 P$

$\pi r^2 (c - s)$

$-\frac{1}{\pi r^2}, -P$

Notice that if the gas is compressed, then the work W done is given by $W = \int_{V_1}^{V_0} (\underline{\hspace{1cm}})\, dV$. If the gas is expanding, the work W done is given by $W = \int_{V_0}^{V_1} (\underline{\hspace{1cm}})\, dV$.

P

P

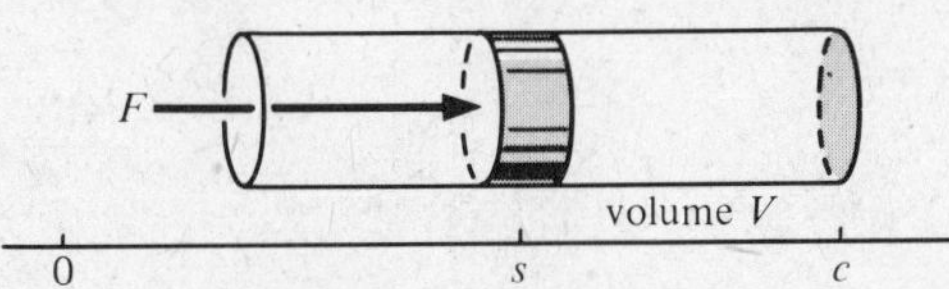

122 The expansion of a gas in a cylinder causes a piston to move, so that the volume of the enclosed gas increases from 10 cubic inches to 20 cubic inches. Find the work done if the relation between the pressure and the volume of the gas is given by $P = 40V^{-1.4}$, where P is measured in pounds per square inch and V in cubic inches. Let the area of the cross section of the cylinder be denoted by A and let s denote the distance from the cylinder head to the piston. The volume V of the gas in the cylinder is given by $V =$ _____, and the force F exerted by the gas is $F =$ _____. The total work W done is given by

As, PA

$$W = \int_{10/A}^{20/A} \frac{\underline{\hspace{1cm}}}{V^{1.4}}\, ds = \int_{10/A}^{20/A} \frac{\underline{\hspace{1cm}}}{A^{1.4} s^{1.4}}\, ds$$

$40A$, $40A$

$s^{-1.4}$

$$= 40A^{-0.4} \int_{10/A}^{20/A} (\_\_\_\_\_\_)\, ds$$

$\dfrac{s^{-0.4}}{-0.4}$

$$= 40A^{-0.4} \left(\_\_\_\_\_\_\right)\Big|_{10/A}^{20/A}$$

9.64

$\approx$ ______ inch-pounds.

123 One cubic foot of air at an initial pressure of 50 pounds per square inch expands adiabatically to a final volume of 3 cubic feet according to the law $P = kV^{-1.4}$, where k is a constant. Find the work done.

(50) (144); $k(1^{-1.4})$; 7200, $7200V^{-1.4}$

$V = 1$ if $P = 50$ pounds per square inch = ________ pounds per square foot. Therefore, (50) (144) = ________, so that $k =$ ______. Hence, $P =$ ________. The work W done is

$V^{-1.4}$

$$W = \int_1^3 7200\,(\_\_\_\_\_\_)\, dV$$

$V^{-0.4}$

$$= \frac{7200}{-0.4}(\_\_\_\_\_\_)\Big|_1^3$$

6400.9

$\approx$ ______ foot-pounds.

124 The piston in a cylinder compresses a gas adiabatically from 50 cubic inches to 25 cubic inches. Assuming that $PV^{1.3} = 100$, where P is the pressure in pounds per square inch and V is the volume in cubic inches, how much work in foot-pounds is done on the gas?

The work W done is

$V^{-1.3}$

$$W = \int_{25}^{50} 100(\_\_\_\_\_\_)\, dV$$

$V^{-0.3}$

$$= \frac{100}{-0.3}(\_\_\_\_\_\_)\Big|_{25}^{50}$$

23.83; 1.99

$\approx$ ______ inch-pounds

$\approx$ ______ foot-pounds.

6.4 Energy

$\frac{1}{2}m\left(\frac{ds}{dt}\right)^2$, $\frac{1}{2}mv^2$

125 If a particle P of mass m is moving along the s axis as a consequence of a variable unopposed force F, always acting in a direction parallel to the s axis, then the kinetic energy K of the particle is defined by $K =$ ________ = ________.

126 If a particle is connected to a spring and the spring is stretched, then work can be done by letting the spring return to its natural length. If the force F acting on P depends only on the position coordinate s of P, then the potential energy V of the particle at the

F — point with coordinate s can be defined by $V = \int_s^{s_0} (\_\_\_\_)\, ds$,

where s_0 is the coordinate of an arbitrary, but fixed, reference point from which the potential energy is calculated.

$K + V$

127 The total energy E of the particle P is defined by $E = \_\_\_\_\_\_$.

128 What velocity must a 10-gram projectile have in order to have the same kinetic energy as a 2000-kilogram car traveling at 24 kilometers per hour?

Let v be the velocity of the projectile in kilometers per hour.

2000 — Then $K = \frac{1}{2}\left(\frac{10}{1000}\right) v^2 = \frac{1}{2}(\_\_\_\_\_\_)(24)^2$, so that

$4800\sqrt{5}$, 10,733.13 — $v = \_\_\_\_\_\_ \approx \_\_\_\_\_\_$ kilometers per hour.

129 Given that the potential energy of a relaxed spring is zero, and that the spring constant is 40 pounds per foot, how far must the spring be stretched so that its potential energy is 25 foot-pounds? If the spring is stretched by s feet, then the work W done will be

F, $40s$ — $W = \int_0^s (\_\_\_\_)\, ds = \int_0^s (\_\_\_\_)\, ds$

$20s^2$, $20s^2$ — $= (\_\_\_\_\_\_)\Big|_0^s = \_\_\_\_\_\_$.

$20s^2$, $\frac{\sqrt{5}}{2}$ — Here, we require that $25 = \_\_\_\_$ or $s = \_\_\_\_$ feet.

130 A 1-ton weight is dropped from a height of h feet and hits the ground with the same kinetic energy as that possessed by a car weighing 1.5 tons and traveling 10 miles per hour. Find h.

$2000h$ — The potential energy of the weight at height h is ______ foot-pounds. The mass of the car is

3000, $\frac{375}{4}$ — $\frac{(1.5)(2000)}{g} = \frac{\_\_\_\_}{32} = \_\_\_\_$ slugs.

$\frac{(10)(5280)}{3600}$ — Therefore, we require that $2000h = \frac{1}{2}\left(\frac{375}{4}\right)\left[\_\_\_\_\_\_\right]^2$,

$\frac{15{,}125}{3000}$, 5.04 — so that $h = \_\_\_\_\_\_ \approx \_\_\_\_\_\_$ feet.

6.5 Force Caused by Fluid Pressure

131 If an admissible planar region R is exposed to a fluid under pressure, then there will be a resulting net force F on R. If the pressure of the fluid is constant over the region R, say with a value of P units of force per square unit of area, then $F = \_\_\_\_$, where A is the area of R.

PA

132 Let R be an admissible planar region placed vertically under the surface of an incompressible liquid of constant density w. Establish a reference vertical axis pointing downward, with its origin at the surface level of the liquid. Let $l(s)$ denote the total length of a horizontal cross section of R at level s units below the surface of the liquid. The pressure at depth s is given by ____ units of force per square unit of area. If R lies entirely between the horizontal lines $s = a$ and $s = b$, then the total force F on R is given by

Answer: ws

$$F = \int_a^b dF = \int_a^b (\_\_\_\_\_\_)\, ds = w \int_a^b (\_\_\_\_\_\_)\, ds.$$

Answer: $wsl(s)$, $sl(s)$

The diagram is

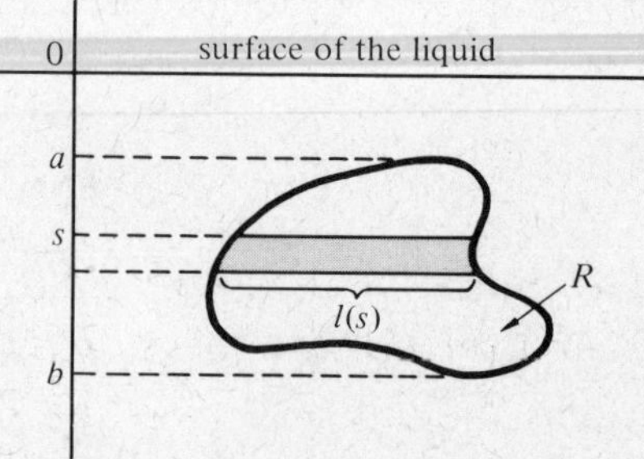

133 Find the force exerted on the semicircular end of a trough if the trough is full of water and the semicircular end has a radius of 3 feet. Assume that the density of water is $w = 62.4$ pounds per cubic foot.

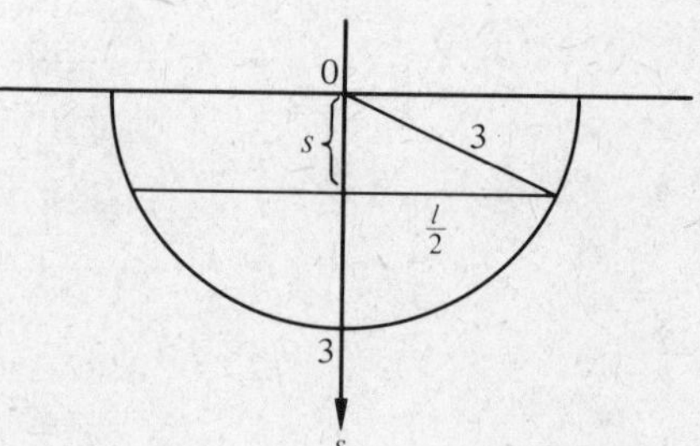

Using the Pythagorean theorem, we have $\left(\frac{l}{2}\right)^2 + s^2 =$ ____ or

Answer: 9

$l =$ ______. The desired force is given by

Answer: $2\sqrt{9 - s^2}$

$$F = w \int_0^3 (\_\_\_\_)\, ds = 62.4 \int_0^3 (\_\_\_\_\_\_)\, ds$$

Answer: sl, $2s\sqrt{9 - s^2}$

$$= 62.4 \left(\_\_\_\_\_\_\right)\Big|_0^3 = 62.4\,(\_\_\_\_)$$

Answer: $-\frac{2}{3}(9 - s^2)^{3/2}$, 18

$=$ ______ pounds.

Answer: 1123.2

134 A water main in the shape of a horizontal cylinder 1.9 meters in radius is half full of water. Find the force on the gate that closes the main.

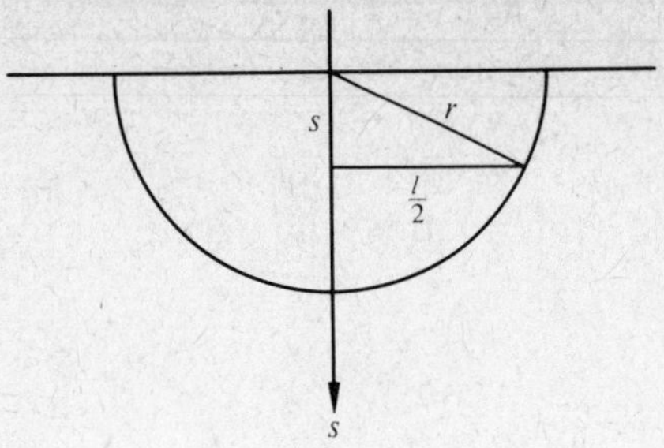

Using the Pythagorean theorem, we have $\left(\frac{l}{2}\right)^2 + s^2 = r^2$, so that

$2\sqrt{r^2 - s^2}$

$l =$ ________________. The force F is given by

$wsl,\ 2s\sqrt{(1.9)^2 - s^2}$

$$F = \int_0^{1.9} (\_\_\_\_)\, ds = 1000 \int_0^{1.9} (\_\_\_\_\_\_\_\_\_\_\_\_\_\_)\, ds$$

$-\frac{2}{3}[(1.9)^2 - s^2]^{3/2}$

$$= 1000 \left(\_\_\_\_\_\_\_\_\_\_\_\_\_\_ \right)\Big|_0^{1.9}$$

$\frac{13{,}718}{3}$

$=$ ________ kilograms.

135 The face of a vertical dam has the shape of an isosceles trapezoid of altitude 16 feet with an upper base of 42 feet and a lower base of 30 feet. Find the total force exerted by the water on the dam when the water is 12 feet deep.

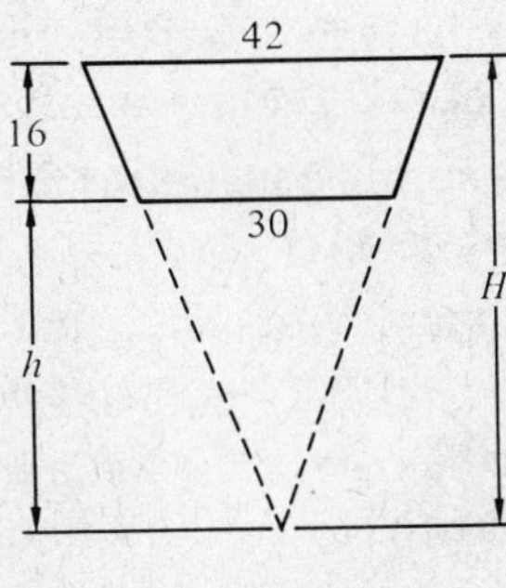

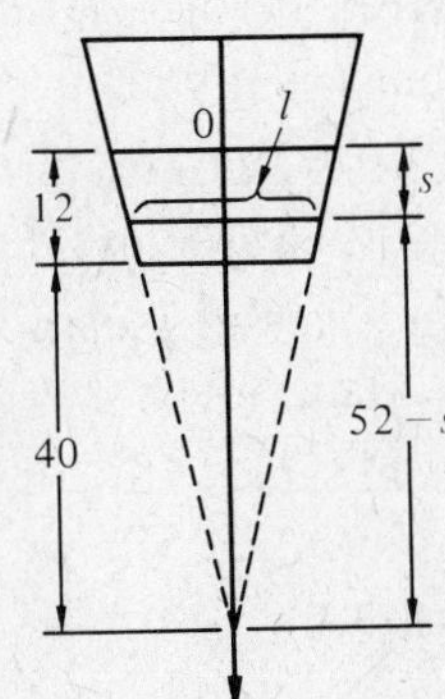

$h + 16$

From the figure, we see that $H =$ ________. Using similar tri-

$h + 16,\ 40$

angles, we have $\frac{h}{30} = \frac{H}{42} = \frac{\quad}{42}$, so that $h =$ ____. By

30

similar triangles again, we have $\frac{l}{52 - s} = \frac{\quad}{40}$. Therefore,

$39 - \frac{3}{4}s$

$l =$ ________. Thus, the total force F is given by

$39s - \frac{3}{4}s^2$

$$F = 62.4 \int_0^{12} sl\, ds = 62.4 \int_0^{12} \left(\_\_\_\_\_\_\_\_ \right) ds$$

$\frac{39s^2}{2} - \frac{s^3}{4}$

$$= 62.4 \left(\_\_\_\_\_\_\_\_ \right)\Big|_0^{12}$$

148, 262.4

$=$ ________________ pounds.

Chapter Test

In Problems 1 through 6, use either the circular disk method, the circular ring method, or the cylindrical shell method to find the volume of the solid generated by revolving the region bounded by the curves about the indicated axis.

1 $y = x^2, x = 0, x = 1$, and $y = 0$ about the x axis.

2 $y = x^2 - 4x + 5$, and $y = x + 1$ about the x axis.

3 $y^2 = 4x$ and $x = 1$ about the line $x = 1$.

4 $x^2 = 4(x - y), x = 0, y = 1$, and $y = 0$ about the y axis.

5 $x^2 = 4(x - y), x = 0, y = 1$, and $y = 0$ about the line $y = 1$.

6 $y = 3 - 2x + x^2$ and $y = 3$ about $y = 3$.

7 Assume that the base of a solid is the circle $x^2 + y^2 = 9$, and on each chord of the circle parallel to the y axis there is erected a square. Find the volume of the resulting solid.

8 Find the length of the curve $9y^2 = 4x^3$ with $y \geqslant 0$ from $x = 0$ to $x = 3$.

9 Find the surface area generated by revolving the curve $a^2y = x^3$ about the x axis between $x = 0$ and $x = a$.

10 The demand equation for a certain commodity is $y = 40(500 + 5x - x^2)$, where x is the demand in thousands of tons and y is the price in dollars per thousand tons. If the current price is \$10,000 per thousand tons, find the consumer's surplus.

11 The rate at which the Zeus Company manufactures machines t weeks after the start of production is $380[1 - 100(t + 10)^{-2}]$ machines per week. How many machines are produced during the tenth week of operation?

12 A cylindrical blood vessel has a radius $R = 0.2$ cm and blood is flowing through this vessel with a velocity $v = 3.0 - 50r^2$ cm/sec at points r cm from the center. Find the rate of flow of the blood.

13 Find the work done in moving a particle under the influence of the force $F(s) = s^2$ from $s = 1$ to $s = 5$. Assume that the distance is in feet and the force is in pounds.

14 A spring obeys Hooke's law. Its spring constant is 2 pounds per inch. How much work is required to stretch it 3 inches from its rest position?

15 Suppose the face of a dam is a square with side equal to 200 feet. What is the maximum total force against the dam, given that the density of the water is 62.4 pounds per cubic foot?

16 A cylindrical tank with a base radius of 5 feet and a height of 20 feet is filled with water. Find the work done in pumping all the water out to the top of the tank. Assume that the density of the water is 62.4 pounds per cubic foot.

Answers

1 $\frac{\pi}{5}$ cubic unit

2 $\frac{117\pi}{5}$ cubic units

3 $\frac{32\pi}{15}$ cubic units

4 $\frac{2\pi}{3}$ cubic units

5 $\frac{16\pi}{5}$ cubic units

6 $\frac{16\pi}{15}$ cubic units

7 144 cubic units

8 $\frac{14}{3}$ units

9 $\frac{1}{27}\pi a^2(10\sqrt{10} - 1)$ square units

10 \$104,166.67 (approx.)

11 280 machines

12 0.25 cm^3/sec (approx.)

13 $\frac{124}{3}$ foot-pounds

14 9 inch-pounds

15 249.6×10^6 pounds

16 980,177 foot-pounds

TABLES

TABLE I Trigonometric Functions

Degrees	Radians	Sin	Tan	Cot	Cos		
0	0	0	0	—	1.000	1.5708	90
1	0.0175	0.0175	0.0175	57.290	0.9998	1.5533	89
2	0.0349	0.0349	0.0349	28.636	0.9994	1.5359	88
3	0.0524	0.0523	0.0523	19.081	0.9986	1.5184	87
4	0.0698	0.0698	0.0699	14.301	0.9976	1.5010	86
5	0.0873	0.0872	0.0875	11.430	0.9962	1.4835	85
6	0.1047	0.1045	0.1051	9.5144	0.9945	1.4661	84
7	0.1222	0.1219	0.1228	8.1443	0.9925	1.4486	83
8	0.1396	0.1392	0.1405	7.1154	0.9903	1.4312	82
9	0.1571	0.1564	0.1584	6.3138	0.9877	1.4137	81
10	0.1745	0.1736	0.1763	5.6713	0.9848	1.3963	80
11	0.1920	0.1908	0.1944	5.1446	0.9816	1.3788	79
12	0.2094	0.2079	0.2126	4.7046	0.9781	1.3614	78
13	0.2269	0.2250	0.2309	4.3315	0.9744	1.3439	77
14	0.2443	0.2419	0.2493	4.0108	0.9703	1.3265	76
15	0.2618	0.2588	0.2679	3.7321	0.9659	1.3090	75
16	0.2793	0.2756	0.2867	3.4874	0.9613	1.2915	74
17	0.2967	0.2924	0.3057	3.2709	0.9563	1.2741	73
18	0.3142	0.3090	0.3249	3.0777	0.9511	1.2566	72
19	0.3316	0.3256	0.3443	2.9042	0.9455	1.2392	71
20	0.3491	0.3420	0.3640	2.7475	0.9397	1.2217	70
21	0.3665	0.3584	0.3839	2.6051	0.9336	1.2043	69
22	0.3840	0.3746	0.4040	2.4751	0.9272	1.1868	68
23	0.4014	0.3907	0.4245	2.3559	0.9205	1.1694	67
24	0.4189	0.4067	0.4452	2.2460	0.9135	1.1519	66
25	0.4363	0.4226	0.4663	2.1445	0.9063	1.1345	65
26	0.4538	0.4384	0.4877	2.0503	0.8988	1.1170	64
27	0.4712	0.4540	0.5095	1.9626	0.8910	1.0996	63
28	0.4887	0.4695	0.5317	1.8807	0.8829	1.0821	62
29	0.5061	0.4848	0.5543	1.8040	0.8746	1.0647	61
30	0.5236	0.5000	0.5774	1.7321	0.8660	1.0472	60
31	0.5411	0.5150	0.6009	1.6643	0.8572	1.0297	59
32	0.5585	0.5299	0.6249	1.6003	0.8480	1.0123	58
33	0.5760	0.5446	0.6494	1.5399	0.8387	0.9948	57
34	0.5934	0.5592	0.6745	1.4826	0.8290	0.9774	56
35	0.6109	0.5736	0.7002	1.4281	0.8192	0.9599	55
36	0.6283	0.5878	0.7265	1.3764	0.8090	0.9425	54
37	0.6458	0.6018	0.7536	1.3270	0.7986	0.9250	53
38	0.6632	0.6157	0.7813	1.2799	0.7880	0.9076	52
39	0.6807	0.6293	0.8098	1.2349	0.7771	0.8901	51
40	0.6981	0.6428	0.8391	1.1918	0.7660	0.8727	50
41	0.7156	0.6561	0.8693	1.1504	0.7547	0.8552	49
42	0.7330	0.6691	0.9004	1.1106	0.7431	0.8378	48
43	0.7505	0.6820	0.9325	1.0724	0.7314	0.8203	47
44	0.7679	0.6947	0.9657	1.0355	0.7193	0.8029	46
45	0.7854	0.7071	1.0000	1.0000	0.7071	0.7854	45
		Cos	Cot	Tan	Sin	Radians	Degrees

TABLE II Natural Logarithms, ln t

t	0.00	0.01	0.02	0.03	0.04	0.05	0.06	0.07	0.08	0.09
1.0	0.0000	0.0100	0.0198	0.0296	0.0392	0.0488	0.0583	0.0677	0.0770	0.0862
1.1	0.0953	0.1044	0.1133	0.1222	0.1310	0.1398	0.1484	0.1570	0.1655	0.1740
1.2	0.1823	0.1906	0.1989	0.2070	0.2151	0.2231	0.2311	0.2390	0.2469	0.2546
1.3	0.2624	0.2700	0.2776	0.2852	0.2927	0.3001	0.3075	0.3148	0.3221	0.3293
1.4	0.3365	0.3436	0.3507	0.3577	0.3646	0.3716	0.3784	0.3853	0.3920	0.3988
1.5	0.4055	0.4121	0.4187	0.4253	0.4318	0.4383	0.4447	0.4511	0.4574	0.4637
1.6	0.4700	0.4762	0.4824	0.4886	0.4947	0.5008	0.5068	0.5128	0.5188	0.5247
1.7	0.5306	0.5365	0.5423	0.5481	0.5539	0.5596	0.5653	0.5710	0.5766	0.5822
1.8	0.5878	0.5933	0.5988	0.6043	0.6098	0.6152	0.6206	0.6259	0.6313	0.6366
1.9	0.6419	0.6471	0.6523	0.6575	0.6627	0.6678	0.6729	0.6780	0.6831	0.6881
2.0	0.6931	0.6981	0.7031	0.7080	0.7130	0.7178	0.7227	0.7275	0.7324	0.7372
2.1	0.7419	0.7467	0.7514	0.7561	0.7608	0.7655	0.7701	0.7747	0.7793	0.7839
2.2	0.7885	0.7930	0.7975	0.8020	0.8065	0.8109	0.8154	0.8198	0.8242	0.8286
2.3	0.8329	0.8372	0.8416	0.8459	0.8502	0.8544	0.8587	0.8629	0.8671	0.8713
2.4	0.8755	0.8796	0.8838	0.8879	0.8920	0.8961	0.9002	0.9042	0.9083	0.9123
2.5	0.9163	0.9203	0.9243	0.9282	0.9322	0.9361	0.9400	0.9439	0.9478	0.9517
2.6	0.9555	0.9594	0.9632	0.9670	0.9708	0.9746	0.9783	0.9821	0.9858	0.9895
2.7	0.9933	0.9969	1.0006	1.0043	1.0080	1.0116	1.0152	0.0188	1.0225	1.0260
2.8	1.0296	1.0332	1.0367	1.0403	1.0438	1.0473	1.0508	1.0543	1.0578	1.0613
2.9	1.0647	1.0682	1.0716	1.0750	1.0784	1.0818	1.0852	1.0886	1.0919	1.0953
3.0	1.0986	1.1019	1.1053	1.1086	1.1119	1.1151	1.1184	1.1217	1.1249	1.1282
3.1	1.1314	1.1346	1.1378	1.1410	1.1442	1.1474	1.1506	1.1537	1.1569	1.1600
3.2	1.1632	1.1663	1.1694	1.1725	1.1756	1.1787	1.1817	1.1848	1.1878	1.1909
3.3	1.1939	1.1970	1.2000	1.2030	1.2060	1.2090	1.2119	1.2149	1.2179	1.2208
3.4	1.2238	1.2267	1.2296	1.2326	1.2355	1.2384	1.2413	1.2442	1.2470	1.2499
3.5	1.2528	1.2556	1.2585	1.2613	1.2641	1.2669	1.2698	1.2726	1.2754	1.2782
3.6	1.2809	1.2837	1.2865	1.2892	1.2920	1.2947	1.2975	1.3002	1.3029	1.3056
3.7	1.3083	1.3110	1.3137	1.3164	1.3191	1.3218	1.3244	1.3271	1.3297	1.3324
3.8	1.3350	1.3376	1.3403	1.3429	1.3455	1.3481	1.3507	1.3533	1.3558	1.3584
3.9	1.3610	1.3635	1.3661	1.3686	1.3712	1.3737	1.3762	1.3788	1.3813	1.3838
4.0	1.3863	1.3888	1.3913	1.3938	1.3962	1.3987	1.4012	1.4036	1.4061	1.4085
4.1	1.4110	1.4134	1.4159	1.4183	1.4207	1.4231	1.4255	1.4279	1.4303	1.4327
4.2	1.4351	1.4375	1.4398	1.4422	1.4446	1.4469	1.4493	1.4516	1.4540	1.4563
4.3	1.4586	1.4609	1.4633	1.4656	1.4679	1.4702	1.4725	1.4748	1.4770	1.4793
4.4	1.4816	1.4839	1.4861	1.4884	1.4907	1.4929	1.4952	1.4974	1.4996	1.5019
4.5	1.5041	1.5063	1.5085	1.5107	1.5129	1.5151	1.5173	1.5195	1.5217	1.5239
4.6	1.5261	1.5282	1.5304	1.5326	1.5347	1.5369	1.5390	1.5412	1.5433	1.5454
4.7	1.5476	1.5497	1.5518	1.5539	1.5560	1.5581	1.5602	1.5623	1.5644	1.5665
4.8	1.5686	1.5707	1.5728	1.5748	1.5769	1.5790	1.5810	1.5831	1.5851	1.5872
4.9	1.5892	1.5913	1.5933	1.5953	1.5974	1.5994	1.6014	1.6034	1.6054	1.6074
5.0	1.6094	1.6114	1.6134	1.6154	1.6174	1.6194	1.6214	1.6233	1.6253	1.6273
5.1	1.6292	1.6312	1.6332	1.6351	1.6371	1.6390	1.6409	1.6429	1.6448	1.6467
5.2	1.6487	1.6506	1.6525	1.6544	1.6563	1.6582	1.6601	1.6620	1.6639	1.6658
5.3	1.6677	1.6696	1.6715	1.6734	1.6752	1.6771	1.6790	1.6808	1.6827	1.6845
5.4	1.6864	1.6882	1.6901	1.6919	1.6938	1.6956	1.6974	1.6993	1.7011	1.7029
5.5	1.7047	1.7066	1.7084	1.7102	1.7120	1.7138	1.7156	1.7174	1.7192	1.7210
5.6	1.7228	1.7246	1.7263	1.7281	1.7299	1.7317	1.7334	1.7352	1.7370	1.7387
5.7	1.7405	1.7422	1.7440	1.7457	1.7475	1.7492	1.7509	1.7527	1.7544	1.7561
5.8	1.7579	1.7596	1.7613	1.7630	1.7647	1.7664	1.7682	1.7699	1.7716	1.7733
5.9	1.7750	1.7766	1.7783	1.7800	1.7817	1.7834	1.7851	1.7867	1.7884	1.7901
6.0	1.7918	1.7934	1.7951	1.7967	1.7984	1.8001	1.8017	1.8034	1.8050	1.8066
6.1	1.8083	1.8099	1.8116	1.8132	1.8148	1.8165	1.8181	1.8197	1.8213	1.8229
6.2	1.8245	1.8262	1.8278	1.8294	1.8310	1.8326	1.8342	1.8358	1.8374	1.8390
6.3	1.8406	1.8421	1.8437	1.8453	1.8469	1.8485	1.8500	1.8516	1.8532	1.8547
6.4	1.8563	1.8579	1.8594	1.8610	1.8625	1.8641	1.8656	1.8672	1.8687	1.8703

TABLE II Natural Logarithms, ln *t* (*continued*)

t	0.00	0.01	0.02	0.03	0.04	0.05	0.06	0.07	0.08	0.09
6.5	1.8718	1.8733	1.8749	1.8764	1.8779	1.8795	1.8810	1.8825	1.8840	1.8856
6.6	1.8871	1.8886	1.8901	1.8916	1.8931	1.8946	1.8961	1.8976	1.8991	1.9006
6.7	1.9021	1.9036	1.9051	1.9066	1.9081	1.9095	1.9110	1.9125	1.9140	1.9155
6.8	1.9169	1.9184	1.9199	1.9213	1.9228	1.9242	1.9257	1.9272	1.9286	1.9301
6.9	1.9315	1.9330	1.9344	1.9359	1.9373	1.9387	1.9402	1.9416	1.9430	1.9445
7.0	1.9459	1.9473	1.9488	1.9502	1.9516	1.9530	1.9544	1.9559	1.9573	1.9587
7.1	1.9601	1.9615	1.9629	1.9643	1.9657	1.9671	1.9685	1.9699	1.9713	1.9727
7.2	1.9741	1.9755	1.9769	1.9782	1.9796	1.9810	1.9824	1.9838	1.9851	1.9865
7.3	1.9879	1.9892	1.9906	1.9920	1.9933	1.9947	1.9961	1.9974	1.9988	2.0001
7.4	2.0015	2.0028	2.0042	2.0055	2.0069	2.0082	2.0096	2.0109	2.0122	2.0136
7.5	2.0149	2.0162	2.0176	2.0189	2.0202	2.0215	2.0229	2.0242	2.0255	2.0268
7.6	2.0282	2.0295	2.0308	2.0321	2.0334	2.0347	2.0360	2.0373	2.0386	2.0399
7.7	2.0412	2.0425	2.0438	2.0451	2.0464	2.0477	2.0490	2.0503	2.0516	2.0528
7.8	2.0541	2.0554	2.0567	2.0580	2.0592	2.0605	2.0618	2.0631	2.0643	2.0665
7.9	2.0669	2.0681	2.0694	2.0707	2.0719	2.0732	2.0744	2.0757	2.0769	2.0782
8.0	2.0794	2.0807	2.0819	2.0832	2.0844	2.0857	2.0869	2.0882	2.0894	2.0906
8.1	2.0919	2.0931	2.0943	2.0956	2.0968	2.0980	2.0992	2.1005	2.1017	2.1029
8.2	2.1041	2.1054	2.1066	2.1078	2.1090	2.1102	2.1114	2.1126	2.1138	2.1150
8.3	2.1163	2.1175	2.1187	2.1199	2.1211	2.1223	2.1235	2.1247	2.1258	2.1270
8.4	2.1282	2.1294	2.1306	2.1318	2.1330	2.1342	2.1353	2.1365	2.1377	2.1389
8.5	2.1401	2.1412	2.1424	2.1436	2.1448	2.1459	2.1471	2.1483	2.1494	2.1506
8.6	2.1518	2.1529	2.1541	2.1552	2.1564	2.1576	2.1587	2.1599	2.1610	2.1622
8.7	2.1633	2.1645	2.1656	2.1668	2.1679	2.1691	2.1702	2.1713	2.1725	2.1736
8.8	2.1748	2.1759	2.1770	2.1782	2.1793	2.1804	2.1815	2.1827	2.1838	2.1849
8.9	2.1861	2.1872	2.1883	2.1894	2.1905	2.1917	2.1928	2.1939	2.1950	2.1961
9.0	2.1972	2.1983	2.1994	2.2006	2.2017	2.2028	2.2039	2.2050	2.2061	2.2072
9.1	2.2083	2.2094	2.2105	2.2116	2.2127	2.2138	2.2148	2.2159	2.2170	2.2181
9.2	2.2192	2.2203	2.2214	2.2225	2.2235	2.2246	2.2257	2.2268	2.2279	2.2289
9.3	2.2300	2.2311	2.2322	2.2332	2.2343	2.2354	2.2364	2.2375	2.2386	2.2396
9.4	2.2407	2.2418	2.2428	2.2439	2.2450	2.2460	2.2471	2.2481	2.2492	2.2502
9.5	2.2513	2.2523	2.2534	2.2544	2.2555	2.2565	2.2576	2.2586	2.2597	2.2607
9.6	2.2618	2.2628	2.2638	2.2649	2.2659	2.2670	2.2680	2.2690	2.2701	2.2711
9.7	2.2721	2.2732	2.2742	2.2752	2.2762	2.2773	2.2783	2.2793	2.2803	2.2814
9.8	2.2824	2.2834	2.2844	2.2854	2.2865	2.2875	2.2885	2.2895	2.2905	2.2915
9.9	2.2925	2.2935	2.2946	2.2956	2.2966	2.2976	2.2986	2.2996	2.3006	2.3016

TABLE III Exponential Functions

x	e^x	e^{-x}
0.00	1.0000	1.0000
0.05	1.0513	0.9512
0.10	1.1052	0.9048
0.15	1.1618	0.8607
0.20	1.2214	0.8187
0.25	1.2840	0.7788
0.30	1.3499	0.7408
0.35	1.4191	0.7047
0.40	1.4918	0.6703
0.45	1.5683	0.6376
0.50	1.6487	0.6065
0.55	1.7333	0.5769
0.60	1.8221	0.5488
0.65	1.9155	0.5220
0.70	2.0138	0.4966
0.75	2.1170	0.4724
0.80	2.2255	0.4493
0.85	2.3396	0.4274
0.90	2.4596	0.4066
0.95	2.5857	0.3867
1.0	2.7183	0.3679
1.1	3.0042	0.3329
1.2	3.3201	0.3012
1.3	3.6693	0.2725
1.4	4.0552	0.2466
1.5	4.4817	0.2231
1.6	4.9530	0.2019
1.7	5.4739	0.1827
1.8	6.0496	0.1653
1.9	6.6859	0.1496
2.0	7.3891	0.1353
2.1	8.1662	0.1225
2.2	9.0250	0.1108
2.3	9.9742	0.1003
2.4	11.023	0.0907
2.5	12.182	0.0821
2.6	13.464	0.0743
2.7	14.880	0.0672
2.8	16.445	0.0608
2.9	18.174	0.0550

x	e^x	e^{-x}
3.0	20.086	0.0498
3.1	22.198	0.0450
3.2	24.533	0.0408
3.3	27.113	0.0369
3.4	29.964	0.0334
3.5	33.115	0.0302
3.6	36.598	0.0273
3.7	40.447	0.0247
3.8	44.701	0.0224
3.9	49.402	0.0202
4.0	54.598	0.0183
4.1	60.340	0.0166
4.2	66.686	0.0150
4.3	73.700	0.0136
4.4	81.451	0.0123
4.5	90.017	0.0111
4.6	99.484	0.0101
4.7	109.95	0.0091
4.8	121.51	0.0082
4.9	134.29	0.0074
5.0	148.41	0.0067
5.1	164.02	0.0061
5.2	181.27	0.0055
5.3	200.34	0.0050
5.4	221.41	0.0045
5.5	244.69	0.0041
5.6	270.43	0.0037
5.7	298.87	0.0033
5.8	330.30	0.0030
5.9	365.04	0.0027
6.0	403.43	0.0025
6.5	665.14	0.0015
7.0	1096.6	0.0009
7.5	1808.0	0.0006
8.0	2981.0	0.0003
8.5	4914.8	0.0002
9.0	8103.1	0.0001
9.5	13,360	0.00007
10.0	22,026	0.00004

TABLE IV Hyperbolic Functions

x	sinh x	cosh x	tanh x
0.0	0.00000	1.0000	0.00000
0.1	0.10017	1.0050	0.09967
0.2	0.20134	1.0201	0.19738
0.3	0.30452	1.0453	0.29131
0.4	0.41075	1.0811	0.37995
0.5	0.52110	1.1276	0.46212
0.6	0.63665	1.1855	0.53705
0.7	0.75858	1.2552	0.60437
0.8	0.88811	1.3374	0.66404
0.9	1.0265	1.4331	0.71630
1.0	1.1752	1.5431	0.76159
1.1	1.3356	1.6685	0.80050
1.2	1.5095	1.8107	0.83365
1.3	1.6984	1.9709	0.86172
1.4	1.9043	2.1509	0.88535
1.5	2.1293	2.3524	0.90515
1.6	2.3756	2.5775	0.92167
1.7	2.6456	2.8283	0.93541
1.8	2.9422	3.1075	0.94681
1.9	3.2682	3.4177	0.95624
2.0	3.6269	3.7622	0.96403
2.1	4.0219	4.1443	0.97045
2.2	4.4571	4.5679	0.97574
2.3	4.9370	5.0372	0.98010
2.4	5.4662	5.5569	0.98367
2.5	6.0502	6.1323	0.98661
2.6	6.6947	6.7690	0.98903
2.7	7.4063	7.4735	0.99101
2.8	8.1919	8.2527	0.99263
2.9	9.0596	9.1146	0.99396

x	sinh x	cosh x	tanh x
3.0	10.018	10.068	0.99505
3.1	11.076	11.122	0.99595
3.2	12.246	12.287	0.99668
3.3	13.538	13.575	0.99728
3.4	14.965	14.999	0.99777
3.5	16.543	16.573	0.99818
3.6	18.285	18.313	0.99851
3.7	20.211	20.236	0.99878
3.8	22.339	22.362	0.99900
3.9	24.691	24.711	0.99918
4.0	27.290	27.308	0.99933
4.1	30.162	30.178	0.99945
4.2	33.336	33.351	0.99955
4.3	36.843	36.857	0.99963
4.4	40.719	40.732	0.99970
4.5	45.003	45.014	0.99975
4.6	49.737	49.747	0.99980
4.7	54.969	54.978	0.99983
4.8	60.751	60.759	0.99986
4.9	67.141	67.149	0.99989
5.0	74.203	74.210	0.99991
5.1	82.008	82.014	0.99993
5.2	90.633	90.639	0.99994
5.3	100.17	100.17	0.99995
5.4	110.70	110.71	0.99996
5.5	122.34	122.35	0.99997
5.6	135.21	135.22	0.99997
5.7	149.43	149.44	0.99998
5.8	165.15	165.15	0.99998
5.9	182.52	182.52	0.99998

TABLE V Common Logarithms, $\log_{10} x$

x	0.00	0.01	0.02	0.03	0.04	0.05	0.06	0.07	0.08	0.09
1.0	.0000	.0043	.0086	.0128	.0170	.0212	.0253	.0294	.0334	.0374
1.1	.0414	.0453	.0492	.0531	.0569	.0607	.0645	.0682	.0719	.0755
1.2	.0792	.0828	.0864	.0899	.0934	.0969	.1004	.1038	.1072	.1106
1.3	.1139	.1173	.1206	.1239	.1271	.1303	.1335	.1367	.1399	.1430
1.4	.1461	.1492	.1523	.1553	.1584	.1614	.1644	.1673	.1703	.1732
1.5	.1761	.1790	.1818	.1847	.1875	.1903	.1931	.1959	.1987	.2014
1.6	.2041	.2068	.2095	.2122	.2148	.2175	.2201	.2227	.2253	.2279
1.7	.2304	.2330	.2355	.2380	.2405	.2430	.2455	.2480	.2504	.2529
1.8	.2553	.2577	.2601	.2625	.2648	.2672	.2695	.2718	.2742	.2765
1.9	.2788	.2810	.2833	.2856	.2878	.2900	.2923	.2945	.2967	.2989
2.0	.3010	.3032	.3054	.3075	.3096	.3118	.3139	.3160	.3181	.3201
2.1	.3222	.3243	.3263	.3284	.3304	.3324	.3345	.3365	.3385	.3404
2.2	.3424	.3444	.3464	.3483	.3502	.3522	.3541	.3560	.3579	.3598
2.3	.3617	.3636	.3655	.3674	.3692	.3711	.3729	.3747	.3766	.3784
2.4	.3802	.3820	.3838	.3856	.3874	.3892	.3909	.3927	.3945	.3962
2.5	.3979	.3997	.4014	.4031	.4048	.4065	.4082	.4099	.4116	.4133
2.6	.4150	.4166	.4183	.4200	.4216	.4232	.4249	.4265	.4281	.4298
2.7	.4314	.4330	.4346	.4362	.4378	.4393	.4409	.4425	.4440	.4456
2.8	.4472	.4487	.4502	.4518	.4533	.4548	.4564	.4579	.4594	.4609
2.9	.4624	.4639	.4654	.4669	.4683	.4698	.4713	.4728	.4742	.4757
3.0	.4771	.4786	.4800	.4814	.4829	.4843	.4857	.4871	.4886	.4900
3.1	.4914	.4928	.4942	.4955	.4969	.4983	.4997	.5011	.5024	.5038
3.2	.5051	.5065	.5079	.5092	.5105	.5119	.5132	.5145	.5159	.5172
3.3	.5185	.5198	.5211	.5224	.5237	.5250	.5263	.5276	.5289	.5302
3.4	.5315	.5328	.5340	.5353	.5366	.5378	.5391	.5403	.5416	.5428
3.5	.5441	.5453	.5465	.5478	.5490	.5502	.5514	.5527	.5539	.5551
3.6	.5563	.5575	.5587	.5599	.5611	.5623	.5635	.5647	.5658	.5670
3.7	.5682	.5694	.5705	.5717	.5729	.5740	.5752	.5763	.5775	.5786
3.8	.5798	.5809	.5821	.5832	.5843	.5855	.5866	.5877	.5888	.5899
3.9	.5911	.5922	.5933	.5944	.5955	.5966	.5977	.5988	.5999	.6010
4.0	.6021	.6031	.6042	.6053	.6064	.6075	.6085	.6096	.6107	.6117
4.1	.6128	.6138	.6149	.6160	.6170	.6180	.6191	.6201	.6212	.6222
4.2	.6232	.6243	.6253	.6263	.6274	.6284	.6294	.6304	.6314	.6325
4.3	.6335	.6345	.6355	.6365	.6375	.6385	.6395	.6405	.6415	.6425
4.4	.6435	.6444	.6454	.6464	.6474	.6484	.6493	.6503	.6513	.6522
4.5	.6532	.6542	.6551	.6561	.6571	.6580	.6590	.6599	.6609	.6618
4.6	.6628	.6637	.6646	.6656	.6665	.6675	.6684	.6693	.6702	.6712
4.7	.6721	.6730	.6739	.6749	.6758	.6767	.6776	.6785	.6794	.6803
4.8	.6812	.6821	.6830	.6839	.6848	.6857	.6866	.6875	.6884	.6893
4.9	.6902	.6911	.6920	.6928	.6937	.6946	.6955	.6964	.6972	.6981
5.0	.6990	.6998	.7007	.7016	.7024	.7033	.7042	.7050	.7059	.7067
5.1	.7076	.7084	.7093	.7101	.7110	.7118	.7126	.7135	.7143	.7152
5.2	.7160	.7168	.7177	.7185	.7193	.7202	.7210	.7218	.7226	.7235
5.3	.7243	.7251	.7259	.7267	.7275	.7284	.7292	.7300	.7308	.7316
5.4	.7324	.7332	.7340	.7348	.7356	.7364	.7372	.7380	.7388	.7396
5.5	.7404	.7412	.7419	.7427	.7435	.7443	.7451	.7459	.7466	.7474
5.6	.7482	.7490	.7497	.7505	.7513	.7520	.7528	.7536	.7543	.7551
5.7	.7559	.7566	.7574	.7582	.7589	.7597	.7604	.7612	.7619	.7627
5.8	.7634	.7642	.7649	.7657	.7664	.7672	.7679	.7686	.7694	.7701
5.9	.7709	.7716	.7723	.7731	.7738	.7745	.7752	.7760	.7767	.7774
6.0	.7782	.7789	.7796	.7803	.7810	.7818	.7825	.7832	.7839	.7846
6.1	.7853	.7860	.7868	.7875	.7882	.7889	.7896	.7903	.7910	.7917
6.2	.7924	.7931	.7938	.7945	.7952	.7959	.7966	.7973	.7980	.7987
6.3	.7993	.8000	.8007	.8014	.8021	.8028	.8035	.8041	.8048	.8055
6.4	.8062	.8069	.8075	.8082	.8089	.8096	.8102	.8109	.8116	.8122

TABLE V **Common Logarithms, $\log_{10} x$ (*continued*)**

x	0.00	0.01	0.02	0.03	0.04	0.05	0.06	0.07	0.08	0.09
6.5	.8129	.8136	.8142	.8149	.8156	.8162	.8169	.8176	.8182	.8189
6.6	.8195	.8202	.8209	.8215	.8222	.8228	.8235	.8241	.8248	.8254
6.7	.8261	.8267	.8274	.8280	.8287	.8293	.8299	.8306	.8312	.8319
6.8	.8325	.8331	.8338	.8344	.8351	.8357	.8363	.8370	.8376	.8382
6.9	.8388	.8395	.8401	.8407	.8414	.8420	.8426	.8432	.8439	.8445
7.0	.8451	.8457	.8463	.8470	.8476	.8482	.8488	.8494	.8500	.8506
7.1	.8513	.8519	.8525	.8531	.8537	.8543	.8549	.8555	.8561	.8567
7.2	.8573	.8579	.8585	.8591	.8597	.8603	.8609	.8615	.8621	.8627
7.3	.8633	.8639	.8645	.8651	.8657	.8663	.8669	.8675	.8681	.8686
7.4	.8692	.8698	.8704	.8710	.8716	.8722	.8727	.8733	.8739	.8745
7.5	.8751	.8756	.8762	.8768	.8774	.8779	.8785	.8791	.8797	.8802
7.6	.8808	.8814	.8820	.8825	.8831	.8837	.8842	.8848	.8854	.8859
7.7	.8865	.8871	.8876	.8882	.8887	.8893	.8899	.8904	.8910	.8915
7.8	.8921	.8927	.8932	.8938	.8943	.8949	.8954	.8960	.8965	.8971
7.9	.8976	.8982	.8987	.8993	.8998	.9004	.9009	.9015	.9020	.9025
8.0	.9031	.9036	.9042	.9047	.9053	.9058	.9063	.9069	.9074	.9079
8.1	.9085	.9090	.9096	.9101	.9106	.9112	.9117	.9122	.9128	.9133
8.2	.9138	.9143	.9149	.9154	.9159	.9165	.9170	.9175	.9180	.9186
8.3	.9191	.9196	.9201	.9206	.9212	.9217	.9222	.9227	.9232	.9238
8.4	.9243	.9248	.9253	.9258	.9263	.9269	.9274	.9279	.9284	.9289
8.5	.9294	.9299	.9304	.9309	.9315	.9320	.9325	.9330	.9335	.9340
8.6	.9345	.9350	.9355	.9360	.9365	.9370	.9375	.9380	.9385	.9390
8.7	.9395	.9400	.9405	.9410	.9415	.9420	.9425	.9430	.9435	.9440
8.8	.9445	.9450	.9455	.9460	.9465	.9469	.9474	.9479	.9484	.9489
8.9	.9494	.9499	.9504	.9509	.9513	.9518	.9523	.9528	.9533	.9538
9.0	.9542	.9547	.9552	.9557	.9562	.9566	.9571	.9567	.9581	.9586
9.1	.9590	.9595	.9600	.9605	.9609	.9614	.9619	.9624	.9628	.9633
9.2	.9638	.9643	.9647	.9652	.9657	.9661	.9666	.9671	.9675	.9680
9.3	.9685	.9689	.9694	.9699	.9703	.9708	.9713	.9717	.9722	.9727
9.4	.9731	.9736	.9741	.9745	.9750	.9754	.9759	.9763	.9768	.9773
9.5	.9777	.9782	.9786	.9791	.9795	.9800	.9805	.9809	.9814	.9818
9.6	.9823	.9827	.9832	.9836	.9841	.9845	.9850	.9854	.9859	.9863
9.7	.9868	.9872	.9877	.9881	.9886	.9890	.9894	.9899	.9903	.9908
9.8	.9912	.9917	.9921	.9926	.9930	.9934	.9939	.9943	.9948	.9952
9.9	.9956	.9961	.9965	.9969	.9974	.9978	.9983	.9987	.9991	.9996

TABLE VI Powers and Roots

Number	Square	Square Root	Cube	Cube Root
1	1	1.000	1	1.000
2	4	1.414	8	1.260
3	9	1.732	27	1.442
4	16	2.000	64	1.587
5	25	2.236	125	1.710
6	36	2.449	216	1.817
7	49	2.646	343	1.913
8	64	2.828	512	2.000
9	81	3.000	729	2.080
10	100	3.162	1,000	2.154
11	121	3.317	1,331	2.224
12	144	3.464	1,728	2.289
13	169	3.606	2,197	2.351
14	196	3.742	2,744	2.410
15	225	3.873	3,375	2.466
16	256	4.000	4,096	2.520
17	289	4.123	4,913	2.571
18	324	4.243	5,832	2.621
19	361	4.359	6,859	2.668
20	400	4.472	8,000	2.714
21	441	4.583	9,261	2.759
22	484	4.690	10,648	2.802
23	529	4.796	12,167	2.844
24	576	4.899	13,824	2.884
25	625	5.000	15,625	2.924
26	676	5.099	17,576	2.962
27	729	5.196	19,683	3.000
28	784	5.292	21,952	3.037
29	841	5.385	24,389	3.072
30	900	5.477	27,000	3.107
31	961	5.568	29,791	3.141
32	1,024	5.657	32,768	3.175
33	1,089	5.745	35,937	3.208
34	1,156	5.831	39,304	3.240
35	1,225	5.916	42,875	3.271
36	1,296	6.000	46,656	3.302
37	1,369	6.083	50,653	3.332
38	1,444	6.164	54,872	3.362
39	1,521	6.245	59,319	3.391
40	1,600	6.325	64,000	3.420
41	1,681	6.403	68,921	3.448
42	1,764	6.481	74,088	3.476
43	1,849	6.557	79,507	3.503
44	1,936	6.633	85,184	3.530
45	2,025	6.708	91,125	3.557
46	2,116	6.782	97,336	3.583
47	2,209	6.856	103,823	3.609
48	2,304	6.928	110,592	3.634
49	2,401	7.000	117,649	3.659
50	2,500	7.071	125,000	3.684
51	2,601	7.141	132,651	3.708
52	2,704	7.211	140,608	3.733
53	2,809	7.280	148,877	3.756
54	2,916	7.348	157,464	3.780
55	3,025	7.416	166,375	3.803
56	3,136	7.483	175,616	3.826
57	3,249	7.550	185,193	3.849
58	3,364	7.616	195,112	3.871
59	3,481	7.681	205,379	3.893
60	3,600	7.746	216,000	3.915
61	3,721	7.810	226,981	3.936
62	3,844	7.874	238,328	3.958
63	3,969	7.937	250,047	3.979
64	4,096	8.000	262,144	4.000
65	4,225	8.062	274,625	4.021
66	4,356	8.124	287,496	4.041
67	4,489	8.185	300,763	4.062
68	4,624	8.246	314,432	4.082
69	4,761	8.307	328,509	4.102
70	4,900	8.367	343,000	4.121
71	5,041	8.426	357,911	4.141
72	5,184	8.485	373,248	4.160
73	5,329	8.544	389,017	4.179
74	5,476	8.602	405,224	4.198
75	5,625	8.660	421,875	4.217
76	5,776	8.718	438,976	4.236
77	5,929	8.775	456,533	4.254
78	6,084	8.832	474,552	4.273
79	6,241	8.888	493,039	4.291
80	6,400	8.944	512,000	4.309
81	6,561	9.000	531,441	4.327
82	6,724	9.055	551,368	4.344
83	6,889	9.110	571,787	4.362
84	7,056	9.165	592,704	4.380
85	7,225	9.220	614,125	4.397
86	7,396	9.274	636,056	4.414
87	7,569	9.327	658,503	4.431
88	7,744	9.381	681,472	4.448
89	7,921	9.434	704,969	4.465
90	8,100	9.487	729,000	4.481
91	8,281	9.539	753,571	4.498
92	8,464	9.592	778,688	4.514
93	8,649	9.644	804,357	4.531
94	8,836	9.695	830,584	4.547
95	9,025	9.747	857,375	4.563
96	9,216	9.798	884,736	4.579
97	9,409	9.849	912,673	4.595
98	9,604	9.899	941,192	4.610
99	9,801	9.950	970,299	4.626
100	10,000	10.000	1,000,000	4.642